Frontiers of Propulsion Science

Frontiers of Propulsion Science

Edited by
Marc G. Millis
NASA Glenn Research Center
Cleveland, Ohio

Eric W. Davis
Institute for Advanced Studies at Austin
Austin, Texas

Volume 227
PROGRESS IN
ASTRONAUTICS AND AERONAUTICS

Frank K. Lu, Editor-in-Chief
University of Texas at Arlington
Arlington, Texas

Published by the
American Institute of Aeronautics and Astronautics, Inc.
1801 Alexander Bell Drive, Reston, Virginia 20191-4344

American Institute of Aeronautics and Astronautics, Inc., Reston, Virginia

2 3 4 5

Copyright © 2009 by the American Institute of Aeronautics and Astronautics, Inc. Printed in the United States of America. All rights reserved. Reproduction or translation of any part of this work beyond that permitted by Sections 107 and 108 of the U.S. Copyright Law without the permission of the copyright owner is unlawful. The code following this statement indicates the copyright owner's consent that copies of articles in this volume may be made for personal or internal use, on condition that the copier pay the per-copy fee ($2.50) plus the per-page fee ($0.50) through the Copyright Clearance Center. Inc., 222 Rosewood Drive, Danvers, Massachusetts 01923. This consent does not extend to other kinds of copying, for which permission requests should be addressed to the publisher. Users should employ the following code when reporting copying from the volume to the Copyright Clearance Center:

978-1-56347-956-4/09 $2.50 + .50

Data and information appearing in this book are for informational purposes only. AIAA is not responsible for any injury or damage resulting from use or reliance, nor does AIAA warrant that use or reliance will be free from privately owned rights.

ISBN 978-1-56347-956-4

THE editors dedicate this book to Dr. Robert L. Forward and Sir Arthur C. Clarke. We honor these individuals for both the advances they made in spaceflight and for the inspirational effect their work had on the rest of us. Each was competent, visionary, and entertaining. Each risked topics beyond the comfort zones of their peers and succeeded in making landmark progress. Their works have become milestones in history. Additionally, their science fiction helped provoke thought on a deeper level and entertained us in the process. Without pioneers like these, this book would not exist. Their vision and passion compels us to carry on, hopefully continuing the progress that will one day allow humanity to travel amongst the stars.

Progress in Astronautics and Aeronautics
Editor-in-Chief
Frank K. Lu
University of Texas at Arlington

Editorial Board

David A. Bearden
The Aerospace Corporation

John D. Binder
viaSolutions

Steven A. Brandt
U.S. Air Force Academy

Jose Camberos
U.S. Air Force Research Laboratory

Richard Curran
Queen's University of Belfast

Sanjay Garg
NASA Glenn Research Center

Eswar Josyula
U.S. Air Force Research Laboratory

Gail A. Klein
Jet Propulsion Laboratory

Konstantinos Kontis
University of Manchester

Richard C. Lind
University of Florida

Ning Qin
University of Sheffield

Oleg A. Yakimenko
U.S. Naval Postgraduate School

Christopher H. Jenkins
Montana State University

Foreword

AS AN aerospace researcher, I appreciate the value and challenge of advancing revolutionary ideas. While my partners and I are opening near-Earth spaceflight to citizens, the creators of this book are extending *far* beyond that—investigating how to enable interstellar flight.

This is the stuff of breakthroughs; those notions that sound crazy at first, but actually lead to enormous improvements in the human condition. I love that word, "breakthrough." It means taking risks to explore what average researchers consider nonsense, and then persevering until you've changed the world. While most aerospace professionals play it safe with yawning improvements in technology, the authors of this book take the risk of seriously considering ideas that presently look impossible... to some. From their own initiative they created this first-ever technical book on star-drive science.

Although the topics of this book might sound like science fiction to the less adventurous—covering ideas like *antigravity*, *space drives*, *warp drives*, and *faster-than-light travel*—these goals are dealt with here as a rigorous scientific inquiry. The authors identify what's already been investigated with comparisons to the foundations of physics, cover some dead-end approaches, and show where next to focus attention to continue systematic, rigorous advancements. This is not light reading. The details and citations are numerous. In short, this book offers the seeds for undiscovered breakthroughs.

It is hoped that this book will inspire students and young professionals. In these pages are the starting materials from which they can begin to make their mark on history. And to help these future pioneers, the editors saw fit to include a chapter on how to conduct such visionary work within typically stodgy establishments. Lessons of prior breakthroughs are contrasted to the demands of bureaucracies, with specific suggestions on how to make such high-risk/high-gain research seem downright prudent.

While I continue to open spaceflight to the masses and NASA reaches back for the moon, it is comforting to know that the scientists and engineers behind this book are looking beyond current activities to answer "what comes after that?"—the breakthroughs that will take us to the stars.

Burt Rutan
Scaled Composites LLC
August 2008

Table of Contents

Preface xvii

Acknowledgments xxvii

I. Understanding the Problem

Chapter 1. Recent History of Breakthrough Propulsion Studies 1
Paul A. Gilster, *Tau Zero Foundation, Raleigh, North Carolina*

Introduction 1
Challenging Initial Stages 2
Reflecting on a Time Line 3
Advanced Propulsion Concepts Emerge 3
Vision-21 and Its Aftermath 7
Wormholes and Warp Drive Physics 8
Quantum Tunneling at FTL Speeds 9
Podkletnov "Gravity Shield" 9
Considering the Quantum Vacuum 10
Institute for Advanced Studies at Austin 11
Breakthrough Propulsion Physics Project 11
Project Greenglow 17
ESA's General Studies Programme and Advanced Concepts Team 19
Broad Strategies for Future Research 20
Continuing Research Projects 21
Faster than Light 22
Conclusions 23
Acknowledgments 23
References 23

Chapter 2. Limits of Interstellar Flight Technology 31
Robert H. Frisbee, *Jet Propulsion Laboratory, California Institute of Technology, Pasadena, California*

Introduction 31
Challenge of Interstellar Missions 32
Fundamentals of Interstellar Flight 43
Rocket-Based Propulsion Options 65
Nonrocket (Beamed-Momentum) Propulsion Options 81
Illustrative Missions 85
Most Critical Technologies and Feasible Engineering Limits 97
Closing Comments 107

Acknowledgments... 109
Appendix: Derivation of the Classical and Relativistic Rocket Equations....... 109
References 123

II. Propulsion without Rockets

Chapter 3. Prerequisites for Space Drive Science 127
Marc G. Millis, *NASA Glenn Research Center, Cleveland, Ohio*

Introduction 127
Methods... 128
Major Objections and Objectives 128
Estimating Potential Benefits.................................. 138
Hypothetical Mechanisms...................................... 149
Next Steps 168
Conclusions 169
Acknowledgments 170
References 170

Chapter 4. Review of Gravity Control Within Newtonian and General Relativistic Physics........................... 175
Eric W. Davis, *Institute for Advanced Studies at Austin, Austin, Texas*

Introduction 175
Gravity Control Within Newtonian Physics 176
Gravity Control Within General Relativity 184
Miscellaneous Gravity Control Concepts 198
Conclusions 221
Acknowledgments 223
References 223

Chapter 5. Gravitational Experiments with Superconductors: History and Lessons 229
George D. Hathaway, *Hathaway Consulting Services, Toronto, Ontario, Canada*

Introduction... 229
Experimental Traps and Pitfalls 231
Historical Outline.. 236
Summary and Future Directions................................ 245
References 246

Chapter 6. Nonviable Mechanical "Antigravity" Devices 249
Marc G. Millis, *NASA Glenn Research Center, Cleveland, Ohio*

Introduction... 249
Oscillation Thrusters 249

Gyroscopic Antigravity 254
Unrelated Devices 259
Conclusions 260
References 260

Chapter 7. Null Findings of Yamashita Electrogravitational Patent .. 263
Kenneth E. Siegenthaler and Timothy J. Lawrence, *U.S. Air Force Academy, Colorado Springs, Colorado*

Nomenclature 263
Introduction 263
Experiment 1 267
Experiment 2 278
Conclusions 290
Acknowledgments 291
References 291

Chapter 8. Force Characterization of Asymmetrical Capacitor Thrusters in Air 293
William M. Miller, *Sandia National Laboratories, Albuquerque, New Mexico*; Paul B. Miller, *East Mountain Charter High School, Sandia Park, New Mexico*; and Timothy J. Drummond, *Sandia National Laboratories, Albuquerque, New Mexico*

Introduction 293
Summary of Theories 294
Overall Experiment Setup 295
Results 297
Discussion of Data as it Relates to Theories 319
Conclusions 325
Acknowledgments 326
References 326

Chapter 9. Experimental Findings of Asymmetrical Capacitor Thrusters for Various Gasses and Pressures 329
Francis X. Canning, *Simply Sparse Technologies, Morgantown, West Virginia*

Introduction 329
Experimental Setup 332
Qualitative Experimental Results 333
Numerical Calculations of Electric Fields 336
Theories Versus Quantitative Experimental Results 337
Vacuum Results 339
Conclusions 339
Acknowledgments 340
References 340

Chapter 10. Propulsive Implications of Photon Momentum in Media 341
Michael R. LaPointe, *NASA Marshall Space Flight Center, Huntsville, Alabama*

Introduction 341
Background 342
Electromagnetic Fields in Dielectric Media................. 345
A Century of Controversy: Theory, Experiment, and Attempts at Resolution 347
Propulsion Concepts Based on Electromagnetic Momentum Exchange 356
Feigel Hypothesis... 366
Conclusions 368
Acknowledgments 368
References 368

Chapter 11. Experimental Results of the Woodward Effect on a Micro-Newton Thrust Balance 373
Nembo Buldrini and Martin Tajmar, *Austrian Research Centers GmbH—ARC, Seibersdorf, Austria*

Introduction 373
Theoretical Considerations 374
Thrust Balance... 376
Setup... 378
Results.. 380
Conclusions 388
Acknowledgments 388
References 388

Chapter 12. Thrusting Against the Quantum Vacuum........... 391
G. Jordan Maclay, *Quantum Fields LLC, Richland Center, Wisconsin*

Introduction 391
Physics of the Quantum Vacuum 393
Measurements of Casimir Forces 399
Space Propulsion Implications 402
Vibrating Mirror Casimir Drive 405
Unresolved Physics....................................... 415
Conclusions 417
Acknowledgments 418
References 419

Chapter 13. Inertial Mass from Stochastic Electrodynamics........ 423
Jean-Luc Cambier, *U.S. Air Force Research Laboratory, Edwards Air Force Base, Edwards, California*

Introduction 423
Background 424

Stochastic Electrodynamics 425
Quantum Electrodynamics 433
The Way Forward 438
Conclusions 441
Appendix A: Relativistic Transformation to a Uniformly Accelerated Frame 442
Appendix B: Estimate of Rest Mass from Classical Interaction with ZPF 445
Appendix C: Boyer's Correlation Function
    and Poynting Vector 446
Appendix D: Unruh–Davies and Classical Oscillator 451
References 452

III. Faster-than-Light Travel

Chapter 14. Relativistic Limits of Spaceflight 455
Brice N. Cassenti, *Rensselaer Polytechnic Institute, Hartford, Connecticut*

Introduction 455
Principle of Special Relativity 456
Paradoxes in Special Relativity 459
Empirical Foundations 463
Relativistic Rockets 465
Conclusions 468
References 469

Chapter 15. Faster-than-Light Approaches in
General Relativity 471
Eric W. Davis, *Institute for Advanced Studies at Austin, Austin, Texas*

Introduction 471
General Relativistic Definition of Exotic Matter and the Energy Conditions 472
Brief Review of Faster-than-Light Spacetimes 484
Make or Break Issues 489
Conclusions 500
Acknowledgments 502
References 502

Chapter 16. Faster-than-Light Implications of Quantum
Entanglement and Nonlocality 509
John G. Cramer, *University of Washington, Seattle, Washington*

Introduction 509
Quantum Entanglement, Nonlocality, and EPR Experiments 510
Quantum No-Signal Theorems 512
Nonlocality vs Special Relativity? 513
Momentum Domain Entanglement and EPR Experiments 514
Coherence-Entanglement Complementarity 518
Nonlocal Communication vs Signaling 520

Superluminal and Retrocausal Nonlocal Communication 522
Paradoxes and Nonlocal Communication 523
Nonlinear Quantum Mechanics and Nonlocal Communication. 524
Issues and Summary 525
Conclusions. 526
Appendix: Glossary and Description of Key Concepts. 526
References 528

IV. Energy Considerations

Chapter 17.   Comparative Space Power Baselines 531
Gary L. Bennett, *Metaspace Enterprises, Emmett, Idaho*

Introduction 531
Non-nuclear Space Power Sources 535
Nuclear Power Sources 537
Conclusions 565
References 565

Chapter 18.   On Extracting Energy from the
Quantum Vacuum 569
Eric W. Davis and H. E. Puthoff, *Institute for Advanced Studies at Austin, Austin, Texas*

Introduction 569
Early Concepts for Extracting Energy and Thermodynamic Considerations 570
Origin of Zero-Point Field Energy 574
Review of Selected Experiments 577
Additional Considerations and Issues 588
Conclusions 597
Acknowledgments 598
References 599

Chapter 19.   Investigating Sonoluminescence as a Means of
Energy Harvesting 605
John D. Wrbanek, Gustave C. Fralick, Susan Y. Wrbanek, and
Nancy R. Hall, *NASA Glenn Research Center, Cleveland, Ohio*

Introduction 605
Approaches. 607
Challenges for Application 609
Sonoluminescence at NASA 610
Indications of High Temperature Generation 617
Energy Harvesting 625
Summary and Future Directions 631
Conclusions 633
Acknowledgments 634
References 634

Chapter 20.   Null Tests of "Free Energy" Claims **639**
Scott R. Little, *EarthTech International, Austin, Texas*

Introduction. 639
Testing of Energy Claims 640
Some Tests of Breakthrough Energy Claims 641
Conclusions 648
References 648

V.   From This Point Forward

**Chapter 21.   General Relativity Computational Tools and
Conventions for Propulsion** **651**
Claudio Maccone

Introduction 651
Recommended Propulsion Computational Conventions 652
Representative Problems in Propulsion Science 655
Review of Existing Computational Tools 655
Conclusions 660
Acknowledgments 661
References 661

Chapter 22.   Prioritizing Pioneering Research **663**
Marc G. Millis, *NASA Glenn Research Center, Cleveland, Ohio*

Introduction 663
Historical Perspectives. 665
Combining Vision and Rigor. 673
Research Project Operating Principles 681
Devising Prioritization Criteria. 686
NASA Breakthrough Propulsion Physics Research Solicitation 689
From Individual to Overall Progress 694
Final Lessons and Comparisons 699
Conclusions 700
Appendix 701
References 714

Subject Index. **719**

Author Index. **737**

Supporting Materials **739**

Preface

"Fortune favors the bold."
—Book X of *The Aeneid* by Virgil (19 B.C.E.)

I. Purpose

THIS book is the first-ever compilation of emerging science relevant to such notions as *space drives, warp drives, gravity control,* and *faster-than-light* travel—the kind of breakthroughs that would revolutionize spaceflight and enable human voyages to other star systems. Although these concepts might sound like science fiction, they are appearing in increasing numbers in reputable scientific journals. The intent of this book is to provide managers, scientists, engineers, and graduate students with enough starting material to comprehend the status of this research and decide for themselves if and how to pursue this topic in more depth.

As with any young topic, it can be difficult to comprehend the potential benefits, the pros and cons of the competing approaches, and then decide what actions are warranted. To that end, the editors have endeavored to collect impartial overviews of the best-known and most relevant approaches. In many cases, dead-end lessons are included to counter recurring claims and offer examples for how to assess such claims. In addition, the methods for dealing with such pioneering topics are included, both from the historical perspective and more specifically from the lessons learned from NASA's Breakthrough Propulsion Physics Project. It is hoped that this volume will give future researchers the foundations to eventually discover the breakthroughs that will allow humanity to thrive beyond Earth.

This research falls within the realm of *science* instead of *technology*, with the distinction that science is about uncovering the laws of nature while technology is about applying that science to build useful devices. Because existing technology is inadequate for traversing astronomical distances between neighboring stars (even if advanced to the limit of its underlying physics), the only way to circumvent these limits is to discover new propulsion science. In addition to their utility for spaceflight, the discovery of any new force-production or energy-exchange principles would lead to a whole new class of technologies. The implications of success are profound.

Objectively, the desired breakthroughs might turn out to be impossible, but progress is not made by conceding defeat. Breakthroughs have a habit of taking pessimists by surprise, but can equally remain elusive. Although no breakthroughs appear imminent, enough progress has been made to provide the groundwork for deeper studies, both with the science itself and with the programmatic methods for tackling such provocative goals. If the history of scientific and technological revolutions is any indication, this topic could one day eclipse familiar aerospace technology and enable humanity to travel to habitable planets around neighboring stars. For now, however, the work is predominantly at stages 1 and 2 of the scientific method; that is, defining the problem and collecting data, with a few approaches already testing hypotheses.

Regardless of whether the breakthroughs are found, this inquiry provides an additional perspective from which to seek answers to the lingering unknowns of our universe. While general science continues to assess cosmological data for its implication to the birth and fate of the universe, a spaceflight focus will cast these observations in different contexts, offering insights that might otherwise be overlooked from the curiosity-driven inquiries alone. Therefore, even if there are no spaceflight breakthroughs to be found, adding the inquiry of spaceflight expands our ability to better understand the universe. The lessons learned in the attempt will advance science in general.

The editors of this volume optimistically look forward to a time when this first-ever technical book on space drives and faster-than-light travel becomes outdated. It will be interesting to look back decades from now to see what future discoveries transpired and then compare them with the directions examined in these chapters. Regardless of the book's inevitable obsolescence, the editors hope that the methods of continued discovery implicit in this book will have lasting impact.

II. Chapter by Chapter

An introduction to each chapter follows, so that the context of each can be better understood. Consider each chapter to be a primer to its topic rather than a definitive last word. There is much that could change with further research, and in some cases *different interim conclusions* are found in different chapters. Such divergent conclusions are a reflection of the embryonic state of this research, where many unresolved issues still exist. In many cases, chapters present ideas that are found to not work. From experience with the NASA Breakthrough Propulsion Physics Project, it was found that dead-end approaches are repeatedly researched because the null results were never distributed. Other chapters describe work that is currently under investigation where more definitive results are imminent. And lastly, much research on this topic has not even begun.

To convey the relative maturity of propulsion studies, a historical perspective is offered in **Chapter 1**. Although the notions for breakthrough spaceflight have been around for quite some time, the actual scientific publications started to accumulate around 1995, with a few examples predating this by many years. From 1996 through 2002, NASA funded work on these subjects through the Breakthrough Propulsion Physics Project. During that same period, several other organizations sponsored work of their own, including British Aerospace Systems. Most of this research now continues under the discretionary time and resources of individual researchers scattered across the globe.

It would not be appropriate to have a book about seeking new propulsion *science* without first articulating the edge of interstellar flight *technology*, both the foreseeable embodiments and their upper theoretical limits. **Chapter 2** provides a thorough overview of the edge of interstellar technology, along with details about optimum trajectory and mission planning. Comparisons between the required mission energies and available terrestrial energy are provided to convey the scale of the challenge. Methods to minimize trip time and vehicle system mass are offered, including assessments of the impact of acceleration and maximum cruise velocity. A number of technological options are described, including their performance predictions, spanning the technology of light sails through antimatter-annihilation rockets.

Presently, the scientific foundations from which to engineer *space drives*—propulsion that uses only the interactions between the spacecraft and its surrounding space—do not exist. To help initiate the systematic search and assessment of possibilities, **Chapter 3** transforms the major objections to the notion of a space drive into a problem statement to guide future research. The major objections are the scarcity of indigenous reaction mass (for momentum conservation), and the lack of known methods to impart net forces against such matter. By examining these issues and the various forms of matter in the universe, and by examining 10 hypothetical space drives, a problem statement is derived. From simple energy analyses, estimates for potential benefits and various analytical approaches are identified. It is found that the very definitions of spacetime and inertial frames warrant deeper research. When viewed in the context of space propulsion rather than general science, these questions present different research paths that still have not been explored.

When it comes to moving spacecraft without using rockets or light sails, one of the most commonly raised approaches is that of manipulating gravity. **Chapter 4** examines several ways of approaching this challenge, from simple Newtonian concepts, General Relativity Theory, semi-classical Quantum Gravity Theory, Quantum Field Theory, and others. Although it is possible, in principle, to create or modify acceleration fields, the present theoretical approaches require large (kilometer scale) and massive (solar-system scale) devices operating at extremely high energy levels (relativistic). The chapter also addresses the cosmological antigravity interpretations of dark energy, to find that this does not lead to obvious propulsion opportunities. The details of a wide variety of specific approaches are discussed, and key issues and unexplored research paths are identified.

Whereas **Chapter 4** dealt with *theoretical* approaches, **Chapter 5** examines recent *experimental* approaches to create gravitational-like effects using superconductors. An introduction to the theoretical basis for the relationship between gravity and superconductors from General Relativity is presented, followed by descriptions of a variety of experimental complications that occur when exploring such notions. These complications are included to better convey the difficulty of accurately deducing how nature works via low-temperature experiments. And finally, an overview is provided of earlier and ongoing attempts to experimentally observe gravitational effects with superconducting and low-temperature devices. Some experiments have turned out to be dead-ends, while others are still under evaluation.

As much as it is important to identify the promising areas of future research, it is also important to clearly state why some approaches will not work, especially when such approaches are repeatedly suggested. This is the case with mechanical devices that purport to create net forces or antigravity-like effects. **Chapter 6** gives examples of very common devices, discusses why they might appear to be breakthroughs, and explains why they are not. It also offers suggestions for tests to provide convincing evidence of their operation, including methods that are written at a level suitable for the independent researchers who frequently propose similar devices. It is hoped that, by providing this information, other researchers who are asked to review such submissions can more quickly and effectively respond.

Chapter 7 presents the findings of the Yamishita electrogravitational patent, which is another null result that warrants reporting. The patent claim involves coupling electrical charge with a rotating mass to produce a gravitational-like force (which was not observed in these independent tests). The chapter also shows how to inexpensively test such claims in a manner that helps educate students on the methods of scientific inquiry. These experiments were conducted as student projects at the U.S. Air Force Academy, in Colorado Springs, Colorado.

A very common device promoted over the Internet is the "Lifter," which has numerous variants (Biefeld–Brown, Asymmetrical Capacitor Thrusters, electrogravitics, etc.), some of which have existed for more than 80 years. This device involves high-voltage capacitors that create thrust by interacting with the surrounding air. Perhaps because of the ease of their construction and the scarcity of rigorous publications on this phenomenon, many jump to the conclusion that the effect is evidence of "antigravity." **Chapter 8** reports on careful measurements of various capacitor configurations, itemizing the resulting correlations. The final conclusion is that the observed effects are consistent with *corona wind*, which can also be referred to as *ion drift*.

In addition to the comprehensive in-air tests of the previous chapter, **Chapter 9** reports on tests in nitrogen and argon at atmospheric pressure and at various partial vacuums. Several other device geometries are examined also as are thorough variations of polarity and ground connections. The combination of grounding options, geometry, and polarity are found to affect thrust. Because the thrust is found to be inversely proportional to the pressure, these findings support the coronal wind conclusion of Chapter 8. More specifically, this chapter concludes that the thrust is from the charged ions leaking across the capacitor that undergo multiple collisions with air, transferring momentum to neutral air molecules in the process.

Whereas the momentum of a photon in vacuum is well understood, the momentum of a photon passing through dielectric media has generated significant debate, beginning with the Abraham–Minkowski controversy of 1908. In particular, the Abraham formulation of the electromagnetic stress tensor in a dielectric medium predicts a photon momentum that differs from the Minkowski formulation, and experimental tests have not yet been able to resolve which perspective is correct. This controversy has entered the realm of breakthrough spaceflight, with more than one device proposed to create net thrust via mechanisms embodied by this ambiguity (Corum, Brito). **Chapter 10** reviews the underlying physics of photon momentum in dielectric media and the potential propulsion implications. The most critical make-or-break issues are identified, as are the conceptual and operational difficulties associated with known experiments. In short, it is found that both the Abraham and Minkowski formulations match experiments, with the caveat that the assumptions and conventions of each formulation must be applied consistently when analyzing the entire system. Because some of these details are subtle, it is easy to reach misleading conclusions if one errantly uses portions of each formulation to analyze a given problem.

James F. Woodward, a science professor at California State University, has been experimenting with a technique to induce net thrust using a particular interpretation of Mach's principle—a principle that deals with the very definition of an inertial frame. **Chapter 11** reports independent tests of this propulsion concept, comparing it with previous experimental claims. Two devices that were developed and tested by Woodward were independently tested using a sensitive thrust balance developed for field-emission electric thrusters. The results do not seem to be in full agreement with the findings claimed by Woodward and collaborators. Nevertheless, the importance of such a discovery is sufficient to recommend continuing experimentation to reach a complete understanding of the phenomenon. Even if this effect is found not to occur, the issues raised by Woodward's approach offer several provocative questions for deeper investigation.

Quantum electrodynamics theory, whose predicted effects have been verified to 1 part in 10 billion, predicts that the lowest energy state of the electromagnetic field still contains energy that can produce forces between nearby surfaces. These *Casimir* forces have been measured and found to agree with predictions. **Chapter 12** examines whether this quantum vacuum might be exploited to propel a spacecraft. Restrictions resulting from the conservation of energy and momentum are discussed. A propulsion system based on an uncharged, conducting, mirror that vibrates asymmetrically in the vacuum produces real photons that create thrust. Even though the thrust is even less than a photon rocket, this action demonstrates that the vacuum can be used for propulsion. Technological improvements, some of which are proposed, may increase the accelerating force. Many questions remain about the supporting theory, and further experiments are needed to probe questions about the quantum vacuum that are far beyond current theory.

Controversial theories exist in the peer-reviewed literature that assert that inertia is a side effect of accelerated motion through quantum vacuum fluctuations. **Chapter 13** examines the constructs of one of these theories to identify the critical make-or-break issues and opportunities for discriminating tests. Although the specific approach examined in this chapter has several shortcomings, the more general notion of the connection between inertia and quantum fields remains open for deeper investigations.

To open the next section on faster-than-light travel, it is appropriate to have a chapter on Special Relativity, the theoretical tool for describing the consequences of hyper-fast motion through spacetime. Experimental observations that support special relativity are described in **Chapter 14**. The chapter outlines the basic groundwork of special relativity and then proceeds to show the paradoxes created if faster-than-light travel were allowed *within* spacetime. Paradoxes are presented and shown to be primarily concerned with our concept of time. Depending on how *time* is defined, it may be possible to resolve all of these paradoxes. The other caveat to faster-than-light travel is the issue of *spacetime itself*. Although faster-than-light motion is clearly a problem *within* spacetime, the situation is different when toying with *manipulating* spacetime. Those manipulations are the subject of the next chapter.

Whereas Special Relativity forbids faster-than-light travel *within* spacetime (or at least sets interesting constraints upon motion), the situation is different in General Relativity where spacetime itself can be manipulated. **Chapter 15** examines the variety of theoretically postulated faster-than-light schemes (warp drives and wormholes in particular) to identify the energy requirements and make-or-break issues. Many of the related issues are provocative topics in their own right in science, and when viewed from the point of view of hyperfast travel, present interesting approaches toward their resolution. It is shown that many of the energy restrictions commonly imposed are *conventions* rather than physical absolutes. The possibility of testing some elements of theory is raised, in particular the use of extreme electric and magnetic field strengths possible with ultrahigh-intensity lasers.

General Relativity is not the only branch of science with curious issues of faster-than-light phenomenon. Quantum Theory has its own set of provocative topics that are commonly, and sometimes errantly, associated with faster-than-light travel. **Chapter 16** reviews such quantum "nonlocality" effects and articulates numerous experiments and their interpretations. One of the prime complexities is having exacting definitions of what is meant by "entanglement" and "nonlocality" and how these compare to the light-speed constraints of Special Relativity. These complexities are explained in the chapter, and it is shown that, in many situations, there are no conflicts. The chapter also shows that the assumptions that lead to the "no-signal theorems" warrant close inspection when assessing their applicability to some contemporary experiments. Under certain circumstances, situations leading to retrocausal paradoxes can be conceived where the effect precedes the cause. These situations are described, but no resolutions exist at this point.

To open the next section about energy implications, **Chapter 17** reviews the past and projected technology that is based on accrued science. Numerous devices are reviewed, including multiple power sources and conversion methods and estimates of their upper performance limits.

The phenomenon of quantum fluctuations and its high-energy density—as calculated from some estimates—has led some to ponder if the quantum vacuum can be tapped as an energy source. **Chapter 18** examines a span of both theoretical and experimental approaches. It concludes that energy conversions between vacuum fluctuations and tangible effects are possible in principle (albeit small), but that *continuous* energy extraction appears problematic on the basis of the current understanding of quantum electrodynamics theory. Specific experimental approaches are described for investigating these issues directly.

Sonoluminescence, which is light generated from acoustic cavitation in fluids, has been associated with claims of energy production. To provide grounding in this topic, **Chapter 19** covers recent experimental methods and findings. Instrumentation techniques that measure optical, radiation, and thermal properties of the phenomenon of sonoluminescence are described. Initial efforts are directed to the generation and the imaging of sonoluminescence in water and solvents. Evidence of high-energy generation in the modification of thin films from sonoluminescence is seen in heavy water but not seen in light water. The attainable energy output was found to be less than the energy input for the experiments discussed. Improvements for realizing fusion processes and energy harvesting are suggested. This phenomenon also serves as an empirical path to further understand the interaction of sound, light, and fluid physics.

In much the same way that it is important to publish the space drive approaches that produced null findings, **Chapter 20** presents experimental findings of a variety of energy conversion devices purported to produce more energy than they consume. Although such concepts can be summarily dismissed for violating thermodynamic laws, the prospect that there might be genuine *unobvious* energy sources has led many to suggest such devices. This chapter details the experimental methods and experiences derived from testing such claims to help other researchers become alert to the complexities and possible pitfalls of related experiments.

To address challenges of breakthrough spaceflight within the sciences of General Relativity and Quantum Field Theory, computer computational tools are required. Calculations in these disciplines are extensive and involve complex notational conventions. The time to do these calculations manually is prohibitive, as is the risk of inducing transcription errors. **Chapter 21** reviews the main tensor calculus capabilities of the three most advanced and commercially available "symbolic manipulator" tools. It also examines different conventions in tensor calculus and suggests conventions that would be useful for space propulsion research.

Based on lessons from history and the NASA Breakthrough Propulsion Physics Project, suggestions are offered in **Chapter 22** for how to comparatively assess the variety of approaches toward revolutionary spaceflight. Perspectives of prior scientific revolutions are examined in the context of seeking spaceflight revolutions. The intent is to identify techniques to improve the management of such visionary research. Methods tested in the course of the Breakthrough Propulsion Physics Project are provided, including details from its one formal cycle of research solicitations. Key recommendations are to rigorously *contrast* existing foundations of science with the visionary goals of breakthrough spaceflight, and to dissect the grand challenges into more approachable, short-term research objectives.

III. Snapshot of the Book

Following the principles of a "traceability map," as discussed in Chapter 22, Fig. 1 shows a relational depiction of the contents of this book. In the far left column of this map, the relevant disciplines of science are listed. In the far right column, the visionary goals are listed and broken down into categories of different approaches. From there, these categories branch out into numerous concepts and devices. In the columns between the foundations of science and these concepts and devices, there are various unknowns, critical issues, and curious effects.

Because of printing limitations, the connecting lines between these various facets cannot be shown, but many of the connections can be inferred by noting the numbers in the parallelograms attached to the blocks. These numbers refer to the chapters in which the specific items are discussed. From this representation it is obvious that a given concept is connected to many issues and that there are

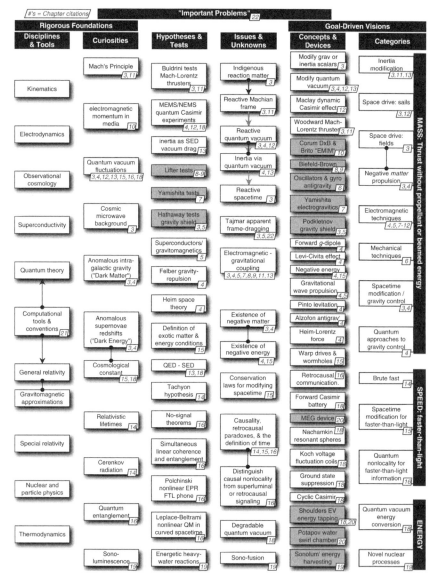

Fig. 1   Relational depiction of this book's contents.

issues that pertain to many concepts. In an expanded version of this map it would be possible to see how all of the items are interconnected. The items having more connections would be those that are more relevant to other facets. Another detail of the figure is that the shaded blocks refer to null tests or to concepts that are shown analytically not to be viable. All of the rest of the subjects are open for further study.

IV. Taking the Next Step

As stated earlier, this book should not be interpreted as the definitive last word on the topic of seeking spaceflight breakthroughs. Instead, it is a primer to get started. A variety of known approaches have been reviewed, including numerous dead ends to avoid. For those researchers who are contemplating investing in this field, we hope that this book provides you with the foundation necessary to make genuine progress.

Marc G. Millis
Eric W. Davis
November 2008

Acknowledgments

THANKS are due to the AIAA Nuclear and Future Flight Propulsion Technical Committee for helping support the development of this book, in particular the participation of the following three authors: Jean-Luc Cambier, Brice N. Cassenti, and Robert H. Frisbee; and for providing the following reviewers: Dana Andrews, Brice Cassenti, Allen Goff, Jochem Hauser, and Pavlos Mikellides. Thanks are due to all the other authors as well; Gary L. Bennett, Nembo Buldrini, Francis X. Canning, John G. Cramer, Timothy J. Drummond, Gustav C. Fralick, Paul A. Gilster, Nancy R. Hall, George D. Hathaway, Michael R. LaPointe, Timothy J. Lawrence, Scott R. Little, G. Jordan Maclay, Claudio Maccone, Paul B. Miller, William H. Miller, H. E. Puthoff, Kenneth E. Siegenthaler, Martin Tajmar, John D. Wrbanek, and Susan Y. Wrbanek; and my co-editor and contributing author, Eric W. Davis.

And what publication is possible without the help of reviewers for their time and insights? Thanks are owed to the following additional reviewers: Peter Badzey, Gary Bennett, Dan Cole, John Cole, Steve Cooke, Roger Dyson, Ron Evans, Larry Ford, Peter Garretson, Bernard Haisch, Nancy Hall, Don Howard, Andrew Ketsdever, Geoff Landis, Mike Lapointe, Carolyn Mercer, William Meyer, William Miller, H. E. Puthoff, Alfonso Rueda, William Saylor, Roy Sullivan, Alexandre Szames, Martin Tajmar, Jeff Wilson, Clive Woods, Jim Woodward, and Ed Zampino.

Marc G. Millis
Eric W. Davis
November 2008

Chapter 1

Recent History of Breakthrough Propulsion Studies

Paul A. Gilster*
Tau Zero Foundation, Raleigh, North Carolina

I. Introduction

THE knowledge that it is impossible to reach even the nearest stars in a single human lifetime using chemical propulsion technologies has launched an active search for alternatives. Interstellar flight is not beyond reach, as Robert Forward understood, a fact that prompted the researcher to conduct a series of studies advocating propulsion methods that worked within the known laws of physics, such as laser-pushed "lightsails" and rockets driven by antimatter reactions. But beginning in the early 1990s, a second track of studies into what can be called "breakthrough propulsion" began to emerge in workshops, papers, and organizational efforts within NASA [Breakthrough Propulsion Physics (BPP) Project], ESA, and private industry (e.g., BAE Systems Project Greenglow). Hoping to uncover new physical principles of motion and energy, all have energized research while, as is common when working at the frontiers of knowledge, encountering ideas that remain controversial.

This chapter examines the recent history of advanced propulsion as a discipline. Its goal is to place the contents of this book in the context of the broader progression of science and technology. The ideas under investigation are remarkable in their range, and engage the community to establish new methodologies for dealing with theoretical and experimental work at the edge of known physics. They include so-called "warp drive" theories involving the manipulation of space-time itself, as well as wormhole concepts that, like the warp drive, offer fast transit without violating Einstein's Special Relativity. Also under study: vacuum fluctuation energy, quantum tunneling, and the coup-

---

Copyright © 2008 by the American Institute of Aeronautics and Astronautics, Inc. All rights reserved.
*Lead journalist.

ling of gravity and electromagnetism. While the theoretical literature spawned by these concepts has grown, so has the appreciation that their study must occur within the rigorous constraints of conventional physics while remaining open to experimental results that may extend our understanding of those principles. Thus breakthrough propulsion research requires a balancing act, insisting upon scientific rigor while examining concepts that are all too easily sensationalized and misunderstood.

The need for continuing review of recent work in this field becomes apparent when surveying the resources available to researchers. In the late 1970s and early 1980s, Robert Forward and astronautical engineer Eugene Mallove published several bibliographies of interstellar studies in the pages of the *Journal of the British Interplanetary Society*. Their 1980 effort demonstrates the range of work being done on these topics even then, including 2700 items filed under seventy subject categories. Forward and Mallove eventually had to abandon their updates due to the labor involved; unfortunately, this valuable service has not been resumed by others and remains unaddressed. Regarding the smaller topic of emerging propulsion physics literature, no such bibliography has yet been initiated. In the absence of such a resource, we can begin by surveying the growth of these studies over the past two decades in the hope of more methodical updates to come. A précis attempting an overview of concepts now under active investigation and noting those likely to be discarded follows.

II. Challenging Initial Stages

The broader questions remain: Has enough progress been made from which starting points for meaningful research exist? And are studies that seek breakthrough physics for propulsion justified if they seem to violate commonly accepted principles of physics such as conservation of momentum? As will be shown in this historical overview and the content of subsequent chapters, enough progress has been made to provide starting points for deeper research. It is too soon, however, to predict whether any breakthroughs actually exist to be discovered. Regarding the issue of possibly violating accepted physics, the real challenge is to find those approaches that do not violate well-established physical laws while also rigorously challenging provisional hypotheses or extending research to where it has not yet been taken. In other words, this field of study does not aim to violate physics, but rather to further extend physics.

An issue related to the appearance of violating physics is whether there is value in pursuing breakthroughs that might not exist. Consider first the value of curiosity-driven physics, for which no application goals guide the pursuit. The value is in what is learned along the way. So too is the case of pursuing breakthroughs that might not exist—the minimum value is what is learned along the way. Granted, if the research does yield the desired breakthroughs the value will be still higher, but even the pessimistic outcome presents the same learning opportunities as research that is unbound by application goals. This caveat remains: For the research to be of value it must adhere to the same high standards as curiosity-driven physics. Unfortunately, given the allure of the grand goal of star flight, where the stakes are higher for humanity, it is common to encounter sensationalistic work where premature claims are made without any rigor to back them up. From the statistics compiled in the course of NASA's BPP Project [1], roughly 10% of the correspondence came from proponents claiming breakthroughs with no credible evidence. This behavior can taint the topic and amplify the concerns of other professionals. Again, to make genuine progress, the emphasis is on the rigor and reliability of the research, rather than on the magnitude of the claims. With that priority understood, this topic does offer opportunities for learning more about the workings of our Universe.

Recent developments, particularly those relating to the presence of unseen dark matter and the dark energy force driving the accelerating expansion of the universe, illustrate how engineering applications ultimately fall back upon theoretical physics for their foundation. But a focus on applications can clarify theory. Seen through the lens of spaceflight, the anomalous galactic rotations that led to the inference of dark matter become significant not only because of their pertinence toward understanding the origin and fate of the universe, but also because they suggest the possibility of an indigenous reaction mass in interstellar space. Dark energy, discovered through anomalous recessional redshifts, raises the question of manipulating gravity. Thus the definition of a problem, the essential first step in scientific methodology, provokes different lines of inquiry depending upon the objective of the study being performed. Multiple approaches raise the chances of resolving underlying issues in physics.

III. Reflecting on a Time Line

To help convey the stage within which this work is set, Table 1 presents a time line of events. The left side of the table lists general advances in physics and books about scientific progress, while the right side lists publications and events that are focused more specifically on practical interstellar flight. A key point is that, while general science continues to evolve and reflect on its progress, the earliest works pointing toward new space propulsion science are just beginning to emerge. There are no publications indicating certain breakthroughs yet, but several theoretical constructs are introduced. Also, the time line reflects the fact that surveys of the options and initial research tasks have commenced. The topic is not, however, at the point where progress is widespread or focal approaches have emerged. The discipline of breakthrough propulsion is still in the early stages of emergence, ripe with opportunities for conducting pioneering work.

It is interesting to note how much time passed before an international conference on General Relativity took place after the appearance of Einstein's theory. It is also interesting to note the timing of the publication of books reflecting on the progress of science and technology, in particular, Kuhn [2], Foster [3], Horgan [4], and Dyson [5].

IV. Advanced Propulsion Concepts Emerge

Although speculations on space travel and the demands of interstellar flight stretch back to the Renaissance and even earlier, the modern study of advanced propulsion may be said to have begun with a lecture given by Eugen Sänger at the Fourth International Astronautical Congress [6]. Sänger's investigations into

Table 1 Time line of pertinent events

General science events	Year	Propulsion goal-specific events
Einstein's General Relativity Theory	1916	
Term "quantum mechanics" coined by Max Born	1924	
Schrödinger develops wave mechanics	1926	
Heisenberg's Uncertainty Principle	1927	
Quantum electrodynamics	1947	Ackeret, relativistic rocket analysis
De Broglie, *The Revolution in Physics*	1953	Sänger, photon rockets studied
	1954	
First international conference on General Relativity	1955	
	1956	
Bondi, "Negative Mass in General Relativity"	1957	
Bohr, *Atomic Physics and Human Knowledge*	1958	
Heisenberg, *Physics and Philosophy*		
	1959	
Maiman develops the first practical laser	1960	
Aharonov-Bohm effect observed	1961	
Kuhn, *Structure of Scientific Revolutions* (paradigm shifts)	1962	
	1963	Forward, "Guidelines to Antigravity"
	1964	
Cosmic microwave background radiation observed	1965	
	1966	
	1967	
	1968	
John Wheeler coins the term "black hole"	1969	
String Theory born	1970	British Aircraft Corporation's Future Concepts Group
	1971	
	1972	

1973	Misner, Thorne, Wheeler, *Gravitation*
1974	
1975	
1976	
1977	
1978	
1979	
1980	
1981	Aspect, experimental test of Bell's inequalities
1982	
1983	
1984	
1985	Foster, *Innovation: The Attacker's Advantage*
1986	
1987	
1988	Morris and Thorne on traversable wormholes; Forward studies negative matter propulsion; Forward, *Future Magic*
1989	Barbour, Mach, "Absolute or Relative Motion"; Matloff, *The Starflight Handbook*; BAe Gravity Control Workshop
1990	NASA GRC Vision-21 Workshop No.1
1991	U.S. AFRL: Tally, on Biefeld-Brown effect
1992	Podkletnov claims "gravity shield"
1993	Milloni, *The Quantum Vacuum*
1994	Alcubierre on warp drive; Haisch studies inertia as ZPE drag; Woodward, propulsion patent no.1; NASA JPL Workshop on quantum and FTL travel

(continued)

Table 1 Time line of pertinent events (continued)

General science events	Year	Propulsion goal-specific events
	1995	Visser, *Lorentzian Wormholes*
		Cramer gives visual clues to search for wormholes
		Krauss, *Physics of Star Trek*
Horgan, *The End of Science*	1996	Millis provides space drive problem definition
Dyson, *Imagined Worlds* (new tools lead to revolutionary discoveries)	1997	NASA, Breakthrough Propulsion Physics (BPP) Project workshop
	1998	Krasnikov tubes
Discovery of accelerated expansion	1999	NASA BPP research solicitation
	2000	
	2001	NASA BPP Project findings presented at Joint Propulsion Conference
	2002	Puthoff on the polarizable vacuum and propulsion (*Journal of the British Interplanetary Society*)
	2003	Hathaway dismisses Podkletnov
	2004	Europen Space agency's Gravity Control study
		Gilster, *Centauri Dreams*
		NASA BPP Project contractor reports published
		Maclay and Forward, gedanken ZPE propulsion
	2005	Matloff, *Deep Space Probes*
	2006	NASA BPP Project final findings published
	2007	BIS warp drive symposium

antimatter rocketry led to a so-called "photon rocket" design that would use the gamma rays produced by the annihilation of electrons and positrons for thrust. Although the idea foundered on the inability to direct the gamma ray exhaust stream, the work led to another Sänger study of interstellar propulsion possibilities in 1956 [7]. Other significant papers of the same era include those of Ackeret [8], von Hoerner [9], and Shepherd [10], all of which examined the issues raised by relativistic rocket flight.

The Air Force Office of Scientific Research began a series of symposia in 1957 that investigated the status of basic and applied research into advanced propulsion. These gatherings, held every two to three years and lasting until the 1970s, resulted in published proceedings whose papers addressed breakthrough propulsion concepts considered futuristic but within the current paradigm of physics. The year 1975 would see NASA's earliest research into these concepts and the publication of its ensuing technical report [11]. A major figure in early studies was Robert Forward, who researched advanced propulsion while working at the Hughes Research Laboratory in Malibu, California. Much of Forward's work was published in the form of research study reports for the Air Force under contract to Franklin B. Mead Jr. at the Propulsion Directorate of the Air Force Research Laboratory at Edwards Air Force Base, California. Forward's popular book, *Future Magic: How Today's Science Fiction Will Become Tomorrow's Reality*, reviewing the breakthrough propulsion concepts he had examined for interstellar flight, appeared in 1988, bringing these studies to a general audience [12].

V. Vision-21 and Its Aftermath

One recent attempt to reassess the prospects for significant advances in spaceflight methods was Vision-21, a volunteer activity within NASA's Lewis Research Center (now Glenn Research Center) in Cleveland, Ohio. While individual researchers studied such questions, largely in their spare time and through occasional workshops with formal support [13–17] an organizational impetus began to emerge at a symposium held at this NASA center in 1990, dubbed "Vision-21: Space Travel for the Next Millennium." Roughly half the contributed papers focused on propulsion, with all papers (other than keynote addresses) presented in poster format to maximize the interaction between speakers and audience [18]. Contributors were charged with examining possible technologies as they might emerge, not just within decades but over the course of the next 1000 years. Although some ideas we would now consider under the heading of breakthrough propulsion did appear both in papers and panel discussion, including the possibility of space-coupling (reacting with spacetime itself to produce propulsive forces) [19], the bulk of these papers reflected current thinking in physics taken to its engineering extreme through concepts such as laser-pushed lightsails, fusion, and antimatter propulsion.

The significance of the conference, however, lay in the scientists behind it. Working under the Vision-21 rubric, a group of researchers at NASA Lewis had met regularly for several years before the meeting, and were encouraged to develop ideas outside NASA's current technological sphere. These scientists and engineers developed methodologies that would later coalesce in the BPP

Project. But in 1990, the arrival of almost 200 guests from other NASA centers as well as universities, research facilities, and businesses flagged the growing interest in building a research process that could tackle concepts at the very edge of physics. These were ideas too speculative to trigger official sponsorship, and thus were in need of an alternate track that could proceed at least informally.

VI. Wormholes and Warp Drive Physics

Vision-21 occurred in the interval between two publications that galvanized the nascent field and suggested fundamentally new areas for investigation. Kip Thorne and (then) graduate students Michael Morris and Uli Yurtsever proposed a novel means of fast transit through engineered wormholes that could be developed by sufficiently advanced technological cultures [20]. The wormhole acts as a hyperspace tunnel through which travelers pass, with no need for faster than light travel because the wormhole connects different regions within the universe, different universes, or even different times. To create and maintain the wormhole geometry, a wormhole with a one-meter radius required an amount of negative energy comparable to the mass of the planet Jupiter [21,22]. Continuing wormhole research examines the stability of such "shortcuts" through spacetime and the power required to keep wormholes open. But one effect of the early papers of Thorne et al. was to put into high relief the understanding that fast interstellar travel would demand propulsion strategies drawn from physical principles only now being investigated.

Breakthrough propulsion began to emerge as a discipline in its own right, given further impetus by the work of theoretical physicist Miguel Alcubierre. Noting that spacetime itself was not bound to the same speed of light constraint as objects within it (a nod to inflationary theories of the early cosmos), the latter argued that the problems of special relativity (and inertia) could be resolved by modifying spacetime so that its volume expanded behind the starship while being compressed in front of it. The vessel itself, moving in a "bubble" of normal space and never surpassing the speed of light, would be carried to its destination, much like a surfer on a wave [23]. With Alcubierre's 1994 paper in place, the stage was set for the broader institutional investigation of issues that had until recently seemed like nothing more than escapist television fare. The physicist's words on the matter set the tone. Describing the effects on ship and crew as, never exceeding the speed of light, they are taken to their destination: "The spaceship will then be able to travel much faster than the speed of light. However, as we have seen, it will always remain on a timelike trajectory, that is, inside its local light-cone: Light itself is also being pushed by the distortion of spacetime. A propulsion mechanism based on such a local distortion of spacetime just begs to be given the familiar name of the 'warp drive' of science fiction" [23]. The science fiction connection has continued to bedevil breakthrough propulsion studies since, though it must be noted that the genre's uses in promoting unfettered looks at imaginative ideas have captivated more than a few physicists. Although rarely accurate at technical prediction, science fiction is known to have inspired careers and even particular research paths [24].

Space drive studies now began to take form. Sergei Krasnikov, seeing that the interior of the Alcubierre "bubble" would be causally disconnected from its exterior, introduced an alternative strategy, the creation of a negative energy "tunnel" at sublight speeds. Reaching their destination, the crew could return through the same tunnel to the point of origin at speeds faster than light [25]. Such emerging ideas grew within a broader context; i.e., the understanding that gravity and electromagnetism are related phenomena within a coupled spacetime [26]. This being the case, the possibility of manipulating gravity or inertia through electromagnetism comes into play [27]. Energy demands placed many such concepts in deep theoretical waters. Both the Thorne and Alcubierre investigations showed that manipulation of spacetime required matter with a negative energy density. The Alcubierre starship would require positive energy to contract spacetime in front of the vessel and negative energy to expand it behind. As it is unknown whether negative energy densities can exist, propulsion studies could only focus on the theoretical, with alarming results. Michael Pfenning and Larry Ford were soon to calculate that the Alcubierre drive apparently demanded more energy than was available in the entire universe to make it function [28]. Subsequent work would reduce that energy requirement considerably [29] but massive amounts of negative energy were still demanded, not only by the Alcubierre warp drive concept but also by wormhole physics, the wormhole's opening drawing on its effects to remain stable. The most current estimates of the energy required for such notions are presented in Chapter 15.

VII. Quantum Tunneling at FTL Speeds

If the warp drive triggered a small burst of media interest in breakthrough propulsion, the attention of scientists and lay people alike was also drawn to the odd phenomenon of photons tunneling through barriers at velocities apparently higher than the speed of light in a vacuum. Experiments conducted in the early 1990s by Aephraim M. Steinberg et al. detected photons tunneling through a mirror barrier at 1.7 times the speed of light [30]. The researchers were quick to note that their results did not violate causality because the effect could not be used to carry information, but explanations of wave packet behavior could not constrain a growing interest in the phenomenon, which would be the subject of subsequent investigations (to be discussed later) by both NASA and the European Space Agency (ESA). Chapter 16 addresses these issues directly.

VIII. Podkletnov "Gravity Shield"

Public interest was also fired by the possibility of manipulating the force of gravity, the ultimate prize being an "antigravity" effect that might produce new spacecraft designs, not to mention radically transforming all other forms of transportation. Two years before Alcubierre's paper appeared, E. Podkletnov and R. Nieminen published the results of their work with a rotating superconductor disk [31]. Podkletnov's disk (made of yttrium barium copper oxide, or YBCO) was cooled to superconducting temperatures by liquid helium. It was then levitated by a magnetic field, with electric current supplied to coils surrounding the disk, and rotated between 3000 and 5000 revolutions per minute. Podkletnov asserted that he had discovered a gravity shielding effect that accounted for a two percent weight reduction in small objects held above the spinning disk. Naively, such an effect would seem to violate conservation of energy, but the apparatus did require an energy input. Podkletnov continued his work, producing a 1996 follow-up paper that was leaked before publication to the *British Sunday Telegraph*, causing unfortunate claims about "... the world's first antigravity device" and a flurry of tabloid-style speculations in the press.

Before Podkletnov's experiments, Ning Li and Douglas Torr had investigated antigravity effects, predicting in 1991 that superconductors could play a role in producing them [32]. Li and Torr were influential in convincing NASA's Marshall Space Flight Center in Huntsville, Alabama to begin its own study, which would attempt to recreate Podkletnov's configuration. The project, which ran from 1995 to 2002, was unable to complete the needed test hardware with the resources then available [33]. The experimental configuration was, however, reproduced in a privately funded test, with results published in 2003. Using equipment 50 times more sensitive than had been available to Podkletnov, the group found no evidence of a gravity-like force [34]. A subsequent Podkletnov claim, that a force beam could be created from the application of high-voltage discharges near superconductors, appeared on the *arXiv* physics preprint site on the Internet in 2001 but remains unsubstantiated by independent researchers [35]. Nonetheless, the publicity attendant to Podkletnov's work infused the mid-1990s with both energy and serious skepticism.

IX. Considering the Quantum Vacuum

A 1994 workshop held at the Jet Propulsion Laboratory shortly before the appearance of the Alcubierre warp drive paper examined the topic of faster-than-light travel from a range of perspectives, noting the experimental approaches that might be used to study the subject [36]. These included resolving the question of the neutrino's rest mass to determine whether or not evidence of imaginary mass existed. Another approach was to measure the speed of light inside the space between closely spaced conductive plates (a "Casimir cavity") to look for evidence of negative energy. Other possibilities were to study cosmic rays above the atmosphere to look for behavior characteristic of tachyons, and to search for astronomical evidence of wormholes, which should show a distinctive visual signature caused by the distortions indicative of negative mass at the entrance [37].

The informal workshop noted a key issue of faster-than-light travel, the apparent violation of causality, concluding that while such violations would be unavoidable in this context, their physical prohibition could not be established. Also, new theories were emerging on the connection between gravity and electromagnetism, already known to be coupled phenomena. Gravity's effects upon spacetime have been confirmed, mass being known to warp the spacetime against which electromagnetism is measured. In 1968, Andrei Sakharov had raised the possibility that gravity could be an induced effect related to changes in the vacuum energy that pervades the universe [38]. A theory published by Bernard Haisch et al. early in 1994 suggested that inertia results from the resistance of a particle's velocity to change as it moves through this vacuum energy background (dubbed zero-point energy or ZPE), whose electromagnetic oscillations remain in the vacuum after all other energy has been removed [39]. Subsequent NASA-sponsored research (1996–1999) on the cosmological implications of quantum vacuum energy produced five papers, two examining the emerging view of inertia [40,41]. A subsequent paper looked at the possibility of engineering such effects [42]. Chapter 13 examines the quantum-vacuum-inertia hypothesis in detail.

X. Institute for Advanced Studies at Austin

In addition to academic and government-sponsored work, privately sponsored organizations also investigate such topics. The Institute for Advanced Studies at Austin, Texas, established in 1985 by H. E. Puthoff, continues to examine ZPE's implications through work on gravitation and the nature of inertia, while probing issues of energy generation and cosmology. The broader theory behind this work sees all matter as reliant on the existence of the ZPE field, with gravity as a necessary consequence. Max Planck first introduced a term for ZPE into the equations for blackbody spectral radiation in 1912. Einstein would recognize the physical reality of ZPE the following year, while Milliken's experiments in 1924 provided spectroscopic evidence (see Chapter 18 for more on the background of ZPE studies). Since then, it has been the subject of investigations into the Casimir effect, the Lamb shift, photon noise in lasers, van der Waal forces, spontaneous atomic emission, and the stability of the ground state of the hydrogen atom from radiative collapse. Active debate continues between cosmologists and quantum field theorists over the energy density of the ZPE, there being no accepted bridge between quantum theory and spacetime physics.

With these subjects in the air, the factors that would lead to the establishment of NASA's BPP Project were now in play. They drew as well on earlier work conducted by the U.S. Air Force and other agencies, which had already examined numerous propulsion concepts that stayed within the realm of classical physics, including nuclear rocketry, laser-beamed lightsails, laser rockets, and antimatter propulsion, but also considered more speculative topics [43–47]. Robert Forward's continuing investigations under Air Force Research Laboratory contracts and in-house Hughes Research Laboratory programs ranged from antigravity studies to antimatter, leading to his keynote address at the Vision-21 conference and providing inspiration to numerous researchers going forward. Seven years of academic research culminated in Matt Visser's graduate-level textbook on wormhole theory [21], while physicists in New York attended a conference called "Practical Robotic Interstellar Flight: Are We Ready?" convened by Edward Belbruno at New York University [48].

XI. Breakthrough Propulsion Physics Project

The year 1996 saw NASA's decision to create a comprehensive strategy for advancing the state of the art in space propulsion. Appearing under the rubric Advanced Space Transportation Program (ASTP), the new plan became the charge of the Marshall Space Flight Center (MSFC), which was given the task of making the new 25-year plan more visionary than previous attempts. ASTP was to examine propulsion concepts ranging from near-term concepts to genuinely revolutionary ideas that could enable interstellar travel. The latter could clearly draw on an established context at NASA's Lewis Research Center, where individuals had pursued such topics through informal sessions and the Vision-21 Conference that grew out of them. Accordingly, MSFC charged the Lewis Center with the task of developing the BPP Project to address those issues at the most visionary end of the ASTP scale. Rather than seeking refinements to existing methods, the BPP Project would specifically target propulsion breakthroughs from physics. Its objective: Near-term, credible, and measurable progress in attaining these results.[†]

Under the leadership of Marc Millis, the BPP Project set about its mission noting the historical pattern common to technological revolutions. When steam ships displaced sail, an existing technology had reached the limits of its physical principles, underlining the need for new methods. The BPP Project operated under the assumption that chemical and solid rocket technologies, as was the case with aircraft supplanting ground transport, were approaching their performance limits. Intended to be application-oriented, the project would tie its discoveries to the unique problem of spaceflight. As opposed to broader and more theoretical research, it would present different lines of inquiry related to the physics of propulsion. The program chose a "Horizon Mission Methodology" that set the "impossible" goal of interstellar travel as its focus, hoping to spur thinking beyond existing solutions into the realm of entirely new possibilities [49].

The major barriers to interstellar travel were quickly identified in the form of three "grand challenges":

1) *Mass:* Discover fundamentally new propulsion methods that either eliminate or drastically reduce the need for propellant. Such methods might create propulsive forces through the manipulation of inertia or gravity, or they might interact with the properties of space itself, or mine the interactions between matter, energy, and spacetime.

2) *Speed:* Discover how to achieve transit speeds that could drastically reduce travel times. This goal entailed looking for ways to move a vehicle at or near the maximum limits for motion through spacetime, or to exploit the motion of spacetime itself (warp drive theory).

3) *Energy:* Discover ways to generate the massive amounts of onboard energy that theory suggested would be required by propulsion systems like these. Propulsion goals were thus seen as closely linked to energy physics.

BPP was to be implemented according to a series of conditional steps, each contingent upon the completion of the previous step. These milestones were organized as follows [50]:

1) *1996:* Determine that sufficient scientific foundations exist to pursue the objective (completed).

2) *1997:* Determine that affordable research candidates exist (completed).

[†]The early history of the project is recounted in a thesis by Szames, A.D., "Histoire et enjeux du projet NASA Breakthrough Propulsion Physics [History and stakes of NASA's BPP Project]," (unpublished). Diplôme d'Etudes Approfondies (DEA) Science Technologie Société, Paris: Conservatoire National des Arts & Métiers (CNAM), 1999.

RECENT HISTORY OF BREAKTHROUGH PROPULSION STUDIES 13

3) *1998:* Devise means to prioritize and select research tasks (completed).
4) *1999/2000:* Solicit and select first round of research tasks (completed).
5) *2001/2002:* Streamline the review and selection process via a research consortium, to be operated by the Ohio Aerospace Institute (terminated when funding was cut).

A. Kickoff Workshops

Because it could not be known whether breakthroughs like these were possible, the BPP Project promised only incremental progress toward achieving them. Using a system of collaborative networking among individual researchers—one connecting the various NASA centers, government laboratories, universities, industries, and individual scientists—the program hoped to create a multidisciplinary team open to collaboration and collateral support. Although the Internet was seen as a primary communications medium (and supporting Web sites established), a brainstorming session in Austin, Texas took place in February of 1997. The program's major event would be a workshop on propulsion breakthroughs held on 12–14 August 1997 in Cleveland, Ohio. Fourteen invited presentations, 30 poster papers, and a series of parallel breakout sessions were designed to generate a list of candidate research tasks, with participants asked to entertain for the duration of the event that the breakthroughs sought were indeed achievable within the context of sound and tangible research approaches. The stated goal: Credible progress toward incredible possibilities [51]. Between the Austin and Cleveland meetings, some 128 ideas on research approaches, many of them redundant, were generated.

Below is a list of the invited presentations in the order presented. References are cited to related or equivalent works:

1) L. Krauss (Case Western Reserve University), "Propellantless Propulsion: The Most Inefficient Way to Fly?" [52]

2) H. E. Puthoff (Institute for Advanced Studies at Austin), "Can the Vacuum Be Engineered for Spaceflight Applications?: Overview of Theory and Experiments" [39,53]

3) R. Chiao (University of California at Berkeley) and A. Steinberg, "Quantum Optical Studies of Tunneling Times and Superluminality" [54]

4) J. Cramer (University of Washington), "Quantum Nonlocality and Possible Superluminal Effects" [55]

5) R. Koczor and D. Noever (Marshall Space Flight Center), "Experiments on the Possible Interaction of Rotating Type II YBCO Ceramic Superconductors and the Local Gravity Field" [33]

6) R. Forward (Forward Unlimited), "Apparent Endless Extraction of Energy from the Vacuum by Cyclic Manipulation of Casimir Cavity Dimensions" [56]

7) B. Haisch (Lockheed) and A. Rueda, "The Zero-Point Field and the NASA Challenge to Create the Space Drive" [39]

8) A. Rueda (California State University) and B. Haisch, "Inertial Mass as Reaction of the Vacuum to Accelerated Motion" [39]

9) D. Cole (IBM Microelectronics), "Calculations on Electromagnetic Zero-Point Contributions to Mass and Perspectives" [57]

10) P. Milonni (Los Alamos), "Casimir Effect: Evidence and Implications" [58]

11) H. Yilmaz (Electro-Optics Technical Center), "The New Theory of Gravitation and the Fifth Test" [59]

12) A. Kheyfets (North Carolina State University) and W. Miller, "Hyper-Fast Interstellar Travel via Modification of Spacetime Geometry" [23,28]

13) F. Tipler III (Tulane University), "Ultrarelativistic Rockets and the Ultimate Future of the Universe" [60]

14) G. Miley (University of Illinois), "Possible Evidence of Anomalous Energy Effects in H/D-Loaded Solids—Low Energy Nuclear Reactions"

Following these presentations, participants divided into six breakout groups, with each of the three "grand challenge" goals addressed by two of the six. From the ensuing discussions, ideas for next-step research tasks were solicited focusing on strategies that would be of short-duration, low-cost, and incremental in their nature.

B. BPP Project Research Tasks

Although the Cleveland conference was to be the BPP Project's first and only workshop, work proceeded actively in other venues. In its short life, the program sponsored five tasks through competitive selection, two in-house tasks and one minor grant. Sixteen peer-reviewed journal articles resulted [61,62]. Summaries of the eight tasks appear below:

1) *Define Space Drive Strategy:* What research paths lead to the creation of a space drive, defined as a device creating propulsion without a propellant? The vehicle must exert external net forces on itself while satisfying conservation of momentum. This task conceived and assessed seven hypothetical space drives and analyzed remaining research approaches including 1) investigating new sources of reaction mass in space; 2) revisiting Mach's principle to consider coupling to surrounding mass via inertial frames; and 3) studying the coupling between gravity, inertia, and controllable electromagnetic phenomena [27]. (Chapter 3 is a follow-on to this initial work.)

2) *Test Schlicher Thruster:* Experiments conducted in-house were performed to investigate claims by Rex Schlicher et al. [62] that a specially terminated coax could create more thrust than could be attributed to photon pressure. No such thrust was observed [63].

3) *Assess Deep Dirac Energy:* Examined the question of additional energy levels and energy transitions in atomic structures [64]. A theoretical assessment demonstrated that some of these states were not possible, but the topic is not considered fully resolved.

4) *Cavendish Test of Superconductor Claims:* This work was performed as a lower-cost way to examine claims of gravity shielding advanced by Podkletnov and Niemenen [31]. Not intended as a replication of Podkletnov's experiments, it was rather designed to test related consequences. Radio frequency radiation used in these experiments was found to couple too strongly to supporting instrumentation to produce a clear resolution of any gravitational effect [65]. No follow-up to

this approach is expected. (Chapter 5 discusses the experimental challenges in general with superconductors and provides a brief history of related work.)

5) *Test Woodward's Transient Inertia:* Experiments were conducted to test James Woodward's claim that transient changes to inertia can be induced by electromagnetic means, with possible relation to Mach's principle [66–67]. The test program was changed when later Woodward publications defined a smaller effect than originally anticipated. No discernible effect appeared in the revised testing. Woodward's experiments and investigations of their theoretical background continue and the issue of transient inertia is considered unresolved. (Chapter 11 discusses subsequent independent testing.)

6) *Test EM Torsion Theory:* These experiments studied a torsion analogy to the coupling between electromagnetism and spacetime [68]. The experiments were later found to be insufficient to resolve the issue correctly, and the possibility of asymmetric interactions that might prove valuable in propulsion [69] is considered unresolved [70].

7) *Explore Quantum Tunneling:* Studied the question of faster-than-light information transfer, as raised by the phenomenon of quantum tunneling, where signals appear to pass through barriers at superluminal speed [71]. The information transfer rate was found to be only apparently superluminal (the entire signal remains light-speed limited), producing no causality violations. The experimental and theoretical work sponsored on this subject examined the special case where energy is added to the barrier. (Chapter 16 examines the faster-than-light implications of quantum mechanics.)

8) *Explore Vacuum Energy:* The quantum vacuum energy, also known as ZPE, draws on Heisenberg's Uncertainty Principle, which implies that an absolute zero electromagnetic energy state is not possible in the space vacuum. The manifestation of such "vacuum energy" in the so-called "Casimir effect" is known to pull two parallel plates together at dimensions on the micron scale. BPP Project-sponsored investigations used micro-electro-mechanical (MEM) rectangular Casimir cavities, showing the possibility of creating net propulsive forces using this energy, although these forces are exceedingly small [72]. The quantum vacuum continues to be the subject of active research into the nature of space itself. (Chapter 12 specifically addresses the notion of using the quantum vacuum as a reactive medium for space propulsion.)

The BPP Project ran for seven years and the funding devoted to this effort was $1.6 million in total over those seven years [50]. In October of 2002, the Advanced Space Transportation Program was reorganized, with all research deemed at less than Technology Readiness Level 3 (at which active research and development including laboratory studies is initiated) being canceled. Thus ended the BPP Project. It was in the very next month that the final report of the President's Commission on the Future of the U.S. Aerospace Industry, which had been charged to assess the health of the aerospace industry in the United States, appeared. Chapter 9 of that report presented a justification for the kind of research that the BPP Project had represented: "In the nearer-term, nuclear fission and plasma sources should be actively pursued for space applications. In the longer-term, breakthrough energy sources that go beyond our current understanding of physical laws, such as nuclear fusion and anti-matter, must be credibly investigated in order for us to practically pursue human exploration of the solar system and beyond. These energy sources should be the topic of a focused basic research effort . . ." [73]. The same report also emphasized the need to explore energy sources such as ZPE.

C. BPP Assessment: Nonviable Concepts

Credible investigation, of course, identifies not only research avenues to be pursued, but those that are unlikely to prove fruitful for energy generation or propulsive effects. Drawing on the BPP Project's research tasks and subsequent investigations, a recent paper clarified those approaches deemed to be nonviable [74]. These include:

1) *Quantum Tunneling as a Faster-than-Light Venue:* Because the effect is only apparent (the entire signal still being limited by the speed of light even if its leading edge appears to exceed that speed), this phenomenon does not demonstrate faster-than-light information transfer and therefore has no broader implications for faster-than-light travel.

2) *Asymmetric Capacitor Lifters (Biefeld-Brown Effect):* Involving the creation of thrust by high-voltage capacitors, the effect (sometimes discussed under the rubric "electrogravitics" or "electrostatic antigravity" was found to be the result of coronal wind [75]. Chapters 8 and 9 describe experimental tests of these devices in more detail, showing that asymmetric capacitor lifters operate according to known physics rather than being evidence of new gravitational physics, as is often claimed.

3) *Hooper Antigravity Coils:* Although said to reduce the weight of objects placed beneath them, these self-canceling electromagnetic coils (based on U.S. Patent 3,610,971 by W. J. Hooper) were shown to produce no such effect [76].

4) *Oscillation Thrusters:* These are mechanical devices claimed to produce a net thrust using the motion of internal components. The apparent thrust seen in such devices is a misinterpretation of mechanical effects. A gyroscopic thruster, for example, can move in a way that seems to defy gravity, but its motion is actually caused by gyroscopic precession, the forces involved being torques around the axes of the gyroscope mounts. No net thrust is thus created. (Chapter 6 describes these forces and how they are commonly misinterpreted.)

5) *Podkletnov Gravity Shield:* Privately funded replication of Podkletnov's experimental configuration has demonstrated no gravity shielding effect over spinning superconductors [34].

The BPP Project's research represents one part of a larger contribution, which is the demonstration that breakthrough physics related to specific space-related tasks can be productively studied. An independent review panel of the Space Transportation Research Program reached this conclusion in 1999, stating:

> The BPP approach was unanimously judged to be well thought out, logically structured and carefully designed to steer clear of the undesirable fringe claims that are widespread on the Internet. The claim that the timing is ripe for a modest Project of this sort was agreed to be justified: Clues do

appear to be emerging within mainstream science of promising new areas of investigations. The team concurred that the 1997 BPP kickoff workshop did identify affordable candidate research tasks which can be respectably pursued, and that the proposed research prioritization criteria were a valid way to select from amongst these (and future) proposals. The Project approach was deemed to be sound: emphasizing near-term progress toward long-term goals; supporting a diversity of approaches having a credible scientific basis; holding workshops to exchange ideas; solicit constructive criticism and assess progress; aiming toward testable concepts [77].

XII. Project Greenglow

While the BPP Project proceeded, a variety of other research projects emerged, each looking at advanced propulsion capabilities from its own angle. Prominent among these investigations is Project Greenglow, whose antecedents actually pre-date the BPP Project by a decade with beginnings in 1986. In some ways similar to the BPP Project although funded privately, Project Greenglow was sponsored by Europe's BAE Systems (created by the merger of British Aerospace and Marconi Electronics Systems) to study a range of approaches to advanced propulsion, maintaining the BPP Project's outlook that such work could best be advanced in this nascent stage by broad examination rather than narrowing to particular channels. Project Greenglow's emphasis was on near-term aeronautics developments, but the similarities with the BPP Project in its research interests were made explicit in an online announcement, although no formal connection existed between the groups:

> Our approach to the research programme will be much along the lines of the newly established NASA Breakthrough Propulsion Physics Programme which has, as its central theme, the goal of developing propellantless propulsion. We hope that there will be elements where our two programmes come together and we will support interaction between the two. Following the lead set by NASA, it is intended that most of the administrative effort and day-to-day communications will be conducted via the *Internet*. Also, the aim will be to keep the research programme as open as possible, within the restrictions imposed by academic ownership and commercial return on investment, as we recognise that there is a great deal of interest throughout the *world* on gravitational research, particularly on the possibility of the control of gravitational fields. We may wish to harness this support in our quest for funding [78].

Project Greenglow's origins actually date as far back as the 1970s, in what was then the British Aircraft Corporation's Future Concepts Group, which was established to investigate new fighter designs and technologies. What aerospace engineer and Project Greenglow consultant John E. Allen calls "speculative propulsion" entered the domain of serious study in 1986 [79]. A 1990 meeting at Preston (Lancashire) highlighted the company's interest in gravitational research, while links to researchers in government and private industry to determine their level of interest were made in the early 1990s. The funding of exploratory programs at several British universities began in 1996, their outgrowth being Project Greenglow itself in 1997. While the project would examine technologies based on conventional physics—solar sails, ion drives, and microwave thrusters—a Greenglow "route map" included research vectors into quantum electrodynamics, ZPE, and quantum gravity. Work on gravitation was performed at Lancaster University and Sheffield University. Project Greenglow director R. A. Evans notes:

> The existence of gravitomagnetism was predicted more than a century ago, but has never been detected. From an engineering point of view, being able to generate artificial gravitomagnetic fields offers the prospect of being able to induce gravity fields. Following on from this would be the control of inertia which would open up the possibility of field propulsion without expelling mass [79].

With superconductivity research also handled at Sheffield, the University of Kingston was charged with the study of engineering design issues, Birmingham with the Casimir force, and Dundee and Strathclyde with microwave thrust technologies.

The project supported a series of incremental research tasks including investigation of the Podkletnov gravity shield claims, vacuum energy extraction, and various approaches to the coupling of electromagnetism and gravity [80]. On the issue of gravity shielding, Project Greenglow commissioned a study performed at Sheffield University that examined the effect of a rotating superconductor on the weight of a nearby test mass. No effect was observed. Left undetermined by the study was the question of whether the methods Podkletnov used in levitating the superconductor disk (three 1 MHz solenoids) could explain the gravitational shielding effects.

But experiment, theory, and design were seen as only a part of the larger Greenglow program. The project also examined published concepts, theories, and patents seemingly germane to advanced propulsion, reaching the conclusion that most of these avenues were impractical due to the weakness of the forces alleged to exist. The project thus turned to a method not dissimilar from the BPP Project's own "horizon mission methodology," creating a hypothetical system that implied the existence of physics breakthroughs and could be used to design revolutionary aircraft. These "what if" concepts led to a series of aerospace vehicle designs; their engineering was explored and consequences on world transportation were analyzed. As opposed to the BPP Project, Project Greenglow emphasized global outcomes of physics breakthroughs rather than the steps needed to achieve those breakthroughs. Assuming a new kind of force interacting with space, researchers used the term "mass dynamic drive" as their breakthrough model, incorporating it ultimately into 20 aerospace projects. A major challenge in such work was noted by one Greenglow consultant:

> The fundamental difficulty is the totally different professional worlds of the advanced physicist and the engineering designer. It is most unlikely that the key would be discovered by accident. So how can the chance be increased significantly? The essence must be to bring together the two worlds of project design and advanced physics in an intense serendipitous initiative [81].

Whether funding for this privately financed project will continue is not known, as is the case for several other potential studies. Reports in *Jane's Defense Weekly* that American aerospace giant Boeing was investigating superconductivity effects related to Podkletnov's work remain unconfirmed [82]. Meanwhile, an article in *Aviation Week and Space Technology* has claimed that a U.S. aerospace company is working on ZPE research in conjunction with the Department of Defense [83]. The private nature of such potential work makes the prospects for disclosure unknown.

XIII. ESA's General Studies Programme and Advanced Concepts Team

The ESA General Studies Programme (GSP) acts as a think tank examining the feasibility of new concepts and methodologies, with studies selected from proposals submitted by ESA staff. GSP's feasibility studies last between 12 and 24 months, with 30 to 50 new studies typically initiated during each cycle. Through its Ariadna program, the GSP serves as an interface between the ESA and university researchers whose work may have space applications. Ariadna allows academics to propose projects for investigations that range from two to six months in duration. An ESA study on the potential use of gravity control for propulsion identified the "Pioneer effect" as well as gravitomagnetic fields in quantum materials as an area for continuing research [84]. The same report recommended ongoing space experiments to search for possible violations of Einstein's Equivalence Principle.

The related ESA Advanced Concepts Team (ACT) is a multidisciplinary research group based at the European Space Research and Technology Centre in Noordwijk, The Netherlands. Created in 2001, ACT includes among its battery of tasks research into fundamental physics in the fields of gravitation, quantum physics and high-energy physics. Within the range of its assessments have come revolutionary concepts that the ACT team has analyzed for potential benefits, ruling out those concepts found to be nonviable. ACT has also developed two mission architectures for missions to the outer Solar System that could be used to test the anomalous acceleration detected in both Pioneer probes (along with the Galileo and Ulysses spacecrafts) that may involve a novel physical effect [85].

ACT's charter sometimes takes it into theoretical realms where fruition might be decades away, the kind of work that a busy agency faced with launch dates on upcoming missions seldom has time to investigate. Of particular relevance to breakthrough propulsion are three studies on what ACT calls "nonissues," topics that upon investigation have been declared to be nonviable:

1) *Specific Coupling Effects Between Electromagnetism and Gravity:* To quote from ACT's description of this work:

> There are ... some tricks which might be realized by nature which would allow the scalar field to be dynamical under some circumstances (e.g., low gravitation, low temperatures). In this case the scalar could mediate an enhanced coupling between gravitation and some type of matter. A suggestion

for a particular strong interaction has recently been put forward. It even postulates a measurable effect on gravity from the earth's magnetic field. Such a strong gravielectric coupling would however open dramatic technological possibilities including the possibility of electromagnetic levitation [86].

The study relies on experimental data from work on gravitational physics at the University of Washington, deriving constraints that yield upper bounds for the strength of such coupling. ACT's study argues that the effects are too weak for technological applications.

2) *Faster-than-Light Communications by Quantum Tunneling:* As did the BPP Project, ACT examined tunneling effects, its particular emphasis being to determine their relevance for space communications. It found no technical applications in the absence of suitable material for a tunneling barrier in the interplanetary medium. ACT also notes that "long tunneling barriers will only allow faster than light communications with long wavelength signal and hence at low data rates. Hence in the end one will have to opt for either high data rates or high transmission speeds" [87].

3) *Quantum Vacuum Effects:* A study involving the universities of Cologne and Grenoble examined energy or momentum extraction from the quantum vacuum for possible propulsive effects, concluding that the inability to amplify these forces makes them unusable for spacecraft applications [88].

XIV. Broad Strategies for Future Research

Looking to the potential of propellantless propulsion via space drive, two demands emerge. The drive must satisfy conservation of momentum, and the vehicle it propels must be able to induce external net forces on itself. The BPP Project assessed seven hypothetical space drives [27]. From this and later work [74], candidates for continuing research emerged as follows (discussed in Chapter 3):

1) *Dark Matter and Dark Energy:* A space vehicle carrying no fuel of its own must use a reaction mass indigenous to the universe around it. Recent astronomical findings related to dark matter and dark energy reveal possible strategies for finding new forms of reaction mass. The anomalous rotation rates of galaxies suggest the presence of huge amounts of dark matter, perhaps accounting for as much as 90% of the visible galaxy. The possible use of dark matter for propulsion will drive future studies, while the dark energy seemingly responsible for the continuing acceleration of the expansion of the cosmos is in its earliest stages of analysis. Dark energy is perhaps linked to an effect similar to antigravity, and is thus of obvious interest for propulsion theory.

2) *Fundamental Force Relationships:* Electromagnetism, gravity, and space-time are known to be coupled phenomena. General relativity suggests that the amount of electromagnetic energy required to produce gravitational effects would be enormous. Continuing work on couplings at the atomic scale and smaller, however, explores the realm of quantum and particle physics in relation to such coupling. Such work deepens our understanding of general relativity's relationship to quantum mechanics, offering the potential of uncovering hitherto unknown propulsive effects.

3) *Frame Dependent Effects:* Mach's principle sees spatial reference frames for acceleration as connected to the surrounding mass in the universe. An absolute reference frame, which grows out of a literal interpretation of Mach's work, includes all matter in the universe (the cosmic microwave background radiation is an observable phenomenon suggestive of such a reference frame). Possible propulsive effects may be investigated that could induce net forces on a spacecraft relative to the surrounding mass. Frame dependent effects depart from the geometric analysis of general relativity and point to a Euclidean interpretation. The latter treatment is often referred to as an "optical analogy," one that views space as an optical medium with an effective index of refraction that is a function of gravitational potential. Consistent with observation, such an approach may prove useful in developing new propulsion concepts [89].

4) *Quantum Vacuum Effects:* Inertia may be interpreted as an electromagnetic drag force experienced by accelerating matter as it moves against the vacuum under some theories. A corollary is that gravity can be interpreted in terms of the distribution of vacuum energy in the presence of matter. These concepts remain one focus of the Institute for Advanced Studies in Austin, Texas, including recent experimental work [90]. The California Institute for Physics and Astrophysics has pursued privately sponsored research into these and related issues. Given that these theories are subject to much debate, it is not surprising that propulsion applications have been only rarely studied [42]. However, the broader issue of quantum vacuum energy as demonstrated through the Casimir effect is more widely in play, even with the understanding that the latter is appreciable only for the smallest cavities (at the micron level). Recent experimental work has focused on microelectromechanical (MEMS) structures in the examination of the production of net forces from this energy [91,92].

XV. Continuing Research Projects

Beyond these broad topics, the following specific projects are among those undergoing further investigation, their status unresolved:

1) *Gravitomagnetic Fields in Rotating Superconductors:* The coupling of electromagnetism, gravity, and spacetime offers ground for continuing interest. Couplings at the macro-level invoke General Relativity and thus imply substantial levels of electromagnetic energy to produce a measurable effect. But the lack of a unified theory incorporating both gravity and quantum mechanics points to the unresolved issues found at the level of particle physics that may have implications for propulsion. Growing out of work done for the ESA on hypothetical gravity control [84], Martin Tajmar and collaborators have used a spinning superconductor ring to detect an apparent frame-dragging force considerably stronger than what would have been predicted by theory. Tajmar's team at the Austrian Research Center facility in Seibersdorf draws on the understanding that below a critical temperature, such effects begin to appear in various materials, including a nonsuperconducting ring of aluminum. Measuring the gravitomagnetic field in the laboratory involves using rings of these materials, with both accelerometers and ring-laser-gyros positioned near the ring. As the temperature drops, the accelerometers and ring-laser-gyros registered signals suggestive of a gravitomagnetic, frame-dragging effect. Recent work by

Tajmar and colleagues describes experimental details of the original observations [93, 94] and compares their work to that of a New Zealand group that had used a spinning lead disk at superconductor temperatures [95]. The dissimilarity in the experimental setups deployed by the two groups leaves the issue of induced acceleration fields open.

2) *Transient Inertia Phenomena:* James Woodward's experiments on oscillatory inertia changes created by electromagnetic means continue in his and other hands (Woodward has actually taken out a patent—U.S. Patent No. 5,280,864—on the use of the technology for propulsion purposes). Woodward's "mass fluctuations" challenge conservation of momentum unless the Mach Principle be invoked to assert that all matter is in some fashion connected to all other mass in the universe, even the most distant. Experimental results have been unable to confirm or refute these theories [96] (see Chapter 11).

3) *Abraham-Minkowski Debate:* The controversy here concerns electromagnetic momentum within dielectric media (the equations for electromagnetic momentum in vacuum are widely accepted). Momentum transfer between matter and electromagnetic fields is described by contradictory equations derived by Max Abraham and Hermann Minkowski. The latter's work suggests that in materials where light travels at slower velocities, electromagnetic fields should exert greater momentum (the inverse of Abraham's findings). Propulsion concepts emerging from the Minkowski formulation have been studied under such terms as "electromagnetic stress tensor propulsion," "Heaviside force," and "Slepian drive," while recent work by Hector Brito suggests a propulsive device shaped around this theory, though the matter is still considered unresolved [97]. The West Virginia-based Institute for Scientific Research has also conducted experiments on the propulsive possibilities involved [98], while a study completed by the U.S. Air Force Academy rules out any net propulsive effects deriving from internal momenta as described by Minkowski [99]. Another recent paper argues that net forces are not possible using these principles [42]. Chapter 10 examines this topic in more detail.

XVI. Faster than Light

Unlike many of these topics, which are susceptible to experimental analysis, the warp drive and wormhole concepts that did so much to reawaken interest in these investigations remain purely theoretical constructs. The magnitude of negative energy required makes creating suitable laboratory experiments unlikely, although some theorists such as Eric Davis continue to study the possibility (see Chapter 15). In a study for the Air Force Research Laboratory, Davis examined using nuclear explosion magnetic compression or ultrahigh-intensity tabletop lasers to create laboratory wormholes [100]. The likelihood of anything but theoretical study of both warp drive and wormholes for the near future is small. There remains, however, the possibility of detecting evidence of wormholes through astronomical data by noting their gravitational lensing effects on distant light [37].

Since Alcubierre's 1994 paper, some 50 publications have appeared addressing one or another aspect of his "warp drive" concept. In identifying key physics obstacles, especially the enormous amounts of negative energy needed to generate and sustain warped spacetime, this work launched international debate that continues in conferences and workshops such as the annual Space Technology and Applications International Forum (STAIF) in Albuquerque, New Mexico. A faster-than-light symposium hosted by the British Interplanetary Society met on 15 November 2007 to discuss warp drive, the potential of vacuum energy, and international standards for the mathematical discussion of these concepts. Both warp drive and wormhole theories will doubtless remain in the speculative realm for the indefinite future, but they do present provocative thought experiments for deeper study of the related physics.

XVII. Conclusions

It is clear from this assessment that much of the most recent work on breakthrough propulsion is because of individual efforts, often with scant funding, scattered among government agencies, research laboratories, universities, and private organizations. Government-sponsored projects like BPP operated over a seven-year period and with a total budget of approximately $1.6 million. This scenario is likely to persist, given that the major space agencies have particular agendas focusing on near-term missions. In the absence of demonstrated success, many of the approaches examined here will continue to receive sparse attention. However, the upsurge in studies within the past 15 years indicates an appreciation of the implications of success. Should we find alternatives to costly chemical propulsion, especially in the area of "propellantless" spacecraft, the entire Solar System would swing into focus as a practical site for a space-based infrastructure that might one day lead to interstellar missions.

The stakes in such work are high, but the issues under investigation often parallel those actively studied in general physics. The coupling of electromagnetism, gravity, and spacetime is obviously of significance well beyond the realm of propulsion, as is the physics of inertial frames, but the perspective provided by a practical, engineering viewpoint may prove beneficial to the theoretical study of these matters. Similarly, an examination of the quantum vacuum may or may not lead to propulsive effects, but using that particular filter may reveal new theoretical possibilities. Propulsion physics, then, may be considered as a discipline useful in its own right, recognizing that some or all of the hoped for technologies may prove impractical or impossible to achieve. Even so, the history of technology offers the prospect that ideas that fail may be replaced by workable concepts not currently understood. It being the business of the future to surprise us, we should not be too quick to rule out such scenarios.

Acknowledgments

The author wishes to thank Marc G. Millis, whose assistance throughout the research and writing of this chapter proved invaluable. Thanks are owed as well to Eric W. Davis for his insightful comments in reviewing the manuscript.

References

[1] Millis, M. G., and Nicholas, T., "Responding to Mechanical Antigravity," NASA TM-2006-214390; AIAA Paper 2006-4913.

[2] Kuhn, T. S., *The Structure of Scientific Revolutions*, Univ. Chicago Press, Chicago, 1962.

[3] Foster, R. N., *Innovation: The Attacker's Advantage*, Summit Books, New York, 1986.
[4] Horgan, J., *The End of Science: Facing the Limits of Knowledge in the Twilight of the Scientific Age*, Addison Wesley, Reading, MA, 1996.
[5] Dyson, F., *Imagined Worlds*, Harvard Univ. Press, Cambridge, MA, 1997.
[6] Sänger, E., "The Theory of Photon Rockets," *Space Flight Problems*, Laubscher, Biel-Bienne, Switzerland, 1953, pp. 32–40.
[7] Sänger, E., "On the Attainability of the Fixed Stars," *Proceedings of the 7th International Astronautical Congress*, Rand Corp., Santa Monica, CA, 17–22 Sept. 1956, pp. 89–133.
[8] Ackeret, J., "On the Theory of Rockets," *Journal of the British Interplanetary Society*, Vol. 6, 1947, pp. 116–123.
[9] von Hoerner, S., "The General Limits of Space Travel," *Interstellar Communication*, A. G. W. Cameron (ed.), New York, Benjamin, 1963, pp. 144–159.
[10] Shepherd, L. R. "Interstellar Flight," *Journal of the British Interplanetary Society*, Vol. 11, 1952, pp. 149–167.
[11] Papailiou, D. D. (ed.), "Frontiers in Propulsion Research: Laser, Matter-Antimatter, Excited Helium, Energy Exchange, Thermonuclear Fusion," TM-33-722 (NASA-CR-142707), Jet Propulsion Lab., Cal Tech, Pasadena, CA, 1975.
[12] Forward, R. L., *Future Magic: How Today's Science Fiction Will Become Tomorrow's Reality*, Avon Books, New York, 1988; updated and republished as *Indistinguishable from Magic*, Baen Books, New York, 1995.
[13] Forward, R. L., "Feasibility of Interstellar Travel: A Review," *Acta Astronautica*, Vol. 14, 1986, pp. 243–252.
[14] Evans, R. A. (ed.), *BAe University Round Table on Gravitational Research*, Rept. FBS 007, British Aerospace Ltd., Preston, U.K., 26–27 March 1990.
[15] Millis, M., and Williamson, G. S., "Experimental Results of Hooper's Gravity-Electromagnetic Coupling Concept," NASA TM-106963, Lewis Research Center, 1995.
[16] Niedra, J., Myers, I., Fralick, C., and Baldwin, R., "Replication of the Apparent Excess Heat Effect in a Light Water-Potassium Carbonate-Nickel Electrolytic Cell," NASA TM-107167, Lewis Research Center, 1996.
[17] Mead, F., Jr., "Exotic Concepts for Future Propulsion and Space Travel," *Advanced Propulsion Concepts, 1989 JPM Specialist Session, (JANNAF)*, CPIA Publ. 528:93–99, 1989.
[18] Landis, G. L. (ed.), *Vision-21: Space Travel for the Next Millennium*. Proceedings, NASA Lewis Research Center, 3–4 April 1990, NASA CP 10059, 1990.
[19] Millis, M. G., "Exploring the Notion of Space Coupling Propulsion," *Vision-21: Space Travel for the Next Millennium*. Proceedings, NASA Lewis Research Center, NASA CP 10059, 3–4 Apr. 1990, pp. 313–322.
[20] Morris, M. S., and Thorne, K. S., "Wormholes in Spacetime and Their Use for Interstellar Travel: A Tool for Teaching General Relativity," *American Journal of Physics*, Vol. 56, 1988, pp. 395–412; also see Morris, M. S., Thorne, K. S., and Yurtsever, U., "Wormholes, Time Machines, and the Weak Energy Condition," *Physical Review Letters*, Vol. 61, 1988, pp. 1446–1449.
[21] Visser, M., *Lorentzian Wormholes: From Einstein to Hawking*, AIP Press, New York, 1995.
[22] Davis, E. W., "Advanced Propulsion Study," Air Force Research Lab., Final Report AFRL-PR-ED-TR-2004-0024, Edwards AFB, CA, April 2004, p. 65.

[23] Alcubierre, M., "The Warp Drive: Hyper-Fast Travel Within General Relativity," *Classical and Quantum Gravity*, Vol. 11, May 1994, pp. L73–L77.
[24] Emme, E. D. (ed.), *Science Fiction and Space Futures Past and Present*, American Astronautical Society Historical Series, Vol. 5, American Astronautical Society, San Diego, CA, 1982.
[25] Krasnikov, S. V., "Hyperfast Interstellar Travel in General Relativity," *Physical Review D*, Vol. 56, 1997, pp. 2100–2108.
[26] Misner, C. W., Thorne, K. W., and Wheeler, J. A., *Gravitation*, W. H. Freeman & Co., New York, 1973.
[27] Millis, M., "Challenge to Create the Space Drive," *Journal of Propulsion and Power*, Vol. 13, 1997, pp. 577–582.
[28] Pfenning, M. J., and Ford, L. H. "The Unphysical Nature of Warp Drive," *Classical and Quantum Gravity*, Vol. 14, 1997, pp. 1743–1751.
[29] van Den Broeck, C., "A 'Warp Drive' with More Reasonable Total Energy Requirements," *Classical and Quantum Gravity*, Vol. 16, 1999, pp. 3973–3979.
[30] Sternberg, A. M., Kwiat, P. G., and Chiao, R. Y., "Measurement of the Single-Photon Tunneling Time," *Physical Review Letters*, Vol. 71, 1993, pp. 708–711.
[31] Podkletnov, E., and Nieminen, R., "A Possibility of Gravitational Force Shielding by Bulk $YBa_2Cu_3O_{7-x}$ Superconductor," *Physica C*, Vol. 203, 1992, pp. 441–444.
[32] Li, N., and Torr, D. G., "Effects of a Gravitomagnetic Field on Pure Superconductors," *Physical Review D*, Vol. 43, 1991, pp. 457–459.
[33] Li, N., Noever, D., Robertson, T., Koczor, R., and Brantley, W., "Static Test for a Gravitational Force Coupled to Type II YBCO Superconductors," *Physica C*, Vol. 281, 1997, pp. 260–267.
[34] Hathaway, G., Cleveland, B., and Bao, Y., "Gravity Modification Experiment Using a Rotating Superconducting Disk and Radio Frequency Fields," *Physica C*, Vol. 385, 2003, pp. 488–500.
[35] Podkletnov, E., and Modanese, G., "Impulse Gravity Generator Based on Charged YBa2Cu3O7−y Superconductor with Composite Crystal Structure," *arXiv:physics/0108005 v2*, August 2001.
[36] Bennett, G. L., Forward, R. L., and Frisbee, R. H., "Report on the NASA/JPL Workshop on Advanced Quantum/Relativity Theory Propulsion." AIAA Paper 1995-2599, 31st ASME/SAE/ASEE Joint Propulsion Conference and Exhibit, San Diego, CA, 10–12 July 1995.
[37] Cramer, J., Forward, R. L., Morris, M., Visser, M., Benford, G., and Landis, G. L., "Natural Wormholes as Gravitational Lenses," *Phyical Review D*, Vol. 51, 1995, pp. 3117–3120.
[38] Sakharov, A., "Vacuum Quantum Fluctuations in Curved Space and the Theory of Gravitation," *Soviet Physics Doklady*, Vol. 12, 1968, pp. 1040–1041.
[39] Haisch, B., Rueda, A., and Puthoff, H. E., "Inertia as a Zero-Point Field Lorentz Force," *Physical Review A*, Vol. 49, No. 2, 1994, pp. 678–694.
[40] Rueda, A., and Haisch, B., "Inertial Mass as Reaction of the Vacuum to Accelerated Motion," *Physics Letters A*, Vol. 240, 1998, pp. 115–126.
[41] Rueda, A., and Haisch, B., "Contribution to Inertial Mass by Reaction of the Vacuum to Accelerated Motion," *Foundations of Physics*, Vol. 28, 1998, pp. 1057–1108.

[42] Puthoff, H. E., Little, S. R., and Ibison, M., "Engineering the Zero-Point Field and Polarizable Vacuum for Interstellar Flight," *Journal of the British Interplanetary Society*, Vol. 55, 2002, pp. 137–144.
[43] Mead, F. B., Jr., *Advanced Propulsion Concepts—Project Outgrowth*, AFRPL-TR-72-31, June 1972.
[44] Cravens, D. L., "Electric Propulsion Study," AL-TR-89-040, Final Report on Contract FO4611-88-C-0014, Air Force Astronautics Lab. (AFSC), August 1990.
[45] Forward, R. L., *21st Century Space Propulsion Study*, AL-TR-90-030, Final Report on Contract FO4611-87-C-0029, Air Force Astronautics Lab. (AFSC), Oct. 1990; see also Forward, R. L., *21st Century Space Propulsion Study* (Addendum), PL-TR-91-3022, Final (Addendum), OLAC Phillips Lab., formally known as Air Force Astronautics Lab. (AFSC), June 1991.
[46] Cravens, D. L., "Electric Propulsion Study," AL-TR-89-040, Air Force Astronautics Laboratory (AFSC), Edwards AFB, CA, 1990.
[47] Talley, R. L., "Twenty-First Century Propulsion Concept," PL-TR-91-3009, Phillips Laboratory, Air Force Systems Command, Edwards AFB, CA, 1991.
[48] "Practical Robotic Interstellar Flight: Are We Ready?" 29 Aug.–1 Sept. 1994, New York Univ. and The United Nations, New York. Sponsored by the Planetary Society.
[49] Anderson, J. L., "Leaps of the Imagination: Interstellar Flight and the Horizon Mission Methodology," *Journal of the British Interplanetary Society*, Vol. 49, 1996, pp. 15–20.
[50] Millis, M., *Breakthrough Propulsion Physics Project: Project Management Methods*, NASA TM-2004-213406, 2004.
[51] Millis, M., *NASA Breakthrough Propulsion Physics Program*, NASA TM-1998-208400, 1998.
[52] Krauss, L. M., *The Physics of Star Trek*, Basic Books, New York, 1995.
[53] Puthoff, H. E., "Gravity as a Zero-Point Fluctuation Force," *Physical Review A*, Vol. 39, 1989, pp. 2333–2342.
[54] Chiao, R. Y., Steinberg, A. M., and Kwiat, P. G., "The Photonic Tunneling Time and the Superluminal Propagation of Wave Packets," *Proceedings of the Adriatico Workshop on Quantum Interferometry*, World Scientific, Singapore, 1994, p. 258.
[55] Cramer, J. G., "The Transactional Interpretation of Quantum Mechanics," *Reviews of Modern Physics*, American Physical Society, Vol. 58, 1986, pp. 647–688.
[56] Forward, R. L., "Extracting Electrical Energy from the Vacuum by Cohesion of Charged Foliated Conductors," *Physical Review B*, Vol. B30, 1984, pp. 1700–1702.
[57] Cole, D., and Puthoff, H. E., "Extracting Energy and Heat from the Vacuum," *Physical Review E*, Vol. 48, 1993, pp. 1562–1565.
[58] Milonni, P. W., *The Quantum Vacuum: An Introduction to Quantum Electrodynamics*, Academic Press, San Diego, CA, 1994.
[59] Yilmaz, H., "Toward a Field Theory of Gravitation," *Il Nuovo Cimento*, Vol. 107B, 1992, pp. 941–960.
[60] Tipler, Frank, *The Physics of Immortality*, Doubleday, New York, 1994.
[61] Millis, M. G., "Prospects for Breakthrough Propulsion from Physics," NASA TM-2004-213082, May 2004.
[62] Schlicher, R. L., Biggs, A. W., and Tedeschi, W. J., "Mechanical Propulsion from Unsymmetrical Magnetic Induction Fields," AIAA Paper 95-2643, Joint Propulsion Conference, San Diego, CA, 1995.

[63] Fralick, G., and Niedra, J., "Experimental Results of Schlicher's Thrusting Antenna," *Joint Propulsion Conference*, AIAA Paper 2001-3657, Salt Lake City, UT, 2001.

[64] Maly, J., and Vavra, J., "Electron Transitions on Deep Dirac Levels I and II," *Fusion Technology*, Vol. 24, 1993, pp. 307–381, and Vol. 27, 1995, pp. 59–70.

[65] Robertson, G. A., "Exploration of Anomalous Gravity Effects by RF-Pumped Magnetized High-T Superconducting Oxides," *Joint Propulsion Conference*, AIAA Paper 2001-3364, Salt Lake City, UT, 2001.

[66] Woodward, J. F., "A New Experimental Approach to Mach's Principle and Relativistic Gravitation," *Foundations of Physics Letters*, Vol. 3, No. 5, 1990, pp. 497–506.

[67] Woodward, J. F., "Measurements of a Machian Transient Mass Fluctuation," *Foundations of Physics Letters*, Vol. 4, 1991, pp. 407–423.

[68] Ringermacher, H., "An Electrodynamic Connection," *Classical and Quantum Gravity*, Vol. 11, 1994, pp. 2383–2394.

[69] Cassenti, B. N., and Ringermacher, H., "The How to of Antigravity," *Joint Propulsion Conference*, AIAA Paper 96-2788, Lake Buena Vista, FL, 1996.

[70] Ringermacher, H., Conradi, M. S., Browning, C. D., and Cassenti, B. N., "Search for Effects of Electric Potentials on Charged Particle Clocks," *Joint Propulsion Conference*, AIAA Paper 2001-3906, Salt Lake City, UT, 2001.

[71] Mojahedi, M., Schamiloglu, E., Hegeler, F., and Malloy, K. J., "Time-Domain Detection of Superluminal Group Velocity for Single Microwave Pulses," *Physical Review E*, Vol. 62, 2000, pp. 5758–5766.

[72] Maclay, J., and Forward, R. L., "A Gedanken Spacecraft that Operates Using the Quantum Vacuum (Dynamic Casimir Effect)," *Foundations of Physics*, Vol. 34, 2004, pp. 477–500.

[73] Walker, R., Peters, F. W., Aldrin, B., Bolen, E., Buffenbarger, R. T., Douglass, J., Fowler, T. K., Hamre, J., Schneider, W., Stevens, R., Tyson, N., and Wood, H., *President's Commission on the Future of the US Aerospace Industry—Final Report: Anyone, Anything, Anytime, Anywhere*, 2002, p. 9-6.

[74] Millis, M., "Assessing Potential Propulsion Breakthroughs," *New Trends in Astrodynamics and Applications*, Belbruno, E. ed., Annals of the New York Academy of Sciences, New York, Vol. 1065, 2005, pp. 441–461.

[75] Tajmar, M., "The Biefeld–Brown Effect: Misinterpretation of Corona Wind Phenomena," *Journal of Propulsion and Power*, Vol. 42, 2004, pp. 315–318.

[76] Millis, M., and Williamson, G. S., "Experimental Results of Hooper's Gravity-Electromagnetic Coupling Concept," NASA TM-106963, Lewis Research Center, 1995.

[77] Merkle, C. (ed.), "Ad Astra per Aspera: Reaching for the Stars," Report of the Independent Review Panel of the NASA Space Transportation Research Program, Jan. 1999.

[78] "What Is Project Greenglow?," Project Greenglow Internet site, URL: http://www.greenglow.co.uk/ [3 January 2008].

[79] Allen, J. E., "Quest for a Novel Force: A Possible Revolution in Aerospace," *Progress in Aerospace Sciences*, Vol. 39, 2003, pp. 1–60.

[80] Szames, A. D., "Le Projet Greenglow pour Controller la Gravite," *Air & Cosmos*, 14 July 2000, pp. 44–45.

[81] Allen, J. E., "Aeronautics-1903; Aerospace-2003; ?? 2103," *Journal of Aerospace Engineering*, Vol. 219 (G3), June 2005, pp. 235–260.

[82] Cook, N., "Anti-Gravity Propulsion Comes 'Out of the Closet,'" *Janes Defense Weekly*, 29 July 2002.
[83] Scott, W., "To the Stars: Zero Point Energy Emerges from Realm of Science Fiction, May Be Key to Deep-Space Travel," *Aviation Week & Space Technology*, March 2004, pp. 50–53.
[84] Bertolami, O., and Tajmar, M., "Gravity Control and Possible Influence on Space Propulsion: A Scientific Study," ESA CR (P) 4365, on Contract ESTEC 15464/01/NL/Sfe, 2002.
[85] Rathke, A., and Izzo, D., "Options for a Non-dedicated Test of the Pioneer Anomaly," *Journal of Spacecraft and Rockets*, ArXiv: astro-ph/0504634, Vol. 43, No. 4, July–Aug. 2006, pp. 806–821.
[86] Rathke, A., "Constraining a Possible Dependence of Newton's Constant on the Earth's Magnetic Field," ESA Advanced Concepts Team, ESTEC, Noordwijk, The Netherlands, Aug. 2004.
[87] ESA Advanced Concepts Team, "Fundamental Physics: Faster-than-light Communications by Tunneling," URL: http://www.esa.int/gsp/ACT/phy/pp/non-issues/faster-than-light.htm [3 January 2008].
[88] Tiggelen, B. A., Rikken, G. L., and Krstic, V., "The Feigel Process: Lorentz Invariance, Regularization, and Experimental Feasibility," Final Report, ACT Web site, URL: http://www.esa.int/gsp/ACT/ariadna/completed.htm [3 January 2008].
[89] Evans, J. K., Nandi, K. K., and Islam, A., "The Optical-Mechanical Analogy in General Relativity: Exact Newtonian Forms for the Equations of Motion of Particles and Photons," *General Relativity and Gravitation*, Vol. 28, 1996, pp. 413–439.
[90] Rueda, A., and Haisch, B., "Gravity and the Quantum Vacuum Hypothesis," *Annals of Physics*, Vol. 14, No. 8, 2005, pp. 479–498.
[91] Maclay, G. J., "Analysis of Zero-Point Electromagnetic Energy and Casimir Forces in Conducting Rectangular Cavities," *Physical Review A*, Vol. 61, 2000, pp. 052110-1 to 052110-18.
[92] Esquivel-Sirvent, R., Villareal, C., and Mochan, W. L., "Casimir Forces in Nanostructures," *Physica Status Solidi (b)*, Vol. 230, 2002, pp. 409–413.
[93] Tajmar, M., Plesescu, F., Seifert, B., and Marhold, K., "Measurement of Gravitomagnetic and Acceleration Fields Around Rotating Superconductors," *Proceedings of the STAIF-2007*, AIP Conference Proceedings 880, 1071, 2007.
[94] Tajmar, M., Plesescu, F., Seifert, B., Schnitzer, R., and Vasiljevich, I., "Search for Frame-Dragging-Like Signals Close to Spinning Superconductors," *Proceedings of the Time and Matter 2007 Conference*, Bled, Slovenia, World Scientific Press, 2007.
[95] Graham, R. D., Hurst, R. B., Thirkettle, R. J., Rowe, C. H., and Butler, P. H., "Experiment to Detect Frame Dragging in a Lead Superconductor," *Physica C*, Vol. 468, Issue 5, March 2008, pp. 383–387.
[96] Cramer, J., Fey, C. W., and Casissi, D. P., "Tests of Mach's Principle with a Mechanical Oscillator," NASA CR-2004-213310, 2004.
[97] Brito, H. H., "Experimental Status of Thrusting by Electromagnetic Inertia Manipulation," *Acta Astronautica*, Vol. 54, pp. 547–558, 2004. Based on paper IAF–01–S.6.02, 52nd International Astronautical Congress, Toulouse, France, 2001.
[98] Corum, J. F., Keech, T. D., Kapin, S. A., Gray, D. A., Pesavento, P. V., Duncan, M. S., and Spadaro, J. F. "The Electromagnetic Stress-Tensor as a Possible Space Drive

Propulsion Concept," AIAA Paper 2001-3654, *37th AIAA/ASME/SAE/ASEE Joint Propulsion Conference*, Salt Lake City, UT, July 2001.

[99] Bulmer, J. S., and Lawrence, T., "Interferometer Examination of the Time Derivative of Electromagnetic Momentum Created by Independent Fields and Applications to Space Travel," USAFA TR 2003–03, United States Air Force Academy, Colorado Springs, CO, 2003.

[100] Davis, E. W., "Teleportation Physics Study," Air Force Research Lab., Final Report AFRL-PR-ED-TR-2003-0034, Air Force Materiel Command, Edwards AFB, CA, 2004.

Chapter 2

Limits of Interstellar Flight Technology

Robert H. Frisbee*
Jet Propulsion Laboratory, California Institute of Technology, Pasadena, California

I. Introduction

ONE of mankind's oldest dreams has been to visit the tiny pinpoints of light visible in the night sky. Much of the science fiction literature in the early 20th century is filled with fantastic stories of galaxy-spanning voyages to distant planetary shores, much as Verne's stories of the latter half of the 19th century dealt with fantastic voyages to the Moon (by cannon/rocket), across Africa (by balloon), and under the sea (by submarine). Elements of classic "space opera" continue today in which the laws of nature, at least as we currently know and understand them, are ignored to varying degrees for the sake of the storyteller's art. In literary criticism this has been called, using a tennis metaphor, "playing with the net down," where all manner of "super-science" technology is available. Here, the writer can make use of faster-than-light (FTL) "warp drives," unlimited power sources (e.g., for powering warp drives or blowing up planets), "deflector shields," traversable wormholes, matter transporters, time machines, and so on. The opposite, "playing with the net up," or "mundane" science fiction [1], deals with stories in which the author explicitly limits the availability of future technologies to relatively near-term extrapolations of existing technologies, with strict adherence to the laws of nature as we currently know and understand them. In this case, the

---

Copyright © 2008 by the American Institute of Aeronautics and Astronautics, Inc. The U.S. Government has a royalty-free license to exercise all rights under the copyright claimed herein for Governmental purposes. All other rights are reserved by the copyright owner.
*Senior Engineer, Propulsion Section.
This work is dedicated to the memory of Robert L. Forward (1932–2002). So much of what we think of when we consider interstellar propulsion concepts and technologies was first conceptualized and evaluated by Bob Forward; our debt to him cannot be overemphasized. Isaac Newton said it best when he said that we see further only because we stand on the shoulders of giants; Bob Forward was one of those giants.

starships travel much slower than the speed of light, relativistic effects are important, power is always at a premium, and physical shielding plates are required to protect against interstellar dust grain impacts.

In the overall context of this book, this chapter takes the "mundane" science fiction approach of asking what types of technologies are required to perform, initially, interstellar precursor missions to distances of hundreds or thousands of astronomical units (AU, 149.6 × 10^6 km), and, ultimately, full interstellar missions with cruise (peak) velocities on the order of 50% of the speed of light, c. Here, we will first identify representative mission requirements (e.g., distance and desired trip time) and then use extrapolations of known science and engineering to develop "paper" spacecraft that might meet the mission requirements.

II. Challenge of Interstellar Missions

Over the last 50 years we have visited most of the major bodies in our solar system, reaching out far beyond the orbit of Pluto with our robotic spacecraft. However, the leap to interstellar distances represents an unprecedented challenge. For comparison, more than 30 years after its launch (on 5 September 1977), the Voyager I spacecraft is about 103.6 AU, or 14.3 light-hours, from the sun, traveling with a solar system escape velocity, $V_{infinity}$, of 17.4 km/s (3.67 AU per year) or 0.006% of the speed of light. Yet this distance, which strains the limits of our technology, represents an almost negligible step toward the light-years that must be traversed to travel to the nearest stars. For example, even though the Voyager spacecraft is one of the fastest vehicles ever built,[†] it would still require almost 75,000 years for it to traverse the 4.3 light-year (LY) distance to our nearest stellar neighbor Alpha Centauri (if it were pointed in that direction). Thus, travel to the stars is not impossible but will, however, represent a major commitment by a civilization simply because of the size and scale of any technology designed to accelerate a vehicle to speeds of a few tenths of the speed of light.

In fact, one of the most daunting aspects of comprehending the scale of an interstellar mission is the sheer size of any transportation system capable of reaching a significant fraction of the speed of light (e.g., 0.1 c or faster). As a point of reference, a "payload" mass of 100 metric tons (MT = 1000 kg), roughly the mass of the space shuttle Orbiter, traveling at 0.5 c has a kinetic energy of 1.29×10^{21} joules (J), including a relativistic mass correction of $1.15 = \{1/[1 - (V/c)^2]\}^{1/2}$. This energy represents about 2.7 years worth of the annual energy production of human civilization (4.85×10^{20} J in 2005). As we will see, adding the required propulsion system to reach interstellar transportation speeds results in vehicle systems with dimensions on the order of planetary diameters, masses of hundreds of billions of tons, and power levels thousands of times that of human civilization (about 15.4 TW in 2005) [2].

[†]The New Horizons Pluto spacecraft started with a much higher velocity leaving Earth than Voyager, but its Solar System escape velocity ($V_{infinity}$) is less than Voyager (e.g., 3 AU/yr versus 3.7 AU/yr). This is because Voyager, initially slower, had the benefit of three gravity assists compared to Pluto's one, so Voyager has the higher final $V_{infinity}$.

A. Interstellar Mission Requirements

As a tool for evaluating the limits of technology for interstellar flight, we can think of a "fast" (0.5 c cruise velocity) interstellar rendezvous mission as our "Vision Mission" [3,4]. This mission represents a "stretch goal" that is intentionally made as difficult as possible so that a simple extrapolation of existing, near-term technologies will not suffice [4]. Ultimately, this gives us a tool to aid in structuring future technology development programs and interstellar precursor missions with a long-range goal of giving us the capability to perform the Vision Mission. As an historical example, we can consider the Apollo lunar landing as the Vision Mission and stretch goal in the early 1960s. This led to development of technologies like large chemical rocket engines, fuel cells, and 0-g cryogenic fluids systems. Similarly, it led to development during the Gemini Program of space operations techniques like rendezvous and docking. Finally, to support the Apollo lunar landings, a number of robotic precursor missions like Ranger, Surveyor, and Lunar Orbiter were flown.

Using our Vision Mission as a guide, we will focus on an evaluation of various propulsion options for interstellar rendezvous missions ranging from 4.3 LY with a 10-year trip time goal, to 40 LY with a 100-year trip time goal. The 40 LY limit was selected to be compatible with the range of future NASA Origins Program telescopes, as will be discussed. Various propulsion candidates are described in more detail that might fulfill the Vision Mission, as well as interstellar precursor missions leading up to implementation of the vision interstellar rendezvous mission. These propulsion candidates [3] include advanced chemical propulsion, electric propulsion, nuclear (fission, fusion, antimatter) propulsion, beamed momentum propulsion (e.g., laser "pushed" sails), electromagnetic catapults, in-situ propellant production concepts (e.g., the Bussard interstellar ramjet), and hybrid systems (e.g., antimatter-catalyzed fission/fusion). Interestingly, because of the high ΔV required for interstellar flyby and rendezvous missions (up to several tenths of the speed of light) and the large dry masses of the candidate propulsion systems, traditional nonpropulsive enhancements (like the gravity assists used by the Voyager spacecraft) are of limited use, resulting in a need for specific impulses (I_{sp}) on the order of 10^4 to 10^7 lb$_f$-s/lb$_m$ (10^2 to 10^5 km/s), depending on the mission overall ΔV [3].

Our ultimate mission goal is the ability to visit scientifically interesting planets circling around other stars. Mission targets, such as planets capable of harboring life and/or habitable by humans, would be identified by the NASA Origins Program. This Program has the long-range goal (by ca. 2040) of detection, remote-sensing spectral analysis, and imaging of potentially habitable planets around stars out to ~40 LY, corresponding to a sphere containing roughly 350 stars [5]. (This total does not include solitary red dwarfs, which are generally incapable of harboring habitable planets.) This will be accomplished by the use of progressively more sophisticated, space-based, observational techniques (e.g., telescopes, interferometers, etc.) to ultimately image Earth-like planets in the potentially habitable region or the "Goldilocks zone"—not too hot, not too cold—around a star. In the near term, however, there are a number of interesting interstellar precursor scientific opportunities that could serve as intermediate, near-term technology demonstration missions on the way to developing propulsion systems capable of achieving the ultimate mission to the stars.

Some examples of these precursor opportunities are shown in Fig. 1 [6]. These include missions to the Heliopause (at ~100 AU), the sun's gravitational lens (550 AU), and the Oort cloud (~2000 AU to ~1 LY, with 1 LY = 63,300 AU). Order-of-magnitude ΔV and corresponding propulsion system I_{sp} requirements for these missions are given in Table 1, where we have assumed that I_{sp}, expressed as exhaust velocity ($V_{exhaust}$) is comparable to the mission ΔV. (This assumption is discussed in more detail in the following.) Figure 2 illustrates the range of ΔV values typically encountered as a function of the trip time desired. Figure 2 also illustrates the difficulty of performing even near-term interstellar precursor missions; for example, advanced nuclear electric propulsion (NEP) requires roughly 50 years to reach the outer edge of the Kuiper belt at 1000 AU. More ambitious missions require the use of fission, fusion, or antimatter reactions as the propellant exhaust, or of photon momentum beamed to the vehicle by, for example, massive lasers in the solar system.

Robotic interstellar missions can be viewed as a natural follow-on to the Origins Program, which will tell us where to send the interstellar spacecraft to provide close-up imaging with a flyby, and detailed in-situ science ("ground truth") with a rendezvous mission. Current emphasis is on a fast interstellar rendezvous mission where the spacecraft stops at its destination. Thus, there is a desire for a high cruise velocity to minimize trip time. For example, to travel 4.3 LY with a 10-year trip time requires an average speed of 0.43 c. However, a high-speed ($\geq 0.1\ c$) flyby is not thought to give significantly more science return than that provided by Origins Program capability in the time frame of interest; in effect, virtually as much imaging capability is provided by advanced telescopes in the solar system as from a rapidly moving spacecraft in a flyby. Also, telescopes in the solar system could monitor the target for an indefinite amount of time; by contrast, a 0.1 c flythrough of, for example, our solar system (with a diameter of 11 light-hours assuming Pluto's orbit) would only allow 110 hours of close-up observation of the target solar system. Thus we see the need for a rendezvous mission, even though this has the effect of doubling the mission ΔV. Finally, for eventual crewed missions to the stars, special emphasis is placed on minimizing trip time with rendezvous assumed to be a given.

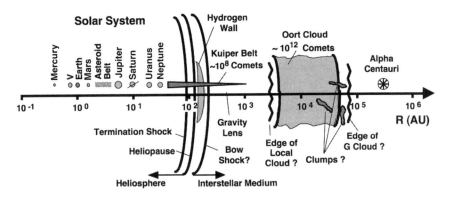

Fig. 1   Scale of the interstellar medium (after Ref. 6).

Table 1 Sample mission, ΔV, and I_{sp} requirements

Mission	Planetary	100–1000 AU	10,000 AU	Slow interstellar flyby	Fast interstellar rendezvous
Mission examples	Inner solar system orbiters Outer solar system flybys Slow outer solar system orbiters	Heliopause (100 AU) Gravity lens (550 AU) Kuiper belt (40–1000 AU)	Oort cloud (2000–60,000 AU)	4.3 LY in 40 yrs	4.3 LY in 10 yrs 40 LY in 100 yrs
Typical ΔV	10 km/s	100 km/s	1000 km/s	$0.1\,c$	$0.5\,c$ for flyby; doubled for rendezvous
Typical I_{sp} (lb_f-s/lb_m) (km/s)	10^3 10^1	10^4 10^2	10^5 10^3	3×10^6 3×10^4 ($0.1\,c$)	1.5×10^7 1.5×10^5 ($0.5\,c$) (ignores relativistic effects)

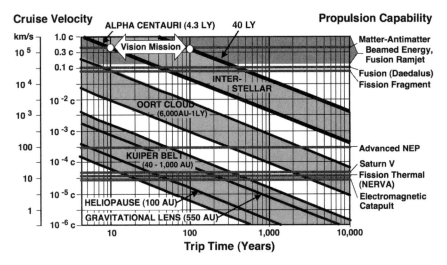

Fig. 2 Mission velocity and trip time.

B. "Impossible" Missions and the Importance of Specific Impulse and Stage Dry Mass

As we will show, the "Rocket Equation" describes the exponential increase in the ratio of the "wet" mass, M_{wet}, of a vehicle (fully loaded with propellant) divided by its "dry" mass without propellant, M_{dry}, as a function of the ratio of velocity change, ΔV, divided by the propulsion system's specific impulse, I_{sp}, or more properly, its exhaust velocity, where $V_{\text{exhaust}} = g_c I_{\text{sp}}$ (i.e., $g_c = 9.80665 \, [\text{m/s}]/[\text{lb}_f\text{-s}/\text{lb}_m]$).

$$M_{\text{wet}}/M_{\text{dry}} = \text{EXP}[\Delta V/(g_c I_{\text{sp}})] = \text{EXP}(\Delta V/V_{\text{exhaust}})$$

Note that in this chapter, we generally give I_{sp} in "Imperial" or foot-pound-second (fps) units of $\text{lb}_f\text{-s}/\text{lb}_m$ because this is how I_{sp} has been (and continues to be) reported in the literature. This also avoids the potential for inappropriately increasing or decreasing the number of significant digits in the value of I_{sp} when converting I_{sp} from $\text{lb}_f\text{-s}/\text{lb}_m$ to m/s; for example, converting a chemical rocket's quoted I_{sp} of 452 $\text{lb}_f\text{-s}/\text{lb}_m$ to 4,433 m/s implies an additional digit of precision not evident in the original literature value.

Historically, I_{sp} was defined as a rocket engine's thrust divided by the rate of propellant mass flowing into the engine. For example, a chemical rocket's thrust would be measured [e.g., in pounds of force (lb_f) or newtons (N)] and the propellant mass flow rate [e.g., pounds of mass per second (lb_m/s) or kilograms per second (kg/s)] derived, for example, from propellant volumetric flow rate (e.g., gallons or liters per second) and the propellant's density. Interestingly, in the metric meter-kilogram-second (mks) system, I_{sp} in N-s/kg reduces directly to V_{exhaust} in m/s because the Newton is a compound unit, where $N = \text{kg-m}/\text{s}^2$. Thus we see that I_{sp} in the "right" units of velocity is simply (and interchangeable) $V_{exhaust}$. To convert I_{sp} in Imperial lb$_f$-s/lb$_m$ (fps) units to velocity, we multiply the value in lb$_f$-s/lb$_m$ by a units conversion factor (g_c) that is numerically equal to the standard acceleration of gravity: 9.80665 [m/s]/ [lb$_f$-s/lb$_m$] = 32.17405 [ft/s]/[lb$_f$-s/lb$_m$]. Also, note that g_c is formally a unit conversion factor and does not depend on the local gravitational field because our unit systems are based on a standard Earth gravity. Finally, one often encounters I_{sp} given in units of "seconds" when it should more properly be expressed as lb$_f$-s/lb$_m$. This is because of the erroneous tendency to "cancel" units of lb$_f$ and lb$_m$ in the lb$_f$-s/lb$_m$ unit, leaving only seconds. In fact, if I_{sp} is defined as thrust divided by propellant *weight* (not mass) flow rate, then I_{sp} can in fact be correctly expressed as seconds. To avoid any confusion, we will generally give I_{sp} in lb$_f$-s/lb$_m$ and $V_{exhaust}$ in velocity units (e.g., km/s or a fraction of the speed of light).

Examples of the increase in vehicle wet mass as ΔV increases (with a given I_{sp} or $V_{exhaust}$) are shown in Fig. 3 for a single-stage vehicle, where we can have an "impossible" mission when ΔV is several times the propulsion system's I_{sp} or $V_{exhaust}$. Thus, a chemical propellant like O_2/H_2, with an I_{sp} on the order of 500 lb$_f$-s/lb$_m$ corresponding to a $V_{exhaust}$ of about 5 km/s, would literally need an ocean's worth of propellant to achieve an interstellar precursor ΔV of about 200 km/s. This is the primary reason why I_{sp} is so strongly emphasized in determining rocket performance. This is also the driver behind the use of multi-stage vehicles so as to reduce the ΔV that must be delivered by each stage, and thus reduce the mass ratio of each stage.

However, there is a second aspect of "impossible" missions that is a consequence of the fact that any real propulsion system has a finite dry mass ($M_{prop\text{-}dry}$) of engines, propellant tanks, structure, etc. For example, we can

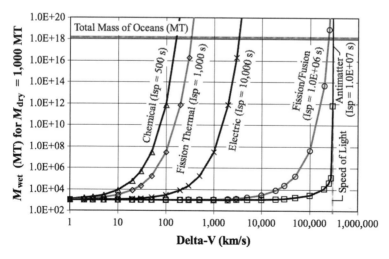

Fig. 3 Increase in vehicle wet mass (M_{wet}) as a function of ΔV and I_{sp} (s = lb$_f$-s/lb$_m$) for representative propulsion systems (based on the Relativistic Rocket Equation, with no loss of propellant). Calculations based on vehicle dry mass (M_{dry}) = 1000 MT.

write the vehicle dry mass as the sum of the payload and propulsion system dry masses:

$$M_{dry} = M_{payload} + M_{dry-prop}$$

Further, as a simplification, we can also divide the propulsion system's dry mass into two components, a fixed mass term, M_{fixed}, independent of the propellant mass, M_p, and a term dependant on propellant mass through a tankage factor, TF:

$$M_{dry-prop} = M_{fixed} + TF \cdot M_p$$

In the extreme limit of large M_p, the payload and propulsion system fixed masses become negligible, and we have

$$TF(max) \sim M_{dry}/M_p$$

This represents the maximum possible value for TF; any nonzero payload or propulsion system fixed masses would tend to lower the allowable tankage fraction of the vehicle. Thus, we can recast the results shown above for a single-stage vehicle and calculate the maximum possible TF based on the Rocket Equation as a function of the propulsion system's I_{sp} and mission ΔV, as shown in Fig. 4.

From this we see that as ΔV grows, the maximum TF allowed by the Rocket Equation decreases and approaches zero at high ΔV. In practical terms, it becomes "impossible" in an engineering sense to build propellant tankage light enough to allow the vehicle to reach a ΔV more than a few times its I_{sp} (or $V_{exhaust}$). For comparison, we have shown the TF for the space shuttle external tank (ET), which holds roughly 700 MT of liquid oxygen (LO_2) and liquid hydrogen (LH_2) at an oxidizer-to-fuel (O/F) ratio of about 6. The ET by itself, an extremely lightweight structure, has a TF of 4%. If we were to include the mass of the space shuttle Orbiter, the "effective" TF would be about 18%, which not surprisingly matches the curved line for the limit of maximum possible TF. (In this example, the mass and ΔV of the solid rocket boosters are omitted.)

We can also see the effect of economy-of-scale by considering the two TF points for the fission/fusion system. These points are based on the total propulsion system masses for the Daedalus two-stage fusion rocket designed by the British Interplanetary Society for a 0.1 c interstellar flyby mission [7]. For this vehicle, the fusion fuels were deuterium (D) and the mass-3 isotope of helium (He^3), with a D/He^3 mass ratio of 2/3. The second stage had 4000 MT of propellant with a total propulsion system only (no payload) TF $\sim M_{dry-prop}/M_p$ of 13.7%. By contrast, the first stage had 46,000 MT of propellant, with a propulsion system TF of only 3.7%. Note that these two points are still considerably below the fission/fusion curves because these data points for TF only consider the dry propulsion system in each stage. By contrast, the curves correspond to the case of considering the total M_{dry} for each stage (i.e., maximum TF $\sim M_{dry}/M_p$) with each stage having $M_{dry} = M_{dry-prop} + M_{payload}$. However, for stage 1, the "payload" mass is all of the fully-loaded stage 2 with its payload.

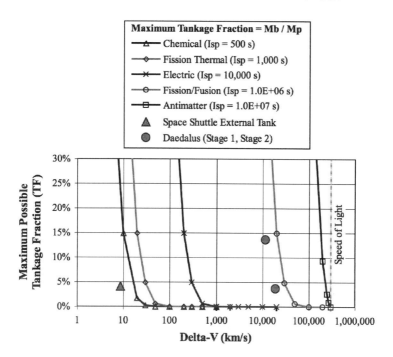

Fig. 4 Decrease in vehicle maximum possible tankage factor (TF $\sim M_{dry}/M_p$) as a function of ΔV and I_{sp} ("s" = lb_f-s/lb_m) for representative propulsion systems (based on the Relativistic Rocket Equation with no loss of propellant).

As a final example of the inherent engineering "impossibility" of using a low-I_{sp} propulsion system for a high-ΔV mission, we can calculate the ratio of $\Delta V/I_{sp}$ (or, more formally, $\Delta V/V_{exhaust}$) as a function of an individual stage's total TF = M_{dry}/M_p. (From the Classical Rocket Equation, $\Delta V/V_{exhaust} = \ln(M_{wet}/M_{dry})$, and $M_{wet} = M_{dry} + M_p$.) As shown in Fig. 5, a total TF of about 15% (representative of many chemical propulsion systems) gives a ΔV only twice the $V_{exhaust}$ of the propulsion system. This is the practical reason why most state-of-the-art propulsion stages are limited to a ΔV of about twice the $V_{exhaust}$ (or corresponding I_{sp}) of their propulsion system; for example, the Space Shuttle has an Earth-to-orbit ΔV of 9 km/s for O_2/H_2 with a $V_{exhaust}$ of 4.5 km/s (I_{sp} = 450 lb_f-s/lb_m). Even if we could—somehow—build a propulsion system with a TF of 0.1%, the achievable ΔV for that system would still only be 6.9 times its $V_{exhaust}$.

Finally, note that we have ignored the effect of using multiple stages to deliver some total ΔV; in practice, this is used to reduce the ΔV per stage and to increase the allowable TF per stage. However, this approach still has its limitations, because the "upper" stages are the "payload" of the "lower" stages. This produces the "pyramid" effect seen in multi-stage rockets: the first stage is very large, with subsequent stages becoming smaller and smaller. For example, if we have a (ΔV per stage)/($V_{exhaust}$) of 2, the Classical Rocket Equation mass ratio of

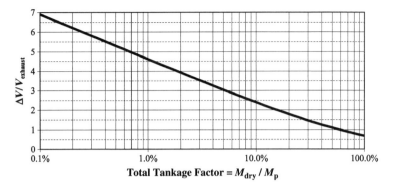

Fig. 5 Ratio of $\Delta V/V_{exhaust}$ as a function of stage TF = M_{dry}/M_p (based on the Classical Rocket Equation).

M_{wet}/M_{dry} per stage is $\exp(2) = 7.38$; thus, a three-stage rocket would have an initial total wet mass of at least $(7.38)^3 = 403$ times the final dry mass of the third stage (including payload), even if stages one and two had a TF = 0.

C. Technology Options Selection Process

Having established the requirements for our fast, interstellar rendezvous Vision Mission (i.e., $V_{cruise} = 0.5\ c$, corresponding to a total mission ΔV of 1 c for a rendezvous mission), we now seek to identify viable propulsion candidates for interstellar missions [3]. To do this, we ask three questions that are used as screening filters,[‡] as shown in Fig. 6. Note that this screening process is used to reduce the large possible number of propulsion options down to a manageable few; in effect, we have used the Horizon Mission methodology [4] where we have intentionally selected a set of mission requirements (from the fast, rendezvous interstellar Vision Mission) that are so demanding that only a limited number of propulsion options are applicable.

1. ΔV Capability

First, we ask the basic question of whether the propulsion system has the capability of providing the required ΔV for the mission. As described previously, the Rocket Equation suggests that it is desirable to have the mission ΔV and propulsion system specific impulse (I_{sp}) or exhaust velocity ($V_{exhaust}$) comparable in size to prevent excessive propellant requirements. (For our Vision Mission case, with $V_{cruise} = 0.5\ c$, this implies that $\Delta V = 0.5\ c = 1.5 \times 10^5$ km/s $\sim V_{exhaust}$, corresponding to $I_{sp} = 15 \times 10^6$ lb$_f$–s/lb$_m$.)

This evaluation criterion quickly eliminates chemical, advanced electric propulsion (EP) and electromagnetic (EM) catapult launchers [8] from consideration

---

[‡]The original "filtering" concept was introduced by H. Harris, Jet Propulsion Laboratory, 1997.

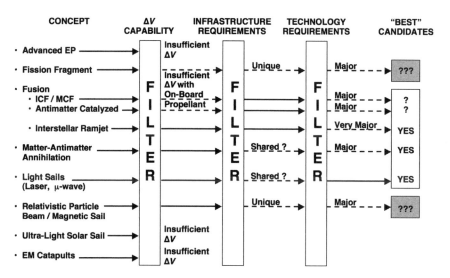

Fig. 6 Fast ($V_{cruise} = 0.5\,c$) interstellar rendezvous mission propulsion option screening process.

for fast interstellar rendezvous missions. Similarly, fission fragment [9,10] and fusion propulsion [3] concepts have a maximum I_{sp} on the order of 1×10^6 lb$_f$–s/lb$_m$ (i.e., $V_{exhaust} = 0.03\,c$), which is too low for a fast rendezvous mission. However, a two-stage fusion rocket has been shown to be a reasonable choice for a slow, $0.1\,c$ interstellar flyby [7]. The one loophole to fusion I_{sp} or $V_{exhaust}$ limitation is the Bussard interstellar fusion ramjet [11] that collects interstellar hydrogen (H) or deuterium (D) for use in a fusion rocket. Because all of the propellant required for the mission is not carried on board the vehicle, the interstellar ramjet effectively "cheats" the Rocket Equation and is capable of supplying unlimited ΔV. Solar sails also "cheat" the I_{sp} limitations of the Rocket Equation; however, even with an ultra-low mass per unit area (areal density), they cannot achieve the required velocities because of the $1/R^2$ drop-off in sunlight intensity (i.e., decrease in photon momentum "push" per unit area) on the sail [3]. Light sails overcome the $1/R^2$ limitation of a solar sail by using laser or microwave power (actually momentum) beaming. However, although microwave light sails (e.g., Starwisp [12]) can be used for interstellar flybys, they cannot be used for rendezvous missions because the long wavelength of microwaves (compared to near-visible laser wavelengths) results in impossibly large optics requirements for focusing the microwaves at interstellar distances. Thus, only laser light sails (with a laser near-visible wavelength on the order of 1 μm) [2,13], matter–antimatter annihilation rockets (with an I_{sp} of 1×10^7 lb$_f$–s/lb$_m$ corresponding to a $V_{exhaust} = 0.33\,c$) [3,14], and fusion ramjets [3] strongly pass the ΔV filter, with fission fragment and fusion propulsion (both with a $V_{exhaust}$ of $0.03\,c$) weakly passing the filter.

2. Infrastructure Requirements

The second evaluation criterion deals with the potential need for a large, possibly space-based supporting infrastructure that is unique for the propulsion concept. The assumption here is that this infrastructure would represent a significant up-front cost that typically would have limited application beyond the interstellar mission.

For example, the fission fragment propulsion concept would require the construction of a unique facility (ground- or space-based) to produce large amounts of short-lived, high-energy, highly-fissionable nuclear fuels such as americium (Am) or curium (Cm) [9]. Similarly, a relativistic particle beam that would "push" a magnetic sail (e.g., MagSail) [15], analogously to the laser-driven light sail, would require an enormous space-based particle beam facility that would have limited applicability beyond in-space transportation.

By contrast, "pure" matter–antimatter annihilation propulsion, where all of the propulsive energy comes from the annihilation reaction, will require major new antiproton production facilities to supply the tons of antimatter required for interstellar missions. However, it must be noted that there are a number of dual-use spin-offs of antiproton research, such as medical applications (e.g., imaging and destruction of cancer tumors in the 1-mm size range) [16], that could justify the infrastructure investment. Similarly, laser (or microwave) light sails would require a major space-based infrastructure consisting of the beam source and the associated optics, but the beamed-energy infrastructure has the unique capability of multiple use as a time-shared power and propulsion source as, for example, a "public utilities in space," with a grid of laser/microwave beams supplying power in space analogous to the electric power and natural gas utilities on Earth.

Thus, the fission fragment and particle beam/magsail concepts strongly fail the infrastructure test. Matter–antimatter and beamed-energy light sail propulsion concepts only weakly fail this test, either because of the potential for multiple in-space or spin-off applications. Therefore, only fusion, matter–antimatter annihilation, and light sail propulsion will be carried on to the third evaluation criterion, technology requirements.

3. Technology Requirements

Our third and final criterion relates to the current technology level and future technology development needs of the various systems. Not surprisingly for an interstellar propulsion system, the technology requirements for all of the three leading candidates will be formidable. Note that *all* of the concepts have numerous uncertainties and major unresolved feasibility issues; there is no clear winner. Rather, the challenge is to identify the approach that has the fewest number of developmental and operational "miracles" required for its implementation. Ironically, the interstellar fusion ramjet has the greatest performance potential but also the greatest number of technology challenges. However, from our perspective today, all three are equally "impossible"; only continued research and analysis will identify which are less "impossible" than others.

III. Fundamentals of Interstellar Flight

In this section, we will develop the governing equations for interstellar missions with regard to the propulsion systems. First, we will outline the derivation of the "Classical" and "Relativistic" Rocket Equation, with particular attention to the case of a matter–antimatter annihilation propulsion system where a large fraction of the initial propellant mass is "lost" during the annihilation process and is thus not available to impart forward momentum to the vehicle. Next, we will describe the equations governing the behavior of laser-pushed sails (light sails), where photon momentum is used to accelerate the sail. We will also briefly discuss some of the issues associated with the Bussard interstellar ramjet. Next, we will discuss the impact of vehicle acceleration (and deceleration at the target), and how too low an acceleration can dramatically impact the total mission flight time. Finally, we will outline an approach to ameliorating the effects on an interstellar vehicle of ultra-high speed (relativistic) impacts with interstellar dust grains.

A. Rocket Equation

We will present the results of the derivation of four versions of the Rocket Equation; details of the derivations are given in the Appendix to this chapter. The first is the Classical Rocket Equation, first derived by Tsiolkovsky [17]. The next is a Relativistic Rocket Equation, in which both the vehicle and the propellant exhaust velocity ($V_{exhaust}$ or equivalently I_{sp}) are moving at a large fraction of the speed of light, such that relativistic corrections must be included. These derivations will follow Forward's method [18]; an alternative mathematical formulation of relativistic systems is discussed in Chapter 14. The next two versions of the Rocket Equation parallel the Classical and Relativistic Rocket Equations, but for the situation where some large fraction of the initial propellant mass or mass-energy is unavailable (e.g., 60% for a proton-antiproton annihilation rocket) for momentum transfer to the vehicle.

1. Classical Rocket Equation

In the Classical Rocket Equation derivation, a rocket with initial mass M ejects a mass of reaction mass (propellant) dm at a constant exhaust velocity (i.e., $V_{exhaust}$ corresponding to I_{sp}) relative to the rocket with a velocity, as shown in Fig. 7. In the center of mass of the system, the resultant rocket velocity is V (i.e., ΔV), and the reaction mass velocity is u. Note that we have shown the situation for both acceleration and deceleration; derivation of the Rocket Equation will be shown for both options. For convenience, we will adopt a sign convention for acceleration and deceleration using expressions like x "$+/-$" y or x "$-/+$" y, where the "upper" sign is for acceleration and the "lower" sign is for deceleration. Thus, the propellant velocity in the center of mass of the system can be written as $u = w \; -/+ \; V$, corresponding to $u = w - V$ for acceleration or $u = w + V$ for deceleration.

To begin the derivation of the Classical Rocket Equation, we first start with the conservation of mass:

$$dM = -dm$$

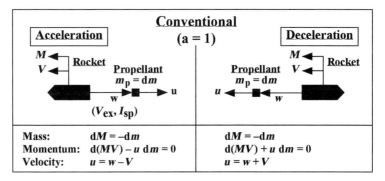

Fig. 7 Classical rocket mass, velocity, and momentum (with no loss of propellant).

The minus sign indicates the decreasing mass of the rocket. Also, the two equations are the same for acceleration or deceleration. Next, we conserve momentum:

$$d(M \cdot V) = +/- u \cdot dm$$

Note the sign difference between the acceleration and deceleration forms of the equation. Finally, we use the classical addition (or subtraction) of velocities:

$$u = w -/+ V$$

Again, note the sign differences. We then use the conservation of mass equation to find dm; this identity will be then be used when we expand the conservation of momentum equation [i.e., $d(MV) = MdV + VdM$]. Thus, expanding the derivatives, combining the above equations, and rearranging gives the derivative form of the equation to be analytically integrated:

$$dM/M = -/+ dV/w$$

If we were performing a numerical integration, we could write: $dM = M[-/+ dV/w]$. Analytical integration gives:

$$\ln(M) = -/+ V/w$$

which, when evaluated for the rocket mass limits of M_{wet} (initial wet mass) and M_{dry} (final "burnout" dry mass), initial and final velocities (V_i and V_f, with $V_f - V_i = \Delta V$), and $w = V_{exhaust} = g_c I_{sp}$, [with $g_c = 9.8$ (m/s)/(lb$_f$-s/lb$_m$)], yields the Classical Rocket Equation:

$$\ln(M_{wet}/M_{dry}) = +/- \Delta V/[g_c I_{sp}]$$
$$M_{wet}/M_{dry} = \exp(+/- \Delta V/[g_c I_{sp}])$$

Note that for acceleration, ΔV is positive ($\Delta V > 0$) since $V_f > V_i$. However, for deceleration, $\Delta V < 0$ since $V_f < V_i$. Thus, both equations give the expected result that $M_{wet} > M_{dry}$ independent of whether the vehicle is accelerating or decelerating.

2. Relativistic Rocket Equation

For the Relativistic Rocket Equation, we follow essentially the same steps as for the Classical Rocket Equation, with the difference being the use of relativistic equations for mass and velocity, and conservation of mass–energy content of the system (as opposed to conservation of mass only in the classical case). Details of the derivation are given in the appendix to this chapter.

First, we start with the conservation of mass–energy; i.e., mass–energy, or mc^2, is conserved. Also, we use relativistic mass, such that:

$$M = M_r/(1 - V^2/c^2)^{1/2} \quad \text{and} \quad m = m_r/(1 - u^2/c^2)^{1/2}$$

where M_r and m_r are the rest masses of the vehicle and propellant, respectively. Note also that the mass–energy conservation equation is the same for acceleration or deceleration. Next, we conserve momentum, for example, $M_r/(1 - V^2/c^2)^{1/2} \cdot V$. Finally, we use the relativistic addition (or subtraction) of velocities:

$$u = (w - /+ V)/(1 - /+ wV/c^2)$$

To begin the derivation, we expand the mass–energy conservation derivative and solve for dm_r. Next, we expand the conservation of momentum derivative, and substitute values for dm_r and u from above. After considerable algebraic manipulation (detailed in the appendix), we obtain the derivative form of the equation to be analytically integrated:

$$dM_r/M_r = -dV/[+/- w(1 - V^2/c^2)] = -c^2/(+/- w) \cdot [dV/(c^2 - V^2)]$$

For numerical integration, we would have:

$$dM_r = M_r\{-dV/[+/- w(1 - V^2/c^2)]\}$$
$$dM_r = M_r\{-c^2/(+/- w) \cdot [dV/(c^2 - V^2)]\}$$

Analytical integration gives:

$$\ell n(M_r) = -c/(+/- 2w) \cdot \ell n[(c + V)/(c - V)]$$
$$= -1/(+/- 2w/c) \cdot \ell n[(1 + V/c)/(1 - V/c)]$$

Evaluating the integrals for the initial and final rest-mass limits of M_{wet} and M_{dry}, initial and final velocities of V_i and V_f, with $V_f - V_i = \Delta V$, and

$w = V_{exhaust} = g_c I_{sp}$, yields the Relativistic Rocket Equation:

$ln(M_{wet}/M_{dry})$
$= c/(+/-2w) \cdot ln\{[(c + V_f)/(c - V_f)]/[(c + V_i)/(c - V_i)]\}$
$= 1/(+/-2w/c) \cdot ln\{[(1 + V_f/c)(1 - V_i/c)]/[(1 + V_i/c)(1 - V_f/c)]\}$
$= 1/(+/-2g_cI_{sp}/c) \cdot ln\{[(1 + V_f/c)(1 - V_i/c)]/[(1 + V_i/c)(1 - V_f/c)]\}$
M_{wet}/M_{dry}
$= \{[(1 + V_f/c)(1 - V_i/c)]/[(1 + V_i/c)(1 - V_f/c)]\}^{[1/(+/-2g_cI_{sp}/c)]}$

Note that we have dropped the subscript "r" denoting rest mass for convenience.

3. Classical Rocket Equation with "Loss" of Propellant

Our goal is to derive a Relativistic Rocket Equation for a matter–antimatter annihilation rocket where a significant fraction of the initial propellant mass, M_p, is converted into forms that are not usable for thrust production (e.g., high-energy gamma ray photons). However, this derivation is complicated by the need to conserve the mass–energy content of the various reactants and annihilation products. As an introduction to this problem, we will first perform a derivation of the Classical Rocket Equation with "loss" of propellant where we only need to conserve mass and momentum; in this case, we assume that a fraction "**a**" of the initial propellant mass is used to produce forward motion of the rocket, as shown in Fig. 8. In effect, a fraction (1 − **a**) of the initial propellant mass (and momentum) is "lost," at least insofar as production of forward momentum is concerned. It is as if a quantity (**a**)M_p is ejected parallel to the direction of motion (to produce some ΔV) and a quantity (1 − **a**)M_p is ejected perpendicular

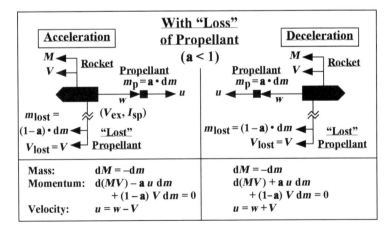

Fig. 8 Classical rocket mass, velocity, and momentum (with "loss" of propellant).

in such a way that $(1 - \mathbf{a})M_p$ does not contribute to the forward ΔV. Finally, note that the usual situation for rocket propulsion is for $\mathbf{a} = 1$; for example, even in a fission or fusion rocket, where some of the rest mass of the nuclear fuel is turned into energy, the amount of mass "lost" (to conversion into energy) is negligible, so the "standard" Classical or Relativistic Rocket Equations (with $\mathbf{a} = 1$) can be used.

For the Classical Rocket Equation with "loss" of propellant, we follow essentially the same steps as for the standard ($\mathbf{a} = 1$) Classical Rocket Equation, with the difference being the use of terms in (**a**) and $(1 - \mathbf{a})$ in the momentum conservation equation. Thus, we again write the equations for conservation of mass and momentum, and addition (or subtraction) of velocities. (Note again the use of the "$+/-$" and "$-/+$" sign convention for acceleration and deceleration.)

$$dM = -dm = -\{(\mathbf{a})dm \text{ [usable]} + (1 - \mathbf{a})dm \text{ [lost]}\}$$

Note that $\mathbf{a} \cdot dm$ represents the "usable" fraction of the (incremental) propellant mass, whereas $(1 - \mathbf{a}) \cdot dm$ represents the fraction "lost" and unavailable for production of velocity (V). For conservation of momentum, we have:

$$d(M \cdot V) = +/- u \cdot \mathbf{a} \, dm \text{ [only } \mathbf{a} \, dm \text{ contributes to } V]$$
$$- V(1 - a) \, dm \text{ [momentum "lost"]}$$

As above, only $u \cdot \mathbf{a} \cdot dm$ contributes momentum toward changing V. Also, in defining w and u, we are only interested in the "bulk" or center-of-mass velocity of the (incremental) propellant mass quantity $\mathbf{a} \cdot dm$ moving parallel to V. Thus, the exhaust velocity ($w = V_{\text{exhaust}}$) or I_{sp} of the rocket engine corresponds to an "effective" I_{sp} or V_{exhaust} that takes into account any inefficiencies in the nozzle in converting the random velocities of the propellant exhaust gas into directed motion (w and u) parallel to V.

Similarly, the (incremental) propellant mass $(1 - \mathbf{a}) \cdot dm$ is ejected at some exhaust velocity (not necessarily w) perpendicular to V (and u) in a symmetric fashion that results in a net cancellation of the perpendicular velocity of the center-of-mass of $(1 - \mathbf{a}) \cdot dm$. However, this unit of mass is still moving forward at the rocket's velocity, V, so the total amount of "lost" momentum is $V \cdot (1 - \mathbf{a}) \cdot dm$. Interestingly, this momentum is totally lost from the system (its value is negative for either acceleration or deceleration), so it also reduces the mass and momentum of the rocket that must be accelerated or decelerated in subsequent differential time steps. However, it also represents a "dead" weight that must be initially loaded into the rocket, which increases M_{wet} over that seen when $\mathbf{a} = 1$.

As before, we expand the conservation of mass equation, solve for dm, and then expand the conservation of momentum equation and substitute dm and u as appropriate to get the equation to be analytically integrated:

$$dM/M = -/+ dV/(\mathbf{a} \, w)$$

If we were performing a numerical integration, we could write:

$$dM = M[-/ + dV/(\mathbf{a}\ w)]$$

Analytical integration gives:

$$\ell n(M) = -/ + V/(\mathbf{a}\ w)$$

and again when evaluated at the integration limits gives:

$$\ell n(M_{\text{wet}}/M_{\text{dry}}) = +/ - \Delta V/(\mathbf{a}\ g_c I_{\text{sp}})$$
$$M_{\text{wet}}/M_{\text{dry}} = \exp[+/ - \Delta V/(\mathbf{a}\ g_c I_{\text{sp}})]$$

Note that the final form of the equation is similar to that of the standard ($\mathbf{a} = 1$) Classical Rocket Equation. The result has been to add **a** to the I_{sp} term, resulting in an "effective" I_{sp} that takes into account the "loss" of propellant. In fact, this result could be anticipated from the classical definition of I_{sp} in terms of thrust (F) and propellant mass flow rate ($\dot{m}$) entering the rocket engine:

$$g_c I_{\text{sp}} = F/\dot{m} = V_{\text{exhaust}}$$

In this equation, $g_c = 9.8$ [m/s]/[lb$_f$-s/lb$_m$], $I_{\text{sp}} = $ lb$_f$-s/lb$_m$, $F = $ Newtons, $\dot{m} = $ kg/s, and $V_{\text{exhaust}} = $ m/s. In the context used here, $\dot{m}$ is the *total* mass of propellant *entering* the rocket engine. However, only a quantity $\mathbf{a} \cdot \dot{m}$ is available to produce thrust, so we would have:

$$g_c I_{\text{sp}} = F/(\mathbf{a}\ \dot{m}\ [\text{Total}])$$

which rearranges to give the same [$\mathbf{a}\ g_c\ I_{\text{sp}}$] term seen in the Rocket Equation above:

$$\mathbf{a}\ g_c I_{\text{sp}} = F/\dot{m}\ [\text{Total}]$$

4. Relativistic Rocket Equation with "Loss" of Propellant

For this derivation, we will combine elements of the derivations of the Relativistic Rocket Equation and Classical Rocket Equation with "loss" of propellant. However, unlike the classical version of this equation, where we conserve only (rest) mass and classical momentum (rest mass multiplied by velocity), the relativistic version requires that we conserve the total amount of mass–energy and relativistic momentum, including that of the "lost" propellant. Thus, because of the need to include the relativistic mass–energy and momentum of the "lost" propellant, the algebra will be more complex, resulting in an equation that must be numerically integrated.

Also, for a matter–antimatter annihilation reaction, we need to bookkeep the rest mass and kinetic energy of the various reactants (protons, antiprotons, electrons, positrons) and annihilation products (charged and neutral pions, gamma ray photons). This is needed to determine both the effective value of "**a**" in the equation (i.e., how much of the mass–energy and momentum are usable for propulsion), as well as to determine the mass–energy and momentum of the various annihilation products. Table A1 in the appendix summarizes mass and energy quantities based on the assumption of a reaction between an atom of hydrogen (H), containing a proton (P^+) and electron (e^-), and an anti-atom of anti-hydrogen containing an antiproton (P^-) and positron (e^+) [14]. The protons and antiprotons annihilate to give neutral and charged pions (π^o, π^+, π^-); the electrons and positrons annihilate to give high-energy photons (gamma rays, γ). The initial annihilation reaction is given as [19]:

$$P^+ + P^- \rightarrow 2.0\pi^o + 1.5\pi^+ + 1.5\pi^- \quad \text{and} \quad e^- + e^+ \rightarrow 2\gamma$$

The neural pions promptly decay into very high-energy gamma rays (each about 355 times more energetic than an electron/positron gamma). The charged pions decay into charged muons (μ) and neutrinos (ν) after traveling about 21 meters at $0.93\ c$ [18,19].

$$2.0\pi^o \rightarrow 4\gamma,\ 1.5\pi^+ \rightarrow 1.5\,\mu^+ + 1.5\,\nu_m,\ 1.5\pi^- \rightarrow 1.5\,\mu^- + 1.5\,\text{anti-}\nu_m$$

To determine the value for **a**, we first determine the fractions (percentages) of the total initial mass–energy contained in the various annihilation species. For the charged pions (π^+, π^-), we can further split the total mass–energy content into rest mass (m_r) and kinetic energy, such that:

$$\text{Rest mass–energy} = m_r c^2$$

$$\text{Kinetic energy} = m_r c^2 \{1/(1 - v^2/c^2)^{1/2} - 1\}$$

$$\text{Total mass–energy} = m_r c^2 /(1 - v^2/c^2)^{1/2}$$

This results in the charged pions having a rest mass that is 22.3% of the total mass–energy content of the reactants (H and anti-H atoms), and a kinetic energy that is about 39.9% of the initial mass-energy. However, even though the total mass–energy content of the charged pions is about 62.2% of the total mass–energy content of the reactants (H and anti-H atoms), this is not the value of **a**. This is because the electromagnet used to deflect the charged pions is not perfectly reflective (some charged pions "leak" through the magnetic field and travel upstream), and for those that are reflected, the resultant beam is not perfectly collimated. This is illustrated in Fig. 9 [14,20], where straight-line particle trajectories are the gamma rays (from neutral pion decay) and non-linear particle trajectories are the charged pions deflected by the magnetic field from a single-loop magnet (shown superimposed on the annihilation region).

Thus, there is an inherent kinetic energy inefficiency in the nozzle that must also be taken into account. Callas [20], using Monte Carlo simulations of particle trajectories in proton/antiproton annihilations, estimated that the magnetic nozzle would be at best 50% kinetic energy efficient, corresponding to a value of **a** = 42.2%. We have arbitrarily assumed a value of **a** = 40% of the initial total mass–energy content of the propellant, corresponding to a

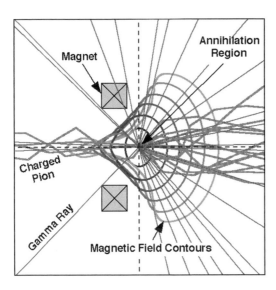

Fig. 9 Proton-antiproton annihilation sample Monte Carlo simulation results.

44.37% nozzle kinetic energy efficiency. These values are summarized in Table A2 in the appendix. Note that the magnetic nozzle does not affect the fraction of rest mass of the charged pions (22.3%); that remains the same. Instead, what we have done is take into account the nozzle inefficiency in converting the kinetic energy of the charged pions, whose 0.9334 c velocity vectors are pointed in random directions, into a directed, collimated stream of charged pions with a 0.6729 c kinetic energy velocity vector pointed directly out the back of the rocket.

Finally, the discussion above has concentrated on the efficiency of the annihilation reaction and electromagnet nozzle in capturing the total mass–energy (rest mass and kinetic energy) of the charged pions. However, there are additional losses in the magnetic nozzle that result in a reduction in overall I_{sp}. For example, although the magnetic nozzle can capture 40% of the reactant's initial mass–energy for thrust production, the "beam" of charged pions leaving the engine are not perfectly collimated (i.e., parallel to V). Using the Monte Carlo particle trajectories, Callas [20] used direct momentum calculations to determine an "effective" I_{sp} or $V_{exhaust}$ of 1×10^8 N-s/kg = m/s = $0.333\ c$ = 1.02×10^7 lb$_f$-s/lb$_m$.

To better understand the distinction between mass–energy content and efficiency (i.e., **a**) and nozzle efficiency (i.e., I_{sp}), it may be useful to consider a chemical rocket engine as an analogy. For example, in the chemical propulsion world there are computer codes that calculate the thermochemical properties (enthalpy, combustion temperature, mole fractions, etc.) of the various reaction products from a given set of propellants (e.g., from hydrazine decomposition or oxygen–hydrogen combustion). Sophisticated versions of these codes can also account for inefficiencies and losses, like incomplete combustion due to finite reaction rates (e.g., kinetics), energy losses to the walls of the reaction chamber, film cooling, and so on. This corresponds to

LIMITS OF INTERSTELLAR FLIGHT TECHNOLOGY   51

the calculation of **a** for the annihilation reaction. For a chemical rocket, the computer program then calculates the isenthalpic expansion of the reaction products through a nozzle with a given expansion ratio and pressure conditions (e.g., expansion into sea-level pressure or vacuum at the nozzle exit). Again, depending on the sophistication of the program, various losses can be calculated, in addition to the inherent limitations based on expansion ratio and nozzle back pressure. This corresponds to the calculation of the magnetic nozzle's I_{sp} [20].

To begin our derivation, we again begin with conservation of mass–energy and momentum, expand the conservation of mass equation, solve for dm_r, and expand the conservation of momentum equation and substitute dm_r and u as appropriate to get the equation to be integrated. This gives us (after considerable algebraic manipulation as described in the appendix):

$$dM_r/M_r = -dV/(1 - V^2/c^2) \cdot [(1 - V^2/c^2 \cdot \mathbf{X}) - /+ wV/c^2(1 - \mathbf{X})]/$$
$$[V \cdot (1 - \mathbf{X}) + /- w \cdot (\mathbf{X} - V^2/c^2)]$$

where we have combined the various terms containing **a** into the quantity **X**:

$$\mathbf{X} = [\mathbf{a}(1 - u^2/c^2)^{-1/2} - (1 - \mathbf{a})/(+/- u) \cdot (1 - V^2/c^2)^{-1/2} \cdot V]/$$
$$[\mathbf{a}(1 - u^2/c^2)^{-1/2} + (1 - \mathbf{a})(1 - V^2/c^2)^{-1/2}]$$

Note that when $\mathbf{a} = 1$, $\mathbf{X} = 1$, and this equation reduces to the form seen for the standard ($\mathbf{a} = 1$) Relativistic Rocket Equation:

$$dM_r/M_r = -dV/(1 - V^2/c^2)/[+/- w] = -dV/[+/- w(1 - V^2/c^2)]$$

Unfortunately, when **a** or **X** are not 1, the various terms in V do not cancel out, leaving us with an equation that must be numerically integrated. For this, we use the equation in the form:

$$dM_r = M_r \cdot \{-dVM_r/(1 - V^2/c^2) \cdot [(1 - V^2/c^2 \cdot \mathbf{X})$$
$$- /+ wV/c^2(1 - \mathbf{X})]/[V \cdot (1 - \mathbf{X}) + /- w \cdot (\mathbf{X} - V^2/c^2)]\}$$

where **X** is defined above and $u = (w -/+ V)/(1 -/+ wV/c^2)$.

5. Comparison of the Rocket Equations

Figure 10 illustrates a comparison of the various Rocket Equations. Not surprisingly, the standard ($\mathbf{a} = 1$) Classical and Relativistic versions do not diverge significantly until the vehicle's final velocity becomes a significant fraction of the speed of light. We also see the dramatic impact that reductions in **a** have to the vehicle mass ratio (M_{wet}/M_{dry}). For example, if an antimatter rocket could have a value of $\mathbf{a} = 1$, its mass ratio would only be 2.2 to accelerate from $V = 0$ to $V = 0.25\,c$. However, with a value of $\mathbf{a} = 0.4$, the mass ratio rises dramatically to 6.4. In practical terms, in order to keep the mass ratio to a

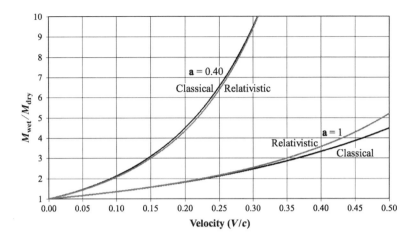

Fig. 10 Comparison of classical and relativistic rocket equations.

reasonable value, a relativistic antimatter rocket, even with an exhaust velocity of 0.3 c, must limit its ΔV per stage to no more than 0.25 c, because of the loss of so much mass–energy from the initial mass–energy content of the propellant (i.e., $1 - \mathbf{a} = 0.6$). Thus, we would need to have a four-stage vehicle for a fast ($V_{\text{cruise}} = 0.5\ c$) interstellar rendezvous mission, with stages 1 and 4 performing the "low" speed acceleration ($0.00 \rightarrow 0.25\ c$ stage 1) or deceleration ($0.25 \rightarrow 0.00\ c$ stage 4), and stages 2 and 3 doing the "high" speed acceleration ($0.25 \rightarrow 0.50\ c$ stage 2) or deceleration ($0.50 \rightarrow 0.25\ c$ stage 3). Finally, note that classical and relativistic acceleration and deceleration are symmetric with respect to each other; for example, the mass ratio is the same when accelerating from $0.00 \rightarrow 0.25\ c$ or when decelerating from $0.25 \rightarrow 0.00\ c$.[§]

Figure 11 illustrates a second effect seen for the Relativistic Rocket Equations where the actual values of V_i and V_f, and not just ΔV, are important. Because of relativistic effects, the mass ratio will be different for a relativistic rocket accelerating from 0.00 to 0.25 c compared to one accelerating from 0.25 to 0.50 c (i.e., both with the same $\Delta V = 0.25\ c$); at higher initial and final speeds, the mass ratio is significantly higher, reflecting the greater impact that relativistic effects have at higher velocities. This effect is completely absent from the Classical Rocket Equations where only ΔV is relevant.

Thus, to have two relativistic stages reach 0.5 c and have the same mass ratio, the first stage would need to have a ΔV greater than 0.25 c, and the second stage a ΔV less than 0.25 c. For example, to accelerate from 0.0 to 0.5 c, stage 1 would have a ΔV of 0.268 c (starting from 0.000 c), stage 2 would have a ΔV of 0.232 c (starting from 0.268 c), and both stages would have the same $M_{\text{wet}}/M_{\text{dry}} = 7.312$. By contrast, classical rocket stages could both have a ΔV of 0.25 c and have the

---

[§]This is exactly true for the standard, $\mathbf{a} = 1$ Relativistic Rocket Equation, as can be see from its analytically integrated form. This appears to also be true when $\mathbf{a} = 0.4$ because the numerical difference between the mass ratio for acceleration and deceleration is of the same order for the numerically-integrated Relativistic Rocket Equations for both $\mathbf{a} = 1$ and $\mathbf{a} = 0.4$.

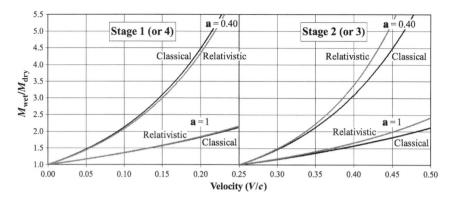

Fig. 11 Comparison of Classical and Relativistic Rocket Equations for a two-stage vehicle with $\Delta V = 0.25\ c$ per stage and $V_i = 0.00\ c$ versus $V_i = 0.25\ c$.

same mass ratio. In effect, relativistic effects make acceleration (or deceleration) from a higher initial velocity "harder" because the relativistic mass increase applies to the total loaded vehicle, so more propellant is used and the mass ratio is higher.

B. Relativistic Light Sail Equations

As discussed above, solar sails are unable to provide the performance (acceleration and peak velocity) needed for interstellar missions. This is primarily driven by the $1/R^2$ drop-off in sunlight intensity, and thus the photon momentum "push" on the sail, as the solar sail moves away from the sun. The classic solution to this problem is to use a laser or microwave source to drive the sail; in this mode, the sail is referred to as a generic "light sail" or laser sail/microwave sail as appropriate to the wavelength of the source. Note that the primary advantage of visible or near-visible laser wavelengths is the ability to focus the laser beam at interstellar distances, given sufficiently large transmitter (laser) and receiver (light sail) optics. In this section we will follow Forward's development of the mathematical relationships used to determine light sail size (based on the assumption of a diffraction-limited optics system), laser power requirements (based on the light sail mass, acceleration, and optical properties such as reflectivity and absorbtivity), light sail mass (based on its mass per unit area or "areal density"), light sail thermally-limited maximum acceleration (based on sail areal density, optical properties including emissivity, and maximum operating temperature) [13].

We will also use Forward's method [13] of using a "two-stage" light sail for an interstellar rendezvous mission, as shown in Fig. 12. In this approach, a large-diameter laser transmitter and light sail first stage is used to accelerate the vehicle (first and second stage) away from Earth. As the vehicle approaches the target star, the first stage and the smaller-diameter second stage (with one-tenth the area of the first stage [13]) separate, and the first stage is used to retro-reflect the laser beam back onto the second stage. This has the effect of

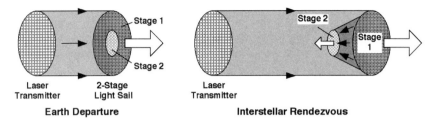

Fig. 12 Laser-driven light sail interstellar rendezvous mission.

slowly accelerating the first stage (because of its larger mass) out of the target solar system, but rapidly decelerating the second stage (because of its smaller mass) and bringing it to a stop in the target solar system. The need to focus the laser onto the first stage at interstellar distances drives the overall size and scale of the mission, as is shown next.

1. Light Sail Diameter

If we assume the limiting (best) case of diffraction-limited optics, the diameter of stage 1 receiving the laser beam, D_r, is determined by the diameter of the laser transmitting optics, D_t, the wavelength of the laser light, λ, and the distance between the transmitter and receiver, L:

$$D_r = 2.440 \lambda L / D_t$$

The receiver's diameter D_r (and the constant 2.440) corresponds to the first minimum (null) in a circular diffraction pattern; the bright central "spot" inside this null ring is called the "Airy Disk" and contains 83.78% of the total beam power (with the remaining power contained in bright diffraction rings outside D_r). For example, with a laser wavelength of 1 μm, a 1000-km diameter transmitter, D_t has a "spot" size (airy disk) D_r of 1000 km at 43 LY; similarly, $D_t = D_r = 316$ km at 4.3 LY.

2. Laser Beam Power

The laser beam power required to push the light sail ($P_{\text{light sail}}$) is directly proportional to the light sail's total mass including payload, $M_{\text{light sail}}$; acceleration, a; and speed of light, c; and inversely proportional to its reflectivity, η, and absorbtivity, α, at the laser beam wavelength¶:

$$P_{\text{light sail}} = M_{\text{light sail}} a c / (2\eta + \alpha)$$

The total laser transmitted beam power, P_{laser}, required to deliver $P_{\text{light sail}}$ to the light sail is dependent on the fraction of the laser beam intercepted by the sail

---

¶The absorbtivity term (α) in the denominator of this equation was omitted in Forward's original derivation [13]; this was corrected by him in later papers.

LIMITS OF INTERSTELLAR FLIGHT TECHNOLOGY

("spot size" efficiency, $\eta_{spot} = 83.78\%$ for a diffraction-limited optics system):

$$P_{laser} = P_{light\ sail} / \eta_{spot}$$

3. Light Sail Mass and Areal Density

The sail itself (without payload) is assumed to consist of three major elements, the sail sheet, sail structure, and sail high-emissivity, ε, coating (to aid in rejecting any power absorbed from the laser input power):

$$M_{sail\ only} = M_{sheet} + M_{structure} + M_{high\text{-}\varepsilon\ coating}$$

Dividing each mass term by the total sail reflecting area ($A_r = \pi D_r^2/4$) gives a corresponding areal density ($\sigma = M/A$, typically expressed in g/m^2):

$$\sigma_{sail\ only} = \sigma_{sheet} + \sigma_{structure} + \sigma_{high\text{-}\varepsilon\ coating}$$

The total light sail mass and corresponding areal density with payload are thus:

$$M_{light\ sail} = M_{sail\ only} + M_{payload}$$
$$\sigma_{light\ sail} = \sigma_{sail\ only} + \sigma_{payload}$$

For our analyses [2], we have diverged slightly from Forward's original assumption that the total light sail areal density (including payload) would be $\sigma_{light\ sail} = 0.1$ g/m^2 independent of sail size [13]. Instead, we have arbitrarily assumed a 100 MT robotic spacecraft payload and fixed the interstellar rendezvous sail-only areal density at $\sigma_{sail\ only} = 0.1$ g/m^2 (independent of sail size). With large sails, this assumption results in a payload areal density that approaches zero; more generally, it allows us to treat the sail by itself and add an arbitrarily large payload for mission analyses. Also, as discussed in more detail below, this allows us to add a dedicated high-emissivity layer of heat-rejecting material to the backside of the sail (the side facing away from the laser) so as to reject any heat absorbed by the sail and thus allow higher laser powers and correspondingly higher acceleration. Thus, if we assume a total sail-only areal density of 0.1 g/m^2 for the interstellar rendezvous light sail, we can then subtract out Forward's values [13] for the sail sheet areal density (0.043 g/m^2 for a 16-nm thick Al—63 Al atoms thick!—sheet) and sail structure areal density (0.030 g/m^2 for a spinning sail), leaving an areal density of 0.027 g/m^2 for the high-emissivity backside coating.

4. Thermally Limited Light Sail Acceleration

One of Forward's important insights was that, although high acceleration is generally desirable to minimize trip time, high acceleration is absolutely essential during the rendezvous portion of a light sail mission to minimize the separation distance between stage 1 and stage 2. This minimizes the optical requirements on stage 1 as it is used as a transmitter to retro-reflect laser beam power onto stage 2

[13]. In principle, this could be achieved by simply increasing the laser power hitting stage 1 of the light sail. Because stage 2 has 1/10 the area and thus mass of stage 1 [13] (assuming $\sigma_{\text{light sail}}$ is the same for both stages), the laser beam retro-reflected from stage 1 back onto stage 2 will cause 10 times higher deceleration in stage 2 than acceleration in stage 1. In effect, the high deceleration of stage 2 allows it to come to rest before the distance between the two stages becomes so great that stage 1 can no longer effectively retro-reflect the laser beam back to stage 2.

However, there is a thermal limit to the amount of laser power (or, more precisely, laser power intensity, i.e., $I = P/A = W/m^2$) that can be used on the sail. In effect, the power per unit area absorbed by the sail must be no greater than what can be radiatively emitted by the sail. (This effect is equally true for solar sails and limits their closest approach to the sun.) Thus, light sail acceleration is ultimately limited by how much laser power (intensity) can be projected onto the sail due to the sail sheet's finite absorbtivity and material temperature limits. This results in a thermally-limited acceleration that is a complex function of sail sheet reflectivity η; absorbtivity, α; emissivity, ε; maximum temperature, $T_{\max}$; and the total light sail areal density including payload, $\sigma_{\text{light sail}}$.

To find the thermally-limited maximum acceleration, $a_{\max}$, we first balance the amount of power or P/A absorbed by the sail with that radiated into space (from both sides of the sail):

$$(P/A) \text{ Absorbed} = (P/A) \text{ Radiated away}$$

$$\alpha(P_{\text{light sail}}/A) = \sigma_{SB}(\varepsilon_{\text{frontside}} + \varepsilon_{\text{backside}})(T_{\max}^4 - T_{\text{sink}}^4)$$

where σ_{SB} is the Stefan-Boltzmann constant (5.67×10^{-8} W/m²K⁴), $\varepsilon_{\text{frontside}}$ and $\varepsilon_{\text{backside}}$ are the emissivities on the front side (facing the laser) and back side (facing away from the laser) of the light sail, respectively, and T_{sink} is the background temperature of space (assumed to be 0 K). (Also, $T_{\max}$ is in Kelvin and $\sigma_{\text{light sail}}$ is converted from g/m² to kg/m² for unit consistency with the other quantities.) Substituting the definition of $P_{\text{light sail}}$ into this equation and solving for the acceleration (and using $M_{\text{light sail}}/A = \sigma_{\text{light sail}}$) gives the thermally-limited acceleration ($a_{\max}$):

$$a_{\max}(m/s^2) = (2\eta + \alpha)\sigma_{SB}(\varepsilon_{\text{frontside}} + \varepsilon_{\text{backside}})(T_{\max}^4 - T_{\text{sink}}^4)/(\sigma_{\text{light sail}} c \alpha)$$

Thus, to achieve high acceleration or deceleration, we need a sail material with low $\sigma_{\text{light sail}}$, high ε, high $T_{\max}$, low α, and high η.

Another very important result of Forward's analyses was the identification of an optimum sail sheet thickness that yields a maximum thermally-limited acceleration [13]. In his analyses, Forward found that for aluminum (Al), the optimum sail thickness was 16 nm; at this thickness, Al has a $\eta = 0.820$, $\alpha = 0.135$, and transmittance $\tau = 0.045$ (with $\eta + \alpha + \tau = 1$). Assuming the emissivity of bare Al ($\varepsilon = 0.06$) on both sides of the sail sheet, and combining this with a conservative maximum operating temperature of 600 K (only 64% of the melting point of Al at 933 K) and a total light sail areal density of 0.1 g/m², results in a thermally limited maximum acceleration of 0.3865 m/s² (0.03944 g). Under these conditions, the maximum laser power intensity hitting the sail is 6.532 kW/m² (4.838 "suns" with one sun = 1.350 kW/m²). However, in order to perform a rendezvous mission, a significantly higher acceleration is required. For this application, Forward assumed the use of an advanced high-emissivity coating on the backside of the sail sheet such that $\varepsilon_{backside} = 0.95$ (and an $\varepsilon_{frontside} = 0.06$ of bare Al); this allowed an increase in the thermally limited sail acceleration to 3.253 m/s² (0.3319 g). For our analyses, we will assume a more conservative $\varepsilon_{backside} = 0.60$ (10 times $\varepsilon_{frontside}$) which results in a thermally limited acceleration of 2.126 m/s² (0.2169 g) and a corresponding laser power intensity on the sail sheet of 35.925 kW/m² (26.611 suns).

Note that a high-ε backside coating was a somewhat arbitrary assumption made by Forward (and this author) in order to have a light sail with a sufficiently high acceleration to perform a rendezvous mission; thus, one significant technology challenge for an interstellar rendezvous light sail is the need to develop a low-σ, high-ε coating capable of producing the required emissivity. Similarly, the need for an extraordinarily thin sail sheet (16 nm), to produce a low $\sigma_{light\ sail}$ for either flyby or rendezvous missions is a major challenge in its own right. One suggested approach to fabricating such an ultra-thin sheet is to use in-space fabrication where Al vapor is sprayed [e.g., by chemical vapor deposition (CVD), technologies] onto a relatively thick substrate (e.g., plastic). The substrate then evaporates (sublimes) or decomposes in sunlight to leave the Al sheet.

Finally, a diffraction-limited optics system produces a power (intensity) distribution across the surface of the light sail with a central peak intensity 4.38 times the average intensity [2]. This raises significant issues with respect to the sail temperature at the center of the sail disk because the total laser beam power hitting the light sail was based on a thermal limit for the *average* intensity across the whole sail, and not the intensity in the center. There are several approaches to ameliorating the additional intensity in the center of the light sail [2]. For example, the simplest approach would be to just accept higher temperature at the center of the light sail. In this case, the 4.38-times greater intensity would increase the sail sheet absolute temperature by a factor of $(4.38)^{1/4} = 1.447$, corresponding to an increase from the nominal 600 K to 868 K. Although this is below the melting point of Al at 933 K, this may still not be feasible because, as Forward notes [13], ultra-thin films can fail by agglomeration (i.e., thin films have a large surface-to-volume ratio, permitting them to reduce their surface energy by forming droplets). For the case of Al, this agglomeration temperature may be as low 725 K [13]. Because of these temperature limitations for Al, other authors have suggested the use of higher-temperature metals and nonmetals [21].

5. Relativistic Effects

For sails moving at 0.5 c, we need to explicitly consider relativistic effects on sail acceleration due to changes in both mass and laser power (redshifting of the wavelength of the laser's photons hitting the sail). Generally, these two relativistic effects result in a reduction in acceleration due to an increase in the (relativistic) mass of the light sail and a reduction in effective power (momentum) of the photons driving the light sail. Also, the increase in wavelength of the redshifted photons can impact the characteristics of the diffraction-limited retroreflection of light from the sail first stage back onto the second stage during the rendezvous maneuver.

The standard equations for relativistic mass increase and wavelength redshifting are:

$$M/M_r = [(1/(1 - V^2/c^2))]^{1/2} = 1.15$$

$$\lambda/\lambda_r = [(1 + V/c)/(1 - V/c)]^{1/2} = 1.73$$

where M = relativistic mass, M_r = rest mass (at $V = 0$), λ = relativistic (redshifted) wavelength, λ_r = wavelength at rest ($V = 0$), and the numerical values are given for $V = 0.5\ c$.

Finally, we have for the photons

$$E = h\nu = hc/\lambda$$

where E = energy (e.g., Joules), h = Planck's constant (6.626×10^{-34} J-s), and ν = frequency (s^{-1} or Hz).

For example, at $V = 0.5\ c$, the light sail's relativistic mass increases by a factor of 1.15; this has a relatively small effect on overall acceleration performance. However, relativistic photon redshift represents a much larger impact than the relativistic mass increase. For example, the wavelength (λ) of laser photons striking the light sail at $0.5\ c$ will be increased (redshifted) by a factor of 1.73, resulting in a corresponding decrease in the photon frequency, momentum, and energy (and power) by a factor of $1/1.73$. Similarly, during the interstellar rendezvous, the photons retro-reflected from stage 1 back onto stage 2 experience a second redshift factor of 1.73; in effect, the photons seen by stage 2 are "double" redshifted, that is, $1/1.73^2 = 1/3.00$. Also, the increased wavelength of photons retro-reflected from stage 1 back onto stage 2 will cause the diffraction-limited spot size on stage 2 to increase by a factor of 1.73. Finally, note that for a constant–power/constant–wavelength transmitter, the overall sail acceleration at $0.5\ c$ will decrease by a factor of $1/(1.15 \cdot 1.73) = 1/2$, as compared to its acceleration at $V = 0$, due to the combined relativistic mass and wavelength increase.** However, in our analyses, we have assumed that laser power is "cheap," and sized the laser system for the worst-case power conditions of acceleration (for a flyby) or deceleration (for stage 2 rendezvous). In part, this is based on the assumption that the laser system can be upgraded as needed relatively easily because it is nearby within our solar system.

An alternative approach would be to vary the wavelength of the laser system to compensate for relativistic redshifting. Generally, this is not easily implemented

---

**Note that relativistic effects were incorrectly ignored in the light sail mission analyses presented in Frisbee and Leifer [3], even though they are discussed in Forward [13]. These errors were corrected in Frisbee [2].

in most lasers, although it is an option for the free electron laser (FEL). Also, transmissions optics (e.g., adaptive optics) are typically optimized for a single wavelength or narrow wavelength band. The issue remains, however, as to the feasibility of large changes in wavelength (e.g., from 1 to $1/3$ μm in the example above) for the laser and its optics, so we have taken the simpler approach of varying laser power as needed.

C. Bussard Interstellar Fusion Ramjet Mission Performance and Momentum Drag

The interstellar fusion ramjet is still highly conceptual; however, it has the potential advantage of providing unlimited range and mission flexibility and so warrants continued study. Here, we will describe the rather unconventional mission performance encountered when dealing with the interstellar fusion ramjet, and describe the impact that the need to minimize momentum drag has on the design of the interstellar hydrogen ram-scoop [22]. Finally, it is important to note that the ram-scoop is not a physical structure, but a magnetic field, so even though the "scoop" has a large frontal cross-sectional area to collect the low-density interstellar fusion fuel, the mass of the scoop consists only of the magnets (and their supporting structure) that project the magnetic field out in front of the vehicle.

1. Interstellar Fusion Ramjet Mission Performance

The mission performance (acceleration, speed, etc.) of an interstellar fusion ramjet has several unique aspects that are not typically encountered in more conventional rocket systems (like an antimatter rocket). For example, the mission can be broken into several steps depending on the propulsion mode being employed. Assuming a constant-thrust (force) fusion engine, there is an initial phase where the vehicle accelerates to the onset of ram-scoop operation, typically $\sim 0.06\ c$, using onboard propellant. During this phase, the mass of the vehicle gradually decreases as onboard propellant is consumed, so the acceleration increases. Because of the low speed at onset of ram-scoop operation, there is a negligible relativistic mass increase of the vehicle. Ram-scoop operation is achieved when the forward speed of the vehicle collects a mass of H or D at the same rate as that consumed and exhausted by the engine. (Interstellar H would be used if a H–H fusion cycle proves feasible; a fallback would be D, as is used in contemporary D–D fusion cycles, although there is less D than H in interstellar space.) In this case, the mass flow rate ($\dot{m}$) of fusion fuel into the ram-scoop equals the $\dot{m}$ of the exhaust of the fusion engine. During this time, the thrust and rest mass remains the same, so acceleration is nearly constant, decreasing slightly due to the relativistic mass increase of the vehicle (e.g., by a factor of 1.15 at 0.5 c).

The unique feature of the interstellar fusion ramjet is that, once it reaches ram-scoop operation, it can continue to accelerate to ultra-relativistic speeds (e.g., $>0.9\ c$) without any mass penalty (other than relativistic mass increase). For speeds above the onset of ram-scoop operation, the ram-scoop magnetic field geometry can be adjusted (e.g., the "scoop" collection area reduced) to maintain the same mass flow ($\dot{m}$) of H (or D) into the fusion engine and thus maintain constant thrust. For a flyby mission, ultra-relativistic speeds can be especially advantageous for distant stellar targets because there is a long distance (and time) over which to accelerate, and a long coast distance that is made more tolerable by a high V_{cruise}. For a rendezvous mission, the same general argument can also be made, especially because the ramscoop can act as a momentum "brake" or "parachute" to slow the vehicle without the need to use the fusion engine.

In the case of deceleration, we simply leave the ram-scoop magnetic field turned on, such that it intercepts interstellar H (or D) and brings the H (or D) to a stop with respect to the vehicle's forward motion. The H (or D) can be vented perpendicular to the vehicle's forward motion so as to produce no sideways ΔV. As long as the vehicle's forward velocity is greater than the fusion engine's V_{exhaust} (e.g., 0.03 c), the intercepted interstellar H/D will act like a decelerating rocket engine with a V_{exhaust} equal to the vehicle's forward speed. During this period, acceleration will increase slightly because relativistic mass is decreasing. Finally, as the forward speed approaches that of the engine's V_{exhaust}, the onboard propellant tanks can be filled from the scooped H (or D). This will cause a significant decrease in deceleration due to the mass of H (or D) propellant added to the vehicle. Finally, the vehicle is turned around and the fusion rocket operated in a normal vehicle-decelerating fashion to bring the vehicle to rest in the target solar system. This will produce an increase in deceleration as onboard propellant is consumed. Note that because the onboard propellant tanks are sized to deliver a ΔV equal to ram-scoop onset (e.g., 0.06 c), there will be residual onboard propellant to supply ΔV (e.g., 0.06 c-$V_{\text{exhaust}} = 0.03\ c$ in this example) for exploration of the target solar system.

2. *Momentum Drag and Its Impact on Ram-Scoop Design*

One major feasibility issue associated with the interstellar fusion ramjet is the same one encountered in supersonic ramjet (scramjet) systems, where the engine thrust must be greater than the ram inlet and engine drag. This suggests a cylindrical-geometry magnetic confinement fusion (MCF) reactor, such as a tandem-mirror reactor, where solenoid magnets confine the plasma radially. Momentum drag would also be an issue for the magnetic scoop, because the H (or D) atoms must be accelerated radially inward toward the centerline of the scoop without being axially decelerated to produce drag. As mentioned above, the ram-scoop is not a physical structure, but a magnetic field; however, it is more complex than a single-coil magnetic nozzle "run backwards." (A magnetic nozzle run this way would "choke" and exclude H at the throat of the scoop.) Another unique requirement of the scoop magnetic field is its immense size (e.g., dimensions of thousands of kilometers), implying powerful magnetic fields.

This suggests that the scoop magnetic field would need to have a very large length-to-diameter (L/D) ratio to minimize the axial contact angle of the ions with the field. An example of this is shown in Fig. 13, which shows a ramscoop with a cone half-angle (i.e., axial contact angle) of 1°. After "reflection" off of the ram-scoop's magnetic "wall," the H (or D) atom is traveling at an angle of 2° (i.e., twice the initial contact angle). For an initial atom speed

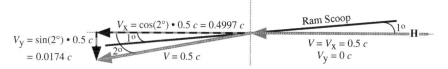

Fig. 13 Ram-scoop geometry for 1% axial momentum drag. (Angles are exaggerated by factor of five for clarity.)

of 0.5 c (relative to the vehicle) corresponding to the maximum cruise velocity, the atoms after reflection have an axial (x-axis) velocity of 0.4997 c, or 0.0003 c slower than their initial speed. This velocity change, 0.0003 c, represents 1% of the I_{sp} (or $V_{exhaust}$) of the fusion rocket engine, so the overall momentum drag is only 1% of the I_{sp} of the engine. In effect, the engine would now have an effective I_{sp} of 0.0297 c because of the momentum drag of the ram-scoop. Note, however that in order to produce only 1% drag from the ram-scoop, the L/D of the ram-scoop is 114.6/1 [i.e., 1/tan(1°) for L/R or 2/tan(1°) for L/D], resulting in the need for a magnetic "wall" 687,500 km long (almost 1.8 times the Earth–moon distance!) for a 6000 km diameter scoop. Needless to say, this represents only one of the many technical challenges facing the interstellar fusion ramjet.

Finally, an alternative approach to ameliorating momentum drag from the ram scoop that has been suggested by B. Cassenti (personal communication, 30 December 2007) would be to use magnetic fields to re-accelerate the scooped atoms beck to their original speed (e.g., 0.5 c) prior to entering the fusion reactor. Energy for this process could come from the fusion reactor and/or from energy produced in the ram-scoop magnetic field due to the atom's initial deflection. Further study is needed to assess the efficiencies and energy flows (i.e., energy sources and sinks) of the various processes to determine the feasibility of this approach.

D. Effects of Acceleration and Cruise Velocity on Trip Time

In this section, we consider first the general problem of acceleration and cruise (coast) velocity and their impact on mission trip time for the case of a fast (0.5 c) interstellar rendezvous mission. Later we will illustrate how acceleration and cruise velocity impact the transportation (propulsion) system mass for several options, such as antimatter rockets and laser sails.

For a given travel distance, trip time will be a function of acceleration as well as cruise velocity. For example, too low an acceleration can adversely impact trip time, because the vehicle spends too much time in the acceleration/deceleration phase and not enough time at peak (cruise) velocity. This problem can be illustrated several ways. First, as shown in Fig. 14, we can consider the case of a rendezvous mission with a given distance and total trip time and find V_{cruise} as a function of acceleration. For this case, we assume a trip time in years numerically equal to twice the distance in LY, so that, ignoring acceleration or deceleration, the average cruise velocity would be 0.5 c. However, in practice, there is some time spent (and distance traversed) during acceleration from Earth and deceleration at the target solar system. (For these analyses, the acceleration and

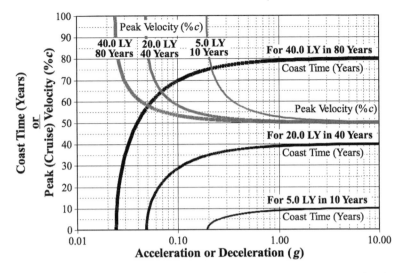

Fig. 14 Coast (cruise) time and peak (cruise) velocity versus acceleration for an interstellar rendezvous mission with equal acceleration and deceleration phases.

deceleration phases are assumed equal.) Thus, a minimum acceleration is needed to reach the target star where the vehicle accelerates to the midpoint in the trajectory, turns around, and immediately begins to decelerate (i.e., there is no time spent coasting); this limiting case requires a peak velocity approaching the speed of light. As acceleration increases, some time is spent coasting, and the peak or cruise velocity approaches $0.5\,c$ as a limiting case for infinite acceleration.

An alternative mission scenario is shown in Fig. 15, where the maximum cruise velocity (V_{max}) is limited to some value ($0.5\,c$ in this case) so as to constrain the overall mission ΔV and thus the vehicle wet mass. In this case, the mission total trip time is a function of acceleration (and deceleration) and distance traversed. Therefore, the trip time at the limit of infinite acceleration is numerically twice the distance because the cruise velocity is limited to $0.5\,c$. In order to minimize the trip time (by maximizing the time spent at peak velocity), the vehicle needs to accelerate (and decelerate) at about $0.01\,g$ ($1\,g = 9.8\,\text{m/s}^2 = 1.03\,\text{LY/yr}^2$) as a minimum. At less than about $0.01\,g$, the vehicle does not even reach the maximum allowed cruise velocity and the total trip time increases dramatically; this effect is seen to become worse as the total travel distance decreases. As shown earlier, higher acceleration is better, but higher acceleration will typically require more power and thus more system mass. Interestingly, there is no significant benefit for acceleration $>1\,g$.

Figure 15 also illustrates the somewhat counter-intuitive result that it can actually be harder to achieve a short trip time for nearby stars (i.e., less distance to travel) than those farther away. This is because at low acceleration and short total distance, the vehicle does not have enough distance to accelerate to a high cruise velocity. For example, performing a 5 LY rendezvous mission at

LIMITS OF INTERSTELLAR FLIGHT TECHNOLOGY

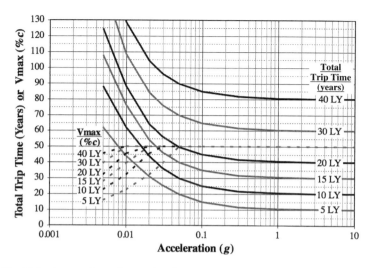

Fig. 15 Trip time and maximum (cruise) velocity, V_{max}, versus acceleration for an interstellar rendezvous mission with equal acceleration and deceleration phases.

an acceleration of 0.01 g takes about 45 years because the vehicle can only accelerate to less than one-half the nominal V_{cruise} (0.5 c) before it has to turn around and begin deceleration. By contrast, at the same 0.01 g, we can travel to 20 LY, 4 times the distance, in about 90 years, twice the time, because the vehicle has more distance to accelerate to roughly 0.45 c before beginning deceleration.

E. Interstellar Dust Impacts

One often-raised issue concerning interstellar missions is the effect of high-speed (e.g., 0.5 c) interstellar dust grain impacts on a vehicle. Typically, the assumption is made that a dedicated dust shield is placed at the front of a vehicle to protect it from relativistic dust impacts [7,14]. The shield thickness will be a function of the vehicle velocity, the distance traveled, and the dust particle number and size (and thus mass) distribution per unit volume in interstellar space [23]. The calculation methodology we have developed previously [14] assumes that the kinetic energy of the dust impact is turned into thermal energy that evaporates (sublimes) the shield material (e.g., graphite). Thus, it is necessary to take into account the energy per unit mass of dust at a given velocity (e.g., $V_{cruise} = 0.5$ c) and then determine the total mass of dust hitting the shield. To do this, we calculate the kinetic energy per unit mass of a dust grain (e.g., J/g), multiply this by the total mass of dust grains encountered along the mission path per unit of shield frontal area (e.g., g/cm^2), and divide by the heat of vaporization of the shield material (e.g., J/g) to determine the mass of shield material (and thus thickness) required per unit area of shield area. A second effect, which might actually be larger, would be to include the effect of additional shield material that could be "spalled" (mechanically broken) off of the surface by the shock of impact.

For light sails, the payload and vehicle systems would need a dust shield, but it is impractical to shield the total sail sheet due to the dust shield's mass. Ironically, it is actually better for the sail sheet to fly perpendicular (face-on) to the velocity vector than fly parallel (edge-on). This is because an edge-on impact could "cut" through more sheet material than a face-on impact. Fortunately, when flying face-on, only a small amount of sail area, $\sim$2.3%, is lost due to dust impacts over even a 40-LY path [2]. This rather modest loss of sail sheet area corresponds to the sum of the cross-sectional areas of all the dust grains hitting the sail. A useful familiar analogy is to imagine the action of a cookie cutter on a sheet of cookie dough, where the area of the "hole" in the sheet (i.e., the cookie) is the same as that of the cookie cutter.

As shown in Fig. 16, the upper end of the size range of the dust particles can be very large compared to the sail sheet thickness. At relativistic speeds, the kinetic energy of the dust grain simply vaporizes the sail sheet material in its direct path. As for the surrounding sail sheet material, there are two effects that tend to limit the size of the sail sheet hole to that of the incoming dust particle. First, at relativistic speeds, the interaction time between the sail sheet and the dust particle is very short (e.g., 16 nm/0.1 $c = 5 \times 10^{-16}$ s), such that the sail sheet material surrounding the dust sail impact region is inertially frozen in place and unable to respond mechanically by tearing or stretching under the force of the dust particle (and be vaporized). Second, the sail sheet is so thin that no significant amount of heat can transfer laterally along the sheet and cause vaporization of additional sail material. In effect, the surrounding material does not have time to tear/stretch or heat significantly, thus localizing the damage to only that area punched out by the dust particle.

Note that this is an area where additional research is needed. For example, an understanding of the physics of dust grain impacts with a shield is not known for relativistic speeds. This research would allow us to quantify whether vaporization or spalling is the dominant effect, and more generally estimate the shield thickness needed. Also, in situ measurements of dust particle number, size, and material during interstellar precursor missions would allow us to benchmark our theoretical particle distribution models.

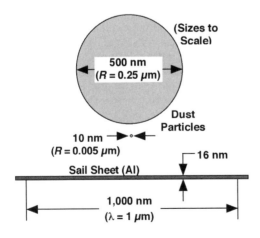

Fig. 16 Comparison of sizes of light sail sheet thickness (16 nm), dust particle diameters (10–500 nm), and laser photon wavelength (1 μm).

LIMITS OF INTERSTELLAR FLIGHT TECHNOLOGY 65

One final observation is that an interstellar rendezvous vehicle is only at high speed in interstellar space. By contrast, flyby mission vehicles encounter thick solar system dust clouds at high speed. Thus, we have the somewhat paradoxical situation that the dust shield for an interstellar flyby could be thicker (and thus heavier) than the one needed for a rendezvous mission.

IV. Rocket-Based Propulsion Options

In this section, we describe some of the general performance limitations of existing propulsion technologies, and consider the potential performance that might be seen for future technologies. Note that in this section, we are addressing "rocket" propulsion systems in the sense that some onboard reaction mass (propellant) is expelled from the vehicle to produce acceleration. Thus, the rocket vehicle has an onboard means of producing the momentum that accelerates it. By contrast, the next section will deal with propulsion technologies that make use of an external source of momentum (e.g., photon pressure from sunlight or a laser). In this scenario, the vehicle possesses no intrinsic onboard momentum source; rather, the external momentum source is manipulated in some fashion to produce the desired vehicle acceleration.

A. Assumptions Used when Projecting Upper Performance Limits

Candidate propulsion systems are categorized by their energy source (e.g., fission, fusion, etc.) and are discussed herein. More detailed discussions can be found in the technical literature (e.g., review articles [24] and individual papers on each topic), from which the various performance parameters listed below were derived.

B. Chemical Propulsion

The highest-I_{sp} SOA chemical propulsion technology is represented by the O_2/H_2 combination with a theoretical (ideal) I_{sp} of 528 lb_f-s/lb_m. In practice, the achievable I_{sp} is approximately 450 lb_f-s/lb_m in such engines as the space shuttle main engine (SSME) or the RL-10 engines in the Centaur upper stage. If they could be produced in high concentrations, more exotic propellant combinations, such as high energy density matter (HEDM) free-radical or "atomic" hydrogen (H), electronic metastable atoms (e.g., triplet helium), or metallic hydrogen have theoretical I_{sp}'s as high as 3,000 lb_f-s/lb_m.

C. Electric Propulsion

Advanced nuclear electric propulsion (NEP) systems have been studied for a variety of interstellar precursor missions (to 1000 AU), although they would have insufficient exhaust velocity (<200 km/s) for interstellar missions using extrapolations of existing electric thruster technologies (e.g., ion, magnetoplasmadynamic, Hall, etc.). Because of the $1/R^2$ decrease in sunlight intensity as one moves away from the Sun, a solar electric propulsion (SEP) vehicle, using solar photovoltaic cells, is not an option. For a high-power NEP vehicle, we could use a uranium-based reactor with either a static thermal-to-electrical power conversion system (with no moving parts; e.g., solid-state thermoelectric power conversion) or dynamic system (with moving parts; e.g., turbines, generators, etc.). An alternative, which is better-suited to small, low-power systems is radioisotope electric propulsion (REP). In this concept, a spacecraft that already has an onboard radioisotope power system [e.g., a plutonium-based radioisotope thermoelectric generator (RTG)] due to the need to operate far from the Sun would be combined with a small, low-power electric propulsion system.

Finally, note that in principle, a high-energy relativistic particle accelerator could be used as an electric thruster with an exhaust velocity approaching that of the speed of light. However, the mass of such a "thruster" would be substantial; one can imagine the low acceleration produced by taking the thrust from the particle accelerators at CERN or the Fermi National Accelerator Laboratory (FermiLab) and dividing this thrust by the mass of these facilities (as "thrusters") along with the mass of their power plants.

D. Fission Thermal and Pulsed Propulsion

Fission thermal and pulsed propulsion systems are characterized by high accelerations (typically on the order of $1g$), but relatively modest I_{sp}'s and corresponding exhaust velocities. In fission thermal propulsion concepts, heat from a fission reactor is transferred to a working fluid (typically H_2) and the high-temperature H_2 gas expelled out a nozzle to produce thrust. SOA performance is represented by the nuclear engine for rocket vehicle applications (NERVA) solid core reactor engine from the early 1970s with an I_{sp} of 825 $lb_f\text{-s}/lb_m$. A modern version of this engine (NERVA derivative), using contemporary technologies in materials, pumps, etc. is projected to have an I_{sp} of 900 $lb_f\text{-s}/lb_m$, with advanced versions potentially reaching 1000 $lb_f\text{-s}/lb_m$. Advanced gas-core reactor engines might have I_{sp}'s as high as 7000 $lb_f\text{-s}/lb_m$, although this would require dedicated external radiators (in addition to regenerative engine cooling from the hydrogen expellant). These systems are at the lower end of applicability to interstellar precursor missions because of their inability to meet the high exhaust velocities and I_{sp}'s needed for these missions (e.g., $V_{exhaust} > 100$ km/s or $I_{sp} > 10,000$ $lb_f\text{-s}/lb_m$).

In the pulsed fission propulsion concept, typified by the Orion vehicle concept (not to be confused with the NASA crew exploration vehicle), a fission "pulse unit" is ejected behind the vehicle and exploded; debris and energy from the explosion impacts a pusher plate that transfers the explosive shock to the vehicle. Orion was originally intended for missions within the solar system, and thus designed with a high thrust and modest I_{sp} around 1800 to 2500 $lb_f\text{-s}/lb_m$. To do this, the pusher plate was intentionally designed to ablate during each pulse to increase the total mass of material ejected as propellant; this had the effect of increasing thrust at the expense of potential I_{sp}. An interstellar version of Orion designed by Freeman Dyson [25] used high-energy fission or fusion pulse units to produce a very high I_{sp} version projected to be capable of reaching a speed of 10,000 km/s (3.3% of the speed of light) in only 10 days at an average acceleration of 1.2 g. At this speed (and with a negligible acceleration time), Alpha Centauri could be reached in 130 years. The vehicle's wet and dry masses were projected to be 400,000 MT and 100,000 MT, respectively, and to use 300,000 1-megaton pulse units. This vehicle also had an ablative pusher plate; however, in this case ablation was used to cool the pusher plate and protect it from the 1-megaton nuclear pulses. For an acceleration period of 10 days, it would be necessary to use 1 pulse unit every 2.9 seconds. Interestingly, these values illustrate the extraordinary efficiency of direct use of fission or fusion energy, as opposed to its use in a thermal rocket engine. For this case, with a mass ratio (M_{wet}/M_{dry}) of 4 and a ΔV of 10,000 km/s, the equivalent I_{sp} is an amazing 736,068 $lb_f\text{-}s/lb_m$, roughly 3/4 the I_{sp} of a "pure" fission fragment or fusion rocket (10^6 $lb_f\text{-}s/lb_m$), as discussed below.

Unfortunately, this would only be a flyby mission; to perform a rendezvous mission we would need an additional "first" stage for acceleration and a "second stage" (the original flyby stage) for deceleration and rendezvous. The first stage (only) would have a wet mass of 1,600,000 MT [18]; the total two-stage vehicle (with a 20,000 MT payload) would have a wet mass of 2,000,000 MT. Keeping the same pulse unit size and delivery rate, it would take 50 days to consume 1,500,000 pulse units and accelerate at an average 0.2 g to a speed of 10,000 km/s (and, 130 years later, 10 days at 1.2 g to decelerate).

However, it should be emphasized that the interstellar version of Orion was at a much less mature level of study than the interplanetary versions. As Dyson himself says, "The starship was like an existence theorem in math. It was to prove that you could do it" [26]. Also, any version of Orion must contend with the issue that the vehicle contains many thousands of nuclear fission (or fusion) pulse units that could potentially be misused; for comparison, the interstellar flyby Orion contains an order-of-magnitude more nuclear pulse units than the total world inventory of nuclear warheads.

E. Fission Fragment Propulsion

Unlike a conventional fission–thermal propulsion system (such as NERVA and others), in which fission energy is used to heat a secondary working fluid (e.g., H), the fission fragment propulsion concept uses fragments of the fission process (e.g., Sr^{90} and Xe^{136} from U^{235} fission) *directly* as the rocket exhaust gas. Because the high-energy fragments are not "cooled" by the addition of any working fluid, this concept has the potential for achieving the ultimate limit of fission energy with an I_{sp} of about 10^6 $lb_f\text{-}s/lb_m$ or an exhaust velocity of 0.03 c. However, as discussed previously, this approach would require the creation of a propulsion-specific "propellant" production infrastructure that might be difficult to justify. Also, as discussed below with the Daedalus fusion propulsion system, which also has an exhaust velocity of 0.03 c, a two-stage fission fragment propulsion vehicle would be needed to reach a cruise velocity of 0.1 c for a flyby mission (and four stages for a corresponding rendezvous mission).

1. Fission Fragment Rocket

In the fission fragment rocket, rods containing the fissionable material are placed in a reactor to control the rate of fission. The rods are made thin to facilitate escape of the fission fragments from the rod. The rods are arranged like the spokes in a rotating bicycle wheel; this allows the rods to be removed periodically from the reactor to quench the reaction rate and cool before being

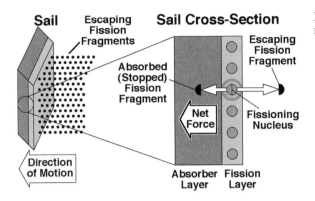

Fig. 17 Fission fragment sail concept.

re-inserted into the reactor. Electromagnetic fields are used to capture, collimate, and direct the charged fragment nuclei out the back of the rocket to produce thrust. As with any propulsion system using electromagnetic fields to control charged particles, we encounter a space-charge limit to the plasma density, with the result that, although the I_{sp} is high, the inherent thrust and thus vehicle acceleration will be low.

2. Fission Fragment "Sail"

In this concept, shown in Fig. 17, a thin sheet (sail) would be doped with a fissionable material near one side of the sail sheet. The fission fragments would escape to space on the side of the sail with the fissionable material; on the other, thicker side, the fragments would be absorbed by the sail sheet to produce thrust. In this case, there would be no moderating reactor to control the rate of fission or to allow periodic cooling. Therefore, it would be necessary to dope the sail such that the rate of fission does not reach a level where waste heat absorbed by the sail exceeds the ability of the sail to radiate the heat to space.

In principle, this system can have a higher acceleration than a corresponding solar sail because the fission fragments produce more momentum than photons. Also, it can be used as a regular solar sail near a star once its fission fuel is depleted. Note, however, that this system throws away at least half of its potential propellant (fission fragment) mass, depending on how well the fragments are absorbed and stopped in the thicker side of the sail. Thus, although it has a potential for high I_{sp}, it will suffer a degradation in performance due to "loss" of propellant.

F. Fusion Propulsion

There are two principal schemes for providing the confinement necessary to sustain a fusion reaction: inertial confinement fusion (ICF), and magnetic confinement fusion (MCF). These confinement schemes result in two very different propulsion system designs. There are literally dozens of different ICF, MCF, and hybrid-ICF/MCF fusion reactor concepts [27]; regardless of the particular reactor type, fusion propulsion concepts fall into two broad I_{sp} categories. At the lower end of I_{sp} (and with correspondingly higher acceleration), fusion rockets are well-suited to future large-scale exploration and exploitation of the solar system (e.g., fusion propulsion could enable four-month round trips to Mars). For these systems, the fusion energy is used to heat a secondary propellant working fluid (such as H) to yield an I_{sp} of about 10^4 to 10^5 $lb_f\text{-}s/lb_m$. For more demanding interstellar missions, the fusion products would be directly exhausted (as with the fission fragment rocket) with a maximum I_{sp} of about 10^6 $lb_f\text{-}s/lb_m$ or an exhaust velocity of 0.03 c.

1. Conventional (All-Onboard Propellants) Fusion Propulsion for Interstellar Missions

In our first example of fusion propulsion for interstellar missions, we will discuss the "conventional" rocket approach in which all of the propellants to be used during the mission are loaded into the vehicle at the start of the mission; this approach is dominated by the strictures of the Rocket Equation because ΔV (>0.1 c) is typically much larger than fusion propulsion's I_{sp} or $V_{exhaust}$ (0.03 c). A second approach, discussed below, seeks to "cheat" the Rocket Equation by continuously refueling the vehicle from interstellar hydrogen.

For a "conventional" fusion rocket to perform a 0.1 c cruise velocity interstellar flyby mission, a two-stage vehicle would be needed, such as the Daedalus two-stage ICF fusion rocket designed by the British Interplanetary Society [7]. Table 2 summarizes the performance of this vehicle in its original two-stage, 0.124 c flyby configuration, where stage 1 had a higher ΔV than stage 2. Also shown is a hypothetical two-stage rendezvous version with the same total vehicle mass, with equal ΔV for both stages.

Finally, we could employ a four-stage vehicle for a rendezvous mission with a cruise velocity of 0.1 c. We would need a new first and second stage to accelerate to 0.124 c, and the original two-stage Daedalus for the deceleration and rendezvous. However, this would result in a very heavy vehicle. For example, if we assume that the first two stages have the same propulsion system TF ($M_{dry\text{-}prop}/M_p$) as the Daedalus' first stage (3.674%), and the same ΔV's and accelerations, then the total vehicle would weigh over 3.5 million MT and the total trip time to 6 LY would be 52.2 years.

2. Bussard Interstellar Ramjet: Extraterrestrial Resource Utilization (ETRU) Fusion Propulsion for Interstellar Missions

Thus far in our discussions of potential interstellar propulsion options, we have seen how strongly I_{sp} (and the corresponding $V_{exhaust}$) impacts performance for flyby and, especially, rendezvous missions. Even direct use of fission and fusion is limited to a $V_{exhaust}$ of 0.03 c, constraining these concepts to rather modest peak velocities and corresponding long trip times. This level of performance is ultimately driven by the Rocket Equation, and its exponential increase in wet mass and propellant mass as ($\Delta V/V_{exhaust}$) increases. One way to "cheat" the Rocket Equation is to *not* carry onboard all the propellant that would be needed for the mission. Instead, we "live off the land," and collect materials along the way that could be used for propellant during the mission. This approach

Table 2 Daedalus two-stage fusion rocket performance to 6.0 LY (Barnard's star)[a]

Mission type	6.0 LY Flyby ($V_{cruise} = 0.124\ c$)			6.0 LY Rendezvous ($V_{cruise} = 0.063\ c$)		
Mission phase	Acceleration	Acceleration	Total mission	Acceleration	Deceleration	Total mission
Total mission trip time (years)	2.050	1.760	50.276	1.998	2.883	96.136
Average acceleration or deceleration (g)	0.0347	0.0279		0.0304	0.0210	
Total mission distance (LY)	0.051	0.183	6.000	0.045	0.242	6.000
Vehicle stage	Stage 1	Stage 2	Total vehicle	Stage 1	Stage 2	Total vehicle
Initial total wet mass, M_{wet} (MT)	54,056	5109	54,056	54,056	7531	54,056
Total propellant mass, M_P (MT)	46,000	4000	50,000	44,835	6551	51,386
Final total dry mass, M_{dry} (MT)	6799	980	4056	9221	980	2670

LIMITS OF INTERSTELLAR FLIGHT TECHNOLOGY

Vehicle mass ratio, M_{wet}/M_{dry}	7.951	5.213	5.862	7.685
Propellant system dry mass, $M_{dry\text{-}prop}$ (MT)	1690	550	1690	530
Propellant system total TF, $M_{dry\text{-}prop}/M_p$	3.674%	13.750%	3.679%	8.090%
Payload mass, M_{PL} (MT)	5109	430	7531	450
Velocity change, ΔV (c)	0.0734	0.0507	0.0626	0.0626
Propellant exhaust velocity, $V_{exhaust}$ (c)	0.03540	0.03070	0.03540	0.03070
Rocket engine thrust, F (N)	7.55E+06	6.63E+05	7.55E+06	6.63E+05
Propellant mass flow rate (kg/s)	0.711	0.0720	0.711	0.0720
"Jet" exhaust power, P_{jet} (TW_{jet})	40.04	3.05	40.04	3.05
Rocket engine run time (years)	2.050	1.760	1.998	2.883
		2240		2670
		430		430
		0.1241		0.1252
		3.810		4.881

[a]Table corrected courtesy of Adam Crowl.

is referred to as extraterrestrial resource utilization (ETRU); for example, chemical propellants needed for a robotic Mars sample return rocket could be produced from CO_2 in the Martian atmosphere and/or water-ice in the soil. We take this same approach in the ultimate ETRU concept, the Bussard interstellar ramjet [11], in which interstellar H would be scooped to provide propellant mass for a fusion propulsion system. Interstellar H would be ionized and then collected by an electromagnetic field. Onset of ramjet operation is at a velocity of about $0.06\ c$. Beyond this ram speed, the scooped H is "free" (as far as the Rocket Equation is concerned), so the vehicle could continue to accelerate even up to ultrarelativistic speeds ($\gg 0.5\ c$). Thus, the only onboard propellant required would be that needed for the initial acceleration to, e.g., $0.06\ c$ (or corresponding deceleration from $0.06\ c$ to rest for a rendezvous), resulting in a significantly lighter vehicle (i.e., $M_{\text{wet}}/M_{\text{dry}} = \exp[0.06/0.03] = 7.4$).

In operation, the Bussard interstellar ramjet would first accelerate up to ram onset speed (e.g., $0.06\ c$) using onboard propellant. At speeds above ram onset, the fusion ramjet is used to accelerate to the final cruise velocity, which is constrained by relativity and *not* the Rocket Equation. To begin a rendezvous mission's deceleration phase, the electromagnetic scoop is turned on but set such that the scooped interstellar H comes to rest (relative to the vehicle) inside the scoop; this is equivalent to allowing the ram scoop to "choke" the gas flow. Momentum drag from the scoop (due to interstellar H coming to rest in the scoop) is actually more efficient than the fusion rocket at speeds above the V_{exhaust} of the rocket ($0.03\ c$). As the speed approaches V_{exhaust}, the scooped H is used to refill the vehicle's propellant tanks for the final fusion rocket deceleration to rest. Because the tanks are sized for a ΔV of $0.06\ c$ (but we only need to decelerate $0.03\ c$), there would be some propellant remaining in the tanks that can then be used for normal fusion propulsion within the target solar system.

However, although the Bussard interstellar ramjet is very attractive for interstellar missions because of its unlimited range and potential for ultrarelativistic speeds, there are several, very major feasibility issues associated with its operation [3]. For example, there do not appear to be any feasible technological solutions to performing pure H–H fusion (short of using the cores of stars). However, D–D fusion appears quite feasible technologically, so it may simply be necessary to collect interstellar D and discard the H (although there is only about 1 D per 10^5 H in interstellar space). A second significant issue is the design of the electromagnetic "scoop," both in terms of its electromagnetic shaping as well as the structural design required to support the magnets that produce the magnetic scoop fields. Finally, a potentially major issue is the need to design a ram-scoop and fusion engine system that minimizes momentum drag on the collected H. This is because, relative to the vehicle, the collected H (or D) represents a decelerating thrust from the momentum of the H (or D) approaching the vehicle (e.g., at $0.5\ c$) versus accelerating thrust from the fusion engine at an exhaust velocity of only $0.03\ c$.

G. Matter–Antimatter Annihilation Propulsion

Matter–antimatter annihilation offers the highest possible physical energy density of any known reaction substance. The ideal energy density ($E/M = c^2$)

of 9×10^{16} J/kg is orders of magnitude greater than chemical (1×10^7 J/kg), fission (8×10^{13} J/kg), or even fusion (3×10^{14} J/kg) reactions. Additionally, the matter–antimatter annihilation reaction proceeds spontaneously, and therefore does not require massive or complicated reactor systems. These properties (high energy density and spontaneous annihilation) make antimatter very attractive for propulsively ambitious space missions like interstellar travel. This section describes those antimatter propulsion concepts in which matter–antimatter annihilation provides all of the propulsive energy. A related concept, in which a small amount of antimatter triggers a much larger microfission/fusion reaction, is discussed below.

1. Overview of Matter–Antimatter Annihilation Propulsion

Not surprisingly, antimatter production, storage, and utilization represent major challenges. Numerous fundamental feasibility issues remain to be addressed, such as scaling up antimatter production rates and efficiencies, storage in a high-density form suitable for propulsion applications, and design and implementation of a complete propulsion system containing all the ancillary systems required to contain the antimatter on the vehicle and ultimately use it in a thruster [14]. Nevertheless, research aimed at addressing these issues is ongoing at a modest level.

As a specific thruster design, we have the "beam-core" antimatter rocket concept [14], in which equal amounts of matter and antimatter (in the form of protons and antiprotons) are combined and annihilate each other to produce high-speed annihilation products (neutral and charged pions). A magnetic nozzle is used to direct the charged pions out of the nozzle to produce an I_{sp} of about 10^7 lb_f-s/lb_m [20] corresponding to an exhaust velocity of 0.3 c.

Note that in this concept, we have emphasized the use of proton–antiproton annihilation because the charged pion annihilation products could be deflected and somewhat collimated to produce a directed stream of "exhaust" (particles with mass and thus momentum) from the engine. In principle, we could have used electrons and anti-electrons (positrons) as our propellants; however, their annihilation products are high-energy photons (gamma rays). This results in two problems with this approach. First, there is no practical way to focus or deflect gamma rays without the use of massive amounts of shielding material, resulting in an extremely low acceleration. Second, although photons have momentum, their momentum is intrinsically low compared to "normal" particles that contain actual mass.

As a simple example of the sort of acceleration we might expect from a "pure" photon rocket, we can consider the "vehicle" depicted in Fig. 18, which consists of a radiation shield to absorb the gamma photons from an annihilation reaction (e.g., 0.511 MeV gammas from electron-positron annihilation) and a radiator to reject the thermal energy produced in the shield by absorbing the gammas. To calculate the acceleration of this vehicle, we will first find the thrust produced by the gammas, and then divide this thrust by the mass of the shield and its radiator. This will yield an upper limit to acceleration because we have ignored the mass of all the other systems on the vehicle; also, by considering only those quantities that scale uniformly with each other, we can present a "generic" derivation and not be concerned with more detailed system mass estimates.

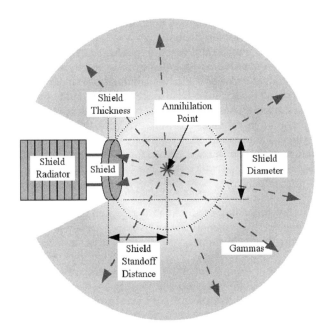

Fig. 18 Simplified electron-positron annihilation gamma photon rocket.

For example, we can arbitrarily select a total annihilation power of 1 TW; the actual value selected will be irrelevant for this simplified case because as power increases, so does thrust, but so also does system mass. For calculation purposes, we assume the shield is 10 m from the annihilation point (standoff distance) and the shield has a 10-m diameter. We will further assume that we wish to absorb 99.9% of the gammas hitting the shield, with 0.1% passing through for an attenuation of 0.001 (i.e., 0.1% of the initial intensity remains after passage through the shield). In this case, the attenuation does weakly affect the calculated acceleration because of the exponential dependence of shield thickness on attenuation, although this effect will be ignored for the purposes of this approximate analysis.

To calculate thrust, we use the total annihilation (i.e., gamma) power (1 TW) and assume that this power is uniformly distributed in a sphere around the annihilation point. We find the fraction of the gamma power incident on the shield by taking the total annihilation power and multiplying this by the ratio of the shield area (perpendicular to the annihilation point) divided by the area of a sphere whose radius equals the shield standoff distance. For our selected geometry, this corresponds to intercepting 6.25% of the total gamma power, or 0.0625 TW. However, our shield doesn't have perfect absorbance (α), instead it only absorbs 99.9% of the gammas, so the total power absorbed is 0.06244 TW. Ultimately, this power will need to be dissipated by the shield's radiator. Also, the total fraction of gammas absorbed by the shield ($0.0625 \cdot 0.999 = 0.06244$) corresponds to the value of "**a**" in the relativistic Rocket Equation with "loss" of propellant, as discussed in the appendix. This can be compared to the proton-antiproton annihilation rocket's **a** of 0.40. Even if we had a hemispherical shield for the electron-positron annihilation rocket, we would always lose at least 50% of the gammas, resulting in a value of **a** < 0.5, with a corresponding mass penalty of additional shield (and radiator) mass.

We can calculate the total gamma photon thrust (F) based on power (P) and the speed of light:

$$F_{photons}[N] = P_{absorbed}[W]/c[m/s]$$
$$= P_{total} \cdot (\text{Area Ratio} = 0.0625) \cdot (\alpha = 0.999)/c$$

Thus, for this (arbitrary) 1 TW system, the total thrust is a surprisingly low 208 N. This is because even a small amount of thrust from photons requires an enormous power due to the factor of the speed of light in the denominator in the equation above.

We have previously described shield thickness (t) calculations for 200 MeV gammas produced by neutral pion decay from proton-antiproton annihilation [14].

$$\text{Attenuation} = \exp(-t/\tau) \text{ or } t = -\tau \ln(\text{Attenuation})$$

where τ is the Shield Attenuation Factor (0.00533 m for tungsten with 200 MeV gammas), and is defined in terms of the shield's density (19.35 g/cm^3 for tungsten) and Mass Attenuation Factor (0.097 cm^2/g for tungsten with 200 MeV gammas).

$$\tau = 1/([\text{Density} = 19.35 \text{ g/cm}^3] \cdot$$
$$[\text{Mass Attenuation Factor} = 0.097 \text{ cm}^2/\text{g}] \cdot [100 \text{ cm/m}])$$

Note that the Mass Attenuation Factor (and thus τ) is specific to a given material and gamma energy. As a first approximation, we will assume that the Mass Attenuation Factor for tungsten scales as a simple ratio of gamma energies:

Mass Attenuation Factor [at 0.511 MeV]
$$= \text{Mass Attenuation Factor [at 200 MeV]} \cdot (200/0.511)$$
$$= 37.96 \text{ cm}^2/\text{g}$$

so that the Shield Attenuation Factor (τ) becomes 1.361×10^{-5} m (at 0.511 MeV). Thus, for tungsten with 0.511 MeV gammas and an Attenuation of 0.001, the shield thickness (t) is:

$$t = -\tau[\text{at 0.511 MeV}] \ln(\text{Attenuation}) = -(1.361 \times 10^{-5} \text{ m}) \ln(0.001)$$
$$= 9.40 \times 10^{-5} \text{ m}$$

With a circular shield diameter of 10 m, a thickness of 9.40×10^{-5} m, and a tungsten density of 19.35 g/cm^3, the shield mass is 1429 kg.

We will assume the same type of advanced main-shield radiator described previously for the proton-antiproton annihilation Antimatter Rocket [14], a two-sided radiator with a mass per thermal Watt radiated (at 1500 K) of

9.677×10^{-7} kg/W$_{thermal}$. With a total thermal power of 0.06244 TW$_{thermal}$, the radiator mass is 60,423 kg.

Finally, we can now rearrange Newton's classic equation (F = MA) to determine the vehicle acceleration (A) based on the total photon thrust (F, 208 N) divided by the vehicle total mass (M, shield plus radiator, 61,852 kg):

$$A = F/M = (208 \text{ N})/(61,852 \text{ kg}) = 0.0336 \text{ m/s}^2 = 0.000343 \text{ gees}$$

With this acceleration, it would take over 1400 years to accelerate to 0.5 c. Thus we see that a "pure" electron-positron annihilation gamma photon rocket is completely impractical if based on brute-force absorption of gamma-ray photon momentum.

Given the requirement for propellants in the form of protons and antiprotons, we find that that actual propellant storage form needs to be that of a condensed phase of molecules (and anti-molecules). This is driven primarily by the density of the propellant and its impact on the propellant tankage mass. For example, we routinely trap, move, and store antiprotons using electromagnetic fields. However, as a charged particle, an antiproton is subject to the space-charge limits imposed on any ion plasma; with current magnet technology, this corresponds to 10^{10} to 10^{12} ions/cm^3 or 10^{-14} to 10^{-12} g/cm^3 for (anti)protons. Thus, it would be wildly impractical to store the tons of antiprotons needed for an interstellar mission as an ion gas; the tanks would be impossibly large and heavy. This then drives us to use ordinary (normal-matter) liquid molecular hydrogen (LH$_2$) and solid molecular anti-hydrogen (anti-SH$_2$). The normal-matter LH$_2$ can be stored as ordinary LH$_2$ at around 20K. However, the anti-SH$_2$ must be stored as magnetically-suspended/levitated pellets at 1K. In this case, the magnetic suspension prevents the antimatter from contacting the propellant tank walls, feed lines, etc. and annihilating. Furthermore, the pellets must be stored at 1K to prevent their sublimation to form gaseous anti-H$_2$, which is not contained by the magnetic field and thus would expand (like any gas) until it reached the tank walls and annihilated [28].

Note that condensed-phase molecular storage implies conversion of antiprotons to anti-atoms (i.e., anti-H, consisting of a hydrogen-like anti-atom containing an antiproton and positron), and subsequent conversion of anti-atoms into antimolecules. This has been seen as a significant challenge due to the need to remove excess energy from the collision partners; for example, an antiproton–positron collision needs to have 13.6 eV (the ionization energy of a hydrogen atom) removed to form an anti-atom, and an anti-H/anti-H collision 4.5 eV (the heat of formation of atomic-H from molecular-H$_2$) to form an antimolecule. However, recent advanced in electromagnetic "cooling" of atoms [29] suggests that this may be more feasible than the laser cooling method previously proposed for this process [28].

Finally, the most serious feasibility issue associated with antimatter propulsion is the sheer difficulty of producing the antimatter. Antiprotons do not exist in nature and currently are produced only by energetic particle collisions conducted at large accelerator facilities [e.g., FermiLab and Brookhaven National Laboratory (BNL) in the United States; CERN in Switzerland; or IHEP in Russia]. This process typically involves accelerating protons to relativistic velocities (very near the speed of light) and slamming them into a metal (e.g., tungsten) target. The high-energy protons are slowed or stopped by collisions with nuclei of the target; the relativistic kinetic energy of the rapidly-moving proton is converted into matter in the form of various subatomic particles, some of which are antiprotons. The antiprotons are then electromagnetically separated from the other particles and stored in electromagnetic rings. Note that antiprotons annihilate spontaneously when brought into contact with normal matter; thus, they must be contained by electromagnetic fields in high vacuums. This greatly complicates the collection, storage, and handling of antimatter. Currently the highest antiproton production/capture/accumulation level (not optimized for rate or efficiency) is of the order of 10 nanograms per year, although planned upgrades to CERN may increase these production rates by a factor of 10 to 100. Additionally, only a much lower level of antiprotons have actually been collected, cooled, and stored after production. Finally, current production/capture/accumulation technology has an energy efficiency of only about 1 part in 10^9 (i.e., 10^9 units of energy are consumed to produce an amount of antimatter that will release 1 unit of energy upon annihilation) [28].

To date, only infinitesimal quantities of antiprotons and anti-atoms have been produced, as shown in Fig. 19. However, it is worth considering the progress of technology in producing "exotic" forms of matter. For example, it has taken little more than a century to go from Dewar's few drops of LH_2 to the 100 tons of LH_2 used in every Space Shuttle launch. One can imagine Dewar's reaction if he were told in 1898 that less than a century later, we would be using hundreds of tons per year for space exploration missions, given the technological challenges he faced in 1898.

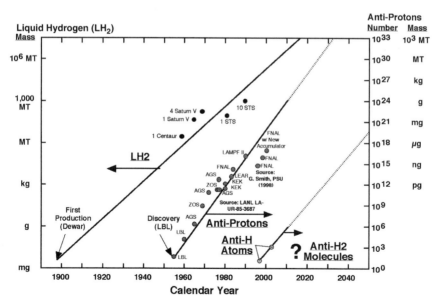

Fig. 19 Comparison of growth in annual production rates of normal-matter LH_2 and antimatter.

2. Matter–Antimatter Annihilation Propulsion Vehicle Sizing

In previous studies,[††] we have developed detailed mass scaling equations for determining the dry mass of a "beam core" antimatter rocket stage [14].[‡‡] As discussed, the Relativistic Rocket Equation with $\mathbf{a} = 0.4$ limits each stage to a ΔV of around 0.25 c. Because the vehicle dry mass is a function of its acceleration and ΔV, we want to determine an "optimum" that produces a reasonable compromise between total vehicle wet mass (and especially antimatter mass), where mass is a function of the acceleration and mass ratio, M_{wet}/M_{dry}, and trip time, where trip time is a function of both acceleration and maximum or cruise velocity, V_{cruise}.

This tradeoff is illustrated in Fig. 20 for the case of our Vision Mission, a 40-LY interstellar rendezvous. Generally, as acceleration increases, we first see a dramatic reduction in trip time; however, for accelerations above about 0.04 g, we see only a modest decrease in trip time and a modest increase in mass. For this system, the curves for 0.05 and 0.06 g accelerations essentially overlap in the vicinity of the "knee" of the curves (roughly, 104-year flight time), with 0.05 g just barely more optimum. At a given acceleration, a much larger effect on mass is seen in increasing the velocity due to the effect of the Rocket Equation's mass ratio. Thus, the best approach seems to use a moderately high acceleration and limit V_{cruise} to a value that keeps M_{wet}/M_{dry} tolerable. For the purposes of the mission analyses presented below, we have somewhat arbitrarily selected an acceleration of 0.05 g and a V_{cruise} that yields a value of $M_{wet}/M_{dry} = 5.0$. For a single-stage vehicle accelerating to some final V_{cruise}, a value of $M_{wet}/M_{dry} = 5.0$ occurs for $V_{cruise} = \Delta V = 0.2186\ c$; for a two-stage vehicle accelerating to some final V_{cruise} with each stage having $M_{wet}/M_{dry} = 5.0$, $V_{cruise} = 0.4173\ c$, with stage 1 accelerating from 0 to 0.2186 c ($\Delta V = 0.2186\ c$) and stage 2 accelerating from 0.2186 c to 0.4173 c ($\Delta V = 0.1987\ c$).

H. Catalyzed Nuclear Fission/Fusion/Antimatter Combinations

Because of the difficulty of "igniting" fusion reactions, several schemes have been proposed to simplify or reduce the energy requirements of the "drivers" of the fusion ignition process. For example, fission reactions can be "catalyzed" by addition of small amounts of antimatter (antiprotons). The energy released from the fission reaction can then be used to ignite a fusion reaction. Because the

---

[††]The antimatter rocket mission analyses presented in Frisbee and Leifer [3] were based on 0th order vehicle mass estimates; a much more detailed set of vehicle system mass and mission performance estimates is given in Frisbee [14].

[‡‡]For the vehicle mass sizing calculations used here, we assumed the case for a 10-fold improvement in superconductor magnets and radiators. Also, because we evaluated several different multi-stage vehicle configurations in which the "top" or final two stages (e.g., stages 1 and 2 in a two-stage vehicle, or stages 3 and 4 in a four-stage vehicle) could be accelerating or decelerating, we assumed that these two stages would have a full dust shield. By contrast, we retained the assumption that "lower" stages (e.g., stages 1 and 2 in a four-stage vehicle) could have a central hole in their dust shields corresponding to the diameter of the upper stage (e.g., stage 3 sitting on "top" of stage 2, or stage 2 sitting on "top" of stage 1).

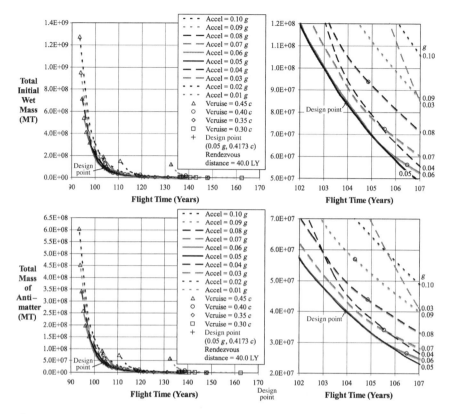

Fig. 20 40-LY interstellar rendezvous antimatter rocket total wet mass (top) and antimatter mass (bottom) as a function of cruise velocity (V_{cruise}).

required amounts of antiprotons are relatively small, it becomes feasible to consider such systems in terms of modest upgrades to existing production facilities (e.g., CERN, FermiLab) and use of existing storage technologies (e.g., as space-charge limited plasmas). However, it should be emphasized that these concepts will have a peak I_{sp} of only 10^6 lb$_f$-s/lb$_m$, because the total energy release is still dependent on the much larger fission and/or fusion reactions. Thus, the overall vehicle may be lighter (because $M_{\text{dry-prop}}$ is lighter), but the propulsion system's exhaust velocity will ultimately be limited to that of the primary energy production mechanism (i.e., fission or fusion $V_{\text{exhaust}} = 0.03\ c$).

1. Muon Catalyzed Fusion

In muon catalyzed fusion, negative muons (μ^-) produced by antiproton annihilation are used to "catalyze" the fusion reaction [30]. The muon is often considered to be a "heavy electron," with a mass 207 times that of the electron. In this concept, the muon replaces the electron in an atom of the fusion fuel (D, T, He^3). The resulting "atom" has a classical Bohr radius 207 times smaller than its electron counterpart; thus, the nuclei are able to approach each other more closely. This in turn enhances the probability of overlap between the wave functions of the nuclei of colliding atoms, and thus increases the probability of fusion. The fusion energy "ionizes" the atom and ejects the muon, which goes on to attach itself to another nucleus, and the process repeats for the lifetime of the muon (2 μs).

2. Antimatter Catalyzed Nuclear Fission/Fusion Combinations

An alternative approach to "conventional" fusion propulsion systems is the Inertial-Confinement Antiproton-catalyzed micro-fission/fusion Nuclear (ICAN) propulsion concept developed by Pennsylvania State University [31]. In this approach to ICF propulsion, a pellet containing uranium (U) fission fuel and deuterium–tritium (D–T) fusion fuel is compressed by lasers, ion beams, etc. At the time of peak compression, the target is bombarded with a small number (10^8 to 10^{11}) of antiprotons to catalyze the uranium fission process. (For comparison, ordinary U fission produces 2 to 3 neutrons per fission; by contrast, antiproton-induced U fission produces ~ 16 neutrons per fission.) The fission energy release then triggers a high-efficiency fusion burn to heat the propellant, resulting in an expanding plasma used to produce thrust. Significantly, unlike "pure" antimatter propulsion concepts that require large amounts of antimatter (because all of the propulsive energy is supplied by matter–antimatter annihilation), this concept uses antimatter in amounts that we can produce today with existing technology and facilities. This technology could enable 100- to 130-day round trip (with 30-day stop-over) piloted Mars missions, 1.5-year round trip (with 30-day stop-over) piloted Jupiter missions, and 3-year one-way robotic Pluto orbiter mission (all with 100 MT payloads).

A recent variation on the ICAN concept is antimatter initiated microfusion (AIMStar) [32], which uses an electromagnetic trap (rather than laser or particle beam implosion) to confine a cloud of antiprotons during the antimatter-induced microfission step. This concept may enable the construction of very small systems (at least as compared to a conventional ICF fusion rocket) because a large ICF-type pellet implosion system is not required.

3. Antimatter-Catalyzed Fission Fragment Sail

A variation on the fission fragment sail concept (discussed above) is the use of antiprotons to induce fission in fissionable material arranged as a sheet, somewhat similar to a solar sail [33]. As with any fission fragment propulsion concept, the I_{sp} and $V_{exhaust}$ are limited to around 0.03 c.

I. Beamed-Energy Propulsion

In beamed-energy propulsion concepts, a remote energy source beams electromagnetic radiation (e.g., sunlight, laser, or microwave energy) to the space vehicle where the energy is used to heat or electromagnetically accelerate onboard propellant to produce thrust. Note that beamed-energy is distinct from the beamed-momentum concepts listed below, because the beamed-energy source only provides energy; the spacecraft still must have onboard propellant

LIMITS OF INTERSTELLAR FLIGHT TECHNOLOGY

and follow the Rocket Equation. By contrast, a beamed-momentum source (which can again be sunlight, laser, or microwave, as well as particle beam) provides actual momentum to the vehicle; in effect, the photon or particle momentum "pushes" the vehicle and no on-board propellant is required.

The primary advantage of beamed-energy systems is the reduction in vehicle mass afforded by taking the energy source off the vehicle. However, the ultimate achievable specific impulse or exhaust velocity is limited by the type of thruster employed on the vehicle. For example, with a thermal engine [e.g., solar thermal propulsion (STP)], where the incoming beamed energy is used to thermally heat a propellant (typically H), materials temperature limits constrain I_{sp} to values similar to NERVA-type engines ($\sim$1000 lb_f-s/lb_m). With an electric thruster [e.g., solar electric propulsion (SEP)], where the incoming beamed energy is first converted into electricity for the thruster's use, the I_{sp} will be representative of the particular type of electric thruster used. However, an SEP system would still be limited by the $1/R^2$ drop-off in sunlight intensity. To overcome this, laser electric propulsion (LEP) uses laser light to provide constant power even far from the Sun. Here, high-performance electric thruster exhaust velocities (<200 km/s) would be adequate for interstellar precursor missions but not for interstellar missions.

V. Nonrocket (Beamed-Momentum) Propulsion Options

In beamed-momentum propulsion, the momentum carried by a stream of particles (e.g., photons or charged particles) is used to push the vehicle; in effect, the stream of particles becomes the "propellant" that supplies the momentum to move the spacecraft. This is in contrast to a beamed-energy system, where the beamed energy (sunlight, laser/microwave beam) provides thermal energy (or indirectly electricity) that is used to energize onboard propellant. Thus, a beamed-momentum propulsion system represents an example of a "propellantless" propulsion system, with both the energy and propellant system taken off the vehicle. As such, it also represents a way to fundamentally "cheat" the Rocket Equation because there is no onboard propellant; in effect, $M_{wet}/M_{dry} = 1$.

Two general types of beamed-momentum systems are considered. The first type uses momentum exchange between photons (solar/laser/microwave sails) and a reflective sheet or "sail." The second type uses momentum exchange between charged particles (from the solar wind or a dedicated particle accelerator) and an electromagnetic field (electromagnetic sails).

A. Assumptions Used when Projecting Upper Performance Limits

Candidate propulsion systems are categorized by their momentum source (e.g., sunlight, laser, etc.) and are discussed below. As with the rocket-based propulsion systems, more detailed discussions can be found in the technical literature.

B. Solar Sails

A solar sail is a propulsion concept that makes use of a flat surface of very thin reflective material supported by a lightweight, deployable structure [34]. Because

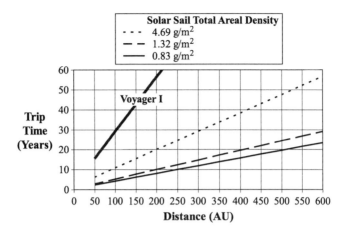

Fig. 21 Representative solar sail trajectories for interstellar precursor missions.

a solar sail uses no propellant, it has an effectively infinite specific impulse; however, the thrust-to-weight (T/W) ratio is very low, 10^{-4} to 10^{-5} for the 9 N/km^2 (5.2 lb$_f$/mile^2) solar pressure at Earth's distance from the sun, resulting in the potential for long trip times in and out of planetary gravity wells.

Solar sails have been considered for interstellar precursor missions; Fig. 21 shows trajectories that make use of intense sunlight pressure near the sun (at 0.25 AU in this example) to "blast" the sail off into a high solar system escape trajectory. Note that the sails shown here represent fairly aggressive technologies in terms of their total areal density (mass, including payload, per unit area of sail sheet, typically in units of g/m^2), as compared to near-term solar sails with areal densities > 10 g/m^2. Overall, solar sail performance is similar to nuclear electric propulsion in terms of achievable solar system escape velocities (V_{infinity}) and thus trip time.

C. Light Sails

One important limitation in solar sails is the $1/R^2$ drop in sunlight intensity as one moves out of the solar system. Nevertheless, as seen above, solar sails could be used for deep space or interstellar precursor missions by first spiraling in close to the sun (e.g., to 0.25 AU) and using the increased sunlight pressure to drive them out of the solar system. It would also be possible to perform interstellar missions with a laser-driven "light" sail. This concept [13] is uniquely suited to interstellar missions since it is one of the few ways that sufficient energy (per unit mass) can be imparted to a vehicle to achieve the high velocities (>0.1 c) required for interstellar missions. This is possible because the spacecraft "engine" (lasers) is left back in Earth's solar system; a somewhat arbitrarily large amount of energy (number of photons per unit of sail mass) could be imparted to the vehicle's propellant (photons) to accelerate the vehicle. In fact, because of the light sail's thermally limited acceleration, input power is ultimately limited by the imperfect reflectivity of the receiver optics; solar or laser light absorbed by the receiver material must be radiated to space as "waste heat" so the maximum power that can be received is a function of the sail's material reflectivity, emissivity, and maximum temperature limits.

Note, however, that for interstellar distances, *very* large optics and laser power levels would be required. For example, a laser operating at 1 μm wavelength requires a transmitter lens with a diameter of 1000 km to illuminate a 1000-km diameter receiver (sail) at 43 LY. Similarly, a very high power level (and ultra-light sail) would be required for reasonable acceleration (typically 0.04 g for flybys to 0.2 g for rendezvous) of the vehicle. For example, the laser power required for a robotic flyby mission to 4.3 LY with a maximum cruise velocity of 0.4 c would be 14 terawatts (TW), which is comparable to the average power produced by all of human civilization (15 TW in 2005). However, any interstellar mission, regardless of the propulsion system, will require high power levels to achieve the high speeds required. Even today we achieve nontrivial propulsion power levels for ambitious space missions; for example, the Saturn V rocket generated a power on liftoff corresponding to about 0.8% of human civilization's total power output in 1969.

The microwave sail (Starwisp) [12] concept is the microwave analog to the laser light sail. This approach has the advantage that the vehicle can be made ultra-lightweight for robotic interstellar mission flybys, thereby reducing both the transmitter power requirements and the size of the transmitter optics (because the microwave sail can be accelerated at high "gs" to its final coast velocity while still relatively near the Earth). In order to achieve this low mass, the "sail" consists of wire mesh with holes in the mesh less than one-half the wavelength of the microwaves. Under these conditions, the sail acts like a solid sheet with respect to the incoming microwave photons. (A related concept, the "perforated" solar light sail, has also been proposed for visible-light sails.) However, microwave sails are only feasible for interstellar flybys, and not rendezvous missions, because the long wavelengths of microwaves (as compared to visible or near-visible light) makes it impractical to build optics large enough to focus a microwave beam onto a sail at interstellar distances.

By contrast, laser light sails have a short enough wavelength to make it possible to focus the beam and enable the use of two-stage light sails for rendezvous missions, as shown in Fig. 22, and even a three-stage light sail [13] for a round trip mission. Ultimately, beamed-momentum light sails represent a *major* development challenge, both because of the extraordinarily demanding technologies, and because of the extraordinarily large scale of the systems. Nevertheless, they do represent one of the few ways to perform interstellar missions with reasonable trip times.

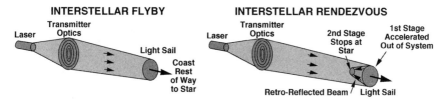

Fig. 22   Beamed-momentum laser light sail missions (after Ref. 13).

D. Electromagnetic Sails

In electromagnetic (EM) sails, charged particles (typically protons) from the solar wind or a charged-particle accelerator are reflected by a magnetic field, analogous to the reflection of solar or laser photons off of a solar/laser sail's reflective sheet. Thus, EM sails are the charged-particle analogs of solar light sails.

1. Electromagnetic Sails

In principle, a solar-wind EM sail could be built using a physical sheet of material, but the momentum per unit area carried by the solar wind is so much less than that from photons that it requires an impossibly lightweight sheet; instead, a (massless) magnetic field, 10s to 100s of kilometers in diameter, substitutes for the solar sail's sheet. Interestingly, EM sails provide many of the same potential benefits and drawbacks of solar sails; for example, sunlight intensity and solar wind density both drop off as the square of the distance ($1/R^2$) from the Sun. (The solar wind maintains a roughly constant velocity of around 300 to 800 km/s throughout the solar system, but the momentum force decreases due to the expansion of the solar wind, and thus increasing dilution of individual particles, at increasing distance from the sun.)

The first proposed EM sail was the magnetic sail, or MagSail, concept [35]. The MagSail consists of a cable of superconducting material, millimeters in diameter, which forms a hoop that is 10s to 100s of kilometers in diameter. The current loop creates a magnetic dipole that diverts the background flow of solar wind. This deflection produces a drag force on the MagSail radially outward from the sun. In addition, proper orientation of the dipole may produce a lift force that could provide thrust perpendicular to the radial drag force.

A newer concept, the mini-magnetospheric plasma propulsion (M2P2) sail [36], uses an artificially generated mini-magnetosphere that is supported by magnets on the spacecraft and inflated by the injection of low-energy plasma into the magnets. (Thus, M2P2 is not, strictly speaking, a true "propellantless" propulsion system; however, the amount of propellant needed to produce the plasma is small, resulting in an effective I_{sp} of 35,000 lb$_f$-s/lb$_m$.) This plasma injection allows the deployment of the magnetic field in space over large distances, comparable to those of the magsail.

2. Particle Beam Driven Electromagnetic Sails and MagOrion

Note that the similarity of EM sails to solar sails extends to their general mission performance; for example, both EM sails and solar sails could be used for the more modest interstellar precursor missions. However, like light sails, EM sails would require a dedicated beamed-momentum source (e.g., charged particle beam) for interstellar missions. Thus, we could use a linear accelerator to fire out high-speed particles (e.g., a relativistic particle beam) or pellets that push against the vehicle [37] as illustrated in Fig. 23. The particles can either hit a physical "pusher plate," or charged particles can push against (and be reflected by) the electromagnetic fields of an EM sail in much the same way that photons are employed in a laser/microwave sail. In principle, a two-stage EM sail could be used for an interstellar rendezvous mission by analogy with the two-stage light sail.

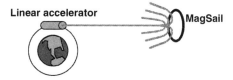

Fig. 23 Relativistic particle beam/magsail concept.

However, one potentially serious problem with the use of a charged ion or neutral particle beam is that the beam can be deflected by stray magnetic fields within a solar system and in interstellar space. Because of the speed of light time delay, it would not be practical to use a feedback control system to correct for the beam deflections. Also, neutral particles would have to be ionized before hitting an EM sail's magnetic fields. Alternatively, neutrals could impact a physical "pusher plate." One advantage of light sails is that their laser/microwave beams are unaffected by magnetic fields, although the beam could be scattered and defocused by interstellar dust clouds.

To avoid the problem of particle beam deflection from a remote source (e.g., from the solar system), we could carry a source of charged particles with the vehicle; for example, we could use the charged particles produced by a nuclear explosion. This would be the EM sail analog of the Orion nuclear pulse concept, with an electromagnetic (rather than physical) "pusher plate," hence the name Magnetic Orion or MagOrion [38]. Interestingly, in the original Orion concept, some of the material of the pusher plate would evaporate (ablate) with each pulse. This serves to both to keep the pusher plate cool, and to add additional "propellant" mass. The result is an increase in thrust, although at the expense of some specific impulse (I_{sp}). By contrast, there is no mass in the MagOrion's magnetic pusher plate, so the effective I_{sp} (15,000 to 45,000 lb$_f$-s/lb$_m$) makes this concept well-suited for interstellar precursor missions.

E. Electromagnetic Catapults

Electromagnetic launchers, such as advanced versions of the rail gun [39] and mass driver (coil gun) [40], could be used to launch small or microspacecraft. However, their demonstrated "muzzle" velocity (<12 km/s) may limit them to solar system exploration missions [8].

VI. Illustrative Missions

Thus far in this discussion, we have described a number of propulsion options that have the potential of performing interstellar precursor missions out to thousands of AU, interstellar flybys, and finally our Vision Mission of a fast (0.5 c V_{cruise}) interstellar rendezvous mission. In this section, we will illustrate the enormous difficulty, but not impossibility, of performing these missions.

A. Next-Step Teething Mission Candidates

Although often forgotten, we have in fact already launched five interstellar precursor spacecraft in the two Voyager, two Pioneer, and the New Horizons Pluto spacecrafts. However, even the fastest (Voyager 1) will barely leave the solar system during its 40-year lifetime. Historically, studies of interstellar missions have considered this 40-year time frame an important mission goal because it represents the "professional lifetime" of scientists and engineers who would launch the spacecraft and then receive the final science data shortly before their retirement. Thus, we tend to see 40- to 50-year mission trip time goals for interstellar precursors and the "easier" interstellar flybys of nearby targets (e.g., Alpha Centauri).

A wide variety of studies have been performed over the years that considered moderately near-term interstellar precursors from hundred to several thousand AU. These missions have included general particle and field measurements of the interstellar medium beyond the Heliopause (as has just recently been possible with the Voyager spacecraft), investigation of Kuiper belt objects, stellar parallax measurements (e.g., with a 100 to 1000 AU triangulation baseline using a telescope on the spacecraft and one at Earth), and astronomy at the Sun's gravitational lens (e.g., 550 AU and beyond). These missions typically require propulsion systems with an I_{sp} of 10,000 lb_f-s/lb_m. Many of these missions could be performed by NEP and its smaller sibling radioisotope electric propulsion (REP), fission fragment propulsion, fusion propulsion, various antimatter-fission-fusion hybrids, and solar sails. From the perspective of the interstellar goal, these missions could provide "engineering" data on the near-sun interstellar environment (e.g., dust grain number and size distribution, density of H and D for fusion ramjets, etc.), as well as highly-autonomous, long-duration engineering test beds for spacecraft systems (e.g., power, communications, etc.).

As next-steps missions beyond 1000 AU we have Oort cloud missions, nominally from 2000 to 20,000 AU for the toroidal inner Oort Cloud, and 20,000 to 60,000 AU (∼1 LY) for the spherical outer Oort Cloud. For reasonable trip times, these missions would need propulsion systems with an I_{sp} of 100,000 lb_f-s/lb_m, which implies fission fragment, fusion, or hybrid antimatter-fission-fusion systems. At this range, first-generation beam-core antimatter rockets and light sails could also begin to be used; this could allow them to be tested in a "small" scale mission before scaling them up to the levels required for a full interstellar mission.

As a "strawman" example, we can imagine an Oort cloud interstellar flyby and rendezvous precursor mission to 0.1 LY (∼6000 AU), with an assumption of a 39.9-year flight time, for a total time from launch to signal return to Earth of 40 years. Table 3 summarizes results for a single-stage light sail and antimatter rocket propulsion system for the flyby mission, and a two-stage light sail and single-stage antimatter rocket propulsion system for the rendezvous mission. Here, the light sail transmitter and receiver (sail) have the same diameter. Also, the acceleration level (gs) fixed by the light sail flyby is used for the antimatter rocket flyby and rendezvous. Note that there is an additional subtlety in calculating the mass ratio (M_{wet}/M_{dry}) for the antimatter rocket rendezvous mission when a single stage is used for acceleration and deceleration. For example, accelerating from rest to 0.002560 c for a flyby mission results in a straight-forward calculation of the antimatter rocket's mass ratio of 1.01868. The corresponding single-stage antimatter rocket rendezvous mission has an acceleration and deceleration to or from the same V_{cruise} (0.002618 c).

Table 3 Comparison of light sail and antimatter rocket for 0.1-LY 39.9-year flyby and rendezvous mission

System	Light sail		Antimatter rocket	
Payload mass (MT)	100		100	
Mission type	Flyby	Rendezvous	Flyby	Rendezvous
Average acceleration (g)				
Stage 1	0.001487	0.0191/0.0316[a]	0.001487	0.001487
Stage 2		0.03290		
V_{cruise} (c)	0.002560	0.002513	0.002560	0.002618
Antimatter rocket mass ratio (M_{wet}/M_{dry})			1.01868	1.03857
Laser sail				
Laser diameter = sail diameter (km)	7.03	47.71		
Sail-only areal density ($\sigma_{sail\ only}$, g/m^2)	0.73	0.100		
Total mass with payload (MT)	102.8	296.7		
Laser power (TW)	0.303	11.210		
Antimatter rocket				
Total wet mass with payload (M_{wet}, MT)			357.1	374.0
Total dry mass with payload (M_{dry}, MT)			350.5	360.2
Total antimatter mass (MT)			3.44	7.29
Total flight time (years)	39.9	39.9	39.9	39.9
Time from launch until signal received at Earth (years)	40.0	40.0	40.0	40.0

[a]Light sail stage 1 acceleration for Earth departure/target rendezvous.

However, the single-stage rendezvous antimatter rocket's mass ratio is not based on the total ΔV (0.005236 c, with M_{wet}/M_{dry} = 1.03862), but rather based on the product of the mass ratio for each step's ΔV (0.005236 c, with M_{wet}/M_{dry} = 1.01910), such that the mass ratio is 1.01910^2 = 1.03857. Finally, the Bussard interstellar fusion ramjet has been far less studied than antimatter rockets or laser light sails; therefore we must defer mission analyses of the fusion ramjet until detailed system studies become available.

These results suggest that an inner Oort cloud mission to 0.1 LY in 40 years would be a reasonable "teething" mission for propulsion systems like light sails and antimatter rockets, capable of ultimately performing our Vision Mission of a fast interstellar rendezvous. For example, if we assume that we selected a flyby mission, as is typically done for interstellar precursors, the light sail and laser transmitter optics are only 7 km in diameter, and the laser system power is a modest 300 GW, or about 7 times the power generated by a Saturn V at liftoff. Also, the sail sheet need not be the high-emissivity version needed for interstellar rendezvous because this mission has a low acceleration, making the fabrication of the sail easier. Similarly, the mass of the antimatter rocket is less than four times the mass of the Space Shuttle Orbiter. However, the issue here would be to achieve antiproton production rates 16 orders-of-magnitude above today's rates (e.g., 100 MT/yr versus 10 ng/yr).

Alternatively, we could choose to perform a rendezvous mission as a means of demonstrating a two-stage light sail mission with all the complexities of pointing and tracking the sail at 0.1 LY, and using the sail's first stage to retro-reflect the laser beam back onto the second stage. Even though a rendezvous mission might not be needed for science reasons, this would give us a chance to exercise all the essential light sail systems required for an interstellar rendezvous.

B. Alpha Centauri (4.3 LY) in 75 Years

The Alpha Centauri star system [41], 4.3 LY away, is a double star (possibly a triple star) system. The central star, Alpha Centauri A, is very similar to our own sun; it is slightly larger and hotter, but both the sun and Alpha Centauri A are in spectral class G2.

The second star, Alpha Centauri B, is smaller and dimmer than our sun; it orbits far enough away from A (11 to 35 AU) that its orbit would not cause any planets in Alpha Centauri A's habitable "Goldilocks" zone to be ejected. For Alpha Centauri A or B, the "safe" zone for immunity from being gravitationally ejected stretches out to about 3 AU for planets in the same plane in which A and B orbit around each other; however, this drops to 0.72 AU for any planets in orbits inclined at right angle to the A/B orbit plane. Figure 24 shows the Alpha Centauri A/B system, with the orbits of the planets of our solar system shown for scale.

Finally, there is a small, cool red dwarf star, Alpha Centauri C, that is quite far from Alpha Centauri A/B (13,000 AU); it may not actually be a gravitationally connected part of the Alpha Centauri system, but rather a wandering star that just happens to be near (in stellar terms) to Alpha Centauri A/B. Interestingly, Alpha Centauri C is slightly closer (4.2 LY) to us than A/B, so C is also called Proxima Centauri.

For this target, we will illustrate a 75-year duration interstellar flyby and rendezvous mission; the results are given in Tables 4 and 5. As discussed previously,

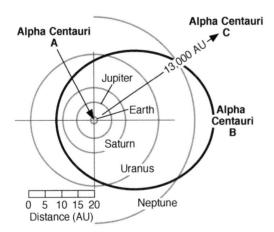

Fig. 24 Alpha Centauri system with orbits of our solar system planets shown to scale.

Table 4   Comparison of Daedalus and Orion for Alpha Centauri 4.3-LY flyby and rendezvous missions

System	Interstellar Orion		Daedalus	
Mission type	Flyby	Rendezvous	Flyby	Rendezvous
Stage 1	Acceleration	Acceleration	Acceleration	Acceleration
Stage 2	Acceleration	Deceleration	Acceleration	Deceleration
V_{cruise} (c)	0.033	0.033	0.1241	0.0626
Total wet mass (M_{wet}, MT)	400,000	2,000,000	54,056	54,056
Total flight time (years)	129.0	129.1	36.6	69.0
Time from launch until signal Received at Earth (years)	133.3	133.4	40.9	73.3

systems like a two-stage fusion rocket (e.g., Daedalus) or an interstellar Orion could perform the flyby mission or rendezvous, but the Orion trip times are significantly longer than those for Daedalus. (Also, the flight times for the Orion are essentially the same for flyby or rendezvous due to Orion's high acceleration.)

As with the 0.1-AU interstellar precursor mission, we can determine the performance of a light sail and antimatter rocket for an Alpha Centauri flyby and rendezvous mission, with an assumption of a 70.7-year flight time, for a total time from launch to signal return to Earth of 75 years. For this mission, with a relatively modest cruise velocity requirement (~0.06 c), we could use a single stage flyby and two-stage rendezvous antimatter rocket. (The light sail always uses a single stage for a flyby, and two stages for a rendezvous mission.)

Here we begin to see the enormous scale of a "true" interstellar mission as compared to an interstellar precursor. For the light sail, the laser powers are 10s to 100s of TW (10^{12} W); for comparison, today all of human civilization operates at an average power level of about 15 TW! Similarly, the amounts of antimatter needed have jumped to 100s to 1000s of MT.

C.   Gliese 876 (15.2 LY) in 75 Years

Gliese 876 [42] is a cool red dwarf (spectral class M3.5), with about 32% of the mass of our Sun. It is the closest star to us known to possess planets. The first planet to be detected around Gliese 876 (Gliese 876 "b" in June 1998) has a mass of about 2.5 Jupiter-masses and orbits its star at 0.21 AU. The next to be detected (Gliese 876 "c" in January 2001) has an even closer orbit, 0.13 AU, and a mass of 0.8 Jupiters. The most recent to be detected (Gliese 876 "d" in June 2005) is a potentially "terrestrial" planet (i.e., rocky as opposed to Jupiter-like gas giant) with a mass of 7.5 Earth-masses (0.024 Jupiter-masses) in a 0.021 AU orbit.

As with the Alpha Centauri mission, the results of a 75-year duration interstellar flyby and rendezvous mission for the light sail and antimatter rocket are summarized in Table 6. Because of the greater distance, flight times for systems like

Table 5 Comparison of light sail and antimatter rocket for 4.3-LY 70.7-year flight time flyby and rendezvous missions

System	Light sail		Antimatter rocket	
Payload mass (MT)	100		100	
Mission type	Flyby	Rendezvous	Flyby	Rendezvous
Average acceleration (gs)				
Stage 1	0.0261	0.0317/0.0326[a]	0.05	0.05
Stage 2		0.1918		0.05
V_{cruise} (c)	0.06182	0.06178	0.06134	0.06187
Antimatter rocket mass ratio ($M_{\text{wet}}/M_{\text{dry}}$)			1.56031	1.56633
Laser sail				
Laser diameter = sail diameter (km)	40.49	312.98		
Sail-only areal density ($\sigma_{\text{sail only}}$, g/m^2)	0.073	0.100		
Total mass with payload (MT)	194.3	8,563.1		
Laser power (TW)	10.682	552.146		
Antimatter rocket				
Total wet mass with payload (M_{wet}, MT)			1,876	6,518
Total dry mass with payload (M_{dry}, MT)			1,202	3,477
Total antimatter mass (MT)			354	1,597
Total flight time (years)	70.7	70.7	70.7	70.7
Time from launch until signal received at Earth (years)	75.0	75.0	75.0	75.0

[a]Light sail stage 1 acceleration for Earth departure/target rendezvous.

Daedalus or the interstellar Orion are probably unacceptable. For example, the flight time for Daedalus for the flyby and rendezvous missions would be 140 and 258 years, respectively; the corresponding value for the interstellar Orion is 471 years (essentially the same for flyby or rendezvous due to Orion's high acceleration). This also illustrates the classic tradeoff of interstellar missions: Do we launch now using a slower vehicle with a longer flight time, or do we wait X years for a faster vehicle to be developed, with a shorter flight time, because by waiting we might actually reach our target sooner? This is a real issue for "slow" ($V_{\text{cruise}} \sim 0.1\ c$) fission or fusion systems; however, it becomes a moot point for relatively "fast" ($V_{\text{cruise}} \sim 0.5\ c$) systems like light sails or antimatter rockets simply because we bump up against the speed-of-light velocity limit. Thus, even with an ultra-relativistic ($V_{\text{cruise}} > 0.9\ c$) Bussard fusion ramjet, we would only have at most a factor of 2 reduction in flight time as compared to a $V_{\text{cruise}} = 0.5\ c$ vehicle.

Table 6 Comparison of light sail and antimatter rocket for 15.2-LY 59.8-year flight time flyby and rendezvous missions

System	Light sail		Antimatter rocket	
Payload mass (MT)	100		100	
Mission type	Flyby	Rendezvous	Flyby	Rendezvous
Average acceleration (g)				
Stage 1 (or 4)	0.0479	0.0476/0.0359[a]	0.05	0.05
Stage 2 (or 3)		0.2054	0.05	0.05
V_{cruise} (c)	0.26618	0.26942	0.26561	0.27953
Antimatter rocket mass ratio (M_{wet}/M_{dry})			2.67912	2.82915
Laser sail				
Laser diameter = sail diameter (km)	128.76	591.56		
Sail-only areal density ($\sigma_{sail\ only}$, g/m^2)	0.073	0.100		
Total mass with payload (MT)	1,053.2	30,332.8		
Laser power (TW)	133.4	3,286.9		
Antimatter rocket				
Total wet mass with payload (M_{wet}, MT)			44,370	1,105,040
Total dry mass with payload (M_{dry}, MT)			11,843	179,273
Total antimatter mass (MT)			17,077	486,025
Total flight time (years)	59.8	59.8	59.8	59.8
Time from launch until signal received at Earth (years)	75.0	75.0	75.0	75.0

[a]Light sail stage 1 acceleration for Earth departure/target rendezvous.

As compared to the Alpha Centauri mission, the greater distance of Gliese 876 has a significant impact on the light sail system, primarily driven by the beam focusing requirements for the longer distances. This has the effect of increasing the laser transmitter optics and light sail diameters, which results in a heavier sail. This heavier sail must also be accelerated to a higher speed in order to cover a longer distance in the same amount of time as the Alpha Centauri mission. For example, the beaming distance for the flyby is longer due to the increased cruise velocity requirement, which in turn drives the laser power level. For the rendezvous, the laser would need to focus the beam onto the sail at Gliese's distance, and provide full power to decelerate stage 2 of the sail at its thermally limited maximum.

For the antimatter rocket, if we attempt to use a single-stage flyby and two-stage rendezvous vehicle, the flight time would exceed our 59.8-year target when we limit the speed to keep M_{wet}/M_{dry} to about 5. Thus, we would have to use a two-stage flyby and four-stage rendezvous vehicle. By using the two-/four-stage combinations, we would keep the ΔV per stage to a much more comfortable level of $M_{wet}/M_{dry} \sim 2.7$. However, even with these limits on V_{cruise} and M_{wet}/M_{dry}, the vehicle mass and especially the mass of antimatter grow to remarkable levels (e.g., half a million tons for the rendezvous mission).

D. Vision Mission: 40 LY

Our Vision Mission to 40 LY was originally selected because a sphere with this radius contains approximately 1,000 stars. In fact, the binary star system 55 or Rho(1) Canceri [43] lies just outside this range at 40.9 LY. This system is especially interesting because the A star of the binary has at least 5 planets, including one inside the star's habitable "Goldilocks" zone. 55 Canceri A is a yellow-orange dwarf (spectral class G8) with about 0.94 times the mass of the Sun. A fifth planet, "f" or "A5," was discovered in November 2007 at a distance of 0.781 AU, within the stars habitable zone. Unfortunately, this planet's mass (45 to 57 Earth-masses) means that it is probably a gas giant; for comparison, it is about half the mass of Saturn (95 Earth-masses). However, looking to our own solar system, 55 Canceri A5 might have a habitable moon analogous to Jupiter and Europa. Finally, star B is a red dwarf about 1,100 AU from star A.

Our Vision Mission has as its goal a "fast" ($V_{cruise} = 0.5\ c$) rendezvous at 40 LY. As shown in Table 7, the light sail system could achieve this, but the antimatter rocket is only able to reach $0.4173\ c$ because of the impact on M_{wet}/M_{dry} of the Relativistic Rocket Equation when $a = 0.4$. As discussed earlier, we must to limit the speed to keep M_{wet}/M_{dry} to about 5.0. Also, stage 2 (or 3) starts at a higher speed than stage 1 (or 4), so the ΔV for stage 2 or 3 ($0.1987\ c$) would be less than that for stage 1 or 4 ($0.2186\ c$) so as to have the same M_{wet}/M_{dry}. Note also that, although staging reduces the ΔV per stage and thus M_{wet}/M_{dry}, each stage has a M_{wet}/M_{dry} of 5.0, so the total four-stage rendezvous vehicle would have a wet mass at least $(5.0)^4 = 625$ times the dry mass of the fourth (final) stage. Thus we see the need for truly prodigious amounts of antimatter (40 million tons!) for the four-stage rendezvous mission. Finally, because of the limitation on V_{cruise}, the trip time is about 20 years longer than for the light sail.

As discussed previously, there are two additional complications introduced when using a light sail for a rendezvous mission. These both involve the problem of using stage 1 to retro-reflect the laser beam back onto stage 2. This has the effect of decelerating stage 2 to a stop, but accelerating stage 1. Also, as stage 2 is decelerating, stage 1 is moving out ahead in front of stage 2, thus increasing the retro-reflection beaming distance between stage 1 and stage 2. The increasing speed of stage 1 exacerbates the already large amount of laser beam redshifting due to the relative speeds between stage 1 and 2. This is at its maximum at the end of the rendezvous maneuver, when stage 2 is at rest and stage 1 is at its maximum velocity, e.g., $0.597\ c$ for this example. This has the result of dramatically increasing the required laser system power by a factor of 3.97 at $0.597\ c$ due to "double" redshifting of the laser's photons seen by stage 2. Finally, redshifting of photons beamed from stage 1 onto stage 2 results in a spot size increase of 1.99 times its classical (nonrelativistic) value. The various relativistic effects are illustrated in Fig. 25 and Table 8.

Table 7 Comparison of light sail and antimatter rocket for 40-LY flyby and rendezvous missions

System	Light sail		Antimatter rocket	
Payload mass (MT)	100		100	
Mission type	Flyby	Rendezvous	Flyby	Rendezvous
Average acceleration (g)				
Stage 1 (or 4)	0.0488	0.0850/0.0388[a]	0.05	0.05
Stage 2 (or 3)		0.1996	0.05	0.05
V_{cruise} (c)	0.5	0.5	0.4173	0.4173
Antimatter rocket mass ratio (M_{wet}/M_{dry})			5.000	5.000
Laser sail				
Laser diameter = sail diameter (km)	239.61	961.95		
Sail-only areal density ($\sigma_{sail\ only}$, g/m^2)	0.073	0.100		
Total mass with payload (MT)	3,400.8	80,044.0		
Laser power (TW)	608.922	17,889.764		
Antimatter rocket				
Total wet mass with payload (M_{wet}, MT)			1,127,632	83,980,666
Total dry mass with payload (M_{dry}, MT)			156,792	7,661,580
Total antimatter mass (MT)			509,691	40,067,520
Total flight time (years)	84.97	84.07	99.9	103.9
Time from launch until signal received at Earth (years)	124.97	124.07	139.9	143.9

[a]Light sail stage 1 acceleration for Earth departure/target rendezvous.

The other potentially serious light sail feasibility issue is the ability of a very lightweight gossamer structure like stage 1 to act as a high-quality optical transmitter for beaming laser light over near-interstellar distances to stage 2. In this example, the stage 1-2 separation distance at the end of the rendezvous maneuver is 0.725 LY, or roughly one-sixth the distance between us and Alpha Centauri.

E. Comparison of Light Sails and Antimatter Rockets for Robotic Interstellar Missions

For our initial mission examples at target distances of less than about 15 LY, the laser light sail and antimatter rocket were at significantly less than their maximum performance as required for the 40-LY Vision Mission. In Fig. 26, we illustrate the performance of these vehicles for 75-year, 40-year, and

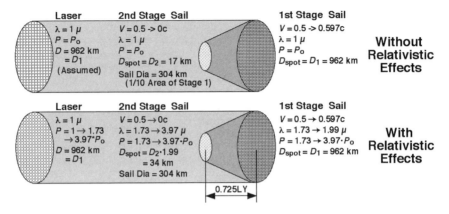

Fig. 25 Comparison between classical and relativistic light sail parameters.

minimum total mission time robotic missions, where the total mission time consists of the flight time plus the time required to return the first data (at the speed of light) back to Earth.

As expected, very "fast" missions, with V_{cruise} approaching 0.5 c, can yield short flight times, but only at the cost of enormous laser powers or quantities of antimatter. (Note the logarithmic power and mass scale in Fig. 26.) Also, for more distant missions, the time required to return the data signal to Earth becomes a significant fraction of the total mission time; e.g., the Vision Mission rendezvous light sail has a flight time of 84 years to 40 LY, plus a 40-year signal travel time back to Earth. However, even at these limits of performance and even if laser power and antimatter were "free," we would ultimately be constrained by such "mundane" engineering factors as the temperature limits of materials (i.e., light sail thermally-limited acceleration) or the Rocket Equation's inexorable mass ratio (i.e., $M_{\text{wet}}/M_{\text{dry}}$ as a function of ΔV per stage).

F. Generational Ships

At the extreme limit of infinite acceleration (i.e., no time spent accelerating) and a cruise velocity of 0.5 c, a robotic mission could only travel as far as 13.33 LY and have the science data returned within the Earth-bound scientist's "professional" lifetime of 40 years, ranging nominally from a person's early 20s to their early 60s. Any robotic mission beyond this distance would require

Table 8 Light sail relativistic effects

Mission step	Earth departure	Rendezvous	
Light sail stage	Stage 1	Stage 1	Stage 2
Velocity (c)	0.00 → 0.50	0.50 → 0.597	0.50 → 0.00
M/M_{rest}	1.00 → 1.15	1.15 → 1.25	1.15 → 1.00
$\lambda/\lambda_{\text{rest}}$	1.00 → 1.73	1.73 → 1.99	1.73 → 3.97

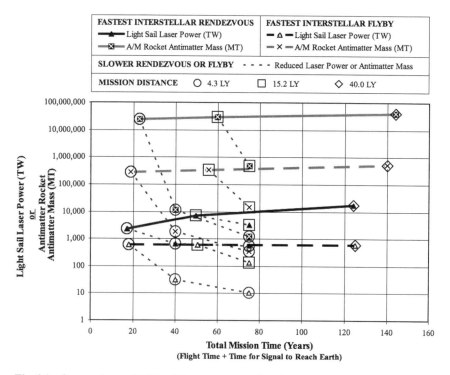

Fig. 26 Comparison of light sail laser power and antimatter rocket antimatter mass as a function of total mission time. Total mission time = flight time plus time for first signal to reach Earth.

multiple generations of humans between launch and final data return. For piloted missions with the crew on board the vehicle, we are faced with two options. First, we can assume a one-way mission, where the primary goal is exploration and not colonization from subsequent generations from the original crew. In this case, we can assume a 40-year time period, but now with the distinction that it is a 40-year flight time, as opposed to flight time plus data return time (at the speed of light) for the robotic mission. In this regard, Forward has explored the ethical issue of sending a crew on a one-way mission in a fictional setting of a piloted light sail mission [44]. Here, the hero is asked about his feelings on embarking on a one-way mission from which there is no "rescue." He replies simply that we are *all* on one-way "missions" through life, with the subtext that the real question is not how long we live, but rather what we do with our lives.

In terms of flight time, if we take the "mundane" science fiction assumption that there are no convenient loopholes for long-duration missions (e.g., no suspended animation/hibernation, no lifespan-extending drugs, etc.), then we are faced with the need for multigeneration human missions. Multigeneration ships would also be needed for colonization [45]. In this case, the mission flight time could be extended indefinitely, but the payloads would become very large due to the need to carry a sufficient human population to prevent inbreeding and/or genetic drift after a few generations. Estimates of the population required have ranged from a low of 50 (a minimum to prevent excessive inbreeding) to around 500 (a level that balances the rate of gain in genetic variation due to mutation with the rate of loss due to genetic drift). A value of 150 to 180 would make it possible to have a stable crew size over 60 to 80 generations (roughly 2,000 years) with each generation having a good selection of potential mates. These numbers could, in principle, be driven even lower with the use of stored/frozen sperm, eggs, or embryos [46].

However, these population estimates were based only on the need to minimize inbreeding and genetic drift effects. Much larger populations may be desirable to overcome the inherent limitations of small populations. For example, the small populations described above did not take into account the psychology or group-dynamics issues of small groups. Also, they made no allowance for the need for a large skills mix of potential colonists. This might be especially important on a new world, where the high-tech imports from the original colony ship would eventually break down, and the colony reduced to a mid-1800s level of technology (e.g., plows pulled by horses rather than tractors). Finally, it may be just as important to have a broad range of social and cultural variation as genetic variation. For example, it is humanity's rapid social/cultural adaptability that has allowed us to survive dramatic changes in environment and fill virtually every ecological niche on Earth over the last 100,000 years, an eye blink in time compared to the slow march of genetic evolution.

Thus, to assess the worst-case mass impact of a large population size, we assumed a vehicle payload of 500,000 MT with a population of 10,000 corresponding to the mass and population of an O'Neill "Island One" (Bernal sphere) space colony [47]. Ironically, this is one time when the enormous mass of the antimatter rocket acts in our favor because is would have such a high dry mass that adding even a large payload (e.g., 500,000 MT) would represent only a modest increase in the dry mass of the total vehicle. For example, adding a 500,000 MT payload to our Vision Mission 40-LY rendezvous four-stage antimatter rocket would only increase its wet mass and antimatter mass by a factor of 32.5 and 32.8, respectively.

Unfortunately, the situation for the two-stage (rendezvous) light sail would be more complicated because of the impact that a large payload has on the stage 2 thermally limited acceleration. Here, adding a large payload to stage 2 (7,267 MT without its 100 MT payload) would decrease its allowable deceleration; we either have to limit the cruise velocity so that stage 2 could be decelerated (from the lower V_{cruise}) without exceeding the thermal limits on stage 1 or 2, or we would have to make the sail larger, so that the payload again would become a small fraction of the mass of the sail. For example, keeping the sail diameter the same as our baseline case (962 km with 100 MT payload), we would have to limit V_{cruise} to $0.055\ c$. Under these conditions, the laser power would grow by roughly a factor of two, from the baseline 17,890 TW to 32,129 TW. However, the flight time would dramatically increase to 737 years. At the other extreme, keeping V_{cruise} at $0.5\ c$ would require an almost seven-fold increase in sail diameter, from the baseline 962 km to 6639 km, and an almost 84-fold increase in laser system power (17,890 TW to 1,495,158 TW). This does, fortunately, keep the flight time to a more reasonable 84.7 years.

Needless to say, these numbers represent prodigious quantities, i.e., "only" 1.3×10^{17} years to produce the 1.3 billion tons of antimatter required for a multi-generation colony ship at today's production rates (10 ng/yr) or a light sail laser system power level "only" 10^5 times the total power output of human civilization. Thus, the physics of the problem says that these missions are not impossible, although one might quite reasonably question their practicality.

VII. Most Critical Technologies and Feasible Engineering Limits

In this section we will discuss several technologies and their potential limits. We will also discuss several other issues, such as societal resources that might be applied to interstellar missions, and the limits of plausibility of performing an interstellar mission as distinct from its physical and technical feasibility.

A. Specific Impulse

In many space propulsion systems it is necessary to trade thrust (which directly impacts vehicle acceleration) for I_{sp} in order to reach an acceptable flight time. This is because for a given total power level, thrust is inversely proportional to I_{sp}. Thus, in order to achieve a fast Earth–Mars round trip time, a fusion propulsion system would be designed to run at an I_{sp} of $\sim$20,000 lb$_f$-s/lb$_m$ [48], even though it is theoretically capable of 10^6 lb$_f$-s/lb$_m$. In this case, the selection of I_{sp} depends on the mission needs in terms of both ΔV and trip time; too high an I_{sp} can result in too low a thrust and acceleration, and thus too long a trip time.

We see the same effect in electric propulsion thrusters. For example, Hall thrusters have an acceptable efficiency ($\sim$50%) in the relatively low I_{sp} range (few thousands of lb$_f$-s/lb$_m$) that is optimum for Earth orbit raising (i.e., ΔV of a few km/s). By contrast, ion thrusters have very poor efficiency in this I_{sp} range. Instead, ion engines have much better efficiencies at I_{sp}'s appropriate for the large ΔV's and long distances of interplanetary missions. In this context, interplanetary distances are long enough that there is an opportunity to use even a low acceleration vehicle to accelerate to a high cruise velocity. Ironically, as far as low-acceleration electric propulsion vehicles are concerned, Mars is *too close* for high-I_{sp} vehicles to build up enough speed for a short flight time; depending on the mission, it may be better to use a lower-I_{sp} option.

Unfortunately, this focus on near-term mission applications, where the ultra-high I_{sp}'s needed for interstellar missions are not required, has hampered development of these systems. Thus, the vast majority of the propulsion technologies discussed here for interstellar precursor and full-up interstellar missions are purely conceptual (e.g., fission fragment, fusion, antimatter); their performance is based on calculations using various theoretical models. Some technologies have had important proof-of-concept experiments performed for them (e.g., in-space deployment of a solar sail-like spinning structure by the Russians, "noncontact" magnetic levitation of normal-matter solid H_2 [49], etc.) but no propulsion technology with performance required for interstellar missions exists. Even for the more modest interstellar precursor mission where I_{sp}'s of 10,000 to 20,000 lb$_f$-s/lb$_m$ are needed, only electric propulsion using ion thruster technology is currently even close, with the power system lagging far behind.

B. Specific Power and Specific Energy Density

We can estimate the ideal energy density (energy per unit mass, E/M) that must be delivered to a propellant by combining the (Classical) Rocket Equation and the definition of kinetic energy (KE) of the propellant (E_p). Thus:

$$KE = 1/2 MV^2 = 1/2 M_p (V_{exhaust})^2 = 1/2 M_p (g_c I_{sp})^2 = E_p$$

Rearranging and solving for $(g_c I_{sp})$ gives:

$$(g_c I_{sp}) = (2E_p/M_p)^{1/2}$$

From the Rocket Equation, we have:

$$\ell n(M_{wet}/M_{dry}) = \Delta V/(g_c I_{sp}) = \Delta V/(2E_p/M_p)^{1/2}$$

We rearrange this equation and solve for the propellant energy density (E_p/M_p) as a function of ΔV and the assumed vehicle mass ratio:

$$(2E_p/M_p)^{1/2} = \Delta V/\ell n(M_{wet}/M_{dry})$$

$$(E_p/M_p) = 1/2[\Delta V/\ell n(M_{wet}/M_{dry})]^2 = 1/2[\Delta V \cdot \ell n(M_{dry}/M_{wet})]^2$$

Thus, the required energy density grows as the *square* of the required ΔV. Figure 27 illustrates this trend for a range of M_{dry}/M_{wet} of 10% (very challenging to build) to 50%, corresponding to $M_{wet}/M_{dry} = 10$ to 2.

From this, we see that a single-stage to low Earth orbit (SSTO–LEO) vehicle is just on the edge of feasibility with chemical (O_2/H_2) propellants, and then only if M_{dry}/M_{wet} is around 10%, a value that represents a major challenge. For a

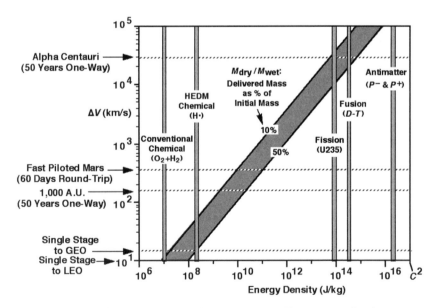

Fig. 27 ΔV capability as a function of propellant energy density.

single-stage to geosynchronous Earth orbit (SSTO–GEO) vehicle, we would need a high energy density matter (HEDM) propellant like free-radical (atomic) H. More ambitious missions require progressively higher and higher energy to be imparted to the propellant, with matter–antimatter annihilation the extreme case. Interestingly, fission and fusion energy densities are very similar, which is why the I_{sp} or $V_{exhaust}$ for fission-fragment and fusion propulsion are essentially the same. Finally, an electron-positron annihilation pair would have an energy density of c^2, but a proton-antiproton pair is less because not all of the initial matter is converted into energy.

We can also calculate the required vehicle specific power, e.g., kg/W, based on the fundamental definition of (classical) rocket engine "jet" or exhaust power as a function of $V_{exhaust}$ ($g_c\, I_{sp}$) and thrust (F):

$$P_{jet} = 1/2 g_c I_{sp} F$$

We combine this with the definition of force as a function of mass and acceleration:

$$F = MA$$

where in this case M is the total mass of the vehicle and A its acceleration. Combining these two equations and solving for specific power as M/P_{jet} (e.g., kg of vehicle/W_{jet}) gives:

$$\text{Specific power} = M/P_{jet} = 1/(1/2 g_c I_{sp} A)$$

We have seen above that the minimum practical acceleration for interstellar missions is 0.01 g; any less than this and the flight times become excessive because so much time is spent in acceleration (or deceleration). By contrast, acceleration at more than 1 g is not needed because the acceleration (or deceleration) times approach zero. Thus, we can "bracket" the range of total vehicle specific power needed for interstellar missions, as shown in Fig. 28.

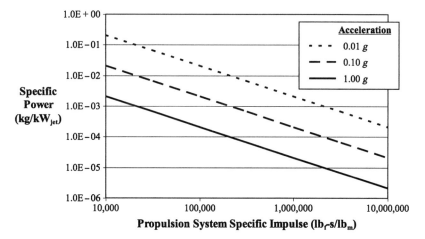

Fig. 28 Total vehicle specific power (kg/kW$_{jet}$) as a function of vehicle acceleration and propulsion system I_{sp}.

Not surprisingly, interstellar fusion vehicles ($I_{sp} \sim$ 1,000,000 lb$_f$-s/lb$_m$) and antimatter rockets ($I_{sp} \sim$ 10,000,000 lb$_f$-s/lb$_m$) will require extraordinarily low specific masses compared to contemporary and near-term vehicles (e.g., NEP on the order of 10 kg/kW$_{jet}$ at megawatt power levels).

C. Critical Nonpropulsion Technologies

Although the focus of this chapter has been on propulsion technology, there are several additional critical technologies that will require major advancements. For example, because of the finite limit of the speed of light, round-trip communication times will be measured in decades. Thus, the vehicle will require extremely advanced autonomy (e.g., software) and avionics (e.g., hardware), which are separate functions in today's spacecraft but must grow to become a single function. More generally, robotic spacecraft will require artificial intelligence on the level of humans if they are to serve as our surrogate explorers at interstellar distances. Similarly, structures technology requires major advancements due to the very large size of the various concepts (e.g., dimensions on the order of thousands of kilometers). In fact, there are two major technology paths in structures; the first involves large, ultra-lightweight, thin-film structures for the light sails; the second involves large, "heavy" structures for systems like the light sail laser array, the antimatter rocket, and the fusion ramjet.

One area that will require significant advancement is reliability because of the long overall mission times. Other critical technologies that will require significant (but not major) advancement include communications systems, power systems, and navigation. Onboard spacecraft power at sub-kilowatt levels could be met by advanced nuclear power systems like radioisotope thermoelectric generators. However, large propulsion system–related, multi-megawatt power systems, like those envisioned for advanced NEP vehicles, may be required for the antimatter rocket and fusion ramjet systems. This includes energy storage systems for startup power, thermal-to-electric power conversion during engine operation, and housekeeping power during coast (for cryogenic refrigeration systems, electromagnetic storage of antimatter, etc.). Various power system options and their limitations are discussed in more detail in Chapter 17. Finally, navigation will require advancements in position knowledge (e.g., advanced optical navigation), timing (e.g., advanced, highly accurate, and stable clocks), and acceleration (changes in position and time). Finally, we will conclude with recommendations for future systems modeling and basic research related to assessing the various elements in interstellar propulsion system.

1. Autonomous Reliability Versus Reparability

Any long-duration space system will require a high level of reliability and system lifetime. With a requirement for systems to operate for decades to centuries, it may be necessary to rethink our traditional assumptions about trading performance for lifetime. For example, instead of pursuing the goal of maximum performance, we may need to design systems for ease of maintenance, repair, or replacement, even if this means sacrificing some level of performance. Also, in the context of a highly intelligent robotic spacecraft, or ultimately a piloted mission, it is possible to imagine a completely autonomous vehicle where replacement parts are manufactured on the vehicle as needed; in effect, the vehicle would have its own "machine shop" and robots to perform the needed work. This also introduces the idea of sacrificing performance for ease of manufacturability in a completely autonomous robotic environment.

2. Minimum Communication System Power

Interestingly, Lesh et al. [50] have already shown that optical (laser) communication over interstellar distances is feasible given modest extrapolations of the technology. This also suggests that programs like SETI (Search for Extraterrestrial Intelligence) may have better success when observations are made at near-visible wavelengths, rather than the current microwave searches.

Figure 29 illustrates the general trends seen in optical and radio communication systems as a function of transmission distance. However, it should be noted that the values shown are from a wide variety of point designs with widely different assumptions [50] about system design (e.g., transmitter and especially receiver architectures), technology levels, efficiencies, and so on. However, we do see the general trend that radio is more power-intensive than optical.

D. Societal Investments in Interstellar Missions

Given the inherent enormous scale of any interstellar mission, one question that can be asked is what resources will a civilization be willing to expend for an interstellar mission? Previously, based on U.S. government spending as a fraction of the U.S. gross national product (GNP), we have estimated that about 10% of the GNP might be available based on a combination of NASA and defense

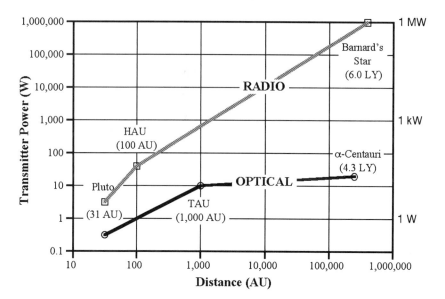

Fig. 29 Radio and optical communication system transmitter power as a function of transmission distance. (Data from Ref. 50.)

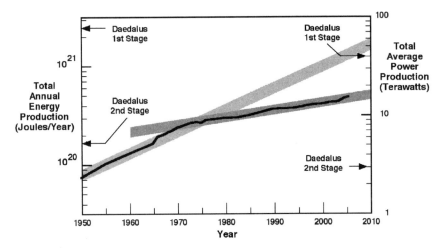

Fig. 30 Comparison of energy and power of Human Civilization and Daedalus fusion rocket.

spending [14]. However, this does not give us a feel for the magnitude of the gap between today's total resources and the requirements of an interstellar mission. Figure 30 attempts to answer this question in terms of the total annual energy and power output of human civilization as compared to the requirements of even a "modest" interstellar capability. Not surprisingly, it would take roughly five years of the total annual energy output of human civilization to equal the energy required for the minimal interstellar capability represented by the Daedalus two-stage $0.1\,c$ flyby fusion vehicle. Finally, as a side note, Fig. 30 also shows the change of slope in annual energy production and consumption due to the energy crisis of the 1970s.

For comparison, the energy content of annihilating the antimatter in the Vision Mission rendezvous four-stage antimatter rocket is capable of vaporizing the entire surface of the Earth to a depth on the order of 100 m. Similarly, the 17.9-PW laser system required for the Vision Mission rendezvous light sail has the capability of delivering the energy equivalent of a 4-megaton nuclear weapon per second into a 0.7-m diameter spot at a distance of 2 AU from the laser system. This suggests that it would be prudent to base an antimatter "factory" or light sail laser in a 1 AU orbit on the side of the Sun opposite Earth, although the light sail laser would have to contend with zodiacal dust scattering if placed in an orbit in the plane of the ecliptic [51]. More generally, the ability of an advanced civilization to destroy itself before reaching interstellar flight capability has been an ongoing issue with estimating the lifetime of a technological civilization for use in the Drake Equation.

E. Physical Feasibility Versus Engineering Feasibility

We mentioned previously the inherent enormous scale of fast interstellar rendezvous missions. Human civilization is certainly capable of committing enormous resources to projects considered important to the civilization. History abounds with examples of amazing engineering feats that would be difficult for us to duplicate, even with 21st century technologies, ranging from the Great Pyramid of Egypt to the Apollo Saturn V at liftoff (with a first-stage rocket power corresponding to 0.8% of the steady-state power level of human civilization in 1969). However, at some point we have to ask about the feasibility of attempting some engineering task, given its inherent "difficulty." As an often-quoted example, we could in principle recreate conditions present during the initial stages of the Big Bang with some incredible extension of current particle accelerator technology; however, to do so would require a particle accelerator ring the diameter of the galaxy!

1. Antimatter Rocket Issues

Thus, there can be situations where the fundamental physics of the problem do not preclude it being done, but the engineering technology required to actually implement a solution are not feasible because of cost, materials, temperatures, etc. For example, we have not identified any physics showstoppers in operating a matter–antimatter annihilation rocket, although the four-stage 40-LY rendezvous vehicle would be enormous in size and require 40-million MT of antimatter. Similarly, there are no inherent physics issues in producing antiprotons because this is already done (albeit at 10s of nanograms per year). Physical feasibility should also extend to the conversion of antiprotons to anti-solid H_2, although the necessary steps have not yet been demonstrated. The problem here is the engineering of an antimatter "factory" capable of producing millions of tons of antimatter per year at very high energy efficiency. In particular, the production efficiency issue may prove to be a potential showstopper, because current production technology has an energy efficiency of only about 1 part in 10^9 (i.e., 10^9 units of energy are consumed to produce an amount of antimatter that will release one unit of energy upon annihilation) [28]. To illustrate the magnitude of the engineering feasibility issue, the annihilation (MC^2) energy of 40-million MT of antimatter (plus an equal amount of matter) corresponds to ~16 million years of current human civilization energy output. At current production efficiencies (10^{-9}), the energy required to produce the antiprotons corresponds to ~16 quadrillion (10^{15}) years of current human civilization energy output. For comparison, this is "only" 601 years of the total energy output of the sun. Forward has identified a number of potential improvements that could be made to various antiproton production facilities around the world, but even at the maximum predicted energy efficiency of these facilities (~0.01%) [28], we would need 160 billion years of current human civilization energy output (2.2 days of the sun's total energy output) for production. Thus, the sheer magnitude of the problem of producing enough energy, in the form of electricity, to make the antimatter is a major issue. In effect, we might very well have to first build a Dyson sphere to collect all of the sun's energy output and convert this into electricity (e.g., for 10 days at 22% sunlight-to-electricity efficiency) before our first interstellar rendezvous mission using an antimatter rocket.

However, although the difficulty of producing large quantities of antiprotons may ultimately prove to be a fatal show-stopper to this approach, it is worth considering the general trends seen in technological advancement. We have already illustrated the tremendous improvement in the production of liquid hydrogen from Dewar's first few drops in the late 1800s to the space shuttle's 100 MT per flight. A similar trend is seen for the production of aluminum metal that emphasizes the importance of technological breakthroughs *and* synergisms with other technologies. For example, in the mid-1850s, aluminum metal was so hard to produce that it was more expensive than gold. Emperor Napoleon III commissioned a set of aluminum dinner cutlery for use by his most honored and important guests; less important guests dined on "mere" silver. And yet, the discovery of the Hall–Heroult electrolysis process barely 30 years later (1886) would drive the cost of aluminum down to the point that the Wright brothers could afford to build an aluminum crankcase engine for their 1903 Flyer. However, the breakthrough of the Hall–Heroult electrolysis process would have been of only modest importance if it had not been coupled with the growth of an electric power industry in the latter third of the 19th century. Thus, the mass production of inexpensive aluminum metal depended on the synergism between a breakthrough discovery and the availability of cheap, reliable, plentiful electric power [52]. Hopefully, between now and our first interstellar missions, there will be an antiproton production breakthroughs and synergisms that echo those of aluminum production.

2. *Laser Light Sail Issues*

In a similar fashion, we have sized the laser light sail assuming diffraction-limited optics on the laser power-beaming system. In practice, it can be very difficult to engineer optics systems that approach this limit of performance. In part, we made the assumption that this would not be an unreasonable goal for the laser transmitter optics based on the recognition that telescope/interferometer systems capable of resolving continent-sized features on planets at 40 LY would have already solved the engineering problems of optics pointing, tracking, and beam jitter that would be required for the laser light sail rendezvous mission. As a numerical example, we can estimate the pointing requirement by assuming a pointing accuracy to within 100 km (e.g., the size of a single pixel when imaging an Earth-sized planet in a 128×128 image array), corresponding to roughly 10% of the sail diameter for the Vision Mission 40-LY rendezvous light sail. The arctangent of the angle of a triangle 100 km high by 40 LY long is 0.26 pico-radians, which is comparable to picking out a 25 cent piece on Mars ($L = 0.6$ AU)! Obviously, we cannot do this today, but this is the class of capability that will be required for the Origins Program to implement space-based observatories like terrestrial planet imager (TPI). Thus, although this is a major challenge, it will have already been solved by the astronomical community before we begin our flights to the stars.

However, a potentially more serious engineering issue with the light sail is its ability to use an ultra-gossamer first-stage (with a "mirror" film 63-atoms thick!) to retro-reflect laser light back onto the second stage during the rendezvous. One approach to solving the potential problem of poor focusing ability in the first stage is to simply increase the diameter of the second stage to capture more of the reflected light, although this will have the effect of decreasing the acceleration of the vehicle as it leaves the solar system. More generally, the size of the second stage is somewhat arbitrary; we assumed a second-stage area 1/10 that of the first stage, although it may be necessary to increase the size of the second stage for nearby targets even when the first stage has diffraction-limited optics capability [2].

Another issue for the light sail is the enormous size (e.g., physical size and power level) of its laser (momentum beaming) system. At the least, we can conduct a "sanity check" on the magnitude of the problem. For example, for a 17.9 PW laser system, the amount of solar power required to run the laser (at 10% efficiency of converting sunlight into laser light) is comparable to "only" the amount of sunlight hitting the surface of the Earth. Thus, the sunlight collector is certainly large, but not of the order of a Dyson sphere.

Second, if we assume that the laser system is composed of individual laser + optics modules that are assembled in space to form the complete 17.9-PW, roughly 1000-km diameter laser transmitter system, we can perform a sanity check on the number of modules required and the optics diameter and laser power per module. For example, if we assume that there are 10^9 modules (comparable to the number of automobiles on Earth), the optics diameter and laser power per module are on the order of a few 10s of meters and 10s of MW, respectively. (For comparison to the familiar automobile engine, 100 horsepower = 133 kW = 0.133 MW.)

Finally, it is possible to estimate the potential mass of the laser transmitter system based on its individual subsystems [2]. Note that this is only an approximation because these mass estimates are based on contemporary and near-term technologies; many of these subsystems may undergo major improvements before we attempt construction of a high-power laser power/momentum beaming system. Using the system sizing estimates given in [2], we obtain the results shown in Figs. 31 and 32.

Thus, the total light sail plus laser system has a mass of 210.8-billion MT, with most of the mass in the radiators, which is not uncommon for any space-based power system. The laser system's mass is roughly 1,000 times the mass of the antimatter rocket. However, this is not an entirely fair comparison, because we have not included the mass of an antimatter "factory," and we have not "amortized" the cost (mass) of the antimatter factory or laser by dividing by the number of missions each would support.

3. Fusion Ramjet Issues

Finally, as discussed earlier, the Bussard interstellar ramjet has both the greatest mission potential and the greatest number of unresolved feasibility issues. For

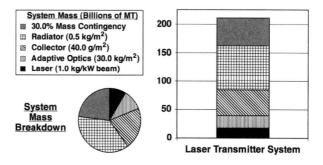

Fig. 31 17.9-PW laser system for the 40-LY 0.5 c interstellar rendezvous light sail mission.

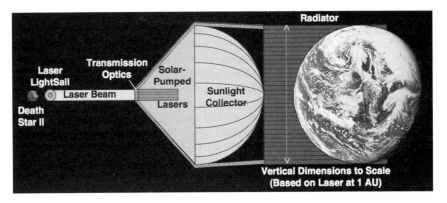

Fig. 32 Laser system size compared to Earth.

example, we know that stars can "burn" H fuel in a fusion reaction, but there do not appear to be any practical engineering solutions to pure hydrogen fusion. Fortunately, an alternative is to use interstellar D, which we currently use in fusion reactors, although this places great demands on the ram-scoop system. Also, the ram-scoop and fusion engines have several unique engineering feasibility challenges. First, we need to design a lightweight (i.e., minimum structure) electromagnet system to produce the required field geometry to collect interstellar H or D [53]. This in itself may be a significant problem given the high L/D required for a drag-free collection system. Similarly, the fusion engine must be designed in such a way that the fusion fuel atoms are never slowed down (relative to the vehicle), otherwise the momentum drag will overcome the rocket exhaust moving at only $0.03\ c$. In engineering terms, this means that the usual techniques of turning or reflecting the fusion plasma in the reactor to increase residence time cannot be used and instead some other approach must be used to confine the plasma in the reactor.

F. Recommendations for Future Work

During the course of our studies we have identified several additional research areas that we recommend be pursued. In several cases, this work would consist of systems-level studies of various elements of the interstellar propulsion systems. For example, the Bussard interstellar fusion ramjet has been studied far less than antimatter rockets or laser light sails. Thus, we would recommend detailed modeling of all aspects of the fusion ramjet, and especially the ram-scoop structure and electromagnetic field, as well as investigation of fusion reactor concepts that would lend themselves to the "drag free" fusion engine requirement of the fusion ramjet.

Beyond system studies, we recommend experimental research on determining "engineering" properties of interest to laser light sails, and especially the optical properties of ultra-thin films (e.g., high-temperature metals, carbon, etc.). Also, there are several fundamental feasibility issues associated with antimatter rockets that need to be demonstrated experimentally, although in several cases normal-matter protons, electrons, hydrogen atoms, and hydrogen molecules can be used instead of antimatter (in appropriate "noncontact" experiments to simulate what would be needed to handle antimatter). For example, although a modest number of anti-H atoms have been produced, the complete process of "noncontact" conversion of H atoms to H_2 molecules to H_2 molecular solid ice needs to be demonstrated. Also, additional work needs to be done in investigating high-efficiency, high-production rate processes for the production of antiprotons, and ultimately conversion into anti-H_2 ice. We would also like to see an estimate of the theoretical ideal limit of antiproton production efficiency in terms of electric energy "in" versus antiproton mass "out," as a point for comparing various current and potential antiproton "factory" efficiencies. Additionally, one of the major inefficiencies of antimatter rockets is the low fraction of proton–antiproton annihilation products with charge and mass (i.e., charged pions) that can be used to produce thrust. Thus, we also recommend research that would seek to increase the fraction of usable annihilation products and thereby reduce the effective "loss" of propellant mass, while still maintaining the extraordinarily high I_{sp} of the antimatter rocket engine. Finally, cross-cutting technologies such as high current density high-temperature superconductors and lightweight radiators that are applicable to many advanced power and propulsion applications should be pursued.

Lastly, impacts with interstellar dust grains at relativistic speeds introduce a requirement for a dust impact shield. However, there are currently no data for particle-shield interactions at the anticipated speeds (e.g., 0.5 c) of interstellar missions, although this has been investigated for micrometeoroid impacts at the more modest speeds of solar system velocities (e.g., several 10s of km/s). Also, interstellar precursors should carry instruments to characterize interstellar dust particle properties (number, size, material distributions) that could serve to benchmark current models based on astronomical observations.

VIII. Closing Comments

The most likely candidates for fast (i.e., $\sim$0.5 c) interstellar rendezvous missions—laser light sails, antimatter rockets, and fusion ramjets—all suffer from either a need for a large infrastructure or a host of major feasibility issues. Thus, there is *no* single concept that is without potentially significant shortcomings and there is no single clear winner. Laser light sails, antimatter rockets, and fusion ramjets are the "best" choices based only on our current level of understanding (i.e., ignorance!).

For example, although fusion propulsion does not require a unique infrastructure (even an He^3 production infrastructure might be shared with terrestrial fusion users), there remain major feasibility issues associated with all of the fusion concepts at the performance levels required for interstellar missions.

Similarly, although matter–antimatter annihilation propulsion has been recognized for decades as the "ultimate" propulsion capability, a major antiproton production infrastructure is required for its implementation. There is however a potential for significant dual-use of antiprotons for non-propulsive uses that might "amortize" and justify construction of an antiproton "factory." Also, there are a number of major feasibility issues associated with the production and storage of high-density forms of antimatter (e.g., solid anti-H_2).

Beamed-momentum light sail systems also require a large space-based infrastructure of beam source and optics systems. However, this infrastructure can lead to a major in-space capability for beamed power to a variety of users, resulting in a "public utility" for power in space. Many of the required technologies for a beamed-momentum light sail system are in the process of being demonstrated (e.g., high-power lasers, adaptive optics, solar sails, etc.); the major issue is scaling to the ultra-thin sail sheets, large sizes, and high powers required for interstellar missions. Finally, although a light sail system may be the most near-term option, it gives the least mission flexibility; every maneuver at interstellar distances must be planned years in advance because of the speed-of-light time lag. Therefore, matter–antimatter annihilation and fusion systems should also be pursued because these systems will provide the highest degree of mission flexibility.

Near-term technology development goals should seek to resolve the major feasibility issues associated with each concept. For example, the mechanical, thermal, and optical properties of ultra-thin film materials is largely unknown and represents a major question that must be answered in order to assess light sails. For matter–antimatter propulsion, experiments with normal matter should seek to duplicate, in a manner appropriate for antimatter, the steps required to go from protons and electrons to atoms, then to molecules, and finally to solid molecular ice. For the fusion ramjet, the H–H (or D–D) fusion reaction, magnetic scoop, and "drag-free" reactor concepts should be computationally modeled. Finally, cross-cutting technologies, such as lightweight radiators or high-temperature, high-field, radiation-resistant superconducting magnets, should be pursued.

Ultimately, the issues that we worry about today may turn out to be a nonproblem given the technological advances that will occur over the next several centuries. We may very well be in the position of Jules Verne trying to predict the engineering required to send men to the moon in his classic *From the Earth to the Moon*. Verne got the space capsule right [54]; it was very similar to the Apollo command module in size, weight, crew (three), and material (aluminum). However, Verne completely missed the mark in designing the launch vehicle; he used a 274-m long cannon (derived from the "high-tech" weapons of the 1860s) instead of a Saturn V rocket (derived from the "high-tech" weapons of the 1960s). Thus, in our case, we may be designing the antimatter rocket like Verne's cannon, and ultimately have a vehicle design that is way too big and heavy, simply because we cannot predict the major technological breakthroughs that will occur over the next several centuries.

We would like to close this chapter with the often-noted observation that interstellar missions are *not* impossible [55]; they are just enormously demanding in terms of size and resources. What is remarkable is that in the brief time we have been a space-faring species (barely a half century), we have already identified three candidate propulsion concepts with the potential for performing the most propulsion-intensive space missions of all, and allowing us to respond to Konstantin Tsiolkovsky's classic challenge [17], "Earth is the cradle of humanity, but one cannot live in a cradle forever."

Acknowledgments

The work described in this chapter was carried out at the Jet Propulsion Laboratory, California Institute of Technology, under a contract with the National Aeronautics and Space Administration (NASA).

The author would like to thank John Cole (former Head of the Revolutionary Propulsion Program at NASA Marshall Spaceflight Center) for providing overall funding support for this task in previous years. Although the Revolutionary Propulsion Program has been cancelled, I hope that it will be reinitiated in coming years to provide an opportunity to continue investigation of advanced propulsion concepts for interstellar missions.

Appendix: Derivation of the Classical and Relativistic Rocket Equations

In this appendix we describe the derivation of four versions of the Rocket Equation. The first is the Classical Rocket Equation, first derived by Tsiolkovsky [17]. The next is a Relativistic Rocket Equation, in which both the vehicle and the propellant exhaust velocity ($V_{exhaust}$ or, equivalently, I_{sp}) are moving at a large fraction of the speed of light, such that relativistic corrections must be included. These derivations will follow Forward's method [18]. The next two versions of the Rocket Equation parallel the Classical and Relativistic Rocket Equations, but for the situation where some large fraction of the initial propellant mass or mass–energy is unavailable (e.g., 60% for a proton–antiproton annihilation rocket) for momentum transfer to the vehicle.

A1. Classical Rocket Equation

In the Classical Rocket Equation derivation, a rocket with initial mass M ejects a mass of reaction mass (propellant) dm at a constant exhaust velocity (i.e., $V_{exhaust}$ corresponding to I_{sp}) relative to the rocket with a velocity, as shown in Fig. A1. (To avoid confusion with the vehicle's velocity "V," $V_{exhaust}$ will be represented by "w" [18].) In the center of mass of the system, the resultant rocket velocity is V (i.e., ΔV), and the reaction mass velocity is u. Note that we have shown the situation for both acceleration and deceleration; derivation of the Rocket Equation will be shown in parallel for both options.

To begin the derivation of the Classical Rocket Equation, we first start with the conservation of mass:

$$\textit{Acceleration: } dM = -dm \qquad \textit{Deceleration: } dM = -dm$$

(The minus sign indicates the decreasing mass of the rocket. Also, the two equations are the same for acceleration or deceleration.)

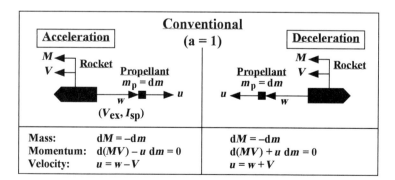

Fig. A1 Classical rocket mass, velocity, and momentum (with no loss of propellant).

Next, we conserve momentum:

$$\text{Acceleration: } d(M \cdot V) = u \cdot dm \qquad \text{Deceleration: } d(M \cdot V) = -u \cdot dm$$

Note the sign difference between the acceleration and deceleration forms of the equation. Finally, we use the classical addition (or subtraction) of velocities:

$$\text{Acceleration: } u = w - V \qquad \text{Deceleration: } u = w + V$$

Again, note the sign differences. We then use the conservation of mass equation to find dm; this identity will be then be used when we expand the conservation of momentum equation (i.e., $d[MV] = MdV + VdM$). Thus, expanding the derivatives, combining the above equations, and rearranging gives:

Acceleration:
$$MdV + VdM = udm$$
$$= -(w - V)dM$$
$$VdM + (w - V)dM = -MdV$$
$$(V + w - V)dM = -MdV$$
$$dM/M = -dV/w$$

Deceleration:
$$MdV + VdM = -udm$$
$$= +(w + V)dM$$
$$VdM - (w + V)dM = -MdV$$
$$(V - w - V)dM = -MdV$$
$$dM/M = +dV/w$$

If we were performing a numerical integration, we could write:

$$\text{Acceleration: } dM = M[-dV/w] \qquad \text{Deceleration: } dM = M[+dV/w]$$

Analytical integration gives:

$$\text{Acceleration: } \ell n(M) = -V/w \qquad \text{Deceleration: } \ell n(M) = +V/w$$

which, when evaluated for the rocket mass limits of M_{wet} (initial wet mass) and M_{dry} (final "burnout" dry mass), initial and final velocities (V_i and V_f, with $V_f - V_i = \Delta V$), and $w = V_{\text{exhaust}} = g_c I_{\text{sp}}$ [with $g_c = 9.8$ (m/s)/(lb$_f$-s/lb$_m$)] yields the Classical Rocket Equation:

Acceleration:
$$\ell n(M_{\text{wet}}/M_{\text{dry}}) = +\Delta V/(g_c I_{\text{sp}})$$
$$M_{\text{wet}}/M_{\text{dry}} = \exp[+\Delta V/(g_c I_{\text{sp}})]$$

Deceleration:
$$\ell n(M_{\text{wet}}/M_{\text{dry}}) = -\Delta V/(g_c I_{\text{sp}})$$
$$M_{\text{wet}}/M_{\text{dry}} = \exp[-\Delta V/(g_c I_{\text{sp}})]$$

Note that for acceleration, ΔV is positive ($\Delta V > 0$) since $V_f > V_i$. However, for deceleration, $\Delta V < 0$ since $V_f < V_i$. Thus, both equations give the expected result that $M_{\text{wet}} > M_{\text{dry}}$ independent of whether the vehicle is accelerating or decelerating.

For convenience, we will adopt a sign convention for acceleration and deceleration using expressions like x "$+/-$" y or x "$-/+$" y, where the "upper" sign is for acceleration and the "lower" sign is for deceleration. Thus, our derivation can be written as

Mass conservation: $\quad dM = -dm$
Momentum conservation: $\quad d(M \cdot V) = +/- u \cdot dm$
Velocity addition: $\quad u = w -/+ V$

LIMITS OF INTERSTELLAR FLIGHT TECHNOLOGY 111

As before, we first expand the conservation of mass equation and solve for dm. Next, we expand the conservation of momentum equation and substitute in values for dm and u found previously:

$$MdV + VdM = +/- udm = -/+ (w -/+ V)dM$$
$$VdM +/- (w -/+ u)dM = -MdV$$
$$(V +/- w - V)dM = -MdV$$

This is rearranged to give forms suitable for analytical or numerical integration:

For analytical integration: $\quad dM/M = -/+ dV/w$
For numerical integration: $\quad dM = dM(-/+ dV/w)$

Analytical integration gives:

Analytical integration: $\quad \ell n(M) = -/+ V/w$
Integration limits: $\quad \ell n(M_{wet}/M_{dry}) = +/- \Delta V/(g_c I_{sp})$
$\quad M_{wet}/M_{dry} = \exp[+/- \Delta V/(g_c I_{sp})]$

A2. Relativistic Rocket Equation

For the Relativistic Rocket Equation, we follow essentially the same steps as for the Classical Rocket Equation, with the difference being the use of relativistic equations for mass and velocity, and conservation of mass–energy content of the system (as opposed to conservation of mass only in the classical case). For example, we use relativistic mass (M and m), such that:

$$M = M_r/(1 - V^2/c^2)^{1/2} \text{ and } m = m_r/(1 - u^2/c^2)^{1/2}$$

where M_r and m_r are the rest masses of the vehicle and propellant, respectively.

First, we start with the conservation of mass–energy; i.e., mass–energy, or mc^2, is conserved. (Note again the use of the "$+/-$" and "$-/+$" sign convention for acceleration and deceleration.)

$$d[M_r/(1 - V^2/c^2)^{1/2}] \cdot c^2 = -d(m_r)/(1 - u^2/c^2)^{1/2} \cdot c^2$$

Note also that the mass–energy conservation equation is the same for acceleration or deceleration.

Next, we conserve momentum[§§]:

$$d[M_r/(1 - V^2/c^2)^{1/2} \cdot V] = +/- d(m_r)/(1 - u^2/c^2)^{1/2} \cdot u$$

[§§]We should note that the equations given above are the correct forms of the mass-energy and momentum conservation equations, respectively, as defined by Forward [18]. In our previous publications [3,14], we had incorrectly included the propellant center-of-mass relativistic velocity correction $(1 - u^2/c^2)^{1/2}$, within the $d(m_r)$ term on the right hand side of the equations; e.g., $d[m_r/(1 - u^2/c^2)^{1/2}]$. Fortunately, we had used an approximation to produce a form of the differential equation that could be analytically integrated, so that the net result for the equation for the analytical integration was the same as that presented.

Finally, we use the relativistic addition (or subtraction) of velocities:

$$u = (w - / + V)/(1 - / + wV/c^2)$$

To begin the derivation, we expand the mass–energy conservation derivative and solve for dm_r:

$$M_r d[(1 - V^2/c^2)^{-1/2}] \cdot c^2 + dM_r \cdot (1 - V^2/c^2)^{-1/2} \cdot c^2$$
$$= -dm_r \cdot (1 - u^2/c^2)^{-1/2} \cdot c^2$$
$$M_r \cdot (1 - V^2/c^2)^{-3/2} V/c^2 dV \cdot c^2 + dM_r \cdot (1 - V^2/c^2)^{-1/2} \cdot c^2$$
$$= -dm_r \cdot (1 - u^2/c^2)^{-1/2} \cdot c^2$$

$$dm_r = -[M_r \cdot (1 - V^2/c^2)^{-3/2} V/c^2 dV + dM_r \cdot (1 - V^2/c^2)^{-1/2}]/(1 - u^2/c^2)^{-1/2}$$

(Again, note that the mass–energy conservation equation is the same for acceleration or deceleration.)

Next, we expand the conservation of momentum derivative, and substitute values for dm_r and u from above:

$$M_r d[V \cdot (1 - V^2/c^2)^{-1/2}] + dM_r \cdot V \cdot (1 - V^2/c^2)^{-1/2}$$
$$= +/ - (dm_r) \cdot (1 - u^2/c^2)^{-1/2} \cdot (u)$$
$$M_r[dV \cdot (1 - V^2/c^2)^{-1/2} + V \cdot (1 - V^2/c^2)^{-3/2}(V/c^2)dV]$$
$$+ dM_r \cdot V \cdot (1 - V^2/c^2)^{-1/2}$$
$$= -/ + [M_r \cdot (1 - V^2/c^2)^{-3/2} V/c^2 dV + dM_r \cdot (1 - V^2/c^2)^{-1/2}]$$
$$\cdot [(w - / + V)/(1 - / + wV/c^2)]$$

Multiplying through by $(1 - V^2/c^2)^{+1/2}$ gives:

$$M_r[dV + (V^2/c^2) \cdot (1 - V^2/c^2)^{-1} dV] + dM_r \cdot V$$
$$= -/ + [M_r \cdot (1 - V^2/c^2)^{-1} V/c^2 dV + dM_r]$$
$$\cdot [(w - / + V)/(1 - / + wV/c^2)]$$

Rearranging gives:

$$dM_r \cdot [V + / - (w - / + V)/(1 - / + wV/c^2)]$$
$$= -dV M_r[1 + (V^2/c^2) \cdot (1 - V^2/c^2)^{-1} + / - V/c^2 \cdot (1 - V^2/c^2)^{-1}$$
$$\cdot (w - / + V)/(1 - / + wV/c^2)]$$

dM_r/M_r

$= -dV[1 + (V^2/c^2) \cdot (1 - V^2/c^2)^{-1} + / - V/c^2 \cdot (1 - V^2/c^2)^{-1}$
$\cdot (w - / + V)/(1 - / + wV/c^2)]/[V + / - (w - / + V)/(1 - / + wV/c^2)]$
$= -dV/(1 - V^2/c^2) \cdot [1 - V^2/c^2 + (V^2/c^2) + / - V/c^2 \cdot (w - / + V)$
$/(1 - / + wV/c^2)]/[V + / - (w - / + V)/(1 - / + wV/c^2)]$
$= -dV/(1 - V^2/c^2) \cdot [1 + / - V/c^2 \cdot (w - / + V)/(1 - / + wV/c^2)]/$
$[V + / - (w - / + V)/(1 - / + wV/c^2)]$

Multiplying the numerator and denominator of the right-hand term by $(1 - / + wV/c^2)$ gives:

$dM_r/M_r = -dV/(1 - V^2/c^2) \cdot [1 - / + wV/c^2 + / - V/c^2$
$\cdot (w - / + V)]/[V(1 - / + wV/c^2) + / - (w - / + V)]$

Expanding and rearranging terms gives forms suitable for analytical or numerical integration:

$dM_r/M_r = -dV/(1 - V^2/c^2) \cdot [1 - / + wV/c^2 + / - wV/c^2$
$- V^2/c^2]/[V - / + wV^2/c^2 + / - w - V)]$
$= -dV/(1 - V^2/c^2) \cdot [1 - V^2/c^2]/[+ / - w(1 - V^2/c^2)]$

For analytical integration: $dM_r/M_r = -dV/[+/ - w(1 - V^2/c^2)]$
$= [-c^2/(+ / - w)] \cdot [dV/(c^2 - V^2)]$

For numerical integration: $dM_r = M_r\{-dV/[+/ - w(1 - V^2/c^2)]\}$
$dM_r = M_r\{[-c^2/(+ / - w)] \cdot [dV/(c^2 - V^2)]\}$

Analytical integration gives:

Analytical integration: $\ln(M_r) = -c^2/(+/-w)[1/(2c)] \cdot \ln[(c + V)/(c - V)]$
$= -c/(+/-2w) \cdot \ln[(c + V)/(c - V)]$
$= -1/(+/-2w/c) \cdot \ln[(1 + V/c)/(1 - V/c)]$

Evaluating the integrals for the initial and final rest-mass limits of M_{wet} and M_{dry}, initial and final velocities of V_i and V_f, with $V_f - V_i = \Delta V$, and $w = V_{exhaust} = g_c I_{sp}$, yields the Relativistic Rocket Equation:

Integration limits:
$$\ln(M_{\text{wet}}/M_{\text{dry}})$$
$$= c/(+/-2w) \cdot \ln\{[(c+V_f)/(c-V_f)]/[(c+V_i)/(c-V_i)]\}$$
$$= 1/(+/-2w/c) \cdot \ln\{[(1+V_f/c)(1-V_i/c)]/[(1+V_i/c)(1-V_f/c)]\}$$
$$= 1/(+/-2g_c I_{\text{sp}}/c) \cdot \ln\{[(1+V_f/c)(1-V_i/c)]/[(1+V_i/c)(1-V_f/c)]\}$$
$$\times M_{\text{wet}}/M_{\text{dry}}$$
$$= \{[(1+V_f/c)(1-V_i/c)]/[(1+V_i/c)(1-V_f/c)]\}^{[1/(+/-2g_c I_{\text{sp}}/c)]}$$

Note that we have dropped the subscript "r" denoting rest mass for convenience.

For the case where the initial (or final) velocity is zero before acceleration (or after deceleration), we have:

$$\ln(M_{\text{wet}}/M_{\text{dry}}) = 1/(+/-2g_c I_{\text{sp}}/c) \cdot \ln[(1+/-V/c)/(1-/+V/c)]$$

where V is the final (or initial) peak velocity after acceleration from $V=0$ (or before deceleration to $V=0$).

A3. Classical Rocket Equation with "Loss" of Propellant

In this section, we perform a derivation of the Classical Rocket Equation with "loss" of propellant where we only need to conserve mass and momentum; the corresponding case for the Relativistic Rocket Equation is complicated by the need to conserve relativistic mass–energy content as well as relativistic momentum. Thus, in the classical case, we assume that a fraction "**a**" of the initial propellant mass is used to produce forward motion of the rocket, as shown in Fig. A2. In effect, a fraction (1 − **a**) of the initial propellant mass (and

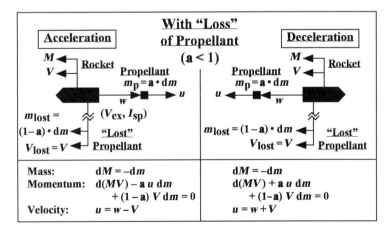

Fig. A2 Classical rocket mass, velocity, and momentum (with "loss" of propellant).

LIMITS OF INTERSTELLAR FLIGHT TECHNOLOGY 115

momentum) is "lost," at least insofar as production of forward momentum is concerned. It is as if a quantity $(\mathbf{a})M_p$ is ejected parallel to the direction of motion (to produce some ΔV) and a quantity $(1 - \mathbf{a})M_p$ is ejected perpendicular in such a way that $(1 - \mathbf{a})M_p$ does not contribute to the forward ΔV. Finally, note that the usual situation for rocket propulsion is for $\mathbf{a} = 1$; for example, even in a fission or fusion rocket, where some of the rest mass of the nuclear fuel is turned into energy, the amount of mass "lost" (to conversion into energy) is negligible, so the "standard" Classical or Relativistic Rocket Equations (with $\mathbf{a} = 1$) can be used.

For the Classical Rocket Equation with "loss" of propellant, we follow essentially the same steps as for the standard ($\mathbf{a} = 1$) Classical Rocket Equation, with the difference being the use of terms in ($\mathbf{a}$) and $(1 - \mathbf{a})$ in the momentum conservation equation. Thus, we again write the equations for conservation of mass and momentum, and addition (or subtraction) of velocities. (Note again the use of the "$+/-$" and "$-/+$" sign convention for acceleration and deceleration.)

Mass conservation: $\quad dM = -dm = -[(\mathbf{a})dm(\text{usable}) + (1 - \mathbf{a})dm(\text{lost})]$

Note that $\mathbf{a} \cdot dm$ represents the "usable" fraction of the (incremental) propellant mass, whereas $(1 - \mathbf{a}) \cdot dm$ represents the fraction "lost" and unavailable for production of velocity (V).

Momentum conservation: $\quad d(M \cdot V) = +/- u \cdot \mathbf{a} \quad dm$

$\quad\quad\quad\quad\quad\quad\quad\quad\quad\quad\quad$ (only $\mathbf{a} \quad dm$ contributes to V)

$\quad\quad\quad\quad\quad\quad\quad\quad\quad\quad\quad - V(1 - \mathbf{a}) \, dm$ (momentum "lost")

As above, only $u \cdot \mathbf{a} \cdot dm$ contributes momentum toward changing V. Also, in defining w and u, we are only interested in the "bulk" or center-of-mass velocity of the (incremental) propellant mass quantity $\mathbf{a} \cdot dm$ moving parallel to V. Thus, the exhaust velocity ($w = V_{\text{exhaust}}$) and I_{sp} of the rocket engine corresponds to an "effective" I_{sp} or V_{exhaust} that takes into account any inefficiencies in the nozzle in converting the random velocities of the propellant exhaust gas into directed motion (w and u) parallel to V.

Similarly, the (incremental) propellant mass $(1 - \mathbf{a}) \cdot dm$ is ejected at some exhaust velocity (not necessarily w) perpendicular to V (and u) in a symmetric fashion that results in a net cancellation of the perpendicular velocity of the center-of-mass of $(1 - \mathbf{a}) \cdot dm$. However, this unit of mass is still moving forward at the rocket's velocity V, so the total amount of "lost" momentum is $V \cdot (1 - \mathbf{a}) \cdot dm$. Interestingly, this momentum is totally lost from the system (its value is negative for either acceleration or deceleration), so it also reduces the mass and momentum of the rocket that must be accelerated or decelerated in subsequent differential time steps. However, it also represents a "dead" weight that must be initially loaded into the rocket, which increases M_{wet} over that seen when $\mathbf{a} = 1$.

Velocity addition: $\quad u = w - / + V$

As above, we first expand the conservation of mass equation and solve for dm. Next, we expand the conservation of momentum equation and substitute in values for dm and u found above:

$$M\mathrm{d}V + V\mathrm{d}M = +/- u\mathbf{a}\ \mathrm{d}m - V(1-\mathbf{a})\mathrm{d}m$$
$$= -/+ \mathbf{a}\ (w-/+V)\mathrm{d}M + V(1-\mathbf{a})\mathrm{d}M$$
$$V\mathrm{d}M +/- \mathbf{a}\ (w-/+V)\mathrm{d}M - V(1-\mathbf{a})\mathrm{d}M = -M\mathrm{d}V$$
$$[V +/- \mathbf{a}\ w - \mathbf{a}\ V - V(1-\mathbf{a})]\mathrm{d}M = -M\mathrm{d}V$$
$$(+/- \mathbf{a}\ w)\mathrm{d}M = -M\mathrm{d}V$$

This is rearranged to give forms suitable for analytical or numerical integration:

For analytical integration: $\mathrm{d}M/M = -/+ \mathrm{d}V/(\mathbf{a}\ w)$
For numerical integration: $\mathrm{d}M = M[-/+ \mathrm{d}V/(\mathbf{a}\ w)]$
Analytical integration: $\ln(M) = -/+ V/(\mathbf{a}\ w)$
Integration limits: $\ln(M_{\mathrm{wet}}/M_{\mathrm{dry}}) = +/- \Delta V/(\mathbf{a}\ g_c I_{\mathrm{sp}})$
$M_{\mathrm{wet}}/M_{\mathrm{dry}} = \exp[+/- \Delta V/(\mathbf{a}\ g_c I_{\mathrm{sp}})]$

Note that the final form of the equation is similar to that of the standard ($\mathbf{a} = 1$) Classical Rocket Equation. The result has been to add $\mathbf{a}$ to the I_{sp} term, resulting in an "effective" I_{sp} that takes into account the "loss" of propellant. In fact, this result could be anticipated from the classical definition of I_{sp} in terms of thrust (F) and propellant mass flow rate ($\dot{m}$) entering the rocket engine:

$$g_c I_{\mathrm{sp}} = F/\dot{m} = V_{\mathrm{exhaust}}$$

In this equation, $g_c = 9.8$ (m/s)/(lb$_f$-s/lb$_m$), where $I_{\mathrm{sp}} = $ lb$_f$-s/lb$_m$, $F = $ Newtons, $\dot{m} = $ kg/s, and $V_{\mathrm{exhaust}} = $ m/s. In the context used here, $\dot{m}$ is the *total* mass of propellant *entering* the rocket engine. However, only a quantity $\mathbf{a} \cdot \dot{m}$ is available to produce thrust, so we would have:

$$g_c I_{\mathrm{sp}} = F/[\mathbf{a}\ \dot{m}\ (\mathrm{Total})]$$

which rearranges to give the same ($\mathbf{a}\ g_c\ I_{\mathrm{sp}}$) term seen in the Rocket Equation above:

$$\mathbf{a}\ g_c I_{\mathrm{sp}} = F/\dot{m}\ (\mathrm{Total})$$

A4. Relativistic Rocket Equation with "Loss" of Propellant

For this derivation, we will combine elements of the derivations of the Relativistic Rocket Equation and Classical Rocket Equation with "loss" of propellant. However, unlike the classical version of this equation, where we conserve only (rest) mass and classical momentum (rest mass multiplied by velocity), the relativistic version requires that we conserve the total amount of mass–energy and relativistic momentum, including that of the "lost" propellant. Thus, because of the need to include the relativistic mass–energy and momentum of the "lost" propellant, the algebra will be more complex, resulting in an equation that must be numerically integrated.

Also, for a matter–antimatter annihilation reaction, we need to bookkeep the rest mass and kinetic energy of the various reactants (protons, antiprotons, electrons, positrons) and annihilation products (charged and neutral pions, gamma ray photons). This is needed to determine both the effective value of "**a**" in the equation (i.e., how much of the mass–energy and momentum are usable for propulsion), as well as to determine the mass–energy and momentum of the various annihilation products. Table A1 summarizes mass and energy quantities based on the assumption of a reaction between an atom of H [containing a proton (P^+) and electron (e^-)] and an anti-atom of anti-hydrogen [containing an antiproton (P^-) and positron (e^+)]. The protons and antiprotons annihilate to give neutral and charged pions ($\pi^°$, π^+, π^-); the electrons and positrons annihilate to give high-energy photons (gamma rays, γ). The initial annihilation reaction is given as [19]:

$$P^+ + P^- \rightarrow 2.0\pi^° + 1.5\pi^+ + 1.5\pi^- \text{ and } e^- + e^+ \rightarrow 2\gamma$$

The neural pions promptly decay into very high-energy gamma rays (each about 355 times more energetic than an electron/positron gamma). The charged pions decay into charged muons (μ) and neutrinos (ν) after traveling about 21 m at 0.93 c [18,19].

$$2.0\pi^° \rightarrow 4\gamma, \quad 1.5\pi^+ \rightarrow 1.5\mu^+ + 1.5\nu_m, \quad 1.5\pi^- \rightarrow 1.5\mu^- + 1.5 \text{ anti-}\nu_m$$

To determine the value for **a**, we first determine the fractions (percentages) of the total initial mass–energy contained in the various annihilation species. For the charged pions ($\pi^\pm$), we can further split the total mass–energy content into rest mass (m_r) and kinetic energy, such that:

$$\text{Rest mass–energy} = m_r c^2$$

$$\text{Kinetic energy} = m_r c^2 [1/(1 - v^2/c^2)^{1/2} - 1]$$

$$\text{Total mass–energy} = m_r c^2 / (1 - v^2/c^2)^{1/2}$$

As shown in Table A1, this results in the charged pions having a rest mass that is 22.3% of the total mass–energy content of the reactants (H and anti-H atoms), and a kinetic energy that is about 39.9% of the initial mass–energy. However, even though the total mass–energy content of the charged pions is about 62% of the total mass–energy content of the reactants (H and anti-H atoms), this is not the value of **a**. This is because the electromagnet used to deflect the charged pions is not perfectly reflective (some charged pions "leak" through the magnetic field and travel upstream), and for those that are reflected, the resultant beam is not perfectly collimated. Thus, there is an inherent kinetic energy inefficiency in the nozzle that must also be accounted for. Callas [20], using Monte Carlo simulations of particle trajectories in proton/antiproton annihilations, estimated that the magnetic nozzle would be at best 50% kinetic energy efficient, corresponding to a value of **a** = 42.2%. We have arbitrarily assumed a value of **a** = 40% of the initial total mass–energy content of the propellant, corresponding to a 44.37%

Table A1 Rest mass, kinetic energy, and total mass–energy content of matter (H atom) + antimatter (anti-H atom) annihilation

Species	Rest mass (MeV)	Fraction of total (%)	Kinetic energy (MeV)	Fraction of total (%)	Effective velocity (c)	Total mass–energy (MeV)	Fraction of total (%)
Initial reactants	*1877.6*	*100.00*	*0.0*	*0.00*		*1877.6*	*100.00*
P^+	938.3	49.97	0	0.00	0.0000	938.3	49.97
e^-	0.5	0.03	0	0.00	0.0000	0.5	0.03
P^-	938.3	49.97	0	0.00	0.0000	938.3	49.97
e^+	0.5	0.03	0	0.00	0.0000	0.5	0.03
Initial products	*688.7*	*36.68*	*1188.7*	*63.32*		*1877.4*	*100.00*
2.0 π^o	269.9	14.38	439.1	23.39	0.9247	709.0	37.76
1.5 π^+	209.4	11.15	374.3	19.94	0.9334	583.7	31.09
1.5 π^-	209.4	11.15	374.3	19.94	0.9334	583.7	31.09
$e^- + e^+ \rightarrow 2\gamma$			1.0	0.05	(1.0000)	1.0	0.05
Decay products	*317.0*	*16.89*	*1559.7*	*83.11*		*1876.7*	*100.00*
2.0 $\pi^o \rightarrow 4\gamma$			709.1	37.78	(1.0000)	709.1	37.78
1.5 $\pi^+ \rightarrow$ 1.5 μ^+	158.5	8.45	288.5	15.37	0.9350	447.0	23.82
1.5 $\pi^+ \rightarrow$ 1.5 ν_m			136.8	7.29	(1.0000)	136.8	7.29
1.5 $\pi^- \rightarrow$ 1.5 μ^-	158.5	8.45	288.5	15.37	0.9350	447.0	23.82
1.5 $\pi^- \rightarrow$ 1.5 anti-ν_m			136.8	7.29	(1.0000)	136.8	7.29

LIMITS OF INTERSTELLAR FLIGHT TECHNOLOGY

Table A2 Hydrogen/anti-hydrogen annihilation mass–energy breakdown to determine "a"

Species	% of Total	Quantity
$2.0\ \pi^\circ \to 4\gamma$	37.76	Total mass–energy
$e^- + e^+ \to 2\gamma$	0.05	Total mass–energy
$1.5\ \pi^\pm$	22.31	Total rest mass
$1.5\ \pi^\pm$	*39.87*	*Total kinetic energy*
$1.5\ \pi^\pm$	62.18	Total mass–energy
$1.5\ \pi^\pm$	22.31	Usable rest mass (100.00% of charged pions)
$1.5\ \pi^\pm$	*17.69*	*Usable kinetic energy (44.37% of charged pions)*
$1.5\ \pi^\pm$	40.00	Total usable mass–energy
$1.5\ \pi^\pm$	0.00	Unusable rest mass (0.00% of charged pions)
$1.5\ \pi^\pm$	*22.18*	*Unusable kinetic energy (55.63% of charged pions)*
$1.5\ \pi^\pm$	22.18	Total Unusable mass–energy
$1.5\ \pi^\pm$	**40.00**	**Net Usable mass–energy (a)**
$\pi^\circ, \pi^\pm, \gamma$	**60.00**	**Net "Lost" mass–energy (1 − a)**

nozzle kinetic energy efficiency. These values are summarized in Table A2. Note that the magnetic nozzle does not affect the fraction of rest mass of the charged pions; that remains the same. Instead, what we have done is take into account the nozzle inefficiency in converting the kinetic energy of the charged pions, with 0.9334 c velocity vectors pointed in random directions, into a directed, collimated stream of charged pions with a 0.6729 c kinetic energy velocity vector pointed directly out the back of the rocket.

Finally, the discussion above has concentrated on the efficiency of the annihilation reaction and electromagnet nozzle in capturing the total mass–energy (rest mass and kinetic energy) of the charged pions. However, there are additional losses in the magnetic nozzle that result in a reduction in overall I_{sp}. For example, although the magnetic nozzle can capture 40% of the reactant's initial mass–energy for thrust production, the "beam" of charged pions leaving the engine are not perfectly collimated (i.e., parallel to V). Using the Monte Carlo particle trajectories, Callas [20] used direct momentum calculations to determine an "effective" I_{sp} or $V_{exhaust}$ of 1×10^8 N-s/kg = m/s = 0.333 c = 1.02×10^7 lb$_f$-s/lb$_m$.

To begin our derivation, we again begin with conservation of mass–energy (and again use the "+/−" and "−/+" nomenclature for acceleration and deceleration):

Mass conservation:

$$d[M_r/(1-V^2/c^2)^{1/2}] \cdot c^2 = -\mathbf{a}\ d(m_r)/(1-u^2/c^2)^{1/2} \cdot c^2$$
$$- (1-\mathbf{a})d(m_r)/(1-V^2/c^2)^{1/2} \cdot c^2$$

Note that unlike the Classical Rocket Equation with "loss" of propellant, where the (incremental) propellant mass dm was independent of velocity, we now have to explicitly include the velocities (u and V) because of the relativistic mass increases. A similar effect is seen for the conservation of momentum:

Momentum conservation:

$$d[M_r/(1 - V^2/c^2)^{1/2} \cdot V] = +/- \mathbf{a} \ d(m_r)/(1 - u^2/c^2)^{1/2} \cdot u$$
$$- (1 - \mathbf{a})d(m_r)/(1 - V^2/c^2)^{1/2} \cdot V$$

As with the classical version, the $(1 - \mathbf{a}) \cdot dm_r$ term is always negative independent of whether the rocket is accelerating or decelerating. Finally, we again have the relativistic addition (or subtraction) of velocities:

Velocity addition:

$$u = (w - / + V)/(1 - / + wV/c^2)$$

To begin the derivation, we expand the mass–energy conservation derivative and solve for dm_r:

$$M_r d[(1 - V^2/c^2)^{-1/2}] \cdot c^2 + dM_r \cdot (1 - V^2/c^2)^{-1/2} \cdot c^2$$
$$= -\mathbf{a} \ d(m_r)(1 - u^2/c^2)^{-1/2} \cdot c^2$$
$$- (1 - \mathbf{a})d(m_r)(1 - V^2/c^2)^{-1/2} \cdot c^2$$

$$M_r \cdot (1 - V^2/c^2)^{-3/2} V/c^2 dV \cdot c^2 + dM_r \cdot (1 - V^2/c^2)^{-1/2} \cdot c^2$$
$$= -d(m_r)[\mathbf{a} \ (1 - u^2/c^2)^{-1/2}$$
$$+ (1 - \mathbf{a})(1 - V^2/c^2)^{-1/2}] \cdot c^2$$

$$dm_r = -[M_r \cdot (1 - V^2/c^2)^{-3/2} V/c^2 dV + dM_r$$
$$\cdot (1 - V^2/c^2)^{-1/2}]/[\mathbf{a} (1 - u^2/c^2)^{-1/2}$$
$$+ (1 - \mathbf{a})(1 - V^2/c^2)^{-1/2}]$$

(Again, note that the mass–energy conservation equation is the same for acceleration or deceleration.)

As before, we expand the conservation of momentum derivative, and substitute values for dm_r and u from above:

$$M_r d[V \cdot (1 - V^2/c^2)^{-1/2}] + dM_r \cdot V \cdot (1 - V^2/c^2)^{-1/2}$$
$$= +/- \mathbf{a} \ d(m_r)(1 - u^2/c^2)^{-1/2} \cdot (u)$$
$$- (1 - \mathbf{a}) \ d(m_r)(1 - V^2/c^2)^{-1/2} \cdot (V)$$
$$= dm_r[+/- \mathbf{a} \ (1 - u^2/c^2)^{-1/2} \cdot (u)$$
$$- (1 - \mathbf{a})(1 - V^2/c^2)^{-1/2} \cdot (V)]$$

LIMITS OF INTERSTELLAR FLIGHT TECHNOLOGY 121

$$M_r[dV \cdot (1 - V^2/c^2)^{-1/2} + V \cdot (1 - V^2/c^2)^{-3/2}(V/c^2)dV]$$
$$+ dM_r \cdot V \cdot (1 - V^2/c^2)^{-1/2}$$
$$= -[M_r \cdot (1 - V^2/c^2)^{-3/2}V/c^2 dV$$
$$+ dM_r \cdot (1 - V^2/c^2)^{-1/2}]/$$
$$[\mathbf{a}(1 - u^2/c^2)^{-1/2} + (1 - \mathbf{a})(1 - V^2/c^2)^{-1/2}]$$
$$\cdot [+/- \mathbf{a}(1 - u^2/c^2)^{-1/2} \cdot (u)$$
$$- (1 - \mathbf{a})(1 - V^2/c^2)^{-1/2} \cdot (V)]$$
$$= -[M_r \cdot (1 - V^2/c^2)^{-3/2}V/c^2 dV$$
$$+ dM_r \cdot (1 - V^2/c^2)^{-1/2}] \cdot (+/- u) \cdot \mathbf{X}$$

where the various terms containing **a** are combined into the quantity **X**:

$$\mathbf{X} = [\mathbf{a}(1 - u^2/c^2)^{-1/2} - (1 - \mathbf{a})/(+/- u) \cdot (1 - V^2/c^2)^{-1/2} \cdot V]/$$
$$[\mathbf{a}(1 - u^2/c^2)^{-1/2} + (1 - \mathbf{a})(1 - V^2/c^2)^{-1/2}]$$

Note that when $\mathbf{a} = 1$, $\mathbf{X} = 1$. Also, this substitution now allows us to continue in a manner analogous to the derivation of the standard ($\mathbf{a} = 1$) Relativistic Rocket Equation. For example, separating terms in dM_r and dV on opposite sides of the equation gives:

$$dM_r[V \cdot (1 - V^2/c^2)^{-1/2} + (1 - V^2/c^2)^{-1/2} \cdot (+/- u \cdot \mathbf{X})]$$
$$= -M_r dV[(1 - V^2/c^2)^{-3/2}V/c^2 \cdot (+/- u \cdot \mathbf{X}) + (1 - V^2/c^2)^{-1/2}$$
$$+ V^2/c^2 \cdot (1 - V^2/c^2)^{-3/2}]$$

We then multiply through by $(1 - V^2/c^2)^{+1/2}$ to get

$$dM_r[V +/- u \cdot \mathbf{X}]$$
$$= -M_r dV[(1 - V^2/c^2)^{-1}V/c^2 \cdot (+/- u \cdot \mathbf{X}) + (1) + V^2/c^2 \cdot (1 - V^2/c^2)^{-1}]$$
$$= -M_r dV/(1 - V^2/c^2) \cdot [V/c^2 \cdot (+/- u \cdot \mathbf{X}) + (1 - V^2/c^2) + V^2/c^2]$$
$$= -M_r dV/(1 - V^2/c^2) \cdot [V/c^2 \cdot (+/- u \cdot \mathbf{X}) + 1]$$
$$dM_r/M_r = -dV/(1 - V^2/c^2) \cdot [V/c^2 \cdot (+/- u \cdot \mathbf{X}) + 1]/[V +/- u \cdot \mathbf{X}]$$

Substituting for $u = (w -/+ V) / (1 -/+ wV/c^2)$ gives

$$dM_r/M_r = -dV/(1 - V^2/c^2) \cdot [V/c^2 \cdot (+/- 1)(w -/+ V)/$$
$$(1 -/+ wV/c^2) \cdot \mathbf{X} + 1]/[V + (+/- 1)(w -/+ V)/$$
$$(1 -/+ wV/c^2) \cdot \mathbf{X}]$$
$$= -dV/(1 - V^2/c^2) \cdot \{1 + [wV/c^2 \cdot (+/- \mathbf{X}) -/+ V^2/c^2$$
$$\cdot (+/- \mathbf{X})]/(1 -/+ wV/c^2)\}/$$
$$\{V + [w \cdot (+/- \mathbf{X}) -/+ V \cdot (+/- \mathbf{X})]/(1 -/+ wV/c^2)\}$$
$$= -dV/(1 - V^2/c^2) \cdot [(1 -/+ wV/c^2) + wV/c^2$$
$$\cdot (+/- \mathbf{X}) -/+ V^2/c^2 \cdot (+/- \mathbf{X})]/$$
$$[V(1 -/+ wV/c^2) + w \cdot (+/- \mathbf{X}) -/+ V \cdot (+/- \mathbf{X})]$$
$$= -dV/(1 - V^2/c^2) \cdot [1 -/+ wV/c^2 +/- wV/c^2 \cdot (\mathbf{X})$$
$$- V^2/c^2 \cdot (\mathbf{X})]/[V -/+ wV^2/c^2 +/- w \cdot (\mathbf{X}) - V \cdot (\mathbf{X})]$$
$$= -dV/(1 - V^2/c^2) \cdot [(1 - V^2/c^2 \cdot \mathbf{X}) -/+ wV/c^2(1 - \mathbf{X})]/$$
$$[V \cdot (1 - \mathbf{X}) +/- w \cdot (\mathbf{X} - V^2/c^2)]$$

Note that when $\mathbf{a} = 1$, $\mathbf{X} = 1$, and this equation reduces to the form seen for the standard ($\mathbf{a} = 1$) Relativistic Rocket Equation:

$$dM_r/M_r = -dV/(1 - V^2/c^2)/[+/- w] = -dV/[+/- w(1 - V^2/c^2)]$$

Unfortunately, when **a** or **X** are not 1, the various terms in V do not cancel out, leaving us with an equation that must be numerically integrated. For this, we use the equation in the form:
For numerical integration:

$$dM_r = M_r \cdot \{-dV/(1 - V^2/c^2) \cdot [(1 - V^2/c^2 \cdot \mathbf{X}) -/+ wV/c^2(1 - \mathbf{X})]/$$
$$[V \cdot (1 - \mathbf{X}) +/- w \cdot (\mathbf{X} - V^2/c^2)]\}$$

where **X** is again defined as

$$\mathbf{X} = [\mathbf{a}(1 - u^2/c^2)^{-1/2} - (1 - \mathbf{a})/(+/- u) \cdot (1 - V^2/c^2)^{-1/2} \cdot V]/$$
$$[\mathbf{a}(1 - u^2/c^2)^{-1/2} + (1 - \mathbf{a})(1 - V^2/c^2)^{-1/2}]$$

and again $u = (w -/+ V) / (1 -/+ wV/c^2)$

To begin the numerical integration, we initialize the calculations with a vehicle initial wet mass (M_{wet}), which can be arbitrarily set equal to 1, and an initial velocity (V_i). We then select a (constant) value for dV and use the above equations to determine dM_r for each calculation step. During each step, we use the values of M_r and V from the previous step to calculate dM_r and "new" values of M_r = (previous M_r – dM_r) and V = (previous V +/– dV), with the +/– reflecting either acceleration or deceleration. We continue the process until we reach the final velocity (V_f) and corresponding vehicle final dry mass (M_{dry}) to obtain the Relativistic Rocket Equation mass ratio M_{wet}/M_{dry} for a given $\Delta V = V_f - V_i$.

References

[1] Williams, S., "Trends," *Asimov's Science Fiction*, Oct./Nov., 2007, pp. 4, 6.

[2] Frisbee, R. H., "Beamed-Momentum LightSails for Interstellar Missions: Mission Applications and Technology Requirements," AIAA Paper 2004-3567, 11–14 July 2004.

[3] Frisbee, R. H., and Leifer, S. D., "Evaluation of Propulsion Options for Interstellar Missions," AIAA Paper 98-3403, 13–15 July 1998.

[4] Anderson, J. L., "Leaps of the Imagination: Interstellar Flight and the Horizon Mission Methodology," *Journal of the British Interplanetary Society*, Vol. 49, 1996, pp. 15–20.

[5] Wilcox, R. M., "Internet STELLAR DATABASE," URL: http://www.stellar-database.com [cited 12 Dec. 2007].

[6] Mewaldt, R., "Interstellar Mission Science Objectives," Presented at the NASA Office of Space Access and Technology (OSAT) Ninth Advanced Space Propulsion Workshop, Pasadena CA, 11–13 March 1998; JPL Internal Document D-15671, R. H. Frisbee (ed.), 1998.

[7] Martin, A. R. (ed.), "Project Daedalus—The Final Report of the BIS Starship Study," *Journal of the British Interplanetary Society*, suppl., 1978.

[8] Jones, R. M., "Electromagnetically Launched Micro Spacecraft for Space Science Missions," AIAA Paper 88-0068, 11 Jan. 1988.

[9] Schnitzler, B. G., Jones, J. L., and Chapline, G. F., "Fission Fragment Rocket Preliminary Feasibility Assessment," Idaho National Engineering Lab. Contract No. DEACO7-76IDO1570 and Lawrence Livermore National Lab. Contract No. W-7405-ENG-88, 1989.

[10] Forward, R. L, "Radioisotope Sails for Deep Space Propulsion and Electrical Power," AIAA Paper 95-2596, 10–12 July 1995.

[11] Bussard, R. W., "Galactic Matter and Interstellar Flight," *Astronautica Acta*, Vol. 6, No. 4, 1960, pp. 179–194.

[12] Forward, R. L., "Starwisp: An Ultra-Light Interstellar Probe," *Journal of Spacecraft and Rockets*, Vol. 22, No. 3, 1985, pp. 345–350.

[13] Forward, R. L., "Roundtrip Interstellar Travel Using Laser-Pushed Lightsails," *Journal of Spacecraft and Rockets*, Vol. 21, No. 2, 1984, pp. 187–195.

[14] Frisbee, R. H., "How to Build an Antimatter Rocket for Interstellar Missions—Systems Level Considerations in Designing Advanced Propulsion Technology Vehicles," AIAA Paper 2003-4696, 20–23 July 2003.

[15] Singer, C. E., "Interstellar Propulsion Using a Pellet Stream for Momentum Transfer," *Journal of the British Interplanetary Society*, Vol. 33, 1980, pp. 107–115;

Zubrin, R. M., and Andrews, D. G., "Magnetic Sails and Interplanetary Travel," *Journal of Spacecraft and Rockets*, Vol. 28, No. 2, 1991, pp. 197–203.

[16] Smith, G. A., "Applications of Trapped Antiprotons," *Hyperfine Interactions*, Vol. 81, No. 1–4, August 1993, pp. 189–196; Lewis, R. A., Smith, G. A., and Howe, S. D., "Antiproton Portable Traps and Medical Applications," *Hyperfine Interactions*, Vol. 109, No. 1–4, Aug. 1997, pp. 155–164.

[17] URANOS Group, "Konstantin E. Tsiolkovsky [Ciolkowski] (1857–1935)," URL: http://www.uranos.eu.org/biogr/ciolke.html [last update Aug. 6, 2001, cited January 22, 2003].

[18] Forward, R. L., "Ad Astra!," *Journal of the British Interplanetary Society*, Vol. 49, 1996, pp. 147–149.

[19] LaPointe, M. R., "Antiproton Powered Propulsion with Magnetically Confined Plasma Engines," *Journal of Propulsion and Power*, Vol. 7, No. 5, Sept.–Oct. 1991, pp. 749–759.

[20] Callas, J. L., *The Application of Monte Carlo Modeling to Matter–Antimatter Annihilation Propulsion Concepts*, Jet Propulsion Laboratory, JPL Internal Document D-6830, CA, 1 Oct. 1989.

[21] Meyer, T. R., McKay, C. P., McKenna, P. M., and Pyror, W. R, "Rapid Delivery of Small Payloads to Mars," AAS Paper 84-172, July 1984, *Proceedings of the Case for Mars II*, pp. 419–431, C. P. McKay (ed.), Vol. 62, Science and Technology Series, American Astronomical Society, 1985; Landis, G. A., "Optics and Materials Considerations for a Laser-Propelled Lightsail," IAA Paper 89-684, 7–12 Oct. 1989; Landis, G. A., "Small Laser-Propelled Interstellar Probe," *Journal of the British Interplanetary Society*, Vol. 50, 1997, pp. 149–154.

[22] Frisbee, R. H., "Impact of Interstellar Vehicle Acceleration and Cruise Velocity on Total Mission Mass and Trip Time," AIAA Paper 2006-5224, 9–12 July 2006.

[23] Draine, B. T., and Lee, H. M., "Optical Properties of Interstellar Graphite and Silicate Grains," *Astrophysical Journal*, Vol. 285, 1 Oct. 1984, pp. 89–108.

[24] Frisbee, R. H., "Advanced Space Propulsion for the 21st Century," *Journal of Propulsion and Power*, Vol. 19, 2003, pp. 1129–1154.

[25] Dyson, F. J., "Interstellar Transport," *Physics Today*, Oct., 1968, pp. 41–45.

[26] Brower, K., *The Starship and the Canoe*, Harper and Row, New York, 1978, p. 149.

[27] Thio, F., "A Summary of the NASA Fusion Propulsion Workshop 2000," AIAA Paper 2001-3669, 8–11 July 2001.

[28] Forward, R. L., "Antiproton Annihilation Propulsion," AIAA Paper 84-1482, 11–13 June 1984; Air Force Rocket Propulsion Laboratory (AFRPL) Technical Report AFRPL TR-85-034, Sept. 1985.

[29] Narevicius, E., Parthey, C. G., Libson, A., Narevicius, J., Chavez, I., Even, U., and Raizen, M. G., "An Atomic Coilgun: Using Pulsed Magnetic Fields to Slow a Supersonic Beam," *New Journal of Physics*, Vol. 9, No. 10, Oct. 2007, pp. 358–366; Narevicius, E., Parthey, C. G., Libson, A., Riedel, M. F., Even, U., and Raizen, M. G., "Towards Magnetic Slowing of Atoms and Molecules," *New Journal of Physics*, Vol. 9, No 4, April 2007, pp. 96–103.

[30] Takahashi, H., "Application of Muon-Catalyzed Fusion, and an Alternative Approach for Space Propulsion," NASA Office of Space Access and Technology (OSAT) Eighth Advanced Space Propulsion Workshop, Pasadena CA, 20–21 May 1997, Proceedings published as JPL Internal Document D-15461, R. H. Frisbee (ed.), 1997.

[31] Lewis, R. A., Smith, G. A., Cardoff, E., Dundore, B., Fulmer, J., Watson, B. J., and Chakrabati, S., "Antiproton-Catalyzed Micro-fission/Fusion Propulsion Systems for Exploration of the Outer Solar System and Beyond," AIAA Paper 96-3069, 1–3 July 1996.

[32] Lewis, R. A., Meyer, K., Smith, G. A., and Howe, S. D., "AIMStar: Antimatter Initiated Microfusion for Precursor Interstellar Missions," AIAA Paper 99-2700, 20–23 June 1999.

[33] Howe, S., and Jackson, G., "Antimatter Driven Sail for Deep Space Missions," Phase I Final Report, Nov. 2002, *NASA Institute for Advanced Concepts (NIAC)*, URL: http://www.niac.usra.edu [cited Aug. 20, 2003].

[34] Garner, C., "Developments and Activities in Solar Sails," AIAA Paper 2001-3234, 8–11 July 2001.

[35] Zubrin, R. M., and Andrews, D. G., "Magnetic Sails and Interplanetary Travel," *Journal of Spacecraft and Rockets*, Vol. 28, No. 2, 1991, pp. 197–203.

[36] Winglee, R., Slough, J., Ziemba, T., and Goodson, A., "Mini-Magnetospheric Plasma Propulsion (M2P2): High Speed Propulsion Sailing the Solar Wind," STAIF Paper CP504, *Space Technology and Applications Forum-2000 (STAIF-2000)*, M. S. El-Genk (ed.), American Institute of Physics, 2000; Winglee, R. M., Slough, J., Ziemba, T., and Goodson, A., "Mini-Magnetospheric Plasma Propulsion: Trapping the Energy of the Solar Wind for Spacecraft Propulsion," *Journal of Geophysical Research*, Vol. 105, 2000, pp. 21,067–21,077.

[37] Singer, C. E., "Interstellar Propulsion Using a Pellet Stream for Momentum Transfer," *Journal of the British Interplanetary Society*, Vol. 33, 1980, pp. 107–115.

[38] Ewig, R., and Andrews, D., "Microfission-Powered Orion Rocket," NASA JPL/MSFC Thirteenth Annual Advanced Space Propulsion Workshop (ASPW 2002), Pasadena CA, 4–6 June 2002; Andrews Space and Technology, "Mini-MagOrion (MMO)," URL: http://www.andrews-space.com/en/corporate/MMO.html [last update 2003, cited June 20, 2003].

[39] Jones, R. M., "Electromagnetically Launched Micro Spacecraft for Space Science Missions," AIAA Paper 88-0068, 11 Jan. 1988, pp. 691–695.

[40] Lipinski, R. J., Beard, S., Boyes, J., Cnare, E. C., Cowan, M., Duggin, B. W., Kaye, R. J., Morgan, R. M., Outka, D., Potter, D., Widner, M. M., and Wong, C., "Space Applications for Contactless Coilguns," *IEEE Transactions on Magnetics*, Vol. 29, No. 1, Jan. 1993.

[41] Sol Company, "Alpha Centauri 3," URL: http://www.solstation.com/stars/alpcent3.htm [cited June 15, 2007].

[42] Sol Company, "Gliese 867/Ross 780," URL: http://www.solstation.com/stars/gl876.htm [cited 15 June 2007].

[43] Sol Company, "55 Rho (1) Canceri 2," URL: http://www.solstation.com/stars2/55cnc2.htm [cited 15 June 2007].

[44] Forward, R. L., *Flight of the Dragonfly*, Baen Books, Riverdale, New York, 1985.

[45] Goddard, R. H., "The Ultimate Migration" (manuscript), 14 Jan. 1918, *The Goddard Biblio Log*, Friends of the Goddard Library, 11 Nov. 1972.

[46] Bowman, L. "Interstellar Travel: A Family Affair," URL: http://news.nationalgeographic.com/news/2002/02/0220_0220_wirelifeinspace.html, 20 Feb. 2002, Scripps Howard News Service, *National Geographic News* [cited 2 Jan. 2008]; Angier, N., "Scientists Reach Out to Distant Worlds," URL: http://query.nytimes.

com/gst/fullpage.html?res=9D05EED81630F936A35750C0A9649C8B63&sec=&spon=&pagewanted=all, 5 March 2002, *The New York Times* [cited 2 Jan. 2008]; Malik, T., "Sex and Society Aboard the First Starships," URL: http://www.space.com/scienceastronomy/generalscience/star_voyage_020319-1.html, 19 March 2002, *Space.Com* [cited 2 Jan. 2008].

[47] O'Neill, G. K., "The Colonization of Space," *Physics Today*, Vol. 27, No. 9, Sept. 1974, pp. 32–40; O'Neill, G. K., "Space Colonies: The High Frontier," *The Futurist*, February 1976.

[48] Orth, C., Klein, G., Sercel, J., Hoffman, N., and Murray, K., "Transport Vehicle for Manned Mars Missions Powered by Inertial Confinement Fusion," AIAA Paper 87-1904, 29 June–2 July 1987.

[49] Paine, C., and Seidel, G., "Brown University Magnetic Levitation/Supercooling Research," NASA Office of Advanced Concepts and Technology (OACT) Third Annual Workshop on Advanced Propulsion Concepts, Pasadena CA, 30–31 Jan. 1992, JPL Internal Document D-9416, R. H. Frisbee (ed.), 1992.

[50] Lesh, J. R., Rugglen, C., and Ceggarone, R., "Space Communications Techniques for Interstellar Missions," *Journal of the British Interplanetary Society*, Vol. 49, 1996, pp. 7–14.

[51] Chalsson, E. J., Quoted in Section 4.3.6, Transcripts of Break Groups, *Proceedings of the Humanity 3000 Workshop, Humans and Space: The Next Thousand Years*, Foundation for the Future, Bellevue, WA, June 2005, p. 182.

[52] Geller, T., "Aluminum: Common Metal, Uncommon Past," *Chemical Heritage Newsmagazine*, Winter 2007/8, Vol. 25, No. 4, URL: http://www.chemheritage.org/pubs/magazine/feature_alum_p1.html [cited Jan. 18, 2008].

[53] Cassenti, B., "The Interstellar Ramjet," AIAA Paper 2004-3568, 11–14 July 2004.

[54] *TRW Space Log*, W. A. Donop Jr. (ed.), Vol. 9, No. 4, Winter 1969–1970.

[55] Forward, R. L., *Indistinguishable from Magic*, Baen Books, Riverdale, New York, 1995.

Chapter 3

Prerequisites for Space Drive Science

Marc G. Millis*
NASA Glenn Research Center, Cleveland, Ohio

I. Introduction

TO CIRCUMVENT the propellant limits of rockets and the maneuvering limits of solar sails, a means to propel spacecraft using only the interactions between the spacecraft and its surrounding space is sought. A general term for such a device is "space drive," which is adopted for convenience from science fiction. (This term first appeared in John Campbell's 1932, *The Electronic Siege* [1].) At present, the scientific foundations from which to engineer a space drive have not been discovered. In fact, the issues, unknowns, and opportunities to seek these discoveries have only recently begun to be articulated [2,3]. To set the stage for further progress, this chapter examines this topic at the level of the first step of the scientific method, specifically defining the problem.

While this chapter focuses on clarifying the problem to guide future research, subsequent chapters deal separately with specific ongoing investigations. These include manipulating gravity, electromagnetic–gravitational couplings, photon momentum in media, inertial modifications, quantum vacuum interactions, and others.

It is important to stress that space drives might be physically impossible, and conversely, that the prerequisite discoveries have just not yet been made. To begin the discussion, the key physics issues, basic energy estimates, hypothetical propulsion concepts, and relevant topics in science are examined. In several instances, provocative conjectures are introduced to demonstrate how space drive goals are distinct from the more general scientific inquiries. The intent is to provide starting points that future researchers can use to specifically address the physics of space drives. Eventually, future research will determine if, and how, space drives are physically possible.

This material is a work of the U.S. Government and is not subject to copyright protection in the United States.
*Propulsion Physicist, Propulsion and Propellants Branch, Research and Technology Directorate.

A further assertion is that adding the challenge of breakthrough spaceflight can enhance the study of the more general lingering unknowns of science. The challenges of spaceflight offer different lines of inquiry, providing insights that might otherwise be overlooked from just curiosity-driven science.

II. Methods

Starting with the major objections to the notion of a space drive, namely conservation of momentum and the scarcity of indigenous reaction mass, specific research objectives are articulated. Next, as both a tool for deeper analysis and to reflect potential benefits, energy comparisons are made between ideal rockets and space drives. This includes demonstrating possible pitfalls when conducting such analyses. Finally, 10 hypothetical space drive concepts are presented to illustrate possibilities and issues. Throughout, relevant physical effects and lingering unknowns are cited. Where possible, the distinctions between general science and the perspectives of propulsion-oriented science are described. The intent is to provide starting points from which to apply scientific progress toward answering the goal of enabling humanity to traverse interstellar space.

Newtonian representations are used predominantly instead of General Relativistic perspectives for several reasons. First, it is easier to introduce space drives using Newtonian treatments. Newtonian terms are more familiar and can be easily presented in the meter-kilogram-sec (MKS) units typical of engineering endeavors; converting General Relativity equations into engineering terms can be more cumbersome. Although General Relativity has broader validity, Newtonian treatments are a valid approximation for the low energy and nonrelativistic situations dealt with here [4]. And finally, Newtonian perspectives allow investigating effects that occur *within* spacetime, while General Relativity deals with spacetime itself. To convey this in terms of an analogy, consider moving an automobile across a landscape. General Relativity allows us to consider how to reshape the landscape so that the automobile (and everything in the vicinity) rolls passively downhill toward the desired destination. The Newtonian perspectives allow us to consider how to move the automobile under its own power relative to the landscape. Both perspectives are intellectually provocative. While the Newtonian perspectives are prevalent here, subsequent chapters (Chapters 4 and 15) examine General Relativity perspectives in more depth.

III. Major Objections and Objectives

A. Conservation of Momentum

The most obvious issue facing the prospect of a space drive is *conservation of momentum*. When accelerating a vehicle there must be an equal and opposite change in momentum imparted to a reaction mass. This encompasses Newton's basic laws of motion. For land vehicles, the reaction mass is the Earth's surface; for rockets, it is their propellant; and for space sails, it is the light that hits the sails. For space drives, the reaction mass is not obvious, but options for further inquiry exist.

Even though one could entertain schemes where momentum conservation does not apply, which would require delving even deeper into undiscovered science, this chapter *imposes* conservation of momentum as a working premise. From that premise, it follows that some form of indigenous reaction mass must be found in space, and that a means to interact with that mass must be discovered to create thrust.

Other phenomena that play a role in conservation of momentum are *inertia* and the *reference frames* upon which momentum conservation is described. Later in this chapter the implications of inertia and the role of reference frames are explored in more depth. For now, the focus is on interacting with reaction masses.

In preparation for examining the possible sources of indigenous reaction mass, first consider the basic equation of momentum Eq. (1), where greater impulse (change in momentum), Δp, can be obtained by either increasing the amount of reaction mass used, m, or increasing the change in velocity, Δv, imparted to that reaction mass. For these introductory exercises, it is sufficient to use non-relativistic equations.

$$\Delta p = m\Delta v \qquad (1)$$

Because the space drive relies on interaction with some form of reaction mass in the *surrounding* space, it is helpful to introduce terms that address the *distribution* of that reaction mass. To this end, Eq. (2) presents a modified version of Eq. (1) where the volume element of the affected space is included and in a manner that still balances the equation. The reaction mass is then expressed as a density, ρ.

To avoid confusion, please be alert to the visual similarity of the terms used to represent the momentum, p, and mass density, ρ, and further between volume, V, and velocity, v. These representations follow common conventions.

$$\Delta p = V\left(\frac{m}{V}\right)\Delta v \quad \Rightarrow \quad \Delta p = V\rho\Delta v \qquad (2)$$

To allow exploring the implications of different momentum transfer schemes, the volume element, V, is factored into the product of an area, A, and distance, r, as shown in Eq. (3).

$$\Delta p = rA\rho\Delta v \qquad (3)$$

It should be stressed that this equation is offered to help identify critical factors, rather than to imply that this would be the exact form of any solutions. For now, this equation merely identifies which factors pertain, and further, if the factors are directly or indirectly proportional to the desired effect. When investigating specific momentum-exchange methods, the equation will likely change. For example, if momentum transfer is accomplished through a collision process, analogous to light hitting a solar sail, then the propulsive effect is directly proportional to the area of the sail, A, and r becomes negligible (r then represents the perpendicular distance to the sail over which the momentum transfer takes place). Conversely, if the momentum exchange is through some intermediary field distributed over space, analogous to how a charged particle is accelerated by an electric field, then the effective distance, r, of that field becomes the dominant term. The equation, as presented, is just a first step toward considering such options and their consequences.

To recap, by starting with a major objection to the notion of a space drive, namely conservation of momentum, key objectives have now been identified. From Eq. (3), it can be seen that a greater change in momentum can be achieved by maximizing any of the following, and that these characteristics would need to be addressed in future space drive research:

- Density of the natural phenomenon in space, ρ, which might constitute a reactions mass (kg/m^3)
- Change in velocity imparted to the reaction mass, Δv (m/s), (having an upper limit of light speed, $c = 3.0 \times 10^8$ m/s)
- Surface area of the spacecraft's space drive effector, A (m^2)
- Effective range of the space drive effect, r (m)

B. Indigenous Reaction Mass

The notion of interacting with some form of indigenous reaction mass in the vicinity of the spacecraft presents the next objection: The vacuum of space appears devoid of tangible matter. There are, however, several phenomena that serve as starting points for deeper inquiry, as shown in Table 1. Consistent with importance of mass density, ρ, as identified by Eq. (3), both the mass density and equivalent energy density for these phenomena are listed. The relation between the mass and energy densities is governed by the familiar $E = mc^2$ relation. The most intriguing of these phenomena are discussed in subsequent paragraphs.

Estimates for the matter in the Universe are still evolving as astronomical observations accumulate. The data shown are based on combinations of textbooks [4,5] and more recent data from the Wilkinson Microwave Anisotropy Probe (WMAP) [6,7]. Table 1 includes the various omegas, Ω_i, (proportions) of the known constituents, starting with the generally accepted assumption that the total mass-density of the Universe satisfies the conditions for a "flat" Universe [5].

Relative to the goal of space drives, the cosmological implications and uncertainties [6] of these values are not of concern. Instead, in the context of spaceflight, the main question is if any of these phenomena can serve as a reaction mass for a spacecraft. This perspective presents different lines of inquiry from the more general cosmological concerns and thereby provides an additional venue through which to learn about these phenomena.

The first issue for space drives is to find phenomena with the highest mass density and, second, to find ways to create forces against these phenomena. At present, none of these phenomena is an obvious candidate for a reaction mass, but some are so inadequately understood that the opportunities for new discoveries are high. These include *Dark Matter*, *Dark Energy*, *quantum fluctuations*, and the deeper meanings of *spacetime* itself. Even if the propulsive study of these phenomena does not lead to a breakthrough, assessing their characteristics

Table 1 Known indigenous space phenomena

Known forms of mass and energy		In terms of mass density, ρ (kg/m³)	In terms of energy density (J/m³)
Total matter in the universe (critical density)	Proportions $\Omega = 1.00$	$9.5 \times 10^{-27\mathrm{a}}$	8.6×10^{-10}
"Dark Energy"	$\Omega_\mathrm{L} = 0.73^\mathrm{b}$	6.9×10^{-27}	6.2×10^{-10}
"Dark Matter"	$\Omega_\mathrm{DM} = 0.22^\mathrm{b}$	2.1×10^{-27}	1.9×10^{-10}
Baryonic matter (normal matter)	$\Omega_\mathrm{B} = 0.04^\mathrm{b}$	3.8×10^{-28}	3.4×10^{-11}
Photons and relativistic matter	$\Omega_\mathrm{rel} = 8.3 \times 10^{-4\mathrm{b}}$	7.9×10^{-31}	7.1×10^{-14}
Cosmic microwave background	$\Omega_\mathrm{CMB} = 10^{-5}$	$10^{-31\mathrm{c}}$	10^{-15}
Quantum vacuum fluctuations			
Inferred as dark energy		$10^{-26\mathrm{d}}$	10^{-9}
Up to nucleon Compton frequency (10^{23} Hz)		10^{18}	$10^{35\mathrm{e}}$
Up to Planck limit (10^{43} Hz)		$10^{98\mathrm{d}}$	10^{113}
Galactic hydrogen		$3.3 \times 10^{-21\mathrm{f}}$	3.0×10^{-4}
Spacetime itself			
In terms of total mass density		$9.5 \times 10^{-27\mathrm{a}}$	8.6×10^{-10}
General Relativity analogy to Young's modulus		5.3×10^{25}	$4.8 \times 10^{42\mathrm{g}}$

Values in this table that are accompanied by a reference citation are from that reference. Other values shown are calculated from the cited values.
[a]From Ref. 5.
[b]From Ref. 6.
[c]From Ref. 4.
[d]Maclay, Chapter 12 in this book.
[e]Davis and Puthoff, Chapter 18 in this book.
[f]From Ref. 19.
[g]See Eq. 6 to 8.

from the perspective of space drives will lead to a more thorough understanding by providing another perspective from which to analyze the data. In other words, even if the propulsive goals are not viable, using a propulsion perspective will improve our understanding of these phenomena.

1. "Dark Energy"

Dark Energy is only a working hypothesis to explain other anomalous observations. In this case, the original phenomenon that led to the Dark Energy hypothesis is the *anomalous redshifts* from the most distant stars. Specifically, the light coming from the most distant Type 1a supernovae (whose distance can be more accurately inferred than other distant objects) is even more redshifted than that attributable to Hubble expansion [8]. These anomalously higher redshifts are interpreted as indications that the Universe's expansion is accelerating [9]. To explain what might be causing this, the term Dark Energy has become the working hypothesis, specifically asserting that an invisible form of energy is causing this expansion [10].

One interpretation is that this dark energy is related to quantum vacuum fluctuations, but the values calculated by astronomical observations and from quantum regimes are mismatched by roughly 120 orders of magnitude (see Chapter 12 in this book). Clearly, this is a subject where more discoveries await. The propulsive implications of the Dark Energy hypothesis, or more generally of the anomalous phenomena that led to the Dark Energy hypothesis, have not yet been explored.

2. "Dark Matter"

Dark Matter is also a working hypothesis for other anomalous phenomena. The first confirmed empirical observations that lead to the Dark Matter hypothesis are the *anomalous rotations of galaxies* [11]. When considering galactic rotation rates and newtonian mechanics, the stars at the outer regions of a galaxy should have been moving away from the center of the galaxy. Instead, galaxies appear to hold together as if some form of invisible matter is adding to the gravitational attraction that binds the galaxy's stars together. More recently, other observations are showing indications of unseen gravitating matter. Specifically, *gravitational lensing* has been observed around clusters of galaxies to a degree beyond that attributable to the visible matter of these galaxies [12].

Whether it is indeed a new form of matter or indicative of other physics yet to be discovered is not yet known. Other theories are being explored such as the MOND hypothesis (modified newtonian dynamics) [13,14]. The utility of such theories or the actual phenomena that led to the Dark Matter hypothesis have not been explored in the context of spaceflight, either as a direct reaction mass or as something indicative of other phenomena.

3. Cosmic Microwave Background Radiation

Considered to be a remnant of the Big Bang, space is filled with photons whose frequency spectrum matches that of blackbody radiation corresponding to a temperature of 2.7K. While cosmological studies focus on the slight anisotropic details of this background [6,7], its prominent isotropic nature makes it intriguing for deep spaceflight. Because the overall spectrum is so isotropic, Doppler shifts due to motion through this background provide reliable measures of velocity relative to the mean rest frame of the Universe. For example, the Earth's motion through this background has been measured as 365 km/s [15]. This cosmic microwave background is thus a natural phenomenon for deep spaceflight navigation. Its prospects as a propulsive media are discussed in Section V of this chapter.

4. Quantum Vacuum Fluctuations

As a consequence of the Uncertainty Principle in quantum physics, the energy state of any system can never be exactly zero. This is true of the electromagnetic spectrum of empty space. Chapter 12 provides an introduction to both the theoretical and experimental details of this phenomenon, along with discussion of its utility as an effective reaction mass for propulsion. Chapter 18 also discusses the energy implications of this phenomenon. Because subsequent chapters in this text provide details, only preliminary information is offered here.

One of the main unknowns of this quantum vacuum energy is its fundamental energy density. Estimates vary widely depending upon the upper cutoff frequency applied to the calculations; Table 1 lists just three possibilities. The first is the estimate based on the assertion that Dark Energy is identical to quantum vacuum energy; the energy density is then calculated from estimates of the total mass of the Universe and other factors [6]. At the other extreme, if the Planck frequency is used as the cutoff value, an energy density of roughly 120 orders of magnitude higher is calculated [Maclay, Chapter 12]. The Planck frequency (10^{43} Hz) is considered the absolute highest possible frequency allowed by the application of the Uncertainty Principle to the structure of space-time itself [16]. An intermediate estimate based on the Compton frequency of the nucleon (10^{23} Hz) is about 80 orders of magnitude less than the Planck limit and about 40 orders of magnitude greater than the Dark Energy estimate. Clearly, this is an area where more discoveries can be made.

5. Virtual Particle Pairs

A collateral phenomenon to quantum vacuum energy is *virtual particle pairs*, where particle–antiparticle pairs continually appear and then disappear from space. The prospects for using this phenomenon for space flight date back to at least 1997 [17].

Regardless of the wide discrepancy in the estimated energy levels of the quantum vacuum background, there is a fundamental limit to the deliberate production of such matter from the vacuum. The creation of mass from energy, is governed by the familiar $E = mc^2$ relation. When considering imparting a velocity, v, to the particles, the more complete form of the energy–mass equation applies [18]:

$$E = mc^2 / \sqrt{1 - (v/c)^2} \qquad (4)$$

where E is the energy required to create a mass, m, with a velocity, v. Considering that this equation represents 100% efficient conversion, this significant energy expenditure is only a lower limit on the energy required to impart momentum to virtual pairs.

6. Galactic Hydrogen

In the voids of our Galaxy, there exist small amounts of hydrogen that have been considered as a possible reaction mass for spaceflight. One concept, the Bussard interstellar ramjet, is discussed in Section V.A of this chapter.

Based on analyses of the mass-to-light ratios of our Galaxy [19], it is estimated that the density of molecular and atomic hydrogen is 3.3×10^{-21} kg/m^3. Other sources assert that some areas may be as high as 10^{-20} kg/m^3 [20].

7. Spacetime Itself

In addition to the lingering mysteries of Dark Energy and Dark Matter, among others, there is the deeper unknown of *the* fundamental phenomenon that underlies the Universe itself: *spacetime*. Spacetime is the background against which all things exist and their interactions are measured. From the Hubble expansion, it appears that spacetime can change dimensions. As evidenced from the lightspeed limit, spacetime has properties that govern the motion of mass and energy. And spacetime even has the property of being an *inertial frame*—specifically, the property of a space that serves as a reference frame for acceleration and rotation. But because it is so ubiquitous, it is difficult to isolate for deeper study.

When faced with the Hubble expansion, an obvious question is: "Expanding relative to what?" As a reflection of the cosmologist perspective, examine the following quote from a general relativity textbook [4].

> It's simplest and most elegant to assume that the observed homogeneity and isotropy extend over the whole universe. In an exactly homogenous model there can be no center and no boundary. In that context it doesn't make sense to talk about the universe expanding into something or from somewhere.

To a propulsion engineer, knowing the components of what he or she is working with is of paramount importance. If there is some underlying structure to spacetime itself (against which spacetime expands) then it must also be considered. Given how much is still not known and considering that the "no center and no boundary" assertions are working assumptions, it is reasonable to consider that additional constituents can be hypothesized.

Spacetime is the background against which both electromagnetic and gravitational fields exist. The factors that govern how these fields behave are well modeled, but *how* they function and if additional factors await discovery are unknown. Considering the Dark Energy and Dark Matter anomalies and the fact that physics on the quantum scale has still not been merged with the large scale physics of General Relativity, it is clear that there is much left to learn.

Consider the lightspeed limit. The evidence shows that lightspeed is always the same in any *local* inertial frame. It is a relation that governs the fundamental correlation between space and time for all electromagnetic phenomena. But when viewed more globally and in the presence of matter, the situation is more complicated. Gravity bends light or spacetime itself, and there is more than one way to model the observations. For example, in the *geometric* versions of General Relativity, the speed of light is used as the defining constant, and consequentially, space and time *curve* in the presence of matter. Within General Relativity there also exist "optical analogies," where space is represented as an optical medium with an effective index of refraction that is a function of gravitational potential [21,22]. In this case, the speed of light is a variable. Although different from the more common geometric interpretation, this interpretation has been shown to be consistent with physical observables, and comparisons between the optical and geometric perspectives have also been published [23,24].

To explain this difference using a simpler analogy, consider the basic equation of constant velocity motion:

$$d = vt \tag{5}$$

where d is the distance covered after a time, t, of moving at velocity, v. From the *geometric* versions of General Relativity, the velocity, v, is held as the universal constant (lightspeed), and thus both space, d, and time, t, have to "curve" when this relation gets distorted by the presence of matter. Conversely, from the *optical analogy* perspective, space, d, is taken to be fixed, and thus both the speed, v, and time, t, must adjust when this relation gets distorted by the presence of matter. Both of these options are available for deeper study, and the choice of which to pursue depends on what problem one is trying to solve.

Little attention is typically focused on the optical analogy because it does not predict any new effects that are not already covered by the more common geometric perspective and because it raises unanswered issues with coordinate systems choices (Matt Visser, personal communication, 2 October 2002). But it is these coordinate system issues that are provocative from the spaceflight point of view, and to further illustrate this issue it is necessary to address Mach's principle.

Mach's principle asserts that an *inertial frame*, specifically the property of a space to be a reference frame for acceleration, is actually created by, and connected to, the surrounding mass in the Universe [25]. A related assertion is that a *literal* interpretation of Mach's Principle implies an *absolute reference frame*, coincident with the mean rest frame for all the matter in the Universe [26]. This is because the matter in the Universe creates the inertial frame and, hence, the location and intensity of that inertial frame tracks with the position and distribution of that matter. Such considerations will evoke issues with coordinate transforms.

It can be argued that, even if such an underlying, machian inertial frame did exist, it could not be detected because inertial forces do not appear until there is a *change* in velocity. Conversely, however, it can be argued that this nondetectability is a consequence of how one chooses to model phenomena. A crucial underlying premise of both *Special* and *General Relativity* is that their very foundations are constructed to provide *frame-independent* representations [4]. This comes from the desire to find representations that are valid everywhere, with universal correctness. But if Nature does possess frame-*dependent* phenomena, it would be difficult to describe these with theories that are limited to frame-*independent* representations.

There have been attempts to incorporate frame-dependent considerations into General Relativity, such as the Brans-Dicke Theory [16,27]. There have also been attempts to incorporate Mach's principle into General Relativity [28], but it remains an unresolved problem in physics. In the Bondi and Samuel paper [28], which deals with the connection between the Lense–Thirring effect (frame dragging) and Mach's principle, no less than 10 variations of Mach's principle are articulated, none of which specify propulsive considerations.

As mentioned previously, Nature does possess at least one frame-*dependent* phenomenon, namely the cosmic microwave background. While curiosity-driven science is focused on understanding the minute variations in this background for its implications to the origins, structures, and fate of the whole Universe [6,7,29], a propulsion engineer sees a fore–aft energy difference when moving through space. Such an effect also provokes conjectures about inducing motion by deliberately creating fore–aft energy differences around a spacecraft. (Hypothetical mechanisms are listed in Section V of this chapter.)

If, like the cosmic microwave background, there is a literal *machian* inertial frame that is *connected* to the surrounding mass of the Universe, then it might serve as a propulsive medium—specifically an effective reaction mass to thrust against. By thrusting against this progenitor inertial frame, the reaction forces are imparted to the surrounding mass of the Universe. At this point, this is sheer conjecture, but considering how little is known about the nature of inertial frames themselves, this notion at least warrants deeper consideration. Continuing with conjectures, the notion that there might be a connection between the cosmic microwave background and such a machian frame also warrants consideration.

Pursuing such conjectures leads to a different line of inquiry than the more common frame-independent, *geometric* form of General Relativity. The propulsion perspective leads to euclidean treatments like the previously mentioned "optical analogies." This is where frame-dependence might be more readily addressed. The difficulties with coordinate transforms could be due to the implicit connection to frame-dependent inertial frames, but again, at this stage, these are just conjectures for future research. Situations such as this are what support the assertion that meaningful and different knowledge will be revealed by the pursuit of spaceflight breakthroughs, which might otherwise be overlooked by curiosity-driven physics.

Proceeding now to provide some initial estimates of the *effective* reaction mass of spacetime itself, two versions are offered. The first is just to use the average mass density of the whole Universe to reflect how much mass contributes to the machian frame. Here the value already cited from Carroll and Ostlie [5] applies. The second version uses an analogy to Young's modulus from Einstein's field equation.

Young's modulus is a way of quantifying the property of materials for how much they distort when under strain. The higher the value, the stiffer the material, or said another way, the less it will distort when subjected to a given force. The units for Young's modulus, Y, are Force/Area, which is equivalent to the units of Energy/Volume, and is defined as [30]:

$$Y = \frac{\text{Stress}}{\text{Strain}} \qquad (6)$$

Compare this to Einstein's Field Equation (shown with MKS units) where the major portions are isolated in parentheses to simplify the analogy [4]:

$$\left(R^{\mu\nu} - \frac{1}{2}g^{\mu\nu}R - \Lambda g^{\mu\nu}\right) = \left(\frac{8\pi G}{c^4}\right)(T^{\mu\nu})$$

$$(\text{Strain}) = \left(\frac{8\pi G}{c^4}\right)(\text{Stress}) \qquad (7)$$

All the terms on the left-hand side represent the *curvature* of spacetime, which is analogous to the *strain* (distortion) resulting from an imposed stress. The right-hand side includes the *energy–momentum stress tensor*, $T^{\mu\nu}$, which is analogous to the *stress* imposed onto spacetime. The scaling factor $(8\pi G/c^4)$ is based on the properties of spacetime and is analogous to the reciprocal of Young's modulus. Continuing with the analogy, it follows that spacetime (where G is Newton's constant and c is lightspeed) has an analogous Young's modulus per:

$$\frac{\text{Stress}}{\text{Strain}} = \left(\frac{c^4}{8\pi G}\right) = 4.8 \times 10^{42} \frac{N}{m^2} \tag{8}$$

This extremely high value (the units are equivalent to energy density) suggests that spacetime is very stiff. This means it takes a significant amount of matter or field energy (represented by the *energy–momentum stress tensor*, $T^{\mu\nu}$) to induce much of a curvature on spacetime. To what degree, if any, this makes spacetime a viable reaction mass is completely unknown. This calculation and perspective is offered to simply provoke deeper inquiries.

8. Summary

At present, none of the indigenous phenomena of space are obvious candidates for a space drive reaction mass, but some are so inadequately understood that the opportunity for new discoveries is high. These include *Dark Matter, Dark Energy, quantum fluctuations*, and the deeper meanings of *spacetime* itself. Even if none of these leads to a desired breakthrough, assessing their characteristics from the perspective of a space drive increases the chances of more thoroughly understanding nature by providing another perspective from which to analyze the data.

In the first step of the scientific method, where the problem is defined, that definition affects how data are subsequently collected and analyzed. For space drive inquiries, some basic initial questions include:

- What are the indigenous phenomena in space?
- Can forces be induced by interacting with indigenous phenomena?
- If so, can *net* forces be created between a device and a phenomenon?
- Are the forces and the amount of accessible reaction mass sufficient to propel a spacecraft?
- If not directly applicable for propulsion, might the phenomenon be indicative of other potentially more useful phenomena? In other words, if the phenomenon is not suitable as a reaction mass, might it be useful to measure other phenomena?

To examine the issues and limitations of interacting with any of these candidates, it is helpful to envision a number of hypothetical space drives and then analyze their viability. Before proceeding to that step, the energy associated with space drives is examined next. These energy analyses highlight pitfalls of prior assessments, give a crude estimation of potential benefits, and provide another means to explore the physics of space drives.

IV. Estimating Potential Benefits

A. Avoiding Pitfalls

The historic tendency when trying to gauge the value of an emerging technology is to compare it using the characteristics of the incumbent technology. Such provisional assessments can be seriously misleading, however, when the emerging technology uses fundamentally different operating principles. For example, the value of steamships is misleading when judged in terms of sails and rigging [31]. Although reduced sail area and rigging lines are indeed a consequence of steamships, the true benefit is that shipping can continue regardless of the wind conditions and with far more maneuvering control. Similarly, the benefits of a breakthrough space drive would likely surpass the operational conventions of rocketry. Issues such as optimizing specific impulse become meaningless if there is no longer any propellant. Three examples are offered next to illustrate the pitfalls of using rocket equations to describe the physics of a space drive.

1. Infinite Specific Impulse Perspective

The first and common misleading practice when describing a hypothetical space drive is to view it as a rocket with an *infinite* specific impulse. This seems reasonable at first because a higher specific impulse leads to less propellant, so an infinite specific impulse should lead to zero propellant. As shown from Eq. (9), however, specific impulse is a measure of the thrust, F, per propellant weight flow rate ($g\, \mathrm{d}m/\mathrm{d}t$) [32]. For a true space drive, the $\mathrm{d}m/\mathrm{d}t$ term would be meaningless, rendering the entire equation inappropriate for assessing propellantless propulsion.

$$I_{sp} = \frac{F}{g\dfrac{\mathrm{d}m}{\mathrm{d}t}} \qquad (9)$$

Furthermore, as shown from Eq. (10), which is based on the energy imparted to the propellant from the rocket's frame of reference [33], an infinite specific impulse, I_{sp}, implies that a propellantless space drive would require infinite energy (substituting $I_{sp} = \infty$).

Conversely, this same equation can be used to conclude that a propellantless space drive would require zero energy if there was no propellant (substituting $m_p = 0$). Neither of these extremes is likely the case. (In the strictest sense when extrapolating, the squared impulse term would dominate the mass term, resulting in infinite energy. Regardless, this equation is not appropriate when contemplating propellantless propulsion.)

$$E = \frac{1}{2} m_p \left(I_{sp} g \right)^2 \qquad (10)$$

2. Modify Rocket Inertia

To illustrate another misleading use of the rocket equations, consider the notion of manipulating the inertia of a rocket. If such a breakthrough were

PREREQUISITES FOR SPACE DRIVE SCIENCE

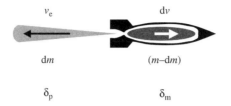

Fig. 1 Rocket in field-free space.

ever achieved, the implications and applications would likely extend beyond rocketry. Even if used on a rocket, there are a number of different ways to envision applying such an effect, each yielding considerably different conclusions; i.e., apply the inertial change 1) to the whole rocket system, 2) just to the expelled propellant, or 3) just to the rocket with its stored propellant. In the latter, a condition of operation is that the propellant resumes its full inertia as it is accelerated out of the rocket. This notion is similar to the science fiction concept of "impulse drive with inertial dampers" as presented in the *Star Trek* television series [34].

It should be noted that, at present, there are no confirmed techniques to affect such a change in inertia, even though experiments are underway [35]. It is important to stress that this is only a hypothetical example to illustrate the sensitivity of the findings to the methods, rather than to suggest that this is a viable breakthrough. Numerous variations on this analysis are possible.

Consider a rocket in field-free space (Fig. 1). To derive the rocket equation, one can start with conservation of momentum, where the rocket expels an increment of propellant, dm, to produce an incremental change in the rocket's velocity, dv.

The standard equation to represent this conservation of momentum has been slightly modified into Eq. (11), where coefficients are inserted to represent hypothetical manipulations of the inertia of the expelled propellant, δ_p, and the rocket, δ_m, where the rocket includes the stored propellant. Values of δ greater than one imply an increase, less than one imply a decrease, and equal to one represents no change. To be clear, when considering such isolated modifications of inertia, the δ_p term *only* affects the propellant as it is ejected from the rocket, and the δ_m term only affects the inertia of the *stored* propellant. In this latter case, it is assumed that the propellant regains its full inertia as it is accelerated out of the rocket.

$$-v_e(\delta_p)\mathrm{d}m = \mathrm{d}v(\delta_m)(m - \mathrm{d}m) \tag{11}$$

Proceeding with the normal steps to derive the rocket equation, it can be shown [36] that the final result for the Δv imparted to the rocket is represented by:

$$\Delta v = v_e \ln\left(\frac{m + m_p}{m}\right) \frac{\delta_p}{\delta_m} \tag{12}$$

Consider now the implications of modifying the inertia of the whole rocket system, which implies equal changes to δ_p and δ_m. In this circumstance there

Table 2 Different ways to modify rocket inertia

Which inertia is modified: From Eq. (12)	Propellant δ_p	Rocket δ_m	Net effect
Unmodified	1	1	Baseline: $\Delta v' = \Delta v$
Whole rocket system	δ	δ	No change: $\Delta v' = \Delta v$
Just rocket with its stored propellant	1	δ	$\Delta v' = 1/\delta \Delta v$
Just ejected propellant	δ	1	$\Delta v' = \delta \Delta v$

is no change at all in Δv. This null finding was one of the observations reported by Tajmar and Bertolami [37]. Alternatively, consider that the inertia of the rocket with its stored propellant is somehow reduced, while the inertia of the expelled propellant regains its full value. In this case the improvement in Δv tracks inversely to δ_m. In other words a δ_m of 0.5, representing a 50% decrease in the rocket's inertia, would yield a 50% increase in Δv. Table 2 summarizes how the different assumptions yield different results.

In addition to the ambiguity and wide span of results when using rocket equations to predict the benefits of modifying inertia, this approach does not provoke the questions needed to further explore such conjectures. For example, the issue of energy conservation is not revealed from the prior equations; although momentum was conserved, energy conservation is not addressed. It is presumed that any benefit must come at some expense, and because energy is a fundamental currency of mechanical transactions, it is reasonable to expect that such a benefit requires an expenditure of energy. These equations do not provide the means to calculate the extra energy required to support this hypothetical change in the rocket's inertia.

3. Gravitationally Altered Launch Pad

Another misleading use of rocket equations is when considering the implication of a *hypothetical* "gravity shield." This idea was provoked from the "gravity-shield" claim [38] that was later found not to be reproducible [39], but this scenario still serves to illustrate key issues. This example highlights issues associated with the *Equivalence Principle*. The Equivalence Principle asserts that *gravitational* mass is identical to *inertial* mass. If the gravitational mass is modified, then the inertial mass would be similarly modified.

Consider placing a launch pad above a hypothetical gravity shield (Fig. 2). A naive assumption would be that the reduced gravity would make it easier for the rocket to ascend as if being launched from a smaller planet.

There is more than one way to interpret this situation if one entertains possibilities beyond the initial "gravity shield" conjecture. In addition to considering that the gravitational field, g, is modified, one can consider that the "gravity shield" is, instead, altering the *mass* of the rocket above the device. In the case where the rocket mass is modified, one can further consider that just its *gravitational* mass is affected or, if the Equivalence Principle is in effect, that both its

PREREQUISITES FOR SPACE DRIVE SCIENCE

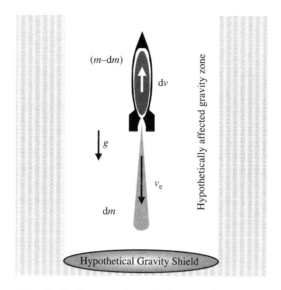

Fig. 2 Rocket over hypothetical gravity shield.

gravitational and *inertial* mass are equally affected. Figure 3 illustrates these assessment options.

To explore these options, start with Eq. (13) for a rocket ascending in a gravitational field [40]. The term on the left represents the mass and acceleration

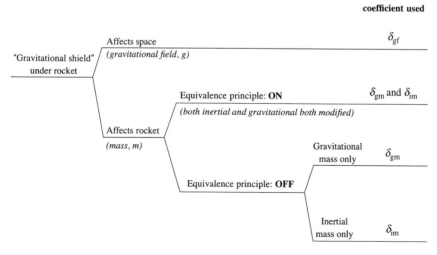

Fig. 3 Analysis options for gravitationally shielded launch pad.

of the rocket; the middle term is the force of gravity; and the right term is the reaction force from the expulsion of an increment of propellant with an exhaust velocity, v_e, relative to the rocket.

$$m\frac{dv}{dt} = -mg - v_e\frac{dm}{dt} \tag{13}$$

To consider the hypothetical modifications, coefficients are inserted next to reflect modifications to the gravitational field, δ_{gf}, and to the rocket's gravitational mass, δ_{gm}, and its inertial mass, δ_{im}. As before, values of δ greater than one imply an increase, less than one a decrease, and equal to one represents no change. Also, the equation is now rearranged to isolate the inertial terms from the gravitational terms:

$$-(\delta_{gm})m(\delta_{gf})g = (\delta_{im})\left[m\frac{dv}{dt} + v_e\frac{dm}{dt}\right] \tag{14}$$

The left side represents the gravitational contributions while the right side represents the inertial contributions. It can be shown that this equation results in the following representation for the Δv of the rocket:

$$\Delta v = -\frac{\delta_{gm}}{\delta_{im}}\delta_{gf}g\Delta t + v_e \ln\left(\frac{m_{initial}}{m_{final}}\right) \tag{15}$$

The increment of time during which propellant is expelled is represented by Δt, and accordingly the two mass terms reflect the *initial* (higher) and *final* (lower) masses of the rocket (including its stored propellant) over this time interval. With the exception of the modification coefficients, this equation is identical to that of a normal rocket ascent in a gravitational field.

Table 3 shows how the different possible interpretations of the hypothetical gravity shield (as outlined in Fig. 3) might affect this situation. If it were assumed that the gravitational field, g, is modified, the result would be as naively expected; it would be the same as launching in a different gravitational environment. If, however, it is assumed that the device affects the mass of the rocket, there are further possibilities. If the Equivalence Principle is in effect, then both the gravitational and inertial mass are equally affected, resulting in no change in the rocket's Δv. If the Equivalence Principle is not in effect, then only the gravitational or inertial masses are affected, resulting in an analogous case to launching in a different gravitational environment.

As before, the energy implications of these conjectures are not addressed in such approaches. These examples illustrate how misleading results become possible if only rocket equations are used to ponder space drive physics. To proceed with a more fundamental basis of comparison, *energy* considerations are examined next.

Table 3 Different analysis results for gravitationally shielded launchpad

Modified term From Eq. (15)	Gravitational field δ_{gf}	Gravitational mass δ_{gm}	Inertial mass δ_{im}	Net effect
Unmodified launch	1	1	1	Baseline: $g' = g$
Gravity modified	δ	1	1	$g' = \delta g$
Rocket inertial and gravitational mass (Equivalence Principle in effect)	1	δ	δ	No change: $g' = g$
Rocket gravitational mass (Equivalence Principle negated)	1	δ	1	$g' = \delta g$
Rocket inertial mass (Equivalence Principle negated)	1	1	δ	$g' = g/\delta$

B. Energy as a Basis of Comparison

Although comparisons built on the incumbent methods might be useful for introductory purposes, a deeper understanding of the benefits and issues are better illustrated by using a more fundamental basis. When considering moving a mass from one place to another, energy is the fundamental currency. Using this basis, three situations will be compared; deep space travel, Earth to orbit, and levitation. These comparisons are presented for three reasons: 1) to continue to illustrate the difference between space drive considerations and the use of rocket equations, 2) to reflect the magnitude of potential benefits compared to rockets, and 3) to provide starting points for deeper inquiry.

1. Deep Space Travel Energy

To compare the energy requirements of a rocket and a hypothetical space drive, the following assumptions are used. To more fully understand the challenges, approaches, and potential benefits of breakthrough propulsion, it would be fruitful to repeat the analysis using different assumptions.

- The space drive is interpreted to be simply a device that converts potential energy (regardless of the source of this energy) into kinetic energy.
- Both the rocket and the space drive are assumed to be 100% efficient with their energy conversions. Absent of any real mechanism, this at least compares the upper performance limits.
- The thrusting duration is assumed to be much shorter than the trip duration, which for interstellar travel is reasonable.
- For the rocket, constant exhaust velocity is assumed.
- Nonrelativistic trip velocity and exhaust velocity are assumed.
- The energy requirements for a rendezvous mission are based on equal Δv's for acceleration and deceleration.

To compare a rocket to another method that does not require propellant, we need an equation for rocket energy where the propellant mass is represented in terms of the vehicle's empty mass and the Δv of the mission—variables shared by the space drive. A common way to calculate the total kinetic energy of a rocket system, including both the rocket and the propellant, is just to calculate the kinetic energy imparted to the propellant from the rocket's frame of reference where the rocket has zero velocity (hence a zero contribution to the total kinetic energy) [32,33]. This is consistent with the previously stated assumptions.

$$E = \frac{1}{2} m_p (v_e)^2 \qquad (16)$$

Next, to convert this into a form where the rocket's propellant mass, m_p, is represented in terms of the exhaust velocity and the mission Δv, we apply the following form of the Rocket Equation, which is an algebraic equivalent of the Tsiolkovski equation:

$$m\left(e^{\left(\frac{\Delta v}{v_e}\right)} - 1\right) = m_p \qquad (17)$$

Substitution of this form of the Rocket Equation into the Kinetic Energy Equation yields this approximation:

$$E = \frac{1}{2}(v_e)^2 \, m\left(e^{\left(\frac{\Delta v}{v_e}\right)} - 1\right) \qquad (18)$$

Before proceeding, a limitation must be communicated. For these introductory exercises, the comparisons are limited to nonrelativistic regimes. For rockets, this implies limiting the exhaust velocity to $\leq 10\%$ lightspeed, where γ is less than about 1%. (γ is the factor by which length and time are affected and is a function of the speed between observers.) The corresponding upper limit to specific impulse follows easily from the equation relating specific impulse to exhaust velocity (where g is the Earth's gravitational acceleration) [32]:

$$v_e = I_{sp} g \qquad (19)$$

Setting the exhaust velocity to 10% of lightspeed (beyond which relativistic effects must be considered), the limiting specific impulse is found to be:

$$(10\%)\left(3.0 \times 10^8 \frac{m}{s}\right) \geq I_{sp}\left(9.8 \frac{m}{s^2}\right) \Rightarrow I_{sp} \leq 3.0 \times 10^6 s \qquad (20)$$

Because a space drive has been defined for this exercise as a device that converts potential energy into kinetic energy, the basic equation of kinetic energy is used to calculate the required energy, where the values of vehicle mass and mission Δv are the same as with the rocket. For these first-step exercises, the source of the stored potential energy need not be specified. The first issue to deal with is the magnitude of energy.

$$E = \frac{1}{2}m(\Delta v)^2 \tag{21}$$

Two things are important to note regarding the energy differences between a rocket and a hypothetical space drive. First, the energy for a rocket is an *exponential* function of Δv, whereas the ideal energy of a space drive is a squared function of Δv. This by itself is significant, but it is important to point out that a rocket and a space drive treat additional maneuvers differently.

For a rocket it is conventional to talk in terms of increases to Δv for additional maneuvers. For example, a rendezvous mission has twice the Δv (accelerate and decelerate) than just a flyby (accelerate). For space drives, however, the additional maneuvers are in terms of additional kinetic energy. To illustrate this difference, consider a mission consisting of multiple maneuvers, n, each having the same incremental change in velocity, Δv. Notice in Eqs. (22) and (23) the location of the term representing the number of repeated maneuvers, n. In the case of the space drive, additional maneuvers scale linearly, while for rockets they scale exponentially:

$$\text{Rocket maneuvers: } E = \frac{1}{2}(v_e)^2 m \left(e^{\left((n)\frac{\Delta v}{v_e}\right)} - 1 \right) \tag{22}$$

$$\text{Hypothetical space drive maneuvers: } E = (n)\frac{1}{2}m(\Delta v)^2 \tag{23}$$

To put this into perspective, consider a mission to send a 5000 kg probe over a distance of 5 light-years in a 50-year time frame. This range is representative of the distance to our nearest neighboring star (4.3 light-years) and the 50-year time frame is chosen because it is short enough to be within the threshold of a human career span, yet long enough to be treated with nonrelativistic equations. This equates to a required trip velocity of 10% light speed. The probe size of 5000 kg is roughly that of the Voyager probe plus the dry mass of the Centaur Upper Stage (4075 kg) that propelled it out of Earth's orbit [41]. The comparison is made for both a flyby mission and a rendezvous mission.

The results are listed in Table 4. The rocket case is calculated for two different specific impulses, one set at the upper nonrelativistic limit previously described, and another set at an actual maximum value achieved during electric propulsion lab tests [42].

Even in the case of the nonrelativistic upper limit to specific impulse— an incredibly high-performance hypothetical rocket—the space drive uses a *factor of 2 to 3 less energy*. When compared to attainable values of specific impulse—values that are still considerably higher than those currently used in spacecraft—the benefits of a space drive are enormous. For just a flyby mission, the gain is *72 orders of magnitude*. When considering a rendezvous mission, the gain is almost *150 orders of magnitude*. Again, though these results are intriguing, they should only be interpreted as the magnitude of

Table 4 Deep spaceflight energy comparisons (5000 kg, 5 Ly, 50 yr)

	Flyby ($n=1$) (Joules)	Compared to space drive Eq. (22)/Eq. (23)	Rendezvous ($n=2$) (Joules)	Compared to space drive Eq. (22)/Eq. (23)
Space drive, Eq. (23)	2.3×10^{18}		4.5×10^{18}	
Theoretical rocket, Eq. (22) $I_{sp} = 3{,}000{,}000$ s	3.8×10^{18}	1.7	1.5×10^{19}	3.3
Actual rocket, Eq. (22) $I_{sp} = 17{,}200$ s[a]	10^{91}	10^{72}	10^{168}	10^{149}

[a]From Ref. 42.

gains *sought* by breakthrough propulsion research. Other assessments and results are possible.

In the case of deep space transport, the energy was previously calculated assuming a constant exhaust velocity for the rocket and thrusting durations that were negligible compared to trip times. Although reasonable assumptions for interstellar flight, it would also be instructive to repeat the energy comparisons with assumptions of constant acceleration, constant thrust, constant power, and when optimized for minimum trip time. To further explore these notions, it would also be instructive to repeat all of these comparisons using the relativistic forms of the equations.

Newtonian equations are not the only way to further explore these notions. From the formalism of General Relativity, there are a variety of transportation concepts that do not require propellant, including: a *gravitational dipole toroid* (inducing an acceleration field from frame-dragging effects) [43], *warp drives* (moving a section of spacetime faster than light) [44], *wormholes* (spacetime shortcuts) [45,46], and *Krasnikov tubes* (creating a faster-than-light geodesic) [47].

To explore these General Relativity formalisms in the context of creating space drives requires the introduction of entirely different energy requirements than with the Newtonian versions explored in this paper. In the General Relativity approach, one must supply enough energy to manipulate all of the surrounding spacetime so that the spacecraft naturally falls in the desired direction. Although such approaches require considerably more energy than the simple Newtonian concepts, they are nonetheless instructive.

2. Earth to Orbit Energy

Consider next the case of lifting an object off the surface of the Earth and placing it into orbit. This requires energy expenditures both for the altitude change and for the speed difference between the Earth's surface and the orbital velocity. Again, the source of this energy is not considered. The point explored first is the amount of energy required. For the hypothetical space drive, this energy expenditure can be represented as:

$$E_{\text{Space drive}} = U + K \tag{24}$$

where U is the potential energy change associated with the altitude change, and K is the kinetic energy change associated with different speeds at the Earth's surface, r_E, and at orbit, r_O. The change in potential energy, which requires expending work to raise a mass in a gravitational field, is represented by:

$$U = \int_{r_E}^{r_O} G \frac{M_E}{r^2} m \, dr \qquad (25)$$

where G is Newton's constant (6.67×10^{-11} m$^3 \cdot$ kg$^{-1} \cdot$ s^{-2}), M_E is the mass of the Earth (5.98×10^{24} kg), r is the distance from the center of the Earth, and m is the mass of the vehicle.

The change in kinetic energy requires solving for the orbital velocity (function of orbital radius, r_O) and the velocity of the Earth's surface (Earth's circumference divided by one rotation period) and can be shown to take this form [36]:

$$K = \frac{1}{2} m \left[\left(G \frac{M_E}{r_O} \right) - \left(\frac{2\pi r_E}{24 \, h} \right)^2 \right] \qquad (26)$$

For the case of using a space drive to place the shuttle orbiter ($m = 9.76 \times 10^4$ kg) into a typical low Earth orbit, ($r_O = 400$ km), the energy required is found to be 3.18×10^{12} Joules.

To assess the required energy for a rocket to accomplish the same mission, the following equation is used [33]:

$$E = \left(\frac{1}{2} F I_{sp} g \right) t \qquad (27)$$

The parenthetical term is the rocket *power*, which is presented in this form for two reasons: to show this additional detail of the Rocket Equation and to introduce the idea of contemplating *power* in addition to just *energy*. While power implications are not further explored here, they constitute a fertile perspective for further study.

Entering the values for the Space Shuttle System, as presented in Table 5, into Eq. (27), the total energy for delivering the Shuttle orbiter via rockets is found to be 1.16×10^{13} Joules.

Comparing this rocket energy value to the hypothetical space drive energy, Eq. (28), where the efficiency of both systems is assumed to be 100%, indicates that the space drive is 3.65 times more energy efficient.

$$\text{Gain} = \frac{\text{Rocket}}{\text{Space drive}} = \frac{1.16 \times 10^{13}}{3.18 \times 10^{12}} = 3.65 \qquad (28)$$

3. Levitation Energy

Levitation is an excellent example to illustrate how contemplating breakthrough propulsion is different from rocketry. Rockets can hover, but not for

Table 5 Data and calculations of Space Shuttle energy to reach low Earth orbit[a]

	Space Shuttle main engines	Solid rocket boosters	Orbital maneuvering system
Number of engines	3	2	2
Thrust each, F	470×10^3 lbs	6 lbs	6×10^3 lbs
	$(2.1 \times 10^6$ N)	$(12.9 \times 10^6$ N)	$(271 \times 10^3$ N)
Specific impulse, I_{sp}	453 s	266 s	313 s
Total power	1.40×10^{10} W	3.36×10^{10} W	8.31×10^8 W
Burn duration, t	514 s	126 s	200 s
Total energy used, E	7.19×10^{12} J	4.24×10^{12} J	1.66×10^{11} J
Total Combined Energy = 1.16×10^{13} Joules			

[a]Values taken from STS-3 Thirds Space Shuttle Mission Press Kit, March 1982, Release #82-29.

very long before they run out of propellant. For an ideal breakthrough, some form of *indefinite* levitation is desirable, but there is no preferred way to represent the energy or power to perform this feat. Because physics defines work (energy) as the product of *force* acting over *distance*, no work is performed if there is no change in altitude. Levitation means hovering with no change in altitude. This zero-energy expenditure is also obtained in the case of indefinitely levitating a permanent magnet over an arrangement of other permanent magnets that are, themselves, supported by the Earth.

Regardless, there are a variety of ways to toy with the notion of indefinite levitation that look beyond this too-good-to-be-true zero-energy requirement. In addition to the potential energy approach to be examined next, here are a variety of other ways to contemplate levitation energy:

1) *Helicopter analogy:* Calculate the energy and power required to sustain a downward flow of reaction mass to keep the vehicle at a fixed altitude.

2) *Normal accelerated motion:* Rather than assess levitation energy directly (where the mass sustains zero velocity in an accelerated frame), calculate the energy or power required to continuously accelerate a mass at 1 g in an inertial frame. In this case, the normal Force × Distance formula can be used, but issues with selecting the integration limits arise.

3) *Escape velocity:* Calculate the kinetic energy for an object that has achieved escape velocity. (This approach actually results in the same value as when calculating the absolute potential energy.)

4) *Thermodynamic:* Treat levitation analogously to keeping a system in a nonequilibrium state, where equilibrium is defined as free-fall motion in a gravitational field and where the stable nonequilibrium condition is defined as levitation at a given height.

5) *Damped oscillation:* Calculating the energy of oscillation about a median hovering height, but where an energy cost is incurred for both the upward and downward excursions, where damping losses are included.

6) *Impulse:* Rather than use the Force × Distance formula, use the "impulse" treatment of Force × Duration. Like the accelerated motion approach, this introduces issues with integration limits.

For now, only one approach is illustrated, specifically the nullification of gravitational potential. Usually Eq. (29) is used to compare gravitational potential energy differences between two relatively short differences in height (the integration limits), but in our situation we are considering this energy in the more absolute sense. This equation can be applied to calculate how much energy it would take to completely remove the object from the gravitational field, as if moving it to infinity. This is more analogous to *nullifying* the effect of gravitational energy. This is also the same amount of energy that is required to stop an object at the levitation height, r, if it were falling in from infinity with an initial velocity of zero (switching signs and the order of integration limits).

$$U = \int_{r_E}^{\infty} G \frac{M_E}{r^2} m \, dr = G \frac{M_E}{r_E} m \quad (29)$$

where G is Newton's Constant (6.67×10^{-11} m$^3 \cdot$ kg$^{-1} \cdot$ s^{-2}), M_E is the mass of the Earth (5.98×10^{24} kg), r is the distance from the center of the Earth, and m is the mass of the vehicle.

Using this potential energy equation, it could conceivably require 62 megaJoules to levitate 1 kg near the Earth's surface. As an aside, this is roughly twice as much as putting 1 kg into low Earth orbit. Again, these assessments are strictly for illustrative purposes rather than suggesting that such breakthroughs are achievable or that they would even take this form if achievable. Some starting point for comparisons is needed, and this is just one version.

To avoid confusion, the potential energy calculated here refers to the energy due to the gravitational field that we want to counteract. How the vehicle achieves this feat, or where it gets the energy to induce this effect, is not specified. At this stage, estimating the energy requirements is a necessary prerequisite.

V. Hypothetical Mechanisms

With conservation of momentum and energy examined, the next step is to consider hypothetical space drive mechanisms. These mechanisms are presented to help identify the key issues needing resolution rather than to imply that any of these are viable propulsion devices. To interpret these concepts at the level that they are intended, consider this section as a brainstorming exercise. By this it is meant that the concepts are intended to show that there are many different ways to approach the challenge of creating a space drive and to provoke deeper inquiry into the many unresolved science questions encountered while considering them.

Ten different hypothetical space drive concepts are offered for this exercise. Table 6 lists these compared with a cursory assessment of their applicability for interacting with the known indigenous space phenomena (taken from Table 1). Note that rough order of magnitude estimates are provided for the mass densities of the phenomena. At this early stage it is sufficient to consider only rough orders of magnitude. Also, many of the phenomena are not known to a high degree of precision. Within Table 6, where a particular propulsion concept intersects with a particular phenomenon, a cursory indication of the applicability or viability is offered. In those cases where a particular option has

Table 6   Hypothetical space drives and candidate phenomena

Candidate phenomena Mass density (kg/m³)[a] Hypothetical concept		Galactic hydrogen 10^{-21}	Dark Energy 10^{-26}	Dark Matter 10^{-27}	Cosmic microwave background 10^{-31}	Quantum vacuum $10^{36\pm62}$	Spacetime $10^{-1\pm26}$	Fields due to charges	Fields due to masses
Interstellar jet	Fig. 4	D[b]	D	D	D	TBD	TBD	N/A	N/A
Differential sail	Fig. 5a	D	TBD	TBD	D	TBD	TBD	N/A	N/A
Induction sail	Fig. 5b	D	TBD	TBD	D	TBD	TBD	N/A	N/A
Diode sail	Fig. 5c	D	TBD	TBD	D	TBD	TBD	N/A	N/A
Inertia modified rocket	Fig. 1	N/A	N/A	N/A	N/A	N/A	Relevant	N/A	Relevant
Oscillating inertia thruster	Fig. 7	N/A	N/A	N/A	N/A	N/A	Relevant[c]	N/A	Relevant
Diametric drive	Fig. 9	N/A	TBD	TBD	N/A	N/A	TBD	D	Relevant[d]
Disjunction drive	Fig. 10	N/A	TBD	TBD	N/A	N/A	N/A	TBD	TBD
Gradient potential drive	Fig. 11	N/A	TBD	TBD	N/A	TBD	TBD	TBD	TBD
Bias drive	Fig. 12	N/A	TBD	TBD	N/A	TBD	TBD	TBD	TBD

D = "Doubtful." The combination of this propulsion concept and phenomenon is cursorily assessed in this chapter and found to be of doubtful viability.
TBD = "To Be Determined." The combination of this propulsion concept and phenomenon has not been sufficiently studied to reliably determine viability.
N/A = "Not Applicable." No obvious way to apply this phenomena to the hypothetical device is recognized.

[a]From Table 1.
[b]From Refs. 48–50.
[c]See Refs. 35, 51, 52.
[d]See Refs. 53–56.

been studied, a reference is cited. The abbreviations used in the table are explained below the table. Note that many possibilities remain unexplored. The analyses of each of the 10 hypothetical mechanisms follow.

A. Hypothetical Interstellar Jet Propulsion

As a first example, consider the well-known Bussard interstellar ramjet [48–50] illustrated in Fig. 4. As the vehicle moves forward, it scoops up the indigenous interstellar hydrogen and channels it into a chamber where the hydrogen nuclei undergo fusion (while in motion) and then release the resulting energy and helium nuclei out the rear of the spacecraft. In addition, some means is required to initially set the spacecraft in motion, such as the method shown in Fig. 4, where a laser sail is used for both the initial push and to add energy to the spacecraft (via the concentrator). Even though there are profound engineering difficulties with this concept [49,50,57], it serves as an example of the idea for using indigenous matter in space as a reaction mass.

In more general terms, consider being able to interact with indigenous matter or energy in space analogous to the way a jet engine captures and accelerates air particles to produce thrust. This would require a sufficient amount of matter to provide a meaningful reaction mass, and require a means to impart forces on that matter.

Considering that tangible matter in space is scarce, a large volume of space would have to be swept to obtain sufficient reaction mass to be practical. To give some sense of the scale of this challenge, Table 7 shows the volume of space that might contain a metric ton (MT) or 1000 kg of matter for the various candidates of indigenous reaction mass. For completeness, equivalent

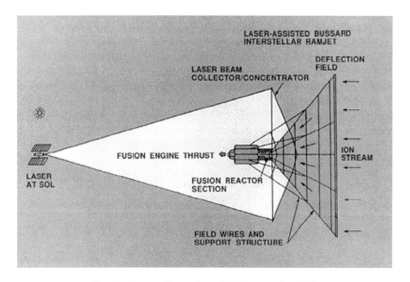

Fig. 4 Bussard ramjet with laser assist [57].

Table 7 Equivalent volumes to contain a metric ton (MT) of matter

Candidate phenomenon	Galactic hydrogen	Dark energy	Dark matter	Cosmic microwave background	Quantum vacuum	Spacetime
Equivalent mass density (kg/m^3)[a]	10^{-21}	10^{-26}	10^{-27}	10^{-31}	$10^{36 \pm 62}$	$10^{-1 \pm 26}$
Volume required to encompass 1 MT (m^3)	10^{24}	10^{29}	10^{30}	10^{33}	$10^{-33 \pm 62}$	$10^{4 \pm 26}$

[a]See Table 1.

mass estimates for the indigenous forms of *energy*, using the familiar $E = mc^2$ relation, are offered.

At this point most of these are impractical even before moving on to the issue of how to induce forces against this matter. Again, as identified in Section III.B, the majority of possibilities are with the unknowns of the quantum vacuum and the nature of spacetime itself. Presently there are no known methods or proposed concepts that apply the analogy of a "jet" to any phenomenon other than the Bussard interstellar ramjet.

B. Hypothetical Sail Drives

The next theme from which to envision space drives is by entertaining the possibility of collisions with some form of indigenous *energy* in space. To put this into a more familiar context, recall the basic principle of a solar sail, where energy (photons) from the Sun impinges on a sail, pushing it away from the Sun. In the context of space drives, energy sources, like the cosmic microwave background radiation, *surround* our spacecraft. To obtain a *net* thrust, more momentum must be imparted to the rear of the spacecraft than to the front.

As an aside, the possible drag force from the cosmic microwave background radiation on a laser-driven light sail was recently assessed in terms of a "terminal velocity." The terminal velocity is achieved when the drag force becomes large enough to cancel the accelerating force from the laser. From this analysis, the terminal velocity was calculated to be 0.99997 of the speed of light [58].

While momentum transfer from photons is familiar, the more general notion of energy density imparting pressure is considered here. Again it is emphasized that these are hypothetical examples to illustrate possibilities and issues, rather than to assert that these are viable mechanisms. In that context, all of the indigenous media listed in Table 1 are considered in terms of an energy density. Some are already in that form, while the mass densities are converted using the familiar $E = mc^2$ relation, as shown in Table 8. From there, a rudimentary estimate of the "pressure" follows directly from the energy density since the units of energy density (J/m^3) are identical to pressure (N/m^2).

To illustrate possibilities for how to induce net forces from such surrounding energy, three variations of hypothetical space drive sails are illustrated in

Table 8 Equivalent "pressure" from indigenous energy densities

Candidate phenomenon	Galactic hydrogen	Dark energy	Dark matter	Cosmic microwave background	Quantum vacuum	Spacetime
(N/m^2) Equivalent "pressure" from energy density[a]	10^{-9}	10^{-9}	10^{-10}	10^{-15}	$10^{52 \pm 61}$	$10^{16 \pm 26}$
(psi) (kPa) (psi and kPa are approximately the same order of magnitude)	10^{-5}	10^{-5}	10^{-6}	10^{-11}	$10^{56 \pm 61}$	$10^{20 \pm 26}$

[a]See Table 1.

Fig. 5a–5c. In these illustrations, the rectangular box represents a cross-sectional element of "sail" and the small arrows represent the impinging energy or photons. The large arrows indicate the direction of acceleration. For introductory purposes, simple force equations are provided within Fig. 5, where F represents the net force, P represents the pressure of the medium acting on the device (Table 8 values), and A represents the area of the device. The first terms on the right of each equation are the *rear* surface pressures; the second terms are the *forward* surface pressures. Their difference yields the net force on the sail, whose equation is shown underneath. In the case of the *induction sail* (Fig. 5b), the additional coefficient, δ, represents the percentage by which the energy impinging on the sail has been locally altered, reciprocally across the front and back of the sail. A δ greater than one implies an increase, less than one a decrease, and equal to one implies no change.

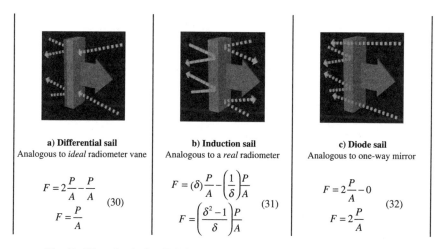

a) Differential sail
Analogous to *ideal* radiometer vane

$$F = 2\frac{P}{A} - \frac{P}{A}$$
$$F = \frac{P}{A}$$ (30)

b) Induction sail
Analogous to a *real* radiometer

$$F = (\delta)\frac{P}{A} - \left(\frac{1}{\delta}\right)\frac{P}{A}$$
$$F = \left(\frac{\delta^2 - 1}{\delta}\right)\frac{P}{A}$$ (31)

c) Diode sail
Analogous to one-way mirror

$$F = 2\frac{P}{A} - 0$$
$$F = 2\frac{P}{A}$$ (32)

Fig. 5 Hypothetical sail drives. (Graphics courtesy of John MacNeil.)

1. Differential Sail

The first version, the *differential sail* (Fig. 5a), assumes that the photons impinging on the rear are reflected (two units of momenta) while the ones on the front are absorbed (imparting one unit of momentum), resulting in a net photon pressure in the forward direction. This concept is analogous to an *ideal* radiometer. Figure 6 shows an actual radiometer. In both the ideal and real situations, the photons striking the black side of the paddles are absorbed, while they are reflected from the white side. The distinction between the ideal and real radiometers depends on whether air or vacuum surrounds these paddles. In the ideal case, as with our differential sail, the paddles are in true vacuum so the forces are only due to photon momenta. This would move the paddles in the direction pointing from white to black, as reflection (white) imparts twice the momentum than absorption (black).

In a real radiometer, some residual air is still present in the bulb. In this case, the black side of the paddle absorbs more energy than the white, and then heats the residual air more than on the white side. As the air molecules collide more energetically on the black side than the white, the paddles are pushed by the air in the direction from black to white. This is exactly opposite to the direction than with the ideal radiometer.

Fig. 6 Crookes radiometer. (Courtesy of Timeline.)

Before proceeding to the next concept (which is similar to a *real* radiometer), a critical limitation of this idea needs to be explained. Although the effects described are entirely possible, they cease once the sail has reached thermal equilibrium with the background. In short, the radiation absorbed will be re-emitted, making the absorbing side, in effect, a delayed reflector. Once in thermal equilibrium, both the front and the back undergo the same momentum transfer. A more detailed explanation of this limitation is presented in Chapter 12, which deals with using vacuum fluctuations as a propulsive medium.

2. Induction Sail

The operation of a real radiometer is analogous to the next hypothetical sail drive concept, the induction sail. This concept assumes that the surrounding medium has an energy density or pressure that can be locally altered in much the same way that the real radiometer's paddles affect the air temperature in the vicinity of the paddles, raising the air temperature near the black side more so than the white side. In Fig. 5b, this difference of the medium's energy density is graphically represented by the line density of the impinging rays. In the hypothetical induction sail, the pressure of the medium has been altered by a factor, δ, behind the sail and by a reciprocal amount in front of the sail, resulting in a net pressure difference across the sail of $(\delta^2 - 1)/\delta$. Note that in the case of $\delta = 1$, which implies no change to the medium, the net force is zero, as expected.

This concept does not suffer the same equilibrium limitation as the differential sail, since this induction sail requires that an energy flow be maintained. The sail is not allowed to reach thermal equilibrium. The kinetic energy acquired by the sail will be from the energy expended to maintain this nonequilibrium state, minus conversion losses.

3. Diode Sail

The next hypothetical sail-drive, the diode sail (Fig. 5c), does not have the problems of removing any absorbed energy, at least in the ideal sense. It uses an analogy to a diode, or one-way mirror, for its operation. The energy impinging on the back is perfectly reflected, while the energy impinging on the front passes through as if the sail were perfectly transparent. This concept is useful to introduce another level of detail, specifically non-ideal conditions. The diode sail equations shown in Fig. 5c represent perfect conditions. Consider, instead, if reflection, absorption, or transparency were not perfect and coefficients had to be inserted into the equation to account for this. In this case the equation would take the form:

$$F = 2(k_R)\frac{P}{A} - (1 - k_T)\frac{P}{A} \qquad (33)$$

where k_R represents a reflectivity coefficient and k_T represents a transparency coefficient. Consider, for example, if both coefficients were only 50%, meaning that only half of the energy impinging the back were reflected, only half from the front passed through, and the rest of the impinging energy was absorbed. The performance would only be a quarter of the ideal situation. At this point, the thermal equilibrium issues raised with the differential sail apply equally to this concept too.

Absent of real sail devices, preliminary upper performance limits can be estimated from the energy density of the various candidate indigenous energy sources, as listed in Table 8. Consider first the performance in the context of using the cosmic microwave background radiation. The energy density of this medium equates to a pressure of only 10^{-15} N/m^2. To put this in context, to produce the same thrust as an ion engine, specifically around 100 mN (the approximate maximum thrust of the NSTAR ion engine used for NASA's Deep Space 1), a perfect differential sail would need a surface area equivalent to a square having sides of 10 million kilometers.

At the other extreme, consider interacting with the quantum vacuum energy. Depending on the upper cutoff frequency used to calculate the energy of this medium, greatly different values result. If the upper cutoff frequency were the Compton frequency of a nucleon (10^{23} Hz), a sail of 1 square meter could produce 10^{35} Newtons. As intriguing as this seems, it is important to stress that these initial assessments are not definitive. While they help bracket the possibilities, the genuine limits of actual interactions remain to be discovered. One example of a more thorough investigation into the possibilities of thrusting against the quantum vacuum energy is given in see Chapter 12, which shows that net thrust is possible in principle but likely to produce even less thrust per power than a photon rocket.

As a closing note to this subsection on sail drive concepts: In 1991 a patent was granted for a "spaceship to harness radiations in interstellar flights," [59] which summarized its operation thus:

> The resultant force of radiations being absorbed from the rear and reflected from the front propels the spacecraft forward.

Per the preceding discussions, the thrust direction is opposite to that expected from its own description unless the device operates analogously to the induction sail. But given the propeller shown at the front of the cylindrical spacecraft in the figures of the patent, it is doubtful any rigorous analysis was conducted to substantiate the claims.

C. Hypothetical Inertial Modifications

1. Inertia Modified Rocket

In Section IV, the notion of modifying the inertia of a rocket or its propellant was raised as a space drive concept (Fig. 1) and can be included again here, in the interest of completeness, as an example of hypothetical mechanisms. As already shown, such notions evoke questions of energy conservation and the deeper implications of the Equivalence Principle. In particular, the question of the energy required to produce the desired change in inertia is both an issue and an approach to deeper inquiries. (For a refresher of the possibilities, refer to Tables 2 and 3.)

For the interested student, a logical next step is to calculate the energy and power required to sustain such a hypothetical inertial modification as the rocket thrusts.

2. Oscillatory Inertia Thruster

A completely different approach that entertains the implications of inertial modification exists in Patent No. 5,280,864 [51], which is based on the transient inertia observations reported by Woodward [35,52]. Chapter 11 examines an experiment of Woodward in more detail. For this chapter, this concept is only examined at an introductory level to illustrate the implications.

It should be emphasized that this concept is not in the category of the "non-viable mechanical 'antigravity' devices" described in Chapter 6. Whereas those concepts involve the oscillatory motion of masses, the focal distinction of the Woodward approach is that the *inertia* of its component masses change, not just their position and velocity. This key difference casts it in an entirely different category.

In the Woodward concept, a device cyclically changes the distance between two masses, while the inertia of each mass is oscillating about its nominal mean, so that the system as a whole shifts its position to the right. One full cycle is shown in Fig. 7, starting with the fully extended position and with both masses at their nominal and equal value. (The operation is described in the caption.)

A critical question behind this concept is, again, conservation of momentum. Here the role played by the inertial frame is nontrivial. Because the inertia of the masses vary, the role played by the inertial frame, in defining the position of the center of mass, is no longer clear. For example, this concept evokes the question: "Is inertia an *intrinsic* property of *only* matter, or does it measure a *relationship* between matter and spacetime?" As such, this concept is an excellent venue through which to revisit deeper meanings of Mach's principle.

The inventor of this concept evokes a fairly literal interpretation of Mach's principle, specifically that an inertial frame is created by, and connected to, the surrounding matter in the Universe. Using this perspective, the reaction forces are imparted to the surrounding mass of the Universe to conserve momentum. This evokes the even deeper question: If the Universe was pushed a little to the left, while the device moved to the right, what would be the background inertial frame for the motion of this Universe/thruster interaction? Again, pursuing such concepts introduces deep, provocative questions of basic physics. Even if the Woodward devices or theories are later dismissed, this approach serves as the foundation for thought experiments about the connection between inertial matter and spacetime, the total matter in the Universe, and characteristics of matter and spacetime that have not yet been discovered.

While the principles of this device are unfamiliar (especially the notion of variations in inertia and the relation this has to a Machian inertial frame), there is a somewhat analogous case that can at least illustrate the notion of using a field for propellantless propulsion. Specifically, consider the concept shown in Fig. 8 where a satellite uses tethers with masses at their tips for orbital maneuvering [60]. By taking advantage of the nonlinear nature of a gravitational well, an orbiting satellite can modify its orbit without expenditure of propellant. If the satellite extends a mass at the end of a tether toward Earth and another such mass away from Earth, the imbalanced reactions will create a net force toward the Earth. This is because the downward force on the near-Earth tether increases

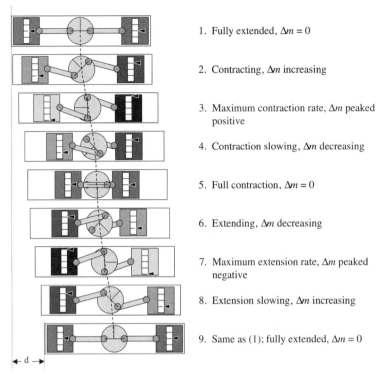

Fig. 7 Inertia oscillation thruster. The first part of the cycle, (1) starts with the fully extended position and with both masses at their nominal and equal value. Note the bar graphs on each mass, whose arrow pointers indicate their current inertia (in mid-position at stage 1). The Δm term refers to the *difference* between the forward (right-side) mass and the rearward (left-side) mass. At the start of the cycle, the masses are equal, so Δm is zero. As the device contracts through stages 1–5, the forward inertial mass increases, peaking to a maximum value at stage 3 of the cycle, and returning to its nominal value as the contraction is completed at stage 5. During this time, the mass at the front is greater than the mass at the rear, so the center position shifts forward. When the device begins to expand, the rearward mass increases, while the forward mass decreases, again returning to their nominal, equal values by the time the extension completes. During the extension, the mass at the rear was higher, and the front lower, shifting the center of the device again to the right.

more than the outward force on the outer tether as the tethers are deployed. By alternately deploying and retracting long tethers at different points during the orbit (apogee and perigee), an orbiting satellite can change its orbital altitude or eccentricity. The result given is a purely classical result, which depends only on the $1/r^2$ force dependence of Newtonian gravity. It is an example of "propellantless" propulsion in that it allows orbital motion to be changed without expenditure of reaction mass, yet it still satisfies Newtonian conservation of

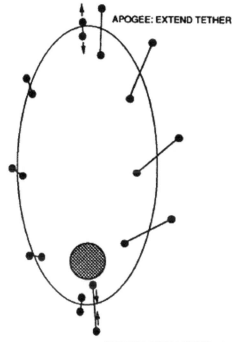

Fig. 8 Taking advantage of nonlinearities: Tether orbital maneuvering [55].

momentum. Here the Earth is the reaction mass and its gravitational field is how it is connected to the spacecraft. A somewhat different form of propulsion against a gravitational field can be derived from General Relativity [61].

D. Hypothetical Field Drives

Gravitational fields accelerate masses and electric fields accelerate charges. To entertain the analogy of using Newtonian *field* interactions for a space drive, it is necessary to assume that there is some means for a vehicle to induce a field around itself that will, in turn, accelerate itself. At first glance, the expectation is that the induced forces would just act between the vehicle's field-inducing device and the rest of the vehicle, like trying to move a car by pushing on it from the inside. In such cases all the forces act *internally* and there would be no net motion of the vehicle. For reference, this issue can be called the "net external force requirement." In addition, field drives will also have to satisfy conservation of momentum, which also appears difficult to satisfy at first glance.

There is more than one way to explore possibilities of field drives to address these requirements. Four "what-if" examples follow: the diametric drive (Fig. 9), disjunction drive (Fig. 10), gradient potential drive (Fig. 11), and the bias drive (Fig. 12). Each of these offers a different approach to convey the issues faced when pondering space drives that are based on Newtonian fields. For completeness, propulsion concepts that are based on Einstein's geometric General Relativity are covered in Chapters 4 and 15.

1. Diametric Drive

The first hypothetical field drive is based on the existing concept of negative mass propulsion [53–56], the basic premise of which is extended here to more general terms. Chapter 4 also examines this concept and provides a basic diagram (see Chapter 4, Fig. 2).

Negative mass is a hypothetical entity where it is assumed that all the characteristics of mass behave with opposite sign. The juxtaposition of negative and positive mass leads to both masses accelerating in the same direction. Because of their mutual characteristics, a repulsive force is created between the positive and negative masses, but because the negative mass also has negative inertia, it accelerates in the direction *opposing* the force. It has been shown that such a scheme does not violate conservation of momentum or energy [54], and is discussed further by Landis [55]. The crucial assumption to the success of this concept is that negative mass has negative *inertia.*

Whereas Chapter 4 discusses the original concept of negative mass propulsion, more general terms are used here to set the stage for covering additional possibilities and issues. Rather than just describing this concept as the interaction of the two *point* masses, consider this concept in terms of a scalar potential, as illustrated in Fig. 9, where only the *x* and *y* axes are plotted to allow the magnitude of the scalar potential to graphically project into the third visual dimension.

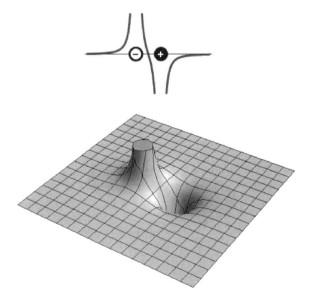

Fig. 9 Hypothetical diametric drive. *Negative mass propulsion* is shown here in a more general form of a scalar potential created by positive and negative point sources. (Singularities truncated for clarity.)

This visual analogy and the tools of vector analysis and scalar potentials make it easier to consider various phenomena of spacetime and matter that exists within spacetime. In the context of negative mass propulsion, Fig. 9 would represent a *gravitational* scalar potential. It could equally represent an electric potential or the density of any of the space phenomena previously discussed. The remaining space drives presented here will all be examined in the context of scalar potentials. It should be noted that scalar potentials are a Newtonian perspective more so than a General Relativistic approach.

To proceed with this broader analogy, it is necessary to more explicitly distinguish factors that govern how the presence of sources affect the scalar potential, and then how the scalar potential affects the matter. In the familiar cases of mass and charge, the cause of a field is the same thing that reacts to a field. Masses create gravitational fields that affect masses and charges create electric fields that affect charges. To delve deeper, consider the analysis by Bondi in 1957 in which the concept of negative mass propulsion was articulated [53]. To explore issues of the Equivalence Principle and others, Bondi made distinctions between the various properties of mass and then assessed the consequences of changes to the signs of each characteristic. He found that all the characteristics must have the same sign to avoid logical inconsistencies, but that either a positive or negative sign could be assigned. Table 9 summarizes these various characteristics, the defining equations, and the signs that define both positive and negative mass.

First, *inertial mass* is the characteristic that sets the defining relationship between force and acceleration. The *active* gravitational mass is the one whose mass *creates* the gravitational field. The last characteristic, the *passive* gravitational mass, is the one that *reacts* to the presence of a gravitational field.

When considering other natural phenomena for their utility to space drives, it is helpful to entertain such distinctions. For example, consider the applicability of electrical fields in the context of this diametric drive. Although opposing charges do exist, placing them next to each other will not result in a unidirectional acceleration. The distinction that renders the notion of the diametric drive incompatible for electric fields is that all charges, regardless if they are positive or negative, have *positive* inertia. The magnitude of a charge's *inertia* is not directly tied to the magnitude of its electrical charge.

Table 9 Dissecting the characteristics of mass

Characteristic	Representative equation	Negative mass	Normal mass
"Inertial mass," m_I (Defines the relation between force and acceleration)	$m_I = \dfrac{F}{a}$	$m_I < 0$	$m_I > 0$
"Active gravitational mass," m_A (Creates a gravitational field)	$m_A = \Phi \dfrac{r}{G}$	$m_A < 0$	$m_A > 0$
"Passive gravitational mass," m_P (Responds to the presence of a gravitational field)	$m_P = \dfrac{F}{-\nabla \Phi}$	$m_P < 0$	$m_P > 0$

F = force, a = acceleration, r = distance from the mass, Φ = gravitational scalar potential, and G = Newton's constant.

To further pursue the general notion of a diametric drive in the context of other phenomena such as Dark Matter, Dark Energy, or spacetime itself, the various properties of these phenomena and their relation with spacetime would need to be carefully distinguished and subsequently analyzed. No such definitions or analyses yet exist to this author's knowledge.

To provide another example of separating the features of mass and then examining the consequences, the disjunction drive is offered next.

2. Disjunction Drive

Another spin-off from the idea of negative mass propulsion is the disjunction drive. The disjunction drive entertains the possibility that the source of a gravitational field, or the "*active* gravitational mass" as Bondi called it, m_A, and that which reacts to a field, which Bondi called the "*passive* gravitational mass," m_P, can be physically separated. By displacing them across space, the passive mass is shifted to a point where the field has a slope, thus producing reaction forces on the passive mass.

Because we have physically separated the active and passive gravitational masses, we must further distinguish separate inertial characteristics for each to proceed to examine the consequences, where m_{IP} is the inertial mass of the Passive mass, and m_{IA} is the inertial mass of the active mass.

Equation (34) and Fig. 10 illustrate the gravitational scalar potential that results from physically separating the active and passive masses a distance, d, apart. By displacing them in space, the passive mass is shifted to a point where

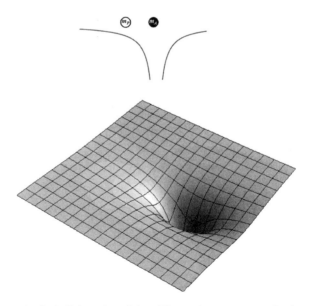

Fig. 10 Hypothetical disjunction drive. The active mass, m_A, is the only source affecting this gravitational potential, whereas passive mass, m_P, is the only thing affected by the gravitational potential. (Singularity truncated for clarity.)

the field has a slope, thus producing reaction forces between the source and the reactant.

$$\Phi = \frac{-Gm_A}{\sqrt{(x-d)^2 + y^2 + z^2}} \tag{34}$$

where Φ is the gravitational scalar potential, G is Newton's constant, m_A is the active mass that creates the gravitational field, d is distance separating the location of the active and passive masses, and x, y, and z, are the usual rectilinear coordinates. This scalar potential is graphically represented in Fig. 10, where only the x and y axes are plotted to allow the potential, Φ, to graphically project into the third visual dimension.

The force, F, on the passive mass, as defined from the gradient of the scalar potential from the active mass, is found to be:

$$F = m_{IP} \frac{Gm_A}{d^2} \tag{35}$$

where, in this case, a positive sign is assigned to denote the orientation of the x-axis pointing in the direction from the passive mass toward the active mass. Also, because the Gm_A/d^2 term represents the gravitational *acceleration* due to the active mass, then the type of mass shown for the passive mass *must* be its inertial mass, m_{IP}, to satisfy the $F = ma$ form of this equation.

Next, imposing Newton's action and reaction requires that this same force is acting on the active mass, but in the opposite direction, and where again the inertial mass property of the active mass, m_{IA}, is the proper form to use in this example, and where a_A is the acceleration of the active mass corresponding to that force:

$$F = -m_{IA} a_A \tag{36}$$

In this case, the negative sign denotes that its direction is opposite to the passive mass's acceleration. By imposing these conditions, momentum conservation is satisfied. The resulting motion, however, will just bring the masses back together because both are accelerating toward each other.

If, however, the distance separating them were fixed by using some hypothetical rigid device of length, d, then the acceleration of both masses would have to be the same. Since the passive mass does not present any gravitational force on the active mass (and the active mass is defined as not having characteristics to respond to a gravitational field), then both masses would accelerate due to the gravitational force of the active mass on the passive mass only, yet tempered by their inertia. In this case, the acceleration of the combined masses ($m_{IP} + m_{IA}$), and taking into account their inertia, would result in the following net acceleration, a, of the system:

$$a(m_{IP} + m_{IA}) = \frac{Gm_A}{d^2} m_{IP} - \frac{Gm_A}{d^2} m_{IA}$$

$$a = \frac{Gm_A}{d^2} \frac{(m_{IP} - m_{IA})}{(m_{IP} + m_{IA})} \tag{37}$$

Note that if the magnitudes of both the active and passive inertial masses have the same value, then there would be no net acceleration.

There are a variety of possibilities that could be posited and analyzed based on these introductory notions. For example, the implications of having different magnitudes of the constituent mass characteristics could be explored, where the issues of conservation of momentum becomes more complex. Additionally, point sources other than just mass could be explored, such as electrical charges and perhaps even provisional characteristics for Dark Matter or Dark Energy. It is hoped that the disjunction drive example will provide future researchers with starting points for how to explore these other options.

3. Gradient Potential Drive

The previous two field-drive conjectures dealt with point sources. This next type of hypothetical field mechanism considers that somehow a small region of space around the spacecraft is affected without the use of point sources, specifically where a localized slope in scalar potential is induced across a spacecraft that then causes forces on the vehicle. It is not yet known if and how such an effect can be created, but this brainstorming exercise is offered to provide some analytical foundations for further inquiries and to address issues that arise.

The first obvious issue is the *net external force requirement*. In general, when pondering the creation of fields to induce forces, the forces act between the field-inducing device and the object that the field acts on. To illustrate this, consider the following equation that represents a gravitational scalar potential for a spacecraft plus an imposed short-range gradient across the vehicle:

$$\Phi = \left(\frac{-Gm}{r}\right) + \left(-xAe^{-r^2}\right) \quad (38)$$

where the first parenthetical term is the contribution from the spacecraft's own mass, m, G is Newton's gravitational constant, and r is the distance in Cartesian coordinates; $r = (x^2 + y^2 + z^2)^{1/2}$. The final parenthetical term is the hypothetical asymmetric effect imposed onto this scalar potential. This induced gradient is represented by a magnitude, A, with a negative slope in the positive x direction, and is localized by a Gaussian distribution (e^{-r^2}) centered at the origin. This localizing equation was arbitrarily chosen for illustration purposes only. Again, this example is only the foundation for future thought experiments where such assumptions can be individually varied.

This scalar potential is shown as a surface plot over an x–y plane in Fig. 11, which is equal to the superposition of the potentials from the vehicle and the induced local gradient.

By calculating the gradient of the scalar potential at the location of the vehicle, specifically the derivative of F with respect to r of the induced pitch effect at $r = 0$, the acceleration for the vehicle is determined to be equal to A, and acts in the positive x direction.

It is not clear where the reaction forces will be imparted. Consider the force on the spacecraft from the asymmetric gradient: $F = mA$. On what does the

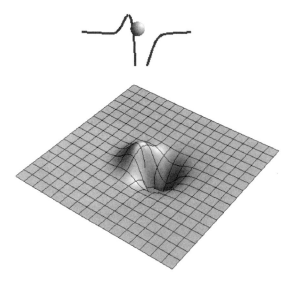

Fig. 11 Hypothetical gradient potential drive. A localized gradient in the gravitational potential along the x-axis is superimposed over that of the spacecraft. (Singularity truncated for clarity.)

counterforce act? Is it the induced gradient itself, or the device that induced that gradient? This option presents a point of departure for subsequent analysis.

The first and more common option is to consider that the forces act between the field-inducing device (which is on the vehicle) and the vehicle, which does not satisfy the net *external* force requirement. Hence, the force would be acting within the components of the vehicle, with no resulting motion.

The second option to get around this situation is suppose that the effect that created the gradient and the effect that this gradient has on the mass of the spacecraft are two, separate interactions. An easier example with which to illustrate this option is the *bias drive*, described next.

4. Bias Drive

The final type of hypothetical field mechanism entertains the possibility that the vehicle creates an asymmetry in the gravitational scalar potential, not by imposing it directly, but rather by altering the properties of spacetime itself. As a second step, this gravitational gradient then acts on the mass of the spacecraft in the usual manner, accelerating it in the positive x direction. Again, it must be emphasized that this hypothetical notion is introduced to illustrate options and issues rather than to assert that such features can be achieved.

To better understand this concept, consider the analogy of a "soap boat." The soap boat is a demonstration of the surface tension of water and how asymmetric changes in that surface tension can propel a toy boat. The demonstration begins with a toy boat floating in clean water. Then, a small quantity of detergent is placed in the water behind the boat and immediately the boat is propelled forward. The reaction mass is the water including the detergent. The force is from the difference in surface tension behind and in front of the boat.

To apply this analogy to our spacecraft, consider that Newton's constant, G, is analogous to the surface tension of the water and that some method has been discovered to asymmetrically alter Newton's gravitational constant over a small region of space. As shown by both versions of Eq. (39) and in both versions of Fig. 12, an asymmetric alteration of Newton's gravitational constant will induce a gradient in the scalar potential, Φ:

Multiplicative Modification

$$\Phi = -\left(\frac{m}{r}\right) G \left(1 + xBe^{-r^2}\right) \tag{39a}$$

Exponential Modification

$$\Phi = -\left(\frac{m}{r}\right) G^{\left(1 + xbe^{-r^2}\right)} \tag{39b}$$

where m is the mass of the spacecraft, G is Newton's gravitational constant, r is the distance in Cartesian coordinates; $r = (x^2 + y^2 + z^2)^{1/2}$, and both B and b represent the magnitudes of two different hypothetical modifications of Newton's constant. In the version shown in Eq. (39a), G is *multiplied* by an asymmetric modifier, similar to the version used in Fig. 11: Gradient potential drive. The "+1" identity term is necessary to return the Newtonian gravitational potential to its original form when the propulsive effects are off ($B = 0$), the x provides the gradient, and the Gaussian distribution (e^{-r^2}) limits the extension of the effect. This localizing equation was arbitrarily chosen for illustration purposes only. In the version shown in Eq. (39b), G is raised to the power of the local asymmetric affect, of identical form, but now where B is replaced with b to simply distinguish it from the prior version.

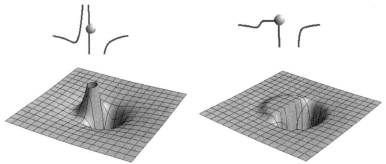

a) Multiplicative modification, B, Eq. (39a) b) Exponential modification, b, Eq. (39b)

Fig. 12 Hypothetical bias drives. To asymmetrically distort gravitational potentials, Newton's gravitational constant, G, is altered to take on a locally confined gradient along the x-axis, which then creates a gradient in the potential at the location of the spacecraft. (Singularities truncated for clarity.)

It is interesting to note from the figure, that the asymmetry posited by the *exponential modifier* [Eq. (39b) in Fig. 12b] has the characteristics of not allowing the scalar potential to exceed zero. At this stage of exploration, it should again be emphasized that these possibilities are introduced to demonstrate multiple approaches as tools for thought experiments, rather than to suggest that these are viable techniques.

When addressing the question of the reaction mass, it is now easier to suggest an alternative perspective. Because Newton's constant is considered a property of *space*, rather than a consequence of the field created by the vehicle, it is easier to consider *space itself* as the reaction mass.

One of the issues not yet raised with any of these propulsion thought experiments is the time rate of change of a scalar potential when responding to changes in the intensity or position of source phenomena. In electrostatics, where net charges can be varied to include a null net value, the consequences of this time delay are easier to study. This delay suggests that the field resulting from a charge is not rigidly connected to that charge, but rather is a combined function of the charge and the properties of the space. When considering the propagation limits for changes in an electrostatic potential, the phenomena of magnetism and the formalisms of Special Relativity come into play.

In the case of *gravitational* scalar potentials, where the source masses cannot be nullified, considering notions of time-delayed potentials (also called "retarded potentials") is less familiar. Although there are theories exploring the time-rate-of-change limitations of gravitational scalar potentials [62], the implications of such approaches for space drives have not been explored.

To return to the question at hand, namely the source of reaction mass, consider if the space itself is the reaction mass. As another analogy, consider the scalar potential to represent air pressure. A gradient would represent a pressure difference where the air is the reaction mass. Next, consider the prior train of thought in the context of Mach's principle, specifically the literal interpretation that leads to the possibility that there is a *progenitor* inertial frame caused by, and coincident with, all the mass in the Universe. Does this mean that forces induced against such a Machian frame might translate to the surrounding matter that creates this frame? At this point these are just conjectures, but the thought experiments offered by these hypothetical propulsion concepts provide yet another venue with which to revisit such fundamental—and still unresolved—questions of physics. If such explorations also lead to the discovery of a propulsive effect, all the better.

5. *Closing Comments on Field Drives*

The most critical issues when considering field drives are *conservation of momentum* and the *net-external force requirement*. It is hoped that the examples and discussions offered here can help future researchers avoid common mistakes and realize how to further investigate the possibilities and issues. The role played by spacetime is not trivial in these examples, nor are the distinct characteristics that make up masses, charges, and any other phenomena to be considered. By considering these various approaches and details, the requisite features needed to provide a space drive can eventually be articulated. Although these questions deal with the same data of nature, it casts them in a different context that can illuminate possibilities that might otherwise be overlooked. The next section begins to compile these unknowns into a problem statement to guide future research.

VI. Next Steps

With the indigenous reaction matter, energy considerations, and hypothetical thrusting mechanisms introduced, this section compiles the critical questions and unexplored options for space drives.

A. Problem Statement

Step 1 of the scientific method is to define the problem. To that end the critical issues from the various space drive concepts have been compiled into the following problem statement. Simply put, a space drive requires some controllable and sustainable means to create asymmetric forces on the vehicle without expelling a reaction mass, and some means to satisfy conservation laws in the process. Regardless of which concept is explored, a mechanism must exist that can affect a property, or the relationship between properties, of matter and spacetime that satisfies these conditions:

Physics issues:

 1) Satisfies conservation of momentum.
 2) Satisfies conservation of energy.
 3) Satisfies the *net external force requirement*, inducing a *unidirectional* acceleration of the vehicle.
 4) The effect must be *sustainable* as the vehicle moves.
 5) The physics proposed for the propulsive mechanism, and for the properties of matter and spacetime used for the propulsive effect, must be completely consistent with empirical observations.

Practical issues:

 6) The effect must be controllable (on/off, throttle-able, and directional).
 7) The effect must be effective enough to propel the vehicle at practical performance levels.

An important detail not elaborated on previously is item 4 in the problem statement—*sustainability*. Sustainability refers to the ability to continue the propulsive effect throughout the vehicle's motion. This implies that the force-inducing effect must work in both an inertial frame and an accelerated frame. This provides further mathematical constraints with which to scrutinize approaches. It also requires that the force-producing field is carried along with, or propagated with, the vehicle, or at least that the effect can be induced again after the vehicle has been set in motion. This is yet another level of detail for future space drive research to explore.

B. Continuing Research

Throughout this chapter, a variety of next steps are mentioned. These are compiled here for the reader's convenience. Addressing any of these in more depth would add to the overall progress in this field of study. This is not a comprehensive list of possibilities, but rather just those raised in the course of this chapter.

1) Explore the anomalies of cosmology from the point of view of spaceflight, in particular by going deeper to revisit the progenitor observations that lead to the working hypotheses of Dark Matter and Dark Energy.

2) Explore the connection between the quantum vacuum and spacetime itself, from the propulsion point of view of reactive media.

3) Continue fundamental research into the nature of spacetime itself, considering the literal interpretation of Mach's principle with its preferential inertial frames, along with other analogues [62–64].

4) Assess the propulsive implications of the "optical analogies" and if those can incorporate Mach's principle and address the issues of preferred reference frames.

5) Explore possible connections between Mach's principle and the cosmic microwave background radiation.

6) Repeat space drive energy analyses for relativistic conditions and with different assumptions from those specified in Sec. IV.B.

7) Analyze space drives from the perspective of *power* calculations (energy expended over time). This approach raises provocative questions from integration limits.

8) Assess sustained levitation using any of the analysis approaches suggested earlier in Sec. IV.B.

9) From Table 6 entitled Hypothetical Space Drives and Candidate Phenomena, pick any cell with the entry "TBD" (To Be Determined), and investigate the correlations further.

10) Repeat *bias drive* analysis, entertaining different localizing equations, different hypothetical scalar potential characteristics, and dissecting the individual characteristics that define things that *affect* the scalar potential and those things *affected by* the scalar potential, including time-rate-of-change factors for the various characteristics. The intent is to look for *actions* with different properties than their *reactions*.

VII. Conclusions

To provide a framework for assessing emerging science in the context of creating a space drive, a problem statement is now offered in this chapter. It is based on considering the key objections to the notion of a space drive: 1) conservation of momentum, 2) scarcity of indigenous reaction mass, and 3) inducing *net* forces relative to a reaction mass. Various properties of space are described, with most prospects residing with the least understood phenomena, specifically the *quantum vacuum*, *Dark Matter*, *Dark Energy*, and *spacetime* itself, which includes the fundamental unknown of the constituents of *inertial frames*. Of these, both the quantum vacuum and spacetime itself appear to have the greatest potential effective reaction mass density, specifically $10^{36 \pm 62}$ kg/m^3 and $10^{-1 \pm 26}$ kg/m^3, respectively. Note the wide uncertainty bands in those values, reflecting how much is still unknown about these phenomena. As starting points for future propulsion research, 10 different hypothetical propulsion concepts are offered, including inertial modification, space sail concepts, and field drive concepts. To address potential benefits and provide starting points for continued analyses, energy calculations are offered in several different contexts, with the most provocative being the case of indefinite levitation in a gravitational field.

Evidence continues to grow that our Universe is more complex than is yet explainable; it therefore may be prudent to consider more than one way to look at this evidence. General Relativity accurately describes the motions of matter at the scale of solar systems, and it can account for the spacetime warping from the presence of extreme energy densities, but it is not yet able to reconcile the anomalous rotation rates of galaxies (dubbed the Dark Matter problem), the anomalous accelerated redshifts (dubbed the Dark Energy problem), completely address Mach's principle, or incorporate quantum mechanics. Although quantum theory accurately describes subatomic electromagnetic energy transitions that cascade to macroscopic effects, it has not yet been converted to also address gravity.

Curiosity-driven physics often seeks a "theory of everything," one overarching theory to explain everything from the quantum level to cosmological scales. Approaches include string theory, quantum gravity, and variations of General Relativity to account for the new large-scale observations. As an alternative perspective that builds on the same physical observations and theoretical foundations, this chapter suggests using the narrower goal of breakthrough propulsion as a problem around which to initiate deeper investigations. This introduces different contexts for investigating Mach's principle, the anomalous rotation rates of galaxies, the anomalous distant redshifts, quantum fluctuations of the vacuum, and the fundamental coupling of electromagnetism, inertia, gravity, and spacetime.

Acknowledgments

The author is grateful to Ed Zampino, William Meyer, Geoff Landis, and Mike LaPointe for discussions that helped distill the ambition of space drives into its key scientific issues. Such individuals, those able to simultaneously entertain imaginative departures and apply analytical rigor, are rare to find and a joy to work with. Extra thanks are also due to William Meyer for his assistance with the three-dimensional plots.

References

[1] Farmer, M., Science Fiction Citations for the Oxford English Dictionary, URL: <http://www.jessesword.com/sf/list> [cited 4 May 2007].

[2] Millis, M. G., "Challenge to Create the Space Drive", *Journal of Propulsion and Power*, Vol. 13, No. 5, 1997, pp. 577–582.

[3] Bertolami, O., and Tajmar, M., *Gravity Control and Possible Influence on Space Propulsion: A Scientific Study*. ESA CR (P) 4365, on Contract ESTEC 15464/01/NL/Sfe, 2002.

[4] Hartle, J. B., *Gravity; An Introduction to Einstein's General Relativity*, Addison Wesley, San Francisco, CA, 2003.

[5] Carroll, B. W., and Ostlie, D. A., *An Introduction to Modern Astrophysics*, 2nd ed., Pearson Education, Saddle River, NJ, 2007.

[6] Robitaille, P.-M., "WMAP: A Radiological Analysis," *Progress in Physics*, Vol. 1, 2007, pp. 3–18.

[7] Spergel, D. N., Bean, R., Doré, O., Nolta, M. R., Bennett, C. L., Dunkley, J., Hinshaw, G., Jarosik, N., Komatsu, E., Page, L., Pereis, H. V., Verde, L., Halpern, M., Hill, R. S., Kogut, A., Limon, M., Meyer, S. S., Odegard, N., Tucker, G. S., Weiland, J. L., Wollack, E., and Wright, E. L., "Three-Year Wilkinson Microwave Anisotropy Probe (WMAP) Observations: Implications for Cosmology," *The Astrophysical Journal Supplement Series*, Vol. 170, No. 2, 2007, pp. 377–408.

[8] Riess, A. G., Kirshner, R. P., Schmidt, B. P., Jha, S., Challis, P., Garnavich, P. M., Esin, A. A., Carpenter, C., Grashius, R., Schild, R. E., Berlind, P. E., Huchra, J. P., Prosser, C. F., Falco, E. E., Benson, P. J., Briceño, C., Brown, W. R., Caldwell, N., Dell'Antonio, I. P., Filippenko, A. V., Goodman, A. A., Grogin, N. A., Groner, T., Hughes, J. P., Green, P. J., Jansen, R. A., Kleyna, J. T., Luu, J. X., Macri, L. M., McLeod, B. A., McLeod, K. K., McNamara, B. R., McLean, B., Milone, A. A. E., Mohr, J. J., Moraru, D., Peng, C., Peters, J., Prestwich, A. H., Stanek, K. Z., Szentgyorgyi, A., and Zhao, P., "BVRI Light Curves for 22 Type Ia Supernovae," *Astronomical Journal*, Vol. 117, 1999, pp. 707–724.

[9] Goldhaber, G., Groom, D. E., Kim, A., Aldering, G., Astier, P., Conley, A., Deustua, S. E., Ellis, R., Fabbro, S., Fruchter, A. S., Goobar, A., Hook, I., Irwin, M., Kim, M., Knop, R. A., Lidman, C., McMahon, R., Nugent, P. E., Pain, R., Panagia, N., Pennypacker, C. R., Perlmutter, S., Ruiz-Lapuente, P., Schaefer, B., Walton, N. A., and York, T., "Timescale Stretch Parameterization of Type Ia Supernova B-Band Light Curves," *The Astrophysical Journal*, Vol. 558, part 1, 2001, pp. 359–368.

[10] Huterer, D., and Turner, M. S., "Prospects for Probing the Dark Energy via Supernova Distance Measurements," *Physical Review D*, Vol. 60, No. 8, 1999, pp. 081301–081306.

[11] Rubin, V. C., and Ford, W. K., Jr., "Rotation of the Andromeda Nebula from a Spectroscopic Survey of Emission Regions," *Astrophysical Journal*, Vol. 159, 1970, pp. 379–403.

[12] Kneib, J. P., Ellis, R. S., Smail, I., Couch, W. J., and Sharples, R. M., "Hubble Space Telescope Observations of the Lensing Cluster Abell 2218," *Astrophysical Journal*, Vol. 471, 1996, pp. 643–656.

[13] Milgrom, M., "Dynamics with a Non-Standard Inertia-Acceleration Relation: An Alternative to Dark Matter in Galactic Systems," *Annals of Physics*, Vol. 229, 1994, pp. 384–415.

[14] Milgrom, M., "MOND – Theoretical Aspects," *New Astronomy Reviews*, Vol. 46, 2002, pp. 741–753.

[15] Rabounski, D., "The Relativistic Effect of the Deviation Between the CMB Temperatures Obtained by the COBE Satellite," *Progress in Physics*, Vol. 1, 2007, pp. 24–26.

[16] Misner, C. W., Thorne, K. S., and Wheeler, J. A., *Gravitation*, W. H. Freeman & Co., New York, 1973.

[17] Rider, T. H., "Fundamental Constraints on Large-Scale Antimatter Rocket Propulsion," *Journal of Propulsion and Power*, Vol. 13, No. 3, 1997, pp. 435–443.

[18] Taylor, E. F., and Wheeler, J. A., *Spacetime Physics*, W.H. Freeman and Co., San Francisco, CA, 1966.

[19] Flynn, C., "On the Mass-to-Light Ratio of the Local Galactic Disc and the Optical Luminosity of the Galaxy," *Monthly Notices of the Royal Astronomical Society*, Vol. 372, 2006, pp. 1149–1160.

[20] Cutnell, J. D., and Johnson, K. W. *Physics*, 3rd ed. Wiley, New York, 1995, p. 441.

[21] Eddington, A. S., *Space, Time and Gravitation*, Cambridge University Press, Cambridge, U.K., 1920.

[22] de Felice, F., "On the Gravitational Field Acting as an Optical Medium," *General Relativity and Gravitation*, Vol. 2, 1971, pp. 347–357.

[23] Evans, J., Nandi, K. K., and Islam, A., "The Optical-Mechanical Analogy in General Relativity: Exact Newtonian Forms for the Equations of Motion of Particles and Photons," *General Relativity and Gravitation*, Vol. 28, 1996, pp. 413–439.

[24] Puthoff, H. E., "Polarizable-Vacuum (PV) Approach to General Relativity," *Foundations of Physics*, Vol. 32, 2002, pp. 927–943.

[25] Mach, E., *The Science of Mechanics*, 5th English ed., Open Court Publishing Co., London, 1942.

[26] Barbour, J. B., and Pfister, H. (eds), *Mach's Principle: From Newton's Bucket to Quantum Gravity (Einstein Studies)*, Birkhäuser, Boston, 1995.

[27] Weinberg, S., *Gravitation and Cosmology: Principles and Applications of the General Theory of Relativity*, Wiley, New York, 1972.

[28] Bondi, H., and Samuel, J., "The Lense-Thirring Effect and Mach's Principle," *Physics Letters A*, Vol. 228, pp. 121–126.

[29] Ciufolini, I., and Wheeler, J. A., *Gravitation and Inertia*, Princeton Univ. Press, Princeton, NJ, 1995.

[30] Symon, K. R., *Mechanics*, 3rd ed., Addison-Wesley, Reading, MA, 1971.

[31] Foster, R. N., *Innovation; The Attacker's Advantage*, Summit Books, New York, 1986.

[32] Seifert, H. (ed.), *Space Technology*, Wiley, New York, 1959.

[33] Berman, A. I., *The Physical Principles of Astronautics: Fundamentals of Dynamical Astronomy and Space Flight*, Wiley, New York, 1961.

[34] Sternbach, R., and Okuda, M., *Star Trek: The Next Generation Technical Manual*, Pocket Books, New York, 1991.

[35] Woodward, J. F., "Flux Capacitors and the Origin of Inertia," *Foundations of Physics*, Vol. 34, 2004, pp. 1475–1514.

[36] Millis, M. G., "Assessing Potential Propulsion Breakthroughs," *New Trends in Astrodynamics and Applications*, Belbruno, E., (ed.), Annals of the New York Academy of Sciences, New York, Vol. 1065, 2005, pp. 441–461.

[37] Tajmar, M., and Bertolami, O., "Hypothetical Gravity Control and Possible Influence on Space Propulsion," *Journal of Propulsion and Power*, Vol. 21, No. 4, July–Aug 2005, pp. 692–696.

[38] Podkletnov, E., and Nieminen, R. "A Possibility of Gravitational Force Shielding by Bulk YBCO Superconductor," *Physica C*, Vol. 203, 1992, pp. 441–444.

[39] Hathaway, G., Cleveland, B., and Bao, Y. "Gravity Modification Experiment Using a Rotating Superconducting Disk and Radio Frequency Fields," *Physica C*, Vol. 385, 2003, pp. 488–500.

[40] Resnick, R., and Halliday, D., *Physics*, 3rd ed., Wiley, New York, 1977.

[41] Boston, M., "Simplistic Propulsion Analysis of a Breakthrough Space Drive for Voyager", *AIP Conference Proceedings No. 504*, El-Genk, M. S., (ed.), American Institute of Physics, New York, 2000, pp. 1075–1078.

[42] Byers, D. C., "An Experimental Investigation of a High Voltage Electron Bombardment Ion Thruster," *Journal of the Electrochemical Society*, Vol. 116, No. 1, 1969, pp. 9–17.
[43] Forward, R. L., "Guidelines to Antigravity," *American Journal of Physics*, Vol. 31, 1963, pp. 166–170.
[44] Alcubierre, M., "The Warp Drive: Hyper-fast Travel Within General Relativity," *Classical and Quantum Gravity*, Vol. 11, 1994, p. L73.
[45] Morris, M. S., and Thorne, K. S., "Wormholes in Spacetime and Their Use for Interstellar Travel: A Tool for Teaching General Relativity," *American Journal of Physics*, Vol. 56, 1988, pp. 395–412.
[46] Visser, M., *Lorentzian Wormholes: From Einstein to Hawking*, AIP Press, New York, 1995.
[47] Krasnikov, S. V., "Hyperfast Interstellar Travel in General Relativity," *Physical Review D*, Vol. 57, 1998, p. 4760.
[48] Bussard, R. W, "Galactic Matter and Interstellar Flight," *Astronautica Acta*, Vol. 6, 1960, pp. 179–194.
[49] Bond, A., "An Analysis of the Potential Performance of the Ram Augmented Interstellar Rocket." *Journal of the British Interplanetary Society*, Vol. 27, 1974, pp. 674–688.
[50] Jackson, A., "Some Considerations on the Antimatter and Fusion Ram Augmented Interstellar Rocket," *Journal of the British Interplanetary Society*, Vol. 33, 1980, pp. 117–120.
[51] Woodward, J. F., "Method for Transiently Altering the Mass of an Object to Facilitate Their Transport or Change Their Stationary Apparent Weights," US Patent # 5,280,864, 1994.
[52] Woodward, J. F., "A New Experimental Approach to Mach's Principle and Relativistic Gravitation," *Foundations of Physics Letters*, Vol. 3, No. 5, 1990, pp. 497–506.
[53] Bondi, H., "Negative Mass in General Relativity," *Reviews of Modern Physics*, Vol. 29, No. 3, July 1957, pp. 423–428.
[54] Forward, R. L., "Negative Matter Propulsion," *Journal of Propulsion and Power*, Vol. 6, No. 1, Jan.–Feb., 1990, pp. 28–37.
[55] Landis, G., "Comment on 'Negative Mass Propulsion'," *Journal of Propulsion and Power*, Vol. 7, No. 2, 1991, p. 304.
[56] Winterberg, F., "On Negative Mass Propulsion," 40th Congress of the International Astronautical Federation, International Astronautical Federation Paper 89 668, Malaga, Spain, Oct. 1989.
[57] Chew, G., Dolyle, M., and Stancati, M., *Interstellar Spaceflight Primer, Final Report on NASA Contract NASW-5067*, Order 167, Science Applications, International Corp., Schaumburg, IL, 2001.
[58] McInnes, C. R., and Brown, J. C., "Terminal Velocity of a Laser-Driven Light Sail," *Journal of Spacecraft*, Vol. 27, No. 1, 1989, pp. 48–52.
[59] Carmouche, W. J., "Spaceship to Harness Radiations in Interstellar Flights," US Patent 5,058,833, 22 Oct 1991.
[60] Landis, G. A., "Reactionless Orbital Propulsion Using a Tether," *Acta Astronautica*, Vol. 26, No. 5, 1992, pp. 307–312; also as National Aeronautics and Space Administration TM-101992, April 1989.

[61] Gueron, E., Maia, C. A. S., and Matsas, G. E. A., "'Swimming' Versus 'Swinging' Effects in Spacetime," *Physical Review D*, Vol. 73, No. 2, id. 024020, January 2006.

[62] Jefimenko, O. D., *Gravitation and Cogravitation; Developing Newton's Theory of Gravitation to Its Physical and Mathematical Conclusions*, Electret Scientific Co., West Virginia, 2006.

[63] Barceló, C., Liberati, S., and Visser, M., "Analogue Gravity," *Living Reviews Relativity*, Vol. 8, No. 12, [online article], URL: http://www.livingreviews.org/lrr-2005-12 cited, 12 Sept. 2007.

[64] Winterberg, F. M., *The Planck Aether Hypothesis: An Attempt For a Finitistic Theory of Elementary Particles*, Verlag Relativistichen Interpretationen, Karlsbad, Germany, 2000.

Chapter 4

Review of Gravity Control Within Newtonian and General Relativistic Physics

Eric W. Davis*

Institute for Advanced Studies at Austin, Austin, Texas

I. Introduction

GRAVITY is a pervasive force that we are unable to affect. Unlike the electromagnetic force, for which an engineering prowess has been achieved, gravity remains an immutable force. The scientific knowledge of gravity has matured to where its behavior is well modeled, both from the Newtonian and the more detailed general relativistic perspectives, but *why* it functions as modeled is still not understood. With the accumulated knowledge, propulsion devices can be engineered that *overcome* the force of gravity through indirect means, but gravity itself cannot yet be modified. Aerodynamic machines fly in air and rockets fly in the vacuum of space. We can even take advantage of gravity, for example, with "gravity assists" for space maneuvers and for converting the Earth's water cycle into electrical energy via hydroelectric dams, but here again without any ability to change gravity itself.

Gravity is weak compared to the other fundamental forces in nature. Considering the forces between an electron and a proton, the gravitational force is 40 orders of magnitude weaker than the electric force. But near the surface of the Earth, where 10^{24} kg of gravitational mass lies underneath, the accumulated strength is significant. On the surface of the Earth, gravity is strong enough to accelerate objects downward at 9.81 m/s^2. Considering the relative weakness of gravity compared to more controllable phenomena, it seems reasonable to consider that gravity might one day become a phenomenon that we can engineer.

Copyright © 2008 by the American Institute of Aeronautics and Astronautics, Inc. All rights reserved.
*Senior Research Physicist.

A colloquial expression for the goal of affecting gravity is "antigravity." This term specifically means the negation or repulsion of the force of gravity. A more general term that encompasses this notion and other possibilities is "gravity control."

This chapter examines a variety of approaches toward the goal of controlling gravity. Both Newtonian and general relativistic perspectives will be explored. The reader should bear in mind that most of these concepts are nowhere near having any form of practicable engineering implementation. The intent is to familiarize the audience with the approaches that already exist, explain their limits and unknowns, and then to identify lingering research objectives that are relevant to the quest of controlling gravity.

The approaches covered include using Newton's Law of Gravity to simply nullify the gravity field of one body acting on another body by using a clever arrangement of masses. Also, the theoretical possibility of antigravity appears in quantum gravity theories, cosmological vacuum (or dark) energy theory, and quantum field theory. We will review additional topics of gravity control that include the production of artificial gravity via ultrahigh-intensity electric or magnetic fields, the production and use of gravitational waves for rocket propulsion, antigravity (self-lifting) forces induced by the Casimir effect or by nonretarded quantum interatomic dispersion forces in a curved spacetime (i.e., in a background gravitational field), and speculative quantum unified field theories that predict unusual new antigravity forces.

II. Gravity Control Within Newtonian Physics

The basic form of Newton's Law of Gravity is given by the standard expression for the gravitational force (F_{grav}) that mutually acts between two masses [1]:

$$F_{grav} = -\frac{Gm_1 m_2}{r^2} \quad (1)$$

where the negative sign indicates that F_{grav} is a mutual force of attraction, G is Newton's universal gravitation constant (6.673×10^{-11} Nm^2/kg^2), m_1 and m_2 are two interacting masses, and r is the radial distance between the two masses (note: we adopt MKS units throughout, unless otherwise indicated). We observe in Eq. (1) that the force of gravity acting on a small test mass becomes stronger when the other (gravitating) mass is larger in magnitude or when the distance between them is very small, or both. Also recall that we can use Eq. (1) and Newton's Second Law of Motion ($F = ma$) to define the magnitude of the gravitational acceleration a_g that acts on a small test mass m due to a larger (gravitating) mass M [1]:

$$a_g = \frac{GM}{r^2} \quad (2)$$

If we choose Earth to be the larger gravitating mass so that $M = M_\oplus$ (5.98×10^{24} kg), then according to Eq. (2), a small test mass m placed near the Earth's surface, whereby $r \approx R_\oplus$ (6.378×10^6 m), will experience a downward gravitational acceleration of $a_g \equiv g = 9.81$ m/s^2.

A. Negating Newtonian Gravity the Hard Way

It is possible to design an antigravity machine that can nullify Earth's gravity field using Newton's Law of Gravity. One way to use Eq. (1) to nullify the Earth's gravitational pull at a particular location would be to locate another planet of equal mass above that location [2,3]. The forces from the two Earth masses will cancel each other out over a broad region between them. Everything within this broad region will be in free fall. This is not a practical solution for aerospace flight, however, because we have no way to manipulate and control another planetary-sized body.

Along similar lines, Forward [2,3] considered using a ball of ultradense compact matter, corresponding to dwarf star or neutron star matter ($\sim 10^{11}$ to 10^{18} kg/m^3), having a diameter of 32 cm and a mass of 4×10^6 tons. This ultradense ball will have a surface gravitational (attractive) force of 1-g. We could then place this small, ultradense ball near the surface of the Earth and its 1-g gravity field will cancel the Earth's 1-g gravity field. All test objects placed in the broad region between the small ultradense ball and the Earth will thus be in free fall. Another option Forward [2–5] suggested would be to shape the compact ultradense matter into a disk that is 45 cm in diameter and 10 cm thick, with the same mass and density as the small ultradense ball. Its gravitational acceleration is $a_g = 4G\rho\tau$, where ρ is the mass density of the disk and τ is its thickness. In this case, the disk will have a force of gravitational attraction that is the same on both sides, and it will be uniform near the center of the disk where the strength of the gravitational force will be 1-g. If this disk were to be placed very close to and above the Earth's surface, then there will be a gravitational force of 2-g above the disk (equal to 1-g due to the Earth's gravity field plus 1-g due to the top-side gravity field of the disk) while underneath the disk near its center, there will be a gravity-free (or free fall) region because the Earth's gravity field underneath is canceled by the gravity field of the disk's bottom side. While these are interesting antigravity machines, they are unfortunately not feasible from an engineering standpoint because we do not yet have the technology or means to create and handle ultradense compact matter.

B. Approaches for Orbital Stability

As another approach, Fig. 1 shows a Six-Mass Compensator designed by Forward [6]. This device approximately nullifies the Earth's residual gravitational tidal forces that act upon all orbiting bodies such as satellites and space stations. In a free-fall orbit around the Earth, the overall downward gravitational acceleration is canceled to first-order by the orbital motion alone. However, there are residual tidal accelerations (or forces) that are very weak ($\sim$micro-g). These tidal forces are not exactly zero because the free-fall motion of orbiting bodies does not entirely cancel the Earth's gravity field.[†] Such weak tidal forces will

---

[†]The only part of a body that is under absolutely zero net gravity force in a free-fall orbit is its center of mass. Those parts of a body located some distance away from the center of mass will experience tidal forces due to gravity gradients (or differential acceleration). Residual tides arise when the distant parts of an orbiting body follow their own orbits that are either higher (lower velocity), lower (higher velocity), or in a plane tangent to the Earth with respect to the orbit of the center of mass. The result is outward-vertical and inward-horizontal tidal forces.

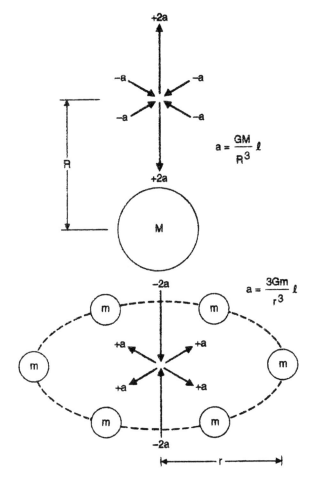

Fig. 1 Orbital tidal patterns of Earth and the Six-Mass Compensator [6].

have a measurable impact upon sensitive lab experiments or manufacturing processes that require an absolute 0-g environment. And the residual tidal forces will grow larger in magnitude as an orbiting body grows larger in size and/or as the size of 0-g sensitive experimental or manufacturing devices grow larger.

Forward's Six-Mass Compensator is designed to reduce the weak horizontal and vertical tidal forces by a factor $\sim 10^6$. The magnitude and sign of the horizontal and vertical tidal accelerations ($a = GM\ell/R^3$) in orbit are illustrated at the top of Fig. 1, where R is the orbital distance ($=R_\oplus + h$, where h is the orbital altitude) from the center of Earth (mass $M = M_\oplus$) and ℓ is the distance from the center of mass of an arbitrary point on the orbiting body. We observe that the outward tidal acceleration in the vertical direction is twice as strong as the uniform inward tidal acceleration (compression) in the horizontal direction. Therefore, the compensator illustrated at the bottom of Fig. 1 generates forces that approximately cancel the residual tidal forces. The compensator is comprised of six equal mass (m) spheres arranged in a ring of radius r. The ring encloses the region to be compensated and the magnitude of the compensating acceleration is $a = 3Gm\ell/r^3$. A solid ring or any number of spherical masses greater than three can be used instead, but Forward estimated that six spherical masses is optimum. Because the compensator is nullifying only the weak residual orbital tidal forces, the compensator spheres can be made of ordinary density matter (e.g., tungsten or lead) having a mass of 100 kg and size of 20 cm. The plane of the ring compensator must always be oriented tangent to the surface of the Earth below. The gravitational tidal pattern of the ring compensator can be fine tuned to match the tidal pattern of the Earth at any altitude by simply adjusting the radius and tilt of the ring compensator. Because the Earth's gravitational tidal forces grow stronger at lower altitudes, the region to be compensated will grow smaller, while the compensation region will grow larger at higher altitudes because the tidal forces will grow weaker.

Unfortunately, this particular antigravity machine offers no propulsion benefit; therefore, we will not consider it further.

C. Newtonian Levitation Energy Estimate

An ideal propulsion breakthrough could take the form of the antigravity-based levitation of an aerospace vehicle within the Earth's atmosphere. Rockets like the Air Force DC-XA can hover above the ground for a time limited by the amount of available rocket fuel [7]. An ideal antigravity propulsion device would allow for the indefinite levitation of a vehicle above the Earth's surface. It is illustrative to estimate the energy required to levitate a 1-kg test mass above the Earth's surface. This will help quantify a potentially key engineering parameter for such a levitation system. Millis [8] provides a generic estimate by considering the amount of energy ($\mathcal{E}_{lev}$) required to nullify the magnitude of the Earth's gravitational potential energy for a test mass m hovering at height h above the Earth's surface:

$$\mathcal{E}_{lev} = \frac{GM_\oplus m}{h} \qquad (3)$$

Equation (3) can also be derived by calculating how much energy is required to completely remove a test mass from the Earth's surface to infinity. Millis points out that this calculation is more in line with the analogy to nullify the effect of gravitational energy. Equation (3) also represents the energy required to stop a test mass at the levitation distance h if it were falling in from infinity with zero initial velocity.

Setting $h \approx R_\oplus$ and $m = 1$ kg in Eq. (3), we get $\mathcal{E}_{lev} = 62.5$ MJ. This is 2.05 times the kinetic energy required to put the test mass into low Earth orbit (LEO). However, this estimate for $\mathcal{E}_{lev}$ will require some adjustment depending on the type of theory and its technological implementation because the operational energetics of a putative antigravity propulsion system must be considered in conjunction with $\mathcal{E}_{lev}$.

180   E. W. DAVIS

D. Negative Matter Propulsion

Negative matter is a hypothetical form of matter whose active and passive gravitational, inertial, and rest masses are opposite in sign to ordinary (positive) matter. Thus, negative matter has negative mass. Negative matter is not to be confused with antimatter, because antimatter is a form of positive matter having positive mass and quantum properties (i.e., spin and electric charge) that are opposite to its positive matter counterparts. Forward [9,10] and others [11–15] studied the physics of negative matter within the context of both Einstein's Special and General Theories of Relativity, and discovered many unusual properties that could provide a form of breakthrough propulsion. These properties are given below [9,10,14].

1) An object with negative mass gravitationally repels all other types of matter, both positive and negative.[‡]

2) The gravitational coupling between a negative mass and a positive mass does not violate the conservation of momentum, because negative matter in motion has negative momentum.

3) The gravitational coupling between a negative mass and a positive mass does not violate the conservation of energy, because negative matter in motion has negative kinetic energy.

4) The total energy required to create negative matter, along with an equal amount of positive matter, is zero, because negative matter has negative rest-energy.

The consequences of these four properties upon a hypothetical negative matter antigravity propulsion system are elucidated in the following sections.

1. Concepts for Negative Matter Propulsion

We can establish the fundamental basis for negative mass propulsion via a thought experiment that demonstrates the negative matter property in item 1 above. If we take a ball of negative matter of mass $(-M)$ that is initially at rest, and place it near a rocket of positive equal mass $(+M)$ that is also initially at rest, then the gravitational field of the ball will repel the rocket, while the gravitational field of the rocket will attract the ball. The result of this unique gravitational coupling is that the rocket and the ball will move off in the same direction with an acceleration that is proportional to the force of gravity between them.[§] Figure 2 illustrates this effect. This mechanism appears to provide an unlimited amount of unidirectional acceleration without requiring either a reaction mass or an energy source (see also, Ref. 15).

Forward [9,10] also provided a Newtonian analysis for the case when the magnitude of $-M$ and $+M$ are unequal. In this case, a consideration of the motion of the center of mass of the combined system becomes important. So,

---

[‡]Positive (ordinary) mass attracts all types of matter, including negative matter.
[§]Otherwise the ball would violate the principle of equivalence which demands that all objects, without exception, acquire the same acceleration in the same gravitational field.

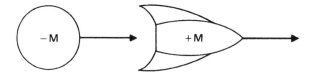

Fig. 2 Negative matter propulsion [9,10].

in addition to the unidirectional acceleration of the combined system, a decrease or increase in the separation of the two objects is superimposed. In a general relativistic analysis, Bonnor and Swaminarayan [12] showed that in the case of a uniformly accelerating pair of gravitationally interacting bodies with opposite masses, the two masses are not quite equal and opposite because they are being measured in an accelerating reference frame. For the two masses to keep a constant separation in the accelerated reference frame, the negative mass should be very slightly larger in magnitude than the positive mass, depending on the initial separation.

The case of physically coupling a particle of negative matter and a (positive matter) rocket together using a stiff spring was analyzed by Forward [9,10]. He designed a propulsion system that has the negative matter particle, of a mass equal in magnitude to the entire rocket, located in an "engine room." In order to move the rocket forward, one must pull a spring attached to the back wall of the engine room and hook it to the negative matter particle. The subsequent inertial reaction of the negative matter particle will cause it to accelerate in the forward direction, thus pulling the rocket forward with an acceleration that is proportional to the strength of the spring. One would unhook the spring in order to stop accelerating and, in order to decelerate the rocket to a stop, one would replace the spring attached to the back wall of the engine room with a spring attached to the front (or forward) wall. Forward also proved that this propulsion system does not violate the laws of conservation of momentum and energy even when the magnitudes of the two masses are not equal.¶

Finally, Forward [9,10] also considered the case of physically coupling a particle of negative matter and a (positive matter) rocket together using electrostatic forces. It is assumed in this case that negative matter comes in an electrically charged variety, whereby the electric charge acts like a "handle" that can be used to push or pull the negative matter particle at a distance using electrostatic forces. It was shown that in both cases when the charges are both equal and opposite, the same results of unidirectional motion of the combined system occurred at an acceleration that is proportional to the strength of the electrostatic force. Again, there was no violation of the laws of conservation of momentum and energy even when the magnitudes of the two masses are not equal.

¶It is the elasticity of the spring that keeps the conservation laws from being violated, because the spring also applies forces to the two objects.

2. Consideration of the Conservation Laws

Items 2 through 4 in the previous list appear to be controversial. In this section we shall examine each of them in turn within the context of Newtonian mechanics. It would appear that negative matter propulsion automatically violates both of the conservation laws of momentum and energy; however, it can be proved that this is not so. In the following it is assumed that the magnitudes of the positive mass and the negative mass are equal, but the results remain unchanged even in the case when they are unequal. When the ball of negative matter and the (positive matter) rocket were both initially at rest (i.e., zero velocity), their individual momentum is zero, and so their total initial momentum is also zero. But when the two objects reach some final velocity v, their total final momentum $\sum \mathcal{P}$ is still zero, because:

$$\sum \mathcal{P} = \mathcal{P}_+ + \mathcal{P}_- \\ = (+M)v + (-M)v \\ = 0 \qquad (4)$$

where $\mathcal{P}_+$ is the final momentum of the rocket and $\mathcal{P}_-$ is the final momentum of the ball. Equation (4) shows that the negative (final) momentum of the ball cancels out the positive (final) momentum of the rocket, and so momentum is conserved. In addition, when the ball and the rocket are both initially at rest, their individual kinetic energy is zero, and so their total initial kinetic energy is also zero. After the two objects reach final velocity v, their total final kinetic energy $\sum \mathcal{KE}$ is still zero, because:

$$\sum \mathcal{KE} = \mathcal{KE}_+ + \mathcal{KE}_- \\ = \frac{1}{2}(+M)v^2 + \frac{1}{2}(-M)v^2 \\ = 0 \qquad (5)$$

where $\mathcal{KE}_+$ is the final kinetic energy of the rocket and $\mathcal{KE}_-$ is the final kinetic energy of the ball. Equation (5) shows that the negative (final) kinetic energy of the ball cancels out the positive (final) kinetic energy of the rocket, and so kinetic energy is also conserved.

Finally, if we could somehow produce negative matter in order to exploit its antigravity propulsion properties, then we must produce it along with equal amounts of positive matter in order to conserve energy. This is exactly what is required for the production of antimatter even though these two types of matter are completely dissimilar. Antimatter production always occurs via the production of particles and antiparticles in pairs. And so, negative matter production must also invoke the production of negative matter particle–positive matter particle pairs. However, unlike the production of antimatter that requires $2mc^2$ (c is the speed of light, 3.0×10^8 m/s) of combined rest-energy to produce each particle–antiparticle pair (each particle of mass, m), the energy cost for producing a negative matter particle–positive matter particle pair is $(+M)c^2 + (-M)c^2 = 0$. This surprising counterintuitive result is possible because the negative rest-energy of the negative matter cancels the positive rest-energy of the equal quantity of positive matter.

Forward [9,10] pointed out that negative rest-energy, negative momentum, and negative kinetic energy are not standard concepts within Newtonian mechanics, but they do not appear to lead to any logical contradictions because they are not (logically) forbidden by Special or General Relativity Theory. If negative matter is forbidden, then the proof must come from some other theory of physics (e.g., quantum field theory, quantum superstring theory, etc.).

3. Make or Break Issues: Negative Matter Propulsion

Can we produce negative matter in the lab and/or does it naturally exist somewhere in the universe? The answer to the first part of this question is no or not yet. The answer to the second part is we do not know. One property that negative matter has in common with antimatter is that, if it does exist naturally somewhere in the universe, then it is just as scarce as antimatter. The cosmological abundance of antimatter is negligible. This conclusion is partly based on the observed minuscule amount of antimatter in cosmic rays. All of the antimatter present in cosmic rays can be accounted for by radioactive decay processes or by nuclear reactions involving ordinary matter. We also do not observe the signatures of electron–positron annihilation, or proton–antiproton annihilation coming from the edges of galaxies, or from places where two galaxies are near each other. As a result, we believe that essentially all of the objects we see in the universe are made of matter and not antimatter. Yet we can produce copious amounts of antimatter in particle accelerator devices. Nearly the entire abundance of antimatter that was created during the Big Bang had been annihilated away prior to or during cosmological nucleosynthesis. Likewise, negative matter could have been produced along with equal amounts of matter during the Big Bang, but we can speculate that some underlying universal quantum symmetry property was broken as the universe cooled, thus causing the negative matter particles to be destroyed by some unknown particle interaction processes. This may account for the absence of naturally occurring negative matter.

Forward [9,10] speculated that the large voids existing between galaxies and superclusters of galaxies are regions of negative matter. Because negative matter repels itself, it cannot gravitationally bind into larger clumps of negative matter. However, because Coulomb electrostatic forces are enormously stronger than gravitational forces, electrically charged bulk negative matter attracts itself and does so much more strongly than ordinary matter. Thus, negative matter can become electrostatically bound into larger clumps [15]. This is because for negative matter, gravity and electric charge exchange roles: The "gravitational charge" of negative matter gravitationally repels like-gravitationally charged particles,** whereas electrostatic forces will attract electrically charged negative matter. So we could then imagine that the large extragalactic voids throughout the

---

**The gravitational charge e_g of a mass M is customarily defined as [14]: $e_g \equiv (-G)^{1/2}M$.

universe are places where negative matter has collected into huge electrically bound clumps. This could explain the mystery of dark matter [16–18]. However, dark matter is not likely to be composed of negative matter because dark matter gravitationally attracts all other matter (including other dark matter), and the total cosmological energy budget requires that dark matter has a positive rest-energy density (and positive pressure).

However, there is nothing to forbid the existence or artificial creation of negative matter. No logical contradiction proofs have succeeded in forbidding it. The most serious known objection to negative matter is the one based on the Principle of Causality or, equivalently, on the basis of the Second Law of Thermodynamics. Terletskii [14] correctly points out that all other objections can be reduced to the Principle of Causality or they can be shown to be linked with it to some degree. Further, because it is recognized that the Principle of Causality is a naturally apparent expression of the Second Law of Thermodynamics, then the objections to negative matter lose all physical justification. Negative matter should exist on an equal par with antimatter and all other exotic elementary quantum matter (artificially created in accelerators) because Murray Gell-Mann's Totalitarian Principle states that in physics "anything which is not prohibited is compulsory" [19]. Furthermore, exotic matter (e.g., antimatter, negative matter, superluminal matter, etc.) is in no way precluded by either relativistic quantum field theory or Einstein's two relativity theories because it is these very theories that suggest their possibility. For example, C. D. Anderson's 1932 discovery of anti-electrons (or positrons) in cosmic rays confirmed Dirac's 1928 relativistic quantum mechanics prediction of antimatter.

III. Gravity Control Within General Relativity

In the sections that follow we describe and summarize the known types of antigravity that can be derived from Einstein's General Theory of Relativity.

A. Antigravity via Gravitomagnetic Forces

1. Gravitomagnetic Foundations

Heaviside [20] (in 1883), Einstein (prior to the 1915 publication of his General Theory of Relativity), Thirring [21,22], and Thirring and Lense [23] showed that General Relativity Theory provides a number of ways to generate non-Newtonian gravitational forces via the splitting of gravitation into electric and magnetic field type components [24]. These forces can be used to counteract the Earth's gravitational field, thus acting as a form of antigravity. General Relativity Theory predicts that a moving source of mass-energy can create forces on a test body that are similar to the usual centrifugal and Coriolis forces, although much smaller in magnitude. These forces create accelerations on a test body that are independent of the mass of the test body, and the forces are indistinguishable from the usual Newtonian gravitational force. We can counteract the Earth's gravitational field by generating these forces in an upward direction at some spot on the Earth.

Forward [25] linearized Einstein's general relativistic field equation and developed a set of dynamic gravitational field relations similar to Maxwell's electromagnetic field relations. The resulting linearized gravitational field relations are a version of Newton's Law of Gravitation that obeys Special Relativity. The linearized gravitational field relations show that there is a unique correspondence between the gravitational field and the electric field. For example, the Newtonian gravitational field of an isolated mass is the gravitational analog to the electric field of an isolated electric charge.

Likewise, there is an analogy to a magnetic field contained within the linearized gravitational field relations. In Maxwellian electrodynamics, a magnetic field is due to the flow of an electric charge or an electric current. In other words, the electric field surrounding an electric charge in motion will appear as a magnetic field to stationary observers. If the observers move along with the charge, they see no relative motion, and so they will only observe the charge's electric field. Thus, the magnetic field is simply an electric field that is looked at in a moving frame of reference. In an analogous fashion, the linearized gravitational field relations show that if a (gravitational) mass is set into motion and forms a mass current, then a new type of gravitational field is created that has no source and no sink. This is called the Lense–Thirring effect, or rotational frame dragging effect, in which rotating bodies literally drag spacetime around themselves.

2. *Forward's Dipole Gravitational Field Generator*

Forward [26,27] used the linearized gravitational field relations plus aspects of the Lense-Thirring effect to develop models for generating antigravity forces. One example of an antigravity generator is based on a system of accelerated masses whose mass flow can be approximated by the electrical current flow in a wire-wound torus. According to Maxwellian electrodynamics, an electric current flowing through a wire that is wrapped around a torus (or ring) causes a magnetic field to form inside the torus. If the current I in the wire increases with time, then the magnetic field, B, inside the torus also increases with time. This time-varying magnetic field in turn creates a dipole electric field, E, as shown in Fig. 3. The magnitude of the electric field at the center of the torus is given by $E = -(\mu_0 N r^2 \dot{I})(4\pi R_t^2)^{-1}$, where μ_0 is the vacuum permeability constant ($4\pi \times 10^{-7}$ H/m), N is the total number of turns of wire wound around

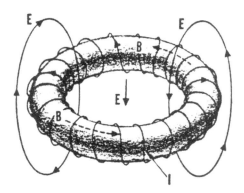

Fig. 3 **Dipole electric field generator [27].**

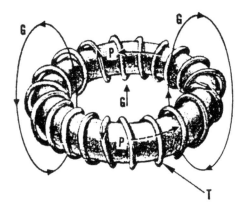

Fig. 4 Dipole gravitational field generator [27].

the torus, $\dot{I}$ is the time rate-of-change of the electric current flowing through the wire, r is the radius of one of the loops of wire, and R_t is the radius of the torus.

In a similar fashion, Forward's antigravity device is a dipole gravitational field generator. As shown in Fig. 4, a mass flow, T, through a pipe wound around a torus induces a Lense–Thirring field, P, to form inside the torus. If the mass flow is accelerated, then the P-field increases with time, and thus a dipole gravitational field, G, is created. The magnitude of the antigravitational field at the center of the torus is given by $G = (\eta_0 N r^2 \dot{T})(4\pi R_t^2)^{-1}$, where η_0 is the vacuum "gravitational permeability" constant ($\equiv 16\pi G/c^2 = 3.73 \times 10^{-26}$ m/kg),[††] N is the total number of turns of pipe wound around the torus, $\dot{T}$ is the time rate-of-change of the mass current flowing through the pipe, r is the radius of one of the loops of pipe, and R_t is the radius of the torus [27]. One should note the striking similarity between the two equations for the dipole electric and dipole gravitational fields.

Using the above equation for G, Forward [26,27] showed that we would need to accelerate matter with the density of a dwarf star through pipes as wide as a football field wound around a torus with kilometer dimensions in order to produce an antigravity field (at the center of the torus) of $G \approx 10^{-10} a_{acc}$, where a_{acc} is the acceleration of the dwarf-star-density matter through the pipes. The tiny factor 10^{-10} is composed of the even smaller η_0, which is the reason why very large devices are required to obtain even a measurable amount of acceleration. To counteract the Earth's gravitational field of 1-g requires an antigravity field of 1-g (vectored upward), and thus the dwarf-star-density material within the pipes must achieve $a_{acc} = 10^{11}$ m/s^2 in order to accomplish this effect.

Forward [5] also identified a configuration comprised of a rotating torus of dense matter that turns inside out like a smoke ring as another type of dipole gravitational field generator. As shown in Fig. 5, an inside-out turning ring of

---

[††]The vacuum "gravitational permittivity" constant is [25]: $\gamma_0 \equiv (4\pi G)^{-1} = 1.19 \times 10^9$ kg · s^2/m^3.

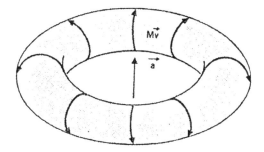

Fig. 5 Inside-out whirling dense matter torus [5].

very dense mass (M) will create an upward force (of acceleration a) in the direction of the constant mass motion (Mv, where v is the mass velocity). This is also a feature of the Lense–Thirring effect. Forward's linearization analysis generalizes all of these effects into the following two key ingredients that are required to produce antigravity forces:

1) Any mass with a velocity and an acceleration exerts many different general relativistic forces on a test mass.

2) These forces act in the direction of the velocity and in the direction of the acceleration of the originating mass.

In summary, these forces are equivalent to gravitational forces, which can be used to cancel the Earth's gravitational field.

One can also view this genre of devices as a gravity catapult machine in which the machine pushes a body away using its general relativistic antigravity forces to impart a change in velocity. A space launch operator on the ground wanting to send a payload up into orbit would just ratchet up the strength of the (upward directed) antigravity field to some value above 1-g, and after pressing the release button the payload accelerates up and away into orbit. These devices could also be placed in Earth orbit, stationed anywhere within the solar system, or even distributed throughout the galaxy in order to establish a network of gravity catapults. Space travelers could begin their trip by being launched from the catapult on the Earth's surface, and when they reach space they would jump through various catapults as needed to reach their destination.

3. Make or Break Issues: Gravitomagnetic Antigravity

We presently do not have the technology to achieve the astronomical mass densities, extreme velocities or accelerations of mass motion, and the large device dimensions required to produce large enough antigravity forces for useful propulsion. The issues we are faced with are: 1) dense materials, and 2) gravitational properties of matter. Forward [27] suggested investigating neutron–neutron interactions. One could cool a gas of thermal neutrons from a nuclear reactor to extremely low temperatures using magnetic confinement or magneto-gravitational traps, and concentrate them into a small region through the interaction of the trap's magnetic field with the magnetic moment of the neutrons. The Fermi energy[‡‡] of the bound neutrons limits the neutron density to $\sim 10^{-3}$ kg/m^3. However, the formation of putative tetraneutrons[§§] or the existence of a superconductive-type phase space condensation will create bosons that do not have this limitation. It turns out that exotic quantum states of matter such as Bose–Einstein (BE) and Fermionic condensates[¶¶] transcend the Fermi energy limit and thus possess highly unusual material properties. BE condensates were first created in 1995 and Fermionic condensates in 2003, but both are still undergoing laboratory exploration.

As for the gravitational properties of matter, we know from electromagnetism that the permeability (μ) of magnetic materials such as iron is anomalously large and nonlinear, which allows for the construction of highly efficient electromagnetic field generators. The gravitational equivalent to the magnetic permeability is a property of matter that is still largely unexplored. A material possessing an anomalously large, very nonlinear gravitational permeability (η) would be useful in the construction of highly efficient, very small scale gravitational field generators. One would expect all materials to have an η that is different from η_0 because the atoms comprising any material have quantum spin. Forward [27] reported that a rough estimate indicates there is a very small difference between η and η_0. It is thus necessary to implement a coordinated theoretical program to determine the value of η for all known forms of matter and an experimental program to find materials that might possess anomalously large or nonlinear properties that can be used to intensify time-varying gravitational fields.

Forward [27] also described an unsuccessful experimental attempt to find materials that have the property of converting time-varying electromagnetic fields into time-varying gravitational fields. This speculative property exploits the fact that the magnetic and inertial moments are combined in an atom via the usual quantum angular and spin momentum coupling. Other theoretical and experimental concepts incorporating the use of rotating magnets to generate gravitomagnetic/antigravity forces will be reviewed in Sect. IV.C.3, while similar concepts using rotating superconductors are reviewed in Chapter 5. Note particularly that Chapter 5 reviews the emerging experimental observations of Martin Tajmar in which an apparent frame-dragging effect is observed near super-cooled rotating rings as measured by ring laser gyros and accelerometers. At the time of this writing these effects were being reported but not yet independently confirmed.

[‡‡]In condensed matter physics, this is the energy of the highest occupied quantum state in a system of fermions at zero absolute temperature.

[§§]These are a hypothetical stable cluster of four neutrons, but recent empirical evidence suggests they exist. Readers should consult the technical literature for more information by using "tetraneutrons" as a search term.

[¶¶]BE or Fermionic condensates are a macroscopic collection of bosons or fermions (electrons, nucleons, or atoms) that collapse into the same quantum state when they form at near-zero absolute temperature.

B. Exact Relativistic Antigravity Propulsion

Felber [28] used the Schwarzschild solution of Einstein's general relativistic field equation to find the exact relativistic motion of a payload in the gravitational field of a mass moving with constant velocity. His analysis gives a relativistically exact (strong gravitational field condition) calculation showing that a mass, which radially approaches or recedes from a payload at a relative velocity of $v_{crit} > c/3^{1/2}$ ($v_{crit} \equiv$ critical velocity), will gravitationally repel the payload as seen by distant inertial observers. In other words, any source mass, no matter how large or small it is or how far away it is from a test body (payload), will produce an antigravity field when moving at any constant velocity above v_{crit}.

The exact relativistic strong-field condition that establishes the lower limit criterion for v_{crit} to induce antigravity repulsion of a payload (as measured by distant inertial observers in the rest frame of the source or in the initial rest frame of the payload) is given by [28]:

$$\gamma^2 > \frac{3}{2}\psi\left[1 - \frac{L^2}{GMr}\left(\frac{\psi}{3} - \frac{GM}{rc^2}\right)\right] \qquad (6)$$

In this expression, $\gamma \equiv (1 - \beta^2)^{-1/2}$ is the standard relativistic Lorentz transformation factor which is a function of the normalized relativistic velocity parameter $\beta = v/c$, $\psi \equiv \psi(r) = 1 - (2GM/rc^2)$ is the g_{00} (or time–time) component of the static Schwarzschild spacetime metric*** of a source (or central) body of mass M, L is the constant specific angular momentum of a ballistic payload of mass m, and r is the radial distance of the approaching/receding payload from M. One can solve the inequality in Eq. (6) for β (or v) under the condition that a payload far from M, such that $r \gg b$ (b is the periapsis distance of the payload from M) and $r \gg GM/c^2$, and find that the payload will become gravitationally repelled by M whenever $\gamma^2 > 3/2$ or $\beta > 3^{-1/2}$. In order to derive an exact solution, Felber considered the case for which $M \gg m$ so that the energy and momentum delivered to the payload has a negligible back-reaction on the source body's motion. He found that a strong gravitational field is not required for antigravity propulsion because a weak-field solution achieves the same results.

Felber discovered another interesting facet about this new relativistic antigravity effect. He found that there is also an antigravity field that repels bodies in the backward direction with a strength that is one-half the strength of the antigravity field in the forward direction. Thus, a stationary body will repel a test body that is radially receding from it at any $v > v_{crit}$. To delineate the propulsion benefit from

---

***A spacetime metric (ds^2) is a Lorentz-invariant distance function between any two points in spacetime that is defined by $ds^2 = g_{\mu\nu}dx^\mu dx^\nu$ where $g_{\mu\nu}$ is the metric tensor, which is a 4×4 matrix that encodes the geometry of spacetime and dx^μ is the infinitesimal coordinate separation between two points. The Greek indices ($\mu, \nu = 0 \ldots 3$) denote spacetime coordinates, $x^0 \ldots x^3$, such that $x^1 \ldots x^3 \equiv$ space coordinates and $x^0 \equiv$ time coordinate. The Schwarzschild metric is $ds^2 = -(1 - 2GM/c^2r)c^2dt^2 + (1 - 2GM/c^2r)^{-1}dr^2 + r^2(d\theta^2 + \sin^2\theta d\varphi^2)$. The corresponding metric tensor is a diagonal matrix: $g_{\mu\nu} = \text{diag}[-(1 - 2GM/c^2r), (1 - 2GM/c^2r)^{-1}, r^2, r^2\sin^2\theta]$.

this technique, Felber determined that the maximum velocity ($v_{\text{pmax-wf}}$) that can be imparted to a payload initially at rest by the weak (gravitational) field of a larger source mass moving toward the payload at constant $v > v_{\text{crit}}$ is $v_{\text{pmax-wf}} \ll c[\beta - (3\beta)^{-1}]$. For the strong-field case, the maximum velocity ($v_{\text{pmax-sf}}$) that can be imparted to the payload (initially at rest) by the larger source mass moving toward the payload at any constant v is $v_{\text{pmax-sf}} = \beta c$. Felber's analysis includes examples using black holes for the large source mass.

This form of antigravity propulsion is not too surprising because Misner et al. [29], Ohanian and Ruffini [30], and Ciufolini and Wheeler [31] report that general relativistic calculations show that the time-independent Kerr (spinning black hole) gravitational field exhibits an inertial frame-dragging effect similar to gravitational repulsive forces in the direction of a moving mass at relativistic velocities. This and Felber's exact solution are among the genre of Lense–Thirring type effects that produce antigravity forces. It is interesting to note that even though General Relativity Theory admits the generation of antigravity forces at relativistic velocities [32], they have not been seen in laboratory experiments because repulsive force terms are second and higher-order in the source mass velocity. To invent a relativistic driver for a captured astronomical body in order to use it to launch payloads into relativistic motion presents a large technical challenge for future experimenters.

C. Negative Energy Induced Antigravity

Negative energy density and negative pressure are acceptable results both mathematically and physically in General Relativity and quantum field theories, and negative energy/pressure manifests as gravitational repulsion (antigravity). Chapter 15 provides a detailed review of the various forms of quantum field theoretic sources of negative energy that are found in nature. For the purpose of the present discussion, we will confine ourselves to a discussion of how negative energy can be used to produce antigravity for the simplest case of counteracting the Earth's gravitational field. To counteract or otherwise reduce gravity merely requires the deployment of a thin spherical shell (bubble) of negative energy around an aerospace vehicle. This particular case study will serve as a useful illustrative comparison with the Newtonian antigravity case discussed in Section II.A.

We are interested only in the slow (nonrelativistic) motion, weak (gravity) field regime that characterizes the physics of the Earth, Sun, other forms of solar system matter, most interstellar matter (excluding compact dense stars and black holes), and small test masses. In this case the time–time component of the Ricci curvature tensor ($R_{\mu\nu}$) is given by $R_{00} \approx G\rho/c^2 \approx (7.41 \times 10^{-28})\rho$ m^{-2}. This is the primary quantity inside the general relativistic field equation[†††] that encodes and measures the curvature of spacetime around a source of mass-energy and characterizes the weak or strong gravity field

---

[†††]The Einstein field equation is: $G_{\mu\nu} \equiv R_{\mu\nu} - (1/2)g_{\mu\nu}R = -(8\pi G/c^4)T_{\mu\nu}$, where $G_{\mu\nu}$ is the Einstein curvature tensor and $R \equiv R_\mu^\mu$ (the trace of $R_{\mu\nu}$) is the Ricci scalar curvature. In simplest terms, this relation states that gravity is a manifestation of the spacetime curvature induced by a source of matter ($T_{\mu\nu}$).

regime for all forms of astronomical mass density (ρ). For example, the Earth's mass density is 5500 kg/m^3 so $R_{00} \approx 4.08 \times 10^{-24}$ m^{-2}, which indicates that an extremely flat space surrounds the Earth and thus the system is within the weak field regime. Gravitational physics in the weak field regime is completely described by the standard Schwarzschild spacetime metric, which leads to the usual Newtonian and post-Newtonian gravitational physics.

Two simple approaches can be used to determine the negative energy density required to counteract the Earth's gravitational field: case a) integrate Einstein's general relativistic field equation, or case b) use an already derived result from General Relativity that gives the repulsive force acceleration in terms of the spacetime metric components. For case a, we have the generalized gravitational Poisson equation from the Einstein field equation:

$$-R_{00}\sqrt{-g_{00}} = \frac{4\pi G}{c^4} Tr(T_{\mu\nu})\sqrt{-g_{00}}$$
$$= \frac{4\pi G}{c^4} T^\mu_\mu \sqrt{-g_{00}} \quad (7)$$
$$= \frac{4\pi G}{c^4} \rho^*_\mathcal{E}$$

where the definition

$$\rho^*_\mathcal{E} \equiv T^\mu_\mu \sqrt{-g_{00}}$$

is used, $\rho_\mathcal{E}^* \equiv$ rest-energy density + compressional potential energy (a.k.a. pressure), $g_{00} \equiv g_{00}(r)$ is the time–time component of the metric tensor $g_{\mu\nu}$, and $Tr(T_{\mu\nu}) \equiv T^\mu_\mu$ is the trace (sum of diagonal matrix elements) of the stress–energy–momentum tensor $T_{\mu\nu}$ (a matrix quantity that encodes the density and flux of a matter source's energy and momentum). We use the identity

$$-R_{00}\sqrt{-g_{00}} \equiv \nabla^2 \sqrt{-g_{00}}$$

to rewrite Eq. (7) as

$$\nabla^2 \sqrt{-g_{00}} = \frac{4\pi G}{c^4} \rho^*_\mathcal{E}, \quad (8)$$

where ∇^2 is the standard Laplace differential operator. The left-hand side of Eq. (8) is the gravitational potential. Now we integrate Eq. (8) once over a region of space exterior to a ball (or thin spherical shell) of rest-energy density to obtain

$$|\nabla \sqrt{-g_{00}(r)}| = \frac{GM}{r^2} \equiv g \quad (acceleration, \text{ m/s}^2) \quad (9)$$

where we use the standard spherically symmetric spacetime (or Schwarzschild) coordinate system (t, r, θ, φ) in which time t, radial space coordinate r, and angular space coordinates (θ, φ) have their usual meaning.

The second approach (case b) can be derived by recalling that in the exterior Schwarzschild spacetime around a central mass M (a ball or thin spherical shell), we have that

$$\sqrt{-g_{00}(r)} = 1 - \frac{GM}{r}$$

Because the definition is given that

$$g \equiv |\nabla\sqrt{-g_{00}(r)}|$$

we then perform the radial derivative of $(1 - GM/r)$ and again arrive at Eq. (9).

Because from Special Relativity we have that $M = \mathcal{E}/c^2$ (for a given rest-energy, $\mathcal{E}$), a negative energy state is identical to a negative mass state (see Sec. II.D) [14]. Thus we can replace the mass M in Eq. (9) with the negative energy density $-\rho_{\mathcal{E}}^* = -\rho c^2 = -Mc^2/V$ by using the volume ($V = 4\pi r^2 \delta r$) of a thin spherical shell of radius r and thickness δr, and rearrange quantities to solve for $\rho_{\mathcal{E}}^*$ to get the final result that we seek:

$$\begin{aligned}\rho_{\mathcal{E}}^* &= \frac{-gc^2}{4\pi G \delta r} \\ &= \frac{-(1.05 \times 10^{27})}{\delta r} \quad (\text{J/m}^3)\end{aligned} \quad (10)$$

where g is now the acceleration due to gravity near the Earth's surface. Equation (10) gives the negative energy density required to generate a repulsive gravitational force that counteracts the Earth's gravity field from the surface all the way up to LEO (as g in LEO is only a few percent smaller than on the surface). Any realistic value that one chooses for the bubble wall thickness δr will give a negative energy density that will always be on the order of the equivalent negative energy density of a dwarf star or neutron star. The technical challenge to implement this kind of antigravity, however, is daunting.

In the next section we discuss the case of a cosmological antigravity that is generated by a form of matter having a positive energy density and negative pressure. Another case of exotic matter having both negative energy density and negative pressure is treated in Chapter 15 in which a special form of antigravity is generated that leads to the creation of faster-than-light spacetimes such as traversable wormholes and warp drives.

D. Cosmological Antigravity

It turns out that there is already a naturally occurring antigravity force that acts throughout the universe. Actually, this force acts upon the entire spacetime structure of the universe, and it is called cosmological inflation. Cosmological inflation causes the universe to expand at an ever accelerating rate. In what follows, we examine the nature of this cosmological antigravity force and its potential breakthrough propulsion application.

1. Pressure as a Source of Gravity

Newtonian gravitation is modified in the case of a relativistic perfect-fluid (where $p \ll \rho_\varepsilon$ cannot be assumed). The stress-energy tensor $T^{\mu\nu}$ for this case is [29]:

$$T^{\mu\nu} = (\rho_\varepsilon + p)U^\mu U^\nu - pg^{\mu\nu} \tag{11}$$

where ρ is the fluid mass density, $\rho_\varepsilon \equiv \rho c^2$ is the fluid rest-energy density (or just energy density), p is the fluid pressure, U^μ is the 4-velocity vector of the fluid, and $g^{\mu\nu}$ is the metric tensor. We can contract the Einstein general relativistic field equation using the identity $g^\mu_\mu = 4$ to obtain $R = (8\pi G/c^4)T$, which is the Ricci curvature scalar. And so Eq. (11) becomes $T = \rho_\varepsilon - 3p$, which is just the trace of $T^{\mu\nu}$. Because $T = \rho_\varepsilon - 3p$, we get a modified gravitational Poisson equation:

$$\nabla^2 \phi = 4\pi G(\rho_\varepsilon + 3p) \tag{12}$$

where ϕ is the gravitational potential. It should be noted that the energy density and pressure are kept as separate terms as opposed to Eqs. (7) and (8) in the previous section. Equation (12) means that a gas of particles all moving at the same speed u has an effective gravitational mass density of $\rho(1 + u^2/c^2)$. Thus, for example, a radiation-dominated fluid generates a gravitational attraction twice as strong as one predicted by Newtonian gravity theory according to Eq. (12).

2. Vacuum Energy of Einstein's Cosmological Constant

A major consequence of the Einstein field equation is that pressure p becomes a source of gravitational effects on an equal footing with the energy density ρ_ε. One consequence of the gravitational effects of pressure is that a negative-pressure equation of state that achieves $\rho_\varepsilon + 3p < 0$ in Eq. (12) will produce gravitational repulsion (i.e., antigravity). The Einstein field equation that includes a cosmological constant Λ is:

$$G^{\mu\nu} + \Lambda g^{\mu\nu} = -\frac{8\pi G}{c^4} T^{\mu\nu} \tag{13}$$

where $G^{\mu\nu}$ is the Einstein curvature tensor. The Λ term, as it appears in Eq. (13), represents the curvature of empty space. Now if one moves this term over to the right-hand side of Eq. (13), which has become widespread practice in recent times, then

$$G^{\mu\nu} = -\left(\frac{8\pi G}{c^4} T^{\mu\nu} + \Lambda g^{\mu\nu}\right) \tag{14}$$

whereby this term now behaves like the stress-energy tensor of the vacuum, $T^{\mu\nu}_{\text{vac}}$, which acts as a gravitational source:

$$T^{\mu\nu}_{\text{vac}} = \frac{\Lambda c^4}{8\pi G} g^{\mu\nu} \tag{15}$$

One should note that the absence of a preferred frame in Special Relativity means that $T^{\mu\nu}_{\text{vac}}$ must be the same (i.e., isotropic or invariant) for all observers. There is only one isotropic tensor of rank 2 that meets this requirement: $\eta^{\mu\nu}$ (the Minkowski flat spacetime metric tensor in locally inertial frames). Therefore, in order for $T^{\mu\nu}_{\text{vac}}$ to remain invariant under Lorentz transformations, the only requirement we can have is that it must be proportional to $\eta^{\mu\nu}$. This generalizes in a straightforward way from inertial coordinates to arbitrary coordinates by replacing $\eta^{\mu\nu}$ with $g^{\mu\nu}$, thus justifying the curved spacetime metric tensor in Eq. (15). By comparing Eq. (15) with the perfect-fluid stress-energy tensor in Eq. (11), we find that the vacuum looks like a perfect fluid with an isotropic pressure p_{vac}, opposite in sign to the energy density ρ_{vac}. Therefore, the vacuum must possess a negative-pressure equation of state (according to the First Law of Thermodynamics):

$$p_{\text{vac}} = -\rho_{\text{vac}} \qquad (16)$$

The vacuum energy density should be constant throughout spacetime, because a gradient would not be Lorentz invariant. So by substituting Eq. (16) into $\rho_{\varepsilon} + 3p$, we see that

$$\begin{aligned}\rho_{\text{vac}} + 3p_{\text{vac}} &= \rho_{\text{vac}} + 3(-\rho_{\text{vac}}) \\ &= -2\rho_{\text{vac}} \\ &< 0\end{aligned} \qquad (17)$$

The vacuum equation of state is therefore manifestly negative. Last, when incorporating ρ_{vac} into the Einstein field equation as a gravitational source term, and comparing its corresponding (Lorentz invariant) stress-energy tensor $\rho_{\text{vac}} g^{\mu\nu}$ with Eq. (15), then the usual identification (or definition) is made that:

$$\rho_{\text{vac}} \equiv \frac{\Lambda c^4}{8\pi G} \qquad (18)$$

Thus the terms "cosmological constant" and "vacuum energy" are essentially interchangeable in this perspective and mean the same thing (whereupon $\rho_{\text{vac}} \equiv \rho_{\Lambda}$), which is seen in the present-day cosmological literature.

By substituting Eq. (18) into Eq. (16), one observes that a positive Λ will act to cause a large-scale repulsion of space (because this gives a negative vacuum pressure), whereas a negative Λ (giving a positive vacuum pressure) will cause a large-scale contraction of space. Because Λ is a constant, the vacuum energy is a constant (i.e., time-independent). This then implies a problem with energy conservation in an expanding universe since we expect that energy density decreases as a given volume of space increases, which is the case for the ordinary matter and cosmic microwave background that is observed in extragalactic space. In other words, the matter and radiation energy densities decay away as the universe expands while the vacuum energy density remains constant.

The cure for this apparent energy conservation problem is the vacuum equation of state given by Eq. (16). A negative pressure is something like tension in a rubber band. It takes work to expand the volume rather than work to compress it. The proof of this is as follows [33]: the energy created in the vacuum by increasing (expanding) space by a volume element dV is $\rho_{vac}dV$, which must be supplied by the work done by the vacuum pressure $-p_{vac}dV$ during the expansion of space, therefore we must have $p_{vac} = -\rho_{vac}$. In other words, the work done by the vacuum pressure maintains the constant vacuum energy density as space expands. Therefore, the vacuum acts as a reservoir of unlimited energy that provides as much energy as needed to inflate any region of space to any given size at constant energy density.

3. Dark Energy

Dark energy is an easily misunderstood form of energy in cosmology. There are two sets of evidence pointing toward the existence of something else beyond the radiation and (ordinary and dark) matter itemized in the overall cosmic energy budget.[‡‡‡] The first comes from a simple budgetary shortfall. The total energy density of the universe is very close to critical. This is expected theoretically and it is observed in the anisotropy pattern of the cosmic microwave background (CMB). Yet, the total matter density inferred from observations is 26% of critical. [Note: 26% total matter density = 4% ordinary (baryonic) matter + 22% dark matter.] The remaining 74% of the energy density in the universe must be in some smooth, unclustered form that is dubbed "dark energy." The second set of evidence is more direct. Given the energy composition of the universe, one can compute a theoretical distance versus redshift diagram. This relation can then be tested observationally.

Riess et al. [34] and Perlmutter et al. [35] reported direct evidence for dark energy from their supernovae observations. Their evidence is based on the difference between the luminosity distance in a universe dominated by dark matter and one dominated by dark energy. They showed that the luminosity distance is larger for objects at high redshifts in a dark energy-dominated universe. Therefore, objects of fixed intrinsic brightness will appear fainter if the universe is composed of dark energy. The two groups measured the apparent magnitudes of a few dozen type Ia supernovae at redshifts $z \leq 0.9$, which are known to be standard distance candles (meaning they have nearly identical absolute magnitudes at any cosmological redshift-distance).[§§§] The supernovae data strongly disfavored (with high confidence) the flat, matter-dominated ($\Omega_m = 1$, $\Omega_\Lambda = 0$) universe and the pure open universe ($\Omega_m = 0.3$, $\Omega_\Lambda = 0$) models.[¶¶¶] After this discovery, a lot

---

[‡‡‡]Dark matter and dark energy are not to be confused. Dark matter is a nonluminous, nonabsorbing, nonbaryonic form of matter that only interacts with all other forms of matter via gravitational and weak nuclear forces. Dark matter has a positive rest-energy density and a nearly negligible positive pressure. Thus, it has no beneficial application for breakthrough propulsion physics.

[§§§]In cosmology, the redshift z serves as a surrogate for distance (in light-years) or look-back time.

[¶¶¶]Ω_m = ratio of energy density contained in matter (as measured today) to the critical energy density; $\Omega_\Lambda \equiv \Omega_{vac}$ = ratio of energy density in a cosmological constant to the critical energy density; $\rho_{cr} \equiv 3H_0^2/8\pi G$ is the critical energy density, where H_0 is the present-day Hubble rate.

of attention was paid to choosing an appropriate name for this new energy. "Quintessence" was one good choice because it expresses the fact that, after cosmological photons, baryons, neutrinos, and dark matter, there is a fifth essence in the universe. More recently, the term dark energy is used more often, with quintessence referring to the subset of models in which the energy density can be associated with a time-dependent scalar field or a time-dependent cosmological vacuum energy.

In analyzing the cosmological modeling results suggested by the type Ia supernovae data, it becomes apparent that the only form of dark energy budgeted for in the models is the cosmological constant. To consider other possibilities, we evaluate the time evolution of the general relativistic conservation law for energy, $\nabla_\mu T^\mu_\nu = \nabla_\mu T^\mu_0 = 0$, where $\nu = 0$ to signify time evolution and ∇_μ is the covariant derivative (or spacetime curvature gradient), in an expanding universe as applied to the cosmological constant [29]:

$$\frac{\partial \rho_\varepsilon}{\partial t} + \frac{\dot{a}}{a}(3\rho_\varepsilon + 3p) = 0 \qquad (19)$$

where a is the scale factor of the universe and $\dot{a}$ is the time derivative of a. Equation (19) is derived using Eq. (11) in the case of a perfect isotropic fluid where there is no gravity and velocities are negligible such that $U^\mu = (1, 0, 0, 0)$, and the energy density and pressure evolve according to the continuity and Euler equations. The only way Eq. (19) can be satisfied with constant energy density is if the pressure is defined by Eq. (16). One might imagine energy with a slightly different pressure and therefore energy evolution. Define the equation of state w:

$$w = \frac{p}{\rho_\varepsilon} \qquad (20)$$

A cosmological constant corresponds to $w_\Lambda \equiv w_{\text{vac}} = -1$, matter (ordinary and dark) to $w_{\text{matter}} \approx 0$, and radiation to $w_{\text{rad}} = 1/3$.**** The earlier Riess and Perlmutter supernovae data (fixing the universe to be flat) showed that values of $w_{\text{de}} > -0.52$ for dark energy are strongly disfavored. In fact, Riess and a team of collaborators (a.k.a. the "Higher-Z Team") recently published new observational data and analysis that include a much larger survey of type Ia supernovae that are at much higher cosmological redshift [36]. The measured spectra of ancient ($z \geq 1$, or up to 10 billion light-years distance or a look-back time of up to 10 billion years ago) and recent ($z \leq 0.1$, or ≤ 1 billion light-years distance or a look-back time of ≤ 1 billion years ago) were compared and showed that there was no evolutionary change in the physics that drives type Ia supernovae explosions and their subsequent spectral luminosity output. This establishes the efficacy of using type Ia supernovae as a standard distance candle for cosmological dark energy surveys. The Higher-Z Team's results also concluded, with

---

****Nonrelativistic (ordinary and dark) matter has a very tiny positive $p \propto T_{\text{emp}}/m$ (T_{emp} is absolute temperature, m is mass), while a relativistic gas (of radiation) has $p = \rho_\varepsilon/3 > 0$.

98% confidence, that $w_{de} = -1.0$, and that this is a perpetual constant (over at least 10 billion years time) [36]. This result falsifies all quintessence models for cosmology. Therefore, a cosmological constant is consistent with the dark energy data to a high degree of precision and statistical confidence whereby we can now state that dark energy is the vacuum energy of Einstein's cosmological constant because $w_{de} = w_\Lambda = -1$ [37,38].

Equation (19) can be integrated to find the evolution of the dark energy density $\rho_{de} = \rho_\Lambda$ as a function of the cosmological scale factor a:

$$\rho_{de} \propto \exp\left\{-3 \int_a \frac{da'}{a'}[1 + w_{de}(a')]\right\} \quad (21)$$

where a' is the dummy integration variable for the scale factor. Because $w_{de} = -1\ (= w_\Lambda)$ is a constant in Eq. (21), we then obtain $\rho_{de} \propto a \exp[-3(1 + w_{de})]$ or $\rho_{de} = \rho_\Lambda \propto a^0$. This is exactly what we expected on the basis of our previous analysis in Sec. III.D.2. For a comparison with this result, one should note that $\rho c^2 \propto a^{-3}$ for (ordinary and dark) matter and $\rho_{rad} \propto a^{-4}$ for radiation such that $\rho c^2 \to 0$ and $\rho_{rad} \to 0$ as $a \to \infty$ while $\rho_{de} = \rho_\Lambda$ remains constant.

4. Cosmological Inflation as a Form of Antigravity

It has already been shown in the literature that an inflationary solution (i.e., an accelerating universe) can solve the cosmological horizon problem. We know that General Relativity Theory ties the expansion of the universe to the energy within it, so we now ask what type of energy can produce acceleration. The answer is found by examining the time–time and space–space components of the Einstein field equation [29]:

$$\left(\frac{\dot{a}}{a}\right)^2 = \frac{8\pi G}{3}\rho_\varepsilon \quad (22)$$

$$\frac{\ddot{a}}{a} + \frac{1}{2}\left(\frac{\dot{a}}{a}\right)^2 = -4\pi G p \quad (23)$$

where $\ddot{a}$ is the second time derivative of the scale factor a. One should note that the left-hand side of Eq. (22) is also the square of the Hubble parameter H, and that Eq. (22) is also known as the first Friedmann equation. Multiplying Eq. (22) by $1/2$ and then subtracting the result from Eq. (23) gives

$$\frac{\ddot{a}}{a} = -\frac{4\pi G}{3}(\rho_\varepsilon + 3p) \quad (24)$$

Equation (24) is known as the second Friedmann equation or the cosmological acceleration equation. Acceleration is defined by $\ddot{a} > 0$, so for this condition to be met we require that $\rho_\varepsilon + 3p < 0$ in Eq. (24). So inflation requires that $p < -\rho_\varepsilon/3$ in order to meet this constraint. Because the energy density is always positive, the pressure must be negative. We saw in the previous section that the accelerated expansion (or cosmological inflation), which causes supernovae to appear very faint, can be caused only by a dark (or vacuum) energy with $p = p_{\text{vac}} < 0$.

5. Breakthrough Propulsion Application of Dark/Vacuum Energy

If we could somehow harness a local amount of dark/vacuum energy, then could we use its negative pressure property to produce an antigravity propulsion effect? To answer this question we can use the experimentally measured value for $\rho_{\text{de}} = \rho_\Lambda \approx 2.4\rho_0 c^2 \approx 10^{-9}$ J/m^3, where ρ_0 is the present-day value of the total cosmological mass density of (ordinary and dark) matter [36,39]. Using this number we can work through the math and estimate that the total amount of dark/vacuum energy contained within the volume of our solar system amounts to the mass equivalent of a small asteroid. This means that its repulsive gravitational influence upon planetary orbital dynamics inside the solar system is completely inconsequential. Only on the extragalactic-to-cosmological scale will its repulsive gravitational property achieve strong enough influence over matter and spacetime. On this basis, we conclude that it is highly unlikely, if not impossible, that one will be able to invent a technology in the near future that can acquire and exploit a near-cosmological amount of dark/vacuum energy to implement a useful antigravity propulsion system. There are a number of unpublished proposals on the Internet that claim to achieve this goal for a faster-than-light "warp drive" propulsion concept using questionable theoretical approaches. However, White and Davis [40] proposed a first-order experiment that is designed to generate artificial dark/vacuum energy in the lab for the purpose of exploring a warp drive propulsion concept that is rooted in D-Brane quantum gravity theory (see Chapter 15).

IV. Miscellaneous Gravity Control Concepts

A. Artificial Gravity via the Levi-Civita Effect

Levi-Civita [41] considered the possibility of generating a static uniform (cylindrically symmetric or solenoidal) magnetic or electric field to create an artificial gravity field in accordance with General Relativity Theory. Levi-Civita's spacetime metric for a static uniform magnetic field was originally conceived by Pauli [42]:

$$ds^2 = (dx^1)^2 + (dx^2)^2 + (dx^3)^2 + \left\{ \frac{(x^1 dx^1 + x^2 dx^2)^2}{a_{\text{mag}}^2 - [(x^1)^2 + (x^2)^2]} \right\} \\ - \left[c_1 \exp\left(\frac{x^3}{a_{\text{mag}}}\right) + c_2 \exp\left(-\frac{x^3}{a_{\text{mag}}}\right) \right]^2 (dx^4)^2 \tag{25}$$

where c_1 and c_2 are integration constants that are determined by appropriate boundary conditions, and $x^1 \ldots x^4$ are Cartesian coordinates ($x^1 \ldots x^3 \equiv$ space,

$x^4 \equiv ct$ for time) with orthographic projection. The important parameter in Eq. (25) is [42]:

$$a_{\text{mag}} = c^2 \left(\frac{4\pi G}{\mu_0}\right)^{-1/2} B^{-1}$$

$$\approx (3.48 \times 10^{18}) B^{-1} \quad (\text{m}) \tag{26}$$

where a_{mag} is the constant radius of curvature of the spacetime geometry given by Eq. (25), μ_0 is the vacuum permeability constant, and B is the magnetic field intensity. The constant radius of curvature for the case of an electric field in Eq. (25) is found by replacing B in Eq. (26) with the electric field intensity E along with an appropriate change in the related vacuum electromagnetic constants (i.e., $B\mu_0^{-1/2} \to E\varepsilon_0^{1/2}$; where ε_0 is the vacuum permittivity constant, 8.8542×10^{-12} F/m). Thus, we have $a_{\text{elec}} \approx (1.04 \times 10^{27}) E^{-1}$ for the case of a static uniform (cylindrically symmetric) electric field that induces artificial gravity.

The physical meaning of the spacetime geometry described by Eq. (25) is difficult to interpret because its mathematical form is arcane. Puthoff et al. [43] recast Eq. (25) into a more transparent form using cylindrical spacetime coordinates $(t, r = a_{\text{field}}, \theta, \varphi, z)$:

$$ds^2 = -\left[c_1 \exp\left(\frac{z}{a_{\text{field}}}\right) + c_2 \exp\left(-\frac{z}{a_{\text{field}}}\right)\right]^2 c^2 dt^2$$
$$+ a_{\text{field}}^2 (d\theta^2 + \sin^2\theta \, d\varphi^2) + dz^2 \tag{27}$$

where a_{field} ($\equiv a_{\text{mag}}$ or a_{elec}) is the constant radius of curvature; time t, z-axis in space, and angular space coordinates (θ, φ) have their usual meaning. We can study Eq. (27) to better understand the spacetime geometry that the magnetic or electric field induces. The spatial part of Eq. (27), $a_{\text{field}}^2 (d\theta^2 + \sin^2\theta \, d\varphi^2) + dz^2$, is recognized as the three-dimensional metric of a "hypercylinder" (denoted by the topological product space $S^2 \times \Re$, where S^2 is the ordinary sphere and $\Re$ is the real line that defines the linear space dimension corresponding to the z-axis). Equation (27) shows that Levi-Civita's spacetime metric is simply a spatial hypercylinder with a position dependent gravitational potential.

This spacetime geometry is interesting from the standpoint that it describes a unique cylindrically shaped "trapped" space having an artificial gravity field that depends on the magnitude of the applied electric or magnetic field intensity. Puthoff et al. [43] showed that the gravitational potential inside this trapped space has the effect of slowing down propagating light beams (i.e., reducing the speed of light) to a minimum value at the center of the trapped space equidistant from the ends of the solenoidal electric or magnetic field. The magnitude of light-speed reduction is a function of the strength of the magnetic or electric field intensity that is applied by a field generator. For example, we require $B \approx 4.93 \times 10^{18}$ Tesla to generate a gravitational field that slows a light beam to $c/2$ at the center of the trapped space, assuming that the magnetic field region generated by the solenoid is 1-m in length. This greatly exceeds the B-field intensity of magnetars ($\sim 10^{11}$ Tesla). The B-field energy density for our example is $B^2/2\mu_0 \sim 10^{43}$ J/m^3, which greatly exceeds the rest-energy density of a neutron star ($\sim 10^{35}$ J/m^3). The equivalent electric field case requires $E \approx 1.47 \times 10^{27}$ V/m with a corresponding energy density $\varepsilon_0 E^2/2 \sim 10^{43}$ J/m^3. For either electric or magnetic field case, the solenoidal field creates a spacetime curvature with $a_{\text{field}} \approx 0.71$ m, which corresponds to an equivalent artificial gravitational acceleration (at the center of the solenoid) of 3.19×10^{16} m/s^2 (or 3.25×10^{15} g).

It is an interesting exercise to compare these magnetic and electric field intensities with the critical quantum electrodynamic (QED) vacuum breakdown field intensities where nonlinear quantum effects begin to appear. This is where intense fields pack enough energy to excite elementary particles out of the vacuum. The critical QED vacuum breakdown electric field intensity is $E_c = 2m_e^2 c^3/\hbar e \approx 10^{18}$ V/m while the critical breakdown magnetic field intensity is $B_c = E_c/c \approx 10^{10}$ Tesla, where E_c is defined by the total rest-energy of an electron–positron pair created from the vacuum divided by the electron's Compton wavelength and its charge, m_e is the electron mass (9.11×10^{-31} kg), e is the electron charge (1.602×10^{-19} C), and $\hbar$ is Planck's reduced constant (1.055×10^{-34} J·s). In comparing these field intensities with those from the previous example, we observe that the 1-m solenoid field intensities exceed the QED vacuum breakdown field intensities by more than eight orders of magnitude, which implies that nonlinear QED effects will appear and potentially influence, for better or for worse, one's attempt to create the trapped space geometry. The nonlinear QED effects will likely cause some kind of quantum back-reaction upon the trapped space geometry, but we are unable to predict the exact effects or their magnitude.

On the other hand, note that the artificial gravitational acceleration of 3.25×10^{15} g from the 1-m solenoid example is very extreme, and certainly not compatible with space travelers who should experience accelerations ~ 1-g over an extended duration. For the purpose of a gravity control model that requires accelerations ~ 1-g, we repeat the example of the 1-m solenoid Levi-Civita effect to create a spacetime curvature equivalent to an artificial gravitational acceleration of 1-g and ascertain the required electric or magnetic field intensities. In a 1-g gravitational (acceleration) field, the speed of light is reduced by 0.42 m/s, which corresponds to $a_{\text{mag}} = a_{\text{elec}} \approx R_\oplus$. Using this result and working the math backwards, we find that the magnetic field intensity required to achieve this effect is $B \approx 5.46 \times 10^{11}$ Tesla (energy density $\sim 10^{29}$ J/m^3) while the corresponding electric field intensity is $E \approx 1.63 \times 10^{20}$ V/m (energy density $\sim 10^{29}$ J/m^3). These field intensities are still larger than B_c and E_c by more than one to two orders of magnitude. Present-day tabletop petaWatt laser technology is beginning to reach E_c- and B_c-field intensities in the laboratory. Advanced pulsed-power technology is not too far behind. It appears reasonable to stay alert to the possibility that laser technology will evolve in the near future to achieve electric or magnetic field intensities on the order of that required to generate a 1-g artificial gravity field.

B. Gravitational Wave Rockets

Can we exploit an intense beam of gravitational wave radiation to propel a spacecraft? It turns out that Einstein's General Theory of Relativity does allow for a beam of gravitational waves to be used as a rocket propellant. Because gravitational waves are ripples in the shape (i.e., curvature) of spacetime, we can use this propellant to attain acceleration simply by ejecting one "hard vacuum" (i.e., the beam of gravitational waves) into another (i.e., the background spacetime). A common example of this propulsion effect is that of a star undergoing asymmetric octupole collapse, which achieves a net velocity change of 100 to 300 km/s via the anisotropic emission of gravitational waves [44].

We elaborate on this further by pointing out an important result from General Relativity: Because a gravitational wave has a definite energy, it therefore is a source that induces its own gravitational field. This induced field is a second-order effect in the $h_{\mu\nu}$, which is tremendously amplified in the case of high-frequency gravitational waves by the very large factor λ^{-2} (note that $\nu \propto 1/\lambda$ from wave optics, where ν is frequency and λ is wavelength) introduced by the terms quadratic in $\partial_\gamma h_{\mu\nu}$ (i.e., terms second-order in $1/\lambda$) that comprise the gravitational wave stress-energy pseudotensor [45].[††††] Therefore, the gravitational wave itself produces the background field upon which it propagates.

1. Gravitational Wave Rocket Based on Photon Rocket Spacetime Metric

Bonnor and Piper [46] performed a very lengthy and rigorous analysis for their study of gravitational wave rocket motion. They obtained the gravitational wave rocket equations of motion directly by solving the Einstein general relativistic field equation using the spacetime metric of a photon rocket [46]:

$$ds^2 = [1 - (f^2 r^2 c^{-4})\sin^2\theta - (2frc^{-2})\cos\theta - (2GMc^{-2}r^{-1})]c^2 dt_{\text{ret}}^2$$
$$+ 2c dt_{\text{ret}} dr - r^2 d\Omega^2 - (2fr^2 c^{-2}\sin\theta) c dt_{\text{ret}} d\theta \quad (28)$$

where the spherical coordinates (r, θ, φ) have their usual meaning, t_{ret} is retarded time, $d\Omega^2 \equiv d\theta^2 + \sin^2\theta \, d\varphi^2$, $M \equiv M(t_{\text{ret}})$ is the mass of a particle at the origin, and $f \equiv f(t_{\text{ret}})$ is its acceleration along the negative direction of the polar axis. Also note that Bonnor and Piper use a metric signature for ds^2 that is opposite of the one used for ds^2 in the previous sections.[‡‡‡‡] Equation (28) is the solution of the Einstein field equation having a stress-energy tensor for the photon fluid that propels the rocket: $T_{\mu\nu} = r^{-2}(2c\dot{M} - 6Mf\cos\theta)\xi_\mu \xi_\nu$, where an overdot signifies differentiation with respect to t_{ret}, and ξ_μ is a null vector

---

[††††]In the linearized general relativistic field equation for gravitational waves, $h_{\mu\nu}$ is the small first-order (weak-field) perturbation metric tensor quantity, x^γ are general spacetime coordinates, and $\partial_\gamma h_{\mu\nu}$ is the ordinary partial-derivative of $h_{\mu\nu}$ with respect to x^γ.

[‡‡‡‡]A metric signature counts how many dimensions of spacetime have a timelike or spacelike character. This is strictly a matter of choice that is established by assigning a negative sign to the timelike component(s) and a positive sign to the spacelike component(s) in ds^2, or vice versa.

(i.e., a zero-length vector in four-dimensional spacetime that is orthogonal to itself) [46]. The first term in parentheses quantifies the loss of mass and the second term quantifies the change of momentum by the rocket because of the anisotropic emission of photons.

However, the photon fluid stress-energy tensor must somehow be canceled out so that one actually solves the vacuum Einstein field equation $R_{\mu\nu} = 0$, because the gravitational waves that propel the rocket are not a physical fluid; instead they are ripples in the shape of spacetime that move through the surrounding background spacetime. Bonner and Piper therefore added terms to Eq. (28) that produce new terms within the resulting vacuum field equation, which cancel out the photon fluid stress-energy tensor in order to arrive at the equations of motion. In order to carry out their program, they found that a gravitational source loses mass by the emission of quadrupole waves and it gains momentum from recoil when it emits quadrupole and octupole waves. Thus, the terms that they added to Eq. (28) are those representing quadrupole and octupole gravitational waves.

As the rocket emits gravitational waves, it will lose mass ΔM in terms of the quadrupole oscillations induced by its internal motions because the conservation of energy guarantees that radiation reaction forces will pull down the internal energy of the rocket at the same rate as gravitational waves carry energy away [46]:

$$\Delta M = -\frac{G}{30c^7} \int_{-\infty}^{t} \tilde{j}^2 \, \mathrm{d}t_{\mathrm{ret}}$$

$$\approx -(1.02 \times 10^{-71}) \int_{-\infty}^{t} \tilde{j}^2 \, \mathrm{d}t_{\mathrm{ret}} \quad (\mathrm{kg})$$

(29)

where $\tilde{j}$ is the third derivative with respect to retarded time of the time-dependent function $j(t_{\mathrm{ret}})$ in the mass quadrupole moment $Q = m\ell^2 j(t_{\mathrm{ret}})$; m is the initial mass of the rocket; ℓ is a linear dimension associated with the rocket; the quantity inside the integral represents the power output in gravitational waves as being roughly the square of the internal motion power flow and has the dimension of [(energy)^2/time]; and the negative sign indicates that mass is being lost from the rocket. The rocket will acquire an acceleration (i.e., thrust) due to the combined rates of change of the quadrupole and octupole moments induced by its internal motions [46]:

$$f = \frac{mG\ell^5}{630c^7}(2\dot{p}\dot{q} - 3p\ddot{q} + 3q\ddot{p})$$

$$\approx (4.84 \times 10^{-73}) \, m\ell^5 (2\dot{p}\dot{q} - 3p\ddot{q} + 3q\ddot{p}) \quad (\mathrm{m/s}^2)$$

(30)

where $\dot{p}$ and $\ddot{p}$ are the first and second retarded time derivatives of $p(x') \equiv \mathrm{d}^2 j(x')/\mathrm{d}x'^2$, $\dot{q}$ and $\ddot{q}$ are the first and second retarded time derivatives of $q(x') \equiv \mathrm{d}^3 k(x')/\mathrm{d}x'^3$, x' is a dummy variable, and $k(t_{\mathrm{ret}})$ is the time-dependent function in the mass octupole moment $O = m\ell^3 k(t_{ret})$. If the rocket starts from rest at time $t_{ret} = 0$, then at a later time $t_{ret} = t_1$, the rocket will acquire a final velocity v_{final}, which is found by integrating Eq. (30) between these time limits [46]:

$$v_{final} = \frac{mG\ell^5}{315c^7} \int_0^{t_1} \dot{p}\dot{q}\, dt_{ret}$$

$$\approx (9.69 \times 10^{-73}) m\ell^5 \int_0^{t_1} \dot{p}\dot{q}\, dt_{ret} \quad (m/s) \tag{31}$$

To put things into a quantitative perspective, we use Eqs. (30) and (31) to make estimates for f and v_{final} in the simplest case of a gravitational wave rocket having a linear dimension (radius or length) of 1-m and an initial mass of 1-kg. In this case, Eq. (30) shows that the magnitude of the combined rates of change of the quadrupole and octupole moments (i.e., the combined quadrupole and octupole oscillations due to the internal motions of the rocket) will need to be $\sim 10^{72}$ in order to generate a propulsive acceleration of 1 m/s^2 and attain a final velocity of 1 m/s (after integrating the acceleration over an appropriate duration of time). If we scale our example rocket up to a realistic linear dimension of 50 m and a mass of 10^5 kg, the magnitude of the combined quadrupole–octupole oscillations will need to be $\sim 10^{58}$ in order to generate the same acceleration and final velocity. These two examples demonstrate how very important the mass and linear dimension scaling in $m\ell^5$ is for determining f and v_{final} given some physically reasonable magnitude for the combined quadrupole–octupole oscillations. We conclude from this that stellar-sized objects having $\ell \geq 10^9$ m ($\geq$ solar radius) and $m \geq 10^{30}$ kg ($\geq$ solar mass) will achieve superior propulsive $f \geq 10^4$ m/s^2 and $v_{final} \geq 10^5$ m/s for combined quadrupole–octupole oscillations of magnitude $\sim 10^2$. This is physically consistent with the gravitational wave propulsion effect produced by a star undergoing asymmetric octupole collapse. In all cases, $|\Delta M| \sim 10^{-71}$ times the internal motion power flow via Eq. (29), and so will be minute.

Baker [47] reviewed a series of patent designs for devices that would produce high-frequency gravitational waves in the lab for an exploratory study of their use in rocket propulsion. The various devices involve active elements comprised of a small mass or a system of small masses (e.g., coils and/or piezoelectric crystals on the order of sub-millimeter and less than gravitational wave wavelength in size) that undergo a rapid change in acceleration (of mass motion or angular momentum), which he defines as the "jerk" (i.e., third-time derivative motion). By applying a series of "rapid jerks" over a picosecond or less to the active elements, using strong electric, magnetic, and electromechanical driving forces, the devices will generate a significant quadrupole moment and thus emit $\geq 10^{12}$ Hz gravitational waves. The upshot is that Baker's devices rely on very fast moving, high-frequency events ($\sim 10^9$ to 10^{12} Hz) and very strong electric, magnetic, or electromechanical driving forces in order to generate very large magnitude quadrupole oscillations that give rise to large-magnitude gravitational wave radiation power output (see also [48], [49]).

2. Producing Gravitons via Quantization of the Coupled Maxwell–Einstein Fields

In the previous section we saw that it will be necessary to violently "shake things up" inside a rocket in order to produce gravitational waves for propulsion. However, it turns out that there are a few alternative methods for generating gravitational waves. These entail the production of quantized gravitational waves, called gravitons, in the laboratory via specialized electromagnetic field or elementary particle interactions. In this and the sections that follow, we review these alternative concepts.

One candidate process uses electrostatic or magnetostatic fields to annihilate an incident photon with production of a graviton. For this case, electrodynamics and gravitation may be written in Hamiltonian form and a quantization carried out in an approximation scheme. For weak gravitational fields we assume that spacetime is almost flat, and that an almost-Minkowski metric tensor is appropriate. Then we can write the spacetime metric tensor $g_{\mu\nu}$ as a Minkowski flat spacetime metric tensor $\eta_{\mu\nu}$ plus a small quantity of the first order:

$$g_{\mu\nu} = \eta_{\mu\nu} + h_{\mu\nu} \tag{32}$$

where $\eta_{\mu\nu} \equiv \mathrm{diag}(-1, 1, 1, 1)$ and $h_{\mu\nu}$ is a small, first-order metric tensor quantity representing a small perturbation in otherwise flat spacetime. We choose spacetime coordinates such that $g_{\mu 0} = \eta_{\mu 0}$, which define a time-orthogonal coordinate system in which the time axis is everywhere orthogonal to the spatial coordinate "grid" (or curves) that form a given spatial hypersurface wherein we perform our calculations. Using these assumptions we can write the Hamiltonian H for the coupled Maxwell–Einstein fields in the approximate form [50]:

$$H = H_G + H_M + \int [(h_{ij}\chi_i\chi_j/8) - 2h_{lk}F_{rl}F_{mk}\delta^{rm}]d^3x \tag{33}$$

which is written in quantum field theory natural units (i.e., $G = c = \hbar = 1$). The first term in Eq. (33) is the gravitational field Hamiltonian H_G, which contains the gravitational field variables (h_{ij}) and momenta (π_{ij}); the second term is the Maxwellian electromagnetic field Hamiltonian H_M, which contains the electromagnetic field variables (vector potential A_i, which defines the electromagnetic field-strength tensor F_{ij}) and canonical momenta χ_i; and the third term is the coupled field interaction term expressed as a three-dimensional (spatial) volume integral. The Latin indices appearing on quantities inside the coupled field interaction term represent spatial coordinates ($x^1 \ldots x^3$), and $\delta^{rm} \equiv \mathrm{diag}(1, 1, 1)$ is the three-dimensional (spatial) unit tensor. Equation (33) describes sets of gravitational and electromagnetic field oscillators. In considering the interaction term, when the theory is quantized, we see from Eq. (33) that it is made up of sums of products, each containing one gravitational field operator and two electromagnetic field operators. This interaction then implies that a photon can decay into another photon and a graviton (see Fig. 6). A careful study of this process (the math is very dense) shows that the interaction matrix elements for it do not vanish unless all three particles propagate in the same direction. However, all three particles are bosons possessing zero rest-mass (note: this is a class of

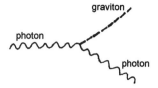

Fig. 6 Decay of photon (*left*) into a photon and graviton (*right*).

elementary particles having integer quantum spin of which photons and gravitons are representative members). Energy and momentum can be strictly conserved only if all particles propagate in the same direction. Therefore, this process cannot occur except, possibly, at energies $\gg 10^{28}$ eV corresponding to decay photon wavelengths $\ll 10^{-34}$ m [note: 1 eV = 1.602×10^{-19} J and the Planck energy is $(\hbar c^5/G)^{1/2} \approx 1.96 \times 10^9$ J $\approx 10^{28}$ eV].

Further study of the interaction term in Eq. (33) shows that graviton production becomes possible if photons are incident on a Coulomb electrostatic or magnetostatic field. The cross section will be very small. For a Coulomb scatterer containing uniform electric or magnetic fields, with linear dimensions that are large compared with the incident photon wavelength, the graviton production cross section (σ) is [50]:

$$\sigma = \frac{8\pi^2 G \ell \mathcal{E}_{\text{Coul}}}{c^4} \quad (34)$$

$$\approx (6.50 \times 10^{-43}) \ell \mathcal{E}_{\text{Coul}} \quad (\text{m}^2)$$

where $\mathcal{E}_{\text{Coul}}$ is the Coulomb scatterer energy and ℓ is the scatterer's linear dimension in the direction of propagation of the photon. Note the absence of Planck's constant in Eq. (34). This is not a totally unexpected result for the very simple reason that the interaction of two boson fields has a classical limit.

An example theoretical estimate for the graviton production cross section is that of a cubic meter of space containing 10^8 J of electric energy, which then gives $\sigma \approx 6.50 \times 10^{-35}$ m^2 according to Eq. (34). This is an exceedingly low cross section. In contrast to this, our Galaxy has a total magnetic field energy density $\approx 2 \times 10^{-13}$ J/m^3 (for a galactic magnetic field $\approx 7 \times 10^{-10}$ Tesla) contained within a galactic volume $\approx 1.80 \times 10^{61}$ m^3 possessing a linear dimension (galactic diameter) $\ell_{\text{Galaxy}} \approx 7.57 \times 10^{20}$ m (or 80,000 light-years) [51,52]. Therefore, the total galactic magnetic field energy is $\approx 3.61 \times 10^{48}$ J giving the result that $\sigma \approx 1.78 \times 10^{27}$ m^2 from Eq. (34), and thus converting approximately 1 part in 10^{13} incident photons into gravitons by this scattering process. These gravitons should be detectable near the Earth with wavelengths ($\lambda_{\text{grav}} \ll \ell_{\text{Galaxy}}$) ranging from extremely low frequencies ($\lambda_{\text{grav}} \approx 0.1$ parsec to 20 AU) to very high frequencies ($\lambda_{\text{grav}} \approx 3$ m to 0.3 cm) [53].§§§§ Thus, an astronomical experiment for detecting gravitons is possible to do with present technology.

---

§§§§1 AU (mean Earth–Sun distance) = 1.50×10^{11} m, 1 parsec = 3.26 light-years, 1 light-year = 9.46×10^{15} m.

To produce very high-to-ultrahigh frequency gravitons in the lab will require the scattering of ultrahigh-intensity petaWatt laser photons (beam intensity $\sim 10^{19}$ to 10^{34} W/m^2) from a Coulomb-electrostatic or magnetostatic field scatterer possessing a total energy $>10^9$ J. Such ultrahigh scatterer field energies can potentially be produced in the laboratory using pulsed power systems. At present the Z-Machine at Sandia National Laboratory achieves a peak magnetostatic field energy output of $\sim$ *several* $\times\ 10^6$ J from the implosion of target hohlraums, which is only 0.1% of what is required to produce a significant number of laboratory gravitons. Gravitons produced by pulsed power systems will necessarily be very high-to-ultrahigh frequency because their wavelengths will be much smaller than the linear dimension of the Coulomb scatterer.

The X-1 machine, a follow-on device to the Z-Machine, is conceptually designed to produce more than one order of magnitude higher field energy ($\sim$ *several* $\times\ 10^7$ J), which is still only 1% of what is required. However, the history of the Z-Machine program has shown that the energy and power output achieved in practice have always been a factor of two or more larger than design estimates, and that further design improvements to the implosion hohlraums led to energy and power outputs that increased by one order of magnitude within 24 months. Therefore, it is expected that the X-1 machine, if it should become operational, could be improved to the point of achieving field energies $\sim 10^8$ to 10^9 J, and that a third generation device will routinely achieve field energies $>10^9$ J within a few years after that.

The small linear dimension of the scattering field ($\ell \leq 5$ m) in a pulsed power system remains a major obstacle to achieving a sizable cross section via Eq. (34) and efficient graviton production rates. However, the role of the ultrahigh-intensity laser is to compensate for this by saturating the scattering field with an ultrahigh-intensity flux of incident laser photons, thus boosting graviton production to significant measurable levels. Because the Coulomb scattering field also polarizes the quantum vacuum zero-point fluctuations (a.k.a. virtual particle pairs), there will then be additional (nonlinear QED) scattering of the incident laser photons by the polarized vacuum resulting in the generation of second-harmonic photons [54]. It is possible that second-harmonic photons will contribute to or otherwise further boost graviton production in this scheme. Additional research will be required to quantify this contribution as well as provide an estimate for the overall graviton production rate and total graviton radiation power achieved.

We now speculate on what a graviton rocket propulsion system might look like based on the previous discussion. But first we point out that ultrahigh-intensity lasers are tabletop devices, whereas pulsed power systems have been miniaturized to fit inside commercial trucks. On this basis, we assume that an ultrahigh-pulsed power (total output energy $>10^9$ J) and ultrahigh-intensity laser system comprises a graviton rocket propulsion system that efficiently generates a high number of gravitons possessing a high radiation power output. The gravitons are high-frequency, highly collimated and possess a well-defined energy spectrum. We imagine the propulsion system to be comprised of many long ($\ell \leq 500$ m) linear arrays composed of individual ultrahigh-pulsed power implosion hohlraum segments, and each of the linear arrays are further arranged to form several cylindrically concentric super-arrays. Affixed at one end of the concentric super-array is a cylindrically concentric array of ultrahigh-intensity lasers. The lasers fire at each of the linear arrays the instant the hohlraums implode, thereby generating an intense avalanche of collimated high-frequency gravitons, which are emitted from the other end of the arrays and ejected out the rear of the rocket. Perhaps it might be better to have the linear arrays arranged to form concentric cones instead of concentric cylinders so that gravitons emitted from each of the linear arrays can merge into a single beam. There is also the benefit of an additional thrust component produced by the outgoing beam of photons generated via the photon conversion process (per Fig. 6). Further work will be required to estimate the necessary physical parameters and photon conversion efficiency in order to define the propulsive performance for this graviton rocket concept.

3. Graviton Production via Particle Accelerators

Novaes and Spehler [55] describe an alternative scheme for producing gravitons in the lab using a laser and an e^+e^- linear collider (e^+ = positively charged positron/antielectron and e^- = negatively charged electron). They evaluated the cross section for the *electron + photon → electron + graviton* reaction in the framework of linearized gravitation. They analyzed this scattering reaction considering an incident photon coming either from a laser beam (assumed flash energy of 2.5 J, same repetition rate as the electron beam pulse frequency, and root-mean-square radius of 20 μm), or from a Compton back-scattering process. They propose that the collision of laser photons of a few eV's, at small scattering angle, with an energetic electron beam is able to produce a highly collimated graviton beam possessing a defined energy spectrum.

Four different e^+e^- linear collider designs were examined by Novaes and Spehler in their study: 1) SLAC (beam energy $\mathcal{E}_{beam} = 50$ GeV and collider flux-luminosity $\mathcal{L} = 5 \times 10^{29}$ cm^{-2}s^{-1}); 2) Palmer-G ($\mathcal{E}_{beam} = 250$ GeV, $\mathcal{L} = 5.85 \times 10^{33}$ cm^{-2}s^{-1}); 3) Palmer-K ($\mathcal{E}_{beam} = 500$ GeV, $\mathcal{L} = 11.1 \times 10^{33}$ cm^{-2}s^{-1}); and 4) VLEPP ($\mathcal{E}_{beam} = 10^3$ GeV, $\mathcal{L} = 10^{33}$ cm^{-2}s^{-1}).¶¶¶¶ The laser photon energy was assumed to be 1 eV for each of these cases. For the direct laser photon scattering process, the estimated total graviton radiation power emitted (for each of the four colliders) ranges from 1.17×10^{-20} eV/s (or 1.87×10^{-39} W, for SLAC) to 1.53×10^{-18} eV/s (or 2.45×10^{-37} W, for VLEPP) [55]. For the back-scattered photon process, the estimated total graviton radiation power emitted (for each of the four colliders) ranges from 1.69×10^{-28} eV/s (or 2.71×10^{-47} W, for SLAC) to 1.32×10^{-23} eV/s (or 2.11×10^{-42} W, for VLEPP) [55]. We expect that the graviton production process outlined in the previous and following sections will yield rates and emitted radiation power many dozens of orders of magnitude higher than those estimated for the laser-e^+e^- collider scheme.

Chen [56] and Chen and Noble [57] reviewed the prospect for high-yield graviton production in the lab using new advanced particle accelerator technology, and confirmed that the most effective way to produce gravitons is through resonant

---

¶¶¶¶ 1 GeV = 10^9 eV.

photon-graviton conversion in a strong external electromagnetic field à la the Gertsenshteĭn effect, which is reviewed in the following section (see also Ref. 58). They propose to achieve the excitation of 10^{11} to 10^{13} V/m plasma waves in an advanced semiconductor or metal crystal channel collider by using either laser wakefield or side-injected laser techniques. In crystal channels, the electrostatic fields are known to be as large as 10^{16} V/m. Ultrahigh center-of-mass energies possibly approaching the Planck energy scale will be made possible by the high-field-gradient acceleration of charged particles along crystal channels that collide within these strong-focusing crystal (atomic scale) channels.

Gravitational synchrotron radiation and gravitational bremsstrahlung radiation are the two resonant excitation mechanisms that were identified by Chen [56]. In the former mechanism, the external electromagnetic field interaction with a charged particle serves as a means to bend its trajectory inside the collider, thus causing the mass of the particle to behave like a gravitational charge. Hence, a gravitational charge undergoing acceleration along a curved path will emit gravitons in the form of gravitational synchrotron radiation. For this case, the estimated total rate of graviton emission is $\sim 10^{-6}$ to 10^3 gravitons per second with a frequency of 70 to 600 kHz using the design parameters of a few conventional high energy storage ring facilities [56]. In the latter mechanism, the collision of e^+e^- beams in a high energy linear collider causes a substantial fraction of beam energy to be lost through bremsstrahlung when particles are bent by the strong collective macroscopic electromagnetic field of the oncoming beam. In this case, the very intense collective field intrinsic to the colliding beams ($\sim 10^4$ Tesla) and the very intense bremsstrahlung that penetrates through such a field gives rise to resonant excitation of gravitational bremsstrahlung radiation. For this case, the estimated total rate of graviton emission is $\sim 10^{-25}$ to 10^{-22} incoherent gravitons per second and 10^{-15} to 10^{-3} coherent gravitons per second using the design parameters of a few conventional high energy storage ring facilities [56]. This result should not be too surprising because high energy accelerators are not designed for optimized gravitational wave production. Unfortunately, Chen does not provide any gravitational synchrotron and bremsstrahlung radiation emission estimates using advanced crystal channel collider technology, but only states that their length scale will have to be enormous (~ 100 km for a single crystal) to achieve graviton yields that are many dozens of orders of magnitude larger than what could be achieved using conventional accelerator technology.

As a final note, it is important to point out that a mobile (~ 1 GeV) electron accelerator less than 3-m in diameter and weighing a few tons was designed by R. R. Wilson in 1952 [59]. Wilson's mobile miniaturized accelerator was designed to consume electrical power of a few kilowatts and achieve a continuous γ-ray beam flux-luminosity $\sim 10^8$ cm^{-2}s^{-1} for a beam projected out to 1-mile distance with an intensity $\sim 10^{14}$ photons per second. A mobile miniaturized accelerator could achieve beam parameters that are many orders of magnitude larger than Wilson's original design using present-day accelerator technology. Therefore it is not out of the question to imagine that a modern-day version of Wilson's mobile miniaturized electron accelerator could be properly scaled and optimized for use as a gravitational wave (graviton) rocket engine. The question still remains whether the emitted graviton beam power will be of large enough magnitude to produce any meaningful rocket propulsion. This is addressed in Sec. IV.B.5.

4. Gertsenshteĭn Effect

In this section we review the genre of schemes for producing gravitational waves that are based upon the Gertsenshteĭn effect. Gertsenshteĭn [60] described a new resonance phenomenon in which a propagating electromagnetic wave can couple its electromagnetic field-strength tensor to that of a transverse background electromagnetic field to give rise to a nontrivial stress-energy tensor, which serves as a source for the linearized Einstein field equation to excite a gravitational wave. In quantum language, this corresponds to a mixing between the propagating photon and a graviton via a Yukawa-type coupling mediated by a virtual photon from the background field. Gertsenshteĭn's embodiment of this was that a monochromatic light wave, of angular frequency ω, propagating through a stationary uniform transverse magnetic field produces an outgoing gravitational wave, also of angular frequency ω, with an amplitude that is proportional to the distance traveled in the background magnetic field. The source of this gravitational wave is the mixed stationary-radiative term of the total electromagnetic field-strength tensor, which is comprised of static and radiative magnetic fields. The Gertsenshteĭn effect is considered to be the primary method for the laboratory generation of gravitational waves.

Gravitational waves produced in this manner are often called Gertsenshteĭn waves (GWs). The production of GWs could not be tested in the lab because the light and magnetic field intensities required to observe the effect were not available in the 1960s. Gertsenshteĭn estimated that only astronomical processes could be capable of producing GWs. The very high light and magnetic field intensities required to produce GWs became available with the advent of tabletop ultrahigh-intensity lasers and ultrahigh-pulsed power technologies in the mid-1990s.

There are several variations of the Gertsenshteĭn effect published in the literature. Pustovoĭt and Gertsenshteĭn [61] proposed generating synchrotron radiation via ultrarelativistic particles moving in a constant magnetic field to produce GWs. A variation of this was the proposal by Sushkov and Khriplovich [62] to generate synchrotron radiation via ultrarelativistic particles moving in the Coulomb electric field of a huge electrical charge. Grishchuk and Sazhin [63] devised a modified Gertsenshteĭn effect in the form of an electromagnetic resonator with a toroidal shape and rectangular cross section in which a periodic electromagnetic field is produced. The periodic electromagnetic field is the time-dependent part of the electromagnetic field stress-energy tensor in the general relativistic field equation, and thus serves as the source of GWs. The result is a standing cylindrical gravitational wave that is produced in the focal region of the resonator. The interference focusing of the gravitational radiation within the resonator occurs via the interference of coherent gravitational waves freely propagating from different parts of the resonator, because gravitational waves cannot be reflected by the material walls of the resonator cavity. Portilla and Lapiedra [64] developed a concept that produces both electromagnetic and gravitational waves by shaking an electric charge in a homogeneous stationary magnetic field while the system is inside either a homogeneous or inhomogeneous dielectric medium. Their work dramatically improved upon the Gertsenshteĭn effect and its ability to produce GWs. And last, Navarro et al. [65] take the previous scheme in another direction by proposing that high-frequency GWs can be produced by irradiating a dielectric film in a resonant cavity.

The early Russian schemes all suffer from extremely low electromagnetic-to-GW conversion efficiency, on the order of or lower than that estimated for the photon decay process described in Sec. IV.B.2, and therefore, extremely low GW luminosity (or power) output. The generic GW luminosity ($\mathcal{L}_{GW}$) produced by the schemes described previously is given by [64]:

$$\mathcal{L}_{GW} = \frac{84\pi GB^2 \mathcal{F}_{EM} \Delta}{5(n^2 - 1)^2 \omega^4} \tag{35}$$

where B is the static magnetic field intensity inside the resonator cavity, $\mathcal{F}_{EM}$ is the incident electromagnetic (light) flux with angular frequency ω, n is the index of refraction of the diffracting dielectric medium inside the resonator, and Δ is a dimensionless parameter that depends on the combined refractive index of the dielectric medium and vacuum pressure inside the resonator. Equation (35) shows how important it is to use ultrahigh field intensities to produce a measurable GW luminosity. A better quantitative measure for the strength of the Gertsenshteĭn effect is given by the ratio of $\mathcal{L}_{GW}$ to incident electromagnetic luminosity, $\mathcal{L}_{EM}$ [65]:

$$\frac{\mathcal{L}_{GW}}{\mathcal{L}_{EM}} = (5 \times 10^{-9}) \left(\frac{B}{1 \text{ Tesla}}\right)^2 \left(\frac{10^{-9} \text{ Torr}}{p}\right)^2 \\ \times \left(\frac{2\pi \times 10^{10} \text{ Hz}}{\omega}\right)^2 \left(\frac{\mathcal{Q}_f}{5,000}\right) \tag{36}$$

where p is the vacuum pressure inside the resonator and $\mathcal{Q}_f$ is the resonator's quality factor (defined as 2π times the ratio of the energy stored in a resonator to the energy lost per cycle). As an illustrative example, we use Eq. (36) to estimate the $\mathcal{L}_{GW}$ produced if the National Institute of Standards and Technology's Synchrotron Ultraviolet Radiation Facility (SURF) were used to provide the incident light flux. The SURF UV beam luminosity is $\mathcal{L}_{EM} \sim 10^6$ W with $\omega \approx 2\pi \times 10^{12}$ Hz, and we assume that the resonator has a magnetic field intensity $B \approx 10$ Tesla with $p \sim 10^{-9}$ Torr and $\mathcal{Q}_f = 5000$ (typical for many resonator designs). Using these parameters, we can expect to generate GWs with $\mathcal{L}_{GW} \approx 50$ μW, which should be within reach of carefully designed high-frequency gravitational wave detectors. It is beyond the scope of the present discussion to review the large number of proposed gravitational wave detectors and their underlying physics.

5. Make or Break Issues: Gravitational Wave Rockets

According to General Relativity, gravitational waves (or gravitons) must propagate at the speed of light. This means that the exhaust velocity (v_e) of a gravitational wave rocket will also be the speed of light, as it is for photon rockets. And the specific impulse (I_{sp}) of a gravitational wave rocket is $I_{sp} = v_e/g = c/g \approx 3.06 \times 10^7$ s, as it also is for photon rockets. This means that gravitational wave rockets will operate at the maximum efficiency possible. Furthermore, the propulsive performance (β) of a rocket propulsion system is measured by its ability to generate thrust (F_{thrust}) for a given unit of exhaust jet power (P_{jet}): $\beta = F_{thrust}/P_{jet} = 2\varepsilon_{ff}/v_e$, where ε_{ff} (≤ 1) is the efficiency involved in the conversion of stored energy into the kinetic energy of the exhaust stream. For conventional chemical rockets (e.g., the space shuttle main engines), $v_e \approx 4.45 \times 10^3$ m/s and $\varepsilon_{ff} \approx 0.5$, hence $\beta \approx 2.25 \times 10^{-4}$ N/W. For gravitational wave rockets, $\beta = 2\varepsilon_{ff}/c \approx (6.67 \times 10^{-9})\varepsilon_{ff}$ N/W, where ε_{ff} will be determined by the particular mechanism chosen to produce gravitational waves. By comparing these two results, we see that gravitational wave rockets require $>10^4$ times the jet power of chemical rockets in order to deliver the same 1 N of thrust.

In the previous sections we learned that there are numerous proposals for producing gravitational waves in the lab. The make or break issues for gravitational wave rockets reside within the ability of any one of the proposals to achieve the production of high-intensity, high-frequency gravitational waves with reasonable efficiency. The electromechanical quadrupole–octupole technique will require a tradeoff between the mass and size scaling of the rocket and the magnitude of the internal quadrupole-octupole oscillations that can be achieved. Our analysis showed that only astronomical-sized objects having physically reasonable internal quadrupole–octupole oscillations can achieve high thrust and high final velocities. If gravitational waves, or gravitons, are produced via electromagnetic-gravitational resonance or resonant photon decay processes, then the make or break issues reside within the ability to achieve the required ultrahigh (external) electromagnetic field intensities. Tabletop ultrahigh-intensity laser technology has reached the threshold where this becomes possible to do. However, producing gravitational waves via conventional and advanced particle accelerator techniques does not appear to be feasible at present because the gravitational beam luminosity and power output is estimated to be too minute for effective rocket propulsion. More theoretical and nascent experimental work needs to be done in all of these areas to identify the best approach for producing gravitational waves at a level that will support a possible rocket propulsion application.

C. Quantum Antigravity Propulsion

Quantum antigravity can be found within the very large genre of quantum gravity theories in which repulsive gravity terms appear as quantum corrections to the classical Newtonian gravitational force law. Generally, one can derive such correction terms by quantizing the Einstein general relativistic field equation or by starting with a particular type of quantum field theory (e.g., supersymmetric field theory, quantized five-dimensional Kaluza–Klein unified field theories, quantum superstrings/D-Brane theory, quantum loops or knots, Yang–Mills theories, etc.) and work backward to find the corresponding gravity theory. The particular mathematical form and quantitative magnitude that quantum correction terms can have depends totally upon the quantization procedure and order of approximation used in a given quantum gravity theory. However, the linearized semi-classical quantum gravity theory is related to Einstein's classical nonlinear General Relativity Theory whereby the former uniquely implies the latter, provided that the graviton, which exchanges the gravitational force between two massive particles or photons, is a pure spin 2 particle. In this theory, the stress-energy tensor of the matter fields is quantized while gravitation (via the Einstein curvature tensor) is still treated classically. Semi-classical quantum gravity is a quantum field theory in curved spacetime that has been successful in reproducing a few of the predictions and many of the foundational precepts of General Relativity Theory.

A particular example of what a quantum antigravity correction term looks like was derived in 1984 by R. L. Forward and this author, with instruction provided by R. P. Feynman and M. Scadron, during a summer quantum gravity seminar sponsored by the Hughes Research Laboratories in Malibu, California. We began by studying the Feynman quantization procedure for the case of single-photon exchange between two charged particles, which tells us about the underlying nature and quantum corrections to the static Coulomb force. From this study we discovered that the same is also true for the case of single-graviton exchange between two massive spin 0 particles in connection with the static Newtonian force. By applying Feynman's quantization procedure [66–68] to the linearized Einstein field equation in the nonrelativistic limit, we derived the following static graviton-exchange potential [$V_{grav}(r)$] for two spin 0 particles undergoing a gravitational interaction:

$$V_{grav}(r) = -\frac{Gm_1m_2}{r} + \frac{4\pi G\hbar^2}{c^2}\delta^3(r) \qquad (37)$$

where m_1 and m_2 are the masses of the interacting particles, r is their radial separation, and $\delta^3(r)$ is the three-dimensional Dirac δ-function with r the position vector of some reference point in space. The first term in $V_{grav}(r)$ is immediately recognized as the attractive Newtonian gravitational potential, while the second quantum correction term is repulsive. Also, the second term is independent of the interacting particle masses and can only be measured for bound quantum s-states because the product of the coefficient $4\pi(G\hbar^2/c^2) \approx 10^{-94}$ with the δ-function gives only a minute physical effect at the atomic scale. The second term happens to be analogous to the usual quantum correction to the Coulomb or nuclear force. If the two particles were to have non-zero quantum spin, then $V_{grav}(r)$ will be modified by additional spin–orbit and spin–spin correction terms. Furthermore, there are additional velocity-dependent corrections to $V_{grav}(r)$ that generate the general relativistic post-Newtonian modifications of the classical equation of motion of a particle in a gravitational field.

But the most important characteristic to observe about the quantum antigravity correction term in $V_{grav}(r)$ is that its magnitude is incredibly minute, only affecting bound quantum s-states. In general, quantum gravity correction terms at any level of approximation, whether gravitationally repulsive or attractive, will have coefficients $\sim G(\hbar^\delta/c^\kappa)$ (for $\delta, \kappa > 1$), and therefore will not have a measurable impact on any macroscopic system that embodies any form of breakthrough propulsion. Because these quantum corrections are so minute, and because there is no single universally accepted quantum gravity theory to work with, investigators have had little reason to look into the potential application of quantum gravity correction terms to breakthrough propulsion physics.

However, this is not the entire story because there are many interesting quantum field theoretic phenomena that exist outside of that which arise in quantum corrections to Newtonian gravity. In what follows, we review the recent discovery of antigravity forces that arise within both QED vacuum fluctuation and nonretarded quantum interatomic dispersion force theories in curved spacetime, as well as two highly speculative concepts that invoke some form of quantum unified field theory.

1. Antigravity via Quantum Vacuum Zero-Point Fluctuation Force

Calloni et al. [69,70] explored the possibility of verifying the equivalence principle for the zero-point energy of QED. They used semi-classical quantum gravity theory to evaluate the net force produced by quantum vacuum zero-point fluctuations (ZPF) acting on a rigid Casimir cavity in a weak gravitational field. Their analysis assumed the rigid Casimir cavity to be a nonisolated system at rest in the Earth's gravitational field, which is modeled using the standard Schwarzschild spacetime metric geometry, so that they could evaluate the regularized (or renormalized) stress-energy tensor, $\langle T_{\text{vac}}^{\mu\nu} \rangle_{\text{ren}}$, of the quantized vacuum electromagnetic field between two plane-parallel ideal metallic plates lying in a horizontal plane. $\langle T_{\text{vac}}^{\mu\nu} \rangle_{\text{ren}}$ encodes the Casimir effect, which has a negative energy density and a negative pressure along the vertical (acceleration) axis between the plates. (See Secs. II and V.B.1 in Chapter 18 for more information about the Casimir effect.) Their results agreed with the equivalence principle because they showed that quantum vacuum ZPF (i.e., virtual quanta) do gravitate because the energy of each ZPF mode is redshifted by the factor $(-g_{00})^{1/2} = [1 - (2GM/c^2 r)]^{1/2}$ even though the modes remain unchanged. In other words, the electromagnetic vacuum state in a weak gravitational field is redshifted. This effect remains true for strong gravitational fields.

The resulting antigravity force (F_{CasGrav}) derived by Calloni et al. is [70]:

$$F_{\text{CasGrav}} = \frac{\pi^2 A \hbar g}{180 c d^3}$$
$$\approx (1.89 \times 10^{-43}) \frac{A}{d^3} \quad (\text{N}) \tag{38}$$

where A is the area of the plates and d is their separation. Equation (38) states that a Casimir device in a weak gravitational field will experience a tiny push in the upward direction (i.e., the opposite direction with respect to the Earth's gravitational acceleration). This is consistent with the interpretation that the negative Casimir energy in a gravitational field will behave like a negative mass (see, e.g., Sec. II.D). F_{CasGrav} is actually the sum of two separate force terms: the first term arises from the Casimir energy encoded in $\langle T_{\text{vac}}^{00} \rangle_{\text{ren}}$, which is interpreted as the Newtonian repulsive force on an object with negative energy, and the second term arises from the pressure along the vertical (acceleration) axis, which is interpreted as the mass contribution of the spatial part of the stress-energy tensor. To evaluate F_{CasGrav} for the case of any gravitating body of interest, one must replace g in Eq. (38) with Eq. (2).

Calloni et al. further point out that a real Casimir cavity is an isolated system in which the actual (total) resulting force is the Newtonian force on the sum of the rest-Casimir energy and rest-mechanical mass whereby the contribution of the vacuum ZPF leads to a gravitational repulsion (F_{CGexp}) on the Casimir device that is given by [70]:

$$F_{\text{CGexp}} = \frac{1}{4} F_{\text{CasGrav}}$$
$$\approx (4.73 \times 10^{-44}) \frac{A}{d^3} \quad (\text{N}) \tag{39}$$

which is the force that should be experimentally tested. Equation (39) takes into consideration that the contribution to the total force on a real cavity resulting from the spatial part of the stress-energy tensor is balanced by the contribution from the mechanical stress-energy tensor. Given that the typical dimensions of a Casimir device are very small, it appears that F_{CGexp} will be very difficult, if not impossible, to measure using present-day lab technology.

However, Calloni et al. propose an experimental device that could significantly magnify the repulsive force up to a measurable scale. Their proposed device is a multilayered series of rigid Casimir cavities with each cavity consisting of two thin metallic disks that are separated by a dielectric material which is inserted to maintain rigidity. They suggest SiO_2 for the dielectric material because it is an efficient dielectric with low absorption over a wide range of frequencies, and it is an inexpensive material that is easy to fabricate into layers. The introduction of the dielectric material is equivalent to enlarging the optical path length by the refractive index n so that the cavity plate separation $d \mapsto nd$. The Casimir effect has been tested down to plate separations ~ 60 nm, while separations ≤ 10 nm is possible with present technology. At ≤ 10 nm distances, dielectric absorption and finite conductivity are expected to decrease the effective Casimir pressure compared to a cavity comprised of perfect mirrors. For example, a plate separation of 6.5 nm corresponds to a decreasing factor ζ of 0.07 for plates made of aluminum. Finite temperature and plate surface roughness could also introduce additional corrections to the Casimir pressure. Calloni et al. propose to magnify the total force by using $N_\ell = 10^6$ layers of rigid cavities with each cavity having a diameter of 35 cm and thickness of 100 nm, for a total device thickness of 10 cm.

All these engineering factors taken together led Calloni et al. to recast F_{CGexp} into the following new form [70]:

$$F_{\text{CGexp}} \approx \zeta N_\ell \frac{\pi^2 A \hbar g}{720 c (nd)^3}$$
$$\approx (4.73 \times 10^{-44}) \frac{\zeta N_\ell A}{(nd)^3} \quad (\text{N}) \tag{40}$$

Calloni et al. also suggest that a feasible experiment will require modulating F_{CGexp} in order to obtain a measurable force. They are investigating the possibility of modulating ζ by varying the temperature in order to induce a periodic transition from conducting state to superconducting state. They estimate that doing this could achieve $\zeta_{\max} \approx 0.5$, and thus produce a force $F_{\text{CGexp}} \sim 10^{-14}$ N at a modulation frequency on the order of 10s of mHz for $d = 5$ nm and $n = 1.46$ (for SiO_2 dielectric). This result is more than two orders of magnitude larger than the force that the VIRGO gravitational wave antenna is expected to detect at several 10s of Hz. If one could fabricate a device consisting of 10^9 layers, then $F_{\text{CGexp}} \sim 10^{-11}$ N. This suggests that cavities made from thin-film deposited surfaces or photonic band-gap materials would be the best approach for fabricating a multilayer Casimir device.

Bimonte et al. [71,72] also derived Eq. (38) for this very same problem by using Green-function techniques in the Schwinger–DeWitt quantum ether prescription for $\langle T_{\text{vac}}^{\mu\nu}\rangle_{\text{ren}}$ in a curved spacetime. They also computed the weak gravitational field-induced correction terms for the Casimir pressure on the plates $\langle T_{\text{vac}}^{00}\rangle_{\text{ren}}$ and the total energy $\mathcal{E}_{\text{CasGrav}}$ stored in the Casimir device, which is given by [71]:

$$\mathcal{E}_{\text{CasGrav}} = -\frac{\pi^2 A \hbar c}{720 d^3}\left(1 + \frac{5}{2}\frac{gd}{c^2}\right) \quad (J) \tag{41}$$

The correction terms for the different (measurable) physical quantities of interest are generally $\sim g/c^2$.

Finally, Calloni et al. point out that the overriding concern with performing an experiment to test F_{CGexp} is whether cavities can be made sufficiently rigid, if the effect of surface roughness and defects can be quantified to improve the force estimate, and if the necessary signal modulation can be achieved in the lab. However, micro- and nano-manufacturing is maturing to the point where rigidity, surface roughness, and close plate separations are becoming routinely controllable. While the numerical estimate for F_{CGexp} is quite feeble, it is still significant because it is at the very low end of the macroscopic scale, and it might be possible to devise advanced methods to magnify the force to a magnitude that benefits a propulsion application. However, the upward force will have to be larger than the weight of the propulsion system in order to achieve levitation. This could be very difficult to do, but this is a concept ripe for further exploration.

2. Antigravity via Nonretarded Quantum Interatomic Dispersion Force

Pinto [73] evaluated the net lifting force produced by nonretarded electrostatic dipole–dipole interactions (i.e., nonretarded van der Waals dispersion forces) acting on a quantum system of polarizable particles in a curved spacetime. The foundation of Pinto's study was the original discovery made by Fermi [74] that classical electrostatic theory must be reformulated in a curved spacetime in order to properly evaluate the effects of gravitation upon the Coulomb electric field of a single charged particle. In this case, the Laplace equation of electrostatics for a single charged particle can be generalized in the presence of a gravitational field and then extended to show that a system of classical charged particles undergoes a gravity-induced self-lifting force. Fermi and other investigators arrived at this counterintuitive result by computing the gravity-induced self-force acting on an isolated electric dipole in a weak gravitational field and showing that the self-force (times dipole size) is exactly equal to the gravitational equivalent of the electrostatic internal energy of the dipole.

The net gravity-induced (electrostatic levitation) self-force (F_{DipGrav}) is given by [73]:

$$F_{\text{DipGrav}} = \left(\frac{q_e^2}{4\pi\varepsilon_0 r}\right)\left(\frac{g}{c^2}\right) \quad (\text{N}) \tag{42}$$

where q_e is the electric charge on a particle and r is the radial distance between two charged particles in the dipole. There is an additional term of order g^2/c^4 in F_{DipGrav} that is neglected because it is negligible in magnitude. Equation (42) states that an electric dipole will experience a push in the upward direction (opposite direction with respect to the Earth's gravitational acceleration), that is, the dipole undergoes self-acceleration in which one charged particle in the dipole appears to be chasing the other charged particle. As an example, for a dipole comprised of two charges (e.g., an electron-proton system) held at fixed r to levitate in the Earth's gravitational field, r would have to be $\sim 10^{-15}$ m (the size of an atomic nucleus). An experiment to test this prediction on such a small scale is too difficult to control or measure.

An energy analysis done by Pinto showed that there is a distance r between two charges (each of rest-mass m_0) in a dipole (of mass $M_{\text{dip}} = 2m_0$) such that their electrostatic potential energy, $U_{\text{dip}} = -q_e^2/4\pi\varepsilon_0 r$, becomes equal to the unrenormalized mass of the system as $r \to \infty$. At this distance, the effective total gravitational mass $M_{\text{dip}} + U_{\text{dip}}/c^2 = 0$ and the self-force alone can support the dipole at rest against its own weight. The self-acceleration of the dipole is such that the acceleration process can continue indefinitely, which poses a problem for energy conservation because the dipole can be left to self-accelerate for an arbitrary period of time and then stopped to harness the resulting kinetic energy. This process could be used to extract unlimited energy from the system. Pinto claims that there is no conflict with energy conservation because the renormalized inertial mass of the accelerating system is $M_{\text{dip-ren}} = M_{\text{dip}} + U_{\text{dip}}/c^2 = 0$ and the total energy of the system is zero at all times regardless of speed. This claim requires re-evaluation because there are subtle boundary conditions involved that might have been overlooked in the analysis.

Fermi's discovery led to a new subfield of research devoted to the study of electrodynamics and dipole and interatomic dispersion forces in a curved spacetime. Pinto's theoretical program extended the result of these studies by considering a system of polarizable atoms and adopting an approach in which the effect of a gravitational field in General Relativity is modeled as an effective optical medium. In other words, the spacetime vacuum is treated as a nonuniform optical medium with a varying index of refraction that defines the components of a flat spacetime metric geometry [43,75]. There is no spacetime curvature due to sources of matter in this model, instead its equivalent general relativistic effects (i.e., gravitation) are produced by varying the vacuum index of refraction, comprised of the vacuum electromagnetic permittivity and permeability parameters, in response to the presence of matter sources. Pinto's lengthy analysis gives the van der Waals dispersion self-force for two polarizable atoms in a curved spacetime (i.e., a weak gravitational field) as [73]:

$$F_{\text{vdWGrav}} = -U_{\text{vdW}}^{(0)} \left(\frac{g}{c^2}\right)$$
$$= -\left(-\frac{6q_e^2 a_0^5}{4\pi\varepsilon_0 r^6}\right)\left(\frac{g}{c^2}\right) \quad (43)$$
$$\approx (2.44 \times 10^{-57}) \frac{q_e^2}{r^6} \quad (N)$$

where a_0 is the Bohr radius (5.292×10^{-11} m), r is the radial distance between two atoms, and $U_{\text{vdW}}^{(0)}$ is the flat spacetime van der Waals (interatomic potential) interaction energy to second-order in quantum perturbation theory. Pinto used Eq. (43) to estimate the gravity-induced self-acceleration (a_{lift}) for the case of two hydrogen atoms in their ground state at $r = 20a_0$, and found that $a_{\text{lift,H}} = F_{\text{vdWGrav}}/2m_H \approx 4 \times 10^{-15}$ m/s² (m_H = mass of hydrogen atom). For the case of two positronium (Ps) atoms, he found that $a_{\text{lift,Ps}} \approx 8 \times 10^{-12}$ m/s².

Pinto's strategy is to dramatically magnify F_{vdWGrav} to a large enough magnitude that it becomes viable for space propulsion applications. He claims that this can be done by manipulating $U_{\text{vdW}}^{(0)}$, which is dependent on the atomic polarizability and strongly affected by the quantum state in which the atoms are prepared. Interatomic forces can also be manipulated by means of external electromagnetic fields that can transform van der Waals forces into a first-order interaction. He evaluated a number of schemes and settled on the following techniques for manipulating dispersion forces: 1) excitation of polarizable atoms to Rydberg states in external time-dependent electric fields, 2) polarizability resonant enhancement by laser radiation, and 3) laser-induced near-zone orientational average of the dispersion force. Also, in order to generate a macroscopic self-lifting force, it will be necessary to apply these techniques to a cluster of trapped atoms because the total self-lifting force acting on the center-of-mass of a trapped gas composed of N_a identical polarizable atoms is N_a^2 times the self-lifting force acting on a single pair of interacting atomic dipoles. Item 1 has a two-part contribution to the magnification of the self-lifting force: a) one part from $\alpha^2(\omega)E^2$ due to the effect of external time-dependent electric fields on atomic polarization, where $\alpha(\omega)$ is the atomic polarizability as a function of the electric field frequency ω and E is the electric field intensity; and b) another part from using highly-excited Rydberg atoms (with principal quantum number $n_p \gg 1$ and Bohr radius $a_n = n_p^2 a_0$) whose polarizability scales as n_p^7. Item 2 leads to a magnification by factors of $\alpha(\omega)/\alpha_0 \approx 10^3$ to 10^5 (α_0 is the static value of the polarizability) via detuning of the laser excitation radiation frequency from the nearest atomic transition resonance of the atoms in the trapped cluster. Item 3 leads to a further magnification due to the effect of the incident laser radiation on the dispersion force being averaged over all directions, which changes the interatomic potential ($\propto 1/r^6$) into a gravity-like $1/r$ potential.

Pinto's study suggests that the combined effect of items 1 to 3 will magnify the self-lifting force to the point where a cluster of trapped atoms will not only hover unsupported in the Earth's gravitational field, but will also generate an additional upward thrust. On the basis of extensive theoretical and empirical studies, along with the typical parameters for laboratory laser and optical atomic matter trap technologies, he estimates that $a_{lift} \geq 1.5\text{-}g$ (in the upward direction). Trapped atom gravimeters can be used to observe this effect in the lab. Pinto also points out that other polarizable systems such as nanoparticles, microspheres, and quantum dots can be used in place of atoms. The trapping of latex spheres into a form of optical matter by means of intense laser radiation has already been demonstrated in the lab. In addition, an analogy to the item 1 to 3 manipulations that produce dramatically enlarged polarizabilities in trapped interacting nanoparticles and microspheres have also been demonstrated in the lab.

Pinto proposes a levitation propulsion thruster in which the combined system of trapped interacting polarizable particles and external confining fields forms a single thruster element comprising a fraction of the mass of the entire vehicle. The reaction of the self-lifting force exerted by this element against the external confining fields results in the transfer of force (thrust) to the entire vehicle. In order to achieve levitation, this requires that the upward thrust per polarizable particle be larger than its own weight if the fraction of the thrusting mass is smaller than the mass of the rest of the vehicle. The propulsive levitation condition is expressed as [73]: $F_{thrust} = (M_{veh} + m_A N)g$ or $F_{thrust}/m_A N g \geq 1$, where F_{thrust} is the total gravity-induced thrust, M_{veh} is the vehicle mass, m_A is the mass of individual polarizable particles, and N is the total number of trapped polarizable particles.

Pinto identified numerous technical challenges that will have to be overcome before this concept can be put to practice. One challenge is that polarizability resonant enhancement also leads to atomic transitions and decay that result in the recoil and evaporation of atoms from inside the trap. Another is the difficulty of maintaining continued confinement of a trapped cluster of polarizable particles in a specific three-dimensional array while the cluster is simultaneously opposing the amplified interatomic forces and producing thrust. The confinement lifetime of trapped polarizable particles is finite and there is the possibility that these particles might be evaporated away or destroyed in a time that is too short to deliver the required thrust to the vehicle. Therefore, a scheme for active repopulation of the trapped cluster will have to be developed. The design of particle cluster traps and associated external confinement fields are of primary importance to determine the effective thrusting time of every polarizable particle. In addition, Rydberg atoms suffer from finite radiative lifetimes and are sensitive to external perturbations, so dispersion force manipulation might lead to the ionization of atoms. Tradeoffs will have to be made between all of the relevant system parameters in order to discover the "sweet spot" that achieves levitation and upward acceleration. These and other yet-to-be-identified technical challenges need to be addressed via further empirical and theoretical studies.

3. Heim's Quantum Theory for Space Propulsion

Heim [76–79], Heim et al. [80], and Dröscher and Heim [81] (a.k.a. the Heim theory group; see also Ref. 82), propose a unification of the quantum field theory of elementary particles with General Relativity via a six- or eight-dimensional spacetime geometry, called Heim space. Heim space is a quantized space comprised of elemental surfaces with orientation (spin) and a size $\sim$ (Planck length)$^2 = (\hbar G/c^3) \sim 10^{-70}$ m^2. Heim space may also comprise several subspaces, each of which are equipped with their own space metric. Combining these subspaces by employing certain selection rules, a set of partial metric tensors is obtained which form a polymetric that represents all of the known fundamental interaction forces. The partial metrics are interpreted as a physical interaction field or elementary particle. However, Heim theory actually predicts two more fundamental interaction fields in addition to the four experimentally known ones.

The six fundamental interactions emerge in our four-dimensional spacetime and represent real physical fields carrying energy. The two additional interaction fields are identified as a gravitophoton interaction (i.e., the conversion of photons into a gravitational-like field) and a quintessence or vacuum interaction (also a conversion of photons into another type of gravitational-like field).***** The gravitophoton interaction is mediated by two hypothetical massless gravitophoton particles, one that is gravitationally attractive and the other that is gravitationally repulsive. The massless quintessence interaction particle mediates a very weak repulsive gravitational-like force that is much smaller in magnitude than the two gravitophoton interactions. Therefore, according to Heim theory, the gravitational force that we experience is comprised of three different fundamental interactions having four different interaction quanta, namely the usual gravitons of Heim-quantized General Relativity Theory, both repulsive and attractive gravitophotons, and the repulsive quintessence (or vacuum) particle. All of the related interaction coupling constants were derived and numerically calculated by the Heim theory group. The gravitophoton and quintessence interactions are Lorentz-invariant, and the laws of momentum and energy conservation are strictly obeyed. However, the speed of light is not the limiting speed for an inertial transformation in Heim theory. The gravitophoton interaction also has the peculiar property of reducing the inertia of a body.

The physical interpretation for the gravitophoton interaction field led the Heim theory group to conclude that this field could be used to both accelerate a body and to cause its transition into a parallel space that possibly admits superluminal motion. Dröscher and Häuser [83–88] propose that these effects act as two separate modes of space propulsion that could lead to gravity control and reduction of inertia at low energy densities. The first propulsion mode is based on the prediction that a Heim–Lorentz force F_{gp} arises as a byproduct of the gravitophoton interaction. This force is structurally similar to the magnetic Lorentz force [83–88]:

$$F_{gp} = -\Lambda_p e \mu_0 (v^T \times B) \quad \text{(N)} \qquad (44)$$

*****The use of the word "quintessence" by the Heim theory group is spurious because it is defined by the cosmological research community as a time-dependent vacuum scalar field. The use of this term is not justified because Heim theory indicates that the gravitophoton vacuum interaction field has no time-dependence and its coupling constant is fixed.

where Λ_p is a dimensionless (highly-nonlinear) function of the probability amplitude of the gravitophoton particle, e is the electron charge, μ_0 is the vacuum permeability constant, $\mathbf{v}^T$ is the circumferential velocity of a rotating body of mass m, and $\mathbf{B}$ is the magnetic field vector. The Heim–Lorentz force is a repulsive gravitational force that is generated by the pair production of gravitophotons (pair production energy is zero). Equation (44) states that an accelerating force is generated by a freely rotating ring-magnet located above a current-loop that produces an external inhomogeneous magnetic field surrounding the ring-magnet, all of which gives rise to gravitophoton pair production via the conversion of photons (that couple the electrons inside the current-loop) to the virtual electrons inside the ring-magnet. Because of the relative strength of their coupling constants and corresponding interaction cross sections, the repulsive gravitophoton gives rise to a propulsive force that acts upon the rotating ring, while the attractive gravitphoton is absorbed by the atomic nuclei comprising the ring material. Furthermore, the virtual electrons are quantum vacuum zero-point fluctuations that surround the atomic nuclei comprising the ring, which are polarized by the externally applied magnetic field. The role of the rotating ring's circumferential speed, the virtual electrons inside it, and the external magnetic field is to catalyze the photon-to-gravitophoton conversion process inside the ring so that the kinetic energy of the ring, under acceleration by the Heim–Lorentz force, is provided by the quantum vacuum and not by the external magnetic field.

This configuration automatically suggested an experiment that Dröscher and Häuser have detailed in their publications whereby an apparatus to test the Heim–Lorentz force prediction will also serve as a prototype "gravitophoton field propulsion" device. Dröscher and Häuser calculated the magnitude of the Heim–Lorentz force for a range of ring rotation rates, external magnetic field strengths, ring size and mass, etc. They predict Heim–Lorentz forces ranging from 7.14×10^{-43} N to 1.45×10^9 N for the case of a 100 kg ring (diameters ranging from 1 m to 6 m) that is embedded in an external magnetic field (ranging from 2 Tesla to 50 Tesla), and rotating with a circumferential speed ranging from 103 m/s to 700 m/s [83–88].

The second mode of gravitophoton field propulsion is highly speculative. Dröscher and Häuser suggest that the gravitophoton interaction can also be exploited to reduce the inertial mass of a spacecraft by a factor of 10^4, which will result in the increased speed of the spacecraft by a corresponding factor of 10^4 without increasing its kinetic energy. (This appears to violate the conservation of energy.) The inertia reduction is what causes the spacecraft to slip into a parallel space where it undergoes superluminal motion. The physical process that allegedly makes this possible is the emission of repulsive gravitophotons by the gravitophoton field propulsion device (using a 60 Tesla magnetic field), which convert the spacecraft's own gravitons into repulsive quintessence particles [83–88]. This process is claimed to reduce the gravitational potential of the spacecraft, thus reducing its inertia. The explanation given by Dröscher and Häuser for the parallel space transition process is somewhat obscure.

The reader should be cautioned that the edifice of Heim's theory is very sophisticated and mathematically very dense. The efficacy of its theoretical foundation has not been proven and its theoretical predictions have not been tested in the lab nor published in the peer-reviewed literature. However, at the time of this writing, the Heim theory group is working to establish their first lab experiment.

4. Alzofon's Antigravity Propulsion

Alzofon [89] began exploring the fundamental nature of the gravitational force during the 1950s–1960s, and began to question whether it could be controlled by artificial means. Many of his contemporaries in the aerospace engineering and sciences community were also exploring this question at a time when several large aerospace industry firms were establishing their own gravity physics research institutes in the United States. The goal of these institutes was to find a physical principle that would lead to the development of a technology for the artificial control of gravity in order to reduce, or otherwise, eliminate the influence of the Earth's gravitational field upon aerospace vehicles. The hope was to revolutionize aerospace transportation by eliminating the need to overcome the Earth's gravity.

Alzofon later began a phenomenological study in which he proposed that a specimen composed of a very pure isotope of aluminum (e.g., Al^{27}) with small spherical iron inclusions (or magnesium plus chromium inclusions), embedded in a static magnetic field, and undergoing excitation by pulsed microwave radiation, would generate a net decrease of the alloy's weight to the point of levitation [90]. This embodiment is said to alter the force of gravity by means of altering nuclear entropy using dynamic nuclear orientation via pulsed polarization of the magnetic moments of nucleons, and its interaction with polarized electron spins. In this mechanism, the paramagnetic electron spin resonance and alignment–disalignment cycle resonance of pulsed dynamic nuclear magnetic moment orientation inside the specimen modifies the interaction between nucleons and spacetime metric fluctuations, thus reducing the specimen's inertial mass. This is hypothesized to result in a corresponding reduction of the Earth's gravitational force acting upon the specimen [90,91].

At present, efforts are underway in private laboratories to expand the empirical envelope of Alzofon's original embodiment to see whether a measurable alteration can be produced in the force of gravity acting on specimens. However, a complete theoretical model for this concept is required in order to provide self-consistent predictions that can be used to guide any empirical studies. No complete, self-consistent theoretical model yet exists for this genre of gravity control devices, and Alzofon's model is a hypothesis at best. However, empirical studies have been known throughout history to guide the way toward the development of mature theoretical models when such is largely absent at the beginning. This is how the laws of gravitation and motion, electrodynamics, optics, quantum theory and many others got their start.

V. Conclusions

In this chapter we reviewed and analyzed a broad number of gravity control concepts that are found within Newtonian gravity theory, General Relativity Theory, semi-classical quantum gravity theory, quantum field theory, nonretarded quantum interatomic dispersion force theory, and speculative quantum unified field theories. We found that plausible mechanisms exist within Newtonian and general relativistic theories whereby one could embody a realistic device that produces gravity control or an antigravity force. However, we discovered that there are daunting technical challenges that arise in each of the proposed embodiments. Mechanical embodiments that produce antigravity forces require kilometer-sized apparatus, astronomical-sized masses and densities, or extreme mass velocities and accelerations. There are other subtleties involved, such as the possibility of different forms of matter having a highly nonlinear gravitational permeability, which could dramatically mitigate such large-scale requirements.

Embodiments of antigravity devices that use negative matter are speculative because negative matter has not been observed in nature or in the laboratory. Dark energy (i.e., the vacuum energy of Einstein's cosmological constant) and dark matter, however, have both been observed. Our analysis showed that dark energy has no useful breakthrough propulsion application even though it has an antigravity property that drives the inflationary expansion of the universe. And it turns out that dark matter has no special antigravity property, so we can ignore it. Negative energy has been produced in the lab in very small quantities. The technologies used for producing negative energy are nascent, and so it will be some time before it can be ascertained whether they are capable of producing the astronomical amounts of negative energy required to generate significant antigravity forces.

Also, embodiments that produce artificial gravity or gravitational waves for propulsion require electromagnetic field intensities approaching the breakdown field intensities of the quantum vacuum, or they require particle accelerator energies that approach or exceed Planck energy. Modern laser technology has reached field intensities approaching vacuum breakdown field intensities while pulsed power technology is beginning to catch up. There is ongoing theoretical and empirical testing of new alternative accelerator technologies designed to achieve Planck energy on a tabletop. It is expected that particle beam luminosity and power output from tabletop accelerators could be more than a dozen orders of magnitude higher than what is achieved by conventional large-scale accelerator technology. This is an ongoing development in experimental particle physics.

Antigravity forces generated by the Casimir effect or by nonretarded quantum interatomic dispersion forces in a curved spacetime (i.e., gravitational field) are very feeble, but there are proposals based on other theoretical and empirical studies which suggest that these forces can be amplified to macroscopic level. However, there are a number of difficult technical challenges to overcome in order to achieve success. There are other antigravity concepts based on speculative quantum unified field theories, but it is beyond the scope of this chapter to review all of them. However, two of the more well-known concepts were reviewed, and our evaluation showed that they are highly speculative and remain subject to theoretical and/or empirical verification.

The field of gravity control is ripe for focused, dedicated research. Laboratory technology continues to evolve and push the envelope on what is possible. Extreme field strengths, power, luminosity, and even mass motion are being achieved by tabletop devices. So the technological means for exploring extreme physical conditions in the lab exist and should be used to explore gravity control.

Acknowledgments

The author thanks the Institute for Advanced Studies at Austin and H. E. Puthoff for supporting this work. Portions of this work previously originated under Air Force Research Laboratory (AFMC) contract F04611-99-C-0025. The author acknowledges the entire body of technical contributions made to gravitational and breakthrough propulsion physics research over four decades by R. L. Forward. Bob Forward was both a personal and professional mentor to this author as well as to many others, and he will not be forgotten. The author also extends many thanks to F. B. Mead, Jr. (AFRL/PRSP, Edwards AFB, CA) and Brig. Gen. S. P. Worden (USAF ret., NASA-Ames) for encouraging and supporting our exploration into this topic. Finally, I thank M. G. Millis (NASA-Glenn), E. H. Allen (Lockheed Martin), J. Häuser (Univ. of Applied Sciences), G. Hathaway (Hathaway Consulting), M. Tajmar (Austrian Res. Centers), R. P. Feynman (CalTech), and M. Scadron (Univ. of Arizona) for many useful discussions.

References

[1] Symon, K. R., *Mechanics*, 3rd ed., Addison-Wesley, Reading, MA, 1971, pp. 10.

[2] Forward, R. L., *Future Magic: How Today's Science Fiction Will Become Tomorrow's Reality*, Avon Books, New York, 1988, pp. 103–132.

[3] Forward, R. L., *Indistinguishable from Magic*, Baen Books, New York, 1995, pp. 148–178.

[4] Forward, R. L., "A New Gravitational Field," *Science Digest*, Vol. 52, 1962, pp. 73–78.

[5] Forward, R. L., "Far Out Physics," *Analog Science Fiction/Science Fact*, Vol. 95, 1975, pp. 147–166.

[6] Forward, R. L., "Flattening Spacetime Near the Earth," *Physical Review D*, Vol. 26, 1982, pp. 735–744.

[7] Davis, E. W., "Advanced Propulsion Study," Air Force Research Lab., Final Report AFRL-PR-ED-TR-2004-0024, Edwards AFB, CA, 2004, pp. 2–5.

[8] Millis, M. G., "Assessing Potential Propulsion Breakthroughs," *New Trends in Astrodynamics and Applications*, Belbruno, E. (ed.), Annals of the New York Academy of Sciences, Vol. 1065, 2005, pp. 441–461.

[9] Forward, R. L., "Negative Matter Propulsion," *Journal of Propulsion and Power*, Vol. 6, No. 1, Jan.–Feb. 1990, pp. 28–37.

[10] Forward, R. L., "Negative Matter Propulsion," *24th AIAA/ASME/SAE/ASEE Joint Propulsion Conference*, AIAA Paper 1988-3168, Boston, MA, July 1988.

[11] Bondi, H., "Negative Mass in General Relativity," *Reviews of Modern Physics*, Vol. 29, No. 3, July 1957, pp. 423–428.

[12] Bonnor, W. B., and Swaminarayan, N. S., "An Exact Solution for Uniformly Accelerated Particles in General Relativity," *Zeitschrift für Physik A Hadrons and Nuclei*, Vol. 177, 1964, pp. 240–256.

[13] Bicak, J., Hoenselaers, C., and Schmidt, B. G., "Solutions of the Einstein Equations for Uniformly Accelerated Particles Without Nodal Singularities," *Proceedings of the Royal Society of London, Series A, Mathematical and Physical Sciences*, Vol. 390, 1983, "I. Freely Falling Particles in External Fields," pp. 397–409; "II. Self-Accelerating Particles," pp. 411–419.

[14] Terletskii, Y. P., *Paradoxes in the Theory of Relativity*, Plenum Press, New York, 1968, Chap. VI.

[15] Landis, G. A., "Comment on 'Negative Matter Propulsion,' " *Journal of Propulsion and Power*, Vol. 7, No. 2, 1991, p. 304.

[16] Clowe, D., Bradac, M., Gonzalez, A., Markevitch, M., Randall, S. W., Jones, C., and Zaritsky, D., "A Direct Empirical Proof of the Existence of Dark Matter," *Astrophysical Journal Letters*, Vol. 648, 2006, pp. L109–L113.

[17] Linder, E. V., "The Universe's Skeleton Sketched," *Nature*, Vol. 445, 2007, p. 273.

[18] Massey, R., et al., "Dark Matter Maps Reveal Cosmic Scaffolding," *Nature*, Vol. 445, 2007, pp. 286–290.

[19] Bilaniuk, O.-M., and Sudarshan, E. C. G., "Particles Beyond the Light Barrier," *Physics Today*, Vol. 5, 1969, pp. 43–51.

[20] Heaviside, O., "A Gravitational and Electromagnetic Analogy," *The Electrician*, Vol. 31, 1893, pp. 281–282, 359.

[21] Thirring, H., "Über die Wirkung Rotierender Ferner Massen in der Einsteinschen Gravitationstheorie," *Physikalische Zeitschrift*, Vol. 19, 1918, pp. 33–39.

[22] Thirring, H., "Berichtigung zu Meiner Arbeit: 'Über die Wirkung rotierender ferner Massen in der Einsteinschen Gravitationstheorie,' " *Physikalische Zeitschrift*, Vol. 22, 1921, pp. 29–30.

[23] Thirring, H., and Lense, J., "Über den Einfluss der Eigenrotation der Zentralkörper auf die Bewegung der Planeten und Monde nach der Einsteinschen Gravitationstheorie," *Physikalische Zeitschrift*, Vol. 19, 1918, pp. 156–163.

[24] Mashhoon, B., Hehl, F. W., and Theiss, D. S., "On the Gravitational Effects of Rotating Masses: The Lense-Thirring Papers Translated," *General Relativity and Gravitation*, Vol. 16, 1984, pp. 711–750.

[25] Forward, R. L., "General Relativity for the Experimentalist," *Proceedings of the Institute of Radio Engineers*, Vol. 49, 1961, pp. 892–904.

[26] Forward, R. L., "Antigravity," *Proceedings of the Institute of Radio Engineers*, Vol. 49, 1961, p. 1442.

[27] Forward, R. L., "Guidelines to Antigravity," *American Journal of Physics*, Vol. 31, 1963, pp. 166–170.

[28] Felber, F. S., "Exact Relativistic 'Antigravity' Propulsion," *Proceedings of the STAIF-2006: 3rd Symposium on New Frontiers and Future Concepts*, El-Genk, M. S. (ed.), AIP Conference Proceedings Vol. 813, AIP Press, New York, 2006, pp. 1374–1381.

[29] Misner, C. W., Thorne, K. S., and Wheeler, J. A., *Gravitation*, W. H. Freeman & Co., New York, 1973.

[30] Ohanian, H., and Ruffini, R., *Gravitation and Spacetime*, 2nd ed., W. W. Norton & Co., New York, 1994.

[31] Ciufolini, I., and Wheeler, J. A., *Gravitation and Inertia*, Princeton Univ. Press, Princeton, NJ, 1995.

[32] Sachs, M., "The Mach Principle and the Origin of Inertia from General Relativity," *Mach's Principle and the Origin of Inertia*, Sachs M., and Roy, A. R. (eds.), C. Roy Keys, Inc., 2003.

[33] Peacock, J. A., *Cosmological Physics*, Cambridge Univ. Press, Cambridge, UK, 1999, pp. 25–26.

[34] Riess, A. G., et al., "Observational Evidence from Supernovae for an Accelerating Universe and a Cosmological Constant," *Astronomical Journal*, Vol. 116, 1998, pp. 1009–1038.

[35] Perlmutter, S., et al., "Measurements of Ω and Λ from 42 High-Redshift Supernovae," *Astrophysical Journal*, Vol. 517, 1999, pp. 565–586.

[36] Schwarzschild, B., "High-Redshift Supernovae Indicate that Dark Energy Has Been Around for 10 Billion Years," *Physics Today*, Vol. 60, 2007, pp. 21–25.

[37] Riess, A. G., et al., "New Hubble Space Telescope Discoveries of Type Ia Supernovae at $z \geq 1$: Narrowing Constraints on the Early Behavior of Dark Energy," *Astrophysical Journal*, Vol. 659, 2007, pp. 98–121.

[38] Astier, P., et al., "The Supernova Legacy Survey: Measurement of Ω_M, Ω_Λ and w from the First Year Data Set," *Astronomy and Astrophysics*, Vol. 447, 2006, pp. 31–48.

[39] Particle Data Group, "Review of Particle Physics," *Journal of Physics*, Vol. 33, 2006, pp. 210–232.

[40] White, H. G., and Davis, E. W., "The Alcubierre Warp Drive in Higher Dimensional Spacetime," *Proceedings of the STAIF-2006: 3rd Symposium on New Frontiers and Future Concepts*, El-Genk, M. S. (ed.), AIP Conference Proceedings, Vol. 813, AIP Press, New York, 2006, pp. 1382–1389.

[41] Levi-Civita, T., *Rendiconti della Reale Accademia dei Lincei*, Series 5, Vol. 26, 1917, p. 519.

[42] Pauli, W., *Theory of Relativity*, Dover, New York, 1981, pp. 171–172.

[43] Puthoff, H. E., Davis, E. W., and Maccone, C., "Levi-Civita Effect in the Polarizable Vacuum (PV) Representation of General Relativity," *General Relativity and Gravitation*, Vol. 37, 2005, pp. 483–489.

[44] Bekenstein, J. D., "Gravitational-Radiation Recoil and Runaway Black Holes," *Astrophysical Journal*, Vol. 183, 1973, pp. 657–664.

[45] Landau, L. D., and Lifshitz, E. M., *The Classical Theory of Fields: Course of Theoretical Physics Vol. 2*, 4th rev. English ed., Butterworth and Heinemann, Oxford, 1998, pp. 347–350.

[46] Bonnor, W. B., and Piper, M. S., "The Gravitational Wave Rocket," *Classical and Quantum Gravity*, Vol. 14, 1997, pp. 2895–2904.

[47] Baker, R. M. L., Jr., "Preliminary Tests of Fundamental Concepts Associated with Gravitational-Wave Spacecraft Propulsion," *AIAA Space 2000 Conference & Exposition*, AIAA Paper 2000-5250, Long Beach, CA, 2000.

[48] Baker, R. M. L., Jr., and Li, F.-Y., "High-Frequency Gravitational Wave (HFGW) Generation by Means of X-ray Lasers and Detection by Coupling Linearized GW to EM Fields," *Proceedings of the STAIF-2005: 2nd Symposium on New Frontiers and Future Concepts*, El-Genk, M. S. (ed.), AIP Conference Proceedings, Vol. 746, AIP Press, New York, 2005, pp. 1271–1281.

[49] Baker, R. M. L., Jr., Li, F.-Y., and Li, R., "Ultra-High-Intensity Lasers for Gravitational Wave Generation and Detection," *Proceedings of the STAIF-2006: 3rd Symposium on New Frontiers and Future Concepts*, El-Genk, M. S. (ed.), AIP Conference Proceedings, Vol. 813, AIP Press, New York, 2006, pp. 1352–1361.

[50] Weber, J., and Hinds, G., "Interaction of Photons and Gravitons," *Physical Review*, Vol. 128, 1962, pp. 2414–2421.

[51] Zombeck, M. V., *Handbook of Space Astronomy & Astrophysics*, 2nd ed., Cambridge Univ. Press, Cambridge, UK, 1990, p. 82.

[52] Cox, A. N. (ed.), *Allen's Astrophysical Quantities*, 4th ed., AIP Press, New York, 1999, p. 571.

[53] Douglass, D. H., and Braginsky, V. B., "Gravitational Radiation Experiments," *General Relativity: An Einstein Centenary Survey*, Hawking, S. W., and Israel, W. (eds.), Cambridge Univ. Press, Cambridge, UK, 1979, p. 98.

[54] Kaplan, A. E., and Ding, Y. J., "Field-gradient-induced Second-harmonic Generation in Magnetized Vacuum," *Physical Review A*, Vol. 62, 2000, p. 043805.

[55] Novaes, S. F., and Spehler, D., "Gravitational Laser Backscattering," *Physical Review D*, Vol. 47, 1993, pp. 2432–2434.

[56] Chen, P., "Resonant Photon-Graviton Conversion in EM Fields: From Earth to Heaven," SLAC-PUB-6666, Stanford Univ., Stanford, CA, 1994.

[57] Chen, P., and Noble, R. J., "Crystal Channel Collider: Ultra-High Energy and Luminosity in the Next Century," *Proceedings of the 7th Workshop on Advanced Accelerator Concepts*, Chattopadhyay, S., McCullough, J., and Dahl, P. (eds.), AIP Conference Proceedings, Vol. 398, AIP Press, New York, 1997, pp. 273–285.

[58] De Logi, W. K., and Mickelson, A. R., "Electrogravitational Conversion Cross Sections in Static Electromagnetic Fields," *Physical Review D*, Vol. 16, 1977, pp. 2915–2927.

[59] Schweber, S. S., "Defending Against Nuclear Weapons: A 1950s Proposal," *Physics Today*, Vol. 60, 2007, pp. 36–41.

[60] Gertsenshteĭn, M. E., "Wave Resonance of Light and Gravitational Waves," *Soviet Physics JETP*, Vol. 14, 1962, pp. 84–85.

[61] Pustovoĭt, V. I., and Gertsenshteĭn, M. E., "Gravitational Radiation by a Relativistic Particle," *Soviet Physics JETP*, Vol. 15, 1962, pp. 116–123.

[62] Sushkov, O. P., and Khriplovich, I. B., "The Gravitational Radiation Emitted by an Ultrarelativistic Charged Particle in an External Electromagnetic Field," *Soviet Physics JETP*, Vol. 39, 1974, p. 1.

[63] Grishchuk, L. P., and Sazhin, M. V., "Excitation and Detection of Standing Gravitational Waves," *Soviet Physics JETP*, Vol. 41, 1976, pp. 787–793.

[64] Portilla, M., and Lapiedra, R., "Generation of High Frequency Gravitational Waves," *Physical Review D*, Vol. 63, 2001, p. 044014.

[65] Navarro, E. A., Portilla, M., and Valdes, J. L., "Test of the Generation of High Frequency Gravitational Waves by Irradiating a Dielectric Film in a Resonant Cavity," *Proceedings of the STAIF-2004: 1st Symposium on New Frontiers and Future Concepts*, El-Genk, M. S. (ed.), AIP Conference Proceedings, Vol. 699, AIP Press, New York, 2004, pp. 1122–1126.

[66] Feynman, R. P., and Hibbs, A. R., *Quantum Mechanics and Path Integrals*, McGraw-Hill, New York, 1965.

[67] Feynman, R. P., Morinigo, F. B., and Wagner, W. G., *Feynman Lectures on Gravitation*, Hatfield, B. (ed.), Addison-Wesley, Reading, MA, 1995.

[68] Zee, A., *Quantum Field Theory in a Nutshell*, Princeton Univ. Press, Princeton, NJ, 2003, Chap. VIII.1.

[69] Calloni, E., Di Fiore, L., Esposito, G., Milano, L., and Rosa, L., "Gravitational Effects on a Rigid Casimir Cavity," *International Journal of Modern Physics A*, Vol. 17, 2002, pp. 804–807.

[70] Calloni, E., DiFiore, L., Esposito, G., Milano, L., and Rosa, L., "Vacuum Fluctuation Force on a Rigid Casimir Cavity in a Gravitational Field," *Physics Letters A*, Vol. 297, 2002, pp. 328–333.

[71] Bimonte, G., Calloni, E., Esposito, G., and Rosa, L. "Energy-Momentum Tensor for a Casimir Apparatus in a Weak Gravitational Field," *Physical Review D*, Vol. 74, 2006, p. 085011.
[72] Bimonte, G., Calloni, E., Esposito, G., and Rosa, L. "Erratum: Energy-Momentum Tensor for a Casimir Apparatus in a Weak Gravitational Field," *Physical Review D*, Vol. 75, 2007, p. 089901(E).
[73] Pinto, F., "Progress in Quantum Vacuum Engineering Propulsion," *Journal of the British Interplanetary Society*, Vol. 59, 2006, pp. 247–256.
[74] Fermi, E., "Sull'elettrostatica di un Campo Gravitazionale Uniforme e sul Peso delle Masse Elettromagnetiche," *Nuovo Cimento*, Vol. 22, 1921, pp. 176–188.
[75] Puthoff, H. E., "Polarizable-Vacuum (PV) Approach to General Relativity," *Foundations of Physics*, Vol. 32, 2002, pp. 927–943.
[76] Heim, B., "Vorschlag eines Weges einer Einheitlichen Beschreibung der Elementarteilchen," *Zeitschrift für Naturforschung*, Vol. A32, 1977, pp. 233–243.
[77] Heim, B., *Elementarstrukturen der Materie: Einheitliche Strukturelle Quantenfeldtheorie der Materie und Gravitation, Band 1*, 3rd ed., Resch Verlag, Innsbruck, 1998.
[78] Heim, B., *Elementarstrukturen der Materie: Einheitliche Strukturelle Quantenfeldtheorie der Materie und Gravitation, Band 2*, 2nd ed., Resch Verlag, Innsbruck, 1996.
[79] Heim, B., "Ein Bild vom Hintergrund der Welt," *Welt der Weltbilder, Imago Mundi, Band 14*, Resch, A. (ed.), Resch Verlag, Innsbruck, 1994.
[80] Heim, B., Dröscher, W., and Resch, A., *Einführung in Burkhard Heim Einheitliche Beschreibung der Welt*, Resch Verlag, Innsbruck, 1998.
[81] Dröscher, W., and Heim, B., *Strukturen der Physikalischen Welt und ihrer Nichtmateriellen Seite*, Resch Verlag, Innsbruck, 1996.
[82] Willigmann, H., *Grundriss der Heimschen Theorie*, Resch Verlag, Innsbruck, 2002.
[83] Dröscher, W., and Häuser, J., "Physical Principles of Advanced Space Propulsion Based on Heim's Field Theory," *38th AIAA/ASME/SAE/ASEE Joint Propulsion Conference*, AIAA Paper 2002-4094, Indianapolis, IN, July 2002.
[84] Dröscher, W., and Häuser, J., "Future Space Propulsion Based on Heim's Field Theory," *39th AIAA/ASME/SAE/ASEE Joint Propulsion Conference*, AIAA Paper 2003-4990, Huntsville, AL, July 2003.
[85] Dröscher, W., and Häuser, J., "Guidelines for a Space Propulsion Device Based on Heim's Quantum Theory," *40th AIAA/ASME/SAE/ASEE Joint Propulsion Conference*, AIAA Paper 2004-3700, Ft. Lauderdale, FL, July 2004.
[86] Dröscher, W., and Häuser, J., "Heim Quantum Theory for Space Propulsion Physics," *Proceedings of the STAIF-2005: 2nd Symposium on New Frontiers and Future Concepts*, El-Genk, M. S. (ed.), AIP Conference Proceedings, Vol. 746, AIP Press, New York, 2005, pp. 1430–1440.
[87] Dröscher, W., and Häuser, J., "Magnet Experiment to Measuring Space Propulsion Heim-Lorentz Force," *41st AIAA/ASME/SAE/ASEE Joint Propulsion Conference*, AIAA Paper 2005-4321, Tucson, AZ, July 2005.
[88] Dröscher, W., and Häuser, J., "Spacetime Physics and Advanced Propulsion Concepts," *42nd AIAA/ASME/SAE/ASEE Joint Propulsion Conference*, AIAA Paper 2006-4608, Sacramento, CA, July 2006.
[89] Alzofon, F. E., "The Origin of the Gravitational Field," *Advances in the Astronautical Sciences*, Jacobs, H. (ed.), Vol. 5, Plenum Press, New York, 1960, pp. 309–319.

[90] Alzofon, F. E., "Anti-Gravity with Present Technology: Implementation and Theoretical Foundation," *17th AIAA/SAE/ASME Joint Propulsion Conference*, AIAA Paper 81-1608, Colorado Springs, CO, July 1981.

[91] Alzofon, F. E., "The Unity of Nature and the Search for a Unified Field Theory," *Physics Essays*, Vol. 6, 1993, pp. 599–608.

Chapter 5

Gravitational Experiments with Superconductors: History and Lessons

George D. Hathaway*

Hathaway Consulting Services, Toronto, Ontario, Canada

I. Introduction

THIS CHAPTER outlines some of the problems inherent in trying to discover and verify new forces, and it provides a historical survey of gravitational experiments using superconductors. The development of superconductors, and in particular high-temperature ceramic superconductors, has facilitated the pursuit of a potential connection between gravity, electromagnetism, and matter in the solid state. The discovery of yttrium barium copper oxide (YBCO) ceramics that are superconducting at liquid nitrogen (LN_2) temperatures has allowed laboratories around the world to fabricate experimentally useful superconductors cooled by relatively inexpensive LN_2 rather than liquid helium (LHe). However, experimenting at cryogenic temperatures necessitates an increased awareness of a host of technical issues that can cause an expected effect to be masked or otherwise interfered with. Hence, this chapter also deals with several of the problems encountered by the experimentalist in the design and execution of gravitational experiments with superconductors. For the theoretician, the possibility of superconductors representing a macroscopic quantum object has provided several avenues for the development of theories connecting gravity and gravitylike forces to matter. The exploration of these concepts is also highlighted in this chapter.

The possible production of laboratory-scale gravitomagnetic fields using high-temperature superconductors has recently been investigated by Li and Torr [1–3] who expanded on earlier work by DeWitt [4] and Ross [5] to include gravitomagnetic fields in the London equations for superconductors.

Copyright © 2008 by the American Institute of Aeronautics and Astronautics, Inc. All rights reserved.
*Director.

Podkletnov and Nieminen [6] published what is arguably the first possible demonstration of an experimental link between high-temperature (LN_2) superconductor effects and gravity, allegedly in the form of a "gravity shield." Although Podkletnov's findings were later found not to be reproducible [7], they suggested that the experimental search for gravity-related forces could be undertaken in the laboratory. In addition, such work prompted theoreticians to consider possible approaches for using superconductors as a special form of condensed matter capable of modifying and/or producing such forces.

Einstein maintained that gravity is a consequence of curved space-time geometry, implying that we need to manipulate space itself to affect gravity. General Relativity introduces a metric tensor theory of gravity, which allows the prediction of gravitational interactions between bodies but does not explain the fundamental physical basis of gravity. Similarly, Maxwell's vector equations allow us to predict the outcomes of electromagnetic interactions but do not explain the fundamental basis for such interactions. It is possible to reformulate the tensor format of General Relativity into a simpler vector format, valid for weak field approximations and nonrelativistic velocities. Using perturbation theory to compute the equations of motion in the simplified General Relativity equations results in terms that have direct analogues in Maxwell's equations, where for example the flow of electric current is replaced by mass flow. Forward [8,9] was one of the first to investigate this analogue. Of particular interest is a term analogous to the Biot–Savart magnetic field, which is generally referred to as the gravitomagnetic field (also sometimes referred to as gravitational frame dragging or the Lense–Thirring effect). A second term is analogous to the electrostatic Coulomb field, and is called the gravitoelectric field. Essentially, the gravitomagnetic field produces a force between currents of flowing matter while the gravitoelectric field produces a force between masses themselves (the conservative Newtonian gravitational field). The term gravitoelectromagnetic field is often used to refer to both the gravitoelectric and gravitomagnetic fields.

In this analogue, gravity is thus comprised of a conservative velocity-independent field and a velocity-dependent field analogous to the electric and magnetic fields in electromagnetic theory. The simplified General Relativity/Maxwell's equations also lead to a Faraday-like law of induction, which can generate Newtonian gravitational fields from time-varying gravitomagnetic fields. Modern experimental attempts to confirm the existence of the gravitomagnetic field include highly accurate laser ranging of the Earth-moon distance, as well as the gravity probe B satellite [10].

In addition to these fundamental considerations, a variety of other approaches have appeared in the literature. Puthoff [11] extended an idea originally articulated by Sakharov [12], which posits that gravity is related to quantum fluctuations of the vacuum, sometimes called zero-point fluctuations. Alzofon [13] presented an engineering approach to interacting with gravity by means of altering nuclear entropy using dynamic nuclear orientation, that is, pulsed polarization of the magnetic moments of nucleons by interaction with polarized electron spins. Hughes analyzed the Kopernicky conjecture, which holds that gravity is nothing other than the slight difference between forces of Coulomb attraction and repulsion [14].

A distinction should be made between gravitational waves, gravitational "force," and anomalous forces. The existence of gravitational waves is a prediction of General Relativity. Research on gravitational waves is generally divided into two more-or-less distinct realms: low frequency (less than a few hundred Hz) and high frequency. The search for low-frequency gravitational waves currently utilizes large, heavy, long metal detectors, interferometers, strain gauges, and accelerometers in an attempt to detect quadrupole gravitational waves from cosmological sources, such as rapidly rotating binary star systems. Several researchers postulate that high-frequency gravitational waves (HFGW) could be produced under certain laboratory conditions in the near future [15]. It is envisioned that such HFGW could be used for communications, astronomy, microscopy, and possibly even propulsion. The production and detection of these waves may be mediated or influenced by superconductors. Some initial debate on such considerations has been articulated by Woods [16,17]. If gravitational wave interaction with laboratory-scale matter results in detectable forces, such forces should be amenable to detection by available laboratory instruments.

Modifying gravity for propulsive purposes resolves itself into roughly two categories: 1) neutralizing or negating the gravitational attraction of a nearby body, and 2) providing a propulsive force or impulse to a spacecraft by manipulating the underlying physical phenomenon which forms the basis of gravity. For additional details, the reader is referred to Chapter 4.

In any experiment designed to produce a gravitylike force or to interact directly with a local gravity field, the researcher typically must look for extremely small deviations from a null result. Most observations to date demonstrate that interactions between gravity and electromagnetic fields, given the densities and strengths available to even the most well-equipped laboratory, are many orders of magnitude smaller than contemporary equipment can detect. Braginski et al. [18] determined that ordinary matter cannot be used to generate measurable gravitational fields in the laboratory. Demonstrating that a new force has been discovered in the laboratory will require the ability to distinguish between true electromagnetic–gravitational interactions and a host of other effects potentially masquerading as these forces.

II. Experimental Traps and Pitfalls

In the course of trying to measure signals from potentially minuscule (if existent) forces, constant vigilance for more prosaic explanations must be maintained. This is true when measuring not only the type of force (e.g., magnetic, electrostatic, mechanical, gravitational), but also the origin of the force. In even the most carefully designed experiments, it is easy to be fooled by a litany of potential effects resulting from experimental procedures that may have been poorly conceived or carried out. This section provides a brief overview of some of those considerations.

For example, the direct determination of the mass of a test sample cooled to cryogenic temperatures typically requires that the test sample be attached to a measurement device (a balance, for example). If the balance is not cooled, the connection between test sample and instrument must accommodate a very great temperature difference, and problems may arise due to condensation and buoyancy. In order to shield the test sample from the effects of condensation, the sample can be enveloped either in a vacuum or a noncondensable (at the operating temperature) gas. Placing the sample in a vacuum mitigates both buoyancy and condensation concerns, but may preclude effective cooling of the sample. Hence a compromise must be struck between the need to isolate the sample from the environment and the speed at which cooling is expected. Balances that operate in vacuum are useful in this regard. It may be possible to weigh the entire apparatus, cryogen, and all. However, this method introduces its own host of problems, mostly resulting from boil-off of the cryogen.

Determination of the mass of a test sample that is near a superconductor requires that the test sample be isolated from the environment as much as possible, to avoid being shaken, buoyed by air or gas currents, or otherwise perturbed. However, if the sample needs to be cooled, the requirement for strict temperature isolation can be somewhat relaxed.

Exacerbating the problems inherent in cryogenic experiments are those introduced by nearby electric and magnetic fields. As the experimenter endeavors to increase the mechanical isolation between the test sample and the local laboratory environment, the sample's susceptibility to electromagnetic and electrostatic influences becomes greater.

The following summary of the some of the more important experimental concerns is extracted from a compilation published in 2006 [19], which in turn expanded on the joint paper by Reiss and Hathaway [20], which discussed laboratory practices for experiments designed to detect gravitylike forces. The following list is organized into general categories, and there will inevitably be some overlap; it is by no means exhaustive.

A. Mechanical Effects

1. Thermal Effects

This includes thermally driven air, gas, or liquid currents, vapor condensation, and thermal expansion issues when a difference in temperature is encountered. Thermally driven currents act to artificially alter the weight of test samples they encounter. It is often exceedingly difficult to calculate the resulting forces, which tend to randomly fluctuate both temporally and spatially. Condensates can alter the mass of the test article, with the most problematic condensate in cryogenic experiments being water vapor from the air and, to a lesser extent, nitrogen condensate when working at LHe temperatures. Experiments must be designed to avoid any part of a cryogenically cooled apparatus being in contact with condensable gasses. When dealing with severe temperature differences, structural apparatus and other test articles must not be subjected to thermal expansion or contraction, unless this effect can be accurately analyzed and factored into the resulting measurements.

2. Buoyancy Effects

Materials display different buoyancy in different media. If a test sample is weighed in air, then weighed in LN_2, its apparent weight will clearly differ and be shape dependent. More subtly, if the sample is cooled to cryogenic temperatures, its buoyancy cross section changes as a function of its temperature, producing a time-varying weight change that could be misinterpreted as a gravitational interaction. Buoyancy differences may also arise in a single medium (such as air) due to stratification and temperature differences in the medium.

3. Seismic/Vibration Effects

Scales, balances, and force sensors generally operate by detecting or displaying the relative distances between fixed and movable masses, and it is necessary to avoid introducing artificial mechanical motion/noise into the measuring system. Vibrations can occur from laboratory and other sources, such as compressors, fans, and local traffic (both pedestrian and vehicular), and are usually transmitted into the measuring apparatus by structural elements. Mitigation is typically achieved either by loosening the structural design through the installation of dampers, or tightening the design by anchoring the support structures to larger, relatively immobile masses. If these are not easily or completely achievable, altering the frequency of the chief modes of vibration by adding or subtracting masses is often helpful. Wires and cables carrying signals and power may appear negligible as structural vibration carriers, but for experiments at cryogenic temperatures the wires may freeze and become rigid. In addition, the pulsing of current through any portion of a wire with curvature will produce a flexure in the wire, leading to displacement or, for alternating currents, oscillatory vibrations.

4. Vacuum Effects

All materials will to more or less extent outgas (release vapor) in vacuum, hence the use of low outgassing materials for any experiment performed in a vacuum or low ambient pressure environment is desirable. Even a low outgassing material may have minute amounts of contaminant (oils, fingerprints, etc.) on its surface, the release of which can be detected by sensitive balances. Other contaminant material on the inside walls of the vacuum chamber housing the sample may be released and coat or otherwise impinge upon the sample. Of more subtle nature is the slight movement of the vacuum chamber and encompassed test apparatus as the vacuum chamber is pumped down. Such small movements can offset previously aligned balances, producing an apparent weight change in a test sample simply due to a slight shifting in the support structures. It is also worth noting the high degree of variation in outgassing among materials that may be used in test apparatus or support structures involved in such experiments. For instance, brass has a high outgassing rate due to the presence of high vapor pressure zinc; if a brass test mass is used in a vacuum environment, its actual weight loss could change over the course of the experiment and mask other experimentally induced effects [21].

5. Liquid Effects

The evaporation or boiling of liquid (usually the cryogen) around a test sample or the suspension system can clearly contribute noise into the measurement system. This is particularly true when the test sample is suspended in, or is in contact with, a liquid cryogen used to cool it below its transition temperature. Boiling of the cryogen when inserting a sample will cause spurious readings and potentially mask effects near the critical temperature until sample and cryogen are in thermal equilibrium. Similarly, lifting the sample from a cryogen produces high rates of evaporation (often in addition to dripping cryogenic fluid), and will cause an apparent weight change as the temperature of the test sample rises back to ambient thermal conditions.

B. Temporal Effects

Certain experiments require a test sample to be cooled in cryogenic gas or liquid while being suspended from the movable arm of a balance. The precise determination of sample temperature as a function of time is critical to ascribing an observed weight change to a superconductive attribute. However, the attachment of a thermocouple to the superconducting sample necessitates feeding wire leads from the movable sample to the fixed laboratory frame on which the temperature readout sits. Movable slip rings are too stiff and unpredictable to be used as rotating/movable joints in most delicate experiments. Testing for the attainment of critical transition temperature via Meissner flux expulsion is often too cumbersome, and would of necessity disturb the balance. If direct temperature determination proves exceedingly difficult, the best path is for the experimenter to simply allow sufficient time for the test sample and any supporting mass to reach the required equilibrium temperature.

C. Electromagnetic Effects

1. Magnetic Coupling

The incorporation of superconductors in gravity experiments necessarily involves magnetic fields. Potential interactions with the magnetic fields arising from current-carrying and signal wires must be taken into account in the vicinity of the experimental apparatus. Residual magnetism can be found in certain "nonmagnetic" metals and alloys, which may interact with the measurement system. The assumption of perfect magnetic shielding by high-mu magnetic shielding alloys is usually not valid, and care must be taken that the proper level of electromagnetic shielding required by the experiment is obtained. Time-varying electromagnetic fields generated by test equipment, power sources, leads, and contacts can interact with nearby conducting bodies and cause often subtle but nevertheless spurious measurements. More dramatically, the sudden release of trapped magnetic fields in superconductors raised above transition temperature can affect, and be affected by, nearby magnetic or conductive structures.

2. Electric Coupling

As in the use of high-mu metals to shield magnetic fields, an over-reliance on a Faraday cage for complete exclusion of DC or quasi-static electric fields is often not realistic, and extreme care should be taken to account for and mitigate any spurious electric fields that may interfere with the test specimen or measurement apparatus.

3. Electromagnetic Coupling

Improper shielding or conductor pairs that are not tightly twisted can cause unwanted cross talk on nearby signal cables, even shielded ones. Switching transients in high-power circuits can easily couple into signal sensing circuits, causing artificial or spurious readings; this is particularly important for circuits incorporated into electronic balances. Using radio frequency (RF) power with improperly matched loads or inappropriately sized matching networks can cause intense electromagnetic interference in nearby circuits.

4. Grounding Issues

The physical layout of signal grounds should be carefully considered to avoid nearby magnetic interference. Shield grounds should be checked for potential ground loop problems by incorporating a single-point grounding connection. Contacts should be checked for loose, corroded, or bi-metal connections which can introduce unwanted contact potentials.

D. Electrostatic and Related Effects

1. Gradient Effects

The operation of an apparatus at high DC voltages relative to a nearby ground will usually produce spatial gradients in the electric field, unless the experiment is specifically designed to produce uniform fields in the vicinity of the test apparatus. Such gradients may induce subtle motion in nearby free bodies whether conductive or not, such as samples suspended from a balance.

2. Charge Pooling and Induced Charges

Even under high vacuum, without sufficient surface preparation and thermal baking, the interior surfaces of vacuum chambers are covered with invisible pools of surface charges, usually in the form of films of polar water molecules. In very sensitive measurement systems, or where the sample mass being weighed is in close proximity to a wall, the electrostatic forces between such "patch charges" on the wall and test mass may interfere with proper weight measurements. This applies whether or not the surfaces are conductive.

3. Ion and Molecular Effects

In high-voltage DC or AC experiments, even in vacuum, strong electric fields can accelerate outgassed or residual ions in the vacuum chamber, which in turn can impart momentum to a suspended test mass. This may occur either because the ions are streaming off of or onto a sample held at high voltage, or because the sample may be in the "ion wind" produced by a nearby source of ions held at high voltage. Such ion wind effects are difficult to calculate, and so generally need to be avoided by adhering to strict chamber cleanliness procedures, proper choice of vacuum compatible materials (including wire insulation), and operation at chamber pressures below those needed to produce significant ion wind effects. High voltages can cause ablation or sputtering of even the best insulation.

4. Charge Leakage

Charges can leak from insulated conductors, either as a quasi-DC corona or (more likely) in bursts called Trichel pulses in high-voltage experiments [22]. These may be negligible to the force-measuring apparatus at relatively low voltages, but at some threshold value they can suddenly become problematic, and often under generally unpredictable conditions. These can cause direct effects by impinging on or interfering with balances, for instance, or indirect effects through, for example, interference with nearby signal wires. Most commonly, they cause weak conduction paths to nearby grounds, which can interfere with high-gain amplifiers, and so on.

E. Experimental Expectations

The preceding is but a small sampling of the experimental artifacts that must be considered when designing and performing any sensitive laboratory experiment, particularly when searching for minuscule effects in support of controversial theories. In addition, there are psychological pitfalls of which investigators who design experiments are often unaware. For instance, *confirmation bias* is the tendency to trust data that support one's hypothesis, and to distrust (or ignore) data that do not support it. After all, if my data are exactly as my theory predicts, then the experiment is a good one and I made no errors! Another is *errors in logic*. An example of one such incorrect syllogism might be "If A is true, then we will observe B. We do observe B, therefore we have proven A is true." This false logic overlooks other explanations for observing B. Put more correctly the logical statement is: "We do observe B, therefore one possible explanation is that A is true, and there may be other explanations." The all too common *neglect of prior probability* flaunts Bayes' theorem, which shows that if the probability of a theory being true is low before the experiment, then a stronger standard of proof is required to support its correctness. These and other considerations must be recognized and added to the usual repertoire of experimental design and protocol that makes up the bulk of the scientific method.

III. Historical Outline

It is instructive to follow the general historical development of the modern search for a link between electromagnetism, matter, and gravity so that future researchers can understand the scope of the investigation so far. This outline focuses on those works related to superconductor experiments and theory. For more general historical events and context, please refer to Chapter 1. The reader is encouraged to pursue the references cited herein to obtain a fuller appreciation of the effort that has been expended in this area of physics.

A. Possibility of Gravitational Effects in the Laboratory

In a 1966 paper, DeWitt [4] proposed a magnetic analogue to the prediction by Shiff and Barnhill [23] who showed that the electric field inside a regular conductor is not zero in the presence of a Newtonian gravitational field. DeWitt calculated the surface current resulting from nonzero magnetic and

Lense–Thirring fields inside a superconductor in the presence of a nearby gravitomagnetic/Lense–Thirring field. The flux of the combination of magnetic and gravitomagnetic fields was found to be quantized. He suggested that the current could be measured in an apparatus that ensured that the current measurement was far from the gravitomagnetic field. In 1983, Dewitt's work was expanded upon by Ross [5], who produced a modified set of London equations as well as weak-field General Relativity equations in a form very similar to Maxwell's equations. The latter were particularly useful to later researchers as they no longer needed to deal with the complete set of tensor General Relativity equations when dealing with practical calculations. These papers laid the theoretical foundations for the later work of Li, Torr, Tajmar, and others, to be discussed below.

Li and Torr combined and extended this prior work in a 1991 publication on the effects of a superconductor on external gravitomagnetic and magnetic fields [1]. Ordinarily, all magnetic fields are excluded from the interior of a type I superconductor due to the Meissner effect. However, by solving the coupled Maxwell, General Relativity, and London equations for the internal magnetic and gravitomagnetic fields of superconductors exposed to external gravitomagnetic and magnetic fields, Li and Torr predicted a small, residual, internal magnetic field in contrast to normal Meissner field expulsion. This in turn produces an internal gravitomagnetic field. These internal fields are related to one another by the Cooper pair mass-to-charge ratio. One year later, the same authors presented a paper in which they used coupled Ginzburg–Landau equations to calculate the relative strengths of the electric and gravitational fields in superconductors in the presence of magnetic and gravitomagnetic fields. They concluded that under certain circumstances, a secondary gravitational field could be induced inside the superconductor and "... provide a basis for the electrical generation of gravitational fields in the laboratory" [2].

Around the same time Eugene Podkletnov, a Russian materials scientist on staff at the Institute of Materials Science at the Tampere University of Technology in Finland, published a paper with R. Nieminen on an apparent gravity shielding experiment using spinning superconductors [6] (Fig. 1). At the heart of his experimental design was the high-speed rotation of a relatively large (14.5 cm diameter × 6 mm thick) YBCO sintered ceramic superconducting disk in the vapors of LHe. The disk was levitated by Meissner flux expulsion over a single, large, toroidal support electromagnet, which was immersed in LHe and powered by a variable-frequency supply between 50 Hz to 10^6 Hz. At the periphery of the disk diameter, two additional but smaller electromagnets, also powered by variable frequency supplies, were positioned. These two "rotational" electromagnets were used to spin the disk in some unspecified manner. A small nonconducting, nonmagnetic test mass was suspended from an analytical balance about 15 mm from the top of the disk. The test mass weight loss was observed to be about 0.3% when stable conditions were achieved.

Not surprisingly, these results were met with much skepticism. Apart from the difficulty in explaining such a relatively large weight loss on purely theoretical grounds, there was considerable doubt about the validity of the experimental observations. The only information on the physical configuration of the experiment was provided in the sketch reproduced in Fig. 1. Assuming this is

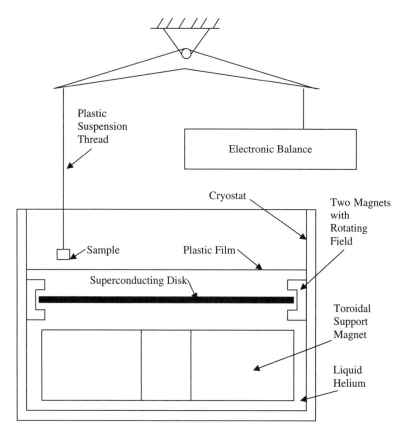

Fig. 1 Podkletnov's spinning disk experiment. (Adapted from Ref. 6.)

representative of the actual experimental setup, one notes that the sample, sample suspension thread, and main balance arm are unprotected from air currents. Concerning the cryostat, for instance, the only thing separating the vapors of LHe from the laboratory atmosphere was a thin plastic film. Ordinarily, so much water vapor and other gases would have condensed on the outer surface of the film as to render it completely opaque, thus making the observation of the disk extremely difficult. If the cryostat was actually designed roughly per the sketch in Fig. 1, the LHe would be boiling so vigorously that it would rupture any film unless adequate He gas escape was provided. As pointed out by de Podesta and Bull [24], thermal currents and buoyancy changes above such a cryostat would be so severe as to render the determination of the weight of a test mass suspended only 15 mm above the disk (and, therefore, only a few mm above a separating film covered with ice) virtually impossible. As such, the design represented in the article appeared to be entirely unsuited for the detailed and delicate measurements reported by Podkletnov. Other important issues, such as how the disk was balanced, how it kept together at high speeds, how much power was used to operate the coils, and what means were employed to prevent the balance from being affected by the magnetic fields from the coils were not addressed in the article.

Nevertheless, Podkletnov and Nieminen's paper provoked some experimentalists to attempt replications of the essence of the experiment. The experimental replications started out using less costly LN_2 with either fixed or rotating permanent magnets, and small (approximately 2 to 3 cm diameter) commercially purchased superconducting disks. Several researchers, including Gonnelli at the Politecnico of Turin (private communication), Woods et al. at the University of Sheffield [25], and this author witnessed very slight apparent weight changes while the disk was passing through its critical temperature, T_c. However, in most cases, the apparent effect was generally no larger than the noise inherent in the system, and further experimentation was not performed. At a subsequent conference in Turin (discussed later in the Sec. III.B), Podkletnov emphasized that unless the exact disk formulation was followed, high-frequency magnetic fields employed (not permanent magnets), and a larger disk spun at lower temperatures, the shielding effect would be extremely small.

After the Podkletnov and Nieminen publication, Li and Torr published another article [3] expanding on their earlier investigations in an attempt to outline the physical mechanism underlying the production of a gravitomagnetic field inside a superconductor. Basic to their evaluation was an assumption of near-zero magnetic permeability in type I superconductors, and the requirement of coherent alignment of lattice ion spins in conjunction with a time-varying applied magnetic vector potential field. With this framework, they determined values for laboratory-scale induced internal gravitomagnetic fields and external gravitoelectric fields, and predicted how these fields could be maximized. After noting an additional assumption about Cooper pair to lattice ion mass ratio, they state "... it appears that the electrically induced gravitoelectric field should be readily detectable in the laboratory."

In 1994, Kowitt [26] showed that the assumption of near-zero permeability inside a superconductor used by Li and Torr in their 1992 and 1993 papers was not valid. Another criticism of the Li and Torr results was from Harris [27]. He argued that Li and Torr's previous results were erroneous because they assumed arbitrary (and extremely small) distances from the lattice ion to the observer, thus producing unreasonably large effects. In fact, Harris pointed out that the correct estimate of the induced gravitoelectric field outside a superconductor is some 20 orders of magnitude smaller than that predicted by Li and Torr. These issues are further discussed in Sec. III.D.

B. Tests of Podkletnov Gravity Shielding Claims

In 1995 Li approached NASA Marshall Space Flight Center to support development of her theory that an artificial gravity field could be measured outside a superconductor. Funding was also secured for replicating the Podkletnov experiment at NASA. After several preliminary experiments at NASA, Koczor et al. failed to see the expected shielding effect in a Podkletnov-like experiment [28]. However, they were using a small commercially available disk levitated above permanent magnets at LN2 temperatures. The following year, after discussions with Podkletnov, Noever and Koczor [29] published the results of their investigations into nonrotating superconductor disks irradiated by radio frequencies from 1 to 15 Mhz and detected a very weak gravity increase. This finding was later shown by the same authors to likely be the result of an instrumentation artifact [30]. Notwithstanding, Koczor commissioned the fabrication of a 27-cm bi-layer disk conforming to Podkletnov's 1997 specifications that had just been published [31], in the hopes of replicating their experiment. Unfortunately, budgetary constraints forced NASA to stop further research in this area few years later, with no further publications forthcoming.

In 1995 this author began preliminary experiments in Toronto after contacting Podkletnov. The approach was to reproduce the original 1992 spinning disk experiment with additional data from Podkletnov and to use a better cryogenic design that allowed the possibility of mechanically spinning the disk. At our Toronto laboratory, manufacturing also began on the large YBCO disks required for the experiment.

Meanwhile, Podkletnov and Vuorinen attempted to publish the results of their own updated spinning disk experiments in the *British Journal of Physics D*, but the paper was withdrawn and later published on the Web [31]. This new experiment involved a large AC-levitated, 27-cm diameter bi-layer sintered YBCO disk spinning at 5000 rpm in two-phase high-frequency RF "rotation" fields, and claimed to show gravitational shielding in the few percent range. The cryostat design used in the updated experiments was somewhat better than the original design in that there was considerably more shielding of the test mass from buoyancy and thermal current effects.

In 1999, a private, unpublished symposium on Podkletnov's work was hosted by R. Gonnelli at the Politecnico in Turin. The state of the Toronto experimental replication was presented, and Gonnelli and others presented their initial findings of a possible tiny weight change in test samples suspended above a disk as it passed through T_c. Podkletnov described how the "gravity shielding" effect was initially discovered. Apparently his group had made large, sputtering target disks of YBCO for single-crystal processing and, to ensure the correct uniformity and porosity, the disks were set into rotation (presumably mechanically) while being levitated over a "supporting solenoid." This allowed quick and complete scanning of the target's surface by means of a small movable test magnet suspended above the rotating disk and connected to an analytical chemical balance. When the smoke from a technician's pipe appeared to rise inexplicably above the apparatus, they considered the possibility of gravitational shielding and substituted a nonmagnetic, nonconducting test mass for the small suspended magnet. It should be noted that the normal rotational speeds for magnetron sputtering targets are in the 10s of rpm, whereas Podkletnov required speeds several orders of magnitude greater than this. Subsequent information from Podkletnov indicated that, to obtain the maximum stable test sample weight loss of about 0.3%, the optimum conditions required operation of the three coils at frequencies of 10^5 Hz and disk rotational speeds of several thousand rpm (E. Podkletnov, private communication).

Although the general consensus at the conference was that the Podkletnov claims were not spurious, only the Toronto group elected to pursue a complete replication.

Our own version of the Podkletnov spinning disk experiment was completed in late 2001, with results published in 2003 [7] (Fig. 2). Here, an external motor

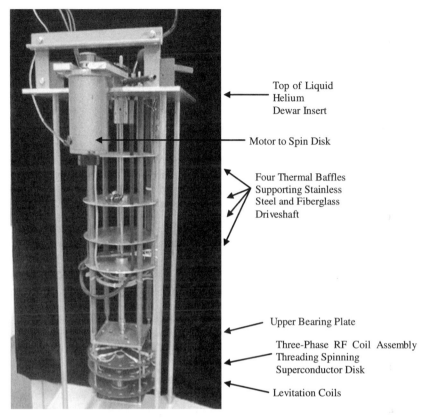

Fig. 2 Apparatus for LHe test of Podkletnov's "gravity shielding" claims [7].

spins the superconducting disk that is at the bottom of a deep LHe dewar. The toroidal disk is threaded by three high-frequency coils and is levitated by three solenoidal coils. The experiment showed a null result. Of potential interest to the reader, the paper contains an overview of the experimental difficulties arising from the nature of the experiment itself, and discusses the inability of Podkletnov to supply critical data on the previous experiments. Unfortunately, neither Podkletnov's 1992 publication nor subsequent discussions with Podkletnov himself allowed a complete understanding of how the original experiment was carried out.

Also of importance, in 1999 Reiss published [32] the results of precise measurements of the weights of superconducting and non-superconducting samples cooled below T_c. He found a slight (~0.5%) weight increase of a high-temperature superconductor for which he was not able to offer a prosaic explanation. He weighed small disk-shaped samples held in a specially-made capsule while dipping it into LN_2. His analysis of possible artifacts is thorough and very useful to other researchers investigating this area. In 2003, he published

[33] an update of his on-going LN_2 experiments with increased precision and artifact reduction. He was still observing weight changes during T_c transition to a repeatable degree not achieved elsewhere.

However, during this same period, Tajmar et al. [34] attempted to replicate several earlier weight-alteration experiments involving high-temperature superconductors but with no success.

C. Podkletnov Force Beam Claims

At the Turin symposium, Podkletnov described further experiments with the so-called high-voltage "gravity beam" (rather than gravity shielding) apparatus. On a small (few cm^2 in area) substrate was grown an array of single crystal whiskers of YBCO. This plate was placed upright in a small LN_2 dewar and electrically attached to a small (~200 KV) van de Graaff machine. A grounded metal annular disk was placed a few centimeters laterally away. The whole assembly was placed in a large bell jar, which was evacuated and backfilled with argon to prevent YBCO degradation by water vapor. When the static machine was operated, a light blue planar "discharge" was seen to pass from the superconductor array to the annulus. At this instant, a pencil standing upright on a table in an adjoining room fell over. The pencil was separated from the experiment by a thick concrete wall.

In 2001, Podkletnov and his collaborator Modanese published a paper on the Web concerning an enhanced version of the gravity beam experiment [35]. This was the first general publication of the high-voltage impulse force experiment. Enough technical description was available to allow assessment of the validity of experimental setup. Unfortunately, many unresolved technical questions cast considerable doubt on whether the experiment actually had been undertaken and, once again, no confirmatory evidence was provided by Podkletnov or Modanese. However, it presented a general summary of the theoretical work by Modanese and had an extensive bibliography.

D. Gravitational Wave Transducers

In 2002, Chiao proposed using superconductors as gravitational wave transducers for RF radiation and attempted to demonstrate this experimentally, apparently without success [36]. Harris provided an explanation for the apparent failure by stating that neither gravitoelectric nor gravitomagnetic fields accompany gravitational waves [37].

In 2005 Woods presented the results of his investigations into the concept of using superconductors to mediate the detection of high-frequency gravitational waves [16,17]. Using Li and Torr's result that the phase velocity of gravitational waves is smaller in superconductors than in ordinary matter by a factor of roughly 300, he reasoned that the refractive index for gravitational waves inside a superconductor might be used to focus or manipulate such waves to increase the possibility of their detection in the laboratory. However, his results are highly dependent on the gravitational wave phase velocity being as predicted by Li and Torr. He also examined the criticisms of Kowitt and Harris and concluded that while Kowitt's concerns may not be tenable, Harris's arguments were harder to refute and the issue might best be resolved through experimental investigation.

E. Tajmar Experiments

In 2001 Tajmar and De Matos published a paper [38] which reviewed previous work and showed that every electromagnetic field is coupled to a gravitoelectric and gravitomagnetic field. According to Tajmar and De Matos, the coupling "... is generally valid and does not require special properties like superconductivity." The authors acknowledged the criticisms of Li and Torr by Kowitt and Harris, and noted that the simple coupling coefficient they derive is exceedingly small. However, they note that it can be increased by using massive ion currents (e.g., a moving/rotating mass or a dense plasma) and alignment of electron and nuclear spins. In a roughly concurrent publication [39], De Matos and Tajmar extended these ideas and used a Barnett effect analogue to show "... any substance (i.e., not necessarily superconductive) set into rotation becomes the seat of a uniform intrinsic gravitomagnetic field..."

In the classic 1950 text on superfluids, London derived an expression for the magnetic field produced by a rotating superconductor or superfluid, which was proportional to the Cooper pair mass-to-charge ratio and the angular velocity [40]. The value of this so-called "London moment" has been measured in the laboratory by Tate et al. [41]. A general expression of the London moment can be used to determine the Cooper pair mass. Surprisingly, the measured Cooper pair mass, which had been predicted to be slightly smaller than twice that of the electron, was actually slightly larger.

In a 2003 paper, Tajmar and De Matos [42] asked if a gravitational effect might explain this mass discrepancy. By applying their previous work to this "Cooper pair mass anomaly," Tajmar and De Matos found that a relatively huge internal gravitomagnetic field would be required to explain the mass anomaly, of a magnitude that could be measured in the laboratory. The gravitomagnetic field was predicted to have the form $B_g = 2\omega\rho^*/\rho$, where ρ^* is the Cooper pair mass density, ρ is the classical bulk mass density of the superconducting material, and ω is the superconductor's rotational speed. To test this hypothesis, Tajmar proposed an experiment "... measuring the torque on a spinning gyroscope produced by the gravitomagnetic field possibly generated by rotating superconductors..." In 2006 Tajmar et al. [43] described the results of just such an experiment (Fig. 3), which involved mechanical spinning of niobium and high-temperature ceramic superconductor rings at LHe temperatures. No external magnetic fields were applied; a sudden angular deceleration of the superconducting rings was used to produce the acceleration required to see the anticipated effect. Tajmar et al. claimed to have found the expected large gravitomagnetic field as detected by nearby sensors such as accelerometers and laser gyroscopes, which matched their theoretical predictions to within a reasonable factor.

By 2007, Tajmar et al. [44] recognized that new data from improved experiments did not match their prior predictions. Nevertheless, an unexplained residual signal persisted that exhibited several unexpected features, including a relatively large coupling constant of 10^{-8} between the observed acceleration

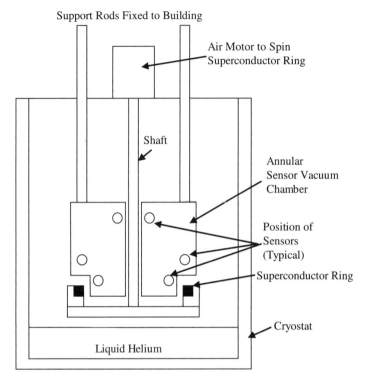

Fig. 3 Tajmar apparatus to search for possible frame-dragging effects. (Adapted from Ref. 44.)

effect and the applied angular velocity. The effect appears to be proportional to angular momentum and inversely proportional to temperature, after passing a critical temperature (which is dependent on the material of the spinning ring and not coincident with the superconducting critical temperature). In addition, the effect is more pronounced in the clockwise rotation direction (as viewed from above), and the effect does not decay as would a dipole field. While Tajmar et al. endeavor to address all possible systematic errors or prosaic explanations, they conclude: "The ... measurements rule out our previous theoretical model which predicted a coupling proportional to the material's Cooper pair and lattice mass density." The residual signal observed in the most recent experiments remains to be explained. At the time of this writing, this most recent reference [44] was still undergoing review prior to its planned publication in the Time and Matter Conference Proceedings, published by World Scientific Press.

In an experiment designed to test Tajmar's results, a team from the University of Canterbury in New Zealand [45] rotated a massive superconducting lead ring near a very large ring laser gyroscope, which allowed measurement of the Sagnac frequency difference between each rotational direction of the lead mass. They measured a variation of Sagnac frequency deviation as a function of superconductor rotational speed approximately 21 times less than that predicted by Tajmar et al. Their data, however, do show a possible indication of an effect similar to the Tajmar observations.

An explicit assumption in the work of Graham et al. [45] was that the putative field was dipolar in nature. They noted that the magnitude of the effect claimed by Tajmar is not dependent on the mass of the superconductor, but that the observed effect would scale roughly as the volume of rotating material, assuming an inverse-cube decay of a dipole field. Thus Graham et al. used a massive lead ring, which was about three times the volume of that used by Tajmar. However, the ring was positioned approximately 20 cm from one leg of the gyroscope compared to the 5-cm distance used in the Tajmar experiment. In addition, Tajmar used fiber optic gyroscopes as detectors, versus the New Zealand's team use of ring laser gyroscopes. Another difference was Tajmar's positioning of the gyroscopes slightly above the spinning ring, whereas the spinning ring used by Graham et al. was in the plane of the gyroscope. Finally, Graham et al. used angular velocities of approximately 100 rad/s, whereas the Tajmar team used angular velocities around 400 rad/s.

F. Other Theoretical Directions

Theoretical work is still being produced on the interactions between the gravitomagnetic field and accelerated matter. A recent paper by Ahmedov [46] investigated the possibility of measuring a voltage due to the action of a gravitomagnetic field on current flowing in orbiting semiconductors as an analogue of the Hall effect resulting from a magnetic field. Rabounski and Borissova [47] developed a nonholonomic metric in which a vertically accelerating mass (not necessarily superconductive) perturbs the "background space" to produce a zone of altered gravity near the mass. They invoke the vibrating balance experiment of Kozyrev [48] to justify their claims.

IV. Summary and Future Directions

If real, a superconductor-mediated interaction between matter and gravity would provide a method for manipulating gravitational fields, with potentially tremendous benefits for future technology. Only a handful of researchers are currently pursuing this avenue, and an increased understanding of the nature of superconductivity is necessary to establish a firm basis from which to proceed. One lingering question is whether there really is a Cooper pair mass anomaly. A second area requiring further investigation is the precise determination of the gravitational wave index of refraction inside a superconductor. Progress in each of these areas will further benefit our understanding of superconductors in general, and may result in new advances that can lead to future flight systems. And lastly, further independent tests of the Tajmar observations would help distinguish if a new effect is being observed, and if so, then better characterize such effects.

In the spirit of the scientific enterprise, experimentalists must ensure that their results, both positive and negative, are readily accessible for other researchers to test. Theoreticians will then have more empirical evidence upon which to further advance models of how Nature functions. And all must ensure that they are cognizant of experimental pitfalls associated with the phenomena they seek to uncover.

References

[1] Li, N., and Torr, D. G., "Effects of a Gravitomagnetic Field on Pure Superconductors," *Physics Review D*, Vol. 43, No. 2, 1991, pp. 457–459.
[2] Li, N., and Torr, D. G., "The Gravitoelectrodynamics of Superconductors: A Theoretical Basis for a Principle of Electrically Induced Gravitation," *Bulletin of the American Physics Society*, Vol. 37, No. 2, 1992, Session G8 paper 10.
[3] Li, N., and Torr, D. G., "Gravitoelectric–Electric Coupling via Superconductivity," *Foundations of Physics Letters*, Vol. 6, No. 4, 1993, pp. 371–383.
[4] DeWitt, B. S., "Superconductors and Gravitational Drag," *Physics Review Letters*, Vol. 16, No. 24, 1966, pp. 1092–1093.
[5] Ross, D. K., "The London Equation for Superconductors in a Gravitational Field," *Journal of Physics A*, Vol. 16, 1983, pp. 1331–1335.
[6] Podkletnov, E., and Nieminen, R., "A Possibility of Gravitational Force Shielding by Bulk $YBa_2Cu_3O_{7-x}$ Superconductor," *Physica C*, Vol. 203, 1992, pp. 441–444.
[7] Hathaway, G., Cleveland, B., and Bao, Y., "Gravity Modification Experiment Using a Rotating Superconducting Disk and Radio Frequency Fields," *Physica C*, Vol. 385, 2003, pp. 488–500.
[8] Forward, R. L., "General Relativity for the Experimentalist," *Proceedings of the IRE*, Vol. 49, 1961, pp. 892–904.
[9] Forward, R. L., "Guidelines to Antigravity," *American Journal of Physics*, Vol. 31, 1963, pp. 166–170.
[10] Fairbank, J. D., Deaver, Jr., B. S., Everitt, C. W. F., and Michelson, P. F. (eds.), *Near Zero: New Frontiers of Physics*, W. H. Freeman and Co., New York, 1988, Chap. VI.
[11] Puthoff, H. E., "Gravity as a Zero-Point Fluctuation Force," *Physics Review A*, Vol. 39, 1989, pp. 2333–2342.
[12] Sakharov, A. D., "Vacuum Quantum Fluctuations in Curved Space and the Theory of Gravitation," *Soviet Physics Doklady*, Vol. 12, 1968, pp. 1040–1041.
[13] Alzofon F. E., "Anti-Gravity with Present Technology: Implementation and Theoretical Foundation," *17th AIAA/SAE/ASME Joint Propulsion Conference*, AIAA Paper 81-1608, Colorado Springs, CO, July 1981.
[14] Hughes, W. L., "On Kopernicky's Conjecture: Gravity Is a Difference Between Electrostatic Attraction and Repulsion," *Galilean Electrodynamics*, Vol. 15, No. 5, 2004, pp. 97–100.
[15] *Gravitational Wave Conference: International High-Frequency Gravitational Waves (HFGW) Working Group*, The Mitre Corporation, McLean, VA, May 2003.
[16] Woods, R. C., "Gravitational Waves and Superconductivity," *Space Technologies and Applications International Forum 2005*, El Genk, M. S. (ed.), American Institute of Physics Conference Proceedings, Melville, NY, 2005.
[17] Woods, R. C., "A Novel Variable Focus Lens for HFGW," *Space Technologies and Applications International Forum 2006*, El Genk, M. S. (ed.), American Institute of Physics Conference Proceedings, Melville, NY, 2006.
[18] Braginski, V. B., Caves, C. M., and Thorne, K. S., "Laboratory Experiments to Test Relativity Gravity," *Physics Review D*, Vol. 15, No. 5, 1977, pp. 2047–2068.

[19] Earthtech International. URL: http://earthtech.org/experiments/Hathaway_Nightmare_List.pdf [Accessed 3 March 2008].
[20] Reiss, H. D., and Hathaway, G. D., "Minimum Experimental Standards in the Laboratory Search for Gravity Effects," *Space Technologies and Applications International Forum 2006*, El Genk, M. S. (ed.), American Institute of Physics Conference Proceedings, Melville, NY, 2006.
[21] O'Hanlon, J. F., *A User's Guide to Vacuum Technology*, 2nd ed., Wiley, New York, 1989, p. 445.
[22] Gallagher, T. J., and Pearmain, A. J., *High Voltage Measurement, Testing and Design*, Wiley, New York, 1983, p. 50.
[23] Schiff, L. I., and Barnhill, M. V., "Gravitation-Induced Electric Field near a Metal," *Physics Review*, Vol. 151, No. 4, 1966, p. 1067.
[24] de Podesta, M., and Bull, M., "Alternative Explanation of 'Gravitational Screening' Experiments," *Physica C*, Vol. 253, 1995, pp. 199–200.
[25] Woods, R. C., Cooke, S. G., Helme, J., and Caldwell, C. H., "Gravity Modification by High-Temperature Superconductors," *37th AIAA/ASME/SAE/ASEE Joint Propulsion Conference*, AIAA Paper 2001-3363, Salt Lake City, UT, July 2001.
[26] Kowitt, M., "Gravitomagnetism and Magnetic Permeability in Superconductors," *Physics Review B*, Vol. 49, 1994, pp. 704–708.
[27] Harris, E. G., "Comments on 'Gravitoelectric-electric Coupling via Superconductivity,'" *Foundations of Physics Letters*, Vol. 12, 1999, p. 201.
[28] Li, N., Noever, D., Robertson, T., Koczor, R., and Brantley, W., "Static Test for a Gravitational Force Coupled to Type II YBCO Superconductors," *Physica C*, Vol. 281, 1997, pp. 260–267.
[29] Noever, D., and Koczor, R., "Radio-Frequency Illuminated Superconductive Disks: Reverse Josephson Effects and Implications for Precise Measuring of Proposed Gravity Effects," *NASA JPL 9th Advanced Space Propulsion Research Workshop & Conference*, Pasadena, CA, 1998.
[30] Noever, D., Koczor, R., Roberson, R., "Superconductor-Mediated Modification of Gravity? AC Motor Experiments with Bulk YBCO Disks in Rotating Magnetic Fields," *34th AIAA/ASME/SAE/ASEE Joint Propulsion Conference*, AIAA Paper 98-3139, Cleveland, OH, July 1998.
[31] Podkletnov, E., "Weak Gravitational Shielding Properties of Composite Bulk YBCO Superconductor below 70K under EM Field," LANL cond-mat/9701074 v. 3, 16 Sept. 1997.
[32] Reiss, H. D., "A Possible Interaction Between Gravity and High Temperature Superconductivity—by a Materials Property?," *15th European Conference on Thermophysical Properties*, Wuerzburg, Germany, Sept. 1999.
[33] Reiss, H. D., "Weight Anomalies Observed During Cool-Down of High Temperature Superconductors," *Physics Essays*, Vol. 16, 2003, pp. 236–253.
[34] Tajmar, M., Hense, K., Marhold, K., and De Matos, C., "Weight Measurements of High-Temperature Superconductors During Phase Transition in Stationary, Nonstationary Condition and Under ELF Radiation," *Space Technologies and Applications International Forum 2005*, El Genk, M. S. (ed.), American Institute of Physics Conference Proceedings, Melville, NY, 2005.
[35] Podkletnov, E., and Modanese, G., "Impulse Gravity Generator Based on Charged YBa2Cu3O7-y Superconductor with Composite Crystal Structure," *arXiv: physics/0108005v2*, Aug. 2001.

[36] Chiao, R., "Superconductors as Quantum Transducers and Antennas for Gravitational and Electromagnetic Radiation," *arXiv: gr-qc/0204012*, Apr. 2002.

[37] Harris, E., "Superconductors as Gravitational Wave Detectors," 69th Annual Meeting Southeastern Section, *American Physics Society*, No. NC.001, 2002.

[38] Tajmar, M., and De Matos, C., "Coupling of Electromagnetism and Gravitation in the Weak Field Approximation," *Journal of Theoretics*, Vol. 3, No. 1, 2001.

[39] De Matos, C., and Tajmar, M., "Gravitomagnetic Barnett Effect," *Indian Journal of Physics*, Vol. 75B, No. 5, 2001, pp. 459–461.

[40] London, F., *Superfluids, Vol. 1*, Wiley, New York, 1950, p. 78.

[41] Tate, J., Cabrera, B., Felch, S., and Anderson, J., "Precise Determination of the Cooper-Pair Mass," *Physics Review of Letters*, Vol. 62, No. 8, 1989 pp. 845–848.

[42] Tajmar, M., and De Matos, C., "Gravitomagnetic Field of a Rotating Superconductor and of a Rotating Superfluid," *Physica C*, Vol. 385, 2003, pp. 551–554.

[43] Tajmar, M., Plesescu, F., Marhold, K., and De Matos, C., "Experimental Detection of the Gravitomagnetic London Moment," *arXiv: gr-qc/0603033*, Mar. 2006.

[44] Tajmar, M., Plesescu, F., Seifert, B., Schnitzer, R., and Vasiljevich, I., "Search for Frame-Dragging-Like Signals Close to Spinning Superconductors," *Proceedings of the Time and Matter 2007 Conference*, Bled, Slovenia, World Scientific Press, 2007.

[45] Graham, R. D., Hurst, R. B., Thirkettle, R. J., Rowe, C. H., and Butler, P. H., "Experiment to Detect Frame Dragging in a Lead Superconductor," *Physica C*, 6 July 2007, submitted for publication.

[46] Ahmedov, B. J., "General Relativistic Galvano-Gravitomagnetic Effect in Current-Carrying Conductors," *arXiv: gr-qc/0701045v1*, Jan. 2007.

[47] Rabounski, D., and Borissova, L., "A Theory of the Podkletnov Effect Based on General Relativity: Anti-Gravity Force Due to the Purturbed Nonholonomic Background of Space," *Progress in Physics*, Vol. 3, 2007, pp. 57–80.

[48] Kozyrev, N. A., "The Vibration Balance and Analysis of Its Operation," *Problems of Research of the Universe*, Vol. 7 in "Astrometry and Celestial Mechanics," Moscow-Leningrad, 1978, pp. 582–584.

Chapter 6

Nonviable Mechanical "Antigravity" Devices

Marc G. Millis*
NASA Glenn Research Center, Cleveland, Ohio

I. Introduction

A QUINTESSENTIAL example of nonviable approaches to create antigravity are the commonly proposed devices that attempt to convert mechanical oscillations or gyroscopic motion into thrusting or gravity-defying effects. When claimed as a breakthrough, such ideas are often summarily dismissed as they are known to violate conservation of momentum. To offer a more effective response, explanations are offered on why these devices *appear* to be breakthroughs, how they operate from the perspective of established physics, and what tests would be required to provide more convincing evidence of how the devices really operate. This also provides a starting point for critically assessing future proposals of this type.

II. Oscillation Thrusters

The *oscillation thruster*, also describable as a *sticktion drive*, *internal drive*, or *slip-stick drive*, is a commonly suggested device that uses the motion of internal masses to create net thrust. One of the most famous oscillation thrusters is the 1959 "Dean drive" described in Patent 2,886,976 [1]. A more recent and simple example is shown in Fig. 1 [2] and further still, Fig. 2 displays an example that uses rotating masses [3]. Although there are many versions, all oscillation thrusters have the following common components: chassis to support a system of masses; conveyor that moves the masses through an asymmetric cycle; and power source for the conveyor.

A crucial feature is that the internal masses go through a cyclic motion where the motion in one direction is quicker than in the other. The result is that the

---

This material is a work of the U.S. Government and is not subject to copyright protection in the United States.
*Propulsion Physicist, Propellant System Branch, Research and Technology Directorate.

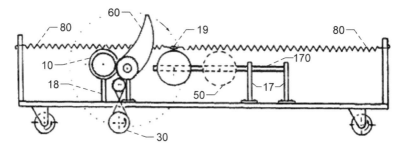

Fig. 1 Linear oscillation thruster. A typical example of an oscillation thruster, specifically from Patent 5,685,196 [2]. As the cam (60) rotates, a mass (50) moves slowly in one direction and is allowed to return quickly in the other. The reaction force from one part of this cycle is sufficient to overcome static friction, while the reaction force is insufficient in the other part of the cycle. This leads to one-directional motion, giving the illusion of net thrust.

whole device moves in surges across the ground, giving the appearance that a net thrust is being produced without the expulsion of a reaction mass.

Because it would constitute a breakthrough to be able to move a vehicle without expelling a reaction mass [4], these devices *appear* to be breakthroughs. Regrettably, such devices are not breakthroughs because they still require a connection to the ground to create net motion. The ground is the reaction mass, and the frictional connection to the ground is a necessary feature to its operation.

More specifically, it is the difference between the static friction (sometimes called *sticktion*) and the dynamic friction between the device and the ground

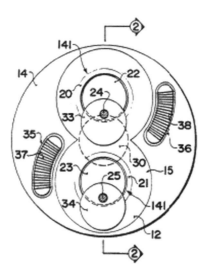

Fig. 2 Rotational oscillation thruster. A typical example of an oscillation thruster that uses rotating masses with convolutions of cycles. The version shown is from Patent 4,631,971 [3].

that is required for their operation. Static friction, the amount of friction encountered when contacting surfaces are not moving relative to one another, is typically greater than the *dynamic* friction when the contacting surfaces are moving relative to one another.

Recall that the device's internal masses move faster in one direction and slower in the other. When the masses move quickly, the device has enough reaction force to overcome the static friction between itself and the ground, and the device slides. When the internal masses return slowly in the other direction, the reaction forces are not enough to overcome the static friction and the device stays in its place. The net effect is that such slip-stick motion causes the device to scoot across the floor.

If the device could be placed into orbit to follow a freefall trajectory absent of any connection with external masses, then the center of mass of the entire system would follow the freefall trajectory without deviation, while the device's external frame and its internal masses would just oscillate with respect to one another. Similarly, if the device were dropped with its thrusting direction pointing either up or down, the rate of fall of the center of mass of the system would be identical regardless if the device were on or off.

To illustrate stick-slip operation, Fig. 3 offers an analogy. The right side of the figure represents half of a cycle where the device does not move, while the left side represents half of the cycle where the device does move. Even though the total impulse ($I = F \times t$), as represented by the *area* of the rectangle, is equal in both phases of the cycle, the force, F, represented by the *width* of the rectangle, and its duration, t, represented by *height* of the rectangle, can vary. When they do, they vary reciprocally so that their product remains constant. (Note that this is

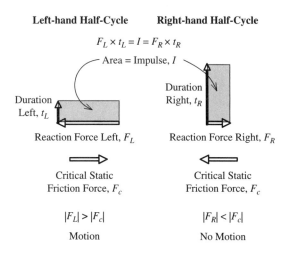

Fig. 3 Equal impulses, different forces. This is a conceptual representation of the $F \times t$ impulse of each half of a generic mechanical oscillator. Although the impulse (area of rectangle) is equal in both halves of the cycle, the force on the right part of the cycle, F_R, is insufficient to overcome the critical friction, F_c, while the force on the left, F_L, is sufficient. The device would move during the left half-cycle and remain stuck to the floor on the right half-cycle.

only a conceptual illustration; a more rigorous analysis would require integrating a variable force over time.) With the impulse the same in both half-cycles, the reaction force is less on the right-hand side than on the left. When compared to the critical static friction force, F_c, which is the amount of force that must be overcome to set the device in motion, it is clear that the reaction force for the right half-cycle is less than this critical force, while on the left it is greater. This means that the device would move during the left-hand half of the cycle, but not during the right-hand half.

More rigorous analyses of a given device can take on a variety of forms, depending on the device in question. Options include using the *impulse* representation ($\int F \cdot dt$) or the *work* representation ($\int F \cdot dl$) over all the phases of the device's cycle. Also, depending on the device, the number of phases could vary. In Fig. 3, there are only two phases, but for the Foster device in Fig. 1, for example, at least three phases would need to be assessed: 1) when the cam (part 60) is displacing the mass (part 50); 2) when the mass returns under the restoring force of spring (part 80); and 3) when the mass comes to a stop back to its initial position.

Because such analyses are time-consuming and are not likely to be understood by many amateurs, it is more effective to suggest rigorous experimental proof. A fitting test is to place the device on a level pendulum stand, as illustrated in Fig. 4, and compare the deflection between the on and off conditions of the device. A sustained net deflection of the pendulum is indicative of genuine thrust. Conversely, if the pendulum oscillates around its null position, which is the expected finding, then the device is not creating net thrust.

To avoid possible spurious effects, it is advised to have the device and its power supply on the pendulum, and to have the tallest pendulum possible (i.e., l in Fig. 4). The reason for containing the power supply with the device is to avoid having the power cords interfere with the free motion of the pendulum.

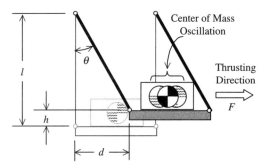

Fig. 4 Basics of a level pendulum test. A pendulum is a simple tool for measuring lateral force and a *level* pendulum keeps the test platform from tilting during operation. If a thrusting device can sustain a deflection of a level pendulum, then there is strong evidence toward the claim of net thrust. It is anticipated, however, that mechanical devices will instead oscillate the pendulum back and forth, with the average position being zero deflection.

The reasons for having a *tall* pendulum is to make its lateral motion (*d* in Fig. 4) more pronounced for a given lateral force, *F*, and to reduce its natural oscillation frequency to be much less than the oscillation frequency of the device. Spurious results are possible if the oscillation frequency of the device and the pendulum are similar. For reference, Eqs. (1) and (2) present common equations for using pendulums to measure lateral thrusting (valid for small deflections, approximating sine $\theta \approx \theta$) [5].

Lateral force as a function of deflection:

$$F = mg \tan \theta \quad (1)$$

Natural frequency of a pendulum:

$$f = \frac{1}{2\pi}\sqrt{\frac{g}{l}} \quad (2)$$

where:

F = lateral force acting to deflect the pendulum (N)
f = pendulum frequency (Hz)
g = gravitational field (9.8 m/s^2)
l = length of the pendulum (m) (Fig. 4)
m = mass at the lower end of the pendulum (other masses taken as negligible) (kg)
θ = deflection angle (radians) (Fig. 4)

A *level* pendulum is recommended instead of a *simple* pendulum to avoid the misleading effects from tilting the base, as illustrated in Fig. 5. Similarly, the reason that an air track is *not* recommended is the misleading effects from

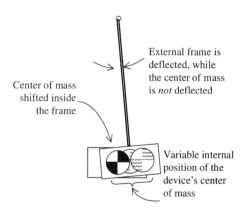

Fig. 5   Pitfall of simple pendulum. With a simple pendulum, a shift in the center of mass of the device (from the motion of its internal masses), gives the false appearance of a deflection when in fact the center of mass of the system is still directly underneath the support.

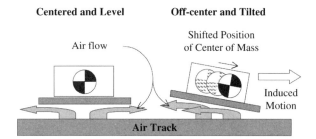

Fig. 6 Pitfalls of air track test. In addition to the fact that air tracks posses some friction, they are not suitable as a testing platform due to the potential for induced tilting. If the center of mass of the device shifts forward, for example, the leading edge of the air platform can tilt downward, creating a preferred channel for the airflow, which in turn creates a propulsive force on the platform.

the tilting of its base, as illustrated in Fig. 6. When the internal masses of the device shift off center, the base can tilt. In the case of the pendulum, this has the effect of inducing an *apparent* deflection of the pendulum. In the case of an air track this has the effect of deflecting more air away from the dipped end, thereby thrusting the device from the reaction to the asymmetric airflow.

Another difficulty when using air tracks is that it is easy for amateur researchers to mistake steady-state motion as evidence of thrust. When activating the device, or when setting an active device on an air track, it is easy to inadvertently impart an initial velocity that will continue to move the device along the track. This motion is not evidence of thrust, but rather of that initial velocity. If the device is actually producing thrust (and the tilting issue described previously is avoided), then the device will *accelerate*, moving faster and faster. Making the distinction between steady-state motion (zero force) and a slightly accelerated motion (thrusting) can be difficult with the naked eye. Instead, reliable measures of velocity versus time are required.

To keep an open yet rigorous mind to the possibility that some physical phenomenon has been overlooked, it should be necessary that future proposals on these types of devices pass a pendulum test and that rigorous supporting data addressing all possible false-positive conclusions be provided. Although a "jerk" effect (time rate change of acceleration) [6] has sometimes been mentioned as a theoretical approach to understand such devices, no physical evidence has been reported to substantiate that such a jerk effect exists, and theoretical difficulties arise because of momentum conservation violations. If successful net-thrust tests are ever produced and if a genuine new effect is found, then science will have to be revised because it would then *appear* as if such devices were violating conservation of momentum.

III. Gyroscopic Antigravity

Another category of commonly purported mechanical breakthroughs consists of a system of gyroscopes. A famous example is from the 1973 demonstration by Eric Laithwaite, where a spinning gyro is shown to rise upward while it is forced

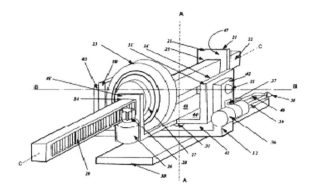

Fig. 7 Laithwaite "propulsion system" patent. Quoting from US Patent 5,860,317: "... the mass of the first gyroscope moves with an associated second movement of the mass of the remainder of the system in substantially the opposite direction, wherein the movement owing to the translation-dominated portion and is larger than the movement owing to the precession-dominated portion of the motion, hence moving the system" [8].

to precess.[†] Although such upward motion is a consequence of conservation of angular momenta, it is easily misinterpreted as an "antigravity" effect. Laithwaite, a professor of applied electricity at the Royal Institution of Great Britain from 1967 to 1975 [7], and Dawson went on to patent a device (Fig. 7) that claims to produce linear force from such torques [8].

Variations on the claim that gyroscopes can create linear thrust are common and typically consist of forcing the axis of a spinning gyroscope to change its orientation in a manner that causes the entire gyroscope to shift upward, thereby creating the impression that an upward "antigravity" force exists [9]. Because a rigorous analysis of this dramatic motion can be difficult, the ambiguities regarding its real physics linger.

Such concepts can be viewed as trying to reverse the effect of a spinning top. Rather than falling immediately over, a spinning top precesses. It appears as if an invisible force is holding it up (Fig. 8). Although the gravitational field, g, which is tilting the top, is *rectilinear*, the manner in which the top reacts to it produces *angular torques*. These torques are working to conserve angular momentum in response to gravity. The notion of reversing this process, of forcing the gyroscopic precessions to induce a linear force to oppose gravity ("antigravity"), seems like reasonable symmetry to expect, but this is not the case.

To illustrate the basic operation of a typical gyroscopic antigravity device, a simple version from Ref. 9 that is based on the Laithwaite demonstration (Fig. 9) will be used at the working example. Laithwaite's demonstration consisted of just a singe 50-lb gyroscope rather than the dual version shown, but

---

[†]Eigenbrot, I. V., email exchange between Dr. Ilya V. Eigenbrot, International and Science Communication Officer, Imperial College London, and Marc Millis, NASA, Glenn Research Center, Cleveland, OH, 15 Sept. 2004.

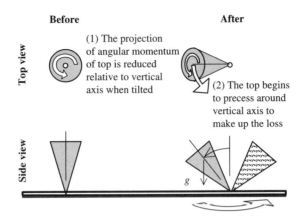

Fig. 8 Dynamics of a spinning top. As gravity, g, pulls downward on a spinning top, the tilting changes the proportion of the top's angular momentum that is projected onto the vertical axis. To conserve angular momentum along this axis, the entire top will begin to precess around. Note that this figure is intended only as a device to help understand the direction of such precessions rather than as a rigorous analysis of the system's dynamics.

the principle of operation is the same. With the gyroscope up to speed, Laithwaite (himself acting as the "main spindle" and "pivot" analogous to Fig. 9) was able to lift the gyroscope by its stem (nonrotating beam that is coincident with the gyroscope's axis), by torquing the stem around horizontally. In Fig. 9, a dual version of the same thing occurs with the stems of two gyroscopes mounted to a "main spindle." When this spindle is rotated, the gyroscope's stems will pivot upward giving the impression of a lifting force. A mechanical stop is included to halt the gyroscope's tilting while it is still aligned to produce this "upward" force.

To understand this in terms of known physics, Fig. 10 presents a visual means to help understand how conservation of angular momenta can lead to such effects. This diagram should be viewed only as a device for understanding the direction of torques from such devices, rather than as a rigorous means to fully convey the system's dynamics. Figure 10 shows the device before and after the main spindle has been torqued. A helpful simplification to comprehend how conservation of angular momentum functions is to consider only one axis at a time, so this illustration concerns itself only with the angular momentum in the axis of the main spindle. To set the initial angular momentum to zero, both the angular momentum of the main spindle and the gyroscope will be set to zero in this view (left or "before" side of Fig. 10). To achieve this zero-momentum initial condition, the position of the gyroscope is set perpendicular to the main spindle so that it projects none of its angular momentum along the axis of the main spindle. Said another way, when viewed from the top (upper left in Fig. 10), there are no apparent rotations. Along the view of this axis, there is zero angular momentum.

In the "after" or right side or Fig. 10, the main spindle is rotating clockwise. At this point it is not necessary to consider what caused this change, as this

NONVIABLE MECHANICAL "ANTIGRAVITY" DEVICES 257

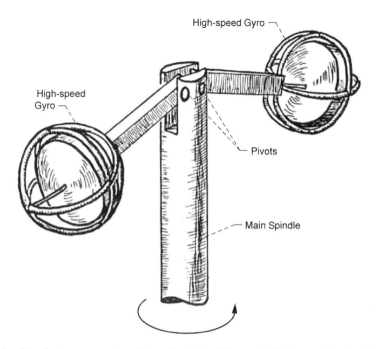

Fig. 9 Classical gyroscopic antigravity claim *"Essential Design of the Laithwaite Engine."* A typical version of a gyroscopic device that purports to exhibit an antigravity effect [9]. With the gyroscopes spinning, the main spindle is torqued. This causes the gyroscopes to flip upward at the pivot points, giving the impression that some antigravity effect is lifting them.

illustration is intended only to help visualize the torque *directions* from angular momenta conservation. Because the rotation of the main spindle has introduced an angular momentum in this view, the stem of the gyroscope has shifted its alignment so that the gyroscope now presents a countering angular momentum so that their sum conserves the initial zero-valued angular momentum. When tilted, the projection of the gyroscope in this view now provides a rotational contribution. Even though it is only an ellipse in this view (i.e., a circle viewed at an angle), this is enough to project a portion of its angular momentum into the plane of rotation of the spindle. When the gyroscopes' stems reach the end of their allowed tilting angle, the torque that induced the tilt will still exist, but it will be acting among the pivot, stem, and the stem's stop. It is not an "upward" force relative to the gravitational field, but rather a torque.

Experimental testing of these gyroscopic devices is not so easy as with the oscillation thrusters. The key difference is that the gyroscopic thrusters require alignment with the Earth's gravitational field to operate. They do not produce *lateral* thrust that can be tested with the pendulum methods previously discussed. Instead, their *weights* must be measured. This is difficult because torques are introduced between the device and the platform on which it is mounted—that is, reaction forces from its internal mechanisms. Furthermore, the frequency of

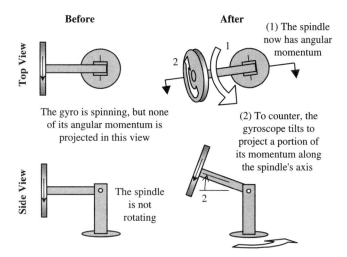

Fig. 10 Picturing conservation of angular momenta. The angular momenta of both situations is identical but achieved differently. On the left, the gyroscope's rotation is aligned perpendicular to the axis of the spindle (zero contribution along the viewed axis) and the spindle is not rotating, hence the total angular momentum is zero. On the right, the spindle has been set into motion, introducing its own angular momentum. To conserve angular momentum, the gyroscope tilts to project an opposing portion of its angular momentum in the axis of the spindle, so that their sum remains zero. Note that this is not a strict representation of the dynamics of this system, but rather a device to help understand the directions that such coupled wheels will move.

vibrations from the gyroscopes might interfere with the operation of any weight-measuring device.

Drop tests, where fall times are measured to detect any changes in gravitational acceleration, are not likely to be a viable testing option for two reasons. First, typical gyroscopic thrusting devices *require* the presence of the gravitational field, which vanishes in free-fall. To illustrate this with a simple experiment, drop a spinning top that is precessing. The moment that it begins its free fall, the precessing will stop. Second, the device is not likely to survive the rapid deceleration at the end of its fall.

One testing option, as illustrated in Fig. 11, is to place two identical devices on each end of a balance beam and look for any tilting of the balance beam when one device is free to operate normally and the other has its stems locked to prevent the upward motion of the gyroscope. Even this simple test can be difficult because it requires two devices and can be vulnerable to spurious effects that could tilt a sensitive balance. Also, the transient impulses from the starting and stopping of the gyroscopes' tilting will induce oscillations in the balance beam.

Again, to keep an open yet rigorous mind to the possibility that there has been some overlooked physical phenomenon, it would be necessary that any experiments explicitly address all the conventional objections and provide convincing evidence to back up the claims. Any test results would have to be

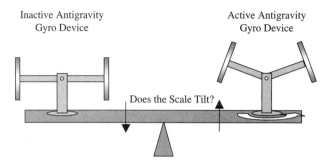

Fig. 11 Testing antigravity gyroscopes. One method of testing the antigravity claims of gyroscopic devices is to place two at either end of a balance beam and to see if there is any difference between an inactive and an operating device. Even though an impulse is expected as the gyroscopes begin their upward tilt, a countering impulse is equally expected when the gyroscopes reach the end of their allowed travel and stop tilting upward. This sequence of impulses is likely to set the balance into oscillation, but a sustained tilting of the balance beam is not expected.

rigorous, impartial, and address all possible causes that might lead to a false-positive conclusion.

One approach to assess such concepts is the "control volume" analysis [10]. To start, a control volume is mathematically constructed around the device. Next the vectors of all the forces and any phenomena that carry momenta that cross the surface of that volume are summed and applied to the center of mass of the system. This includes external masses, fluids, magnetic fields, gravitational fields, electromagnetic radiation, and so on. Motions within the control volume need not be considered, including mechanical devices, fluid flow, internally generated fields, and so on. If nothing crosses the surface that can carry momentum, then there is no net acceleration.

The *orientation* of this control volume, however, is not fixed. The internal motion of gyroscope can alter which direction a device will point without affecting its rectilinear motion. Such is the case of "reaction wheels" discussed next.

IV. Unrelated Devices

It is necessary to make clear distinctions between these gyroscopic *antigravity* devices and similar-sounding devices, specifically "reaction wheels," spinning superconductors, rotating masses in general relativity, and gyroscopic variants of the previously described oscillation thrusters.

A reaction/momentum wheel or torque wheel is an established device used in satellites to change the satellite's pointing direction [11]. Even though the angular momentum of the entire system remains constant, the angular orientation and momentum of the external structure and its internal masses can be changed with respect to one another. For example, by rotating an *internal* wheel clockwise, the *external* cage (i.e., satellite body) will rotate counterclockwise. This is a simple and effective way to change the pointing direction of satellites, but such devices cannot be used to change the *position* of the *center of mass* of the system.

Another similar-sounding claim is the gravity shielding claim involving spinning superconductors [12]. These superconductor claims were later shown not to be reproducible [13]. Further claims of gravitomagnetic effects using spinning superconductors [14] are still under review at the time of this writing. These are separate effects that should not be confused with claims of gyroscopic antigravity.

The concept of using rotating masses to induce forces does have a treatment within General Relativity that can lead to acceleration fields, but it is through the very feeble effect of *frame-dragging*. In 1963, Robert Forward calculated the induced acceleration field from a rapidly rotating ultra-dense toroid [15]. The magnitude of the induced effect is impractically trivial compared to the configurations needed to produce the effect. This example does serve, however, as a theoretical treatise on the subject and is discussed in more detail in Chapter 4.

When gyroscopic devices constrain their motions to a single plane instead of the multi-direction axes versions that are the focus of this section, then they are just a variation of the *oscillation thrusters* discussed previously. The version shown in Fig. 2 is a classic example.

V. Conclusions

Claimed breakthroughs that are based on errant interpretations of mechanical forces are common. Two devices in particular involve oscillating masses that claim net thrust and gyroscopic devices that claim antigravity effects. The oscillation thrusters are misinterpretations of differential friction, while the gyroscopic devices misinterpret torques as linear thrust. To help reduce the burden on reviewers and to give would-be submitters the tools to assess these ideas on their own, examples of these devices, their operating principles, and testing criteria have been discussed.

References

[1] Dean, N. L., "System for Converting Rotary Motion into Unidirectional Motion," U.S. Patent 2,886,976, 19 May 1959.

[2] Foster, Sr., R. E., "Inertial Propulsion Plus/Device and Engine," U.S. Patent 5,685,196, 11 Nov. 1997.

[3] Thornson, B. R., "Apparatus for Developing a Propulsion Force," U.S. Patent 4,631,971, 30 Dec. 1986.

[4] Millis, M. G., "Assessing Potential Propulsion Breakthroughs," *New Trends in Astrodynamics and Applications*, Belbruno, E. (ed.), Annals of the New York Academy of Sciences, New York, Vol. 1065, pp. 441–461.

[5] Resnick, R., and Halliday, D., *Physics*, 3rd ed, Wiley, New York, 1977, pp. 123, 314.

[6] Davis, "The Fourth Law of Motion," *Analog Science Fiction, Science Fact*, 1962, May, pp. 83–105.

[7] *The Royal Institution–Heritage: Ri People*, Royal Institution of Great Britain, London, UK, 2006; URL: http://www.rigb.org/rimain/heritage/ripeople.jsp [cited 1 June 2006].

[8] Laithwaite, E., and Dawson, W., "Propulsion System," U.S. Patent 5,860,317, 19 Jan. 1999.
[9] Childress, D. H., *The Anti-Gravity Handbook*, Adventures Unlimited Press, Stelle, IL, 1985, pp. 18, 20.
[10] Hill, P. G., and Peterson, C. R., *Mechanics and Thermodynamics of Propulsion*, Addison-Wesley, Reading, MA, 1965.
[11] Bayard, D. S. "An Optimization Result With Application to Optimal Spacecraft Reaction Wheel Orientation Design," *Proceedings of the American Control Conference 2001*, Vol. 2, Arlington, VA, June 25–27, 2001, pp. 1473–1478.
[12] Podkletnov E., and Nieminen, R., "A Possibility of Gravitational Force Shielding by Bulk YBCO Superconductor," *Physica C*, Vol. 203, 1992, pp. 441–444.
[13] Hathaway, G., Cleveland, B., and Bao, Y., "Gravity Modification Experiment Using a Rotating Superconducting Disk and Radio Frequency Fields," *Physica C*, Vol. 385, 2003, pp. 488–500.
[14] Tajmar, M., Plesescu, F., Seifert, B., Schnitzer, R., and Vasiljevich, I., "Search for Frame-Dragging-Like Signals Close to Spinning Superconductors," *Proceedings of the Time and Matter 2007 Conference*, Bled, Slovenia, World Scientific Press, 2007.
[15] Forward, R. L., "Guidelines to Antigravity." *American Journal of Physics*, Vol. 31, 1963, pp. 166–170.

Chapter 7

Null Findings of Yamashita Electrogravitational Patent

Kenneth E. Siegenthaler* and Timothy J. Lawrence[†]

U.S. Air Force Academy, Colorado Springs, Colorado

Nomenclature

α = significance level (0.001)
Δ_0 = test value (zero)
μ = mean, general case
$\mu_{Neg-Unc}$ = difference in means for negative/uncharged scenario
$\mu_{Pos-Unc}$ = difference in means for positive/uncharged scenario
$\bar{d}$ = difference in means, general case
H_a = alternate hypothesis
H_0 = null hypothesis
n = number of observations
$n - 1$ = degrees of freedom
s = standard deviation, general case
s_d = difference in standard deviations
t = test statistic
$t_{\alpha,n-1}$ = rejection region

I. Introduction

A. Theoretical Background

ELECTROGRAVITATIONAL theory holds that moving charges are responsible for the mysterious phenomenon we know as gravity. According

---

This material is a work of the U.S. Government and is not subject to copyright protection in the United States.
*Professor of Astronautics, Department of Astronautics.
†Director, Space Systems Research Center.

to Nils Rognerud, author of the paper "Free Fall of Elementary Particles: On Moving Bodies and Their Electromagnetic Forces," gravity as we know it is "... simply a pseudo-force, produced by the special non-shieldable dielectric effect which is produced by the relativistic motions of orbital electrons of ordinary matter" [1]. Much like a moving charge will induce a magnetic field, it will also induce a gravitational field of a much smaller magnitude [1].

Essentially, according to electrogravitational theory, all gravity is produced by the motion of orbital electrons in the atoms that comprise all matter. This supposed dielectric effect is also additive, unlike the magnetic forces of randomly oriented atoms that cancel each other out [1]. The obvious implication of this statement is that more massive objects, hence possessing more orbital electrons than less massive objects, will produce a greater electrogravitational force. This is a simple explanation of why the Earth possesses a stronger gravitational field than the moon, why the sun possesses a stronger gravitational field than the Earth, and so on.

Figure 1 is a simple illustration depicting the manner in which a charged particle induces an electrogravitational field. Notice that the produced field is normal to the direction of motion.

Figure 2 depicts how an atom produces an electrogravitational field around itself. Again, this field is additive; it becomes stronger with the presence of additional atoms [2].

Electrogravitational theory is much more involved than the brief description provided above. However, this elementary description could be applied to artificially inducing an electrogravitational field.

B. Experimental Background

By virtue of the assertion that moving particles will induce an electrogravitational field, a charged, rotating cylinder should produce an electrogravity

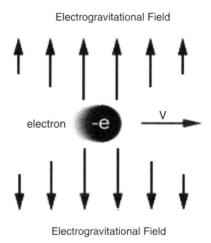

Fig. 1 **Particle inducing an electrogravitational field.**

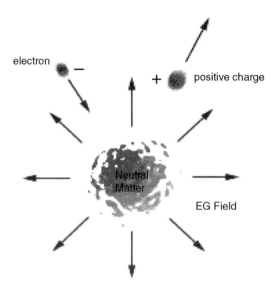

Fig. 2   Atom inducing an electrogravitational field.

field normal to the plane of rotation. If the cylinder is sufficiently charged and rotating rapidly enough, it should alter its weight equivalent in measurable ways. Yamashita attempted this in his 1991 experiment [2]. The result he obtained, an 11-g reduction in weight equivalent of a 1300-g device, is significant [2]. Unfortunately, no one has been able to reproduce his experiment to date. Figure 3 is an illustration of the device that Yamashita had envisioned for the artificial induction of an electrogravitational field.

In order to test the validity of his hypothesis, Yamashita constructed a device similar to Fig. 3 that rotated at approximately 50 revolutions per second (3000 revolutions per minute) [2]. It was comprised of four main parts: a base plate, an electrode, a rotor, and an electric motor [2]. All parts, except the motor, were machined from aluminum. The electrode was 130 mm in diameter and 5-mm thick. It was coated on the interior with a dielectric coating, although Yamashita makes no mention as to what chemicals he used to achieve that purpose. The rotor was 127 mm in diameter, 5-mm thick, and 60-mm high. It also was coated presumably with the same dielectric coating, except on the outside. The machine, at rest, weight equivalent was 1300 g.

In order to test his machine, Yamashita placed it on a scale with a resolution of 1 g. He tested the machine, uncharged, and rotated it to its maximum speed. The difference in weight equivalent between the machine at rest and the machine at this speed was less than 1 g. Yamashita concluded that there was, in fact, a difference in weight equivalent that his scale could not detect, and he attributed this to the rotor's interaction with the surrounding air.

Yamashita applied a charge to the device by bringing into contact for one minute the charged Van de Graaf generator's spherical electrode and machine's

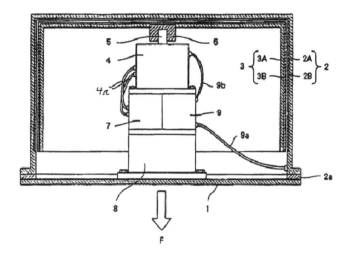

Fig. 3 Yamashita's device.

electrode. He applied a current of 0.5 A to the motor. As the rotor accelerated, the scale read increasingly lower weight equivalent until at top speed, it read a weight equivalent of 1289 g. This represents a decrease of 11 g, or a 1% decrease in the machine's weight equivalent.

Yamashita then reversed the polarity of the rotor. To do this, he attached a spherical electrode to the positive terminal of the generator. He brought the sphere and the machine's electrode into contact, although his patent does not indicate for how long. With the polarity reversed, the machine increased its

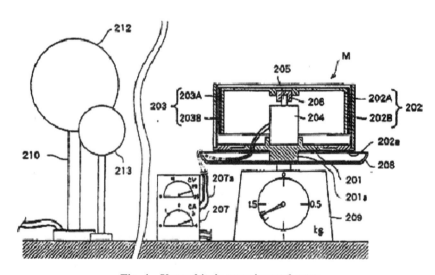

Fig. 4 Yamashita's experimental setup.

weight equivalent by 4 g at top speed. From this experiment, Yamashita concluded that "horizontal rotation of a charged body generates a vertical force," "when the polarity of the charges supplied to the rotating body is reversed, the direction of the generated vertical force is also reversed," "the faster the body is rotated, the stronger is the generated vertical force," and "the direction and strength of the generated vertical force does not depend on the direction of the body" [2]. Yamashita arrived at the last conclusion after his machine produced a force when oriented at an angle. Figure 4 depicts Yamashita's experimental setup.

No one has been able to reproduce Yamashita's experiments to date. A 1% decrease in the weight equivalent of an object is very significant, and it was the aim of the experiments conducted at the United States Air Force Academy to validate Yamashita's claims and to determine whether or not electrogravitational theory is valid.

II. Experiment 1

A. Methods

Efforts were made to follow Yamashita's experiment as closely as possible. His patent, although somewhat vague, gave enough information to conduct an experiment reasonably true to his original.

The first step in attempting to replicate his experiment was to construct a machine reasonably similar to the one he used. If electrogravitation is indeed a real phenomenon, minor differences between the two machines theoretically should not matter in terms of producing results.

In choosing an electric motor to power the rotor, a Global Super Cobalt 400 27T motor was selected. This motor was originally intended to power radio-controlled aircraft, and it was for this reason that it was chosen to power the replica device. The motor boasts a stall current of 64 amps and is capable of reaching 19,500 RPM without a load. Compared to most other electric motors, the Global motor is especially powerful and should have no trouble spinning the rotor at speeds higher than those attained by Yamashita's device. Higher speeds, in theory, should induce a larger electrogravitational field that is easier to measure.

A preliminary design was made using Autodesk Inventor Version 7.0. The replica machine was designed specifically for the Global electric motor and, therefore, it was not exactly the same as Yamashita's device. Figure 5 depicts the replica's electrode.

Figures 6, 7, and 8 depict the rotor, base plate, and motor mounts, respectively.

Yamashita's machine used an unspecified dielectric material to insulate the electrode from the rotor. Dielectric materials come in many different forms, ranging from gels to baked-on coatings to solid sheets. For the purposes of the replica machine, the surfaces that required a dielectric coating were prepared using one coat of gray automotive primer. To act as the dielectric, four coats of blue enamel paint were applied over the primer. The particular enamel used contained the chemical Xylene, which is known to have a dielectric constant of 2.5 at 25°C [3]. With the dielectric applied, the replica machine was assembled. A specially made nylon washer was used to further insulate the

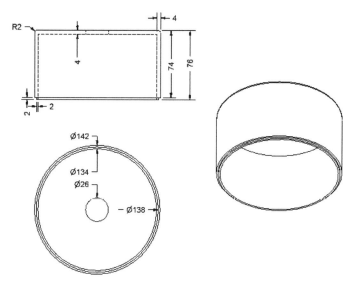

Fig. 5 Replica machine electrode.

rotor from the motor shaft. Finally, to dampen vibrations, foam padding was applied to the base of the machine.

Figure 9 depicts the appearance of the fully assembled machine. With the machine fully assembled, a mathematical relationship could be developed

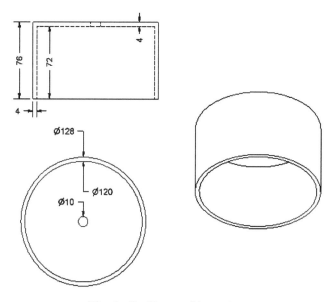

Fig. 6 Replica machine rotor.

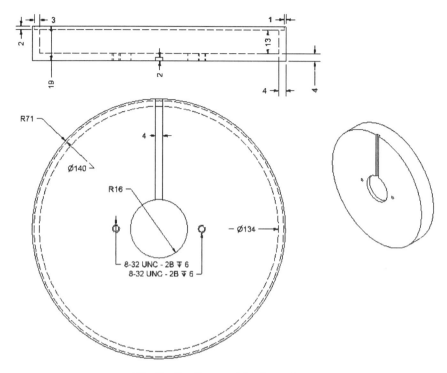

Fig. 7 Replica machine base plate.

between input current and rotor speed. To power the rotor, an Energy Concepts, Inc. Model 20600B high current power supply was used. Due to the high power requirements of the Global electric motor, a regular power supply could not provide sufficient power to turn the rotor, let alone attain the required rotation rate. On the Model 20600B power supply, the 0-24 volt setting was selected. The amps setting was then selected to display the supplied current. A vertical line was drawn on the rotor for purposes of speed calibration. A Power Instruments Digistrobe Model M64 strobe light (calibrated up to 10,000 RPM) was used in conjunction with the vertical calibration line to obtain the rotor speed. Input current was varied, and the rotor speed associated with that current was recorded. Table 1 shows input currents and the resulting rotor speed.

With Table 1, it is relatively easy to develop a function that relates input current to rotor speed using Microsoft Excel. Figure 10 is the Excel graph of Table 1, with rotor speed function shown.

The data points fit well on the third-order polynomial trendline developed by Excel. Equation (1) is the relationship between input current and rotational speed.

$$rpm = 75.341i^3 - 686.26i^2 + 2229.2i - 2047.1 \tag{1}$$

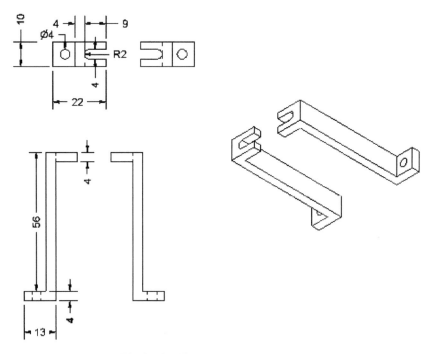

Fig. 8 Replica machine motor mount.

Fig. 9 Replica machine.

NULL FINDINGS OF YAMASHITA

Table 1 Input current and rotor speed

Input current (amps)	Rotor speed (RPM)
3	500
4	700
4.7	1,130
5	1,330
5.8	2,500

This equation is also shown on the graph. Knowledge of this equation was necessary to the experiment. It would not be possible to use a strobe light to monitor the rotor speeds during the test, because the electrode would effectively cover the rotor. It should be noted that the rotor was always spun in a counterclockwise direction; the Global motor is not designed to rotate in the other direction.

To test the replica machine, a thorough test plan was followed. This plan required the use of the Model 20600B power supply to spin the rotor. It also required the use of a Van de Graaf generator to charge the machine. For this purpose, a Wabash Instrument Corp. Winsco Model N-100V generator was used. This particular generator is capable of developing charges up to 250 kv [4]. Finally, to record the possible weight equivalent to change, an Ohaus I-10 FE-7000 precision scale was used, which is capable of supporting up to 25 kg, with a resolution of 1 g. Figure 11 is a photograph of the entire test apparatus; Fig. 12 is a photograph of the machine, the scale, and the power supply.

First, it was necessary to determine whether or not the high levels of static electricity from the Van de Graaf generator had any adverse effects on the scale. To accomplish this, the device was charged for one minute and rotated at full speed in close proximity to the scale. If the reading on the scale remained constant throughout this process, then it would be assumed that there was no interference.

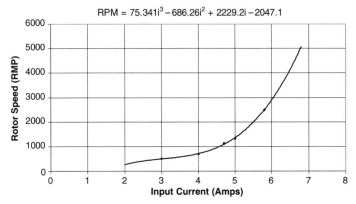

Fig. 10 Rotor speed graph and function.

Fig. 11 Test apparatus.

After establishing that the scale was not affected by the presence of static electricity, the generator was switched off and the machine discharged. The uncharged machine was then placed on the scale. Double-sided tape was used to hold the underside of the machine to the scale. This ensured that the machine did not vibrate across the smooth surface of the scale. The wires were placed in such a way that they did not affect the readings on the scale (Fig. 12). Throughout the course of the experiment, the machine would not leave the scale.

Uncharged, the weight equivalent to of the machine was taken at 500 RPM intervals up to its top speed of 5000 RPM. The purpose of this was to ensure

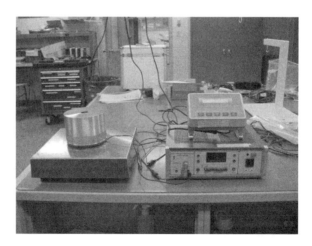

Fig. 12 Detail of machine, scale, and power supply.

that the machine's vibrations did not cause erratic readings, as well as provide a basis for comparison for charged operation. After this test, the machine was allowed to come to a complete stop.

The next test involved charged operation. The machine was charged for one minute from the Van de Graaf generator. A wire was used to connect the spherical electrode of the generator to the electrode on the machine. After the machine was fully charged, the Van de Graaf generator was switched off and grounded to ensure it carried no residual charge that might have caused interference. As was the procedure of the previous test, the weight equivalent to of the machine was recorded at 500 RPM intervals, from 0 to 5000 RPM. At these intervals, the machine's weight equivalent was recorded.

Afterwards, the machine was allowed to stop and was discharged. The Van de Graaf generator was again used to charge the machine. This time, however, contact was made through a wire from the generator's positive electrode on its base to the machine's electrode for one minute. The previous test was again conducted, this time with the opposite charge, and the weight equivalent values were recorded.

The next series of tests involved testing, at full speed, various levels of charge and their effects on the machine's weight equivalent. Using the same procedures previously described for charging, the machine was charged in 10-s intervals, ranging from 10 to 60 s, from the negative spherical electrode of the generator. At each of these charge intervals, the machine operated at full speed, and its weight equivalents were recorded. Between each charge interval, the machine was allowed to come to a complete stop and be fully discharged. For example, the first step of this test involved charging the machine for 10 seconds. It was then tested, stopped, and discharged. The second step involved charging the machine for 20 s and repeating the same process. The third step involved charging the machine for 30 s, and so on, until the machine received one full minute of charging. Finally, this same test was conducted with the machine receiving a positive charge from the electrode on the base of the generator.

It was hoped that by following these procedures that Yamashita's claim would be validated. Furthermore, it was hoped that by using more thorough procedures, electrogravitational theory could either be validated or rejected.

B. Results and Discussion

Although not conclusive by any means, the results of this experiment were nonetheless interesting. By operating the fully charged machine in close proximity to the scale, it was determined that the static electricity had no adverse effect on the other equipment, especially the scale.

The first test involved the uncharged machine at various speeds. Table 2 contains the input currents, associated speeds, and weight equivalents of the machine.

As the rotor was accelerating, the scale fluctuated at ± 2 g above and below the central value. When the rotor reached a steady speed, the fluctuation in some cases disappeared; in others was ± 1 g about the central value. In these cases, it was the central value that was recorded as the machine's weight

Table 2 Uncharged rotation, speed varied

Supply current (A)	Rotor speed (RPM)	Machine weight equivalent to (g)
0.0	0	1315
3.0	500	1315
4.6	1000	1315
5.1	1500	1315
5.5	2000	1315
5.8	2500	1316
6.0	3000	1316
6.3	3500	1316
6.5	4000	1316
6.6	4500	1316
6.8	5000	1316

equivalent. These fluctuations for the most part seemed to be caused by the natural vibrations in the machine, although the rotor's interaction with the surrounding air plays a part as well.

The next test involved varying rotor speed with a full negative charge. Table 3 contains the test data from this scenario, as well as the differences in weight equivalent between the present and the uncharged test cases.

The next test involved varying rotor speed with a full positive charge. The data for this scenario are listed in Table 4, again showing the change in weight equivalent from the uncharged test case.

Figure 13 compares the percent of weight equivalent to deviation during the rotation experiments of an uncharged, negatively charged and positively charged cylinder of Experiment 1.

Table 3 Negative charge, speed varied

Rotor speed (RPM)	Machine weight equivalent to (g)	Weight equivalent to change (g)
0	1315	0
500	1315	0
1000	1315	0
1500	1315	0
2000	1315	0
2500	1315	−1
3000	1315	−1
3500	1314	−2
4000	1314	−2
4500	1314	−2
5000	1314	−2

NULL FINDINGS OF YAMASHITA

Table 4 Positive charge, speed varied

Rotor speed (RPM)	Machine weight equivalent to (g)	Weight equivalent to change (g)
0	1315	0
500	1315	0
1000	1315	0
1500	1316	+1
2000	1316	+1
2500	1316	0
3000	1316	0
3500	1316	0
4000	1316	0
4500	1317	+1
5000	1317	+1

The next test involved varying negative charge at full rotational speed. Table 5 lists the resulting data.

After obtaining the data for the 60 s charge time at full speed, the machine broke. An inaccessible screw holding the motor to the motor bracket came loose, and with no way to adequately tighten it, the machine would vibrate violently above 500 RPM. At this point, enough data had been taken; the constant speed, varying positive charge test was scrubbed.

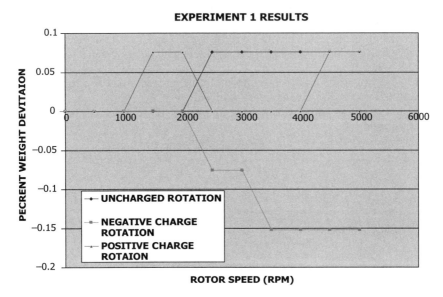

Fig. 13 Comparison of the percent weight equivalent to deviation results for Experiment 1.

Table 5 Constant speed, varied negative charge

Charge time (s)	Machine weight equivalent to (g)	Weight equivalent to difference (g)
0	1315	0
10	1315	−1
20	1315	−1
30	1315	−1
40	1314	−2
50	1314	−2
60	1314	−2

Enough data had been acquired to statistically test whether or not this experiment offered proof that electrogravity was indeed a real phenomenon. Even though there was an apparent decrease in weight equivalent through negatively charged rotation and an increase through positively charged rotation (as Yamashita had predicted), it may not have been a measurable enough of a change to reach a definite conclusion.

The varied rotation speed with one minute charge will be of statistical interest. The uncharged rotation will be used as a basis for separate comparison between the negative- and the positive-charged cases. These data will be considered paired data, because they consists of two observations on the same unit, that unit being speed of rotation. Consequently, a t-test will be used. The null hypothesis, in this case, is that there is no difference between the mean weight equivalent of the uncharged scenario and that of whatever it is being compared to, either the positive or negative. On the contrary, the alternative hypothesis holds that there is a difference in the means. The t-test will be used to either reject or accept the null hypothesis. Rejection of the null means that it is statistically sound to accept electrogravitational theory based on the experimental data. Accepting the null signifies the opposite, or the experimental data are statistically insufficient to prove the existence of an electrogravitational force.

Table 6 shows all pertinent data for statistical analysis of the negative-charged case.

The two hypotheses in this case are as follows:

$$H_0: \mu_{Neg-Unc} = 0$$

$$H_a: \mu_{Neg-Unc} < 0 \tag{2}$$

Table 6 Negative charge statistical data

	Uncharged	Negative	Difference
μ	1315.5455	1314.636364	−0.90909
s	0.522233	0.504524979	0.94388

For this analysis, a significance level of $\alpha = 0.001$ will be used. Furthermore, there are a total of $n = 11$ observations. For this test, the test statistic will be as follows:

$$t = \frac{\bar{d} - \Delta_0}{S_d/\sqrt{n}} \quad (3)$$

This test statistic will be compared to $t_{0.001,11-1} = 4.144$. If the following relationship is found to be true, then H_0 will be rejected.

$$t < -t_{\alpha, n-1} \quad (4)$$

Stepping through all of the math for the comparison of the uncharged and negative-charged data, the following is found to be true:

$$t = \frac{-0.90909 - 0}{0.94388/\sqrt{11}} = -3.194 \quad (5)$$

This particular test value, when compared to $t_{0.001,11-1} = 4.144$, makes Eq. (4) false. Therefore, it is necessary to fail to reject the null hypothesis. The experimental data for the negative case are insufficient to prove the existence of an electrogravitational force.

The same test must be performed between the uncharged and the positive-charged scenarios. Table 7 contains all the necessary data for this comparison.

Unlike the previous case, this case requires a different set of hypotheses. H_a must be the opposite for this case, because positive rotation tended to cause the machine to gain weight equivalent rather than lose weight equivalent. The set of null hypotheses required for this test are as follows:

$$H_0: \mu_{Pos-Unc} = 0$$
$$H_a: \mu_{Pos-Unc} > 0 \quad (6)$$

Likewise, the test relationship shown in Eq. (4) must be changed. The test relationship now becomes:

$$t > t_{\alpha, n-1} \quad (7)$$

Table 7 Positive charge statistical data

	Uncharged	Positive	Difference
μ	1315.5455	1315.909091	0.363636
s	0.522233	0.70064905	0.8202

Using Eq. (5) to obtain the test statistic,

$$t = (0.363636 - 0)/(0.8202/\sqrt{11}) = 1.470 \qquad (8)$$

Using $t_{0.001, 11-1} = 4.144$ as in the previous case, Eq. (7) is not satisfied. Therefore H_0 must be accepted in this case also. There is not enough evidence from the experimental data from the positive case that electrogravity is a real force. Statistical analysis of the varied charge test is not necessary. The analysis performed on the initial data proves that the experimental data are not sufficient to confirm the existence of an electrogravitational force.

III. Experiment 2

A. Methods

After the results obtained in Experiment 1, another attempt was made to replicate the results and make improvements. However, funding issues precluded the implementation of a key improvement, which was conducting the experiment in a vacuum environment. After several inquiries to different electric motor manufacturers, it was found that a vacuum rated motor capable of spinning the rotor at the desired RPM would cost roughly $1000. Additionally, the manufacturers would not be able to supply a motor in a fashion that would meet the timetable requirements. Many of the motors were out of stock or would have to be custom-made for the purposes of this experiment. This postponed the arrival of a motor by several months, precluding their use even if funding were available.

Another objective of the second experiment was to investigate the reason why the previous experiment's machine failed at high speeds. The rotor used previously was plagued with balancing issues. These were first believed to have come from painting the rotor; however, further investigation concluded that the rotor used in Experiment 1 was in fact out of balance without the paint and that the hole drilled for the motor shaft interface was out of center. Additionally, the motor mount previously designed was determined to be insufficient to keep the machine stable at high RPM, especially given the imbalanced rotor. This conclusion was reached through inspection of the motor mount when fully assembled and attached to the base plate and motor. The motor mount was comprised of only two brackets, attached to the motor and base plate with a total of four screws as illustrated in Fig. 8. The motor mount was also attached to the motor with hot glue. The motor used was also questioned, as the slender shaft of the motor may have contributed additional instabilities of the experiment at high speeds.

Another discrepancy found was that the electrode did not fully enclose the rotor—there was a hole at the top that provided access to a screw that held the rotor to the motor interface. Whether or not this hole affected experimental results has not been determined, but it can be hypothesized that such a hole could introduce a larger possibility for aerodynamic effects as opposed to a completely enclosed and sealed device. The electrode issued in Experiment 1 and Yamashita's device can be seen in Figs. 3 to 5. From Yamashita's drawings, it is seen that neither iteration included a hole in the electrode component.

The experiment was conducted as closely as possible to Yamashita's experiment; however, the ambiguity of the European patent application led to educated guesses on a few aspects during the design of the machine. In addition, accuracy limitations of the manufacturing tools at the Air Force Academy introduced errors to the components, more specifically the rotor. These ambiguities and limitations led to many iterations of the machine while in the construction phase of the experiment. The initial design for the machine can be seen in Fig. 14, but the final iteration of the machine differed greatly from this drawing. (The final version of the machine is shown later.)

An enlarged view of the new motor mount design can be seen in Fig. 15. It is easily seen that this design would be much more stable than the two-bracket design from the previous machine; however, implementation of this motor mount was not possible due to issues with the motors, detailed later.

Given the criticality of accuracy in manufacturing the components, the tools used were integral in allowing the components to meet exacting standards; however, it was determined that the facilities at the Air Force Academy were not capable of machining within such tolerances. The limitations present were a great hindrance, especially in manufacturing the rotor. The rotor that was manufactured for Experiment 2 was out of balance in addition to being out of round. The out of roundness of the rotor was mainly due to the thinness of the aluminum. When compared to Experiment 1 rotor (mounted on Experiment 2's motor), the vibrations experienced by the new rotor were much greater; therefore Experiment 1's rotor was reused in this new experiment. In addition it was shortened to reduce the mass moment of inertia values. The same thickness was kept to maintain the durability of the rotor.

After analysis of the previous experiment's motor, it was hypothesized that the motor shaft was bent, contributing to the vibrations experienced. Three different motors were tested with the old rotor and each experienced similar vibration. Although specific dimensions of the motor and motor shaft were not provided in Yamashita's patent application, further investigation of his drawings led to

Fig. 14   Initial design of the machine.

Fig. 15 Motor mount.

the conclusion that a motor with a very thick shaft was used in his experiment; therefore, a shaft diameter of about 0.25 in. was decided upon. Furthermore a thicker shaft would bolster the stability of the spinning rotor given how imbalanced it was. Hobby shops did not provide motors with a shaft thickness in the range of 0.25 in. therefore more creative means were employed to find a desirable motor. A 1.30 VAC brushless electric fan motor with a shaft thickness of 6 mm (slightly under 0.25 in.) was found at a Goodwill store. To power the motor, an Energy Concepts, Inc. Model 20600B high current power supply was used.

The fan cover and blades were removed leaving only half of the fan casing, the motor, speed control, and stand. The integrated speed control of the fan was set to maximum for the entire experiment. A picture of that assembly can be seen in Fig. 16.

Because of the peculiar way the motor was mounted to the fan assembly, it would have been difficult and time consuming to design and build a new mount for the fan motor. Therefore, it was decided that the base plate would be mounted on top of the fan motor with the shaft protruding from the bottom of the base plate. This is illustrated in Fig. 17.

Because the design had gone through several iterations at this point, different sets of holes had been placed on the base plate. After initial construction it was determined that these should covered with aluminum tape to mitigate airflow through this section.

Additionally, the motor mount interface posed a problem as it presented more opportunities for inaccuracies to be introduced. Experiments is rotor-to-motor interface was held to the motor shaft by friction fit, but after many uses this interface became loose. Another method of interfacing the motor to the rotor was improvised by using a small hand drill chuck to clamp onto the motor shaft. Running the motor at maximum speed with the chuck attached produced minimal vibrations. Further detail as to how these components were interfaced is illustrated in Fig. 18.

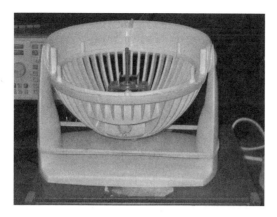

Fig. 16 Modified fan.

A reference line was drawn on the chuck. Then, using a strobe light rated to over 18,000 RPM and the Model 20600B placed on the 130 VAC setting, the maximum rotational speed of the motor was determined to be roughly 3390 rpm. Though this was much slower than the previous maximum speed, it was sufficient to meet the specifications detailed in Yamashita's experiment. The method of determining the rotational speed of the rotor in relation to the current supplied was identical to the previous experiment and is further discussed later in the paper. The rotor-to-motor interface was attached to the chuck via screw. Because the previous experiment's interface produced minimal vibrations when attached to the motor, it was also reused for this experiment; however it was redrilled and tapped to hold a screw that fit into the chuck. The interface was attached to the rotor via two countersunk screws that completed the assembly for attaching the rotor to the motor shaft. Although much more complex, when fully assembled the rotor spun with less vibration than the previous machine.

Fig. 17 Base plate mounted on fan.

Fig. 18 Drill chuck on motor shaft.

Unfortunately, at high speeds these vibrations were still apparent. The final assembly is illustrated in Fig. 19.

The vibrations were mitigated through manually balancing the rotor. Manual balancing was necessary because machine shops offering professional balancing services were not able to fit the rotor on their balancing machines as the shaft size for the motor was too thin. A shop was found which was able to balance rotors with smaller shaft sizes (specifically turbo chargers for cars), but this development occurred too late in the manufacturing process to meet deadline requirements.

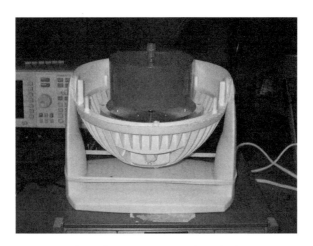

Fig. 19 Rotor mounted on shaft.

NULL FINDINGS OF YAMASHITA

Given that the hole drilled for the interface was off center, a micrometer was used to determine the point on the rotor with the smallest radius. Additional mass was added to this point by attaching solder to the inside of the rotor with aluminum tape. This addition greatly minimized the vibrations of the rotor, especially at high speeds. Fine tuning of the rotor's balance was achieved through an iterative process of adding a mass (section of 18-gauge wire attached with duct tape), spinning up the motor to its maximum velocity, judging whether the addition decreased the total vibration of the machine, and moving the mass to a new location. Locations with the least vibration were marked, compared, and the optimum location of the additional mass was determined through observation. After several masses were added to the rotor (all on the inside surface), it became increasingly difficult to judge the differences in severity of vibration. At this point it was considered that the rotor was balanced to the maximum extent possible given the method implemented.

Once the rotor was balanced, it was painted and retested to see if the paint had any noticeable affects. Because Yamashita's patent application did not specify the exact dielectric layer used, Vanguard Class F Red VSP-E-208 Insulating Enamel was utilized for this experiment to insulate the inner surface of the electrode and the outer surface of the rotor. This insulating enamel is specifically designed to insulate electrical components. To ensure an even coating on the outside of the rotor, the paint was applied as the rotor was spinning. When applied the paint did not have any adverse affects on the balance of the rotor. Once the paint was applied, a calibration curve that related rotor speed to applied current was developed using the strobe light and by making a reference line on the rotor. The results can be seen in Table 8.

From the data in Table 8 a linear regression between each point was derived using the TREND function in Microsoft Excel. This function determined the RPM associated with intermediate levels of current. The TREND function was used between each point because the regression lines that Excel produced did

Table 8 Relation of amps to RPM

Amps	RPM
0.20	621
0.21	1983
0.22	2860
0.23	3015
0.25	3165
0.30	3277
0.35	3330
0.40	3350
0.45	3368
0.50	3377
0.60	3385
0.74	3390

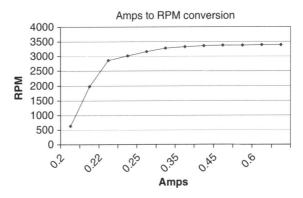

Fig. 20 Graph of amp to RPM conversion.

not match well with the data. Figure 20 is a graph of the data points from the Table 8.

A relation between current and RPM was required because the rotor would not be visible once the electrode was attached.

When connecting the electrode to the assembly, styrofoam was used as an insulative layer to separate the electrode from the base plate. Nylon screws were used to hold the two components together. The final assembly of the device can be seen in Fig. 21. Though the internal components of the machine were exposed, the aerodynamic effects generated by the spinning rotor were minimal, as detailed in the results section of this paper.

To measure the weight equivalent to change, a Mettler PM6100 scale was used. Provided by the chemistry department, this scale had a resolution and range of 0.01 g and 6100 g, respectively. The scale was grounded with a 28-gauge wire in order to protect the equipment from static discharge. This wire

Fig. 21 Machine fully assembled.

Fig. 22   Wire used to ground the scale.

was oriented in such a way that it would not affect the weight equivalent of the machine and is illustrated in Fig. 22.

A Wabash Instrument Corp. Winsco Model N-100V Van de Graf generator was provided by the physics department and used to charge the machine. The Van de Graf generator's spherical electrode was attached to the machine's electrode via wire. Although Yamashita's patent application depicted charging the machine's electrode by directly touching it with the Van de Graf generator's electrode, the sensitivity of the scale utilized precluded the implementation of that procedure. Instead a wire was used to connect the generator's electrode to the machine's. The procedures for the experiment are as follows.

With the power supply and the charge supply off, the weight equivalent of the machine was measured. The wire for the motor was oriented in such a way that they would not affect the weight equivalent of the machine. This was also true for the wire used to charge the machine. These are illustrated in Figs. 23 and 24. The

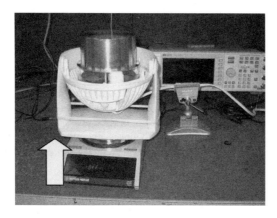

Fig. 23   Clamp holding motor cord.

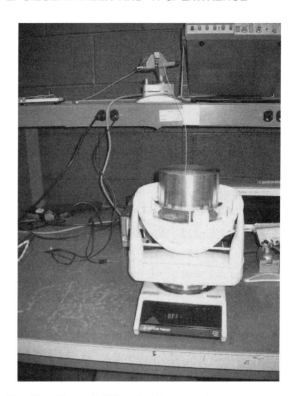

Fig. 24 Clamp holding the charging wire.

mass measurement was made several times while shaking the wires connected to the motor and power supply, and mass differences less than 0.1 g were observed. Given the results of shaking the wires, they would need to be kept still during experimentation.

The power supply was then turned on, and the rotor was accelerated rapidly. From the acceleration test it was seen that the machine's mass would fluctuate by less than ± 0.1 g if maximum current were applied at rest. Further testing indicated that a current increase of 0 to 0.74 amps in 60 seconds produced (relatively slow acceleration) smaller fluctuations in mass. It was then decided that the rotor would be allowed to run at determined amp levels for 10 seconds before proceeding to further accelerate the rotor. Increasing the current by 0.05 amps and allowing the rotor to spin at the amp level for 10 seconds, the mass readings were recorded.

To check whether the electrode held charge, the machine was connected to the Van de Graf generator via wire, which can be seen in Fig. 25. The generator was then turned on and allowed to charge the electrode for one minute. After one minute the generator was disconnected from the electrode and turned off. The device illustrated in Fig. 26 was then connected to the ground socket in a wall outlet and brought within close proximity of the electrode. A spark was observed, verifying that the electrode had been charged by the generator.

Fig. 25 Wire connecting the Van de Graf generator to the electrode.

Then, with the rotor at rest, the Van de Graf generator was turned on and reconnected to the machine's electrode. After one minute of charging the electrode, the wire was disconnected from the machine's electrode and the Van de Graf generator was taken out of proximity and turned off.

The rotor was then accelerated to its top speed, taking readings of the scale at intervals of 0.05 amps, remaining at each level for 10 s. These procedures were then repeated with the positive terminal of the Van de Graf generator charging the electrode. A third test was also performed with the generator attached and continually charging the machine at a low rate while the rotor accelerated.

B. Results and Discussion

When uncharged, the difference in mass between 0 RPM and 3390 RPM was less than 0.02 g. This difference can be considered to be caused by an interaction

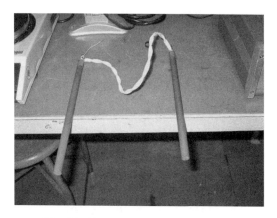

Fig. 26 Device used to check charge.

Table 9 Uncharged operation

Amps	RPM	Mass (g)	Change (g)
0	0	2585.01	0
0.21	1983	2585.02	0.01
0.24	3087.764	2585.01	0
0.29	3254.6	2585	−0.01
0.34	3319.4	2585.01	0
0.39	3346	2585.02	0.01
0.44	3364.4	2585.01	0
0.49	3375.2	2585.01	0
0.54	3380.2	2585.01	0
0.59	3384.643	2585.00	−0.01
0.64	3386.429	2585.01	0
0.69	3388.215	2585.02	0.01
0.74	3390	2585.01	0
0	0	2584.99	−0.02

between the rotor rotation and the surrounding air as the machine was not fully enclosed. Additionally, the slight vibrations that resulted from the machine's operation may have caused these fluctuations given the scale's sensitivity. Table 9 depicts the mass fluctuations when accelerating the uncharged electrode.

It is seen from the data that the machine's mass stays fairly constant while accelerating, indicating very smooth operation; the small fluctuations that do occur may have been caused by slight interference from either air or small vibrations from the machine itself.

When charging the electrode using the negative terminal of the Van de Graf generator, the scale produced the mass changes illustrated in Table 10.

Table 10 Negatively charged operation

Amps	RPM	Mass	Change
0	0	2584.99	0
0.21	1983	2584.99	0
0.24	3087.764	2584.99	0
0.29	3254.6	2584.98	−0.01
0.34	3319.4	2584.98	−0.01
0.39	3346	2584.99	0
0.44	3364.4	2584.97	−0.02
0.49	3375.2	2584.97	−0.02
0.54	3380.2	2584.97	−0.02
0.59	3384.643	2584.99	0
0.64	3386.429	2584.99	0
0.69	3388.215	2585.01	0.02
0.74	3390	2584.99	0
0	0	2585.01	0.02

From Table 10 it is seen that no significant mass changes were registered by the scale. Additionally, the very slight mass fluctuations that did occur were not indicative of a weight equivalent loss pattern that the machine should have been experiencing with negative charge. Though these results are indicative that the theory behind Yamashita's device does not hold true, an inaccuracy of the machine's dimensions may have been the cause of such results.

A protruding screw, which attached the rotor to the drill chuck, caused a gap of about 2 cm between the top of the rotor to the ceiling of the electrode. This is not in accordance with the smaller gap evident from Yamashita's drawings. In addition, the mass of the machine was much greater than the mass of either Berrettini's [5] or Yamashita's [2]. This was mainly due to the additional mass incurred by integrating the fan into the entire assembly of the machine.

When connected to the positive terminal, the machine was accidentally nudged. The scale was allowed to settle and it settled on a new value of 2585.45 g. While connected to the positive terminal, the machine produced the results presented in Table 11.

Figure 27 compares the percent of weight equivalent deviation during the rotation experiments of an uncharged, negative, charged, and positive charged cylinder of Experiment 2.

It is also seen in this case that the machine neither provided a significant mass change or a general increase in mass difference. When continually charged by the Van de Graf generator, the machine yielded similar results for both the positive-charged and negative-charged case. To put this in perspective for the application of this concept as a form of space travel, based upon the results we obtained for the most precise mass displacement measurements conducted in Experiment 2, it would take 830 years for a 500 kg spacecraft to travel from Earth to Mars [6–9].

Table 11   Positively charged operation

Amps	RPM	Mass	Change
0	0	2585.45	0
0.21	1983	2585.45	0
0.24	3087.764	2585.46	0.01
0.29	3254.6	2585.46	0.01
0.34	3319.4	2585.45	0
0.39	3346	2585.44	−0.01
0.44	3364.4	2585.43	−0.02
0.49	3375.2	2585.42	−0.03
0.54	3380.2	2585.45	0
0.59	3384.643	2585.43	−0.02
0.64	3386.429	2585.45	0
0.69	3388.215	2585.44	−0.01
0.74	3390	2585.45	0
0	0	2584.43	−0.02

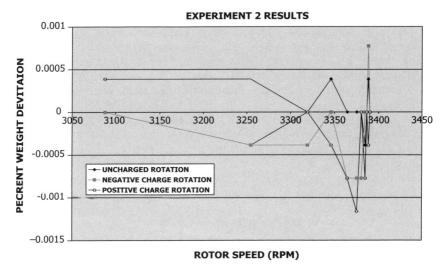

Fig. 27 Comparison of the percent weight equivalent to deviation results for Experiment 2.

IV. Conclusions

The experiments conducted at the United States Air Force Academy in no way confirmed the existence of an electrogravitational force. Because of the equipment and techniques used, the differences in weight equivalent actually seen were too statistically insignificant to prove anything. The replica device, however, did not show any evidence contrary to Yamashita's claims. The first claim, that a horizontal rotating body produces a vertical force, could not be disproved. The spinning rotor and its effect on the replica device's weight equivalent did not disprove the first claim by virtue of the fact that the data needed to prove or disprove this claim were statistically insignificant. Another one of Yamashita's claims, that reversing the polarity should reverse the direction of the force, could not be disproved for the same reason. The device seemed to generally decrease its weight equivalent when given a negative charge, and increase its weight equivalent with a positive charge; again, however, this weight equivalent change was not statistically conclusive. The final claim that could not be rebuked is that the magnitude of the force generated increases with the speed of rotation of the charged body. One can see from the tables that this did tend to happen, although again the results were statistically insignificant. Yamashita's fourth claim, that a force could be produced in any direction, was not observed in this experiment.

In order to be able to prove or disprove Yamashita's claims, it would be absolutely necessary to know exactly what he did in his experiment. Unfortunately, the ambiguity of his patent left a lot of room for guesswork. A number of assumptions were made in the absence of information from Yamashita's patent and these assumptions may be partly responsible for the fact that this experiment did not produce any conclusive results.

These experiments neither denied nor confirmed the existence of an electro-gravitational force. Conclusive proof or disproof would require better data than there collected from these experiments.

However, we feel these experiments were successful because the finding that these concepts were not significant and worthy of further research in the field of breakthrough propulsion physics for NASA allowed resources to be allocated to investigate other concepts.

Acknowledgments

This work would not be possible without all of the hard work and dedication of the cadets. Cadets helped in the design, fabrication, testing, and analysis of all of these experimental campaigns. We acknowledge John Bulmer, Christopher Schlagheck, and Vinny Berrettini for their outstanding work. Without them, this chapter would not have been possible.

References

[1] Rognerud, N., "Free Fall of Elementary Particles: On Moving Bodies and Their Electromagnetic Forces" Technical Paper, Rognerud Research and Development, Concord, CA, *Physics Review D*, 1994, submitted for publication.

[2] Yamashita, H., and Takayuki, Y., *Machine for Acceleration in a Gravitational Field*, European Patent 0486243A2, 1991.

[3] Xylene: Material Safety Data Sheet, CHEM-SUPPLY Pty Ltd, Gillman, South Africa, Sept. 1999.

[4] Schlagheck, C., "Weight Equivalent to Loss Effects Associated with a Rotating Charged Cylinder," AIAA Student Paper Conference, Parks College, St Louis, Missouri, 17–19 April 2002.

[5] Berrettini, V., "The Coupling of Mass and Charge as a Propulsive Mechanism," AIAA Student Paper Conference, University of Minnesota, Minneapolis, Minnesota, 28–30 April 2004.

[6] Tajmar, M. and Jacinto de Matos, C., "Induction and Amplification of Non-Newtonian Gravitational Fields," AIAA Paper 2001-3911, AIAA Joint Propulsion Conference, Salt Lake City, Utah, 9–11 July 2001.

[7] Larson, W. J., and Wertz, J. R. (eds), *Space Mission Analysis and Design*, Microcosm Press, El Segundo, CA, 1999.

[8] Humble, R. W., Henry, G. N., and Larson, W. J., *Space Propulsion Analysis and Design*, McGraw-Hill Company, New York, 1995.

[9] Angelo, J. A. Jr., and Buden, D., *Space Nuclear Power*, Orbit Book Company, Malabar, FL, 1985.

Chapter 8

Force Characterization of Asymmetrical Capacitor Thrusters in Air

William M. Miller*
Sandia National Laboratories, Albuquerque, New Mexico

Paul B. Miller[†]
East Mountain Charter High School, Sandia Park, New Mexico.

and

Timothy J. Drummond[‡]
Sandia National Laboratories, Albuquerque, New Mexico

I. Introduction

FOR some time there have been claims of significant forces generated by voltage-excited asymmetric capacitors (colloquially known as "lifters"), along with theories explaining them. In fact, reports concerning anomalous force production by asymmetric capacitors have existed for more than 80 years. However, few of these were published in any readily obtainable form, fewer still contained significant amounts of quantitative data, and many involved extensive speculation as to the origin of this force. The authors believe that this lack of data, combined with fanciful speculation, may have led most researchers to ignore what might be an important phenomenon. Our purpose was to remedy this situation by making careful measurements and letting the weight of the evidence speak for itself as to the physics underlying the effect.

This chapter reports the results and conclusions of this experimental program. While applying both uniform and sinusoidal voltages to asymmetric capacitors of

---

This material is a work of the U.S. Government and is not subject to copyright protection in the United States.
*Distinguished Member of Technical Staff, System Integration and Technology Department.
[†]Student; previously at East Mountain Christian Academy, Tijeras, New Mexico.
[‡]Principal Member of Technical Staff, Technology and Integration Department.

varying geometry, forces were measured to within ±100 nN. In most cases, voltage, current, and force were measured and absorbed power was calculated.

The force (direction and magnitude) was found to be polarity independent for a uniform applied voltage. Sinusoidal excitation led to constant forces of similar functional form (force as a function of voltage) of slightly lower magnitude. Forces were observed well below the breakdown threshold in air. Frequency dependencies were insignificant. A uniform magnetic field caused no quantitative effect on the force. Geometrical variations in the asymmetry produced only small variation in the measured force. These results are shown to be consistent with a force produced by a corona wind originating at the capacitor electrode associated with the higher electric field.

Early work on forces generated by high-voltage-excited asymmetric capacitors was carried out from 1923 to 1926 by Dr. Paul Alfred Biefeld and his student, Thomas Townsend Brown, both of Dennison University. Contemporary publications [1] by Brown are sketchy, but refer to the effect he observed as being related to "controlling gravitation" and "influencing the gravitational field" with an "apparatus requiring electrical energy." Many of the details involving this can be found in a number of British and U.S. patents [2–6] dating from the late 1920s through the early 1960s. The British patent clearly refers to "controlling gravitation" and "influencing the gravitational field" with an "apparatus requiring electrical energy." However, the later U.S. patents for different devices utilizing an asymmetric electrical field make no such claims beyond the production of mechanical forces. Brown did, however, claim in one patent [6] the production of forces in a vacuum far from potentially interacting walls. He was also issued one French patent on his devices [7].

Some work was done under the auspices of the U.S. government from the mid 1980s through 1990 to better understand this phenomenon. A report describing the forces on an asymmetric capacitor held under vacuum in a chamber that was large compared to the size of the device under test, did report the observation of small forces [8]. Another report detailed possible relationships between electromagnetic fields and gravity [9].

More recent work consists primarily of that by amateur experimenters [10].[§] Carefully controlled and published experiments continue to be rare. Bahder and Fazi reported on theoretical and experimental work they performed at the Army Research Laboratory [11]. Canning et al. reported on tests conducted under high vacuum conditions, in a NASA report [12]; a more recent summary by Canning appears in the next chapter of this book.

II. Summary of Theories

A number of theories have been put forth to explain the motive force of asymmetric capacitors. Not all devices conform to the triangular wire and plate arrangement widely described on the Internet. The common feature is that all lifters are asymmetric capacitors. Brown patented both the wire and rectangular plate geometry as well as the geometry of a sphere above, on the axis of, and

---

[§]Naudin's Web site is arguably the most thorough compilation.

normal to a circular plate. An assessment by the Office of Naval Research (ONR) of Brown's work in 1952 came to the conclusion that, where Brown had documented results, those results could be readily explained in the context of a corona wind [13]. The ONR study dismissed Brown's claim that results were linked to lunar tides. More recently, studies have inconclusively tried to link asymmetric capacitor phenomena to the Earth's global electric or magnetic circuits [14,15]. Without invoking a "fifth force" in nature such interactions are orders of magnitude too small to have a significant impact on lifter phenomena. It has been hypothesized by T. Musha that a strong localized electric field generates a local gravitational field and that the mass associated with the electric field energy is canceled by a negative mass associated with the induced gravitational field. Musha asserts from his experiments that the effect of solar and lunar tides is real. Musha's results do not explain the observed directionality of the nominal Biefeld–Brown effect nor its independence of the polarity of the applied field [16]. More credible assertions that the inertial mass of an object can be manipulated electromagnetically have been put forth by Woodward and Brito in separate approaches [17,18]. These approaches do not, however, claim to explain the Biefeld–Brown effect. An Air Force study postulated that by assuming a five-dimensional continuum, an electrogravitic coupling could be derived to explain the Biefeld–Brown effect. This is a purely speculative approach that may appear to predict the desired result. Such modeling is fraught with danger as any such model will be severely limited by constraints deriving from both cosmology and particle physics [19].

III. Overall Experiment Setup

In reviewing the literature of the previous 80 years, most reports described the (uncalibrated) high voltage needed to lift a several-gram object against the force of gravity. While this indicated the presence of an impressively large force, there were few reports of accurate measurements of both the force and the voltage required to generate it. In some cases, the force measurements were of a very indirect nature, relating to the velocity of a spinning object against the retarding forces of friction. We set out to remedy the continued lack of direct, detailed, and precise measurements of this phenomenon with the hope of discerning among the many theories that have been proposed as to its physical origin.

We built several asymmetric capacitors each consisting of a thin conducting wire (0.0042 in. uninsulated beryllium-copper wire; Alloy 25, Little Falls Alloys, Inc. Paterson) separated from and parallel to a thin conducting plate (0.0016 in. uninsulated aluminum foil; Reynolds 657). The particular shape and size of the device depended on the particular experiment. Direct force measurements (uniform and sinusoidal voltage excitation) were made on traditional triangular-shaped devices (which enhanced their stability), while geometrical variations were made on linear devices to be described later. Descriptions of the measurements made on each device are provided in Sec. IV "Specific Setup" and the results of these measurements are provided in Sec. IV "Results."

Force measurements were made by suspending the various asymmetric capacitors directly from the pan hook of a Mettler M3 microbalance on a thin

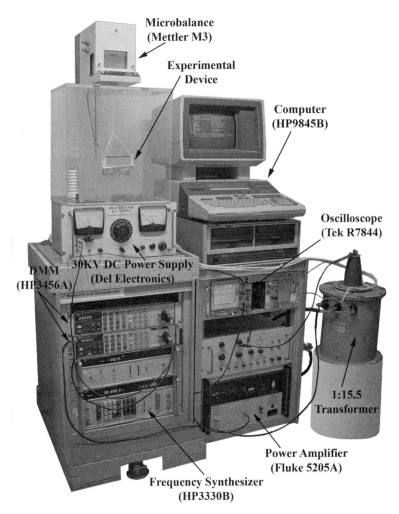

Fig. 1 Complete experimental setup. Note that the device under test has been enhanced to clearly show the wires (*black*) and the supporting lines (*white*).

insulating polypropylene line (thickness of 0.0044 in.). This self-calibrating, zeroing balance can be read to ± 1 μg (approximately ± 10 nN). Even though the entire experimental setup (Fig. 1) was placed on a vibration isolation mount supported by a poured concrete basement foundation, vibration and other motions induced by operating the capacitors allowed accurate reading to be made to no better than ± 10 μg ($\sim \pm 100$ nN). As the balance had a maximum capacity of 3 g (with the balance pan in place) and somewhat more than this with the pan removed (as it was for these experiments), the mass of the capacitor that we used was limited. The force of the effect was measured as the difference between the forces on the unexcited capacitor (the force due to gravity on its mass) and that of the excited capacitor.

Exciting voltages were provided by high voltage uniform and sinusoidal power supplies (described under the "Specific Setup" subsections). The voltages applied to and the currents flowing into the capacitors were measured across a noninductive resistive voltage divider and a noninductive current viewing resistor, respectively. The resulting voltages were measured by a pair of calibrated zeroing digital multimeters (HP3456A, Hewlett Packard Co.) to $\pm 1\,\mu V$. The relationship between the voltage divider readings and the actual voltage applied was determined by calibration using a high voltage probe (Fluke 80K-40). The current viewing resistor was calibrated with the use of an impedance bridge (Model 1656, General Radio). All measurements were performed at an altitude of 2304 m with an approximate air pressure of 76.5 kPa at a temperature between 4°C and 16°C.

Force, voltage, and current measurements were made as follows:

1) A measurement of the weight of the unexcited device was made with the microbalance. Even though the device was in a cage shielded from external air currents and on a vibration-damped table, some small variation in the measurement always occurred. The maximum and minimum of this range was recorded. These variations tended to be small when the force measured was small, and larger at greater forces.

2) With all high voltage controls set to zero, measurements were also made of the applied voltage and the unexcited current, providing a zero reference for later comparison.

3) The high voltage excitation was then applied.

4) The device was allowed to stabilize and a reading of the maximum and minimum force on the microbalance was made.

5) The net force due to the effect was taken as the difference between the force on the excited device under test and that of the gravitational force on the unexcited device.

6) Thirty readings each of the excitation voltage and the current flowing into the device were made. The average and variance were then recorded.

7) This process was repeated from step 3 through step 6 for different excitation voltages until a complete set of data was obtained. Typically, readings were made starting at the lowest and working to the highest voltage.

IV. Results

A. DC Measurements

1. Specific Setup and Measurement Technique

The high voltage DC exciting voltage was provided by a Del Electronics Corp. power supply capable of generating up to 30,000 volts at 10 mA (Fig. 2). Therefore, for safety's sake, the entire experiment was carried out inside a $\frac{1}{2}$ in. thick Plexiglas cage supported on top of this power supply that itself was supported on the vibration isolation table mentioned earlier. The device under test was suspended from the Mettler M3 microbalance in such a way that when measuring from any conducting extremity on the device to the insulating cage walls there was 11 cm to the bottom, 25 cm to the top, 14 cm to either side, 10 cm to the front, and 12 cm to the back of the cage. As the device under test

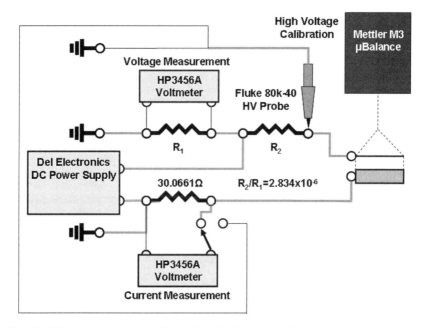

Fig. 2 DC experiment setup. The resistor bridge was calibrated using the Fluke HV probe using the pair of voltmeters. During data acquisition, the second voltmeter was used to measure the voltage across the current viewing resistor.

had approximately 3 cm between the wire and the plate, the cage walls were at least several times this characteristic distance away from the test device to minimize wall interaction concerns.

This initial set of measurements was performed on a single triangular test device built out of 2 mm × 2.5 mm rectangular cross-section balsa wood, each side of which was 200 mm long. The aluminum plate was 40 mm tall and the wire was held 30 mm above this plate and in its plane by balsa wood supports. To minimize corona discharge between the plate and the wire, the aluminum was rolled over the upper balsa wood supporting stick (Fig. 3). A representative picture of a later, but similar device is shown in Fig. 4. The device was supported on three fine strands of polypropylene line attached to each of its three corners. These lines were attached together above the center of the device and a single strand of line then continued upward to the attachment on the balance. The device was leveled horizontally so that the force generated up along the rotational symmetry axis of the triangular lifter was directed vertically. Actual force, voltage, and current measurements were carried out as described in Section III.

2. Results

Some initial qualitative experiments were performed with the symmetry axis of the triangular device held horizontally. In this case the device generated a force along the axis of the device directed from the plate toward the wire. Likewise,

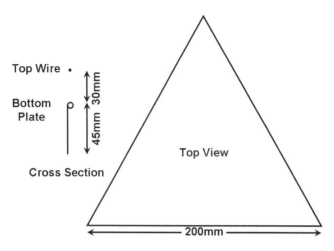

Fig. 3   Original design of asymmetric capacitor.

when the device was suspended with its axis vertical, the force was again directed along the axis from the plate to the wire. Quantitative measurements were made of both the force developed by the device and the current flowing between the wire and the plate as a function of the applied voltage. Figure 5 shows that the force follows a power law dependence on voltage of approximately:

$$F = k_0 V^4 \tag{1}$$

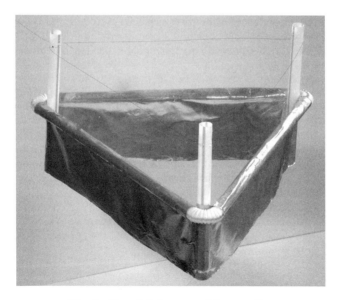

Fig. 4   Completed asymmetric capacitor.

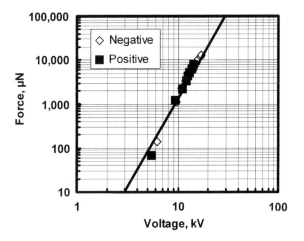

Fig. 5 Measured force as a function of applied DC voltage. Note that the force is identical for both polarities.

where F is the measured force, k_0 is a proportionality constant, and V is the applied voltage. It is exceptionally important to note that the force data for both the positive and negative applied wire polarities were found to be identical within the resolution of our measurements. We observed no polarity dependence for any of the devices tested in this study. Another important point is that we were able to make measurements at higher negative than positive voltages. This is consistent with the fact that negative corona discharges are more stable than positive corona discharges.

The current dependence of the force was also discovered to follow a power law as shown in Fig. 6. This is clearly shown by plotting the force against the current raised to the 2/3 power. This functional dependence is given as:

$$F = k_1 I^{(2/3)} \qquad (2)$$

where again F is the measured force, k_1 is a proportionality constant, and I is the measured current. This relationship describes the data quite well and yields a correlation coefficient between the data points and the power law of 0.998.

Again, both positive and negative polarities yield the same forces. There is no polarity dependence observed. Extension of the best fit line shows an important point; it passes through zero. This clearly implies that current must flow for a force to be generated.

In order to understand the possible nature of any electrical conduction between the wire and the plate, the applied voltage and the current delivered to the device was plotted on a log-log graph. Figure 7 shows that the current follows a power law with a slope of approximately six:

$$I = k_2 V^6 \qquad (3)$$

where k_2 is a proportionality constant.

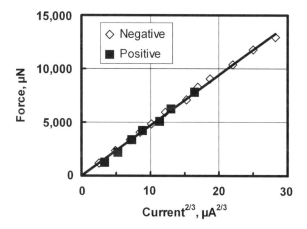

Fig. 6 Measured force as a function of the measured current raised to the 2/3 power.

Note that the functional dependence of the current-voltage curve was approximately independent of polarity and of small changes in the device geometry. The "large 45-mm device" had a side length of 200 mm, a plate height of approximately 46 mm with a wire height of approximately 31 mm. The "large 40-mm device" had a side length of 200 mm, a plate height of 39 mm and a wire height of 30 mm, while the "small 45-mm device" had a side length of 135 mm, a plate height of 45 mm, and a wire height of 30 mm.

In order to better understand the role of current flow in the production of the force, we took the "small" device and coated its wire with glyptol, a paintable conformal insulator, and allowed it to dry. Such a coating would, we presumed, suppress the formation of corona discharge in the immediate vicinity of the wire. Thus, we might expect a decrease in the overall force developed.

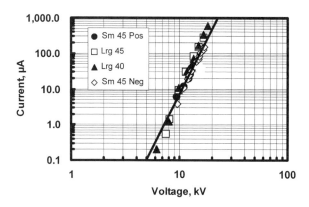

Fig. 7 Plot of the current flowing into the device as a function of the applied voltage.

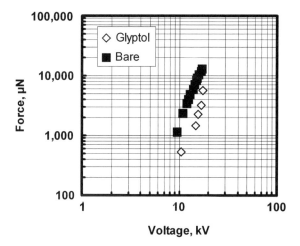

Fig. 8 Comparison of the force developed on identical devices, one with the wire bare and the other with the wire covered in a thin layer of glyptol insulation. Both measurements were taken with a positive polarity on the wire.

This effect can be seen in Fig. 8, which shows the measured force as a function of the applied voltage for capacitors with bare and glyptol-coated wires. Both measurements were made with the wire at positive polarity. Note that at the highest voltage, the force-voltage curve of the insulated wire is approaching that of the bare wire. We interpret this as an increase in the number of points at which the glyptol has broken down along the length of the wire allowing a relatively more uniform corona to form.

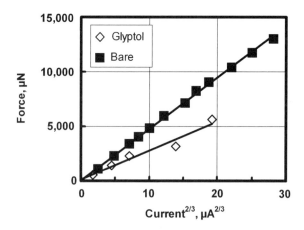

Fig. 9 Comparison of the force developed with bare and glyptol-insulated wires as a function of the current to the 2/3 power. Note that again the extrapolated best linear fit passes through zero for both wire types.

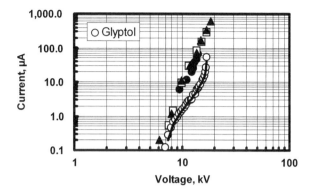

Fig. 10 Comparison of the current voltage relationship of bare versus glyptol-insulated wires. The glyptol-insulated wire shows currents that are significantly below those of the bare wire for intermediate voltages, but approaches the bare wire results at both lower and higher voltages.

We expected that the force-current curve would be related to the nature of the atmosphere only and that the force developed would be proportional to the current that flowed, even if higher voltages were required to bring this about. As can be seen in Fig. 9, this is not the case. The glyptol-insulated wire produced a force much lower than that of the bare wire for a given measured current. However, in both cases the extrapolated best linear fit passes through zero, again confirming that current must flow for the force to be generated.

Finally, the current-voltage plot of Fig. 10 compares the various positive bare wire results with that of the positive glyptol insulated wire. Unlike the bare wire results that follow a single power law dependence, the insulated wire results show a marked degree of curvature. At the lower voltages the bare and insulated wires have similar values and slopes (although the insulated wire data are slightly below that of the bare wire). At intermediate values of voltage, the insulated wire data are significantly below that of the bare wire. Finally, at the highest voltages used, the insulated wire data turn upward to approach the values for the bare wire (much as was seen in the force-voltage curve of Fig. 8).

3. Discussion

The DC force data provide several indications of the nature of the force generated by an asymmetric capacitor.

1) The magnitude and direction of the force are independent of the polarity of the excitation voltage.

2) This force was also not related to the orientation of the device with respect to the gravitational field of the Earth. It appeared to be solely related to geometry of the asymmetric capacitor. Specifically, the force was always directed along the symmetry axis of that capacitor in the direction from the plate to the wire.

3) Within the excitation regime studied (relatively high voltages and high currents), the force exhibits a power law dependence with respect to voltage

(the force is proportional to the square of the voltage) and current (the force is proportional to the current raised to the 2/3 power). We will later show that these data were collected in what we will identify as the Fowler–Nordheim regime.

4) Current must flow for there to be a force. This was shown to be true no matter the condition of the wire (bare or insulated). One can conclude, therefore, that the effect, while requiring a voltage to be applied to generate a current flow, is not so much an electric field phenomenon as a current flow phenomenon. Even though these DC data do not show a force at very low voltage, the presence of a current at lower voltages indicates that such a force should be detected. Measurements at low voltages (described later) will certainly show this to be true in the case of uniform excitation down to a few hundred volts.

5) Devices with minor changes in the geometry behave in similar ways. This will be explored further in the section on geometrical variations.

B. Sinusoidal Excitation Measurements

The fact that the force developed in the device was shown to be independent of the polarity of the applied voltage led us to speculate that the device would also develop a force under sinusoidal voltage excitation. In fact, Canning et al. [12] and others have speculated that even with uniform voltages applied, the leakage of current does not occur in a steady manner, but rather occurs as pulses, implying that neither DC nor AC voltage excitation is truly uniform.

1. Specific Setup

We tested for the presence of electrical pulses under the application of uniform negative DC voltage through the use of a 10× high voltage probe (Tektronix PG122). This was placed on a PVC stand allowing the probe to be placed securely and at a fixed distance between the wire and the plate of the device under test. The output of the probe was fed into an oscilloscope (Tektronix R7844 dual beam oscilloscope) and photographed with a digital camera. The uniform voltage had to be raised above 8.3 kV before any effects were seen. (The results of this will be described in detail.) However, the presence of these pulses (described as Trichel pulses) encouraged us to proceed with the planned measurements of the force on the device under sinusoidal high voltage excitation.

In order to make these measurements, an entirely different excitation setup was needed. As we were interested in how any measured force changed with voltage and with frequency, and as the voltage required might be in excess of 20 kV, we required a high voltage, wide frequency range amplifier. We were unable to find any off-the-shelf equipment meeting our needs. Therefore, we chose to adopt a technique utilizing a low voltage, high precision frequency source (Hewlett Packard 3330B frequency synthesizer) feeding a high voltage amplifier (Fluke 5205 high voltage wideband amplifier with a maximum output of 1200 V at its 100× fixed gain), which in turn fed a custom designed and built high voltage final transformer (modified Energy Systems pulse transformer). As this transformer had an original input to output ratio of 1:7, we could not hope to achieve the final maximum voltage that we required, and so rewrapped both its primary and secondary windings to match our input device and our required output voltage. We made measurements on the original pulse transformer core (14 turn primary bifilar winding, 102 × 2 turn balanced secondary winding) to help in the selection of the final design. Oscilloscope measurements of the original transformer showed that frequencies from 100 Hz to 50 kHz were handled with no distortion or diminution of the transformer ratio. For our final design we chose a primary winding of 93 turns of 28-gauge copper wire per side (bifilar wound), and a secondary winding of 2809 turns (per side of the balanced secondary). While the 93 turns of the primary winding would not provide a sufficient load on the high voltage amplifier at lower frequencies, selection of a greater number of turns would have made wrapping the secondary impractical.

The four coils needed were wound onto PVC mandrels and then insulated with high voltage Formvar varnish. Given our use of a bifilar primary and balanced secondary, the designed output ratio was 1:15.1 and the measured ratio was 1:15.5. Insulation was provided by 4.5 gallons of Shell Aeroshell 30-weight oil, dewatered and degassed at 108°C under a modest vacuum to achieve a suitable breakdown voltage.

Figure 11 shows the calibration curve for this transformer measured as a function of frequency at low input voltage. One DMM (mentioned earlier) was used to measure the ratio of input to output voltage. For frequencies below approximately 5 kHz, a flat voltage multiplication of 15.5 was measured. For use at higher frequencies, we prepared a detailed calibration curve.

As the high voltage amplifier would not properly load into the transformer, we were forced to place a 360 Ω resistor in series between the amplifier output and the transformer input. While this decreased the input voltage to the transformer and hence the output voltage (and was not ideal), it did allow data to be acquired.

The complete experimental setup for the AC measurements using the modified transformer is given in Fig. 12. The Tektronix oscilloscope was used to measure

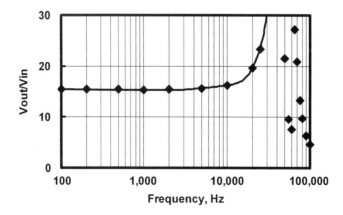

Fig. 11 Transformer voltage input-to-output ratio as a function of frequency.

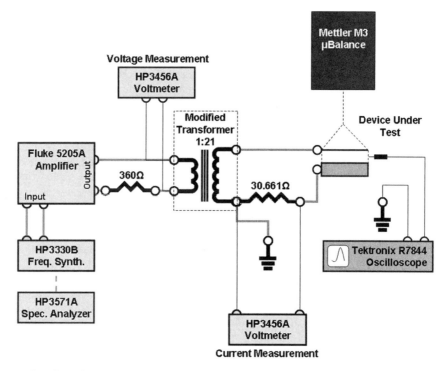

Fig. 12 AC experiment setup. Oscilloscope was used to measure Trichel pulses.

the Trichel pulses while the device was under DC excitation only. We arranged to apply either AC or DC excitation by reconnecting the DC signal source at the test points to the right hand of the modified transformer.

Voltage measurements were made utilizing the root-mean-square (RMS) feature of the digital multimeter (DMM). Current measurements proved to be impossible due to ground loops. Due to a lack of time, the automation of the measurements using the HP9845B was not completed; the equipment was run manually. Each combination of voltage and frequency was measured 30 times with the average reported in this chapter. In addition, we recorded the variance of the observations. The error bars for these measurements were typically less than 50 nN.

2. Results

In this case the wire was kept at a high negative voltage and the plate was grounded. A single pulse is shown in Fig. 13. This clearly shows the double peak characteristic of Trichel pulses. Figure 14 shows a series of oscilloscope pictures obtained as the voltage between the wire and the plate was increased. In this case the probe was kept approximately halfway between the wire and the plate of the device under test. Additional data (not shown here) were taken with the probe

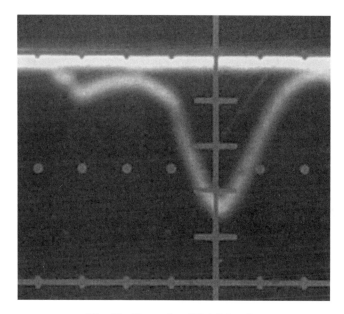

Fig. 13 Example of Trichel pulse.

in various other positions. Similar results were found, although the magnitude of the DC offset of these pulses (at higher voltages) did change.

Note that when the voltage reached a certain threshold value (just below 8336 V for this geometry), infrequent negative-going pulses began to form. As the voltage was increased, the pulse rate also increased. When the voltage reached 8690 V the pulse rate was higher still and the DC voltage level increased.

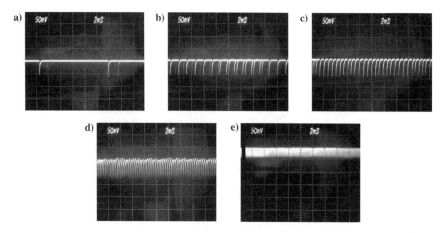

Fig. 14 Onset of Trichel pulses. Voltage was raised to a) 8336 V, b) 8690 V, c) 9323 V, d) 10,054 V, and e) 11,022 V, respectively.

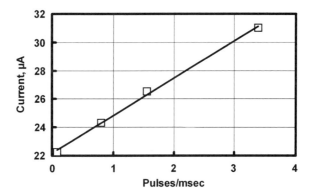

Fig. 15 Number of Trichel pulses per millisecond is linearly correlated with the measured current.

Above this voltage the DC component continued to increase, while the pulse rate became uncountable as individual pulses overlapped.

If the current is solely related to the formation of the Trichel pulses, then the pulse rate should be proportional to the measured current. In fact, a plot of these two variables (Fig. 15) shows this to be the case. Had it increased more rapidly, one might argue that generic leakage currents were superimposed onto the current flow from the Trichel pulses. This is consistent with the physics of Trichel pulses wherein the charge per pulse decreases with frequency due to space charge effects.

These previous data indicate that even in the case of uniform excitation, a regular time-periodic response can be obtained. These data caused us to pursue direct AC measurements. The equipment (described in the previous subsection) was then constructed to take these additional data.

Force and voltage measurements were made at 100 Hz, 200 Hz, 500 Hz, 1.0 kHz, 1.2 kHz, 2 kHz, 4 kHz, 5 kHz, 10 kHz, 20 kHz, 38 kHz, 50 kHz, and 100 kHz with the majority of these measurements taken at 1.2 kHz, 2 kHz, 4 kHz, and 38 kHz. Initially there appeared to be some form of resonance at 38 kHz, so data were first taken at this frequency. The thought was that if a maximum of power would be absorbed by the device, a maximum force should be observed. This did not prove to be true.

The data at 38 kHz (Fig. 16) show the same functional dependence that we observed in the case of DC excitation. However, rather than being of large magnitude as we had expected, the force was quite low. Compare the magnitude of the force at 38 kHz with that measured for DC excitation and for AC excitation at lower frequencies. The force is proportional to the square of the voltage. Note that the scatter of the value of the force is large at the lower applied voltages, but tightens as 1 kV is approached. We believe this to be due to the difficulty of measuring forces in the region below 100 nN.

The force was measured at a series of voltages as the frequency was varied in logarithmic steps from 100 Hz to 100 kHz. Only a relatively small change in force was observed as a function of the frequency over this range.

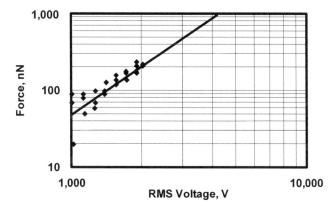

Fig. 16 Force measured at 38 kHz excitation. Note that the force curve follows the same power law as it did for DC excitation.

The data in Fig. 17 indicate that we can expect very little change in force as a function of frequency. Our equipment was such that voltages above 1 kV could only be obtained at frequencies in the 1 to 10 kHz range. Therefore, additional data were obtained in this range and are presented below.

Figure 17 shows data taken at 1.2 kHz, 2 kHz, and 4 kHz. In this case the force again behaves like a power law as before, following a voltage cubed relationship at the higher voltages. However, at the lower voltages it follows a voltage squared relationship.

These data clearly show several things: 1) the lifter is still capable of producing its force, even with an AC voltage applied to it, and 2) it appears that the force of the lifter is the same no matter if the excitation by a sinusoidal (AC) voltage or a DC voltage.

3. Discussion

These data elucidate several additional important facts about the force arising from an asymmetric capacitor.

1) As expected from the polarity independence of force under DC excitation, AC excitation also results in the production of significant forces. In fact, when compared, the DC and AC forces are nearly identical (for the higher voltages where such a comparison is possible).

2) The presence of Trichel pulses and their relationship to the force produced under DC excitation shows that even for such an excitation, nonconstant effects are taking place.

3) It is clear that measurable but small forces continue to be produced even at extremely small voltages (down into the range of hundreds of volts). There is no reason to believe that there is any limit in how low this can go, provided that current is flowing. Practically, devices based on the phenomena of force produced from asymmetric capacitors should concentrate on methods to increase current flow at lower applied voltages.

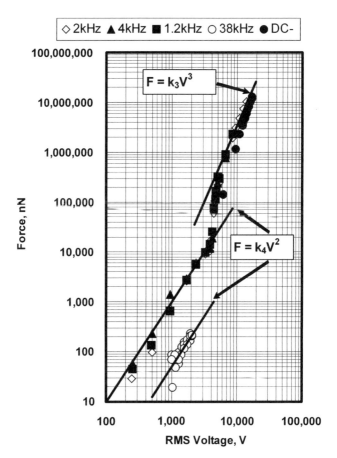

Fig. 17 Force on device as a function of frequency. Note that the AC and DC forces are nearly identical at the higher voltages.

C. Geometric Variations

Understanding the physics behind the force generation in an asymmetrical capacitor should be facilitated by measurements of the force with different asymmetries and hence in different geometrical configurations. However, such variations are not easily accomplished using the standard triangular device that had been used in the pervious experiments. A linear arrangement (to be described below) was selected to allow easy variation of the wire-to-plate separation and the plate depth, thus varying the magnitude of the asymmetry.

1. Specific Setup

The linear arrangement was formed from two vertical hollow, cylindrical plastic supports, each 130-mm tall. In the center of these, another smaller hollow plastic support, slightly longer than 60 mm, was placed such that this support could rotate. On this was placed a 60-mm wide thin aluminum foil plate such that this foil could be rolled around the horizontal support. The lower edge of the plate was supported by a stiff copper wire of 65 mm in length, which was centered and held rigid by slits in the edge of the vertical supports. A thin 60-mm long wire was supported above the horizontal support via a series of holes placed in the vertical supports and spaced every 5 mm above the horizontal pieces, beginning at a 10 mm separation from the top of the aluminum foil plate. These arrangements allowed the wire to be adjusted to any of 10 positions above the plate (10 mm through 55 mm). Rigidity in the structure was ensured by two thin plastic lines at the top and bottom of the arrangement that kept tension on the vertical supports. The entire arrangement was hung from a triangular plastic line support and then from the center of the Mettler M3 balance, much like the triangular device. The plate height could be adjusted continuously from 10 mm through 55 mm. Thus the depth of the plate and the height of the wire could be varied without changing the underlying shape of the device under test. For practical reasons, extra polyethylene weights were symmetrically hung on the bottom of the device to provide a downward force on which the device could exert force. (Recall that the balance can measure weights up to 3 g.) Weighting the device so that the total device weighed nearly this amount guaranteed the maximum total measurement range. This also served to stabilize the arrangement.

The data were taken using positive DC excitation on the wire via the setup previously described (Fig. 2). The maximum voltage that could be used was related to the minimum separation between the wire and the plate. For example, with a 10-mm separation, a maximum of 13 kV could be applied, while at greater than 20 mm, the full 30 kV capability of the high voltage DC power supply could be used. The measurement procedure was identical to that described in the DC measurements section (Section IV.A). However, as this linear design was somewhat unstable against horizontal motion, higher voltages tended to move the device back and forth more readily, thus leading to greater variations in the vertical forces measured.

2. Results

As shown in Fig. 18, the force developed on the device with the plate depth set to 55 mm varies with the wire height and voltage. As is expected from the previous DC force measurements, the force increases with applied voltage. These data also show that for a given voltage, the force generally decreases as the wire height increases. At very small separations between the wire and the plate, with the highest voltages applied, the force decreased. There are a number of modest maxima shown. For 10 to 18 kV, the maximum force is generated at the minimum wire height measured, with a second, lower, local maxima between 45 mm and 50 mm. At higher voltages, these second maxima move toward lower wire heights, until at 30 kV, it is below 20 mm. At such high voltages data could not be taken at the smallest wire heights as breakdown would occur.

Most data taken at 15-mm wire height had to be discarded as these data were taken with a poor connection to the aluminum plate. Additional data

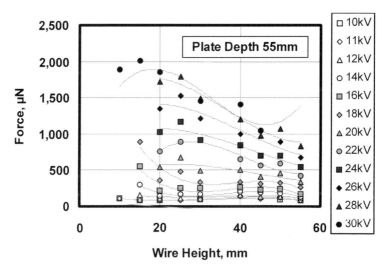

Fig. 18 Force developed on the linear device as a function of wire height and voltage. Note that there are gentle local maxima.

taken at 6 kV were also not shown, as they were not visible on a graph of this scale. Nevertheless they behaved in a way similar to the other data between 10 and 18 kV. There were no unexpected results discovered, save the gentle force maxima. While these may be useful in an engineering design utilizing this effect, they were not of further use in understanding the physics of the phenomena.

Likewise, we studied the effect of increasing the plate depth while using a fixed wire height. The wire height was set at 25 mm to allow a full range of voltages to be used. Our thought was that the greater plate depth might increase the overall asymmetry of the system, thereby increasing the force. This experiment was intended to answer a fundamental question about the force, specifically, if increasing asymmetry was indeed related to the force production. Figure 19 clearly shows that such an effect was not observed. Even when the data were looked at very closely, there appeared to be no correlation between plate depth and force. Plate depths from 17 mm to 55 mm were studied. These were selected to be of sufficient range (greater than a factor of three) and an appropriate magnitude (slightly less than to much greater than the characteristic wire height) that some effect, if it were present, should have been manifested. Please note that while the lowest force was recorded with the 17-mm plate depth, the second lowest was found with the 55-mm plate depth. At best there might be a very slight maxima at a plate depth of 45 mm; however, given the scatter in the data, this experiment appears to have produced a null result.

As was the case in previous DC measurements, the force was found to vary as a power law of both the voltage and the current. Again force varied as the voltage cubed (Fig. 19) and as the current to the 2/3 power (Fig. 20). Furthermore, this did not change appreciably as the plate depth was varied.

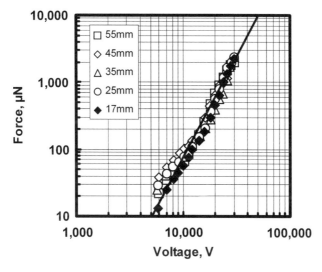

Fig. 19 Force as a function of voltage for a variety of plate depths (in mm). The wire is fixed 25 mm from the plate. The force increases as the cube of the applied voltage. Note that the variation in plate depth over the range studied does not result in significant changes in the force developed.

The force power law as a function of voltage did appear to change as the wire height was varied, as seen in Fig. 21. While the overall data at higher voltages appear to follow the voltage cubed relationship described earlier, the very smallest wire separations may have a slightly steeper slope. This also appeared to change with the wire separation (being steeper with decreasing wire separations).

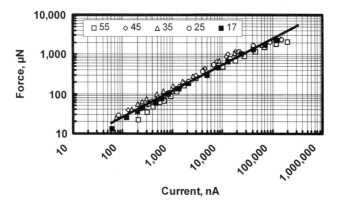

Fig. 20 Force as a function of current for a variety of plate depths (in mm). The wire is fixed 25 mm from the plate The force increases as the current raised to the 2/3 power.

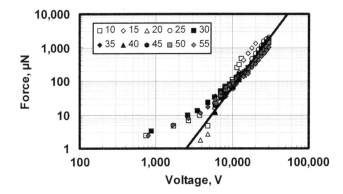

Fig. 21 Force measured as a function of voltage for various wire separations (in mm) and with a fixed plate depth of 55 mm. Note that the data do not all follow a single power law dependence. A line showing a cubic dependence is indicated. For low voltages the force is proportional to the voltage and at higher voltages it is proportional to the cube of the voltage.

At the lowest voltages, the force is approximately proportional to the voltage, with not all of the data behaving in this way.

The relationship of the force with current again behaved approximately as the 2/3 power as previously shown in Fig. 20. There might be very slight variations with wire height (slightly higher slope) but this is difficult to confirm. Note the wide scatter in the data.

One of the advantages of measuring the force, voltage, and current relationships with the linear asymmetric capacitor was the complete absence of corners found in the triangular device. This simpler geometry also allowed us to investigate the different current regimes accessed as the voltage was increased. As we saw in the earlier section on DC behavior, it is the current flow that is most well-correlated with the production of the force. Thus, understanding the origin of the current is most likely to help in understanding the origin of the force, especially as the geometry was changed. The current and voltage data, collected as described above, were replotted with 1000 divided by the voltage plotted on the y-axis and with the natural logarithm of the ratio of the current to the voltage squared plotted on the x-axis. Such a plot should be linear if Fowler–Nordheim currents are flowing. Fowler–Nordheim tunneling, or field emission, is a form of quantum tunneling in which current passes through an insulating barrier (in this case, air) in the presence of a high electric field.

In this experiment, only data taken at the largest combined wire separation (55 mm) and plate depth (55 mm) were plotted so that the form of graph might clearly be seen. This representation of the data (Fig. 22) allowed three regimes to be identified. At the lowest voltages 1000/V had its highest value. This regime is seen to the far right-hand side of the graph. These data exhibit a straight line relationship from the lowest voltages to approximately 6000 V (1000/V ≈ 0.17 1/V). This first region we identify as the "non-self-sustaining current regime." Note that 6000 V is approximately the voltage at which we first saw

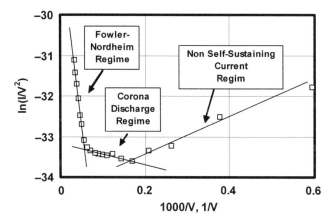

Fig. 22 Fowler–Nordheim plot of linear device (55-mm wire separation and 55-mm plate height). Note the three distinct conduction regimes highlighted by the three lines.

the onset of the largest DC forces and where Trichel pulses were first observed, although smaller forces within this regime could and were measured in both the AC excitation and this experiment. The non-self-sustaining regime is dominated by leakage currents that are not energetic enough to result in current multiplication.

At somewhat higher voltages (from approximately 6000 V to 14,000 V with corresponding values of $1000/V$ of from 0.17 to 0.07 1/V), another linear region was found. We termed this the "corona discharge regime." This is best described as resulting from ion multiplicative effects. Even though a slight positive slope can be seen for this regime, we believe that the actual slope here is approximately zero due to the additive nature of all of the current and the inability to distinguish one particular current type within each regime. The greatest effect tends to dominate these regimes. In this case the third regime tends to add to and cause this slight slope.

The third and final regime at the highest voltages (above 14,000 V from $1000/V$ less than 0.07 1/V) we have identified as the Fowler–Nordheim regime. These very large currents dominate all others. This region is composed of both Fowler–Nordheim current and corona current although, being much larger, the Fowler–Nordheim current dominates. Interpretation of the data relies heavily on these observations and will be discussed at length in Section V. The three regimes found may help to explain the variation in the power law dependence seen in Fig. 21. Note that the behavior changed at approximately the voltages where the Fowler–Nordheim and the corona discharge regimes meet.

All data for the varying wire height (with the fixed plate depth of 55 mm) are shown in Fig. 23. The same behavior exhibited with the 55-mm–55-mm data can be seen at all other wire heights, although less clearly as there is much overlap.

Finally, several sets of data are replotted in Fig. 24 to show a fact of potential engineering significance, that of the force-to-power relationship for this linear

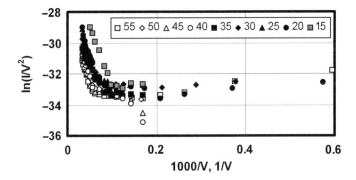

Fig. 23 Fowler–Nordheim plot for 55-mm plate linear device as a function of wire height. Note how the curves move upward with decreasing wire height.

system. Note that the force increases as approximately the square root of the power. While we cannot claim to have an optimized system it is interesting to note (empirically) that

$$F = C\sqrt{P} \quad (4)$$

Again we define the force as F (in μN), the power as P (in μW) and the constant as C, which is 1 (in units of $\mu N/\sqrt{\mu W}$). We have every reason to expect that the force will be proportional to the length of the asymmetric capacitor. Given the length of this device (0.06 m) and for this geometry (wire height of 25 mm) and plate depths from 17 mm through 55 mm, the force per unit length will follow a formula similar to Eq. (4):

$$f = c\sqrt{P} \quad (5)$$

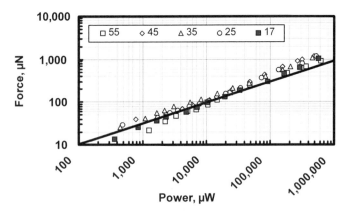

Fig. 24 Force as a function of power for a variety of plate depths (in mm). The wire is fixed 25 mm from the plate. The force increases as the square root of the power.

Where f is the force per unit length (in $\mu N/m$), again P is the applied power in μW and c is the specific force per unit length per unit power of 16.7 $\mu N/(m \cdot \sqrt{\mu W})$.

3. Discussion

These studies of the effect of geometry on the force developed in asymmetric capacitors were surprising in that no large changes were noted with relatively large changes in wire spacing and plate height, and hence in asymmetry. Minor maxima were found as a function of wire height that became more pronounced at the higher voltages. Again, interpretation of these results should be made in terms of the flow of the current (given that no force was measured unless current was observed to have flowed, as in Fig. 9) and not just on the basis of the voltage applied. However, as Fig. 22 has shown, the values of the voltage applied do help delineate the various regimes of current production.

At very small wire-plate separations, the total force generated was always modest. This was almost certainly due to the current leakage between the wire and the plate and the very limited voltages that could be applied without leading to arcing of the system. When such breakdowns did occur, they did not damage the device under test.

From a purely practical point of view, reasonably large forces can be generated by the simple plate-and-wire arrangement. Clearly, the more efficient generation of large currents with more modest voltages would raise the efficiency of this system.

D. Magnetic Field Measurements

There has been speculation that the force generated by an asymmetric capacitor is related to its interaction with the Earth's magnetic field, which has a magnitude on the order of 5×10^{-5} T. Both the magnitude and direction of this field vary depending on the location on the Earth. In our location, just to the east of Albuquerque, New Mexico, at approximately 35°00′50″ north latitude, 106°18′50″ west longitude, and an altitude of approximately 2304 m above mean sea level, the magnetic field had an approximate total magnitude of 5.0676×10^{-5} T and a direction of approximately 9°57′ east of north and a dip from the horizontal of 62°24′ down into the Earth on the date of the calculation (30 Nov. 2003) [20].

1. Specific Setup

Previously mentioned measurements of the lifter's force were always made in the Earth's magnetic field. In order to discern if this (or any) magnetic field gave rise to the observed force, a much larger artificial magnetic field was created and the measurements made in it. To create a uniform magnetic field we designed, built, and operated the experiment in a Helmholtz coil, which has a property of producing a large, relatively uniform magnetic field of known magnitude and direction near its center (Fig. 25). Each of the 2 coils had four windings 8-layers deep for a total of 32 turns of 6-gauge copper wire. The diameter of the inner coil was 75.6 cm and that of the outer coil was 80.0 cm. We separated

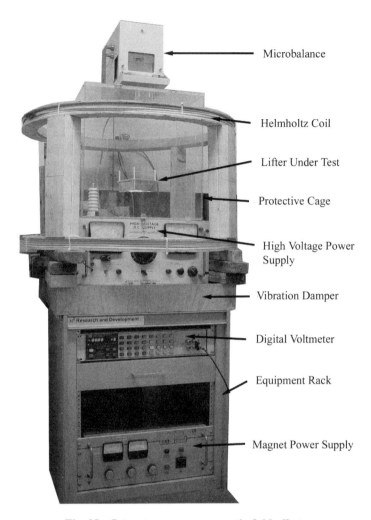

Fig. 25　Setup to measure magnetic field effects.

the 2 coils by 38.1 cm. The coils were energized with a Sorensen DCR40-60A power supply that could produce 40 V at 60 A, although they only needed about 13 V and 60 A to produce a magnetic field approximately 100 times that of the Earth (specifically 5.0×10^{-3} T). A current shunt was used to measure the current flowing through the Helmholtz coil and hence the magnetic field produced. The calibration constant was calculated to be 0.2063 T/V.

It was possible to arrange the coils to produce both a horizontal and a vertical magnetic field. Specifically, the horizontal component was directed approximately upward, and the vertical component was directed downward. However, a failure of the power supply precluded our taking data in the horizontal direction.

2. Results

Force measurements were first made with an unpowered triangular asymmetric capacitor (as described in Sec. IV.A on DC measurements) with the Helmholtz coils de-energized. With the device unpowered and the Helmholtz coils energized, no effect was observed within the experimental resolution of ± 0.00025 mN. This indicates that the artificial magnetic field did not affect the force measurement of the Mettler microbalance to within these limits. Measurements were then made with the device powered to produce a force, but without the Helmholtz coils energized. This provided a background measurement against which to compare the force once the coils were energized. Finally a measurement was made with the device powered and the coils energized. This force measurement was then compared to the one made with no applied external magnetic field.

Table 1 summarizes the results with the external magnetic field pointing vertically downward. Two experiments were performed, one with an insulated wire (coated with glyptol to decrease the formation of any leakage or corona currents) and one left bare. The first number under the column titled "Net Force Change" for each experiment represents the downward force on the unpowered device. The other values in this column represent the net upward force that was measured both with and without the external magnetic field. The final column in the table (Range of Effect) represents the net effect when normalized to the field that was applied compared to the Earth's magnetic field. Note that these are very small numbers. If there was a significant magnetic field effect one would have expected the net force change to have increased proportionally to the applied magnetic field.

3. Discussion

Note that in both cases, the range of the measured effect was very small, equivalent to the measurement errors experienced. The applied magnetic fields (5.034×10^{-3} T and 5.044×10^{-3} T) were nearly 113 times that of the Earth's vertical magnetic field component of 4.4577×10^{-5} T. If the majority of the force were to arise from interactions with the Earth's magnetic field, then one would expect that a force of 2.03 mN or 7.07 mN would have increased to 100 times that in the artificial field provided. They did not. We conclude that within experimental error, magnetic fields have little or nothing to do with the generation of the forces observed.

V. Discussion of Data as it Relates to Theories

The objective of a theory of asymmetric capacitors (or lifters) is to elucidate the mechanism by which a motional force is generated by the application of a sufficiently high voltage. In all lifter experiments to date for which precision measurements of current were made, it has been found that there exists a monotonic dependence of force upon current flowing through the lifter. This is true for both static and sinusoidally time-varying applied potentials. The language here is carefully chosen to avoid the confusion inherent in describing potentials as DC or AC, terms which refer explicitly to the time variation of current. As discussed

Table 1 Changes in measured force with magnetic field[a]

Measurement		Applied external magnetic field (T)	Applied voltage (V)	Net force change (mN)	Range of effect (%)
	No B, no E			21.15515	
	No B with E		14,121	2.03218	
	B and E	0.005034	14,204	2.03218	
Glyptol	B and E maximum		14,204	2.02238	0.0043
	B and E minimum		14,204	2.04199	−0.0043
	No B, no E			19.23957	
	No B with E		14,149	7.06790	
	B and E	0.005044	14,251	7.26893	
Bare	B and E maximum		15,915	7.22971	−0.0197
	B and E minimum		15,915	7.30816	−0.0293

[a]Measurements were made with both insulated and bare wires.

below, the lifter will be considered to be a leaky capacitor. There are two possible current paths: surface leakage over the dielectric supports separating the wire from the plate, and free space currents consisting of ions and electrons. Herein we will assume that only the free space currents generate a force. Under this assumption we also observe that, even under static applied potentials, the current is never truly a DC current as would be characteristic of a linear resistive circuit (ignoring thermal noise). The current is time varying at high frequencies, driven by inhomogeneity in the lifter components and the complexities of free space transport of charged particles [21]. At best, the current may be characterized by a root-mean-square (rms) average current. Even in the case of a sinusoidally varying applied potential, the current should not be assumed to be a simple sinusoidal response. It is also important to emphasize that force must be dependent on a current flow. Under a static applied potential, with no current flow, no work is done by the system that could be transduced into a force capable of moving the lifter a finite distance. Whatever mechanism is at work, it is assumed to obey the symmetry of conservation of energy.

The classical application, which most closely relates to the phenomena observed in lifters, is found in the electrical description of electrostatic precipitation [22]. In a lifter the common configuration is that of a wire parallel to the edge of a rectangular plate and occupying the same plane. A common geometry used in the study of precipitator phenomena is that of a wire in the plane of a perpendicular bisector of a plate (foil) along its major axis. The wire is held parallel to the plane of the plate. A precipitator is designed to have a single high electric field electrode. This is accomplished by designing the large electrode to be free of sharp edges. In discussions of lifter phenomena, the wire is often assumed to be a unique high electric field anode although by nature of the design, often complicated by crude assembly methods, this assumption is of questionable validity. Precipitators are designed to operate with a self-sustaining corona discharge plasma localized in the near neighborhood of the wire electrode. The polarity of a corona discharge is always referenced to the polarity of the high field electrode, in this case the wire. In air, both positive and negative corona discharges are possible. In a positive corona discharge a current of positive ions is ejected from the plasma, to be collected by the plate, while both electrons and negative ions are attracted to the wire where they are collected or neutralized [23]. The situation for a negative corona discharge is more complex [24]. Both electrons and negative ions are ejected from the plasma surrounding the wire. At low currents electrons leaving the plasma form more negative ions by attachment to electronegative molecules, oxygen in particular. This forms a negative ion sheath around the plasma sheath that, at higher currents, causes instability in the plasma causing it to periodically collapse. The resulting pulsed current mode is described as a Trichel pulse mode. The frequency of the Trichel pulses may increase with current from a few Hz to hundreds of MHz. While Trichel pulses have been implicated in the force generation mechanism, they occur only for negative corona discharges, although comparable lifter force has been observed under conditions of both positive and negative discharge.

A corona discharge, either positive or negative, is sustained by an avalanche multiplication process in which electrons are accelerated by the high electric fields near the wire to produce ions via impact and frequently generating more electrons to participate in the ionization process. The avalanche process is self-sustaining only above a certain threshold field. The electron source seeding the avalanche derives from natural ionization events driven by UV light or other background radiation. Below the self-sustaining threshold the seed source is constant, although some current amplification does occur from non–self-sustaining avalanches. The authors are not aware of any critical studies of the current characteristics in this subcorona regime. In the corona regime the current is quadratic in voltage according to the law

$$I = AV(V - V_o) \qquad (6)$$

where I is the current, V is the applied voltage, and A and V_o are constants dependent of the geometry of the system. V_o is the corona onset voltage. This equation is known as the Townsend equation [25] and relates directly to Paschen's law and its critical parameter pd (p being the pressure and d the gap distance). For pd values less than about 1000 torr-cm the Townsend equation holds true. The current is carried entirely by ions. At sufficiently high voltages the corona is typically observed to collapse into a spark discharge in which a direct electron current flows between electrodes limited only by the power supply. This is generally a destructive event. In the early literature on precipitators, an intermediate regime was often reported in which currents were observed to follow a power law, $I \propto BV^n$, with $4 < n < 6$. This behavior was not interpreted at the time. In the present work, it is found that this is a Fowler–Nordheim electron emission current flowing parallel with a reduced ion current. In lifters, the geometry allows for substantial Fowler–Nordheim currents in conjunction with either positive or negative discharges. In the case of a positive corona, the electrons must be sourced from the base of the lifter and not the wire. Because the net force is essentially the same for positive and negative corona discharges for which a substantial Fowler–Nordheim current flows, it must be substantially an electron current thereby contributing negligibly to the force.

Fowler–Nordheim emission is a field-assisted tunneling current that is qualitatively described by

$$j = A(y)(\mathscr{E}^2/\Phi) \exp[-B(y)\Phi^{3/2}/\mathscr{E}] \qquad (7)$$

where j is the current density, $\mathscr{E}$ is the electric field, Φ is the work function of the electron emitter, and $A(y)$ and $B(y)$ are slowly varying functions of $y = 3.79 \times 10^{-5}\sqrt{\mathscr{E}/\Phi}$. Because Φ is so strongly dependent on the geometry, smoothness, and cleanliness of the emitter, it is found to have an effective value typically 10 to 100 times greater than its nominal value. With the approximation that at the emitter surface the electric field is proportional to the applied voltage and the substitution $I = j \times$ (*effective emission area*), $A(y)/\Phi$ and $-B(y)\Phi^{3/2}$ are subsumed, with these proportionality constants, into unknown constants A^* and B^*. Equation (7) is then rearranged as

$$\ln(I/V^2) = B^*/V + \ln(A^*) \qquad (8)$$

which yields linear graphs of Fowler–Nordheim $I-V$ data. On the surface this seems like a large number of approximations. However, as in a parallel diode array, the bulk of the current flows through surfaces having comparable effective Φ and surface fields. Locally the conditions required for Eq. (6) are satisfied and dominate the total current. By emphasizing that a Fowler–Nordheim current emitted from an extended source is spatially inhomogeneous it is possible to understand that a Fowler–Nordheim emission current can coexist with a corona ion current as spatially distinct phenomena.

The assumption that lifters "fly" by generation of a unipolar ion current at the wire electrode is realistic. That the force is independent of the polarity of the applied bias and massive ions may account for virtually all of the current. Lifter forces resulting from a constant ion current have been analyzed by Bahder and Fazi under two assumptions: ballistic transport of ions from wire to skirt (ionic wind) and nonballistic transport were the ions assume a finite mobility determined by collisions with neutral air molecules (ionic drift) [11]. In many references Bahder and Fazi's "ionic drift" is referred to as ionic wind, ion-neutral wind, or corona wind. The ion-neutral wind explicitly calls out the assumed transfer of momentum from ions to neutral air molecules. Bahder and Fazi found that ballistic transport resulted in forces orders of magnitude smaller than those observed in lifters. The magnitude of the forces estimated to result from a corona wind (ionic drift) matched nicely with observed lift forces. The limitation of this analysis, however, was that they did not recognize that the lifter was operating in a corona regime and could not identify a source of ions sufficient to drive the lifter. In contrast, Christenson and Moller, deliberately investigating ionic drift propulsions in the 1960s, designed a circular lifter-like device that produced air velocities of up to 7 ft/s (2.1 m/s) with the device fixed on a stand [26]. The only significant difference from modern circular lifter designs is that a dense array of point emitters, directed toward the skirt of the device, was used in place of a wire. The $I-V$ response of their device was verified to obey the Townsend equation. The air velocity was proportional to the square root of the current flowing through the device.

Published results for corona plasmas generated by sinusoidal high voltage excitation are far less common than for constant excitation. They are of interest primarily as a loss mechanism pertinent to high voltage power transmission lines [27]. The salient features are that a full cycle of applied potential results in a positive corona for $V > V_o$, a nonplasma interphase for $V_o > V > -V_o$, a negative corona for $V < -V_o$, followed by a final nonplasma interphase. It is important to note that for either positive or negative coronas, there is a plasma sheath around the wire containing both positive and negative ions. Some of the extant literature seem to imply that space charge produced by a sinusoidally excited plasma simply oscillates in space outside the plasma region in the nonplasma interphase in favor of a net current [28,29]. However, the field strength decreases monotonically with distance from the wire. There should be a net current of alternating positive and negative ions. Ions leaving the wire during their generative corona and drifting outward will see a weaker return field as the potential swings toward the opposite polarity. Evidence for such a net current is evident in the comparison of the AC and DC lifter data already discussed.

A second-order effect that has been considered in precipitators derives from the induced polarization of the ambient gasses. Such forces are variously referred to as electrohydrodynamic (EHD), dielectrophoretic, thermodynamic, or Kelvin forces. Any electric field will induce a dipole in a polarizable atom or molecule. In a spatially uniform electric field dipoles will simply align with the field. It is generally assumed that air subject to a high electric field constitutes a dilute incompressible fluid of linearly polarizable dipoles. In a spatially nonuniform electric field, dipoles will migrate to the regions of the highest electric fields driven by the gradient in the electric field intensity, irrespective of the polarity of the field. The magnitude of the force is proportional to the polarizability (corresponding macroscopically to the permittivity of the working fluid) of the dipolar molecules and $\nabla(\mathscr{E} \cdot \mathscr{E})$, the gradient of the dot product of the electric field [30]. Common atmospheric gases have low polarizabilities. Ions produced in the corona plasma and plasma reaction products such as ozone (O_3) can have polarizabilities orders of magnitude larger than neutral N_2 or O_2. In wire-plate test configurations Kelvin forces are found to be responsible for circulating flows with momentum transfer to the air approximately an order of magnitude smaller than that transferred by the Coulomb force [31]. Because $\nabla(\mathscr{E} \cdot \mathscr{E})$ is independent of the polarity of the field, we have considered that these Kelvin forces may account for the motional force acting on a lifter. The counter argument has been that the Kelvin force ultimately results in a static situation where the increasing gas pressure in the high field region balances the Kelvin force. Coupled with an ionic flow, a recirculating flow is obtained, which reinforces the corona wind in the plate perpendicular bisector plane containing the wire.

All of the atmospheric pressure results accumulated in this work, as well as those reported in the open literature can be interpreted in the context of an ion drift thrust mechanism using Eqs. (6) and (7). Claims for esoteric thrust mechanisms will not be considered here and it is asserted that all published atmospheric results can be interpreted solely in the context of a corona ion drift wind as asserted by Tajmar [32]. Tajmar was also able to exclude the existence of any significant exotic force by enclosing a lifter in a sealed grounded enclosure and demonstrating that the lifter was not able to impose a net force on the surrounding air in the enclosure or on the enclosure itself. A unique contribution of the present work has been to demonstrate that there are two distinct plasma regimes that can provide significant lift: the pure corona regime and the regime where an ion current flows in parallel with a much larger electron current sourced by Fowler–Nordheim emission. These regimes with be distinguished as the corona wind and corona Fowler–Nordheim regimes.

If we consider a pure corona wind as providing the motive force for lifters, one can make the following positive assertions as being true of both lifters and explicit corona wind studies. A corona wind will always flow from the high field (plasma source) to the low field region of the device. In traditional lifters, including ours, the high field region is typically a wire. For a corona wind the current will be proportional to V^2. Negative coronas will be stable to higher voltages than positive coronas. Only negative coronas generate Trichel pulses and substantially more ozone than positive coronas. Wire vibrations may occur that are driven by simple electromechanical coupling of the wire and the plate in the presence of a plasma sheath [33]. No new physics are required to explain these phenomena.

A new observation for lifters is that currents in excess of the corona current are apparently due to Fowler–Nordheim emission. These currents appear on a log-log plot as being proportional to V^n, $4 < n < 6$. Over a larger range of validity they are described by the Fowler–Nordheim Eq. (7). The Fowler–Nordheim currents are assumed to be pure electron currents that do not contribute significantly to the total force. They do, however, reduce the rate of ion production comparably in both positive and negative discharges. In the triangular lifters studied here the propensity for Fowler–Nordheim emission was strong enough to dominate at all voltages. The onset of Fowler–Nordheim current flow apparently seeded the corona and no pure corona wind regime was observed. In the linear geometry lifters the upper edge of the foil had a much larger radius and a corona wind regime was easily observed prior to the onset of Fowler–Nordheim emission.

VI. Conclusions

Taken together, the experimental data on asymmetric capacitors data exhibit the following:

1) The magnitude of the force is independent of the polarity of the applied electrical excitation (AC and DC data).

2) The direction of the force is independent of the polarity of the applied electrical excitation (AC and DC data).

3) The force can be generated both in line with and perpendicular to the Earth's gravitational field (DC data).

4) Current must flow for the force to be generated (DC data).

5) Electrical insulation of the wire leads to both a decrease in the current flow and a decrease in the force generated (glyptol data).

6) The magnitude of the force is unaffected by magnetic fields up to 5×10^{-3} T (magnetic field data).

7) Even with a constant applied (DC) excitation, non-constant current flow can be observed (DC data).

8) Forces of nearly identical magnitude were observed for both DC and AC excitation.

9) Forces have been measured with small excitation voltages (DC data) and very small excitation voltages (down to 100s of volts) (AC data).

10) Minor changes in the geometry of the asymmetric capacitors lead to minor changes in the forces generated (geometry data).

11) Large changes in geometry lead to only modest changes in the forces generated (geometry data).

12) All parameters that correlate with force production obey power laws (voltage, current, electrical power) (DC, AC, geometry data).

13) The variation of the force with voltage follows a fourth power dependence for DC excitation.

14) The variation of the force with voltage follows a third power dependence at high voltages and a second power dependence for low voltages for AC excitation.

15) The force generated is proportional to the square root of the power delivered to the device.

All of these data, when taken together, can be explained by the ion drift theory as presented earlier. The complete description of the force is somewhat more complicated and requires the consideration of second-order effects including that of the dielectrophoretic (Kelvin) effect.

Acknowledgments

Sandia is a multiprogram laboratory operated by Sandia Corporation, a Lockheed Martin Company, for the United States Department of Energy's National Nuclear Security Administration under Contract DE-AC04-94AL85000.

References

[1] Brown, T. T., "How I Control Gravitation," *Science and Invention*, Aug. 1929.
[2] Brown, T. T., "A Method of and an Apparatus or Machine for Producing Force or Motion," British Patent 300,311, 15 Nov. 1928.
[3] Brown, T. T., "Electrokinetic Apparatus," U.S. Patent 2,949,550, 16 Aug. 1960.
[4] Brown, T. T., "Electrokinetic Transducer," U.S. Patent 3,018,394, 23 Jan. 1962.
[5] Brown, T. T., "Electrokinetic Generator," U.S. Patent 3,022,430, 20 Feb. 1962.
[6] Brown, T. T., "Electrokinetic Apparatus," U.S. Patent 3,187,206, 1 June 1965.
[7] Brown, T. T., "Appareils et Procédés Électrocinétiques," (English translation "Electrokinetic Apparatuses and Processes"), French Patent 1,207,519, 21 Aug. 1961.
[8] Talley, R. L., "21st Century Propulsion Concept," Air Force Astronautics Lab. Rept. AFAL-TR-88-031, AD-A197 537, Edwards Air Force Base, CA, April 1988.
[9] Cravens, D. L., "Electric Propulsion Study," Air Force Astronautics Lab. Rept. AL-TR-89-040, AD-A227 121, Edwards AFB, CA, Aug. 1990.
[10] URL: http://jnaudin.free.fr/lifters/main.htm [cited 2 March 2007].
[11] Bahder, T. B., and Fazi, C., "Force on an Asymmetric Capacitor," Army Research Lab. Rept., ARL-TR-3005, Adelphi, MD, June 2003.
[12] Canning, F. X., Melcher, C., and Winet, E., "Asymmetrical Capacitors for Propulsion," NASA Glenn Research Center, NASA/CR-2004-213312, Oct. 2004.
[13] Cady, W. M., "Thomas Townsend Brown: Electro-Gravity Device," Office of Naval Research File 24-185, Pasadena, CA, Sept. 1952.
[14] Buehler, D. R., "Exploratory Research on the Phenomenon of the Movement of High Voltage Capacitors," *Journal of Space Mixing*, Vol. 2, 2004, pp. 1–22.
[15] Stephenson, G. V., "The Biefeld Brown Effect and the Global Electric Circuit," *Space Technology and Applications International Forum—STAIF 2005*, El-Genk, M. S. (ed.), American Institute of Physics, 2005, pp. 1249–1255.
[16] Musha, T., "The Possibility of Strong Coupling Between Electricity and Gravitation," *Infinite Energy Magazine*, Issue 53, Jan.–Feb. 2004, pp. 61–64.
[17] Woodward, J. F., "Flux Capacitors and the Origin of Inertia," *Foundations of Physics*, Vol. 34, No. 10, 2004, pp. 1475–1514.
[18] Brito, H. H., "Experimental Status of Thrusting by Electromagnetic Inertia Manipulation," *Acta Astronautica*, Vol. 54, 2004, pp. 547–558.
[19] Randall, L., *Warped Passages: Unraveling the Mysteries of the Universe's Hidden Dimensions*, HarperCollins Publishers, New York, 2005.

[20] National Geophysical Data Center, National Oceanic and Atmospheric Administration, "Magnetic Field Calculator", URL: http://www.ngdc.noaa.gov/seg/geomag/jsp/Downstruts/calcIGRFWMMM [cited 10 Feb. 2007].
[21] Carreno, F., and Bernabeu, E., "On Wire-to-Plane Positive Corona Discharge," *Journal of Physics D*, Vol. 27, 1994, pp. 2135–2144.
[22] White, H. J., *Industrial Electrostatic Precipitation*, Addison-Wesley, Reading, MA, 1963, Ch. 4.
[23] Chen, J., and Davidson, J. H., "Electron Density and Energy Distributions in the Positive DC Corona: Interpretation for Corona-Enhanced Chemical Reactions," *Plasma Chemistry and Plasma Processing*, Vol. 22, No. 2, 2002, pp. 199–224.
[24] Chen, J., and Davidson, J. H., "Model of the Negative DC Corona Plasma: Comparison to the Positive DC Corona Plasma," *Plasma Chemistry and Plasma Processing*, Vol. 23, No. 1, 2003, pp. 83–102.
[25] White, H. J., *Industrial Electrostatic Precipitation*, Addison-Wesley, Reading, MA, 1963, Ch. 4.
[26] Christenson, E. A., and Moller, P. S., "Ion-Neutral Propulsion in Atmospheric Media," *AIAA Journal*, Vol. 5, No. 10, 1967, pp. 1768–1773.
[27] Abdel-Salam, M., and Shamloul, D., "Computation of Ion-flow Fields of AC Coronating Wires by Charge Simulation Techniques," *IEEE Transactions on Electrical Insulation*, Vol. 27, No. 2, 1992, pp. 352–361.
[28] Zhang, C. H., and MacAlpine, J. M. K., "A Phase-Related Investigation of AC Corona in Air," *IEEE Transactions on Dielectrics and Electrical Insulation*, Vol. 10, No. 2, 2003, pp. 312–319.
[29] MacAlpine, J. M. K., and Zhang, C. H., "The Effect of Humidity on the Charge/Phase-Angle Patterns of AC Corona Pulses in Air," *IEEE Transactions on Dielectrics and Electrical Insulation*, Vol. 10, No. 2, 2003, pp. 506–513.
[30] Melcher, J. R., *Continuum Electromechanics*, MIT Press, Cambridge, MA, 1981, Ch. 3.
[31] Yabe, A., Mori, Y., and Hijikata, K., "EHD Study of the Corona Wind Between Wire and Plate Electrodes," *AIAA Journal*, Vol. 16, No. 4, 1978, pp. 340–345.
[32] Tajmar, M., "Biefeld-Brown Effect: Misinterpretation of Corona Wind Phenomena," *AIAA Journal*, Vol. 42, No. 2, 2004, pp. 315–318.
[33] Kawasaki, M., and Adachi, T., "Mechanism and Preventive Method of Self-Excited Vibration of Corona Wire in an Electrostatic Precipitator," *Journal of Electrostatics*, Vol. 36, 1996, pp. 235–252.

Chapter 9

Experimental Findings of Asymmetrical Capacitor Thrusters for Various Gasses and Pressures

Francis X. Canning*
Simply Sparse® Technologies, Morgantown, West Virginia

I. Introduction

ASYMMETRICAL capacitor thrusters have been proposed as a source of propulsion. For over 80 years it has been known that a thrust results when a high voltage is placed across an asymmetrical capacitor, when that voltage causes a leakage current to flow. Chapter 8 provides experimental results using the classical "lifter shape" in air, while this chapter provides results for several geometries that are more "capacitor like" and that have greatly varying amounts of asymmetry. Measurements are made in air, nitrogen, and argon at atmospheric pressure and at various partial vacuums. The thrust these devices produce has been measured for various voltages, polarities, and ground configurations and their radiation in the very high frequency (VHF) range has been recorded. A number of possible explanations for the source of the thrust are considered. Several of these are different from those considered in the previous chapter. However, we also consider a model that assumes the thrust is due to electrostatic forces interacting with the leakage current flowing across the capacitor. It further assumes that this current involves charged ions which undergo multiple collisions with air. These collisions transfer momentum. All of the measured data were found to be consistent with this model.

The force produced by asymmetrical capacitor thrusters (ACT) was first observed in 1922. A graduate student, T. T. Brown, working under his advisor, Dr. Paul Biefeld, noticed a force on a device when a high voltage was applied, an effect sometimes called the Biefeld–Brown effect. T. T. Brown received a patent in Great Britain for the use of this effect in 1928 [1]. More recently, this effect has been used to produce devices commonly called "lifters." Lifters are generally

---

Copyright © 2008 by the American Institute of Aeronautics and Astronautics, Inc. All rights reserved.
*Independent consultant.

light-weight devices that have a high voltage supplied by attached wires. They have generated much interest in hobbyists as they lift off of the ground in a way that appears magical to the casual observer. One such device was patented in 1964 [2].

A common feature of these devices is that they apply a high voltage to an asymmetrical capacitor. Some of these devices are called asymmetrical capacitor thrusters. Not only are they asymmetrical, but they generally also have sharp edges and/or sharp corners. One normally does not think of a capacitor as consuming power in its charged state. However, the combination of sharp features and high voltage tends to produce a small leakage current causing power consumption. Potentials in the range of 50,000 to 100,000 volts are commonly used.

This report describes some recent experiments [3,4] that were designed to explain some of the confusing lore about how these devices function. For example, some previous reports suggest that they always created a force toward the side of the capacitor with the sharper physical features, as does Chapter 8. Other reports state that the direction of the force changes when the polarity of the excitation is changed. That conclusion is hard to reconcile with other reports that observe that these devices function with both a DC and an AC voltage applied. We control an additional factor in our experiments and as a result provide a reasonable explanation for all of these observations.

The specific designs that we tested were chosen because they both generated a relatively strong force and had features that would help in determining the mechanism that produced the thrust. There have been several reports on tests of such devices [3–6]. However, this chapter concentrates on reviewing the interpretation of the tests reported in Refs. 3 and 4, in which this author participated. New features of our experimental data are 1) the use of devices with both weak and strong asymmetry, 2) simultaneous control of both the polarity and ground location, and 3) simultaneous measurements of both the current into the capacitor and the current out of the capacitor.

A. Lifter Geometries

A typical lifter is made from materials such as aluminum foil and wire. The wire is near the aluminum foil (Fig. 1) and is on the upper side. The wire may

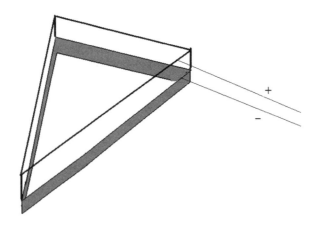

Fig. 1 Typical lifter.

be considered a sharper surface than the edge of the aluminum foil. The two wires are charged at different potentials. For illustration, the top wire is shown as plus, while a negative polarity is just as common.

Lifters seem to most often have only metallic surfaces, although dielectric material may be used. Even lifters that are several feet in size weigh very little, often less than an ounce. The previous chapter presented results for lifter geometries, while we will only present results for asymmetrical capacitor thruster (ACT) geometries.

B. Asymmetrical Capacitor Thruster Geometries

ACTs are similar in design to lifters, but tend to be more recognizable as a capacitor with a significant asymmetry. Nevertheless, one side is more discontinuous than the other. For example, one design uses a disk and a cylinder, where both bodies of revolution share the same axis of rotation.

For a disk and a (hollow) cylinder, the disk is considered to have the sharper features. While each has a surface with an edge, we believe the presence of the other side of the cylinder softens the discontinuity (more accurately, an abrupt change) in the resulting electric fields. A numerical calculation for a two-dimensional version of a disk and cylinder was performed in Ref. 3 to verify this (see Section III). It was found that the electric field strength at the edge of the disk when the cylinder was grounded was approximately twice the electric field strength at the cylinder when the disk was grounded. This numerical calculation for the two-dimensional case supports our arguments about sharper features for the three-dimensional case.

One interesting feature that was used on some designs consisted of adding short individual wires. These wires were obtained from a window screen. The wires that were parallel to an edge were removed so that the wires that remained all pointed in the same direction. These created a stronger "discontinuity" (Section III clarifies what this means). Thus, we were able to compare the performance of designs that did and did not have these sharper features.

Four devices were examined in detail. Device 1 is shown in Fig. 2. This is the capacitor shape that others have tested in the past. It is expected that the disk functions as having a sharper discontinuity than the cylinder. Others have discussed the possible effect of dielectric materials. Thus, we created device 2 which is shown in Fig. 3.

It is important to understand the effects of greater and of less asymmetry. Devices 1 and 2 may be considered to be less asymmetric than a standard

Fig. 2   Test device 1.

Fig. 3 Test device 2.

lifter design. This is because a wire (Fig. 1) is nearly a one-dimensional object while a disk or cylinder (Figs. 2 and 3) has surface area. Figure 3 one can see dielectric material on the left half and a hollow copper cylinder on the right half of the device.

For comparison we created devices 3 and 4, which have a much larger asymmetry than both devices 1 and 2 and standard lifters (Fig. 1). As compared to device 1, devices 3 and 4 make the disk end even sharper and the cylinder less sharp. As shown in Fig. 4, device 3 has very fine wires on the disk. These fine wires add to the discontinuity there. Also, the end of the cylinder nearest to the disk has an added rounded collar, making it a smoother surface.

The final device tested was device 4 that was similar to device 3, with one added feature. Wires pointing away from the disk were added to the rear end of the cylinder (the end most distant from the disk). When sharp features (such as these wires) are placed in a region with a varying electric potential, they generally cause that potential to vary rapidly near that sharp feature. This "near discontinuity" generally causes a large local electric field. (Because of its similarity to device 3, device 4 is not illustrated.)

II. Experimental Setup

Each of the four devices was tested both in a vacuum chamber and in a conducting box having the same dimensions as the vacuum chamber. Two polarities are possible; one has the disk positive and the cylinder negative and the other reversing that polarity. Fortunately, in building the apparatus, it was realized that the ground for the box might be kept at the same potential as either the disk or the cylinder, for either polarity.

Fig. 4 Device 3.

FINDINGS OF ASYMMETRICAL CAPACITOR THRUSTERS

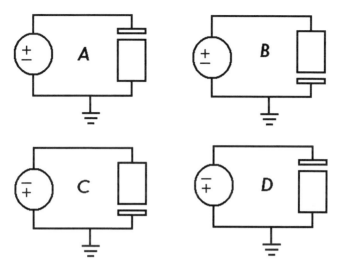

Fig. 5 Four wiring circuits.

The four possible combinations of polarity and ground are shown in Fig. 5 as circuits A through D. Observing all four cases proved very illuminating for understanding the physics of these devices. The resulting data provided an explanation why previous anecdotal information from a variety of sources appeared to be contradictory. That is, the location of the ground affected the results whereas other researchers typically did not specify whether the ground was floating or fixed, or where it was attached.

We also measured two different currents by measuring the current through each lead. We expected that these two currents could be significantly different, because there are three current paths. The first two travel from one side of the capacitor to the other, either through air or through the supporting structure. The third path goes through the "air" (or other gas or vacuum) to surfaces at some distance from the ACT.

III. Qualitative Experimental Results

The results of the experiments in Ref. 3 are summarized in Sections III.A and III.B. Section III.C gives interpretations of those results. These tests were performed using a direct (nonalternating) current.

A. Tests Performed in Air at Atmospheric Pressure

1) Devices 1 and 2 always produced a force on the ACT toward the nongrounded (charged) surface.

2) Devices 3 and 4 always produced a force on the ACT directed from the cylinder toward the disk.

3) Devices 1 and 2 produced a larger force when the disk was the nongrounded surface.

4) The polarity ($+$ Vs $-$) had only a small effect on the magnitude of the force produced (for ground constant).

5) When the cylinder was grounded, devices 3 and 4 produced a larger force than devices 1 and 2.

6) When the cylinder was grounded, device 4 produced more thrust than device 3.

7) The current to the live (nongrounded) side was always larger than the current from the grounded side to ground.

8) When the box containing the apparatus was opened, the hair on one's arm stood on end.

Notice that when we say the polarity had only a small effect on the magnitude of the force produced, we are assuming that when the polarity is changed, the position of the ground does not change. That is, if one changed the polarity in circuit A, one would then have circuit D. Similarly, if one changed the polarity in circuit B, one would get circuit C. However, previous reports generally do not mention the ground and we do not even know if the ground was fixed or floating. Thus, if they change the polarity of circuit A, we do not know if their configuration was circuit B or D, or something else. Thus, it is not surprising that they reported varying results. For example, we observed that for the less asymmetric devices (devices 1 and 2), changing both polarity and the ground (e.g., going from circuit A to B) changed the direction of the force.

B. Tests Performed in a Partial Vacuum in Air and in other Gasses

1) At pressures such as 300 torr in air, the results were similar to atmospheric pressure but forces were weaker.

2) Also, similar results as for air were found in argon and nitrogen, but with somewhat smaller forces.

3) In air, the current flowed in bursts, and VHF radiation was observed.

4) In argon and nitrogen, the current did not flow in bursts and the VHF radiation was absent.

5) These bursts in the current were observed on an oscilloscope.

6) In a significant vacuum, with one exception, no force was observed although experimental sensitivity was low. The exception was a momentary force that occurred at below $1/10,000$ of a torr when a significant spark (arcing) was observed. This occurred the first time a voltage was applied after the vacuum chamber had been closed and the pressure reduced.

C. Interpretations of Results

For the tests performed in air, the air became significantly ionized. This allowed a second current path from the live side to ground through air bypassing the grounded side of the ACT. It is possible that the wires on the cylinder of device 4 reduced the ionization as compared to device 3, and resulted in the larger force observed. However, that is not clear because tests were not done that controlled the ionization, such as by flushing the air.

FINDINGS OF ASYMMETRICAL CAPACITOR THRUSTERS   335

The two devices that were only somewhat asymmetrical (devices 1 and 2) produced a force in a direction that was determined by the location of the ground. This is due to the live side having a large voltage gradient at its surface because the voltage changes abruptly to the ambient value (approximately ground). A large voltage gradient gives a large electric field, which causes ionization. The ions have the same charge (whether plus or minus) as the nearby ACT surface, so they are repelled by it. This produces a force on the ACT which in this case is directed from the grounded side toward the live side.

The two devices that were highly asymmetric devices (devices 3 and 4) always produced a force in a direction that was determined by the asymmetry. That is, the force was always directed from the cylinder toward the disk, regardless of the polarity or the ground location. A reasonable explanation for this is possible because the air was clearly ionized in the box containing the ACT when it was in use. The ionization was displayed in two ways. First, the current into the live side of the ACT was larger than the current from the grounded side as measured through the grounding wire. This significant current difference showed that current was flowing through the air, indicating that it must be ionized. Second, the air was found to make arm hairs stand up straight, again showing the air was charged. This suggests that near the ACT, the ambient voltage could be significantly different from ground due to the charges in the air. Thus, there could be a significant voltage gradient between the grounded side of the ACT and the air around it.

The voltage gradient near the grounded side is expected to be significant, even though it likely is smaller than the voltage gradient near the surface of the live (non-ground) side of the ACT. Thus, for a large enough asymmetry, the asymmetry would be the determining factor in the net force on the ACT. That is, charged particles may be created on both sides of the ACT, and they would be repelled from their respective nearby side of the ACT. The charged particles near each side produce forces in opposite directions. Thus, it is quite reasonable that for a strong enough asymmetry, the net force is always directed toward the sharper side (e.g., the one with the wires).

The current was found to flow continuously when an ACT was operated in argon and in nitrogen, but to flow in bursts when operated in air. This was measured both by a time resolved measurement of each current and indirectly by observing the VHF radiation that resulted from the bursts of current when the ACT was used in air. These bursts are a known phenomenon called "Trichel pulses." Because a force was produced in argon and in nitrogen where these bursts do not occur, it appears that the mechanism of the force is not inherently linked to these Trichel pulses.

The only time a noticeable force was created in a strong vacuum occurred the first time a voltage was applied after the chamber had been closed. A significant arcing was associated with this momentary force. Thus, it is possible that some material was removed from one part of the ACT when that arcing occurred. It is reasonable that moisture due to humid air may have deposited on the ACT while the vacuum chamber was open. Thus, this event may have been due to some material (moisture or otherwise) being ejected from the ACT.

IV. Numerical Calculations of Electric Fields

It is well known that high voltages combined with sharp surfaces cause high electric fields. Those experienced in working with high voltages are very familiar with this. The theoretical reason is very simple. An electric field is the rate of change of the voltage as one measures the voltage at different spatial locations, and the voltage changes rapidly near discontinuities in conducting objects. Another way to see this is to consider a piece of wire charged to a potential of one volt. The charges on the wire tend to repel each other, and are concentrated (denser) at the ends of the wire. Thus, as one moves from an end of the wire off the wire, the voltage changes rapidly (a large electric field) because of the large number of nearby charges. If one moves away from the center of the wire, the voltage changes slowly (a smaller electric field). A numerical simulation of this was given for a lifter type of geometry in Ref. 3. A simulation was also given for an ACT type of geometry, and we describe its results below.

The numerical results presented in Fig. 6 show that in parts a and b, the largest electric field occurs on the charged (not grounded) plate. In part c, where the plates are charged to equal strength (but different signs), the strongest field occurs on the upper (more discontinuous) plate. This supports our arguments (Section III.C) that the strength of the difference in the "discontinuity" versus the strengths of the difference of the voltage on a plate from the ambient voltage determines which electric field will be stronger, and thus the direction of the net force produced.

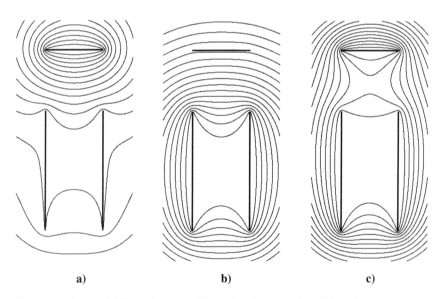

Fig. 6 Equipotential lines for a two-dimensional conducting object. In a), the upper plate is at one volt and the lower plates are at ground. In b), the upper plate is at ground and the lower plates are at one volt. In c), the upper plate is at plus one volt and the lower plates are at minus one volt.

V. Theories Versus Quantitative Experimental Results

It has been shown that all of the qualitative observations can be explained by a model using a flow of ions. It remains to consider other theories and see if they are plausible, and whether an ion flow model can predict the magnitude of the force created. Several possible theories for the mechanism that creates the force (or thrust) are now considered. Except for Section V.C, these are different from the mechanisms considered in Chapter 8.

A. Ablative Material

Due to the high voltages and high electric fields that are present, material might be removed from the disk or cylinder during continuous operation. The possibility that continuously ejected material could provide the force was examined in Ref. 3. However, because these devices were operated for a large number of hours with no visible physical change, an upper limit on the amount of mass that might have been removed was easy to calculate. Using ejection velocities due to thermal effects it was found that any force created by this would be significantly less than that observed, so this effect is not a significant possible mechanism.

B. Electrostatic Forces Involving Image Charges

This concept is simpler to analyze for a lifter than for our test geometry. Consider the possibility that a lifter's charges might interact with image charges due to a ground plane. For example, a concrete floor might have metal reinforcement that produces a current that can be described by image charges. One might assume a perfectly conducting plane under the lifter. (A perfect conductor would produce the strongest forces. A more realistic surface would produce significantly weaker forces.) The lifter creates approximately a charged dipole, which interacts with its image dipole. A simple calculation shows the resulting force is many orders of magnitude too weak. Thus, this cannot be the mechanism that produces the force. Adding dielectric material (as is sometimes done) would increase the force significantly. However, there is always an attractive force between a charge and its image, so this effect would pull the lifter down rather than make it rise. Thus, with or without dielectric material, such an electrostatic force would be in the wrong direction to explain why a lifter lifts.

C. Ion Drift Causing Momentum Transfer to Air

A high electric field near a charged plate will cause charged particles to be ejected. They will be charged with the same sign as that nearby plate (i.e., both plus or both minus), and will be repelled by it. If charged particles at one side of the ACT are accelerated, move to the other side, and decelerate to move with the ACT again, no net force is produced. The net change in momentum of such a particle is zero. For a system consisting of these particles and the ACT, the average force on these particles would be zero so the average force on the ACT would also be zero. The only way a continuous force could be created on an ACT due to these particles would be if these particles transferred momentum to something else. Of course, particles moving through air will have collisions and thus transfer momentum to that air. This is analogous to a propeller moving through air and transferring momentum.

Charged particles would have a large number of collisions traveling from one side of an ACT to the other, and as a result would always be moving much slower than the thermal velocity at sea level and room temperature. Thus, their collision rate would be approximately unchanged due to their motion. With this assumption, it is simple to compute the force produced for a given voltage, distance across the ACT, and current. Further, it may be assumed that all of the charged particles flow in one direction. With these assumptions, the force that would be expected on an ACT was computed in Ref. 3. All of the forces measured in Ref. 3 were found to be smaller than the result of a computed force using an equation that we reproduce below [Eq. (3)]. However, for device 4 with the cylinder grounded and the disk positively charged the result was close at 77% of the computed value. This is in good agreement as Eq. (3) gives an upper limit to the possible force. If not all charges flow in the same direction, there will be a partial cancellation reducing the net force to a value smaller than that given by Eq. (3).

The energy efficiencies, forces produced, and other properties of an ACT or a lifter may be understood from this ion drift model. The calculation of the force produced seems straightforward. One first finds how long it takes a charge to move across the gap between the charged surfaces. The current, multiplied by this time, gives the charge in the gap. The applied voltage divided by the size of the gap gives an average electric field. The electric field times the charge gives a force. This seems simple, but there is one pitfall. As a charged particle moves across the gap, it has multiple collisions with air (or whatever gas is present). With these collisions, it takes orders of magnitude longer for a charge to cross the gap than it would in a vacuum. If this effect is missed, then one would calculate a force that is orders of magnitude too small.

The simple model used in Ref. 3 assumed that in air at standard conditions, there are 10^{10} collisions per second. It was assumed that charged particles also had this number of collisions with air, and that on average each collision caused them to lose all of their momentum across the gap. With these assumptions, for an ion of mass m, the distance traveled between collisions would be

$$d_0 = (1/2) \, a \, t_0^2 = (1/2) \, a \, [10^{-10} \text{ sec}]^2 \text{ where } a = F/m = [eV/d]/m \quad (1)$$

In this equation, e is the charge on the ion, V is the voltage applied, and a is the acceleration. The total time, t, to travel a distance, d, across the gap and the resulting force are:

$$t = 10^{-10} \text{ sec} \, (d/d_0) = 2d \, (10^{10}/\text{sec})/a = 2d^2(10^{10}/\text{sec}) \, m/(eV) \quad (2)$$

$$F = I \cdot t \cdot V/d = 2d \, (10^{10}/\text{sec}) \, mI/e \quad (3)$$

When a 100 kilovolts produces a current of less than one milliamp, the computed force is significantly smaller than a Newton.

This model is very approximate, and the resulting force is often a few times smaller than it predicts. Equation (3) shows that if the current is fixed, then the force produced is proportional to the size of the gap, which is the distance the charges must flow. We observed that device 3 produced significantly less force than device 4. There is a likely explanation for this. The air may have been less ionized for device 4 and therefore the ions were nearly (but not completely) all produced only on the disk side. For our highest-thrust device, device 4 with circuit A, the measured force was 77% of the force computed from Eq. (3). This strongly suggests that this ion drift model explains the origin of the force.

VI. Vacuum Results

Measurements were also made in a vacuum chamber containing dry air, argon or nitrogen at various pressures. All of those results suggested that the force produced by an ACT goes to zero as the pressure is reduced. Some measurements in dry air are shown in Fig. 7.

Figure 7 shows that the forces produced by each ACT decrease as the pressure was decreased. For each device/circuit combination shown, the same voltage was used at each pressure. The devices were attached to an axle that rotated, and the torque they produced is shown on the vertical axis. This has the advantage that they are moving through fresh (less ionized) air and the disadvantage that the force measurements are less accurate than for stationary devices. In all of our measurements, no force was measured for very low pressures.

As the pressure is reduced, the optimal design parameters change. That is, one might (possibly) design an ACT that is optimized for maximum thrust at a given pressure. Regardless, however, the trend shown in Fig. 7 would still apply.

VII. Conclusions

Data from some recent tests performed on ACTs were reviewed and a number of mechanisms were considered for how their thrust is produced.

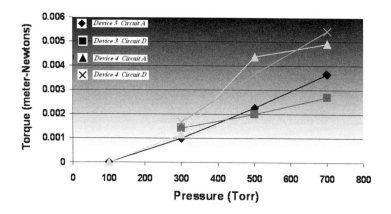

Fig. 7 Operation in dry air at reduced pressures.

These mechanisms were considered theoretically and in light of test results. Only one mechanism seems plausible, and it relies on standard elementary physics. This model consists of ions drifting from one electrode to the other under electrostatic forces. They collide with air as they move, slowing them down and increasing the time that each contributes to the force. Each collision transfers momentum to the surrounding air, much as a propeller does. This model was found to be consistent with all of the observations that were made. This included how, for certain designs, the direction of the force changed depending on which side of the ACT was grounded. It also predicted how, for other designs, the direction of the force did not change depending on which side of the ACT was grounded. Furthermore, it approximately predicted the magnitude of the force (thrust) that was measured. This model also predicted that the direction of the thrust was independent of the polarity of the applied voltage. Furthermore, it predicted which designs were most efficient at producing a force. The force we measured always decreased to zero as the pressure was reduced. In spite of previous speculation about possible new physical principles being responsible for the thrust produced by ACTs and lifters, we find no evidence to support such a conclusion. On the contrary, a multitude of details about their operation is fully explained by a very simple theory that uses only electrostatic forces and the transfer of momentum by multiple collisions.

Acknowledgments

The author thanks Jonathan Campbell at NASA Marshall and Marc G. Millis at NASA Glenn for introducing him to this subject. Also, he thanks his coauthors of Refs. 3 and 4 for performing much of the work that is reviewed here.

References

[1] Brown, T. T., "A Method of and an Apparatus or Machine for Producing Force or Motion," British Patent 300,311, Nov. 1928.
[2] deSeveresky, A. P., "Ioncraft," U.S. Patent 3,130,945, 1964.
[3] Canning, F. X., Melcher, C., and Winet, E., "Asymmetrical Capacitors for Propulsion," NASA Glenn Research Center, NASA TM CR-2004-213312, Oct. 2004.
[4] Canning, F. X., Campbell, J., Melcher, C., Winet, E., and Knudsen, S. R., "Asymmetrical Capacitors for Propulsion," *Proceedings of the 53rd JANNAF Propulsion Conference*, Monterey, CA, Dec. 2005.
[5] Talley, R. L., "Twenty-First Century Propulsion Concept," NASA TM CR-2004-213312, 2004.
[6] Bahder, T. B., and Fazi, C., "Force on an Asymmetrical Capacitor—Final Report, Aug.–Dec. 2002," ARL-TR-3005, NTIS Order Number ADA416740, 2002.

Chapter 10

Propulsive Implications of Photon Momentum in Media

Michael R. LaPointe*
NASA Marshall Space Flight Center, Huntsville, Alabama

I. Introduction

IN 1908, Hermann Minkowski published derivations of the electromagnetic momentum density and Maxwell stress tensor, the latter having an asymmetric form in macroscopic media [1]. Dissatisfied with this inherent asymmetry, Max Abraham soon thereafter derived a symmetric form for the stress tensor in a material medium [2]. From these differing derivations was born the Abraham–Minkowski controversy, the resolution of which continues to be argued nearly a century after their publications. Put succinctly, Abraham's formulation predicts that the momentum density of an electromagnetic field (photon) will decrease when traversing media with an index of refraction greater than unity; Minkowski's derivation predicts that the photon momentum density will increase in such media. The results from a limited number of experiments have been used to argue in support of both derivations. The potential implications for breakthrough propulsion arise from the asymmetric form of the stress tensor, which suggests that net linear momentum may be imparted to a material medium via photon drag effects or electromagnetic interactions with the vacuum. A sustained motion of this self-contained system would allow propulsion without the expenditure of propellant or the application of external forces. Concepts proposing to take advantage of this effect have appeared in the literature, and experimental efforts have been undertaken to determine whether such ideas may be feasible. This chapter reviews the origins and evolution of the Abraham–Minkowski controversy, and discusses the theories, interpretations, and experimental efforts designed to resolve the controversy. Propulsion concepts and related experiments

---

This material is a work of the U.S. Government and is not subject to copyright protection in the United States.
*Project Manager, Science Research and Technologies Project Office.

seeking to exploit net material momentum are described, and potential theoretical and experimental issues are identified. The chapter concludes with a summary of recent developments and prospects for this class of potential breakthrough propulsion devices.

II. Background

Maxwell's derivation of the basic equations used to describe the propagation of electromagnetic fields in vacuum was a crowning achievement of 19th-century physics. This theoretical framework, built upon the experimental efforts of Faraday, Ampere, Henry, and others, has stood the test of time unchanged and serves as the underpinning of all large-scale electromagnetic devices in operation today. In their differential form, Maxwell's equations are written:

$$\nabla \cdot \mathbf{B} = 0 \quad \nabla \times \mathbf{E} = -\frac{\partial B}{\partial t}$$
$$\nabla \cdot \mathbf{E} = \frac{\rho}{\varepsilon_0} \quad \nabla \times \mathbf{B} = \mu_0 \mathbf{J} + \frac{1}{c^2}\frac{\partial \mathbf{E}}{\partial t} \tag{1}$$

where $\mathbf{B}$ is the magnetic field vector, $\mathbf{E}$ is the electric field vector, ε_0 is the permittivity of free space, μ_0 is the permeability of free space, and c is the speed of light in vacuum, equal to $(\mu_0 \varepsilon_0)^{-1/2}$. The terms ρ and $\mathbf{J}$ represent the charge and current density source terms, respectively, which give rise to the free space electromagnetic fields.

Following standard practice [3,4], this set of equations is often recast in terms of the potential functions V and $\mathbf{A}$,

$$\mathbf{B} = \nabla \times \mathbf{A} \quad \mathbf{E} = -\nabla V - \frac{\partial \mathbf{A}}{\partial t} \tag{2}$$

which satisfy the Lorentz condition,

$$\nabla \cdot \mathbf{A} + \frac{1}{c^2}\frac{\partial V}{\partial t} = 0 \tag{3}$$

Defining the electromagnetic field tensor, F, with components

$$F = \begin{pmatrix} 0 & B_z & -B_y & -\frac{iE_x}{c} \\ -B_z & 0 & B_x & -\frac{iE_y}{c} \\ B_y & -B_x & 0 & \frac{-iE_z}{c} \\ \frac{iE_x}{c} & \frac{iE_y}{c} & \frac{iE_z}{c} & 0 \end{pmatrix} \tag{4}$$

allows Maxwell's equations to be written in a more compact form:

$$\nabla \cdot \mathbf{E} = \frac{\rho}{\varepsilon_0}$$
$$\nabla \times \mathbf{B} = \mu_0 \mathbf{J} + \frac{1}{c^2}\frac{\partial \mathbf{E}}{\partial t} \quad \Rightarrow \quad \partial_\mu F_{\mu\nu} = \mu_0 J_\nu$$

(5)

$$\nabla \cdot \mathbf{B} = 0$$
$$\nabla \times \mathbf{E} = -\frac{\partial \mathbf{B}}{\partial t} \quad \Rightarrow \quad \partial_\mu F_{\nu\alpha} + \partial_\nu F_{\sigma\mu} + \partial_\sigma F_{\mu\nu} = 0$$

where the current and charge densities are related through the continuity equation:

$$\nabla \cdot \mathbf{J} + \frac{\partial \rho}{\partial t} = 0 \quad \Rightarrow \quad \partial_\mu J_\mu = 0 \qquad (6)$$

Equation (2) expressing the potentials V and $\mathbf{A}$ can then be written as:

$$\left(\nabla^2 - \frac{1}{c^2}\frac{\partial^2}{\partial t^2}\right)\mathbf{A} = -\mu_0 \mathbf{J}$$
$$\left(\nabla^2 - \frac{1}{c^2}\frac{\partial^2}{\partial t^2}\right)V = -\frac{\rho}{\varepsilon_0} \quad \Rightarrow \quad \left(\nabla^2 - \frac{1}{c^2}\frac{\partial^2}{\partial t^2}\right)A_\mu = -\mu_0 J_\mu \qquad (7)$$

with the Lorentz condition, Eq. (3), taking the form $\partial_\mu A_\mu = 0$.

Of particular interest to this chapter, electromagnetic fields carry both energy and momentum. Using the simplified notation above, we can define the quantity, $F_{\mu\nu} J_\nu$, whose three spatial components relate to the force density, $(\rho \mathbf{E} + \mathbf{J} \times \mathbf{B})$, and whose fourth component describes the power input per unit volume, $(\frac{i}{c} \mathbf{E} \cdot \mathbf{J})$, to the particles that make up the current and charge density source terms. This quantity can be written in terms of the divergence of the symmetric, second-rank electromagnetic stress–energy–momentum tensor, $T_{\mu\alpha}$,

$$F_{\mu\nu} J_\nu = \partial_\alpha T_{\mu\alpha} \qquad (8)$$

where

$$T_{\mu\alpha} = \frac{1}{\mu_0}\left[F_{\mu\nu} F_{\nu\alpha} + \tfrac{1}{4}\delta_{\mu\alpha} F_{\sigma\tau} F_{\sigma\tau}\right] \qquad (9)$$

and $\delta_{\mu\alpha}$ is the Kronecker delta function ($\delta_{\mu\alpha} = 1$ for $\mu = \alpha$, zero otherwise).

The elements of $T_{\mu\alpha}$ are given by:

$$T = \begin{pmatrix} T_{11} & T_{12} & T_{13} & -icg_x \\ T_{21} & T_{22} & T_{23} & -icg_y \\ T_{31} & T_{32} & T_{33} & -icg_z \\ -\frac{i}{c}S_x & -\frac{i}{c}S_y & -\frac{i}{c}S_z & u_{EM} \end{pmatrix} \quad (10)$$

where

$$T_{ij} = \varepsilon_0 E_i E_j + \frac{1}{\mu_0} B_i B_j - \frac{1}{2}\delta_{ij}\left(\varepsilon_0 E^2 + \frac{1}{\mu_0}B^2\right) \quad (i,j \le 3) \quad (11)$$

are the three-dimensional elements of the electromagnetic Maxwell stress tensor;

$$\mathbf{S} = \frac{1}{\mu_0}\mathbf{E} \times \mathbf{B} \quad (12)$$

is the Poynting vector, which describes the time rate of energy flow per unit area;

$$\mathbf{g} = \varepsilon_0 \mathbf{E} \times \mathbf{B} = \frac{1}{c^2}\mathbf{S} \quad (13)$$

is the momentum density carried by the electromagnetic field; and

$$u_{EM} = \frac{1}{2}\varepsilon_0 E^2 + \frac{1}{2\mu_0}B^2 \quad (14)$$

is the electromagnetic energy density, the total electromagnetic field energy per unit volume. The conservation laws of energy and momentum are contained within Eq. (8). For example, setting $\mu = 4$ yields

$$\begin{aligned} F_{4\nu}J_\nu &= \partial_\alpha T_{4\alpha} \\ \frac{i}{c}\mathbf{E}\cdot\mathbf{J} &= \partial_1 T_{41} + \partial_2 T_{42} + \partial_3 T_{43} + \partial_4 T_{44} \\ &= \nabla\cdot\left(-\frac{i}{c}\mathbf{S}\right) + \frac{1}{ic}\frac{\partial u_{EM}}{\partial t} \\ \mathbf{E}\cdot\mathbf{J} &= -\nabla\cdot\mathbf{S} - \frac{\partial u_{EM}}{\partial t} \end{aligned} \quad (15)$$

which is a statement for the conservation of energy. Similarly, summing Eq. (8) over spatial indices 1 to 3 yields an expression for the conservation of momentum for the electromagnetic field:

$$\frac{\partial g_j}{\partial t} - \sum_{i=1}^{3} \nabla_i T_{ij} = -f_j^L \tag{16}$$

where $\mathbf{f}^L$ represents the Lorentz force density, $\mathbf{f}^L = \rho\mathbf{E} + \mathbf{J} \times \mathbf{B}$, acting on the charge distribution.

Maxwell's equations have been amply verified for electromagnetic wave propagation in free space, and are commonly accepted pillars of contemporary physics. In the early part of the 20th century, two well-known figures in theoretical physics, Hermann Minkowksi and Max Abraham, separately sought to extend Maxwell's equations to encompass electromagnetic wave propagation in regions where matter is present. Their results, and the controversies that have followed, are described in the following sections.

III. Electromagnetic Fields in Dielectric Media

Within a dielectric material, Maxwell's equations take the form:

$$\begin{array}{ll} \nabla \cdot \mathbf{B} = 0 & \nabla \times \mathbf{E} = -\dfrac{\partial \mathbf{B}}{\partial t} \\ \nabla \cdot \mathbf{D} = \rho & \nabla \times \mathbf{H} = \mathbf{J} + \dfrac{\partial \mathbf{D}}{\partial t} \end{array} \tag{17}$$

where $\mathbf{D}$ is the displacement field, $\mathbf{H}$ is the magnetic field intensity, and the remaining symbols retain their usual meanings. For linear dielectrics, the constitutive relations between electric and magnetic fields in vacuum and within the dielectric material are given by:

$$\mathbf{D} = \varepsilon\mathbf{E} \quad \mathbf{B} = \mu\mathbf{H} \tag{18}$$

Here ε is the static permittivity of the dielectric, equal to $(\varepsilon_0 + \chi_0)$ where χ_0 is the static dielectric susceptibility, and μ is the static magnetic permeability, equal to $\mu_0(1 + \chi_m)$, where χ_m is the static magnetic susceptibility. The static dielectric and magnetic susceptibilities may depend on the electric and magnetic fields, respectively. The continuity equation takes the modified form:

$$\nabla \cdot \mathbf{J} + \frac{\partial}{\partial t}(\nabla \cdot \mathbf{D}) = 0 \tag{19}$$

Although Maxwell's equations in matter are generally accepted in the above form, the corresponding value of the electromagnetic stress–energy–momentum tensor has been a continuing source of contention. While there have been many competing derivations over the intervening decades [5], the two primary protagonists at the heart of the debate are Minkowski and Abraham, whose apparently competing formulations are briefly described as follows.

A. Minkowski's Formulation

In 1908, Minkowski published his derivation of the stress–energy–momentum tensor for electromagnetic wave propagation in linearly responsive but spatially stationary matter [1]. In Minkowski's formulation, the spatial components of the electromagnetic stress tensor become:

$$T_{ij}^M = E_i D_j + B_i H_j - \frac{1}{2}\delta_{ij}(\mathbf{E}\cdot\mathbf{D} + \mathbf{H}\cdot\mathbf{B}) \quad (i,j \leq 3) \tag{20}$$

where the superscript M denotes the Minkowski form of the stress tensor. The electromagnetic momentum density chosen by Minkowski is:

$$\mathbf{g}^M = \mathbf{D} \times \mathbf{B} \tag{21}$$

and the conservation of linear momentum for the electromagnetic field takes the form:

$$\frac{\partial g_j^M}{\partial t} - \sum_{i=1}^{3} \nabla_i T_{ij}^M = 0 \tag{22}$$

Using the Minkowski form of the momentum density, it can be shown that the momentum flux density for an electromagnetic wave in a stationary material medium increases proportionally to the index of refraction, n, of the material. Assuming for illustration a plane electromagnetic wave traveling in the z-direction, and ignoring losses and dispersion within the material, Maxwell's equations provide the simple solution, $\mathbf{B} = n(\mathbf{E}/c)$, where n is the index of refraction of the material, $n = \sqrt{\mu\varepsilon}c$. Using Eqs. (14), (18), and (21), and substituting the solution for $\mathbf{B}$, yields:

$$g_z^M = \varepsilon EB = \frac{\varepsilon n E^2}{c} = n\frac{\mu_{EM}}{c} \tag{23}$$

Compared with an electromagnetic wave propagating in vacuum, the Minkowski momentum density of an electromagnetic wave propagating within a material medium increases by a factor proportional to the index of refraction of the material.

B. Abraham's Formulation

Dissatisfied with the asymmetric form of Minkowski's energy–momentum tensor, Abraham developed, and in 1910 published, a symmetric form of the electromagnetic stress tensor in spatially stationary matter [2]:

$$T_{ij}^A = \frac{1}{2}(E_i D_j + E_j D_i + B_i H_j + B_j H_i) - \frac{1}{2}\delta_{ij}(\mathbf{E}\cdot\mathbf{D} + \mathbf{H}\cdot\mathbf{B}) \quad (i,j \leq 3) \tag{24}$$

with the superscript A denoting Abraham's form for the stress tensor. The electromagnetic momentum density in this formulation is given by:

$$\mathbf{g}^A = \frac{1}{c^2}(\mathbf{E} \times \mathbf{H}) \qquad (25)$$

The conservation of linear momentum then takes the form:

$$\frac{\partial g_j^A}{\partial t} - \sum_{i=1}^{3} \nabla_i T_{ij}^a = -f_j^A \qquad (26)$$

with the so-called Abraham force density, $\mathbf{f}^A$, given by:

$$\mathbf{f}^A = \varepsilon_0 \chi \frac{\partial}{\partial t}(\mathbf{E} \times \mathbf{B}) \qquad (27)$$

Using the Abraham formulation for momentum density, and again assuming a simple plane wave solution to Maxwell's equations, yields:

$$g_z^A = \frac{EH}{c^2} = \frac{EB}{c^2\mu} = \frac{nE^2}{(c^2\mu)c} = \frac{nE^2\varepsilon}{n^2 c} = \frac{1}{n}\frac{\mu_{EM}}{c} \qquad (28)$$

In Abraham's formulation, the momentum flux density for an electromagnetic wave traversing a stationary medium is inversely proportional to the index of refraction of the material medium, and hence *decreases* from its value in free space. This is clearly at odds with the results derived using Minkowksi's formulation, corresponding to a factor of n^2 difference in momentum flux density between the two approaches. These apparently disparate conclusions by two pre-eminent physicists ignited a smoldering debate, the resolution of which remains in contention nearly a century later. Why these results are important, and in particular why they are important for those seeking to develop breakthrough propulsion capabilities, are discussed in the following section.

IV. A Century of Controversy: Theory, Experiment, and Attempts at Resolution

The momentum densities formulated by Minkowski and Abraham agree for electromagnetic waves propagating through free space, where the index of refraction is unity. The dispute arises for electromagnetic waves passing through material media. In Minkowski's derivation, the electromagnetic wave (photon) momentum increases by a factor of n compared to its vacuum value, while in the Abraham derivation the photon momentum decreases by same factor. In each case, a corresponding change in the momentum of the material balances the change in photon momentum, such that total momentum is conserved. Theoretical and experimental attempts to resolve this apparent inconsistency between the two approaches have yielded considerable insight into the nature of the electromagnetic stress energy momentum tensor, together with a better if not yet complete understanding of the nature of photon propagation in dielectric media. The following paragraphs highlight, in roughly chronological order, some of the analytic and experimental results published over the past century. For a more complete history, the reader is referred to Penfield and Haus [5], Brevik [6], Bowyer [7], and the references therein.

A. Early Years: 1930 to 1960

1. Theoretical Developments

Early discussions on the correct form of the energy–momentum tensor in matter typically centered on the conservation of energy and of linear and angular momentum. In 1935, Halpern demonstrated that electromagnetic waves described by a symmetric energy–momentum tensor will (as required) propagate with the velocity of light, provided the four-dimensional divergence and the diagonal sum of the energy–momentum tensor vanish [8]. Applying this theorem, he concluded that angular momentum is not conserved within the Minkowski formulation, and the asymmetric tensor form is therefore unsatisfactory. Angular momentum can be conserved in Abraham's symmetric tensor derivation, but requires that wave propagation within the medium will exert a force on the medium. Halpern argues that conservation of angular momentum takes precedence, and that Abraham's form for the electromagnetic energy–momentum tensor is preferred.

This generally accepted interpretation was reinforced in a 1953 paper by Balasz [9], who provided a simple thought experiment in support of this stance. The *gedanken* consists of two enclosures in uniform motion with no external forces acting on them. In one enclosure an electromagnetic wave packet passes through a perfect, nondispersive dielectric rod in which the wave velocity is lower than that in vacuum. An identical dielectric rod and wave packet are in the other enclosure, but in this case the wave does not pass through the rod. Started simultaneously, the electromagnetic wave traveling through vacuum will be at a different point than an otherwise identical wave passing more slowly through the dielectric medium; this means that the equivalent mass associated with the wave energy will also be at a different location with respect to time. Because no external forces act on the enclosures, the dielectric rod in which the slower wave packet is traveling must move to keep the positions of the centers of mass of each enclosure displaced by the same amount. This in turn implies that the dielectric rod received momentum from the wave packet, and because the total system momentum is conserved, the momentum of the electromagnetic wave inside the dielectric must be different than the momentum of an otherwise identical electromagnetic wave traveling through vacuum. Balasz then asks which of the two tensor formulations can simultaneously satisfy the conservation of momentum and center of mass requirements, and after an extended bit of algebra shows that only the symmetric form of the energy–momentum tensor satisfies both conditions.

2. Jones and Richards Experiment

In 1954, the results of several carefully designed experiments with sufficient accuracy to address the Abraham–Minkowski controversy were reported by Jones and Richards [10], who immersed various metallic reflectors in air and in dielectric media and measured the radiation pressure exerted on each. The dielectric media ranged from water (index of refraction, $n = 1.33$) to carbon disulphide ($n = 1.61$); a 30-W tungsten lamp was used as the radiation source, with short time constants employed to minimize heating and other spurious forces. Their analysis of the experimental results demonstrated that the ratio of radiation pressure on the reflector in a dielectric medium to that in air was proportional to the refractive index of the dielectric material—an experimental confirmation that the *Minkowski* form of the electromagnetic stress–energy–momentum tensor was correct. However, despite its flagrant challenge to the prevailing theories supporting Abraham's tensor formulation, the work was not well known and received little attention at the time.

B. Intermezzo: 1960 to 1990

1. Simplest Cases Theory

In 1968, Shockley [11] employed a "simplest case" technique to once more argue in favor of Abraham's electromagnetic energy–momentum tensor. Employing center of mass considerations, Shockley's thought experiment consisted of a closed coaxial system, with an inner conducting rod and outer conducting cylinder electrically connected through a resistor. Power is provided by a battery. Shockley concludes, upon analysis of equivalent mass flows and force densities within the system, that the volume integrated form of Abraham's momentum density yields the proper value for total system momentum. In an author's note added in proof, he credits the above cited work of Balasz with using a similar simplest method approach for discerning the correct form of the electromagnetic tensor, but points out that Balasz did not employ the same force density arguments and as such, his arguments were less well-founded. Nevertheless, and in spite of Jones and Richards experimental results, the prevailing theoretical arguments continued to favor Abraham's symmetric form for the electromagnetic stress tensor.

2. Ashkin and Dziedzic Experiment

In 1973, Ashkin and Dziedzic [12] published the results of their experimental efforts to measure the radiation pressure on an air–water interface due to the passage of focused laser light. Their underlying assumption is that the momentum of light will change upon entering a dielectric medium from free space, due to Fresnel reflection and via interactions with the medium. Assuming p_0 is the momentum of light in free space and p is the momentum in the dielectric medium, the net change in momentum is given by $\{p_0(1+R)\} - \{p(1-R)\}$, where n is the index of refraction of the medium and R is the Fresnel reflection coefficient. The difference in momentum must be balanced by a mechanical force on the medium, and they show that if the momentum in the medium is given by $p = (n/c) \cdot u_{EM}$, then light entering the dielectric will exert a net outward force at the surface. If, instead, the electromagnetic field momentum within the dielectric medium is given by $p = u_{EM}/(nc)$, then the light will exert a net inward force on the interface surface. These momentum expressions are equal to the Minkowski (Eq. 23) and Abraham (Eq. 28) terms, respectively, and hence the experiment could provide a discriminator between the competing concepts. The experimental setup is discussed in detail in Ashkin and Dziedzic; the results of their experiments demonstrated a net *outward* force on the free surface of the liquid, in agreement with the predictions of the Minkowski form for the momentum. These observations, which agree with the previously reported results of Jones and Richards, caught the attention of theorists and spurred an effort to better understand the nature of electromagnetic fields and forces within dielectric media.

3. Gordon's Analysis

In the same year that Ashkin and Dziedzic published the results of their experiments, Gordon [13] published an analysis in which the concept of *pseudomomentum* was used to delineate between the Minkowski and Abraham forms for momentum. Like Ashkin and Dziedzic, Gordon worked for Bell Telephone Laboratories at the time and was keenly aware of the experimental efforts of his colleagues. Gordon's analysis showed that for nondispersive dielectric media, Abraham's tensor form correctly represents the momentum density of electromagnetic fields, but the Minkowski form is not precluded if one correctly takes into account the total radiation pressure acting on an object embedded in the dielectric. For a broad range of conditions, Minkowski's momentum density (here termed a pseudomomentum) can be expressed as the relative dielectric constant of the medium multiplied by Abraham's form for the momentum density. Gordon argues that the pseudomomentum form can be used to evaluate the radiation pressure on objects embedded in dielectric media, if it is recognized that the radiation pressure is actually a combination of the mechanical force exerted directly by the field on the object, and the force exerted on the object by mechanical pressures induced in the dielectric medium by the presence of the electromagnetic field. Within this framework, Gordon analyzes the results of the Ashkin and Dziedzic experiment, as well as that of Jones and Richards, and concludes that theory and experiment are in complete accord: Abraham's momentum density is correct for nondispersive dielectric media, and the pseudomomentum formulation embodied by Minkowski's momentum density correctly determines the radiation pressure on objects embedded within such a medium. He concludes with a cautionary note that laboratory experiments designed to measure the true nature of electromagnetic momentum in dielectric media may not be feasible.

4. Walker, Lahoz, and Walker Experiment

Undeterred, in 1975 Walker et al. [14] published a set of experimental results which, unlike previous experiments, supported the *Abraham* form for the force density. The premise of the experiment was that, by measuring the motion of a material body placed in a time-varying electromagnetic field, the total momentum can be separated into ordinary mechanical momentum and electromagnetic field momentum. The electromagnetic momentum can then be compared to the Abraham and Minkowski predictions.† In the Walker et al. experiment, a barium titanate disk with a thin aluminum coating was suspended by a fine wire between the poles of a strong electromagnet, forming a torsion pendulum. A time varying voltage was applied across the disk, providing a slowly varying (quasi-stationary) orthogonal electric field. The resulting torsion pendulum oscillations were readily measured, and were shown to be consistent with predictions made using the Abraham form of the force density, to within an experimental accuracy of $\pm 10\%$. The apparently unambiguous measurement of a repeatable macroscopic effect explicitly due to the Abraham force term appeared to rule out the Minkowski formulation by several standard deviations, and lent strong experimental support to the Abraham side of the long simmering theoretical debate.

5. Brevik's Analysis

However, as the reader has probably anticipated, this result did not close the door on the controversy but instead opened a window of opportunity to gain deeper physical insight into the varying electromagnetic tensor formulations. In 1979, Brevik [6] published a detailed review article that outlined and interpreted the experimental results of the preceding decades. Regarding the Walker et al., experiment, Brevik notes that despite some shortcomings (low frequencies, nonmagnetic test body, and constant magnetic field), the experiment does provide direct experimental evidence for the existence of the Abraham force term and the Minkowski tensor appears inadequate as a description for quasi-static fields. However, Brevik goes on to note that Minkowski's tensor still has use at optical (and higher) frequencies, where the Abraham term is unobservable; in these instances, the relative simplicity of the Minkowski tensor is often more convenient. Brevik also discusses the 1954 experimental results of Jones and Richards, which he terms one of the most important experiments in phenomenological electrodynamics. Following alternative derivations for the radiation pressure, Brevik shows that either the Abraham or the Minkowski formulation can account equally well for the experimental results; that the Abraham form in this case would agree with the Minkowski results quoted by Jones and Richards is due to the vanishing of the Abraham force term at high optical frequencies. Rather than eliminating one form of the electromagnetic field tensor, the experiment can instead be used to show the close relationship between the two tensor forms at optical frequencies. Regarding the 1973 results of Ashkin and Dziedzic, which again appear to support the Minkowski interpretation,

---

†An unpublished but apparently similar experiment to detect the Abraham force density was reported by R. P. James in his Stanford University dissertation. The diligent reader is referred to James, R. P., *Force on Permeable Matter in Time-Varying Fields*, Ph.D. dissertation, Dept. of Electrical Engineering, Stanford University, 1968. See also James, R. P., "A 'Simplest Case' Experiment Resolving the Abraham–Minkowski Controversy on Electromagnetic Momentum in Matter," [abstract] *Proceedings Natural Academy Science*, Vol. 61, Nov. 1968, pp. 1149–1150. The results of James's experiment are briefly discussed in Brevik [6].

Brevik provides a lengthy analysis of the experiment and demonstrates that the Abraham force term under these reported conditions would not contribute to the radiation pressure acting on the liquid interface. As such, the Abraham and Minkowski surface force expressions would be identical, and the experiment does not actually discriminate between the two. Brevik discusses and dissects a number of other experiments as well, and concludes his review by noting that, in most cases, it is sufficiently accurate and often simplest to use the divergence-free form of the Minkowski tensor.

6. Photon Drag Experiment

Soon after Brevik published his review, Gibson et al. [15] published the results of an experiment in which radiation pressure was measured by means of the photon drag effect. This experiment brought to bear a new type of potential discriminator through the direct measurement of photon momentum density. A photon traversing a semiconductor material can transfer momentum to the electrons in the valence or conduction bands of the material, generating an electric field. The high refractive index of semiconductor materials provides significantly different predictions between the Minkowski and Abraham momentum densities [Eqs. (23) and (28)]; by measuring the electric field due to photon drag, the correct tensor description can be established. Experimental results were obtained using silicon and n-type and p-type germanium semiconductors and photon wavelengths up to 1.2 mm; the results were shown to be correctly described by the Minkowski formulation. In 1986, Brevik [16] analyzed the results of the experiment and, while confirming that the Minkowski tensor provided the most straightforward explanation of the results, showed that other tensor formulations were equally able to predict the experimental findings. In concluding, Brevik notes that the photon drag experiment shows the usefulness of Minkowski's tensor in the high-frequency range, from optical to at least millimeter frequencies, but at much lower frequency limits, the results of Walker et al. show that Minkowski's tensor cannot be used and the Abraham form provides the correct description of experimental results. While intermittent discussions continued to appear in print, this resolution appears to have been generally accepted.

C. Recent Developments: 1990 to Today

1. Theoretical Advancements

The concept of electromagnetic pseudomomentum previously developed by Gordon was significantly expanded upon in a 1991 paper by Nelson [17], who used a general Lagrangian approach to develop momentum and pseudomomentum conservation laws for electromagnetic fields interacting with a dispersive, deforming dielectric medium. Upon quantizing the electromagnetic wave energy, Nelson sums the momentum and pseudomomentum terms to find a new conserved quantity, dubbed the wave-momentum, which enters into wave-vector conservation and phase-matching relationships. In 2004, Garrison and Chiao [18] published a quantization scheme for electromagnetic fields in a weakly dispersive, transparent dielectric, and used it to develop canonical and kinetic forms for the electromagnetic wave momentum. The canonical form defines a unique operator that generates spatial transformations in a uniform media; its physical significance has been established via energy and momentum conservation during spontaneous emission, Cerenkov and Doppler effects, and phase matching in nonlinear optical processes. Two choices arise for the kinetic momentum operator, corresponding to the classical Minkowski and Abraham momentum terms. Based on a review of prior experiments (outlined previously), they conclude that the Abraham form is required to correctly describe closed system experiments in which acceleration of the dielectric is allowed, at least for classical, low-frequency electromagnetic fields.

In 2004 and 2005, Mansuripur published a series of papers dealing with radiation pressure and linear momentum related to the once again percolating Abraham–Minkowski discussion. In his first paper [19], Mansuripur applies the classical Lorentz force to evaluate the interaction of electromagnetic fields with the charges and currents in conducting and dielectric materials. Following a cursory review of plane wave reflection from perfect conductors, he next develops a detailed analysis of the momentum density of an electromagnetic plane wave in an isotropic and homogeneous dielectric medium. Mansuripur finds that the correct expression is actually given by an average of the Minkowski and Abraham momentum densities; however, he notes that this result is limited due to the effects of multiple plane wave interference that arises in most practical applications. In such instances, the resulting force on the dielectric medium must be calculated by a direct evaluation of the charge and current interactions with the electromagnetic field components. He shows that the electric component of the radiation field acts on the bound charges within the medium, while the magnetic field component exerts a force on the bound currents. Sample calculations for various configurations are provided, and are shown to be in good agreement with prior Minkowski–Abraham related experiments.

The following year Mansuripur extended his analysis to a detailed study of radiation forces on a solid dielectric wedge immersed in a transparent medium [20]. Again using the Lorentz law, he shows that the linear momentum of the electromagnetic field within the dielectric is given by an equal contribution of both the Minkowski and Abraham momentum terms. For a dielectric surrounded by free space, he finds that the Lorentz force acting on the bound charges and currents exactly equals the time rate of change of the incident electromagnetic momentum. For a dielectric immersed in the transparent liquid medium, the time rate of change of the incident momentum is equal to the force exerted on the wedge plus that experienced by the surrounding fluid. In his third paper published that same year, Mansuripur derives an expression for the radiation pressure of a quasimonochromatic optical plane wave incident on the flat surface of a semi-infinite dielectric medium [21]. He shows that the combined mechanical momentum acquired by the dielectric, plus the rate of flow of electromagnetic momentum within the dielectric (i.e., the Abraham momentum), can together account for the total optical momentum transferred to the dielectric. In this same paper, he then addresses the results of prior photon drag experiments, which showed that the charge carriers within the semiconductor acquired a momentum equal to the Minkowski value. Again using the Lorentz force law, he constructs a model of a thin absorbing dielectric layer (similar to a semiconductor band gap) and finds that an incident captured photon of initial energy $h\nu$

transfers an amount of momentum nhv/c to the medium. The latter value is equivalent to the Minkowski momentum of the photon in the dielectric medium, which is greater than the total incident momentum of the photon. This, in turn, means the dielectric must recoil an amount equal to the difference between the incident photon's initial momentum and the Minkowski momentum value picked up by the excited charge carrier within the dielectric. Thus, by using the Lorentz force law and paying careful attention to boundary conditions, Mansuripur accounts for a broad range of previously published experimental results.

Nearly concurrent with Mansuripur's third paper, Loudon et al. [22] published their own detailed analysis of momentum transfer via the photon drag effect. They constructed a two-component optical system model, with the semiconductor material described by real phase and group refractive indices, and the charge carriers modeled by an extinction coefficient. Using the Lorentz force approach, they calculated the force on the charge carriers and semiconductor host material and found that the momentum transfer to the charge carriers per photon was equal to the experimentally determined Minkowski value, nhv/c, where n is the (phase) refractive index, v is the optical frequency, and h is Planck's constant. The transfer of momentum to the host material is fixed by conservation of momentum considerations, and requires the total momentum transfer per photon to have the value $(hv/c)[(n^2 + 1)/2n]$. Loudon et al. then take their analysis one step further, and calculate the time dependent Lorentz force when the incident light is a narrow-band, single photon pulse. For pulse widths significantly shorter than the attenuation length in the material, the momentum transfer to the semiconductor can be split into surface and bulk contributions, and they find that the total bulk momentum transfer (to charge carriers plus host) is equal to the Abraham momentum value of hv/nc. As did Mansuripur, they argue there is no conflict in these results, as they apply to different field and boundary conditions. They further note that there is really no uniquely correct expression for the transfer of momentum from light to matter, as both the Minkowski and Abraham values can, in principle, be observed by appropriate measurements.

2. Bose–Einstein Condensate Experiment

As theorists continued to hone in on a consensus interpretation, a new experimental approach to measure electromagnetic momentum transfer made its debut appearance. In 2005, Campbell et al. [23] published their measurements of photon recoil momentum from a Bose–Einstein condensate of rubidium atoms. While photon recoil experiments in dilute gas are not new, the use of a Bose–Einstein condensate allowed Campbell et al. to significantly increase the density of the medium, and to observe how altering the index of refraction changes the atomic recoil frequency. The intent of the experiment was to distinguish between two competing theories for the correct amount of atomic recoil momentum expected for photon propagation through a dispersive medium. If an atom absorbs a photon, and no motion is left within the system, the recoil momentum should be $\hbar k$, where k is the wave number of the photon. If instead the atom interacts via dipole moment oscillations with the other atoms in the medium, it will recoil with a momentum of $n\hbar k$, where n is the refractive index of the dilute medium. The results of the carefully conceived and executed experiment demonstrated that the atoms do indeed recoil with a momentum of $n\hbar k$. While not explicitly developed as a test of the Abraham or Minkowski formulations, the experiment demonstrated that, at least under these conditions, the Minkowski formulation predicted the correct momentum transfer. The results sent a new ripple through the theoretical community, which quickly rose to the challenge.

3. Leonhardt's Analysis

Soon after the experimental results of Campbell et al. were published, Leonhardt [24] calculated the energy and momentum balance in quantum dielectrics such as Bose–Einstein condensates. He found that, in the nonrelativistic limit, the total momentum density, **g**, can be expressed as:

$$\mathbf{g} = \rho\hbar\,\nabla\varphi + \mathbf{D} \times \mathbf{B} \tag{29}$$

where ρ is the number density and φ is the phase of the condensate. This result shows that Minkowski momentum is imprinted onto the phase of the quantum dielectric, which agrees with the experimental results of Campbell et al. However, Leonhardt goes on to show that the total momentum density can also be written:

$$\mathbf{g} = \rho m \mathbf{u} + \frac{\mathbf{E} \times \mathbf{H}}{c^2} \tag{30}$$

where m is the atomic mass and **u** is the flow velocity of the dielectric media, defined as

$$\mathbf{u} = \frac{1}{m}\left[\hbar\,\nabla\varphi + \left(\varepsilon - \frac{1}{\mu}\right)\frac{\mathbf{E} \times \mathbf{B}}{\rho}\right] \tag{31}$$

Leonhardt notes that the mechanical momentum, $m\mathbf{u}$, differs from the canonical momentum, $\hbar\,\nabla\varphi$, by the Rontgen term $(\varepsilon - 1/\mu)\mathbf{E} \times \mathbf{B}/\rho$ arising from interactions of the electromagnetic field with the moving electric and magnetic dipoles of the dielectric medium. Leonhardt's interpretation is that variations of the Minkowski momentum are imprinted onto the phase, and the Abraham momentum drives the flow, of the quantum dielectric. Leonhardt derives relativistic covariant formulations of Eqs. (29) through (31), and concludes that the Abraham–Minkowski controversy actually has its roots in the Rontgen interaction.

4. Photonic Band Gaps as Future Testbeds

In 2006 Scalora et al. [25] published the results of detailed derivations for the Minkowski momentum density and the Lorentz force density assuming a dispersive medium. They evaluated the interactions between short optical pulses incident on 1) dielectric substrates of finite length, 2) on micron-size multilayer structures in free space and embedded within a dielectric medium, and 3) on a negative index of refraction material (with negative values for ε and μ). They used a numerical approach to solve the vector Maxwell equations and found that for all the considered cases, the conservation of linear momentum could be satisfied solely using the Poynting vector and the Abraham form for the momentum density. They further surmise that neither the Minkowski momentum density nor the average momentum density advocated by Mansuripur could reproduce the Lorentz force in any of the circumstances considered. Of additional interest, they calculate that a narrow band optical pulse of 600-fs duration and 1-MW/cm^2 peak power, incident upon a multilayer photonic bandgap structure with a mass of 10^{-5} g, may produce accelerations up to 10^8 m/s^2. Although the very short interaction times will limit sample displacement and velocity, they suggest that such experiments could be used to test basic electromagnetic phenomena and momentum transfer to macroscopic media.

D. Summary

Thus stands the famous Minkowski–Abraham debate as of this writing. As Gordon [13] notes, in a statement attributed to Blount, "The argument has not, it is true, been carried on at high volume, but the list of disputants is very distinguished." That the debate shows no signs of abating is evidenced by the growth of recent popular articles appearing in print [26–28], and it appears that a final resolution, if indeed one exists, may await further experiment and analysis. Other electromagnetic tensor formulations, of which there are several, have not been discussed in the abbreviated and incomplete history given here; the reader is referred to the excellent reviews of Penfield and Haus [5] and Brevik [6] for a sampling of these alternative formulations. As we draw this section to a close, perhaps the most succinct and useful summary of the Abraham–Minkowski controversy is provided by Pfeifer et al. [29], who state that "Any choice of the electromagnetic energy–momentum tensor is equally valid provided the corresponding material counterpart is also taken into consideration, as it is only the total energy–momentum tensor which is uniquely defined." It is with these comforting words of conciliation that we move from our review of this century-old debate, and step through the looking glass into proposed propulsive applications.

V. Propulsion Concepts Based on Electromagnetic Momentum Exchange

The subtle interplay between electromagnetic and mechanical momentum has over the years inspired several creative and contentious attempts to produce unidirectional forces and propulsive motion. Unlike typical low thrust photon propulsion schemes in which internally generated electromagnetic waves are beamed away from a spacecraft, or externally generated photons or microwaves are reflected by a spacecraft, the concepts described in this section attempt to exchange mechanical and electromagnetic momentum within a closed system. Time varying electric and magnetic fields are generated within a fixed volume, and an interchange between electromagnetic field momentum and spacecraft momentum is postulated to generate mechanical motion within the prescribed field boundaries. However, the subtleties inherent in the generation and arrangement of the fields, together with the time-varying interactions between fields and sources, can quickly obfuscate the underlying physics, and great care must be exercised to conserve total system energy and momentum. A cautionary note along these lines appeared in print over half a century ago, and serves as a useful introduction to this class of propellantless propulsion concepts.

A. Slepian's Electromagnetic Space Ship

One of the most widely referenced publications on the possible generation of unbalanced electromagnetic forces for propulsion is a short 1949 essay by Joseph Slepian, a distinguished engineer who penned a number of brief "electrical essays" designed to amuse and educate the readership of the *Electrical Engineering* journal. In this particular essay [30], Slepian presented a "means of propulsion which does not require any material medium upon which the propelling thrust is exerted," a prescient description of modern breakthrough propulsion physics efforts. With the typical good humor displayed in his other essays, Slepian proceeds to walk the reader through the various components that constitute his hypothetical spacecraft, and ends the succinct tutorial with two questions: 1) is there an unbalanced force acting on the material system of the spaceship, and 2) can this unbalanced force be used to propel the craft? His answer to the first question was yes; his answer to the second, no. Per the instructive nature of his essays, both answers were explained in a short follow-up publication entitled "Answer to Previous Essay," which appeared the following month [31]. Because the first essay is often cited in support of electromagnetic tensor propulsion concepts, the salient points of Slepian's enjoyable discourse and self-rebuttal are both outlined below.

Slepian begins the construction of his hypothetical space drive using elementary principles from classical electromagnetic theory. The reader is first introduced to the standard Lorentz force: a current-conducting wire situated perpendicular to a magnetic field will feel a force in a direction perpendicular to both the current and magnetic field. The second step of the argument reasserts Newton's classical Laws of Motion, and states that there is an equal but opposite force acting on the poles of the (permanent) magnet that corresponds to the force experienced by the current distribution. The third step in the construction of the electromagnetic space ship replaces the permanent magnet with an electromagnet in the form of separated solenoid coils that provide an equivalent magnetic field. The force on the current distribution running perpendicular to the area between the separated coils is countered by an equal but opposite force acting on the solenoid coils, just as occurred for the poles of the permanent magnet. Slepian then notes that the same forces will arise if an alternating current is used in place of a direct current in the solenoid windings, if the current distribution in the space between the coils also changes direction. In this case the force

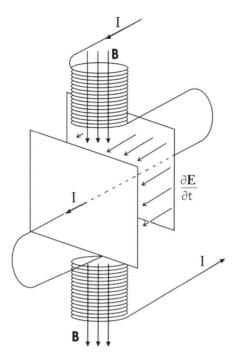

Fig. 1   Slepian's conceptual electromagnetic momentum generator [30].

on the current distribution remains unidirectional, with an equal and opposite unidirectional counterforce acting on the electromagnet.

At this point Slepian breaks the continuous current-carrying wire with two parallel plates (Fig. 1), and introduces the concept of displacement current. Due to the gap between the plates, the conduction current flowing along the wire is no longer continuous; however, the current flowing along the wire is maintained by the displacement current between the plates, which is exactly equal to the conduction current. The displacement current between the plates is given by $\varepsilon_0(\partial \mathbf{E}/\partial t)$, which appears in the set of Maxwell's equations [Eq. (1)]. Slepian then shows that there will be an unbalanced force due to the displacement current appearing in the space between the plates and the magnetic field, equal to:

$$\mathbf{F}_1 = \int \left(\varepsilon_0 \frac{\partial \mathbf{E}}{\partial t} \times \mathbf{B} \right) dV \qquad (32)$$

where V is the volume between the plates. By attaching the device to the side of a spacecraft, the unbalanced force generated on the solenoid coils by the interaction of the oscillating magnetic field with the displacement current would move the craft in a direction perpendicular to both, providing propulsion without propellant. Thus ends Slepian's first essay, with an admonishment to the reader that while in principle this unbalanced force exists, it cannot be used to propel the ship. As with his other essays, Slepian provides his reasoning in a separate article published the following month, allowing the reader time to think about the puzzle prior to giving the answer.

In his second essay, Slepian observes that the above calculations, while correct, do not proceed quite far enough. According to Maxwell's equations, the time-changing magnetic field generates an electric field between the plates; the interaction between the magnetic field and induced electric field provides a force that acts on the parallel plates:

$$\mathbf{F}_2 = \int \left(\varepsilon_0 \mathbf{E} \times \frac{\partial \mathbf{B}}{\partial t} \right) dV \tag{33}$$

The time average of this force acting on the charged plates is just equal and opposite to the time average of the force acting on the solenoid magnets, resulting in no net (time-averaged) motion of the device. How then was Slepian able to assert a positive answer to the question posed in his first essay regarding net unbalanced forces acting on the system? A careful consideration of the oscillating electric and magnetic fields in the space between the plates shows that they are in time quadrature, hence a non-zero *instantaneous* resultant force does indeed exist. However, this instantaneous force alternates in direction, hence the net, time-averaged force acting on the system will be zero. As Slepian remarks, "Yes, there will be an unbalanced net force on the space ship [as calculated], but it will be an alternating force." Hence, no net motion will result.

In his closing remarks, Slepian discusses the conservation of momentum and energy for his hypothetical space drive. Combining the expressions for the forces above, the (instantaneous) unbalanced force on the material system is given by:

$$\mathbf{F} = \mathbf{F}_1 + \mathbf{F}_2 = \int \left(\varepsilon_0 \frac{\partial \mathbf{E}}{\partial t} \times \mathbf{B} \right) dV + \int \left(\varepsilon_0 \mathbf{E} \times \frac{\partial \mathbf{B}}{\partial t} \right) dV$$
$$= \int \frac{\partial}{\partial t} (\varepsilon_0 \mathbf{E} \times \mathbf{B}) \, dV \tag{34}$$

where the quantity in brackets under the last integral is recognized as the momentum density of electromagnetic fields in free space [Eq. (13)]. The electromagnetic energy density in the free space between the plates is then given by Eq. (14). In the hypothetical space drive, the electric and magnetic fields are caused to oscillate at high frequency, resulting in a rapidly changing momentum density between the plates. By conservation of momentum, there will be an equal but opposite rate of change in the momentum of the material system. This again corresponds to an instantaneous unbalanced force acting on the system, whose magnitude is given by Eq. (34). However, as noted above, this force alternates in direction and its time average over an oscillation period still sums to zero.

While "Slepian drives" still appear in the literature, it is clear that Slepian meant his electrical essay to be an informative and entertaining mental diversion for his readers, with no pretext that such a drive would actually work. The problem he posed and the solution he discussed serve to illuminate the difficulties inherent in untangling and understanding the various field interactions that underlie this class of breakthrough propulsion concepts. Far from advocating that such a system would work, his essays instead provide a clarion call that all such systems must remain consistent with the well-established laws of momentum and energy conservation.

B. Electromagnetic Stress-Tensors and Space Drive Propulsion

More recently, Corum et al. [32,33] have sought to extend Slepian's hypothetical electromagnetic space drive to include the effects of nonstationary boundary conditions. Using the definition of the displacement field, $\mathbf{D} = \varepsilon_0 \mathbf{E}$, Corum writes the electromagnetic momentum density force appearing in Eq. (34) as:

$$\mathbf{f}_H = \frac{\partial(\mathbf{D} \times \mathbf{B})}{\partial t} \tag{35}$$

and renames it the Heaviside force density, $\mathbf{f}_H$, due to its appearance in the derivation by Heaviside of Maxwell's stress tensor [34]. In agreement with Slepian, Corum notes that the time average of this force density, integrated over a stationary bounding volume, will provide zero net force. Corum et al. then speculate on the possibility of rectifying this force using a temporally discontinuous action, for example by some appropriate modulation of the boundary conditions. Corum et al. argue that if this is possible, then the average of the integrated force will no longer vanish and unidirectional motion may indeed take place.

Among the experiments Corum et al. cite in evidence of a possible unidirectional force component is the early 20th-century work of R. Hartley, inventor of the Hartley oscillator. Hartley observed a sustained, residual mechanical offset on the plates of a capacitor charged with alternating current [35]. In the 1950s, Manley and Rowe expanded on these results and concluded that time-varying and nonlinear elements may both contribute to the observed mechanical offset [36,37]. Corum et al. propose that a similar effect could occur in the Slepian electromagnetic device, perhaps resulting in a small but sustained unidirectional force on a spacecraft so equipped. In fiscal year 2000 and for a few years following, Corum et al. received funding to conduct a further investigation of this effect. The intent of the research program was to experimentally investigate whether the Heaviside force was real and observable, and Corum et al. set up a program to try and rectify the force on a capacitor to cause unidirectional movement of a wheeled device. The device chosen was a modification of what is historically known as the ampere rail motor, a concept similar to a modern rail gun in which a direct current is passed along conducting rails and across a conductive load; the current through the load interacts with the magnetic fields produced by the current in the rails to provide Lorentz acceleration of the device along the rails. The experiment sought to replicate this using high frequency alternating currents through a capacitor placed in series with the load, in which the hypothesized Heaviside force would produce an offset force, resulting in small but perceptible movement of the device.

A concurrent experimental effort was set up to repeat the 1981 experiments of Lahoz and Graham [38], who used a torsion pendulum to try and observe the momentum density produced by static electric and magnetic field distributions. As reported in 2001, both experimental efforts were proceeding to hardware fabrication in anticipation of near-term testing; however, to date there has been no further publications regarding their status or results, and it appears that this activity is no longer being funded.

In 2002, the U.S. Air Force (USAF) Academy undertook an independent investigation of the Heaviside force conjecture [39]. The effort employed both theory and experiment to investigate the possibility of producing a unidirectional Heaviside force as proposed by Corum et al. The theoretical analysis showed that using the electromagnetic force density per se would not lead to any discernable net thrust, but that the time derivative of the electromagnetic momentum density, as represented by Eq. (35), might, in the words of the report, cause an acceleration of space and hence a discernable force. The USAF Academy research team used a laser interferometer arranged with one beam passing through an apparatus held under vacuum, in which perpendicular electric and magnetic fields would be generated; the other beam traversed a similar space devoid of electromagnetic fields. The beams are recombined at a photodetector, and the signal searched for any effect due to beam traversal of the crossed electric and magnetic fields versus traversal when the fields are absent. The thought was that the crossed and oscillating fields might somehow produce an acceleration of space, which would then be picked up as a change in the interference pattern of the recombined beams. As reported, the laser interferometer did not find any acceleration of space. The team redesigned its experiment to look instead for possible oscillations on a parallel plate capacitor using parallel applied and induced electric fields, corresponding to possible linear electromagnetic momentum transfer. However, it does not appear that sufficient funding remained to carry out the revised set of experiments, and no results of the modified USAF Academy experiment have to date been published. An independent analysis of the original USAF Academy approach was undertaken by the Instituto Universitario Aeronautico in Argentina, which determined that no net forces would be produced under this arrangement, and that no acceleration of space or any matter within that space would occur, confirming the null results of the experiment [40].

Further critiques of the conceptual use of the Heaviside force for propulsion were published by Brito and Elaskar [41], who developed their own concept for propellantless propulsion via the manipulation of electromagnetic fields (see Sec. IV.C). Brito and Elaskar argue that the volume integrated force in the vacuum space between the parallel plates of the system is identically equal to zero, hence any rectification procedure applied to the instantaneous force to generate unidirectional acceleration is meaningless. They argue that this renders moot the possibility of temporally modulating the boundary conditions to provide a non-zero time average force, unless polarized matter is present. However, even modulating the boundary conditions on the dielectric will not circumvent the problem because the momentum of the complete system must still be conserved, resulting in a zero net force. Brito and Elaskar thus conclude that Slepian's instructional construct, and Corum et al.'s extension to include a conjectured unbalanced Heaviside force, are not capable of producing unidirectional acceleration of the system.

C. Electromagnetic Inertia Manipulation (EMIM) Propulsion

In the late 1990s, Brito began publishing his work on theoretical and experimental efforts to use Minkowski's form of the stress tensor as a basis for propellantless breakthrough propulsion [41–46]. Dubbed *electromagnetic inertia manipulation*, or EMIM, the fundamental concept is that the total space-time momentum of a material spacecraft, plus any electromagnetic field it generates, must be a conserved quantity. By considering four-dimensional spacetime instead of three-dimensional space, Brito argues that the four-velocity of the spacecraft can be changed by a proper manipulation of the electromagnetic portion of the total mass tensor (ship plus fields), analogous to an ice skater changing (angular) velocity by changing the moment of inertia. The EMIM system is comprised of the spaceship plus the extended electromagnetic fields it generates; the conjecture is that a net displacement force on the spacecraft would result from momentum exchange with the electromagnetic field. A conceptual schematic of the electromagnetic momentum generating device is shown in Fig. 2.

Brito proposed a set of related hypotheses to describe how such a system might operate, with the formulations based on electromagnetic field momentum, mass tensors, and electromagnetic force densities. While the mass tensor formulation appears most often in his later publications, each approach is briefly described.

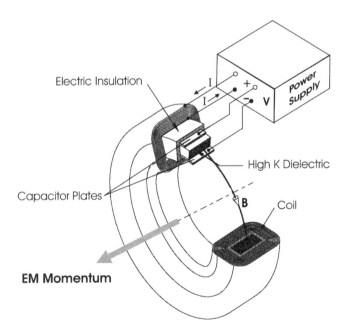

Fig. 2 Brito's conceptual EMIM generator [46]. (Reproduced with permission of the American Institute of Aeronautics and Astronautics.)

1. Electromagnetic Field Momentum

The basis for this derivation directly relates to the original Abraham–Minkowski stress tensor controversy, and Brito re-examines the genesis of Eqs. (21) and (24) to determine whether the non-zero momentum densities apparently arising in Minkowski's formulation can lead to a non-zero total electromagnetic momentum for the coupled ship-field system. Brito invokes the results of Furry [47] to show that the total electromagnetic momentum, **G**, for a static distribution of electric and magnetic fields in a material medium, whose magnitudes rapidly decay to zero at infinity, can in general be expressed as:

$$\mathbf{G} = \int \mathbf{g} dV = -\int \mathbf{x} \left[\nabla \left(\frac{1}{c'}\right)^2 \cdot \mathbf{S} + \left(\frac{1}{c'}\right)^2 \nabla \cdot \mathbf{S} \right] dV \qquad (36)$$

where **x** is a position vector, **S** is the Poynting vector, and c' is the speed of light in the medium, $(\mu\varepsilon)^{-1/2}$. Brito notes the quantity in brackets is simply the divergence of the electromagnetic momentum density, **g**; if this quantity is not divergence free everywhere, then it may be possible to produce a non-zero total electromagnetic momentum. This leads him to adopt Minkowski's formulation for the electromagnetic momentum density (Eq. 21); Abraham's formulation would result in zero total electromagnetic momentum. Referring to Fig. 3, he shows that the quantity $\nabla \cdot \mathbf{S}$ vanishes, which leaves the gradient in the velocity of light within the integral region as the only remaining possibility for generating non-zero total electromagnetic momentum. Such gradients might arise across the

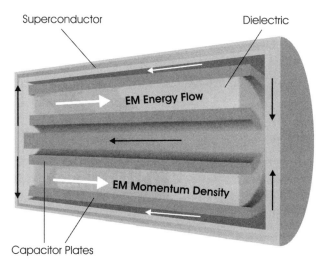

Fig. 3 Electromagnetic energy flow and momentum density in EMIM device [46]. (Reproduced with permission of the American Institute of Aeronautics and Astronautics.)

free surfaces of a finite dielectric material, and Brito argues that the device depicted in Fig. 3 provides just such an arrangement.

Invoking the conservation of total mechanical and electromagnetic momentum, Brito conjectures that the time rate of change of the volume integrated electromagnetic momentum produced by the EMIM device will be balanced by the time rate of change of the mechanical momentum of the physical device, producing a force on the device and providing net thrust. Noting that the electromagnetic fields fall rapidly to zero outside of the dielectric boundaries, he shows that the net thrust, **T**, acting on the device is approximately equivalent to:

$$\mathbf{T} \cong -\int \frac{\partial \mathbf{g}}{\partial t} dV = -\int \frac{\partial}{\partial t} (\mathbf{D} \times \mathbf{B}) dV \qquad (37)$$

where the integration is taken over the material volume of the dielectric medium. In this approach, the possibility of generating net thrust explicitly hinges upon the correctness of the Minkowski interpretation of the electromagnetic momentum density. While similar to Corum's development using the Heaviside force density (Eq. 35), Brito's analysis is restricted to integration volumes bounded by the dielectric medium, with the discontinuity in the speed of light at the material boundaries responsible for the generation of electromagnetic momentum and the offsetting mechanical momentum required for total momentum conservation.

2. Mass-Inertia Tensor

In this formulation, Brito considers the spacecraft mass and the equivalent mass of the electromagnetic field generated by the EMIM device aboard the craft as a single point mass located at the center of mass of the total system. He then derives a mass-inertia tensor, M, given by:

$$M = \left(m_0 + m_{EM}^*\right) I + \left(\mathbf{p}_{EM} \wedge \mathbf{v}\right)/c^2 \qquad (38)$$

where m_0 represents the material rest mass, m_{EM}^* is the equivalent mass of the electromagnetic field in the spacecraft frame of reference, $\mathbf{p}_{EM}$ is the four-momentum of the electromagnetic field, $\mathbf{v}$ is the four-velocity of the point center of mass, $\wedge$ is the wedge product, and I is the identity tensor. For a closed system completely encompassing both material spacecraft and electromagnetic field, the conservation of momentum is written:

$$dM \cdot \mathbf{v} + M \cdot d\mathbf{v} = 0 \qquad (39)$$

Combining Eqs. (38) and (39), Brito concludes that any change in the electromagnetic field momentum will be exactly balanced by a change in the mechanical (spacecraft) momentum. Restated, the formulation argues that momentum can be exchanged between matter and field within the closed system bounded by the electromagnetic field distribution, with the amount of momentum that can be transferred to the spacecraft limited by the amount of electromagnetic momentum contained within the closed system.

3. Electromagnetic Momentum Density

The third approach considered by Brito is again based on Minkowski's formulation for the electromagnetic momentum density, with both the displacement and magnetic fields undergoing harmonic excitation. Within this framework, Brito develops an expression for the time-averaged electromagnetic thrust on a closed system. Related to the EMIM generator shown in Fig. 2, the expression becomes:

$$\langle T \rangle = \frac{\varepsilon_r \omega N I V d}{2c^2} \sin\varphi \qquad (40)$$

where the applied voltage is $V \cdot \sin(\omega t)$, the applied current is $I \cdot \sin(\omega t + \varphi)$, ε_r is the relative permittivity of the medium, N is the number of turns in the magnet coil, and d is the distance between capacitor plates. This equation provides a set of tunable physical parameters that can be used to experimentally evaluate the time averaged force acting on the proposed EMIM device.

Armed with the above hypotheses, Brito performed a series of experiments to validate the electromagnetic inertia manipulation concept represented in Figs. 2 and 3. These experiments, and the current status of the EMIM effort, are discussed below.

4. EMIM Force Experiments

Beginning in the mid-1990s, an experimental embodiment of the conceptual EMIM thruster shown in Fig. 2 was built, tested, and refined in an effort to demonstrate sustained thrust via electromagnetic inertia manipulation. Barium titanate ($BaTiO_3$) ceramic with a high relative permittivity was used as the dielectric medium. Periodic current and voltage supplied to the magnet coils and capacitor plates, respectively, provided the magnetic and electric displacement fields required to generate electromagnetic momentum within the dielectric, with the fields falling rapidly to zero outside the device. Acting as a resonant mass, the device was placed atop a thin vertical cantilever beam (resonant blade), which in turn rested on a vibration-free table. Piezoelectric strain transducers were used to measure the displacement of the cantilever beam and infer the net force acting on the system. A common power supply providing 100 VAC at 30 kHz energized three, 900-turn, parallel-mounted toroidal coils connected in series with three parallel mounted, 10 nF, 8-mm-wide annular capacitors. With this experimental configuration, Brito calculated that a peak electromagnetic momentum of 5×10^{-11} N should be generated by the system, capable of providing a time averaged thrust of around 5×10^{-6} N. A later upgrade of the power supply to 200 VAC at 30 kHz doubled these estimates to around 10^{-10} N and 10^{-5} N, respectively. Additional details on the experimental configurations are included in Brito [43–45] and Brito and Elaskar [46].

Operating first with the 100-VAC supply, Brito measured the displacement of the resonant mass under conditions in which either the coils or the capacitors were energized, in which case no thrust is expected to occur, and when both the coils and capacitors were energized, in which case a displacement was expected to occur via electromagnetic momentum interchange with the mechanical system.

They note that uncertainties arising from mechanical and electrical interference prevent a clean interpretation of the results; nevertheless, the claim is made that the predicted force appears at a level sufficiently above the experimental background noise to demonstrate that a net propulsive effect does indeed occur. The reader is referred to the experimental results presented by Brito and Elaskar [46] to draw their own conclusion as to the validity of this claim.

As might be expected, there are a number of extraneous factors that must be accounted for in the experimental setup and interpretation of results. Among these are electrostatic and magnetic field interactions with laboratory surroundings, interactions with self-induced magnetic fields, differential heating of thruster or test components, ground and air vibrations, geomagnetic field interactions, coronal discharges, ion wind, and myriad other nuanced effects that must be tracked and eliminated as potential sources for any observed net motion. Brito discusses and dismisses a number of these potential sources of uncertainty as negligible [46], but the possibility of unaccounted interactions clearly remains. Such effects are exacerbated by the rapidly pulsed nature of the device, as well as the extremely low levels of anticipated net thrust. While it will be difficult to account for all such interactions, extraordinary claims require extraordinary proof, and a diligent effort is needed to better understand and eliminate any remaining sources of uncertainty before claims of experimental proof are fully justified.

D. Summary

The work of Slepian, Corum, Brito, and others outlined in this section serves to illuminate the challenges inherent in this approach to breakthrough propulsion. The analysis of potential interactions between electromagnetic field and material medium is complex, and sources of experimental noise are often exceedingly difficult to discern. It is an enormous and perhaps unwarranted extrapolation to move from still-debated forms of the electromagnetic stress tensor to conceptual systems designed to generate net motion. Nevertheless, such endeavors do provide additional understanding of electromagnetic field–matter interactions, and help clarify areas for potential future research. One such area with possible implications for breakthrough propulsion is the Feigel hypothesis, briefly discussed below.

VI. Feigel Hypothesis

In 2004, a new and controversial physical interpretation of the momentum transfer between matter and field was published by Feigel [48], who applies a relativistic Lagrangian formulation to the interaction of light and matter in a dielectric medium. He derives the equations of motion and conservation laws, and finds that in the case of a liquid dielectric, the electromagnetic field interactions will cause the medium to move. Returning to the Minkowski–Abraham controversy, he concludes that Abraham's expression represents the momentum of the field, but because the measured momentum also includes a matter contribution, its value corresponds more closely to the Minkowski formulation. In a novel piece of analysis, Feigel predicts that *zero-point vacuum fluctuations* contribute to the motion of a dielectric media in crossed electric and magnetic fields. This astonishing result raises the possibility that momentum can be transferred from the zero-point fluctuations to matter, with the direction and magnitude of the force controlled by externally applied electric and magnetic fields. Feigel estimates that for a relative dielectric constant of 1.5, material density of 10^3 kg/m^3, and externally applied orthogonal electric and magnetic fields of 10^5 V/m and 17 T, respectively, one could achieve a small but measurable velocity of around 50 nm/s due to the contribution of high frequency vacuum modes.

Following this publication, there appeared a number of comments and replies [49,50], culminating with a 2006 paper by van Tiggelen et al. [51] in which they reanalyze Feigel's results using a Lorentz-invariant description with field regularization techniques to cope with vacuum divergence. They find that for an infinite medium, there is no transfer of momentum via Feigel's proposed mechanism. However, they note that a vanishingly small momentum transfer may occur in the squeezed vacuum states provided by inwardly moving parallel plates in a Casimir geometry. For this exercise they assumed the Casimir plates start at a large separation distance, with a magneto-electric material slab at rest in the space between them. The crossed electric and magnetic fields are switched on, and the plates are slowly brought closer together to a finite separation distance (this would occur naturally due to the Casimir force acting to push the plates together). The inward motion of the plates thus converts vacuum energy into kinetic energy, but there is no transfer of momentum to the system because the plates are moving with opposite momentum. However, the electromagnetic momentum density in the space between the plates is changing with the plate separation distance; this change must be compensated by a change in the momentum density of the magneto-electric material in order for the total system momentum to remain zero. Using similar material properties, van Tiggelen et al. calculate that the medium would acquire a momentum density nearly 20 orders of magnitude smaller than that originally predicted by Feigel, which is immeasurable.

However, in a companion report to the European Space Agency (ESA), van Tiggelen et al. [52] calculate that the Feigel effect for a small magneto-electric object immersed in an isotropic, monochromatic radiation field could be set into motion if the external fields are suddenly switched on. They note that experimental verification in this instance would be facilitated by the ease with which monochromatic radiation fields can be generated and modulated. For a field intensity of 10 kW/cm^2, they predict a velocity approaching 10^{-5} cm/s could be achieved for an object made of FeGaO$_3$. Experimental verification could be achieved using a 10-μg crystal of FeGaO$_3$ mounted at the end of a piezo-resistive cantilever. Immersed in a 10-kW/cm^2 radiation field, the crystal would acquire a momentum of approximately 10^{-10} g-cm/s. Using a DC magnetic field and periodic switching of the radiation field could provide an oscillatory force on the mass sample, of sufficient magnitude to be measured experimentally. Recently, van Tiggelen, Rikken and Krzic proposed to perform these experimental measurements, which could help determine whether the controversial effect predicted by Feigel actually exists. As of this writing, no additional information is available on the status of their proposed experimental effort.

VII. Conclusions

The still-debated physical mechanism responsible for the interchange of electromagnetic and mechanical momentum in a dielectric medium retains an allure for breakthrough propulsion research, offering a potentially self-contained method of propulsion without the expenditure of propellant. However, purported demonstrations of unidirectional motion from a rectified harmonic force or by exchanging mechanical momentum with the momentum of an extended electromagnetic field remain to date elusive and unsubstantiated. Experimental efforts are often susceptible to the extraneous noise and unwanted motion inherent in most laboratory environments, while theoretical arguments may easily get lost in the subtle accounting of electric and magnetic field interactions. We are reminded once again of Slepian's decades-old admonishment that all such proposed systems must remain consistent with the conservation of energy and momentum, indispensable principles in the continuing search for new breakthrough propulsion concepts.

Acknowledgments

The author thanks C. Mercer, M. Millis, T. LaPointe, and J. Wilson for several useful comments and corrections. Portions of this research were conducted under the auspices of the Science Research and Technology Projects Office at the NASA Marshall Space Flight Center, whose support is gratefully acknowledged.

References

[1] Minkowski, H., "Die Grundgleichungen fur die Elecktromagnetischen Vorgange in Bewegten Korpen," *Nachrichten von der Gesellschaft der Wissenschaften zu Gottingen*, Vol. 53, 1908, pp. 53–111.

[2] Abraham, M., "Sull'Elettrodinamica di Minkowski," *Rendiconti Circolo Matematico di Palermo*, Vol. 30, 1910, pp. 33–46.

[3] Jackson, J. D., *Classical Electrodynamics*, 2nd ed., Wiley, New York, 1975, Ch. 6.

[4] Cook, D. M., *The Theory of the Electromagnetic Field*, Prentice-Hall, Saddle River, NJ, 1975, Chs. 6, 12, and 15.

[5] Penfield, P., Jr., and Haus, H. A., *Electrodynamics of Moving Media*, MIT Press, Cambridge, MA, 1967.

[6] Brevik, I., "Experiments in Phenomenological Electrodynamics and the Electromagnetic Energy–Momentum Tensor," *Physics Reports*, Vol. 53, No. 3, 1979, pp. 133–201.

[7] Bowyer, P., "The Momentum of Light in Media: the Abraham–Minkowksi Controversy," dissertation, School of Physics and Astronomy, Southampton, UK, Jan. 2005. URL: http://peter.mapledesign.co.uk/writings/physics/2005_dissertation_The_Abraham-Minkowski_Controversy.pdf.

[8] Halpern, O., "A Theorem Connecting the Energy Momentum Tensor with the Velocity of Propagation of Waves," *Physical Review*, Vol. 48, Sept. 1935, pp. 431–433.

[9] Balasz, N. L., "The Energy–Momentum Tensor of the Electromagnetic Field Inside Matter," *Physical Review*, Vol. 91, No. 2, July 1953, pp. 408–411.

[10] Jones, R. V., and Richards, J. C., "The Pressure of Radiation in a Refracting Medium," *Proceedings of the Royal Society London, Series A*, Vol. 221, 1954, pp. 480–498.
[11] Shockley, W., "A 'Try Simplest Cases' Resolution of the Abraham–Minkowski Controversy on Electromagnetic Momentum in Matter," *Proceedings of the National Academy Science*, Vol. 60, 1968, pp. 807–813.
[12] Ashkin, A., and Dziedzic, J. M., "Radiation Pressure on a Free Liquid Surface," *Physical Review Letters*, Vol. 30, No. 4, 1973, pp. 139–142.
[13] Gordon, J. P., "Radiation Forces and Momenta in Dielectric Media," *Physics Review A*, Vol. 6, No. 1, 1973, pp. 14–21.
[14] Walker, G. B., Lahoz, D. G., and Walker, G., "Measurement of the Abraham Force in a Barium Titanate Specimen," *Canadian Journal of Physics*, Vol. 53, 1975, pp. 2577–2568.
[15] Gibson, A. F., Kimmitt, M. F., Koohan, A. O., Evans, D. E., and Levy, G. F., "A Study of Radiation Pressure in a Refractive Medium by the Photon Drag Effect," *Proceedings of the Royal Society of London A*, Vol. 370, 1980, pp. 303–311.
[16] Brevik, I., "Photon Drag Experiment and the Electromagnetic Momentum in Matter," *Physics Review B*, Vol. 33, No. 2, 1986, pp. 1058–1062.
[17] Nelson, D. F., "Momentum, Pseudomomentum, and Wave Momentum: Toward Resolving the Minkowski–Abraham Controversy," *Physics Review A*, Vol. 44, No. 6, 1991, pp. 3985–3996.
[18] Garrison, J. C., and Chiao, R. Y., "Canonical and Kinetic Forms of the Electromagnetic Momentum in an Ad Hoc Quantization Scheme for a Dispersive Dielectric," *Physics Review A*, Vol. 70, No. 5, 2004, p. 053862(8).
[19] Mansuripur, M., "Radiation Pressure and the Linear Momentum of the Electromagnetic Field," *Optics Express*, Vol. 12, No. 22, pp. 5375–5401.
[20] Mansuripur, M., "Radiation Pressure on a Dielectric Wedge," *Optics Express*, Vol. 13, No. 6, 2005, pp. 2064–2074.
[21] Mansuripur, M., "Radiation Pressure and the Linear Momentum of Light in Dispersive Dielectric Media," *Optics Express*, Vol. 13, No. 6, 2005, pp. 2245–2250.
[22] Loudon, R., Barnett, S. M., and Baxter, C., "Radiation Pressure and Momentum Transfer in Dielectrics: The Photon Drag Effect," *Physics Review A*, Vol. 71, No. 6, 2005, p. 063802(11).
[23] Campbell, G. K., Leanhardt, A. E., Mun, J., Boyd, M., Streed, E. W., Ketterle, W., and Pritchard, D. E., "Photon Recoil Momentum in Dispersive Media," *Physical Review Letters*, Vol. 94, No.17, 2005, 170403(4).
[24] Leonhardt, U., "Energy-Momentum Balance in Quantum Dielectrics," *Physics Review A*, Vol. 73, No. 3, 2005, p. 032108(12).
[25] Scalora, M., D'Aguanno, G., Mattiucci, N., Bloemer, M. J., Centini, M., Sibilia, C., and Haus, J. W., "Radiation Pressure of Light Pulses and Conservation of Linear Momentum in Dispersive Media," *Physics Review E*, Vol. 73, No. 5, 2006, p. 056604(12).
[26] Cho, A., "Momentum from Nothing," *Physics Review Focus*, 2004, URL: http://focus.aps.org/story/v13/st3.
[27] Leonhardt, U., "Momentum in an Uncertain Light," *Nature*, Vol. 44, No. 14, 2006, pp. 823–824.
[28] Buchanan, M., "Minkowski, Abraham and the Photon Momentum," *Nature Physics*, Vol. 3, Feb. 2007, p. 73.

[29] Pfiefer, R. N., Nieminen, T. A., Heckenberg, N. R., and Rubinsztein-Dunlop, H., "Two Controversies in Classical Electromagnetism," *Proceedings of the SPIE: Optical Trapping and Optical Micromanipulation III*, Dholakia K., and Spalding, G. (eds.), Vol. 6326, 2006, 62260H(10).

[30] Slepian, J., "Electrical Essay: Electromagnetic Space-Ship," *Electrical Engineering*, Feb. 1949, pp. 145–146.

[31] Slepian, J., "Answer to Previous Essay," *Electrical Engineering*, Mar. 1949, p. 245.

[32] Corum, J. F., Dering, J. P., Pesavento, P., and Donne, A., "EM Stress–Tensor Space Drive," *Space Technologies and Applications International Forum 1999*, El Genk, M. S. (ed.), American Institute of Physics Conference Proceedings CP458, Springer-Verlag, New York, 1999, pp. 1027–1033.

[33] Corum, J. F., Keech, T. D., Kapin, S. A., Gray, D. A., Pesavento, P. V., Duncan, M. S., and Spadaro, J. F., "The Electromagnetic Stress-Tensor as a Possible Space Drive Propulsion Concept," *37th AIAA/ASME/SAE/ASEE Joint Propulsion Conference*, AIAA Paper 2001-3654, Salt Lake City, UT, July 2001.

[34] Heaviside, O., *Electromagnetic Theory: Volumes 1, 2, and 3*, Dover Pub., New York, 1950.

[35] Hartley, R., "Oscillations in Systems with Nonlinear Reactance," *Bell Systems Telephone Journal*, Vol. 15, No. 3, 1936, pp. 424–440.

[36] Manley, J. M., and Rowe, H. E., "Some General Properties of Nonlinear Elements—Part I: General Energy Relations," *Proceedings of the Institute Radio Engineering*, Vol. 44, No. 7, 1956, pp. 904–913.

[37] Rowe, H. E., "Some General Properties of Nonlinear Elements—Part II: Small Signal Theory," *Proceedings of the Institute Radio Engineering*, Vol. 46, No. 6, 1958, pp. 850–856.

[38] Lahoz, D. G., and Graham, G. M., "Experimental Decision on the Electromagnetic Momentum Expression for Magnetic Media," *Journal of Physics A*, Vol., 15, 1982, pp. 303–318.

[39] Seigenthaler, K. E., and Lawrence, T. J., "Breakthrough Propulsion Physics Research at the United States Air Force Academy," *42nd AIAA/ASME/SAE/ASEE Joint Propulsion Conference*, AIAA Paper 2006-4912, Sacramento, CA, July 2006.

[40] Eguinlian, A., and Gallo, B., "Critical Analysis of the Results Presented by the United States Air Force Academy in Report USAFA TR 2003-03," Instituto Universitario Aeronautico Technical Report DMA-008/04, Cordoba, Argentina, Mar. 2004.

[41] Brito, H. H., and Elaskar, S. A., "Overview of Theories and Experiments on Electromagnetic Inertia Manipulation Propulsion," *Space Technologies and Applications International Forum 2005*, El Genk, M. S. (ed.), American Institute of Physics Conference Proceedings CP746, Springer-Verlag, New York, 2005, pp. 1395–1402.

[42] Brito, H. H., "A Propulsion-Mass Tensor Coupling in Relativistic Rocket Motion," *Space Technologies and Applications International Forum 1998*, El Genk, M. S. (ed.), American Institute of Physics Conference Proceedings CP420, Springer-Verlag, New York, 1998, pp. 1509–1515.

[43] Brito, H. H., "Propellantless Propulsion by Electromagnetic Inertia Manipulation: Theory and Experiment," *Space Technologies and Applications International Forum 1999*, El Genk, M. S. (ed.), American Institute of Physics Conference Proceedings CP458, Springer-Verlag, New York, 1999, pp. 994–1004.

[44] Brito, H. H., "Candidate In-Orbit Experiment to Test the Electromagnetic Inertia Manipulation Concept," *Space Technologies and Applications International Forum 2000*, edited by El Genk, M. S. (ed.), American Institute of Physics Conference Proceedings CP458, Springer-Verlag, New York, 2000, pp. 1032–1038.

[45] Brito, H. H., "Research on Achieving Thrust by EM Inertia Manipulation," *37th AIAA/ASME/SAE/ASEE Joint Propulsion Conference*, AIAA Paper 2001-3656, Salt Lake City, UT, July 2001.

[46] Brito, H. H., and Elaskar, S. A., "Direct Experimental Evidence of Electromagnetic Inertia Manipulation Thrusting," *Journal of Propulsion and Power*, Vol. 23, No. 2, Mar.-Apr. 2007, pp. 487–494; also published as *39th AIAA/ASME/SAE/ASEE Joint Propulsion Conference*, AIAA Paper 2003-4989, Huntsville, AL, Jul. 2003.

[47] Furry, W. H., "Examples of Momentum Distributions in the Electromagnetic Field and in Matter," *American Journal Physics*, Vol. 37, No. 6, 1969, pp. 621–636.

[48] Feigel, A., "Quantum Vacuum Contribution to the Momentum of Dielectric Media," *Physical Review Letters*, Vol. 92, No. 2, 2004, p. 020404(4).

[49] Schutzhold, R., and Plunien, G., "Comment on 'Quantum Vacuum Contribution to the Momentum of Dielectric Media,'" *Physical Review Letters*, Vol. 93, No. 26, 2004, 268901(1); Feigel replies in *Physical Review Letters*, Vol. 93, No. 26, 2004, p. 268902(1).

[50] van Tiggelen, B. A., and Rikken, G. L., "Comment on 'Quantum Vacuum Contribution to the Momentum of Dielectric Media,'" *Physical Review Letters*, Vol. 93, No. 26, 2004, p. 268903(1); Feigel replies in *Physical Review Letters*, Vol. 93, No. 26, 2004, p. 268904(1).

[51] van Tiggelen, B. A., Rikken, G. L., and Krstic, V., "Momentum Transfer from Quantum Vacuum to Magetoelectric Matter," *Physical Review Letters*, Vol. 96, No. 13, 2006, p. 130402(4).

[52] van Tiggelen, B. A., Rikken, G. L., and Krstic, V., "The Feigel Process: Lorentz Invariance, Regularization, and Experimental Feasibility. Final Report of a Two Month Study Funded by the ESA Advanced Concepts Team," European Space Agency Internal Report, Ariadna Grant No. 18806/04/NL/MV, 2005.

Chapter 11

Experimental Results of the Woodward Effect on a Micro-Newton Thrust Balance

Nembo Buldrini* and Martin Tajmar[†]

Austrian Research Centers GmbH—ARC, Seibersdorf, Austria

I. Introduction

THE need for efficient space propulsion systems has stimulated researchers to investigate new pathways in modern physics for a theory or a phenomenon that could be exploited for this purpose. Nevertheless, even if some intriguing developments exist in this field, there is lack of experimental evidence: either the experiment gives inconclusive results, or the required test is not affordable because of our limited current technological capability. In any case, there are no wholly accepted experimental proofs that demonstrate such advanced propulsion systems are feasible. In 1990, James F. Woodward [1] started to get interesting results in his experiments to demonstrate that the inertia of a body is the result of Mach's principle; that is, it arises from the gravitational interaction with distant matter in the universe. Since then, he and other investigators have been obtaining remarkable results that are giving, step-by-step, a more complete view of the phenomenon, suggesting that what is detected is a genuine effect. In this paper, we tested some of these devices on a μN thrust balance in a large vacuum chamber. Our results are in general about an order of magnitude below Woodward's past claims. In one configuration, a net thrust effect was observed but still needs further investigation.

Copyright © 2008 by Austrian Research Centers GmbH—ARC. Published by the American Institute of Aeronautics and Astronautics, Inc., with permission.
*Research Scientist, Space Propulsion and Advanced Concepts.
[†]Head, Space Propulsion and Advanced Concepts.

II. Theoretical Considerations

The focus of this paper is on experimental evaluation; therefore, we will present just a brief description of the theoretical background, encouraging the reader to consider the full theoretical aspects that have already been extensively and rigorously treated [1–10]. In 1953, Dennis Sciama showed that inertial reaction forces can be viewed as the reaction to the radiative action of chiefly distant cosmic matter on accelerating local objects [11]. From the case of electrodynamics we know that radiation reaction includes third, as well as second, time-derivatives in the equations of motion, and therefore some peculiar phenomena not encountered in normal mechanics can take place. Starting from this assumption, Woodward derived a series of equations that show what happens to the mass-energy of an object "immersed" in the gravitational (or inertial) field of cosmic matter when it is accelerated by some external force. Interestingly, it turns out that the expression for the local source of the inertial (or gravitational) field has time-dependent terms that can be engineered to be surprisingly large. They are, however, only non-zero for sources that change their state of internal energy as they are accelerated.

Following the formalism adopted in Ref. 5, proper matter density fluctuation arising from Machian "gravinertial" interactions may be written as:

$$\delta \rho_0(t) \approx \frac{1}{4\pi G}\left[\frac{1}{\rho_0 c^2}\frac{\partial^2 E_0}{\partial t^2} - \left(\frac{1}{\rho_0 c^2}\right)^2 \left(\frac{\partial E_0}{\partial t}\right)^2\right] \quad (1)$$

where ρ_0 is body density (kg · m^{-3}), E_0 is the internal energy of the body (J), and G is Newton's gravitational constant ($= 6.67 \cdot 10^{-11}$ m^3 · kg^{-1} · s^{-2}). Capacitors excited by an alternating voltage are one possible system where large, rapid fluctuations in E_o can easily be affected and should produce periodic mass fluctuations δm_0 that might be used to generate stationary forces. Mass fluctuation is just the integral of $\delta \rho_0(t)$ over the volume of the capacitor, and the corresponding integral of the time derivatives of E_0, because $\partial E_0/\partial t$ is the power density will be:

$$\delta m_0 \approx \frac{1}{4\pi G}\left[\frac{1}{\rho_0 c^2}\frac{\partial P}{\partial t} - \left(\frac{1}{\rho_0 c^2}\right)^2 \frac{P^2}{V}\right] \quad (2)$$

where P is the instantaneous power delivered to the capacitor and V its volume. The predicted mass fluctuation can be computed using Eq. (2) which, after differentiation of $P = P_0 \sin(2\omega t)$ and ignoring the second term on the right-hand side, reads:

$$\delta m_0(t) \approx \frac{\omega P_0}{2\pi G \rho_0 c^2} \cos(2\omega t) \quad (3)$$

A simple way to get unidirectional force is to make the capacitor oscillate at a rate that it is accelerated in one direction when the mass undergoes a positive fluctuation, and in the other direction when the fluctuation is negative. Several devices have been built that used piezoelectric materials to exert an oscillating force on a capacitor subject to mass fluctuations [2,6,7]. A more reliable and convenient method to apply this force is the use of a magnetic field perpendicular to the displacement current in the capacitor [3–5,8,10,11]. If the phase between the current in the coil and the voltage applied to the capacitor is 90 degrees, then a Lorentz force from the displacement current and the field of the coil will act on the dielectric with the correct timing to give rise to a unidirectional force. This can be described with the following formalism:

$$F_B = i_d \times B \times L \tag{4}$$

where i_d is the displacement current in the capacitor (A), B the magnetic flux (T), and L the distance between the capacitor plates (m). Considering that i_d and B are orthogonal by construction and have the same frequency, the force may be expressed as follows:

$$F_B \approx B i_d L \cos \varphi \tag{5}$$

The total force acting on the supports of the device is the sum of the Lorentz force and the counterbalancing lattice forces:

$$F_{\text{tot}} = -(F_B + F_{\text{lat}}) \tag{6}$$

which, in the absence of mass fluctuations in the dielectric, turns out to be zero. However, when mass fluctuations are taken into account, the time-average of F_{tot} no longer vanishes in stationary circumstances if the phase relationship among F_B, i_d, and δm_o is such that F_B acts in phase with the mass fluctuation; thus we can write:

$$\langle F_{\text{tot}} \rangle \approx -2 \left(\frac{\delta m_0}{m_0} \right) F_B \sin \varphi \tag{7}$$

where the phase angle φ is the one between the voltage applied to the capacitor and the current in the inductor. The factor of two arises because the mass fluctuation peaks with reversed sign when the lattice restoring forces act during each cycle.

Although several experiments have claimed to have detected the existence of this mass fluctuation effect, the agreement of the results with the theory is not yet completely fulfilled. Usually the predictions underestimate observations by a factor of two to several hundreds, depending on the theoretical model adopted. This mismatching can arise from several factors, such as the interactions at atomic level involved in the production of such mass fluctuations that could reveal themselves to be more complex than expected. Another theoretical approach has been proposed by H. Brito and S. A. Elaskar based on purely electromagnetic considerations [12,13]. Electromagnetic fields possess momentum carried by the crossed E and B fields. Although it could be shown that a torque can be generated by such fields [14], the generation of linear momentum is highly controversial and called "hidden momentum" throughout the literature [15] (see also Chapter 10). Brito and Elaskar followed the "hidden momentum" hypothesis and built a thruster very similar to Woodward's device. Their theory predicted higher thrust values compared to Woodward's Machian approach.

III. Thrust Balance

The balance used in this work [16] has been especially developed to measure the thrust produced by In-FEEP (indium field emission electric propulsion) thrusters, designed and manufactured at the Austrian Research Centers. The force produced by a single thruster of this sort falls in the micro-Newton range, reaching milli-Newtons for clustered systems.

The balance consists of a symmetric arm (a hollow rectangular aluminum profile with a side length of 20 mm and a total length of about 60 cm), which is free to rotate by means of two flexural pivots. The pivots have been selected because of their low friction, high linearity, and negligible hysteresis. In addition they are able to support large loads (in the order of a few kilograms). The selected flexural pivot is a G-10 pivot from C-Flex, Inc. When the thruster, mounted to one end of the arm, is switched on, it will exert a force on the arm, causing it to shift its position. The deflection of the balance is a linear function of the applied force. Additionally, the measurable thrust range is adjustable, changing either the position of the sensor or the thruster (i.e., the distance to the center of rotation). Moreover, the flexural pivot can be substituted by a pivot with a different torsional spring rate.

Different sensors have been evaluated to measure the deflection of the thrust balance. The sensor selected is a D64 fiber optic displacement sensor (Philtec, Inc.), which works by measuring the reflection of light from the target surface. A big advantage of this sensor is that the part inside the vacuum chamber consists only of optical fibers, thus the signal cannot be influenced by electromagnetic interference (Fig. 1).

The typical response curve of the sensor allows the operation of the sensor in either the so-called "near side" or "far side." The near side offers a higher sensitivity at the cost of a decreased linear measurement range (this limits the measurable thrust range and makes the initial setup of the balance more complicated). The far side has a lesser sensitivity but a much larger measurement range.

a) Vacuum chamber   b) Thrust balance

Fig. 1 The vacuum chamber and thrust balance mounted inside the chamber for an In-FEEP test.

In order to simplify the initial setup of the balance and to enable use of both operating modes of the sensor, a movable sensor mount was designed. A vacuum-compatible stepper motor driving a high precision stage allows the movement of the sensor head. By following the response curve of the sensor one can set the working point either in the near or far side of the sensor.

Because the flexural pivot used has low friction, the motion of the balance is almost without damping. To reduce oscillations of the arm, a damping system had to be included in the design. Possibilities for this included oil dampers, electromagnetic dampers, and damping by electrostatic forces. The damping system selected for the balance was the latter. The signal from the displacement sensor is recorded into the computer and is differentiated, amplified, and fed into the voltage command input of a high-voltage power supply connected to a special comb assembly. This works as a velocity proportional damping system. The level of damping can be conveniently set in the control software.

A typical response of the balance is shown in Fig. 2. When a calibration force of 25 μN is applied by the electrostatic comb actuator, the arm moves to a new position. The oscillations are gradually reduced by the damping system. After switching off the voltage supply for the electrodes, the balance returns to its original position. Note that when the calibration force is switched on again, the balance is damped more strongly because the level of damping has been changed in the control software. This is not the highest possible level, however; further tests are needed to show the optimum damping. In general it is preferred to reach a steady state as quickly as possible to maximize measuring time.

Figure 3 displays the response of the balance to the force produced by a real In-FEEP thruster. The black trace represents the thrust calculated using the electrical values of the thruster; in this case the force was about 10 μN. In this

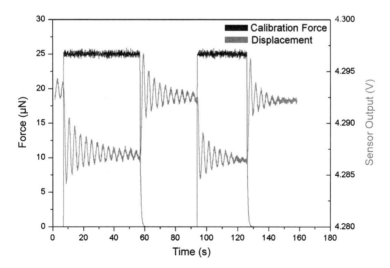

Fig. 2 Response of the balance to a calibration force.

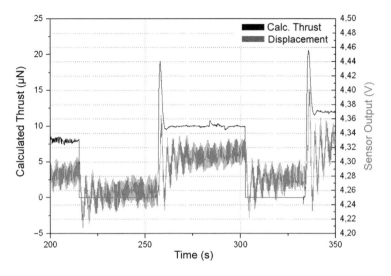

Fig. 3 Balance response to In-FEEP firing.

run the displacement sensor has been used only in the range with less sensitivity to simplify the setup of the balance in the vacuum chamber. In the high sensitivity range, resolutions of less than 1 µN are achievable.

IV. Setup

Figure 4 shows the typical balance assembly used to test mass fluctuation devices. This is a preliminary layout where all the electrical connections are

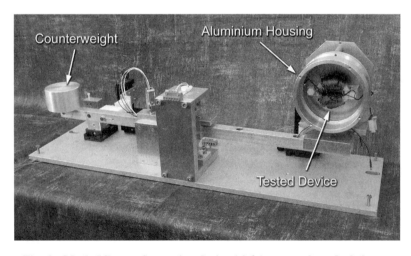

Fig. 4 Mach-5C mass fluctuation device (*right*) mounted on the balance.

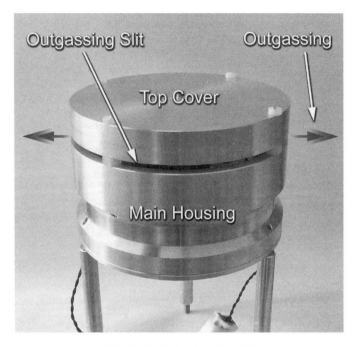

Fig. 5 Radial outgassing slit.

not yet implemented. The counterweight is located on the left-hand side of the beam. The white Teflon blocks next to it constitute part of the damping system actuator. The device to be tested is mounted on the other side of the balance arm. The article is enclosed in an aluminum housing (122-mm external diameter, 54-mm height) that provides EM radiation shielding and mechanical coupling with the thrust balance. Between the top cover and the main housing, a space of several millimeters is retained (Fig. 5) to allow for proper outgassing of the device (the tests are performed at a pressure of 10^{-6} mbar). This outgassing slit solution was selected to avoid other problems during the test of the article as well: an abrupt outgassing episode of some extent is expected every time the device is connected to the power supply (as result of the increase of temperature), and spurious thrust signals resulting from gas emission are probable and must be avoided. The slit provides a radial outlet, so that even if the outgassing is not uniformly distributed over the rim, the disturbance will be minimized as the thrust vector will be perpendicular to the sensitive thrust axis of the balance.

The devices are driven by two Carvin DCM2500 audio amplifiers, a model similar to those used by Woodward [3] and March [8], in order to provide a true replication of the original experiment. One amplifier drives the coil and the other one the capacitors. Stepup transformers are needed to elevate the voltage level from the output of the amplifiers. For the tests with Mach-6C and its modified version Mach-6CP (see experimental results), an additional

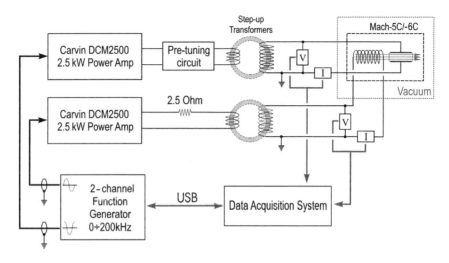

Fig. 6 Experiment setup with independent capacitor/coil driving.

pre-tuning circuit has been inserted in the capacitor line, between the amplifier and the stepup transformers. This circuit permits higher capacitor voltages, which should translate in higher thrust outputs. The signal generator used to pilot the amplifiers is a two-channel wave synthesis generator connected via USB to the computer. It is able to generate various waveforms with a maximum frequency of 200 kHz. The phase shift of each signal can be adjusted via software from 0° to 180° continuously. Having the two signals supplied independently makes it possible to check the dependence of the thrust produced on the relative phase shift, thus eliminating some possible spurious results. Much attention has to be paid in feeding the audio frequency power to the device. The cables have to be robust enough to carry high power levels. Consequently, mechanical interference with the balance cannot be excluded. The experiment setup is shown in Fig. 6.

Dedicated Labview software is used to manage the operation of the device and the acquisition of the data. For each shot, the start and duration of both signals can be set independently.

V. Results

Three different devices, two of them (named Mach-5C and Mach-6C) were provided and already tested by Woodward himself and have been characterized. All these devices are based on Eq. (7), where the unidirectional force is supposed to be achieved rectifying the mass fluctuation of the capacitor dielectric by means of the Lorentz force. The general structure of these devices consists of two or more capacitors surrounded by a coil, which generates a magnetic field perpendicular to the electric field established in the capacitors' dielectric.

A. Mach-5C

The device named Mach-5C is comprised of two 2.2 nF 15 kV Vishay Cera-Mite capacitors (Z5U dielectric, barium titanate, $\varepsilon_r \approx 8{,}500$) glued between the halves of a toroidal inductor core and connected in parallel (measured total capacitance, 4.95 nF). The magnetic flux through the capacitors needed to produce thrust is generated by two coils wound over the capacitors. The measured inductance of the two coils connected in series is 680 μH. In previous measurement sessions with this device, Woodward and Vandeventer [10] recorded a thrust of about 50 μN.

Since the first runs with the Mach-5C device, it was obvious that a pronounced thermal drift signature was present in the arm displacement plot (Fig. 7), recognizable from the fact that the trace does not return promptly back toward zero, as should occur when the device is not energized. Further investigation identified that the cause of the drift was due to a thermomechanical bending of the power supplying wires that go through the balance to the device. Nevertheless, even if such a spurious effect is present, it would not be difficult to see a real thrust signal superimposed on it.

Figure 8 shows the comparison between two signals: one obtained with a 90° phase shift between the capacitor voltage and the coil current, which should correspond to the maximum thrust production, and the other obtained with a 180° phase shift, which should result in a zero thrust trace, per Eq. (7). Mach-5C was energized with 1.6 kV to the capacitors and 2.2 A to the coil; the frequency of the driving signal was 43 kHz. The coil was energized for 4 seconds starting at 115 sec in the plot. After the initial second, the capacitor was energized for 4 sec. The total sequence lasted 5 seconds. The predicted thrust at 90° was 5 μN and not 50 μN, because the voltage going to the capacitors was 1.6 kV and not 3.5 kV, and the frequency was 43 kHz instead of 60 kHz as

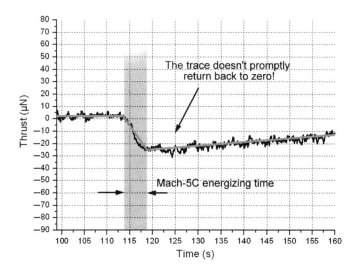

Fig. 7 Typical trace obtained with the Mach-5C device setup.

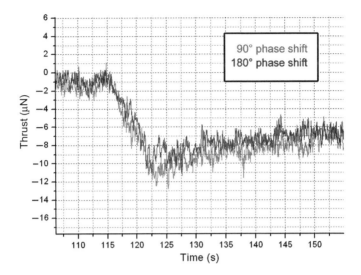

Fig. 8 90° and 180° phase shift superimposed plots.

used by Woodward in his 50-μN test. From the graph it is possible to see a small difference of several μN between the curves, but it is difficult to say if it is due to a genuine effect, as the initial thermal conditions were not monitored. That could lead to different thermal behavior of the wires, confusing the actual results.

Using the lower frequency configuration we can safely say that the thrust produced from the Mach-5C device must be less than 2 μN compared to the 5 μN extrapolated from Woodward's results. However, we have to note that, due to the heating of the capacitors, the phase between coil and capacitor is changing so that if a Machian thrust is present, the maximum time that it acts on the balance is only tenths of a second. Still, a trace on the balance should be possible to see, as we will see in a later section where we tested the balance response with short calibration pulses. As the 2-μN thrust limit is, however, close to the 5 μN estimate, at best we can say that the results are inconclusive from this experiment.

B. Mach-6C

More controllable conditions were achieved with the Mach-6C device (Fig. 9), also built by Woodward, who equipped it with an embedded thermistor. Tests by Woodward claim a thrust generated with this device in the 150-μN thrust range [17]. Monitoring the temperature allows operation of the device far away from the temperature range where the capacitor dielectric can get damaged; moreover, it permits to begin each test with the same starting conditions. The Mach-6C assembly comprises eight capacitors (Y5U dielectric, barium titanate, $\varepsilon_r \approx 5000$), connected in parallel (measured total capacitance, 4.20 nF) and a coil wound in toroidal configuration over them (measured inductance,

Fig. 9 Mach-6C.

1.17 mH). A test from Woodward assessed that with Mach-6C, it was possible to reach higher voltages (and thus higher thrusts) without incurring dielectric aging, a phenomenon that was noted to reduce the performance of the "5" series. We upgraded the driver setup so that it was possible to reach the required higher voltages values at the capacitor leads (in excess of 3.2 kVp). The operation frequency ranged between 50 kHz and 55 kHz. Figure 10a depicts the typical trace of one Mach-6C run. The thermal drift has been virtually eliminated by rearranging the wiring and reducing the firing time to about 2 sec. It appears evident that no appreciable thrust signal is present in the balance arm

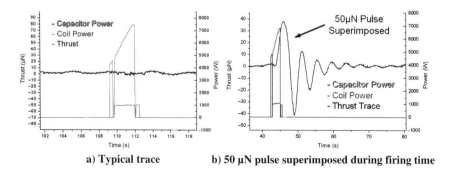

a) Typical trace   b) 50 µN pulse superimposed during firing time

Fig. 10 Mach-6C measurements.

displacement trace within the balance resolution of 1 μN. Compared with Woodward's results of 150 μN, we should have definitely seen a trace.

In order to check if the balance was working under the same conditions as described above, Fig. 10b shows a trace achieved with similar working parameters but with a 50 μN superimposed calibration pulse generated independently by the electrostatic actuator (the duration of the pulse coincides with the running time of the device). The idea behind this test was to show that, if there were actually a propulsive effect coming from the device, it would not be hidden by something occurring during the firing time (counter effects from the wires, EMI, etc.). One problem encountered during experimentation with Mach devices was the difficulty to keep the phase relationship stable between the capacitor voltage and the coil current. Even if the firing time is very short, in fact, the power levels supplied to the device are relatively high and this causes the heating up of the several parts. As mentioned in the previous section, heating directly translates into a phase shift of the applied voltage to the capacitors relative to the phase of the current flowing into the coil. Furthermore, the voltage applied to the capacitor varies as we are receding from the condition of resonance in the circuit comprising the capacitor itself and the driving circuitry. That said, we can assume that ideal thrust conditions only occur during the first tenths of a second. Therefore, in order to evaluate the behavior of our balance at different thrust duration values, a series of short pulses was generated using the calibration actuator, as shown in Fig. 11. The first pulse is the usual calibration and is performed before every device run to provide the actual scaling factor. It consist of a single pulse (usually corresponding to a thrust of 50 μN) generated while keeping the actuator switched on for the time needed by the damping system to bring the balance arm to a static position. The pulses in Fig. 11 were generated with a software-controlled timing procedure. From this

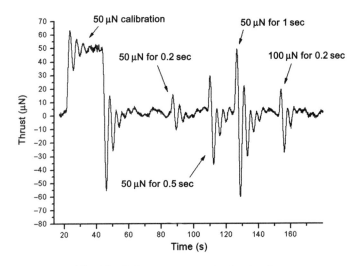

Fig. 11 Balance response to short pulses.

graph we can observe that it is possible to easily discriminate 50 μN pulses as short as 0.2 sec.

In summary, even short pulses of 50 μN and 100 μN should have been clearly identifiable with thrust peaks of up to 30 μN. In our experimental setup we can say that if the Machian thrust effect is present, it must be at least one order of magnitude less than in previous claims.

C. Mach-6CP

The Mach-6CP was tested in a different configuration with the power cables fed through the outgassing slit instead of the bottom of the housing as for the Mach-6C tests (Fig. 12). As we will see, that could have caused an interference with the balance. The graphs in Fig. 13 show the results obtained with a new version of Mach-6C, also provided by Woodward. The same device has been potted (thus the "P" in the model number) in epoxy paste and enveloped in a mild steel housing. This encapsulation provides an additional shield against electromagnetic emission and eliminates the occurring of a possible corona effect on the capacitor leads. It also decreases the cooling time between one run and the next. A thrust of 100 μN to 200 μN was recorded in Woodward's facilities (in vacuum).

One standard procedure to discriminate an authentic effect from a false positive is to run the coil and the capacitors separately. If the effect is genuine, in fact, it has to be present only when the coil and capacitors are working together, so that the magnetic field generated by the coil can act on the displacing ions in the

Fig. 12 Mach-6CP.

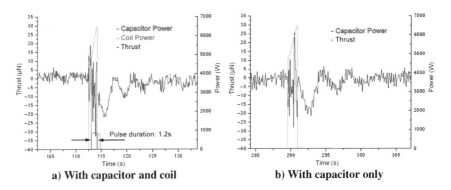

Fig. 13 Mach-6CP measurements.

capacitor dielectric lattice, producing the Lorentz force necessary to rectify the mass fluctuations and produce net thrust. Figure 13 shows the comparison between a shot with the capacitors and coil together, and a shot with only the capacitors in operation. A thrust signature continues to be recorded when the capacitors are working alone. The real cause of this pulse still must be assessed. One possibility could be the different arrangement of the feeding cables (in the "P" version, they have to exit the housing from the outgassing slit), together with an inadequacy of their electromagnetic shielding. In fact, if some electric fields are leaking out, they can interact with the fixed portions of the balance or with the vacuum chamber wall, giving rise to a net force on the device during operation time.

Actually, some electromagnetic field measurements were pursued (with the device operating in air) and a leak in the cable shielding near to the aluminum housing was found: an electric field strength of 2.5 kV/m was detected several centimeters away from the housing, that can perhaps account for the observed effect. The wires were then shielded properly so that a field strength of just 59 V/m was measured. Unfortunately, it has not been possible to date to perform experiments with this last modified version, so we cannot yet confirm the real nature of the effect. There is another explanation that can account for the thrust effect recorded: because the capacitor leads are asymmetric, they generate a magnetic field that can interact with the dielectric of the capacitor itself and generate a thrust according to the mass fluctuation effect. However, the fact that the thrust magnitude is not varying between the two circumstances (in both cases the maximum swing of the balance arm is indicating about 22 µN) suggests that the origin of the recorded effect is not of Machian nature. Otherwise the magnetic field generated by the coil would have added to the produced thrust and a difference between the plots should have been seen.

D. 2-MHz Breadboard Device

A breadboard device has been built and tested here in Austrian Research Centers to explore force production at higher frequencies and using different dielectric materials. It consists of high-power ceramic capacitors, designed for high frequency duty (dielectric: titanium dioxide, $\varepsilon_r \approx 120$) and a coil connected in series, driven at resonance (resonance frequency: 2 MHz) by a compact class-E type amplifier. Although the dielectric relative permittivity of titanium dioxide is quite low compared to the materials employed in the capacitors of the Mach-5C/6C devices, the dissipation factor is one order of magnitude lower, thus allowing their use at higher frequency without incurring severe overheating. Driving the coil and the capacitor in a series (or parallel) configuration has two desirable advantages:

1) The voltage on the capacitor and the current in the coil always have a relative phase shift of 90°, thus, via Eq. (7), maximum thrust is obtainable.

2) At resonance, high voltage (or current) values are built up inside the device leading to high-reactive (not dissipative) power levels on the order of several kVA with a relatively low driving power, related only to the dissipation factor of the circuit.

The device, as shown in Fig. 14, was operated in air on a commercial electronic balance (Sartorius MC-1). Voltages in excess of 2 kVp were obtained with currents on the order of 4 to 5 A. An expected thrust varying from 1 mN to 6 mN was calculated, depending on the theoretical approach used [9], but no real thrust was detected within the sensibility of the used electronic balance (0.1 mN). Allowing for the Machian effects to be a real phenomenon, a possible explanation of the failure of this device could be that the dielectric used and its geometry (very thin layer) is not suitable for producing mass fluctuations. Accordingly, a further investigation using different type of capacitors seems to be reasonable.

Fig. 14 2-MHz breadboard device.

The kind of device here described appears to be an optimal breadboard to continue Machian mass fluctuation testing because of its insensitiveness to capacitor/coil phase shifts and its very compact layout. This last feature makes this design suitable for self-contained setup testing, where device, amplifier, and power source (battery) are placed together on the thrust-measuring apparatus, thus eliminating spurious signals coming from the feeding cables.

VI. Conclusions

The aim of the reported experimental activity was to evaluate the propulsive capabilities of devices based on the Machian mass fluctuation effect. Two test devices, built and already tested by Woodward, have been characterized using a highly sensitive μN thrust balance usually employed for electric ion propulsion. Our results are, for the most part, not in agreement with the data collected by Woodward, even if some interesting thrust plots were recorded during the last test with the Mach-6CP device that suggest further investigation be pursued. In general, we find that any Machian effect eventually produced by such devices is likely more than an order of magnitude below Woodward's past claims.

A device operating at higher frequencies and with different capacitors has been designed and built at the Austrian Research Centers. No thrust has been detected within the sensibility of the electronic balance used, although the minimum thrust measurable with this instrument was well below the magnitude predicted by the theoretical models. However, it must be taken into account that the geometry and the dielectric material of the capacitors may influence the expression of the effect in a way that is still not clear.

Recently, Woodward built a thrust balance similar to the one at the Austrian Research Centers and initial tests seem to be in general agreement with our analysis of the magnitude of the effect [17], though Woodward claims to see a small, real effect. The thrust values of Mach-6CP-like devices are presently under investigation on this thrust balance.

Acknowledgments

The main credit goes to J. Woodward. Without his contributions and the continuing inspiration of his work during the last 10 years, this paper wouldn't have been realized. Many thanks go out to P. March, A. Palfreyman, P. Vandeventer, T. Mahood, and other colleagues for providing useful discussions. We also acknowledge the help of K. Marhold and B. Seifert for designing the original μN thrust balance.

References

[1] Woodward, J. F., "A New Experimental Approach to Mach's Principle and Relativistic Gravitation," *Foundation of Physics Letters*, Vol. 3, No. 5, 1990, pp. 497–506.

[2] Woodward, J. F., "A Laboratory Test of Mach's Principle and Strong-Field Relativistic Gravity," *Foundations of Physics Letters*, Vol. 9, No. 3, 1996, pp. 247–293.

[3] Woodward, J. F., "Flux Capacitors and the Origin of Inertia," *Foundations of Physics*, Vol. 34, No. 10, 2004, pp. 1475–1514.

[4] Woodward, J. F., "Life Imitating 'Art': Flux Capacitors and Our Future in Spacetime," *Proceedings of Space Technology and Applications International Forum (STAIF-2004)*, El-Genk, M. S. (ed.), AIP Conference Proceedings 699, Melville, NY, 2004, pp. 1127–1137.

[5] Woodward, J. F., "Tweaking Flux Capacitors," *Proceedings of Space Technology and Applications International Forum (STAIF-2005)*, El-Genk, M. S. (ed.), AIP Conference 746, Melville, NY, 2005, pp. 1345–1352.

[6] Mahood, T. L., "A Torsion Pendulum Investigation of Transient Machian Effects," M.S. Thesis, California State University – Fullerton (CSUF), 1999.

[7] Mahood, T. L., March, P., and Woodward, J. F., "Rapid Spacetime Transport and Machian Mass Fluctuations: Theory and Experiments," *Proceedings of 37th AIAA/ASME/SAE Joint Propulsion Conference*, AIAA Paper 2001–3907, 9 July 2001.

[8] March, P., "Woodward Effect Experimental Verifications," *Space Technology and Applications International Forum (STAIF-2004)*, El-Genk, M. S. (ed.), AIP Conference Proceedings 699, Melville, NY, 2004, pp. 1138–1145.

[9] March, P., and Palfreyman, A., "The Woodward Effect: Math Modeling and Continued Experimental Verifications at 2 to 4 MHz," *Proceedings of Space Technology and Applications International Forum (STAIF-2006)*, El-Genk, M. S. (ed.), American Institute of Physics, Melville, NY, 2006.

[10] Woodward, J. F., and Vandeventer, P., "Mach's Principle, Flux Capacitors, and Propulsion," *Proceedings of Space Technology and Applications International Forum (STAIF-2006)*, El-Genk, M. S. (ed.), American Institute of Physics, Melville, NY, 2006.

[11] Sciama, D., "On the Origin of Inertia," *Monthly Notices of the Royal Astronomical Society*, 113, 1953, pp. 34–42.

[12] Brito, H. H., and Elaskar, S. A., "Direct Experimental Evidence of Electromagnetic Inertia Manipulation," *39th AIAA/ASME/SAE/ASEE Joint Propulsion Conference and Exhibit*, AIAA Paper 2003–4989, Huntsville, AB, July 2003.

[13] Brito, H. H., and Elaskar, S. A., "Overview of Theories and Experiments on Electromagnetic Inertia Manipulation Propulsion," *Proceedings of Space Technology and Applications International Forum (STAIF-2005)*, El-Genk, M. S. (ed.), AIP Conference Proceedings 746, Melville, NY, 2005, pp. 1395–1402.

[14] Graham, G. H., and Lahoz, D. G., "Observation of Static Electromagnetic Angular Momentum in Vacuo," *Nature*, Vol. 285, 1980, pp. 154–155.

[15] Hnizdo, V., "Hidden Momentum and the Electromagnetic Mass of a Charge and Current Carrying Body," *American Journal of Physics*, Vol. 65, No. 1, 1997, pp. 55–65.

[16] Marhold, K., and Tajmar, M., "Micronewton Thrust Balance for Indium FEEP Thrusters," *41st AIAA/ASME/SAE Joint Propulsion Conference*, AIAA Paper 2005-4387, Tucson, AZ, July 2005.

[17] Woodward, J. F., "Mach's Principle and Propulsion: Experimental Results," *Proceedings of Space Technology and Applications International Forum (STAIF-2007)*, El-Genk, M. S. (ed.), American Institute of Physics, Melville, NY, 2007.

Chapter 12

Thrusting Against the Quantum Vacuum

G. Jordan Maclay*
Quantum Fields LLC, Richland Center, Wisconsin

I. Introduction

THIS chapter addresses the question of how the properties of the quantum vacuum might be exploited to propel a spacecraft. Quantum electrodynamics (QED), the theory of the interaction of light and matter that has made predictions verified to 1 in 10 billion [1], predicts that the quantum vacuum, which is the lowest state of the electromagnetic field, contains a fluctuating virtual photon field. This fluctuating field is predicted to produce vacuum forces between nearby surfaces [1-3]. Recently these Casimir forces have been measured and found to agree with predictions [4-9]. If this virtual radiation pressure could be utilized for propulsion, the goal of propellantless propulsion would be achieved. Restrictions due to the conservation of energy and momentum are discussed. A propulsion system based on an uncharged, conducting mirror that vibrates asymmetrically in the vacuum is described. By the dynamic Casimir effect, the mirror produces real photons that impart momentum and result in a net acceleration. The acceleration is very small, but demonstrates that the vacuum can be utilized in propulsion. Technological improvements, some of which are proposed, may be used to increase the accelerating force. Many questions remain about the supporting theory, and experiments are needed to probe questions about the quantum vacuum that are far beyond current theory.

Rockets employing chemical or ionic propellants require the transport of prohibitively large quantities of propellant. If the properties of the quantum vacuum could somehow be utilized in the production of thrust, that would provide a decided advantage because the vacuum is everywhere. At this embryonic stage, in the exceedingly brief history of interstellar spacecraft, we are trying

---

Copyright © 2008 by G. Jordon Maclay. Published by the American Institute of Aeronautics and Astronautics, Inc., with permission.
*Professor Emeritus, University of Illinois.

to distinguish between what appears possible and what appears impossible within the context of our current understanding of quantum physics and the fundamental laws of physics, particularly conservation of momentum and energy. Science fiction writers have written about the use of the quantum vacuum to power spacecraft for decades but no research has validated this suggestion. Arthur C. Clark, who proposed geosynchronous communications satellites in 1945, described a "quantum ramjet drive" in 1985 in *Songs of Distant Earth*, and observed "If vacuum fluctuations can be harnessed for propulsion by anyone besides science-fiction writers, the purely engineering problems of interstellar flight would be solved."[†] Australian science fiction writer Ken Ingle described, with my fanciful suggestions, the Casimir vacuum drive in his soon to be published book *First Contact*.

In the last 10 years great progress has been made experimentally in measuring Casimir forces, which arise between closely spaced surfaces due to the quantum fluctuations of the electromagnetic field, the quantum vacuum. Although the forces tend to be small, practical applications of vacuum forces have recently appeared in microelectromechancial systems (MEMS) devices [10–12].

QED predicts the behavior of the quantum vacuum, including vacuum forces and the presence of a vast energy in empty space due to a fluctuating electromagnetic field. Unfortunately we do not yet have a proven method to propel a spacecraft by harnessing the vast energy of vacuum fluctuations that QED predicts, and therefore this chapter focuses on general considerations about momentum transfer between the quantum vacuum and a spacecraft. The spacecraft proposed in this paper is described as a "gedanken spacecraft" because its design is intended not as an engineering guide but just to illustrate possibilities. Indeed, based on our current understanding of quantum vacuum physics, one could reasonably argue that the gedanken spacecraft could be propelled more effectively by simply oscillating a charged mirror that would emit electromagnetic radiation or simply using a flashlight or laser to generate photons. Although the performance of the vacuum-powered gedanken spacecraft as presented is disappointing and is no more practical than a spacewarp [13], the discussion illustrates many important ideas about the quantum vacuum, and it suggests the potential role of quantum vacuum phenomena in a macroscopic system like space travel. In fact, with a breakthrough in materials, methods, or fundamental understanding, this approach could become practical, and we might be able to realize the dream of space travel as presented in science fiction. Physicists have explored various means of locomotion depending on the density of the medium and the size of the moving object. It would be interesting to find an optimum method for moving in the quantum vacuum. Unfortunately we currently have no simple way to mathematically explore these various simple possibilities.

[†]Clarke, A. C., Personal Communication. See the Acknowledgments in *The Songs of Distant Earth*. Numerous science fiction writers, including Clarke, Asimov, and Sheffield have based spacecraft on the quantum vacuum.

II. Physics of the Quantum Vacuum
A. Historical Background

Quantum mechanics is one of the great scientific achievements of the twentieth century. It provides models that describe many properties of atoms and molecules, such as the optical spectra and transition probabilities. In its original form, as developed by Schrodinger, Heisenberg, Bohr, and others, quantum mechanics is a nonrelativistic theory that makes the ad hoc assumption that light is emitted and absorbed by atoms in bundles, called photons. The electromagnetic field, however, is treated as an ordinary classical field that obeys Maxwell's relativistic equations, not as a quantized field. Dirac, Heisenberg, Jordan, Dyson, and others began formulation of a relativistic form of quantum mechanics, and made efforts to quantize the electromagnetic field. This quantized field theory of particles and light theory developed over the next few decades with numerous successful predictions. In 1948, Willis Lamb tested a crucial prediction of the field theory, that the 2s and 2p levels of a hydrogen atom would have precisely the same energy. Lamb sent a beam of hydrogen atoms through a cavity exposed to radio frequency radiation, and determined that the 2s and 2p energy levels were in fact split by an energy equivalent to 1000 MHz [14]. Within days, Hans Bethe of Cornell realized the problem and published the solution: The theoretical calculation did not consider the effects of the quantum vacuum on the energy levels of the hydrogen atoms [15]. This ushered in the modern formulation of QED of Feynmann, Schwinger, and Tomonaga [1].

In QED, particles and light are both treated as quantized fields that are fully relativistic. Because the electromagnetic field is quantized, there may be 0, 1, 2, 3 or any number of photons present and, because the fields are relativistic, they can be readily transformed to coordinate systems that are translated or moving uniformly (Lorentz transformations). The entire formalism of QED can be written in tensors that ensure the proper transformation properties of all observables under a Lorentz transformation.

The pervasive and dynamic role of the vacuum state in QED was unexpected to many physicists. The lowest state of the quantized electromagnetic field, which is referred to as the quantum vacuum, was predicted to be filled with photons and electron–positron pairs that appear and disappear continuously and so rapidly that no direct measurement of their presence is possible. Yet these so called "virtual particles" affect measurable properties of atoms, such as the energy levels, magnetic moments, and transition probabilities.

In retrospect, when arguing from nonrelativistic quantum mechanics and the uncertainly principle, it was clear that the vacuum would contain a fluctuating electromagnetic field once the field was treated as a quantized field. The field variables, E_ω and B_ω, representing the electric and magnetic field at a frequency ω, are directly analogous to P_ω, and Q_ω, the position and momentum of a harmonic oscillator of frequency ω. The ground state of the harmonic oscillator has to obey Heisenberg's Uncertainty Principle: $\Delta P_\omega \Delta Q_\omega \geq \hbar$, where ΔP_ω is the uncertainty in the momentum and ΔQ_ω is the uncertainty in the position. In the lowest state, the oscillator is still vibrating, with an energy $\frac{1}{2}\hbar\omega$. If it were not vibrating but was motionless, then the uncertainty in its momentum would be zero. If we knew the approximate position of the oscillator, then this state would violate the Uncertainly Principle. The energy of the nth excited state of the oscillator is $(n+\frac{1}{2})\hbar\omega$.

Similarly quantized electric and magnetic fields cannot vanish, but must, in their lowest state, fluctuate. This isotropic residual fluctuating electromagnetic field, which is present everywhere at zero Kelvin temperature with all electromagnetic sources removed, is often called the zero-point electromagnetic field.

Quantum fluctuations occur in the particle fields as well as the electromagnetic field, so the quantum vacuum is filled with virtual electron–positron pairs, as well as virtual photons. Before Lamb's Nobel Prize-winning measurement, most physicists felt comfortable ignoring the effects of the quantum fluctuations, assuming that they just shifted the energy but did not have measurable consequences. It turns out that quantum fluctuations affect virtually all physical processes, including the mass, charge, and magnetic moment of all particles; the lifetimes of excited atoms or particles; scattering cross sections; and the energy levels of atoms. QED, which accounts for all the vacuum processes, has made experimental predictions of magnetic moments and energy levels that have been verified by experiment to 1 in 10 billion, the most accurate predictions of any scientific theory [1,16].

B. Energy in the Quantum Vacuum

Zero-point field energy density is a simple and inexorable consequence of quantum theory and the uncertainty principle, but it brings puzzling inconsistencies with another well-verified theory, General Relativity. The energy in the quantum vacuum at absolute zero, which is the lowest energy state of the electromagnetic field, is due to the presence of virtual photons of energy $\frac{1}{2}\hbar\omega_n$ of all possible frequencies:

$$E_0 = \frac{1}{2}\sum_{n=0}^{n_{max}} \hbar\omega_n \qquad (1)$$

Usually a cutoff is used for the high frequencies, such as the frequency corresponding to the Planck length of 10^{-34}m which gives an enormous energy density (about 10^{114} J/m^3 or, in terms of mass, 10^{95} g/cm^3). From the perspective of General Relativity, this enormous energy density seems to make no physical sense, and that is why the effects of the quantum fluctuations were neglected for decades. Indeed such a large energy would, according to the General Theory of Relativity, have a disastrous effect on the metric of spacetime. For an infinite flat universe, this vacuum energy density would imply an outward zero-point pressure that would rip the universe apart [17]. Astronomical data, on the other hand, indicate that any such cosmological constant must be ~ 4 eV/mm^3, or 10^{-29} g/cm^3 when expressed as mass [18]. The discrepancy here between theory and observation is about 120 orders of magnitude, and is arguably the greatest quantitative discrepancy between theory and observation in the history of science [19,20]! There are numerous approaches to solve this "cosmological constant problem," such as renormalization, supersymmetry, string theory, and quintessence, but as yet this remains an unsolved problem.

Gradually, the belief has developed that only changes in the energy density give observable effects [21].

Each virtual photon of frequency ω and wave vector $\vec{k}$ ($k = 2\pi/\lambda$) has associated with it a momentum $\hbar \vec{k}$. Because photons are in random directions, the mean momentum of the vacuum fluctuations vanishes, but just as there are fluctuations in the electric and magnetic fields consistent with the uncertainty principle, there are fluctuations in the root mean square momentum. At finite temperatures, real photons begin to appear in the quantum vacuum, but their contribution to the total energy is much smaller than that of the virtual photons.

C. Casimir Forces Predicted in 1948

About the same time as Lamb's experiment, Heindrick Casimir, director of research at Phillips Laboratories in the Netherlands, found some disagreements between the experiment and his model for precipitation of phosphors used in the manufacture of fluorescent light bulbs. Better agreement between theory and experiment could be obtained if the van der Waal's force between two neutral, polarizable atoms somehow fell off more rapidly at larger distances than had been supposed. A co-worker suggested that this might be related to the finite speed of light, which prompted Casimir and Polder to reanalyze the van der Waals interaction. They found that including the retardation effects caused the interaction to vary as r^{-7} rather than r^{-6} at large intermolecular separations r, which gave agreement with the experiment.

Intrigued by the simplicity of the result, Casimir sought a deeper understanding. A conversation with Bohr led him to an interpretation in terms of zero-point energy, and the realization that, by simply considering the changes in vacuum energy arising from the presence of surfaces in the vacuum, forces due to the vacuum fluctuation would appear [22]. To understand this result, consider how inserting two parallel surfaces into the vacuum causes the allowed modes of the electromagnetic field to change. This change in the modes that are present occurs since the electromagnetic field must meet the appropriate boundary conditions at each surface. Thus surfaces alter the modes of oscillation, and therefore the surfaces alter the energy density corresponding to the lowest state of the electromagnetic field. In actual practice, the modes with frequencies above the plasma frequency do not appear to be significantly affected by the metal surfaces because the metal becomes transparent to radiation above this frequency. In order to avoid dealing with infinite quantities, the usual approach is to compute the finite change ΔE_0 in the energy of the vacuum due to the presence of the surfaces:

$$\Delta E_0 = E[\text{energy in empty space}] \quad (2)$$
$$- E_S[\text{energy in space with surfaces present}]$$

where the definition of each term is given in brackets. This equation can be expressed as a sum over the corresponding modes:

$$\Delta E_0 = \frac{1}{2}\sum_{n=0}^{n_{\max}} \hbar\omega_n - \frac{1}{2}\sum_{m=0}^{\text{surfaces}} \hbar\omega'_m \tag{3}$$

The quantity ΔE_0 can be computed for various geometries. The forces F due to the quantum vacuum are obtained by computing the change in the vacuum energy for a small change in the geometry and differentiating. For example, consider a hollow conducting rectangular cavity with sides a_1, a_2, a_3. Let $en(a_1, a_2, a_3)$ be the change in the vacuum energy due to the cavity, then the force F_1 on the side perpendicular to a_1 is:

$$F_1 = -\frac{\delta en}{\delta a_1} \tag{4}$$

where δen represents the infinitesimal change in energy corresponding to an infinitesimal change in the dimension δa_1. This equation also represents the conservation of energy when the wall perpendicular to a_1 is moved infinitesimally:

$$\delta en = -F_1 \delta a_1 \tag{5}$$

Thus, if we can calculate the vacuum energy density as a function of the dimensions of the cavity, we can compute derivatives that give the forces on the surfaces. For uncharged, perfectly conducting, parallel plates with a very large area A, very close to each other (separation of d), the tangential component of the electric field must vanish at the surface, and wavelengths longer than twice the plate separation d are excluded. With the appropriate boundary conditions, we can compute the change in vacuum energy and use Eq. (5) to predict an attractive (or negative) force between the plates:

$$F = -\frac{K}{d^4} \tag{6}$$

where

$$K = \frac{\pi^2 \hbar c}{240} \tag{7}$$

This force F, commonly called the Casimir force, arises from the *change in vacuum energy density* E_{pp} from the free field vacuum density that occurs between the parallel plates [21,23]:

$$E_{pp}(d) = \frac{-K}{3d^3} \tag{8}$$

Two decades after Casimir's initial predictions, a method was developed to compute the Casimir force in terms of the local stress-energy tensor using quantum electrodynamics [24]. Many innovations have followed. Vacuum forces have been computed for other geometries besides the classic parallel plate geometry, such as a rectangular cavity, cube, sphere, cylinder, and wedge. For a cube or sphere, the Casimir forces are outward or repulsive. For a rectangular cavity, the Casimir forces on the different faces may be inward, outward, or zero depending on the ratio of the sides. Situations arise in which there are inward forces on some faces and outward forces on other forces [25]. It is difficult to understand these unusual results intuitively.

The application of Casimir forces in space propulsion is motivated more clearly by the interpretation of the parallel plate Casimir force as arising from radiation pressure, the transfer of momentum from the virtual photons in the vacuum to the surfaces [3]. It is this virtual radiation pressure that we propose to explore as a possible driving force to generate net forces on an object, ultimately to propel a spacecraft. There are very significant advantages if it is possible to use virtual radiation pressure: no propellant may be required, and there is always something to push.

D. Dynamic Casimir Effect

In the dynamic Casimir effect the parallel plates are imagined to move rapidly, which can lead to an excited state of the vacuum between the plates, meaning the creation of real photons [26]. To understand this process from a physical perspective, imagine that in a real moving conductor, the surface charges must constantly rearrange themselves to cancel out the transverse electric field at all positions. This rapid acceleration of charge can lead to radiation. This effect, generally referred to as the dynamic or adiabatic Casimir effect, has been reviewed but not yet observed experimentally [21,23,27]. The vacuum field exerts a force on the moving mirror that tends to damp the motion. Energy conservation requires the existence of a radiation reaction force working against the motion of the mirror [28]. This dissipative force may be understood as the mechanical effect of the emission of radiation induced by the motion of the mirror. This force of radiation reaction can be used to accelerate the mirror, or a spacecraft attached to the vibrating mirror, as discussed in Section V.

The Hamiltonian is quadratic in the field operators, and formally analogous to the Hamiltonian describing photon pair creation by parametric interaction of a classical pump wave of frequency ω_0, with a nonlinear medium [29]. Pairs of photons with frequencies $\omega_1 + \omega_2 = \omega_0$, are created out of the vacuum state. Furthermore, the photons have the same polarization, and the components of the corresponding wave vectors $\vec{k}_1$ and $\vec{k}_2$ taken along the mirror surface must add to zero because of the translational symmetry:

$$\vec{k}_1 \cdot \hat{x} + \vec{k}_2 \cdot \hat{x} = \omega_1 \sin \theta_1 + \omega_2 \sin \theta_2 = 0 \qquad (9)$$

This last equation relates the angles of emission of the photon pairs with respect to the unit vector $\hat{x}$, which is normal to the surface. It is interesting that the photons emitted by the dynamic Casimir effect are entangled photons. This analysis in terms of the analogous effective Hamiltonian is illuminating but not complete for perfect mirrors, because no consistent effective Hamiltonian can be constructed in this case with the idealized and pathological boundary conditions. More realistic results are obtained assuming that the mirrors are transparent above a plasma frequency.

The dynamic Casimir effect was studied for a single, perfectly reflecting mirror with arbitrary nonrelativistic motion and a scalar field in three dimensions in 1982 by Ford and Vilenkin [30]. They obtained expressions for the vacuum radiation pressure on the mirror. In 2001, Barton extended the analysis using a one-dimensional scalar field to a moving body with a finite refractive index [31]. The vacuum radiation pressure and the radiated spectrum for a nonrelativistic, perfectly reflecting, infinite, plane mirror was computed by Neto and Machado for the electromagnetic field in three dimensions, and shown to obey the fluctuation-dissipation theorem from linear response theory [28,32]. This theorem shows the fluctuations for stationary body yield information about the mean force experienced by the body in nonuniform motion. Jaekel and Reynaud computed shifts in the mass of the mirror for a scalar field in two dimensions [33]. The mirror mass is not constant, but rightfully a quantum variable because of the coupling of the mirror to the fields by the radiation pressure. A detailed analysis was done by Barton and Calogeracos in 1995 for a dispersive mirror in one dimension that includes radiative shift in the mass of the mirror and the radiative reaction force [34]. This model can be generalized to an infinitesimally thin mirror with finite surface conductivity and a normally incident electromagnetic field.

E. Alternative Theories of Casimir Forces

The experimental verification of Casimir's prediction is often cited as proof of the reality of the vacuum energy density of quantum field theory. Yet, as Casimir himself observed, other interpretations are possible:

> The action of this force [between parallel plates] has been shown by clever experiments and I think we can claim the existence of the electromagnetic zero-point energy without a doubt. But one can also take a more modest point of view. Inside a metal there are forces of cohesion and if you take two metal plates and press them together these forces of cohesion begin to act. On the other hand you can start with one piece and split it. Then you have first to break chemical bonds and next to overcome van der Waals forces of classical type and if you separate the two pieces even further there remains a curious little tail. The Casimir force, *sit venia verbo*, is the last but also the most elegant trace of cohesion energy [35].

Several approaches to computing electromagnetic Casimir forces have been developed that are not based on the zero-point vacuum fluctuations directly. In the special case of the vacuum electromagnetic field with dielectric or conductive boundaries, various approaches suggest that Casimir forces can be regarded as macroscopic manifestations of many-body retarded van der Waals forces, at least in simple geometries with isolated atoms [1,36]. Casimir effects have also been derived and interpreted in terms of source fields in both conventional [1] and unconventional [37] quantum electrodynamics, in which the fluctuations appear within materials instead of outside of the materials. Lifshitz provided a detailed computation of the Casimir force between planar surfaces by assuming that stochastic fluctuations occur in the tails of the wavefunctions of atoms that leak into the regions outside the surface These fluctuating tails can induce dipole moments in atoms in a nearby surface, which leads to a net retarded dipole-induced dipole force between the planar surfaces [38]. These various approaches that are alternatives to conventional QED always postulate the existence of fluctuations in potentials, wave functions, or electromagnetic fields, and give results consistent with QED formulations in the few cases of simple geometries that have been computed [16].

It should be pointed out that all QED calculations must routinely include the effects of the vacuum fluctuations in order to obtain the correct results. For example, the spontaneous emission from excited atoms depends on transitions induced by the vacuum field.

F. Limitations of Current Theoretical Calculations of Vacuum Forces

The parallel plate geometry (and the approximately equivalent sphere–flat plate geometry or sphere–almost flat plate geometry) is essentially the only geometry for which experimental measurements have been conducted and the only geometry for which the vacuum forces between two separate surfaces (assumed to be infinite) have been computed. In the calculations with spheres, the radius of curvature of the sphere is very large compared to the separation; therefore locally, the geometry is a parallel plate geometry. Vacuum forces are known to exist in other experimental configurations between separate surfaces, but rigorous calculations based on QED are very difficult and have yet to be completed [39]. Because it is experimentally possible to measure forces between various separate surfaces with the improvement in experimental techniques, theoreticians may soon see the need for such computations.

Calculations of vacuum stresses for a variety of geometric shapes, such as spheres, cylinders, rectangular parallelepipeds, and wedges are reviewed in Refs. 2, 21, and 23. In general, calculations of vacuum forces become very complex when the surfaces are curved, particularly with right angles. Divergences in energy appear, and there are disagreements about the proper way to deal with these divergences [40]. The material properties, such as the dielectric constant and plasma frequency of the metal and the surface roughness also affect the vacuum forces, and are often not treated realistically in theoretical calculations. Indeed, in the Lizshitz formulation, the Casimir forces depend on the permittivity and permeability as a function of the frequency over the entire frequency range. Because this information is not generally available, approximations have to be made. In addition, usually only a spatial average of the force for a given area for the ground state of the quantum vacuum field is computed, and material properties such as binding energies are ignored, a procedure which Barton has recently questioned [25,31,41].

III. Measurements of Casimir Forces

It was not until about 1998, that the parallel plate Casimir force was measured accurately [4,9]. Corrections for finite conductivity and surface roughness have been developed for the parallel plate geometry, and the agreement between theory and experiment is now at about the 1% level for separations of about

0.1 to 0.7 μm [5]. In actual practice, the measurements are most commonly made with one surface curved and the other surface flat, using the proximity force theorem to account for the curvature. This experimental approach eliminates the difficulties of trying to maintain parallelism at submicron separations. Mohideen and collaborators have made the most accurate measurements to date in this manner, using an atomic force microscope (AFM) that has a metallized sphere about 250 μm in diameter attached to the end of a cantilever about 200-μm long, capable of measuring picoNewton forces. The deflection of the sphere is measured optically as it is moved close to a flat metallized surface [4]. The more difficult measurement between two parallel plates has been made by Bressi et al. who obtained results that are consistent with theory [6]. Measurements of the force between two parallel surfaces, each with a small (1 nm) sinusoidal modulation in surface height, have shown that there is a lateral force as well as the usual normal force when the modulations of the opposing surfaces are not in phase [7]. Recent measurements have confirmed the predictions, including effects of finite conductivity, surface roughness, and temperature, uncertainty in dielectric functions, to the 1 to 2% level for separations from 65 to 300 nm [8].

A. Forces on Conducting Surfaces

Parallel plate Casimir forces go inversely as the fourth power of the separation between the plates. The Casimir force per unit area between perfectly conducting plates is equivalent to about 1 atm pressure at a separation of 10 nm, and so is a candidate for actuation of MEMS. The relative strength of Casimir, gravitational, and electrostatic forces for parallel, conducting surfaces is shown in Fig. 1 [42].

In 1995 the first analysis of a dynamic MEMS structure that used vacuum forces was presented by Serry et al. [42]. They consider an idealized MEMS component resembling the original Casimir model of two parallel plates, except that one of the plates is connected to a stationary surface by a linear restoring force (spring) and can move along the direction normal to the plate surfaces. The model demonstrates that the Casimir effect could be used to actuate a switch, and might be responsible in part for the "stiction" phenomenon in which micromachined membranes are found to latch onto nearby surfaces during the fabrication process. If the movable surface is vibrating, then an "anharmonic Casimir oscillator" (ACO) results. Other MEMs structures using Casimir forces, such as pistons and interdigitated combs, have been proposed [43].

In MEMS, surfaces may come into close proximity with each other, particularly during processes of etching sacrificial layers in the fabrication process. To explore stiction in common MEMS configurations, Serry et al. computed the deflection of membrane strips and the conditions under which they would collapse into nearby surfaces [44]. Measurements were done by Buks and Roukes on cantilever beams to investigate the role of Casimir forces in stiction [45]. An experimental realization of the ACO in a nanometer-scale MEMS system was recently reported by Chan et al. [10]. In this experiment the Casimir attraction between a 500-μm square plate suspended by torsional rods and a gold-coated sphere of radius 100 μm was observed as a sharp increase in the tilt angle of the plate as the sphere–plate separation was reduced from 300 nm to 75.7 nm. This "quantum mechanical actuation" of the plate suggests "new possibilities for novel actuation schemes in MEMS based on the Casimir force" [10]. In a refinement of this experiment, a

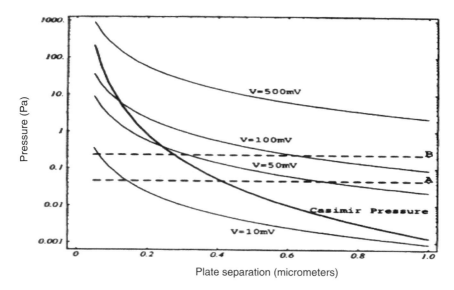

Fig. 1 Casimir, electrostatic, and gravitational pressures. A comparison of the attractive pressures due to the Casimir effect and applied electrostatic voltage (V) between two flat parallel plates of conductors in vacuum. Also shown are the gravitational pressure on 2-μm thick (*dashed line A*) and 10-μm thick (*line B*) silicon membranes [42].

novel proximity sensor was demonstrated in which the plate was slightly oscillated with an AC signal, and the deflection amplitude observed with its rapid inverse fourth power behavior gave an indication of the precise location of the nearby sphere [11]. A measurement using a similar torsion oscillator was recently reported using gold on the sphere and chromium on the plate [46].

B. Forces on Semiconductor Surfaces

One of the potentially most important configurations from the technological viewpoint involves vacuum forces on semiconductor surfaces. The Casimir force for a conducting material depends approximately on the plasma frequency, beyond which the material tends to act like a transparent medium. For parallel plates separated by a distance, d, the usual Casimir force is reduced by a factor of approximately $C(a) = [1 + (8\lambda_p/3\pi d)]^{-1}$, where λ_p is the wavelength corresponding to the plasma frequency of the material [47]. Because the plasma frequency is proportional to the carrier density, it is possible to tune the plasma frequency in a semiconductor, for example, by illumination, by temperature, or by the application of a voltage bias. In principle it should be possible to build a Casimir switch that is activated by light, a device that would be useful in optical switching systems. A very interesting measurement of the Casimir force between a flat surface of borosilicate glass and a surface covered with a film of amorphous silicon was done in 1979 by Arnold et al. [48]. They observed an increase in the Casimir force when the semiconductor was exposed to light. This experiment has yet to be repeated with modern methods and materials. As a first step, Chen et al. have used an AFM to measure the force between a single Si crystal and a 200-μm diameter gold coated sphere, and found good agreement with theory using the Lizshitz formalism [49].

IV. Space Propulsion Implications

A. General Considerations

Conservation of energy and momentum place severe restrictions on what mechanisms may be utilized to propel spacecraft. For example, if a spacecraft is accelerating due to an interaction with the quantum vacuum, then it has to be removing energy from the quantum vacuum. Further the increase in kinetic energy must be equal or less than the decrease in energy in the quantum vacuum. Some general constraints on using the vacuum for space travel, as well as methods of altering the metric of spacetime for space travel, are outlined in the paper by Puthoff et al. [50]. In the analysis of any proposed approaches, we need to consider the momentum and energy of the field plus any objects in the field. Consider, for example, a spacecraft mechanism such as a sail, that alters the normally isotropic quantum vacuum energy density in a local region surrounding the spacecraft. Let $E(\omega, \vec{r}, \vec{r}_S)$ be the change in the vacuum energy density as a function of the frequency ω, the position $\vec{r}$ measured with respect to the center of the sail given by $\vec{r}_S$, which is measured with respect to some fixed location. If, in actual fact, this function $E(\omega, \vec{r}, \vec{r}_S)$ does not depend on $\vec{r}_S$ but has the same shape no matter where the sail is located, then the change in vacuum energy due to the presence of the sail is constant. By the conservation of energy, the sail is moving at a constant velocity, and cannot experience a force due to its interaction with the quantum vacuum. In conclusion, if the change in vacuum energy does not depend on the position of the spacecraft, then the energy and momentum are constant.

B. Sails in the Vacuum

A variety of sail concepts have been proposed [51]. As we mentioned earlier, we can view the vacuum as a source of radiation pressure from virtual photons. The challenge is to design surfaces that alter the symmetry of the free vacuum and produce a net force. Consider, for example, a sail made of two different materials on opposite sides that absorb electromagnetic radiation differently. Can we expect a net force on the sail? A simple classical analysis as shown in Fig. 2 suggests the answer to this question.

For a given frequency, assume the radiation energy density is proportional to $cf(\omega, T)$, the net momentum transfer ΔP_ω to the top surface is

$$\Delta P_\omega = A_\omega f(\omega, T) + E_\omega f(\omega, T) + 2R_\omega f(\omega, T) \tag{10}$$

where A_ω is the absorptivity, E_ω is the emissivity, R_ω the reflectivity, and T the temperature. For a body in thermodynamic equilibrium, $A_\omega = E_\omega$, and by definition, $1 = A_\omega + R_\omega$. Using these restrictions it follows that $\Delta P_\omega = 2f(\omega, T)$, which is independent of the material properties. Therefore, even if the individual A_ω, E_ω, and R_ω are different for the other side, the same relations hold and the

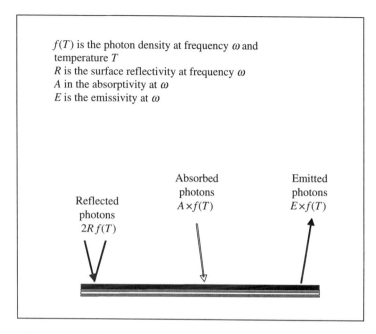

Fig. 2 Schematic of the momentum transfer from zero-point electromagnetic radiation to a sail made from different materials at temperature T on the top and bottom.

force on the opposite side of the sail just cancels this force, and there is no net acceleration. This conclusion holds at every frequency. We assumed the temperature of the sail is the same on both sides because of the intimate contact. If the radiation spectrum corresponds to that at zero temperature, then $f(\omega, 0)$ describes the zero-point field, and both sides of the sail would be at zero Kelvin. On the other hand, if one made a sail in which a temperature gradient was maintained across the sail, a net force might occur, and it would be a function of the energy required to maintain the temperature difference.

There is a complication to this analysis: What happens if the sail is moving? If the radiation density is due solely to the quantum vacuum at zero temperature $[cf(\omega, 0) = \hbar\omega^3/2\pi^2 c^3]$, then the spectral energy density the sail sees does not change with motion. The invariance of the spectrum of the zero-point fluctuations with uniform motion is a special property of the zero-point quantum vacuum. Without this property, one could distinguish a unique rest frame for the universe, violating the intent of special relativity. On the other hand, the thermal fields of real photons do not have this unique invariance. Hence uniform motion in a thermal field results in a Doppler shifted spectrum. For a sail, this means that the spectral energy density is different on the opposite sides of the sail and, provided the integral of the forces over all frequencies were different for the two sides of the sail, it would be possible to obtain a net, thermally generated, force. When one considers the restrictions on the frequency dependence of dielectric coefficients due to causality, it is uncertain if one can generate a net force with this method. The possibility remains unresolved.

Einstein considered this situation for an atom moving in an isotropic thermal field, and showed that the increase in the atom's kinetic energy upon absorption and emission of radiation is balanced by the drag force if the thermal field follows the usual Planck spectrum. Similarly if the atom is moving in the zero-point vacuum field, there is no net force on the atom. But if the spectrum does not have this form, net forces are possible.

By inserting surfaces into the vacuum, we can alter the spectrum of the vacuum fluctuations, which results in net forces. Indeed, wherever there is an inhomogeneous vacuum energy density, there will be a net force on a polarizable particle given by $\frac{1}{2}\alpha\vec{\nabla}\langle E(x)^2\rangle$ [16]. From a propulsion viewpoint, this suggests the possibility of using vacuum forces to eject particles to generate a propulsive force, an approach that may not offer any distinct advantage over more conventional methods.

Friction due to the quantum vacuum has been predicted to exist between two parallel infinite plates that have finite conductivity. The friction arises because of the motion of charges in the surface of the metal moving to maintain the boundary conditions.

In conclusion, for a sail to accelerate due to the quantum vacuum, the sail must be removing energy from the vacuum. This prompts the question, by what processes can one remove energy from the vacuum?

C. Inertia Control by Altering Vacuum Energy Density

One can make use of the negative vacuum energy density that arises in a parallel plate structure in an alternative approach to propulsion. Based on the principles of General Relativity, one would expect the changes in vacuum energy ΔE to correspond to a change in mass $\Delta E/c^2$. Thus the negative energy density of the vacuum in a parallel plate structure should result in the equivalent of a negative mass object. Proposals have been made to test this hypothesis, stating that a negative vacuum energy leads to a reduction in mass by constructing stacks of capacitors; however, the predicted effect is just beyond current measurement capability From the theoretical perspective, one can estimate the positive and negative mass contributions for a parallel plate capacitor made from plates that are only one hydrogen atom thick and one atom apart, and still the total energy is positive. Thus it appears that a parallel-plate Casimir cavity will always have a net positive energy density and cannot be used to create a zero or negative mass spacecraft, or initiate a wormhole [17]. Nevertheless, the negative vacuum energy density may, with more effective approaches, be of use in reducing inertia, which reduces kinetic energy and the amount of work required to accelerate a spacecraft to a given velocity [50].

There is another variant of a negative mass drive that deserves mention. A system of charges that has a negative electrostatic potential energy ΔE would also, by the principles of General Relativity, be expected to have a negative associated mass $\Delta E/c^2$. Thus, in a gravitational field, there would be a levitating force. Pinto has explored this possibility, and suggests the effect may be amplified and made measurable if highly polarizable hydrogen atoms in a magnetic trap are exposed to isotropic laser radiation [52]. Although these technically challenging enhancements may amplify the basic levitating force, it still appears that the effective reduction in mass will be quite small compared to the total mass.

Negative energy drives and the hypothesis linking inertia to vacuum fluctuations [53] are discussed more fully in Chapters 3, 4, 13, and 15.

D. Dynamic Systems

Dynamic systems, in which something moves and interacts with the quantum vacuum, may have the possibility of extracting energy from the vacuum. Hence they may be able to accelerate a spacecraft. The movement might be a macroscopic physical motion, a piezoelectrically driven surface, or the motion of electrons within a semiconductor, possibly altering the plasma frequency or the dielectric constant. We discuss one possible dynamic system next.

V. Vibrating Mirror Casimir Drive

It is possible to conceive of a vacuum spacecraft that operates by pushing on the quantum vacuum with a vibrating mirror [54]. With a suitable trajectory, the motion of a mirror in vacuum can excite the quantized vacuum electromagnetic field with the creation of real photons.

A. Simple Model Based on Energy and Momentum Conservation

The important physical features of using the dynamic Casimir effect to accelerate a spacecraft can be seen in a simplified, heuristic model. Assume that the spacecraft has an energy source, such as a battery, that powers a motor that vibrates a mirror or a system of mirrors in a suitable manner to generate radiation. We will assume that there are no internal losses in the motor or energy source. We assume that at the initial time t_i, the mirrors are at rest. Then the mirrors are accelerated by the motor in a suitable manner to generate a net radiative reaction on the mirror, and at the final time t_f, the mirrors are no longer vibrating, and the spacecraft has attained a non-zero momentum. We can apply the First Law of Thermodynamics to the system of the energy source, motor, and mirror at times t_i, and t_f:

$$\Delta Q = \Delta U + \Delta W \quad (11)$$

where ΔU represents the change in the internal energy in the energy source, $-\Delta W$ represents the work done on the mirrors moving against the vacuum, and ΔQ represents any heat transferred between the system and the environment. We will assume that we have a thermally isolated system and $\Delta Q = 0$ so

$$0 = \Delta U + \Delta W \quad (12)$$

By the conservation of energy, the energy ΔU extracted from the battery goes into work done on the moving mirror $-\Delta W$. Because the mirror has zero vibrational and kinetic energy and zero potential energy at the beginning and the end of the acceleration period, and is assumed to operate with no mechanical friction, all work done on the mirror goes into the energy of the emitted radiation ΔR, and the kinetic energy of the spacecraft of mass M

$$\Delta W = \Delta R + \frac{M(\Delta V)^2}{2} \tag{13}$$

Thus the energy of the radiation emitted due to the dynamic Casimir effect equals

$$\Delta R = -\Delta U - \frac{M(\Delta V)^2}{2} > 0 \tag{14}$$

The frequency of the emitted photons depends on the Fourier components of the motion of the mirror. We assume that the radiant energy can be expressed as a sum of energies of n_i photons each with frequency ω_i:

$$\Delta R = \sum_i n_i \hbar \omega_i \tag{15}$$

The number of photons emitted depends on the cosine of the angle the photon momentum makes with the normal to the surface, as Neto and Machado [28] show. In this simplified calculation, we will assume that all photons are emitted normally from one side of the accelerating surface. This assumption is not valid, but it allows us to obtain a best-case scenario and illustrates the main physical features. If all photons are emitted normally from one surface, then the photon momentum transfer ΔP is

$$\Delta P = \sum_i n_i \frac{\hbar \omega_i}{c} = \frac{\Delta R}{c} \tag{16}$$

where c is the speed of light. Using Eq. (14), we obtain the result

$$\Delta P = \frac{-\Delta U}{c} - \frac{M(\Delta V)^2}{2c} \tag{17}$$

In a nonrelativistic approximation $\Delta P = M \Delta V$ and the change in velocity ΔV of the spacecraft is to second order in $\Delta U / Mc^2$:

$$\frac{\Delta V}{c} = \frac{-\Delta U}{Mc^2} + \left(\frac{\Delta U}{Mc^2}\right)^2 \tag{18}$$

This represents a maximum change in velocity attainable by use of the dynamic Casimir effect (or by the emission of electromagnetic radiation generated by more conventional means) when the energy ΔU is expended. The ratio $\Delta U / Mc^2$ is expected to be a small number, and we can neglect the second term in Eq. (18). As a point of reference, for a chemical fuel, the ratio of the heat of formation to the mass energy is approximately 10^{-10}. With this approximation, we find the maximum value of $\Delta V/c$ equals $\Delta U/Mc^2$, the energy obtained from the energy source divided by the rest mass energy of the spacecraft. It follows that the kinetic energy of the motion of the spacecraft E_{ke} can be expressed as:

$$E_{\text{ke}} = \frac{M(\Delta V)^2}{2} = \Delta U \frac{\Delta U}{2Mc^2} \qquad (19)$$

This result for the upper limit on the spacecraft kinetic energy shows that the conversion of potential energy ΔU from the battery into kinetic energy of the spacecraft is an inefficient process because $\Delta U/Mc^2$ is a small factor. Almost all of the energy ΔU has gone into photon energy. This inefficiency follows because the ratio of momentum to energy for the photon is $1/c$.

In our derivation, the internal energy of the system is used to create and emit photons from some unspecified process; no massive particles are ejected from the spacecraft (propellantless propulsion). We have neglected: 1) the change in the mass of the spacecraft as the stored energy is converted into radiation, 2) radiative mass shifts, 3) complexities related to high energy vacuum fluctuations and divergences, and 4) all dissipative forces in the system used to make the mirror vibrate. These assumptions are consistent with a heuristic nonrelativistic approximation.

In this simplified model, we have not made any estimates about the rate of photon emission and how long it would take to reach the maximum velocity. For configurations considered in the literature, rates of photon emission from the dynamic Casimir effect are estimated to be very low, typically 10^{-5} photons/sec or about 300 photons/yr [28]. Also we will have to vibrate the mirror asymmetrically so that more photons are emitted from one side than the other. In the derivation, however, we never made any assumptions about the mechanism by which photons were generated, so the derivation holds quite generally, whether we simply use a battery and a perfect lightbulb or a vibrating charged surface.

1. Static Casimir Effect as Energy Source

The general analysis in the preceding section can be taken one step further to suggest a spacecraft whose operation is totally based on quantum vacuum properties. The vibrating motor in the spacecraft could be powered by energy removed from the quantum vacuum using an arrangement of perfectly conducting, uncharged, parallel plates. Detailed considerations about extracting energy from the quantum vacuum are discussed in Refs. 55–58, and presented in Chapter 18. The Casimir energy $U_C(x)$ at zero degrees Kelvin between plates of area A, separated by a distance x is:

$$U_C(x) = -\frac{\pi^2}{720} \frac{\hbar c A}{x^3} \qquad (20)$$

If we allow the plates to move from a large initial separation a to a very small final separation b then the change in the vacuum energy between the plates is approximately:

$$U_C(x) = U_C(b) - U_C(a) \tag{21}$$

Substituting Eq. (20) gives the result

$$\Delta U_C \approx -\frac{\pi^2}{720}\frac{\hbar c A}{b^3} \tag{22}$$

The attractive Casimir force has done work on the plates, and, in principle, we can build a device to extract this energy with a suitable, reversible, isothermal process, and use it to accelerate the mirrors. We neglect any dissipative forces in this device, and assume all of the energy ΔU_C can be utilized. Thus the maximum value of $\Delta V_C/c$ obtainable using the energy from the Casimir force "battery" is:

$$\frac{\Delta V_C}{c} = \frac{\pi^2}{720}\frac{1}{Mc^2}\frac{\hbar c A}{b^3} \tag{23}$$

We can make an upper bound for this velocity by making further assumptions about the composition of the plates. Assume that the plate of thickness L is made of a material with a rectangular lattice that has a mean spacing of d, and that the mass associated with each lattice site is m. Then the mass of one plate is:

$$M_P = AL\frac{m}{d^3} \tag{24}$$

The density approximation is good for materials with a cubic lattice, and within an order of magnitude of the correct density for other materials.

In principle it is possible to make one of the plates in the battery the same as the plate accelerated to produce radiation by the dynamic Casimir effect. As the average distance between the plates is decreased, the extracted energy is used to accelerate the plates over very small amplitudes. If we assume we need to employ two plates in our spacecraft, and that the assembly to vibrate the plates has negligible mass, then the total mass of the spacecraft is $M = 2M_P$ and we obtain an upper limit on the increase in velocity:

$$\frac{\Delta V_C}{c} = \frac{\pi^2}{1400}\frac{\hbar}{Lmc}\frac{d^3}{b^3} \tag{25}$$

The final velocity is proportional to the Compton wavelength ($\hbar/mc$) of the lattice mass m divided by the plate thickness L. Assume that the final spacing between the plates is one lattice constant ($d = b$), that the lattice mass m equals the mass of a proton m_p, and that the plate thickness F is one Bohr radius $a_0 = \hbar^2/m_e e^2$, then we obtain (α is the fine structure constant with approximate value of 1/137):

$$\frac{\Delta V_C}{c} = \frac{\pi^2}{1400} \frac{\alpha m_e}{m_p} \quad (26)$$

(A real plate constructed with current technology might easily be three orders of magnitude thicker.) Substituting numerical values we find:

$$\frac{\Delta V_C}{c} = \frac{\pi^2}{1400} \frac{1}{137} \frac{1}{1800} = 2.78 \times 10^{-8} \quad (27)$$

This best-case scenario corresponds to a disappointing final velocity of about 8 m/s, about 10^3 times smaller than for a large chemical rocket. As anticipated, the spacecraft is very slow despite the unrealistically favorable assumptions made in the calculation, yet this simple gedanken experiment does demonstrate that it may be possible to base the operation of a spacecraft entirely on the properties of the quantum vacuum. Using an additional energy source can result in higher terminal velocity.

B. Model for Propulsion Using Vibrating Mirrors

Assume we have a flat, perfectly reflecting, mirror whose equilibrium position is $x = 0$. At a time t where $t_i < t < t_f$ the location of the mirror is given by $x(t)$. Neto has given an expression for the force per unit area $F(t)$ on such a mirror [32]:

$$F(t) = \lim_{\delta x \to 0} \frac{\hbar c}{30 \pi^2} \left(\frac{1}{\delta x} \frac{d^4 x(t)}{c^4 dt^4} - \frac{d^5 x(t)}{c^5 dt^5} \right) \quad (28)$$

where δx represents the distance above the mirror at which the stress-energy tensor is evaluated. The second term represents the dissipative force that is related to the creation of traveling wave photons, in agreement with its interpretation as a radiative reaction. In computing the force due to the radiation from the mirror's motion, the effect of the radiative reaction on $x(t)$ is neglected in the non-relativistic approximation. The divergent first term can be understood in several ways. Physically, it is a dispersive force that arises from the scattering of low frequency evanescent waves. The divergence can be related to the unphysical nature of the perfect conductor boundary conditions. Forcing the field to vanish on the surface requires its conjugate momentum to be unbounded. Thus the average of the stress-energy tensor $\langle T_{\mu\nu} \rangle$ is singular at the surface for the same reason that single-particle quantum mechanics would require a position eigenstate to have infinite energy [30]. This divergent term can be lumped into a mass renormalization, and therefore disappears from the dynamical equations when they are expressed in terms of the observed mass of the body [31,32]. We will not discuss this term further in this calculation, although we will return to the general idea of radiative mass shift in our discussion. We will assume that diffraction effects are small for our finite plates.

The total energy radiated per unit plate area E can be expressed

$$E = -\int_{t_1}^{t_2} dt F(t) \frac{dx(t)}{dt} \qquad (29)$$

Substituting Eq. (28) for $F(t)$ we find

$$E = \frac{\hbar}{30\pi^2} \frac{1}{c^4} \int_{t_1}^{t_2} dt \left(\frac{d^3 x(t)}{dt^3}\right)^2 \qquad (30)$$

The total impulse I per unit plate area can also be computed as the integral of the force per unit area over time:

$$I = \int_{t_1}^{t_2} dt\, F(t) = -\frac{\hbar}{30\pi^2} \frac{1}{c^4} \left(\frac{d^4 x(t)}{dt^4}\bigg|_{t_2} - \frac{d^4 x(t)}{dt^4}\bigg|_{t_1}\right) \qquad (31)$$

The total impulse I equals the mass of the system M per unit area times the change in velocity ΔV in a nonrelativistic approximation:

$$I = M\Delta V \qquad (32)$$

We want to specify a trajectory for the mirror that will give a net impulse. One of the trajectories that has been analyzed is that of the harmonic oscillator [30,59]. In this case, the mirror motion is in a cycle and we can compute the energy radiated per cycle per unit area and the impulse per cycle per unit area. For a harmonic oscillator of frequency Ω and period $T = 2\pi/\Omega$, there is only one Fourier component of the motion, so the total energy of each pair of photons emitted is $\hbar\Omega = \hbar(\omega_1 + \omega_2)$. For a harmonically oscillating mirror the displacement is

$$x_{ho}(t) = X_0 \sin \Omega t \qquad (33)$$

A computation based on Eqs. (30) and (31) shows there will be a net power radiated in a cycle, however, the dissipative force for the harmonic oscillator F_{ho} will average to zero over the entire cycle as shown in Fig. 3, so there will be no net impulse.

In order to secure a net impulse, we need a modified mirror cycle. One such model cycle can be readily constructed by using the harmonic function $x_{ho}(t)$ over the first and last quadrants of the cycle, where the force F_{ho} is positive, and a cubic function $x_c(t)$ over the middle two quadrants where F_{ho} is negative:

$$x_c(t) = \frac{X_0}{2} \frac{(\Omega t - \pi)^3}{(\pi/2)^3} - \frac{3X_0}{2} \frac{(\Omega t - \pi)}{\pi/2} \qquad (34)$$

The coefficients for the cubic polynomial are chosen so that at $\Omega t = \pi/2, 3\pi/2$ the displacement and the first derivatives of $x_c(t)$ and $x_{ho}(t)$ are equal. As can be

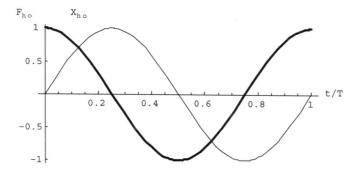

Fig. 3 The displacement x_{ho} and the radiative reaction F_{ho} (*bold line*) for a harmonically oscillating mirror plotted as a function of the normalized time. For convenience F_{ho} and x_{ho} are normalized to one.

seen from Fig. 4, the cubic function $x_c(t)$ matches $x_{ho}(t)$ quite closely in the interval $0.25 < t/T < 0.75$. Of course the higher order derivatives do not match, and that is precisely why the force differs.

The similarity in displacement and the difference in the resulting force is striking. For the mirror displacement $x_m(t)$ in our model we choose:

$$x_m(t) = x_{ho}(t) \quad \text{for } 0 \leq t/T \leq 0.25; \quad 0.75 \leq t/T \leq 1 \quad (35)$$

$$x_m(t) = x_c(t) \quad \text{for } 0.25 < t/T < 0.75 \quad (36)$$

Figure 5 shows $x_m(t)$ plotted with the corresponding force per unit area $F_m(t)$ obtained from Eq. (28). The force $F_m(t)$ is positive in the first and last quarter of the cycle, and vanishes in the middle, where the trajectory is described by

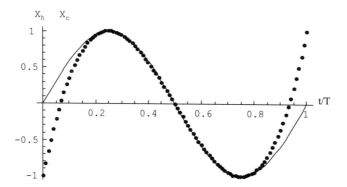

Fig. 4 The normalized displacement $x_{ho}(t)$ for a harmonically oscillating mirror (*solid line*) and the cubic function $x_{c(t)}$, (*bold dotted line*).

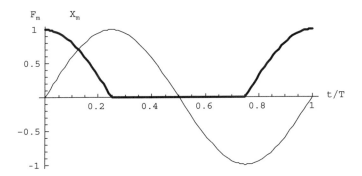

Fig. 5 The normalized displacement $x_m(t)$ and the corresponding normalized radiative force $F_m(t)$ (*bold line*), are shown as functions of the time. The force is positive in the first and last quarters, and zero in the middle quarters of the cycle.

the cubic. The energy radiated per area per cycle for our model trajectory can be obtained from Eq. (30):

$$E_m = -\frac{\hbar c}{60\pi} X_0^2 \left(\frac{\Omega}{c}\right)^5 \tag{37}$$

The total impulse per area per cycle for our model I_m trajectory is

$$I_m = -\frac{\hbar}{15\pi^2} X_0 \left(\frac{\Omega}{c}\right)^4 \tag{38}$$

The impulse is first order in $\hbar$ and is therefore typically a small quantum effect. Thus for our model cycle, the change in velocity per second is $\Delta V/dt$:

$$\Delta V_m/dt = \frac{I_m \Omega}{M} \tag{39}$$

where M is the mass per unit plate area of the spacecraft, and we assume the plate is the only significant mass in the gedanken spacecraft. In order to estimate ΔV_m, we can make some further assumptions regarding the mass of the plate per unit area. As before, we can make a very favorable assumption regarding the mass per unit area of the plates $M = m_p/a_0^2$, which yields the change in velocity per second:

$$\Delta V_m/dt = -\frac{\hbar}{15\pi^2} X_0 \left(\frac{\Omega}{c}\right)^4 \Omega \frac{a_0}{m_p} \tag{40}$$

If we substitute reasonable numerical values [59,60], a frequency of $\Omega = 3 \times 10^{10}$ s^{-1} and an oscillation amplitude of $X_0 = 10^{-9}$ m, we find that

$\Delta V_m/dt$ is approximately 3×10^{-20} m/s² per unit area, not a very impressive acceleration. Physically, one would imagine the surface of the mirror vibrating with an amplitude of just one nanometer. This conservative limitation in the amplitude arises because the maximum velocity of the boundary is proportional to the elastic deformation, which cannot exceed about 10^{-2} for typical materials. The energy radiated per area $E_m\Omega$ is about 10^{-25} W/m². There are a number of methods to increase these values by many orders of magnitude, as discussed below.

The efficiency of the conversion of energy expended per cycle in our model E_m into kinetic energy of the spacecraft $E_{ke} = \frac{1}{2}M(\Delta V_m)^2 = I_m^2/2M$ is given in the nonrelativistic approximation by the ratio:

$$\frac{E_{ke}}{E_m} = \frac{\hbar}{Mc} \frac{1}{\pi} \left(\frac{\Omega}{c}\right)^3 \qquad (41)$$

With our assumptions, the approximate value of this ratio is 10^{-26}, making this conversion an incredibly inefficient process.

C. Methods to Increase Acceleration

The dynamic Casimir effect has yet to be verified experimentally, hence there have been a number of interesting proposals describing methods designed to maximize the effect so it can be measured. In 1994, Law predicted a resonant response of the vacuum to an oscillating mirror in a one-dimensional cavity [61]. The behavior of cavities formed from two parallel mirrors that can move relative to each other is qualitatively different from that of single plates. For example, it is possible to create particles in a cavity with plates separating at constant velocity [62]. The very interesting proposal by Lambrecht et al. concludes that if the mechanical oscillation frequency is equal to an odd integer multiple of the fundamental optical resonance frequency, then the rate of photon emission from a vibrating cavity formed with walls that are partially transmitting with reflectivity, r, is enhanced by a factor equal to the finesse $f = 1/\ln(1/r)$ of the cavity [59,60]. For semiconducting cavities with frequencies in the GHz range, the finesse can be 10^9, giving our gedanken spacecraft an acceleration of 3×10^{-11} m/s² based on Eq. (40). Plunien et al. have shown that the resonant photon emission from a vibrating cavity is further increased if the temperature is raised [63]. For a 1-cm cavity, the enhancement is about 10^3 for a temperature of 290K. Assume that one has a gedanken spacecraft with a vibrating cavity operating at about 290K providing a 10^{12} total increase in the emission rate. This would result in an acceleration per unit area of the plates of 3×10^{-8} m/s², a radiated power of about 10^{-13} W/m² and an efficiency E_{ke}/E_m of about 10^{-14}. After 10 years of operation, the gedanken spacecraft velocity would be approximately 10 m/s, which is about three orders of magnitude less than the current speed of Voyager (17 km/s), obtained after a gravity assist maneuver around Jupiter to increase the velocity. The burn-out velocity for Voyager at launch in 1977 was 7.1 km/s [64].

The numerical results for the model obviously depend very strongly on the assumptions made. For the plate mass/area and system to vibrate the plate, we have made unrealistically favorable estimates; for the oscillation frequency and amplitude, we have made conservative estimates. It is possible that new materials, with the ability to sustain larger strains, could make possible an amplitude of oscillation orders of magnitude larger than 1 nm. Perhaps the use of nanomaterials, such as carbon nanotubes that support 5% strain [65], or "super" alloys [66], would allow a much larger effective deformation. If the amplitude was 1 mm instead of 1 nm, the gedanken spacecraft, or some modification with improved coupling to the vacuum, might warrant practical consideration.

Eberlein has shown that density fluctuations in a dielectric medium would also result in the emission of photons by the dynamic Casimir effect [67]. This approach may ultimately be more practical with large area dielectric surfaces driven electrically at high frequencies. More theoretical development is needed to evaluate the utility of this method. Other solid state approaches may also be of value with further technological developments. For example, one can envision making sheets of charge that are accelerated in MOS type structures. Yablonovitch has pointed out that the zero-point electromagnetic field transmitted through a window, whose index of refraction is falling with time, shows the same phase shift as if it were reflected from an accelerating mirror [68]. To simulate an accelerating mirror, he suggested utilizing the sudden change in refractive index that occurs when a gas is photoionized or the sudden creation of electron hole pairs in a semiconductor, which can reduce the index of refraction from ~ 3.5 to 0 in a very short time. Using subpicosecond optical pulses, the phase modulation can suddenly sweep up low-frequency waves by many octaves. By lateral synchronization, the moving plasma front with its large change in the index of refraction can, in effect, act as a moving mirror exceeding the speed of light. Therefore, one can regard such a gas or semiconductor slab as an observational window on accelerating fields, with accelerations as high as $\sim 10^{20}$ m/s [68]. Accelerations of this magnitude will have very high frequency Fourier components. Equation (40) shows that the impulse goes as the fourth power and the efficiency as the third power of the Fourier component for an harmonic oscillator, which suggests that with superhigh accelerations, an optimum time dependence of the field and an optimum shape of the wavefronts, one might be able to secure much higher fluxes of photons and a much higher impulse per second, with a higher conversion efficiency. On the other hand, a preliminary calculation by Lozovik et al. to investigate this approach suggests that the accelerated plasma method may have limitations in producing photons [69].

Ford and Svaiter have shown that it may be possible to focus the fluctuating vacuum electromagnetic field [70]. This capability might be utilized to create regions of higher energy density. This might be of use in a cavity in order to increase the flux of radiated photons. There may be also enhancements due to nature of the index of refraction for real materials. For example, Ford has computed the force between a dielectric sphere, whose dielectric function is described by the Drude model (based on a simple approximate model for free electrons in a metal), and a perfectly reflecting wall, with the conclusion that certain large components of the Casimir force no longer cancel. He predicts a dominant oscillatory contribution to the force, in effect developing a model for the amplification of vacuum fluctuations. Barton has shown that for materials with a fixed index of refraction, the force for a one-dimensional scalar field goes as $[(n-1)/n]^2$, which suggests the possibility that one might be able to enhance the force by selection of a material with a small index [31].

Another, albeit improbable, approach to a vacuum-facilitated gedanken spacecraft is to consider the possibility of adjusting the radiative mass shift, which we have neglected until this point. There is a very small radiative shift of the mirror due to its interaction with the vacuum, akin to the Lamb shift for an atom [34,59]. To measure a vacuum mass shift, a proposal was made recently to measure the inertial mass shift in a multilayer Casimir cavity, which consists of 106 layers of metal 100-nm thick and 35 cm in diameter, alternating with films of silicon dioxide 5-nm thick [71]. The mass shift is anticipated to arise from the decrease in the vacuum energy between the parallel plates. A calculation shows that the mass shift for the proposed cavity is at or just beyond the current limit of detectability.

It appears that if quantum vacuum engineering of spacecraft is to become practical, and the dreams of science fiction writers are to be realized, we may need to develop new methods to be able to manipulate changes in vacuum energy densities that are near to the same order of magnitude as mass energy densities. Then we would anticipate being able to shift inertial masses by a significant amount. Because mass shifts in computations are often formally infinite, perhaps such developments are not forbidden, and, with more understanding (and serendipity!), may be controllable. With large mass shifts one might be able to build a structure that had a small or zero inertial mass, which could be readily accelerated.

VI. Unresolved Physics

Casimir effects are typically small and difficult to measure. In fact, measurements have only been made for the simplest geometries, such as the parallel plate geometry or the sphere plate geometry. No measurements have ever been made that validate the dynamic Casimir effect. Assuming the effect is verified experimentally, then the question is: Is it possible to amplify these effects and bring them into the useful range? This is certainly one of the challenges of vacuum engineering. Experiments are needed that explore some of the issues that are beyond the present calculational ability of QED; for example, the effect of complex geometries on vacuum forces, or the effect of massive fields or dense, moving nuclear matter on the quantum vacuum. Is it possible to make a stable vacuum field that has a large variation in energy density? What is the effect of changes in vacuum energy on the index of refraction? Can energy density gradients be found on a length scale that is useful to humans? We need to greatly increase our knowledge of the quantum vacuum. The development of a very sensitive small probe that provided a frequency decomposition of the local vacuum energy density would certainly be *very* helpful.

From the status of current research in Casimir forces, it is clear that we are at the tip of the iceberg describing the properties of the quantum vacuum for real systems with real material properties. For example, there is no general agreement regarding the calculations of static vacuum forces for geometries other than infinite parallel plates of ideal or real metals at a temperature of absolute zero.

Non-zero temperature corrections for flat, real metals are uncertain [72,73]. There are fundamental disagreements about the computation of vacuum forces for spheres or rectangular cavities, and about how to handle real material properties and curvature in these and other geometries [31]. Indeed, it is very difficult to calculate Casimir forces for these simple geometries and to relate the calculations to an experiment. Calculations have yet to be done for more complex and potentially interesting geometries. The usual problems in QED—such as divergences due to unrealistic boundary conditions, to curvature, to interfaces with different dielectric coefficients—abound [74].

A. Possible Discriminating Tests

As discussed, to make practical spacecraft based on engineering the quantum vacuum, it seems we need to find new boundary conditions for the vacuum that can alter the vacuum energy density by orders of magnitude more than with our current boundary conditions, which are primarily metal or dielectric surfaces. Perhaps the use of new materials (e.g., those with a negative index of refraction, or an ultra-high electrical carrier density, either steady state or transient), or novel superconducting materials may open the door to new Casimir phenomena. Recently the use of negative index materials was proposed to make a repulsive Casimir force [75]. With significantly increased funding of research, some breakthroughs might be possible.

There are some important experiments that can help our understanding of vacuum energy and Casimir forces that may lead to significant improvements in our engineering capability. Experiments to verify the adiabatic Casimir effect have been suggested in the literature. This is an important theoretical issue that has ramifications in different fields, including astrophysics and elementary particle physics. Within the next five years, some clever experimental approaches will probably be developed to explore the adiabatic Casimir effect. Experiments measuring the Casimir forces for semiconductor surfaces would be helpful in the development of new applications of vacuum forces and to demonstrate that it is possible to alter the Casimir force by altering the carrier density. The measurement of Casimir forces and energies for different geometry and composition objects, such as rectangular cavities or spheres, would provide some badly needed answers for theoretical modeling. Measurements of Casimir forces between separate, nonplanar surfaces are also needed. There may be surfaces that have larger forces than the classic parallel plates, but we have yet to measure anything but parallel plates.

Tests to determine if negative vacuum energy yields a negative force in a gravitation field have been proposed. This is a fundamental question. Order-of-magnitude estimates suggest that a negative mass object is not possible, but that the total mass may be reduced a measurable amount by contributions from negative vacuum energy.

B. Application Implications

Because Casimir forces require small separations, they have been utilized in micro- and nano-electromechanical systems (MEMS/NEMS). As microfabrication technology develops, nanodevices with shorter working distances and more varied materials will be developed, resulting in more applications for Casimir forces. An excellent recent survey by Capasso et al. discusses a variety of devices, including the anharmonic Casimir oscillator and a surface with a Pd film whose electrical properties change in the presence of hydrogen, which could lead to a measurable change in the Casimir force [12]. Also reported were repulsive Casimir forces for certain planar material combinations as predicted by the Lifshitz formulation, for example, a gold-covered planar surface and a silica-covered surface when ethanol is the medium in between. An ultrasensitive magnetometer was proposed using the repulsive Casimir force to levitate a disk with a magnetic dipole in alcohol. Also proposed was a demonstration of quantum torque, analogous to the Casimir force in origin [76]. In this system, the birefringent material is floated by the repulsive Casimir force above a fixed birefringent plate. Steady progress in the application of Casimir forces in MEMS is to be expected.

VII. Conclusions

One objective was to illustrate some of the unique properties of the quantum vacuum and how they might be utilized in propulsion of a spacecraft. We have outlined some of the considerations for the use of vacuum energy to propel a spacecraft and pointed out some directions in which some helpful discoveries may lie.

We have demonstrated that it is possible in principle to propel a spacecraft using the dissipative force an accelerated mirror experiences when photons are generated from the quantum vacuum. Further, we have shown that one could, in principle, utilize energy from the vacuum fluctuations to operate the vibrating mirror assembly required. The application of the dynamic Casimir effect and the static Casimir effect may be regarded as a proof of principle, with the hope that the proven feasibility will stimulate more practical approaches exploiting known or as yet unknown features of the quantum vacuum. A model gedanken spacecraft with a single vibrating mirror was proposed that showed a very unimpressive acceleration due to the dynamic Casimir effect of about 3×10^{-20} m/s^2 with a very inefficient conversion of total energy expended into spacecraft kinetic energy. Employing a set of vibrating mirrors to form a parallel plate cavity and raising the cavity temperature to about 290K increased the output by a factor of the finesse of the cavity times approximately 10^3, yielding an acceleration of about 3×10^{-8} m/s^2 and a conversion efficiency of about 10^{-14}. To put this into perspective, after 10 years at this acceleration, the spacecraft would have only attained a 10 m/s velocity. At least we have computationally suggested (pending experimental verification of the phenomena predicted by QED), that it is possible to use the quantum vacuum to propel a spacecraft. This represents progress.

The vacuum effects that we computed scale with Planck's constant and are therefore very small. In order to have a practical spacecraft based on quantum vacuum properties, it would be preferable that the vacuum effects scale as $\hbar^0$, meaning that the effects are essentially independent of Plank's constant and consequently may be much larger. By itself this requirement does not guarantee a large enough magnitude, but it certainly helps [59]. New methods of modifying the quantum vacuum boundary conditions may be needed to generate the large changes in free-field vacuum energy or momentum required if "vacuum engineering" as proposed in this paper is ever to be practical. For example, the vacuum energy density difference between parallel plates and the region outside them in free space is simply not large enough in magnitude for our engineering purposes. Energy densities, positive or negative, that are orders of magnitude greater are required. Such energy density regions may be possible, at least in some cases. For example, a region appeared in the one-dimensional dynamic system in which the energy density was below that of the Casimir parallel plate region [61]. Similarly, more effective ways of transferring momentum to the quantum vacuum than using photons generated with the adiabatic Casimir effect are probably necessary if a spacecraft is to be propelled using the vacuum.

Although the results of our calculations using the vibrating mirrors to propel a gedanken rocket are very unimpressive, it is important to not take our conclusions regarding the final velocity in our simplified models too seriously. The choice of numerical parameters can easily affect the final result by four orders of magnitude. The real significance is that a method has been described that illustrates the possibility of propelling a rocket by coupling to the vacuum. It is possible that there may be vastly improved methods of coupling. There are numerous potential ways in which the ground state of the vacuum electromagnetic field might be engineered for use in technological applications, a few of which we have mentioned here. As the technology to fabricate small devices improves, and as the theoretical capability of calculating quantum vacuum effects increases, and as experiments help us understand the issues, it will be interesting to see which possibilities prove to be useful and which remain curiosities.

Optical applications of vacuum engineering, such as fabricating lasers in cavities to control spontaneous emission, have not been mentioned. Likewise, the current astrophysical conundrums about dark energy and the cosmological constant, which may relate to vacuum energy, were not discussed because we are not very good at predicting the development of technology. In the 1980s we thought artificial intelligence was going to revolutionize the world, but it did not. In the 1960s, manufacturers were hard put to think of any reason why an individual would want a home computer and today we wonder how we ever survived without them. Circa 1900 an article was published in *Scientific American* proving that it was impossible to send a rocket, using a conventional propellant, to the moon. The result was based on the seemingly innocuous assumption of a single-stage rocket. It is hoped that a paper on vacuum propulsion written 100 years from now, will also find amusing our failure to perceive the key issues and see clearly how quantum propulsion should be done.

Acknowledgments

I gratefully acknowledge helpful comments and suggestions from Dan Cole, and I thank Gabriel Barton, Carlos Villarreal Lujan, Claudia Eberlein, Peter Milonni, and Paulo Neto for discussions about their work. I would like to express my appreciation to the NASA Breakthrough Propulsion Physics Program and Marc G. Millis for the support of earlier versions of this work. I am sad to report that Robert L. Forward died while we were working on the Casimir drive.

References

[1] Milonni, P. W., *The Quantum Vacuum: An Introduction to Quantum Electrodynamics*, Academic Press, San Diego, CA, 1994.

[2] Milton, K. A., *The Casimir Effect*, USA World Scientific, NJ, 2001.

[3] Milonni, P., Cook, R., and Groggin, M., "Radiation Pressure from the Vacuum: Physical Interpretation of the Casimir Force," *Physical Review*, Vol. A38, 1988, p. 1621.

[4] Mohideen, U., and Anushree, R., "Precision Measurement of the Casimir Force from 0.1 to 0.9 Micron," *Physical Review Letters*, Vol. 81, 1998, p. 4549.

[5] Klimchitskaya, G., Roy, A., Mohideen, U., and Mostepanenko, V., "Complete Roughness and Conductivity Corrections for Casimir Force Measurement," *Physical Review A*, Vol. 60, 1999, pp. 3487–3495.

[6] Bressi, G., Carugno, G., Onofrio, R., and Ruoso, G., "Measurement of the Casimir Force between Parallel Metallic Plates," *Physical Review Letters*, Vol. 88, 2002, p. 041804.

[7] Chen F., Mohideen, U., Klimchitskaya, G. L., and Mostepanenko, V. M., "Demonstration of the Lateral Casimir Force," *Physical Review Letters*, Vol. 88, 2002, p. 101801.

[8] Chen, F., Mohideen, U., Klimchitskaya, G. L., and Mostepanenko, V. M., "Theory Confronts Experiment in the Casimir Force Measurements: Quantification of Errors and Precision," *Physical Review A*, Vol. 69, 2004, pp. 022117 1–11.

[9] Lamoreaux, S., "Measurement of the Casimir Force Between Conducting Plates," *Physical Review Letters*, Vol. 78, 1997, p. 5.

[10] Chan, H. B., Aksyuk, V., Kleiman, R., Bishop, D., and Capasso, F., "Quantum Mechanical Actuation of Microelectromechanical Systems by the Casimir Force," *Science*, Vol. 291, 2001, p. 1941.

[11] Chan, H. B., Aksyuk, V. A., Kleiman, R. N., Bishop, D. J., and Capasso, F., "Nonlinear Micromechanical Casimir Oscillator," *Physical Review Letters*, Vol. 87, 2001, p. 211801.

[12] Capasso, F., Munday, J., Iannuzzi, D., and Chan, H., "Casimir Forces and Quantum Electrodynamical Torques: Physics and Nanomechanics," *IEEE Journal of Selected Topics in Quantum Electronics*, Vol. 13, 2007, pp. 400–414.

[13] Morris M. S., and Thorne, K. S., "Wormholes in Spacetime and Their Use for Interstellar Travel: A Tool for Teaching General Relativity," *American Journal of Physics*, Vol. 56, 1988, pp. 395–412.

[14] Lamb, W., and Retherford, R., "Fine Structure of the Hydrogen Atom by a Microwave Method," *Physics Review*, Vol. 72, 1947, p. 241.

[15] Bethe, H., "The Electromagnetic Shift of Energy Levels," *Physics Review*, Vol. 72, 1948, p. 339.

[16] Maclay, J., Fearn, H., and Milloni, P., "Of Some Theoretical Significance: Implications of Casimir Effects," *European Journal of Physics*, Vol. 22, 2001, p. 463.

[17] Visser, M., *Lorentzian Wormholes: From Einstein to Hawking*, AIP Press, New York, 1996, pp. 81–87.

[18] Ostriker, J., and Steinhardt, P., "The Quintessential Universe," *Scientific American*, Vol. 284, 2001, pp. 46–53.

[19] Weinberg, S., "The Cosmological Constant Problem," *Reviews of Modern Physics*, Vol. 61, 1989, p. 1.

[20] Adler, R. J., Casey, B., and Jacob., O. C., "Vacuum Catastrophe: An Elementary Exposition of the Cosmological Constant Problem," *American Journal of Physics*, Vol. 63, 1995, pp. 620–626.

[21] Plunien, G., Müller, B., and Greiner, W., "The Casimir Effect," *Physics Reports*, Vol. 134, 1986, p. 87.

[22] Casimir, H. B. G., "On the Attraction Between Two Perfectly Conducting Plates," *Proceedings of the Koninklijke Nederlandse Akademie van Wetenschappen*, Vol. 51, 1948, pp. 793–795.

[23] Bordag, M., Mohideen, U., and Mostepanenko, V., "New Developments in the Casimir Effect," *Physics Reports*, Vol. 353, 2001, p. 1.

[24] Brown L. S., and Maclay, G. J., "Vacuum Stress Between Conducting Plates: An Image Solution," *Physical Review*, Vol. 184, 1969, pp. 1272–1279.

[25] Maclay, J., "Analysis of Zero-Point Energy and Casimir Forces in Conducting Rectangular Cavities," *Physical Review A*, Vol. 61, 2000, p. 052110.

[26] Moore, G. T., "Quantum Theory of the Electromagnetic Field in a Variable-Length One-Dimensional Cavity," *Journal of Mathematical Physics*, Vol. 11, 1970, p. 2679.

[27] Birrell, N., and Davies, P. C. W., *Quantum Fields in Curved Space*, Cambridge University Press, Cambridge, U.K., 1984, p. 48, 102.

[28] Neto P., Maia, A., and Machado, L. A. S., "Quantum Radiation Generated by a Moving Mirror in Free Space," *Physics Review A*, Vol. 54, 1996, p. 3420.

[29] Jaekel M., and Reynaud, S., "Motional Casimir Force," *Journal of Physics* (France) I, Vol. 2, 1992, p. 149.

[30] Ford, L., and Vilenkin, A., "Quantum Radiation by Moving Mirrors," *Physics Review D*, Vol. 25, 1982, p. 2569; 36.

[31] Barton, G., "Perturbative Casimir Energies of Dispersive Spheres, Cubes and Cylinders," *Journal of Physics A*, Vol. 34, 2001, p. 4083.

[32] Neto, P., "Vacuum Radiation Pressure on Moving Mirrors," *Journal of Physics A*, Vol. 27, 1994, p. 2167.

[33] Jaekel, M., and Reynaud, S., "Quantum Fluctuations of Mass for a Mirror in Vacuum," *Physical Letters A*, Vol. 180, 1993, p. 9.

[34] Barton, G., and Calogeracos, A., "On the Quantum Dynamics of a Dispersive Mirror. I. Radiative Shifts, Radiation, and Radiative Reaction," *Annual of Physics*, Vol. 238, 1995, p. 227.

[35] Casimir, H. B. G., "Some Remarks on the History of the So-called Casimir Effect," *The Casimir Effect, 50 Years Later*, Bordag, M. (ed.), World Scientific, Singapore, 1999, pp. 3–9.

[36] Power E. A., and Thirunamachandran, T., "Zero-Point Energy Differences and Many-Body Dispersion Forces," *Physics Review A*, Vol. 50, 1994, pp. 3929–3939.

[37] Schwinger, J., DeRaad, Jr., L. L., and Milton, K. A., "Casimir Effect in Dielectrics," *Annuals of Physics (NY)*, Vol. 115, 1978, pp. 1–23.

[38] Lifshitz, E. M., "The Theory of Molecular Attractive Forces Between Solids," *Soviet Physics JETP*, Vol. 2, 1956, pp. 73–83.

[39] Maclay, J., and Hammer, J., "Vacuum Forces in Microcavities," *Seventh International Conference on Squeezed States and Uncertainty Relations*, Boston, MA, 4–6 June 2001. URL: http://www.physics.umd.edu/robot 32.

[40] Deutsch D., and Candelas, P., "Boundary Effects in Quantum Field Theory," *Physiscal Reviews D*, Vol. 20, 1979, pp. 20, 3063–3080.

[41] Barton, G., "Perturbative Casimir Energies of Spheres: Re-orienting an Agenda," *International Conference of Squeezed States and Uncertainty Relations*, Boston, May 2000. URL: http://www.physics.umd.edu/robot.
[42] Serry, M., Walliser, D., and Maclay, J., "The Anharmonic Casimir Oscillator," *Journal of Microelectromechanical Systems*, Vol. 4, 1995, p. 193.
[43] Maclay, J., "A Design Manual for Micromachines Using Casimir Forces: Preliminary Considerations," *Proceedings of STAIF-00 (Space Technology and Applications International Forum)*, El-Genk, M. S. (ed.): American Institute of Physics, New York, 2000.
[44] Serry, M., Walliser, D., and Maclay, J., "The Role of the Casimir Effect in the Static Defection and Stiction of Membrane Strips in Microelectromechanical Systems (MEMS)," *Journal of Applied Physics*, Vol. 84, 1998, p. 2501.
[45] Buks, E., and Roukes, M.L., "Stiction, Adhesion, and the Casimir Effect in Micromechanical Systems," *Physical Review B*, Vol. 63, 2001, p. 033402.
[46] Decca, R., Lopez, D., Fishbach, E., and Krause, D., "Measurement of the Casimir Force Between Dissimilar Metals," *Physical Reviews*, Vol. 91, 2003, p. 050402.
[47] Lambrecht A., and Reynaud, S., Comment on "Demonstration of the Casimir Force in the 0.6 to 6 μm Range," *Physical Review Letters*, Vol. 84, 2000, p. 5672.
[48] Arnold, W., Hunklinger, S., and Dransfeld, K., "Influence of Optical Absorption on the van der Waals Interaction Between Solids," *Physical Review B*, Vol. 19, 1979, p. 6049.
[49] Chen, F., Mohideen, U., Klimchitskaya, G. L., and Mostepanenko, V. M., "Investigation of the Casimir Force Between Metal and Semiconductor Test Bodies," *Physical Review A, Rapid Communication*, Vol. 72, 2005, pp. 020101–020104.
[50] Puthoff, H. E., Little, S. R., and Ibison, M., "Engineering the Zero-Point Field and PolarizableVacuum for Interstellar Flight," *Journal of the British International Society*, Vol. 55, 2002, pp. 137–144.
[51] Millis, M. G., "NASA Breakthrough Propulsion Physics Program," *Acta Astronautica*, Vol. 44, No. 2–4, 1999, pp. 175–182; "Challenge to Create the Space Drive," *Journal of Propulsion and Power*, Vol. 13, No. 5, 1997, pp. 577–582; and "Prospects for Breakthrough Propulsion from Physics," NASA TM-2004-213082, May 2004.
[52] Pinto, F., "Progress in Quantum Vacuum Engineering Propulsion," *Journal of the British International Society*, Vol. 39, 2006, pp. 247–256.
[53] Haisch, B., Rueda, A., and Puthoff, H. E., "Inertia as a Zero-Point-Field Lorentz Force," *Physical Review A*, Vol. 49, 1994, pp. 678–694.
[54] Maclay, J., and Forward, R. W., "A Gedanken Spacecraft that Operates Using the Quantum Vacuum (Dynamic Casimir Effect)," LANL Arxiv Physics/0303108, *Foundation of Physics*, Vol. 34, 2004, p. 477.
[55] Forward, R. L., "Extracting Electrical Energy from the Vacuum by Cohesion of Charged Foliated Conductors," *Physics Review B*, Vol. 30, 1984, p. 1700.
[56] Cole, D. C., and Puthoff, H. E., "Extracting Energy and Heat from the Vacuum," *Physics Review E*, Vol. 48, 1993, p. 1562.
[57] Rueda, A., "Survey and Examination of an Electromagnetic Vacuum Accelerating Effect and its Astrophysical Consequences," *Space Science Reviews*, Vol. 53, 1990, pp. 223–345.
[58] Cole, D. C., "Possible Thermodynamic Violations and Astrophysical Issues for Secular Acceleration of Electrodynamic Particles in the Vacuum," *Physics Review E*, Vol. 51, 1995, p. 1663.

[59] Lambrecht, A., Jaekel, M., and Reynaud, S., "Motion Induces Radiation from a Vibrating Cavity," *Physics Review Letters*, Vol. 77, 1996, p. 615.
[60] Dodonov, V., and Klimov, A., "Generation and Detection of Photons in a Cavity with a Resonantly Oscillating Boundary," *Physics Review A*, Vol. 53, 1996, p. 2664.
[61] Law, C. K., "Resonance Response of the Vacuum to an Oscillating Boundary," *Physics Review Letters*, Vol. 73, 1994, p. 1931.
[62] Villarreal, C., Hacyan, S., Jauregui, R., "Creation of Particles and Squeezed States Between Moving Conductors," *Physics Review A*, Vol. 52, 1995, p. 594.
[63] Plunien, G., Schützhold, R., and Soff, G., "Dynamical Casimir Effect at Finite Temperature," *Physics Review Letters*, Vol. 84, 2000, p. 1882.
[64] Boston, M., "Simplistic Propulsion Analysis of a Breakthrough Space Drive for Voyager," *Proc. STAIF-00 (Space Technology and Applications International Forum)*, January 2000, Albuquerque, NM, El-Genk, M.S. (ed.) American Institute of Physics, New York, 2000.
[65] Bernholc, J., Brenner, C., Nardelli, M., Meunier, V., Roland, C., "Mechanical and Electrical Properties of Nanotubes," *Annual Review of Material Research*, Vol. 32, 2002, pp. 347–375.
[66] Saito, T., Furuta, T., Hurang, J.-H., Kuramoto, S., Nishino, K., Suzuki, N., Chen, R., Yamada, A., Ito, K., Seno, Y., Nonaka, T., Ikehata, H., Nagasako, N., Iwamoto, C., Ikuhara, Y., and Sakuma, T., "Multifunctional Alloys Obtained via a Dislocation-Free Plastic Deformation Mechanism," *Science*, Vol. 300, 2003, p. 464. My thanks to Steven Hansën for this reference.
[67] Eberlein, C., "Quantum Radiation from Density Variations in Dielectrics," *Journal of Physics A: Mathematics and General*, Vol. 32, 1999, p. 2583.
[68] Yablonovitch, E., "Accelerating Reference Frame for Electromagnetic Waves in a Rapidly Growing Plasma: Unruh-Davies-Fulling-DeWitt Radiation and the Nonadiabatic Casimir Effect," *Physics Review Letters*, Vol. 62, 1989, p. 1742.
[69] Lozovik, Y., Tsvetus, V., and Vinogradovc, E., "Femtosecond Parametric Excitation of Electromagnetic Field in a Cavity," *JEPT Letters*, Vol. 61, 1995, p. 723.
[70] Ford, L. H., and Svaiter, N. F., "Focusing Vacuum Fluctuations," *Physics Review A*, Vol. 62, 2000, p. 062105.
[71] Calonni, E., DiFiore, L., Esposito, G., Milano, L., and Rosa, L., "Vacuum Fluctuation Force on a Rigid 38 Casimir Cavity in a Gravitational Field," *Physics Letters A*, Vol. 297, 2002, p. 328.
[72] Bezerra, V., Klimchitskaya, G., and Romero, C., "Surface Impedance and the Casimir Force," *Physics Review A*, Vol. 65, 2001, p. 012111.
[73] Genet, C., Lambrecht, A., and Reynaud, S., "Temperature Dependence of the Casimir Force Between Metallic Mirrors," *Physics Review A*, Vol. 62, 2000, p. 012110.
[74] DeWitt, B., *Physics in the Making*, Sarlemijn A., and Sparnaay M. (eds.), Elsevier, Amsterdam, 1989, pp. 247–272.
[75] Leonhardt, V., and Philkin, T., "Quantum Levitation by Left-Handed Materials," *New Journal of Physics*, Vol. 9, 2007, p. 254.
[76] Enk, S., "Casimir Torque Between Dielectrics," *Physical Review A*, Vol. 52, 1995, p. 2569.

Chapter 13

Inertial Mass from Stochastic Electrodynamics

Jean-Luc Cambier*
U.S. Air Force Research Laboratory, Edwards Air Force Base, Edwards, California

I. Introduction

THE origin of inertia, the resistance of an object to a change in its velocity (i.e., acceleration), has been the subject of various hypotheses throughout the development of modern physics. This resistance implies that a force must be applied to provide acceleration to an object (Newton's law) and the constant of proportionality is the so-called (inertial) "mass." One of the most famous conjectures regarding inertia is the Mach principle, which postulates that interaction with the entire universe is responsible for this property of matter. Such arguments of a quasi-philosophical nature played an important role in the development and refinement of the important physical concepts of causality, inertial frames, invariant measures, and eventually the equivalence principle between inertial mass and gravitational mass. While this led to formidable advances in our understanding of the fundamental properties of the universe, Einstein's work did nothing to explain why mass seemed to be an intrinsic property of matter in the first place. The advent of quantum mechanics and quantum field theory led to further insight, but also more unexplained phenomena, such as the existence of fields in a non-vanishing energy ground-state, thus potentially filling the universe with so-called "zero-point energy" (Chapters 1 and 18), while interaction with this field on all scales could lead to effective particle masses that are seemingly infinite. The origin of mass is, therefore, still an open field of research. If mass is strictly the result of the interaction with background fields, then one may hope to be able someday to modify these fields and change the mass. This could have obvious implications for propulsion provided this manipulation can be done in a practical and efficient way. Mass reduction would then allow a

---

This material is a work of the U.S. Government and is not subject to copyright protection in the United States.
*Senior Research Scientist, Propulsion Directorate–Aerophysics Branch.

higher acceleration for a given applied force, and near-vanishing of the mass would allow travel at near the speed of light.

This type of interaction can also proceed when the background field is a random, purely classical field. This is the basis for the stochastic electrodynamics (SED) model, and in particular the hypothesis by Haisch, Rueda, and Puthoff [1] that inertial mass can be the result of such interaction with a fluctuating field. In this chapter, this hypothesis is examined and compared with the modern relativistic, quantum electro-dynamics (QED) theory. The intent is to give the reader an explanation of the premise of this theory and to identify the most critical and unresolved issues for further research. It should be pointed out that any strengths or weaknesses identified here pertain only to this particular hypothesis rather than the general theory of stochastic electrodynamics, or the general notion that inertia might be fundamentally tied to quantum fluctuations. It is hoped that the analysis offered here will alert future researchers to the details that need to be clearly addressed to proceed along these lines.

II. Background

SED is a theory of the interaction of point-like charged particles with fluctuating electromagnetic fields of the vacuum, also called zero-point fields (ZPF), which Haisch, Rueda, and initially Puthoff, among others, have been proposing for a number of years [1–7] as an explanation for the origin of inertial mass. In the first version of the theory, Haisch, et al. [1] (hereafter called HRP) postulate that the coupling of the charged particle with such fields leads to a oscillating motion, that is, an example of "zitterbewegung" (literally, "jittering motion") [8]. SED is intimately tied to the decades-old problem of the electromagnetic vacuum and radiation spectrum. In 1910, Einstein and Hopf [9] used a classical oscillator model to demonstrate (in the nonrelativistic approximation) that the oscillator coupled to a background radiation field, was subject to a retarding force proportional to the velocity, $F = -Rv$, where[†]

$$R = \frac{4\pi^2 e^2}{5mc^2}\left[\rho(\omega_0) - \frac{\omega_0}{3}\frac{d\rho}{d\omega_0}\right] \qquad (1)$$

and $\rho(\omega_0)$ is the energy density of the field at the resonant frequency ω_0 of the oscillator. When the background field consists *only* of the vacuum field (the thermal component being absent), the energy density is proportional to ω^3, and the bracketed term on the right side of Eq. (1) vanishes; thus, in the absence of thermal radiation ($T \equiv 0$) there is no frictional force. It has been shown independently by Marshall [10] and Boyer [11] that the ω^3 scaling law is the required form for the Lorentz invariance, and, therefore, the vacuum field does not exert any force in any inertial frame; this is an important and necessary result, because observers in inertial frames cannot be, by definition, subject to forces and accelerations.

[†]Units throughout are ESU (see Ref. 12); to convert to SI units, replace $e^2 \rightarrow e^2/4\pi\varepsilon_0$.

The main objective of the HRP thesis was to generalize the Einstein and Hopf results to consider noninertial frames, and attempt to obtain another retarding force proportional to the *acceleration*, $F = -m_i a$. By doing so, the inertial mass (the coefficient m_i) would be explained in terms of a coupling to the fluctuating background field of the vacuum. The HRP analysis is almost entirely based on the work by Boyer [11,13–15], where the classical oscillator is also modeled by the linear motion of a point charged particle ("parton," in the terminology of high-energy particle physics). After averaging over the high-frequency oscillations and in the limit of low acceleration and nonrelativistic motion, the Lorentz force is identified as a retarding force (after selecting the appropriate sign) proportional to the acceleration. The inertial mass is given in Ref. 1 as:

$$m_i = \left[\frac{\hbar \omega_\Lambda}{c^2}\right](\Gamma \omega_\Lambda) \qquad (2)$$

where ω_Λ is the upper limit[‡] to the integration of all electromagnetic modes of the ZPF and Γ is a damping constant. Equation (2) appears to indicate that the mass is *completely* obtained as a result of the interaction with the ZPF, and is the basis of the claims of inertia as the result of a drag force from the vacuum. There are, however, a number of issues related to the actual computations performed to obtain the result in Eq. (2) above, and to its interpretation. This chapter is organized as follows; first the HRP theory, as mostly described in the seminal paper [1], is summarized, with some key aspects and assumptions highlighted; second, the theory is compared with the conventional approach of quantum field theory in order to bring to light the similarities and differences; third, we discuss more recent results, as well as a way forward, that is, what should be done to achieve further progress; finally, we provide a brief conclusion.

III. Stochastic Electrodynamics

We principally discuss here the original model used by HRP, which is an extension of the classical oscillator model in the presence of an oscillating electromagnetic, with radiation reaction[§]:

$$m_0 \ddot{r} = -m_0 \omega_0^2 r + \frac{2e^2}{3c^3} \dddot{r} + e E_0 \qquad (3)$$

Here, r is the amplitude of oscillation along the direction of the applied field, which we can choose as the $\hat{y}$ direction without loss of generality. This equation of motion leads to a damping oscillating motion, with

$$\Gamma = \frac{2e^2}{3 m_0 c^3} \qquad (4)$$

[‡]The cutoff can be expressed in terms of energy, frequency or mass: $\Lambda = \hbar \omega_\Lambda = m_\Lambda c^2$.

[§]See Sec. 17.8 of Ref. 12. This model describes the motion of a point-like, charged mass subject to an applied electric field and a restoring force (harmonic potential). The $\dddot{r}$ term is the radiation reaction force.

being the well-known classical damping coefficient (the so-called Abraham–Lorentz constant), which yields the natural line width of spectral lines. The electric field is also along the $\hat{y}$ direction, so that the vector notation can be ignored. When the field frequency is $\omega \approx \omega_0$, the system is, of course, driven to resonance. However, if the field is randomly fluctuating in time, its Fourier transform contains many different frequencies. In this case, the equation of motion becomes a Langevin equation, where the driving term is described by a random variable $\breve{E}$, with a probability distribution function (PDF) such that $\langle \breve{E} \rangle \equiv 0$, but such that its variance is nontrivial, that is, $\langle \breve{E}^2 \rangle \neq 0$. The averaging procedure (brackets $\langle \ \rangle$) is over many cycles of the fluctuating quantity, which is considered equivalent to the PDF average (ergodic hypothesis). HRP consider this fluctuating field to be the ZPF, which contains a wide range of frequencies extending, presumably, up to the Planck frequency:

$$\omega_P = \left[\frac{c^5}{\hbar G}\right]^{1/2} \approx 1.8 \times 10^{43} \text{ rad/s} \tag{5}$$

Therefore, unless the spectrum of the fluctuating field decays exponentially fast in the high-frequency region, extremely high driving frequencies will be present. These can easily be averaged over time scales of interest. An exponentially decreasing function of frequency is the spectral density of a thermal (Planck) distribution. However, the complete energy density function of the electromagnetic field

$$\rho(\omega)d\omega = \frac{\omega^2 d\omega}{\pi^2 c^3} \cdot \left[\frac{\hbar\omega}{e^{\hbar\omega/kT} - 1} + \frac{\hbar\omega}{2}\right] \tag{6}$$

contains the nonthermal, vacuum contribution $\hbar\omega/2$. Integration over all frequencies for this nonthermal contribution leads to a divergent integral ("UV catastrophe"); the Planck frequency [Eq. (5)] acts as an upper limit to regularize the integral, although for any practical purposes, this cutoff frequency is so large that it can be considered as virtually infinite.

The reference frame $\mathbf{S}$ of the oscillator (characterized by the equilibrium position $\vec{R}$) is assumed to have a uniform acceleration a with respect to the laboratory frame,[¶] $\mathbf{K}$. We choose the direction of motion of the frame to be the $\hat{x}$-direction. At any given time, one can also define an inertial frame $\mathbf{K}'$, which coincides with $\mathbf{S}$ but is related to the laboratory frame, by a Lorentz transformation, the parameters of which are time-dependent (see Fig. 1).

Accounting for the uniform acceleration, the oscillator dynamics are given by:

$$m_0 \ddot{r} = -m_0 \omega_0^2 r + \frac{2e^2}{3c^3}\left[\dddot{r} - \frac{a^2}{c^2}\dot{r}\right] + eE_0 \tag{7}$$

where the time derivative is with respect to the proper time in the $\mathbf{S}$ frame, $\dot{r} = dr/d\tau$. HRP determine first a solution of Eq. (7) for the velocity, then separately compute the ensemble average of the net Lorentz force

---

[¶]In Ref. 1, the authors almost exactly replicate the analysis and notation of Boyer [14].

INERTIAL MASS FROM STOCHASTIC ELECTRODYNAMICS

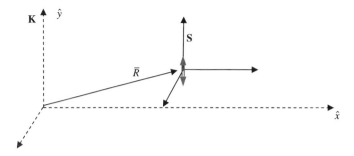

Fig. 1 Schematic of relative frame transformation, where S is the frame attached to the center of motion of the oscillator, and K is the laboratory frame. The oscillating motion (y-direction) is transverse to the relative velocity of the two frames (x-direction). The S frame is uniformly accelerated ("hyperbolic motion") with respect to the laboratory frame.

acting on the system from this initial solution. From the harmonic oscillator approximation, the magnitude of the velocity is $v \approx eE_0/m_0\omega$, which can be recast in the form:

$$\frac{v^2}{c^2} \cong \alpha \cdot \left(\frac{\hbar\omega}{m_0 c^2}\right)^2 \approx \alpha \cdot \left(\frac{\lambda_c}{\lambda}\right)^2 \tag{8}$$

where we have used the energy density $E_\omega^2 \approx \hbar\omega/\lambda^3$ for a single mode of the ZPF, and the following definitions:

$$\text{Fine-structure constant: } \alpha = \frac{e^2}{\hbar c} \tag{9a}$$

$$\text{Compton wavelength: } \lambda_c = \frac{\hbar}{m_0 c} \tag{9b}$$

Let us first assume that m_0 is the observed rest mass of the electron, m_e. Because the vacuum modes extend to very short wavelengths, presumably down to the Planck length—almost 20 orders of magnitude smaller than the characteristic length in Eq. (9b) for the electron—the term in parentheses on the right side of Eq. (8) can be very large.** Thus, the charged particle motion would be ultrarelativistic and the model oscillator described in Eq. (7) would not be accurate, *unless* the cutoff is arbitrarily limited by the physically observed energy scale, that is, $\lambda \geq \lambda_c$, or the particle mass m_0 is very large. We should also point out that Eq. (7) is taken directly from Boyer [14], and is valid in the limit of small oscillating amplitude. A derivation of this equation (notably the reaction force) from a covariant formulation is given in Appendix A to this chapter.

Equation (7) has two fundamental parameters: m_0 and ω_0. The nature of these parameters is not clearly specified in the HRP model, except that m_0 is considered the "bare" mass scale of the oscillator,[††] while ω_0 is considered essentially

---

**In HRP [1], p. 49, it is explicitly mentioned that only very high frequencies contribute to the inertial mass.

[††]This connotation of "bare" mass comes from quantum field theory, as explained in Section IV.

irrelevant by the authors. For a harmonic oscillator, ω_0 describes the properties of the restoring force (stiffness), when combined with the oscillator mass m_0. Because there are no details given regarding the model and the nature of this restoring force, these remain as free, adjustable parameters.

Equation (7) is solved by applying Fourier transforms. Again following Boyer [14]:

$$r(\tau) = (2\pi)^{-1/2} \int_{-\infty}^{\infty} d\Omega \cdot \eta(\Omega) \cdot e^{-i\Omega t} \tag{10a}$$

$$E_0(\tau) = (2\pi)^{-1/2} \int_{-\infty}^{\infty} d\Omega \cdot \Sigma(\Omega) \cdot e^{-i\Omega t} \tag{10b}$$

which, when applied to Eq. (7), yields the solution:

$$\eta(\Omega) = \frac{e}{m_0} \frac{\Sigma(\Omega)}{\omega_0^2 - \Omega^2 - i\Gamma\Omega(\Omega^2 + a^2/c^2)} \tag{11}$$

where Γ is given by Eq. (4). The oscillating velocity is given by its time-Fourier component:

$$\tilde{v} = -i\frac{e}{m_0} \frac{\Omega \cdot \Sigma}{\omega_0^2 - \Omega^2 - i\Gamma\Omega(\Omega^2 + a^2/c^2)} \tag{12}$$

Note that Eq. (10b) is a Fourier transform in the time/frequency domain only; it describes the electric field at the center of the oscillator, that is, E_0 ($r \equiv 0,\tau$) in the S frame. There is no dependence in the position of the charge itself, that is, dependence on r, whether in the initial dynamical Eq. (7), where the electric field is evaluated at E_0 ($r \equiv 0,\tau$), or in the fluctuating velocity. This is valid as long as the spatial dependence of the electromagnetic fields is negligible within the amplitude of motion of the oscillator (small oscillator approximation). Again, because the model considers wavelengths of the fluctuating vacuum field as small as the Planck length, this approximation is debatable.

The Lorentz force is obtained by taking the product of the velocity with the magnetic field. For the latter, HRP use the following Fourier transformation[‡‡]:

$$E_0(0,\tau) \approx \gamma(1-\beta) \int d^3k \cdot \tilde{H}_{ZP} \cdot e^{i(kR - \omega t)} \tag{13}$$

with $\tilde{H}_{ZP} = \hbar\omega/2\pi^2$. The electric field in Eq. (13) is a simplified version of Eq. (10) of HRP, assuming a single polarization in the $\hat{y}$-direction. The magnetic field for this mode is pointing in the $\hat{z}$-direction ($\pm$) and is related to Eq. (13) by a factor $1/c$, which can be ignored when using the natural system. Note that Eq. (13) includes a Lorentz transformation (of parameters $\beta = V/c$ and $\gamma = 1/\sqrt{1-\beta^2}$) from an inertial frame (**K**) where the vacuum mode energy density is defined ($\tilde{H}_{ZP}$) to the inertial frame (**K**′) that is instantaneously at rest with the uniformly accelerated frame **S** that contains the

---

[‡‡]To avoid lengthy descriptions, we have simplified Eq. (8) from HRP by explicitly considering the z-component only and a single field polarization.

oscillator. This explains why the phase contains a term proportional to R (distance to the center of the oscillator) and the time t (and not the proper time). At this point, we should point out that the velocity Eq. (12) depends only on the square of the acceleration; any effect of acceleration (expanding in powers of a^2) from the solution of the oscillator equation of motion would therefore be independent of its sign, and could never contribute to a force proportional to the acceleration, $F = -m_i \cdot a$. In fact, there is no reason to keep the term proportional to a^2/c^2 in Eq. (7) if the only intent is to derive an expression up to the first order in the acceleration. Close examination of the derivation presented in HRP shows that the linear term in a is obtained *only* from the phase of the Fourier transform in Eq. (13), that is, from the expansion of the distance:[§§] $R \approx \frac{1}{2}a \cdot t^2$. Thus, the force computed in the HRP model arises from a dephasing between the oscillating velocity and the oscillating magnetic field. After some further algebraic manipulation and introducing a cutoff frequency (an adjustable parameter), HRP arrive at an expression for the force:

$$F = -(\Gamma \omega_\Lambda) \frac{\hbar \omega_\Lambda}{2\pi c^2} a \qquad (14)$$

and the term in front of the acceleration is thereby identified as the inertial mass. Radiative damping actually plays no role in the HRP model, and it is somewhat misleading to express the solution in Eq. (14) in terms of the damping constant. In fact, it is easy to see from HRP that the Γ term in Eq. (14) is obtained only from the leading prefactor e/m_0 of Eq. (12) and the charge e of the expression for the Lorentz force: $ev \times B$. The inertial mass in Eq. (14) can be expressed as:

$$m_i \cong \frac{2}{3} \alpha \frac{m_\Lambda^2}{m_0} \qquad (15)$$

In the absence of any other *natural* cutoff scale, one should have $m_\Lambda \cong m_P = (\hbar c/G)^{1/2}$, the Planck mass. Note that in order to obtain the known mass of the electron from Eq. (15), that is, $m_i \equiv m_e$, the bare mass m_0 must be of the order of 10^{12} kg (10^{56} GeV/c^2)—more than 20 orders of magnitude *greater* than the Planck mass. One could argue that this is of no significance, because the bare mass is a free parameter, but this is well above any fundamental mass scale imaginable.[¶¶]

In the HRP model, there is also no provision for computing the mass in an inertial frame; the so-called inertial mass can be computed only when the oscillator is accelerated. One must therefore assume that the "rest" mass, obtained in

---

[§§]There are some discrepancies in Eqs. (13) and (1) regarding the expression for the distance R—see Appendix A—although these are of no significance for the final results.

[¶¶]One could make a reverse argument and consider m_0 to be the Planck mass and determine the cutoff energy scale from Eq. (15). This would lead to $m_\Lambda \approx 10^{18}$ eV/c^2, an intermediate scale well above the projected scale of super-symmetry breaking and well below that of grand-unified theory (GUT), see, for example, Ref. [17]. However, because Eq. (15) is based on a classical model of electrodynamics, there is very little relevance to the current state-of-the-art in quantum field theory.

an inertial frame when $a \equiv 0$, is also the same. However, it is physically counter-intuitive, within the context of this *classical* model, to imagine how the "parton" with an initial bare rest mass of $m_0 c^2$ can *lose* energy when subjected to the high-frequency chaotic motion induced by the ZPF in order to reach the observed value of $m_e c^2 \approx 0.512$ MeV. In fact, simple arguments regarding the coupling of a classical oscillator with a fluctuating field yield a resulting *rest* mass with a similar scaling as the HRP result, albeit with a slightly different coefficient*** (see Appendix B of this chapter). Therefore, the mass "correction" due to the coupling of the oscillator with the vacuum field is always positive, and can be extremely large if the cutoff energy is also large.

If the cutoff frequency is set instead to a value such that, with a proper combination of the bare mass the result matches the observed electron mass,

$$m_e c^2 \cong \frac{2}{3} \alpha \frac{(\hbar \omega_c)^2}{m_0 c^2} \qquad (16)$$

a new adjustable parameter is being introduced, which is equivalent to the "size" of the oscillator. Indeed, a cutoff frequency ω_c indicates that all ZPF fluctuations with wavelengths much smaller than the oscillator size c/ω_c do not contribute to a net force due to the averaging process over random phases. However, the oscillator size is now introduced as an ad hoc parameter in the model, making its usefulness questionable. The charged parton being coupled to the ZPF modes, one can argue that any wavelength shorter than its fundamental length scale (i.e., the "extent" of the parton), the Compton length $\lambda_{c,0} = \hbar/m_0 c$ would not effectively contribute to the jittering motion. Thus, the cutoff scale becomes the same as the original, that is, $m_\Lambda \approx m_0$; any attempt to "explain" the electron mass scale from a more fundamental scale (such as Planck mass) is, within the context of this model, highly problematic. Note also that any high-mass parton constituting an electron would, when annihilating with an "anti-parton" of a positron, result in much higher energy gamma rays than observed.††† Therefore, the model is faced with a conundrum: either there exists a rest mass that is excessively high and unphysical, or a particle size is introduced ad hoc to match observations and the model essentially predicts and explains nothing.‡‡‡

One should point out that HRP did consider the case of $m_0 < 0$ (their Appendix B); in that case, one could still start with a large (negative) bare mass and obtain a small (observed) finite mass after accounting for the kinetic energy of zitterbewegung, while a positive inertial mass can be obtained by choosing the

---

***As mentioned earlier, this scaling relies on a non-relativistic approximation, which breaks down when the cutoff energy is much larger than the physical rest-mass energy and a different, linear scaling is obtained when taking into account relativistic effects. Although HRP have not considered the relativistic case, one can reasonable assume that a similar change in scaling could be obtained in that model.

†††This discrepancy could not be explained away by hand-waving arguments that the jitter motion would be reabsorbed by the ZPF, because the problem is with the rest mass energy, $m_0 c^2$, and not the kinetic energy of zitterbewegung.

‡‡‡It is somewhat ironic that in a response [6] to comments by Woodward and Mahood, Haisch and Rueda make a critique of gravitational theory of inertia (Mach's principle) as containing a circular argument, while the authors appear unable to recognize the circularity of their own model.

opposite sign in the solution for the Lorentz force. A negative (unobservable) bare mass is not a priori impossible in quantum field theory (QFT) because the mass parameter is related to the shape of a field potential and a cosmological phase transition. For example, one could consider in a classical model a negative bare mass as arising from a strong (nonperturbative) gravitational solution [16,18], albeit of doubtful relevance to physical space. However, this only raises additional and even more fundamental issues to the problem, with little perspective of resolution within the scope of classical theory.

Haisch and Rueda and others have attempted to develop the SED theory further [2–7], but some of the problems mentioned above have persisted. Abandoning an explicit model such as the damped oscillator, Haisch and Rueda considered the interaction between the ZPF and a particle of a given size (volume V_0) and with a given dimensionless factor, $\eta(\omega)$. The inertial mass is now:

$$m_i = \frac{V_0}{c^2} \int \eta(\omega) \frac{\hbar \omega^3}{2\pi^2 c^3} d\omega \tag{17}$$

Therefore, the model difficulties regarding ultraviolet divergences are swept under the rug by assuming a form factor $\eta(\omega)$. This approach would be viable if the particle had a structure,[§§§] which could be determined experimentally; for example, the structure of the proton in terms of partons (later to be identified as the quarks of the standard model) was a primary object of experimental study during the 1970s. However, as far as we know [19] the electron has no structure up to a scale of approximately 50 GeV; this makes the upper limit of the integral in Eq. (17) much larger than the observed rest-mass of the electron. Thus once again, we are faced with a disparity of scales, that is, one must artificially set the free parameters of the theory to match observed values.

Among recent work, one must mention the use of a covariant approach to compute the inertial mass [3] and a "quantized" version [4]. However, the connection between rest mass and inertial mass is still not being made. For example, Sunahata [4] claims that the particle, seen from the laboratory frame, contains both a mechanical momentum $\vec{p}_0 = \gamma m_0 \vec{v}$ and a ZPF momentum, the latter becoming a function of time as the particle is being accelerated. This separation of variables is not consistent. The total momentum after acceleration can be written as:

$$\vec{P}(t + \Delta t) = \vec{p}_0 + \vec{p}_{zp}(t + \Delta t) = \vec{p}_0 + \vec{p}_{zp}(t) + m_i \int \vec{a} \, dt \tag{18}$$

However, the velocity has now changed, and obviously the "mechanical" momentum should now be $\vec{p}_0 = \gamma m_0 (\vec{v} + \vec{a} \Delta t)$. Therefore there cannot be a separation between a "mechanical" momentum (with an unknown mass, m_0)

[§§§]Replacing a form factor (i.e., distribution of internal constituents) by a frequency-dependent permittivity makes no difference; the latter also implies the existence of a substructure with characteristic time and length scales.

and a ZPF contribution. There are other problems as well. In the HRP model, the linear term in the acceleration $\vec{a}$ arises from the expansion of a phase difference between the electric and magnetic field Fourier components. In the more recent work by Rueda and Haisch [3] and Sunahata [4], the effect of the acceleration arises from the Lorentz transformation of the field *amplitudes*. Indeed, by switching to a semi-quantized version of the SED, the vacuum expectation values of the ZPF involve the product of one field with the Hermitian of another (in order to have a nonvanishing contribution $<0|a \cdot a^+|0>$ of the creation and destruction operators). This results in a phase factor $\exp[i\Theta(k) - i\Theta(k')]$ which, when combined with a delta-function $\delta(\vec{k} - \vec{k}')$, can be eliminated. The mechanism by which the acceleration modifies the ZPF Poynting vector (of which the only relevant component is in the x-direction, i.e., $E_y B_z$) is very suspect. To summarize, Rueda and Haisch obtain field amplitudes of the form:

$$\vec{E}''_{\omega\lambda} = \begin{pmatrix} 0 \\ \gamma(1 - \beta\hat{k}_x) \\ 0 \end{pmatrix} \cdot \tilde{H}_{ZP}(\omega), \quad \vec{B}''_{\omega\lambda} = \begin{pmatrix} 0 \\ 0 \\ \gamma(\hat{k}_x - \beta) \end{pmatrix} \cdot \tilde{H}_{ZP}(\omega) \quad (19)$$

Integration over the wave vectors keeps only terms of even order in k_x, yielding a different expression for the Poynting vector in the new frame of reference, which is interpreted as a resistance to the acceleration. However, Eq. (19) can be obtained for *any* Lorentz transformation. Thus, the model would predict a reaction force for a particle in inertial motion, which is of course unphysical. In fact, the value of the Poynting vector of the vacuum fields was already computed by Boyer [13] as a two-point correlation function, and was found to be identically zero. Thus, the results of Rueda and Haisch [3] are in contradiction with Boyer's. Our own calculations (see Appendix C of this chapter) verify Boyer's results and the identically null Poynting vector.

The recent work by Sunahata [4] has also been hailed as a major step forward by using a quantized version of the ZPF. However, this is somewhat misleading, because the nonvanishing expectation value of the ZPF is itself a result of quantum theory, used in a classical model. Sunahata compared classical evaluations of the ZPF energy density and ZPF Lorentz force with a quantized version of the two-point correlation functions; this is not a full quantum treatment, because the matter fields would also need to be quantized, and an ad hoc symmetrization of the operator ordering was introduced (instead of time ordering). In order to better appreciate the similarities and differences between classical and quantized versions, it is worth examining the state-of-the-art in conventional theory, that is, a fully quantized, relativistic theory: quantum electrodynamics or QED.

Fig. 2   External lines in QED.

Fig. 3 Propagator lines in QED.

IV. Quantum Electrodynamics

In quantum field theory (QFT), where matter and fields are both described at the quantum level ("second quantization"), interactions with fluctuations of the vacuum field are a fundamental aspect of the theory. While a complete and rigorous description of QFT can be found in various monographs [20,21], it is worth presenting here a simplistic summary of the essential elements of the theory. We consider only the theory for quantum electrodynamics (QED), which describes the coupling of charges to electromagnetic fields at this fundamental level. There are some very simple rules to observe, which are related to the so-called Feynman diagrams (Figs. 2–8):

1) Any observable state (whether field or matter) is always physical and can be described as an "external" line, as shown below. The fact that the state is physical implies that momentum and energy conservation are applied, and that all masses are the physical masses observed; these states are said to be "on-shell" (Fig. 2).

2) Virtual states can be generated and propagated from one point to another; these are called "propagator" and their momentum and energy can take a range of values (the arrows shown on propagator lines in Fig. 3 are actually immaterial).

3) The coupling between a charge and a field is described by a three-leg vertex (in QED), and is associated with one unit of charge, or coupling constant: conservation of momentum-energy is applied at each vertex (Fig. 4).

4) Only topologically different diagrams need to be considered.

For example, Fig. 5 describes to the lowest order (Born approximation) the scattering of two particles (A and B) by the exchange of a virtual photon, resulting in two outgoing particles A' and B'. The combination of the delta functions at each vertex ensures that energy and momentum are conserved between the initial and final products.

Given the QFT rules described above, one can also construct the diagram in Fig. 6. In this case, a charged particle (electron, top solid line) is created at

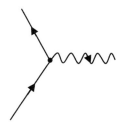

Fig. 4 Vertex diagram (coupling) in QED.

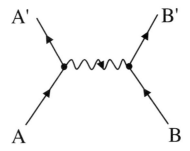

Fig. 5   Lowest-order fermion scattering in QED.

point O along with an anti-particle (positron, bottom solid line), while the interaction between the two is mediated by a photon. All particles in this diagram are virtual, that is, are not required to have positive energy or mass. In fact, one can consider this process as the temporary excitation of an electron from below the Fermi level of a condensed state, thus leaving a positive "hole" inside the condensed state. This process is therefore a vacuum fluctuation, and is a QFT generalization of the Heisenberg principle. Note that because there are no external lines in Fig. 6, that is, there are no observables; although these fluctuations can occur at all times, they are inconsequential.¶¶¶ For such vacuum processes to have an observable effect there must be external lines associated with the diagram. Consider now the diagram obtained by splitting the bottom solid line in Fig. 6 into two external lines (Fig. 7).

This time, the process describes the interaction of an electron (incoming line at left) with the vacuum field, during which a virtual electron (e*) is created, and after coupling to the vacuum field again, a real ("on-shell") electron is produced again. Although Fig. 7 appears benign, it does have important consequences, leading to a *mass renormalization* of the electron. Note that the process is of order e^2, the square of the electron charge (there are two vertices), or $e^2/\hbar c$, the fine-structure constant. More complicated diagrams can be

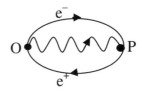

Fig. 6   Vacuum fluctuation diagram in QED.

¶¶¶This distinction is critical to the prospects of "extracting" energy from the vacuum: In QED, while the *fluctuations* of the vacuum (i.e., interaction with temporary fields) can have a profound and measurable effect, there is no known mechanism (i.e., diagram) that leads to net energy extraction. Even the impact on gravity requires, at the fundamental level, external lines corresponding to gravity field (gravitons), and the need for a quantum theory of gravity.

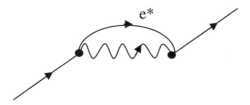

Fig. 7 Electron self-energy diagram.

constructed as well; for example there can be repeated emission/absorption of the virtual photons, leading to an expansion in powers of α, or there can be an interaction with another particle while the electron is in a virtual ("off-shell") state, as shown in Fig. 8. When the particle at the top of Fig. 8 is an atomic nucleus, the diagram describes the Lamb shift of spectral lines. Because $\alpha \ll 1$, one can construct a formal expansion in powers of α, that is, an expansion in the number of vertices. Using this perturbative approach, quantum electromagnetic properties have been calculated to an exquisite degree of precision, making QED one of the most accurate and successful theories known to date.

However, there are some subtle conceptual difficulties with QFT. Note that the self-energy diagram (Fig. 7) contains a *loop*, and because the virtual particles can have any value of momentum and energy, there is an integration over the momentum-energy values corresponding to each loop. The problem is that divergences may occur during this integration. In particular, the diagram (Fig. 7) is UV-divergent and a cutoff must be introduced. This is basically the same problem as in the classical model of ZPF interaction****; i.e., all vacuum modes can interact with the particle, up to the Planck frequency.

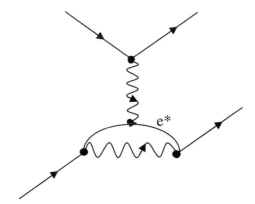

Fig. 8 Renormalized coupling diagram (Lamb shift).

****An important difference is that in QFT the divergences are only logarithmic.

The solution to this problem, as distressing it may appear, is a balance between two mass parameters. In the initial QFT formulation, the "mass" of particles appears as a specific term in the field Lagrangian, for example, $-m\bar{\psi}\psi$, where ψ is a Dirac field describing an electron. The self-energy diagram contributes a mass correction δm (as well as a rescaling of the wave function or field amplitude, which does not need to be considered here), proportional to $\ln \Lambda$, where Λ is the cutoff energy used to regularize the loop integral. Thus, there is a so-called *counter-term* $\delta m \cdot \bar{\psi}\psi$ in the Lagrangian, and the overall contribution to the Lagrangian should be a combination of a "bare" mass m_0 (which is not observable) and the counter-term, such that the observable mass is $m_e = m_0 + \delta m$. Two important remarks are in order:

1) There is an infinite number of possible self-energy diagrams of the type shown in Fig. 7, where additional vertices and loops can be added, yielding an expansion in powers of α (according to the number of vertices V) as well as an expansion in the number of loops (L), with each loop corresponding to an integration over momentum and energy. It would seem that one would need a similar infinite number of counter-terms in the Lagrangian; however, this is not the case for well-behaved theories. This is an important criterion in distinguishing candidate theories, and a QFT is said to be *renormalizable* if there is a finite number of such counter-terms.

2) Although the bare parameters seem arbitrary and a new circular argument may be at hand (that is, the bare mass being adjusted such that the complete result matches observations), the bare mass can actually be a product of a higher-level theory. For example, "masses" can be generated[††††] from a QFT with a higher degree of symmetry (Higgs mechanism). The bare mass would then have a scale comparable to the energy scale at which this breakdown of symmetry occurs (similar to a phase transition in condensed matter), but the so-called "radiative corrections" (Fig. 7) would bring the observable mass to a different value. In this case, the cutoff energy scale Λ would presumably be of the same order as this symmetry breakdown scale. Furthermore, it is important to point out that the QED results are based only on the method of perturbation, the only approach currently available to perform this type of calculation. In addition to the mass, the other parameters of the theory (in particular α, the coupling strength) also are being renormalized.[‡‡‡‡] The overall behavior of the parameters follows a set of nonlinear transformations—the "renormalization group" [23]—as function of energy scale. This behavior plays a critical role in allowing the unification of all forces of nature.

The result from QED is a counter-term—[20, Sec. 7.42]:

$$\delta m \cong \frac{3\alpha}{4\pi} m \left(\ln \frac{\Lambda^2}{m^2} + \frac{1}{2} \right) \approx \underbrace{\frac{3\alpha}{2\pi} m \cdot \ln \frac{\Lambda}{m}}_{\Lambda \to \infty} \qquad (20)$$

where Λ is the energy cutoff and m is the physical, observable particle mass.

[††††]Even negative masses.
[‡‡‡‡]See also the discussion in [32] Sec. 11.10.

Thus, although there are similarities of concept between QED and SED, there are also important differences:

1) QED is a more complete theory, which is fully quantized, relativistic, well developed, and highly accurate. SED was until recently a classical model based on some questionable assumptions and mathematical solutions.

2) Both SED and QED rely on the concept of "bare" parameters, which become renormalized by interaction with the vacuum field (virtual particles). The radiative correction in QED leads to an additive term ($m_e = m_0 + \delta m$), while SED predicts a *multiplicative* correction (for example, $m_e \approx m_P^2/m_0$).

3) QED predicts logarithmic divergences, while SED predicts much more severe quadratic divergences (m_P^2) in the nonrelativistic case and possibly linear in the relativistic case.

4) The radiative corrections in QED yield renormalization of the rest mass, but SED predicts the mass only in the case of acceleration (that is, inertial mass), because the calculations in HRP depend on the existence of a noninertial frame.

These points are debated further in the next section. Item 4 listed above raises another question: does QED also predict a mass shift for accelerated particles? The answer is yes, although the effect is exceedingly small. QFT can be extended to fields at finite temperature; formally, this is done by extending the theory from Minkowski space to Euclidean space with an imaginary time variable β now associated with $1/kT$. The Feynman diagram rules are modified accordingly and include the known quantum distribution functions at finite temperature, that is, Bose and Fermi statistics, for the propagators [24]. An important aspect of finite-temperature QFT is the fact that the vacuum now contributes *external lines* in the Feynman diagrams. These lines are only present because the vacuum now has a thermal distribution of photons, that is, the number operator ($n_\omega = a_\omega^+ a_\omega$) operating on the vacuum state $|0>$ is non-zero. The presence of real (thermal) vacuum photons plays a key role in removing infrared divergences when computing scattering cross-sections (see [25]). The self-energy diagram at finite-temperature also has a temperature-dependent contribution [25–27] of the order $\delta m \approx (T^2/m)$.

Davies and Unruh [28,29] showed the equivalence between acceleration and finite-temperature by examining the correlation functions (i.e. propagators in QFT) in an accelerated frame. The result is that the quantum vacuum in an accelerated frame obtains a thermal distribution with a temperature:

$$kT_U = \hbar a/2\pi c \qquad (21)$$

Therefore, the complete formalism and results of finite-temperature QFT are directly applicable to the case of the accelerated frame, by simply using the Unruh-Davies temperature in Eq. (21) in the results. Note the scale of the temperature in Eq. (21): It would take an acceleration of 2.5×10^{20} m/s^2 to yield a temperature of 1 K; for this reason, the principal area where this effect has been studied is black-hole physics. Obviously, the corrections due to the Unruh–Davies effect are exceeding small, but could potentially be observed in future laboratory experiments [30–32]. Finite-temperature QFT effects have significant impact mostly in the early times after the big bang, when the universe was still at very high temperature.

It should also be pointed out that the effect of a uniform acceleration on the classical vacuum electromagnetic field was also computed by Boyer [13]. This leads to a simple correction to the rest mass that is proportional to the Unruh–Davies temperature, that is, to the acceleration (see Appendix B). The finite-temperature QED renormalization yields, again, to a logarithmic correction, and one could simply use the Unruh–Davies temperature in the final result to obtain the effect of a uniform acceleration. Again, the impact is completely negligible except for exceedingly large acceleration (i.e., near back-hole horizon events).

V. The Way Forward

The SED model of vacuum-induced inertia can be summarized by the following observations:

1) The HRP model of SED is based on a classical oscillator model with radiation reaction force and with uniform acceleration. The principal result of the model is a resistance force that is claimed to be proportional to the acceleration, with the constant being identified as the inertial mass. Higher-order terms (i.e., $a^2\tau^2/c^2$) are not computed, but would likely remain in a more complete solution, thus potentially leading to deviations from Newton's Law.

2) It is claimed in HRP that the SED is essentially a relativistic effect and HRP carried a complex and cumbersome relativistic treatment throughout most of the calculations, yet a nonrelativistic approximation is essential to the validity of the model and is used throughout. In fact, the analysis could have been carried completely with a nonrelativistic approximation, because the inertia effect is obviously observed at low accelerations. Similarly, the examination of the mathematical treatment shows that the radiation dampening of the oscillator, the term proportional to Γ in the denominator of Eq. (12) is also essentially irrelevant, and that the result is obtained as a result of phase difference between the Fourier components of the velocity and magnetic fields expressed in different frames of reference.

3) Another argument [3] for the generation of inertial mass based on a non-zero expectation value of the Poynting vector of the vacuum field is at odds with prior calculations by Boyer [13], which show a vanishing contribution (see Appendix C of this chapter).

4) The inertial mass depends on a bare mass m_0 and a cutoff frequency ω_Λ, which are both free parameters in the theory. Depending on the choice of this cutoff frequency, the model either appears to yield unphysical results or has no real predictive capability.

5) The model used a small oscillator approximation (no field variation in the transverse direction) while the integration over ZPF modes includes very short wavelengths. The model also decouples the oscillating motion from the acceleration in the longitudinal direction due to the Lorentz force, essentially a nonrelativistic approximation that is not valid for the high frequency modes.

Further extensions of the model "removed" some of the problems of the original classical oscillator model of HRP by keeping the form factor adjustable;

however, disparity of mass scales remains a problem that prevents SED from making any mass predictions from any more fundamental scale.

The SED theory, whether the original classical oscillator model or more recent versions, lacks rigor and contains some errors as well as unnecessary complications that obscure much of the physics and the analysis. Nevertheless, there remains the potential for some advances in the theory, which can be fruitful. The original intent of the theory, i.e., determination of the inertial mass as a result of coupling to vacuum energy fluctuations, may have been of interest, but this objective appears to be lost in circular and philosophical (or meta-physical) arguments. Classical models of electron self-energy have been developed for many decades,[§§§§] even after the discovery of quantum mechanics and quantum field theory. Although classical models can still bring some interesting insight into the physics of a problem, as shown for example in the work by Boyer, one must be aware of the fundamental limitations of the approach. In particular, classical theory breaks down at small length and time scales, that is, when their combination becomes of the order of the Planck constant.[¶¶¶¶] By contrast, QFT and in particular QED is a fully relativistic, quantum theory that is extremely well validated. One can argue that, conceptually, there are still some remaining difficulties with QFT, and that it is prudent to re-examine the differences between SED and QED (listed at the end of Section IV) before drawing conclusions on ways to improve SED.

1) *QED is a more complete and accurate theory*: This is certainly the case, because SED relies mostly on classical and nonrelativistic approximations that are not quite valid. Thus, one can reasonably assume that with continued progress in calculating SED effects, such as quantization and inclusion of relativistic effects, SED would become better developed and more accurate. However, it begs the question as to the fundamental difference between the two theories: QED already includes full quantization and relativistic effects, and already accounts for the coupling to vacuum fluctuations, doing so with high precision (e.g., Lamb shift, magnetic moment). What then would a quantized, relativistic version of SED offer? If such a theory yields the same result as QED, what would be the difference, conceptually or computationally? If it predicts different results, either SED is *in*validated (e.g., does not reproduce experimental data), or it predicts *additional* results that QED cannot. For example, the treatment by HRP is obtained as an expansion in terms of the acceleration; higher-order terms (a^2) are also present but not yet computed. Would such deviations from Newton's Law be measurable?

2) *Treatment of bare parameters:* Pending a grand unification theory or validated string theory, the bare mass of the particles in QFT have a certain degree of arbitrariness[*****]; however, there are a number of important effects

---

[§§§§]The reader is urged to consult the book by P. Milonni [33] for a review on past and current models.

[¶¶¶¶]Classical physics can be formally recovered from quantum mechanics in the limit $h \to 0$.

[*****]One can argue that some masses can be explained by a Higgs mechanism, but it mostly defers the question to another level (i.e., why the symmetry breaking of a higher-level QFT would occur at a particular scale) until the ultimate theory can be obtained.

in QFT that make the problem conceptually less difficult; for example, the fact that mass parameters can be associated with phase transitions; divergences are only logarithmic (see items); fundamental symmetries (e.g., SUSY) can be at hand to cancel divergences. As a classical model, it is much more difficult to interpret the incongruity of bare parameters; for example, there is no mechanism for a classical particle to start with a very high rest-mass and end up with a lower total energy when adding high-frequency oscillating motions, and the concept of negative mass in classical theory presents a number of difficulties. Therefore, it would appear that either one needs to have a complete revolution in physics by reverting back to a reinterpreted classical view of nature, or SED needs to graduate and become a quantum theory. There appears to be some minor progress in that direction [4], but in that case, one is then led back to the issues mentioned in item 1.

3) *Nature of divergences:* One could argue that the exact nature of divergences is not important, because these divergences are removed by setting appropriate values of the bare parameters of the theory. However, a logarithmic divergence is much more appealing, especially while attempting to construct a more fundamental theory. The renormalization of the bare parameters (mass, charge, etc.) of a theory has become an essential aspect of modern QFT through the use of renormalization group (RNG) theory [23]. Here, RNG is used explicitly to predict at what energy scale the coupling parameters become similar and when symmetry breaking of a higher-level theory may occur. By contrast, SED cannot make such predictions, does not offer any explanations for the cutoff values or form factors, and, generally speaking, yields much more severe divergences. Again, one can argue that SED is still in its infancy, and that further developments along the lines of quantization may yield more information. Although this is possible, one is led to wonder at what point SED would become relevant, and what would then be the difference from QED, leading us back to items 1 and 2.

4) *True effect of acceleration:* It is puzzling that although the effect of acceleration on the vacuum has already been examined by Unruh and Davies, it is not being used by Rueda and Haisch in the development of SED. One could use simple (even simplistic) arguments combining the classical models and the basic effect of a transformed vacuum field spectrum as a result of acceleration (Appendix D of this chapter) to examine the impact of accelerated motion on mass and self-energy. Because the acceleration essentially brings a thermal component[†††††] to the vacuum field, one could simply use the complete (thermal + zero-point) energy spectrum in the interaction with a classical oscillator. The effect can also be computed easily in QED, using well-known results from finite-temperature QFT, as described earlier.

We suggest that further development of SED theory would require the following:

1) Computing with SED some effects of the vacuum fluctuations that can be compared with *multi-loop* QED and known data (e.g., Lamb shift, electron magnetic moment).

[†††††]P. Milonni ([33], p. 64) describes it as a "promotion" of the quantum modes to the level of thermal fluctuations. This effect has also been shown within the context of SED [15].

2) Computing higher-order effects (a^2) that are not predicted by conventional theory and determine if experimental validation of those effects is possible.

3) Examining fundamental issues regarding the connection with the Unruh–Davies effect and similarities and differences with QED, in a clear and convincing manner.

There is, of course, the problem of experimental validation. Any theory must eventually be able to satisfactorily reproduce experimental results, explain observations that have otherwise remained beyond the capabilities of competing models, and/or predict new observable effects. In this aspect, QED has a significant advantage over SED. It may be that, to distinguish between QED and SED, one may be led to a new physical regime, notably in the levels of acceleration. The recent development of ultra-intense, ultra-short lasers may provide an avenue toward high acceleration levels in a laboratory space, and could be used if SED can be made to predict effects to higher orders in acceleration, for example. However, the problem may not be so simple because the laser field itself creates a zitterbewegung and an effective mass that can drown measurable effects from the ZPF. Another approach is to seek the high accelerations near astrophysical objects, but data interpretation is also subject to a lot of uncertainty. Therefore, there does not appear to be an easy solution to the problem of experimental validation,[‡‡‡‡‡] especially if one must be able to make the clear distinction between expected QED effects and SED effects. Nevertheless, it is absolutely critical for SED to be computable (i.e., does not rely on ad hoc parameters that are adjustable), and predictable (i.e., able to make quantifiable predictions for feasible experiments).

VI. Conclusions

Of course the manipulation of inertial mass opens up many interesting possibilities for space propulsion and access to space. As always, the benefits have to be weighed against the costs; for example, one must determine whether it is advantageous to expend a certain amount of energy in reducing the inertial mass of a spacecraft, versus the energy needed to provide the kinetic energy to that same spacecraft. Assuming the SED theory to be correct, such control of the inertial mass must inevitably come from manipulation of the vacuum field. To this day, there is no definite mechanism that allows us to perform this on macroscopic scales. Even if one were to induce changes in the vacuum field at precise locations and times, the Heisenberg principle implies that such changes are bound to be microscopic, either on the energy or time scale. Thus, manipulation of the inertial mass would require related, major advances in our understanding of the quantum vacuum in general, and extreme cooperation on the part of the laws of physics; that is, the discovery of methods for field manipulation on macroscopic scales. Although we are

---

[‡‡‡‡‡]We must emphasize that commonly cited experiments that are based on the interaction with vacuum fluctuations, such as Casimir force and related cavities, do not necessarily provide a validation of SED, since current state-of-the-art quantum theories are readily used to predict such effects. There must be a clear and measurable difference between QED and SED predictions.

now within the realm of fiction, one may suspect that this is likely to be either absolutely impossible (i.e., being limited to microscales) or extremely costly (requiring, for example, to generate copious amounts of anti-Higgs or other particles with similar effects). Nevertheless, it is much too early to have a definite opinion on the subject, and it is best to wait until further evidence can be obtained, from current efforts at unified theories or other, less conventional approaches. One may be able to learn from surprising effects in condensed-matter, which has often been an important application of QFT (see, for example, Ref. 34).

The potential of SED for explaining fundamental particle properties is somewhat doubtful. The current state of the theory is not especially promising, being afflicted by questionable mathematical approaches and physical models, and caught in circular arguments regarding particle size, cutoff, form factors, etc. Significant advances are still possible, and guidelines have been provided. Until a number of significant steps have been accomplished (i.e., comparison with QED, computation of known effects, prediction of new ones), SED will remain at the fringes of scientific research with little prospect for major impact. On the other hand, SED could potentially play a role in other areas, such as plasma physics (e.g., noise generation, turbulence, chaotic heating), astrophysics, and condensed matter.

Appendix A: Relativistic Transformation to a Uniformly Accelerated Frame

Let us start with covariant formulation of the radiation reaction force [22]:

$$F^\mu = \frac{2e^2}{3c} \frac{d^2 u^\mu}{ds^2} - R \frac{u^\mu}{c^2} \tag{A1}$$

with

$$R = -\frac{2e^2}{3c} \frac{du_\nu}{ds} \frac{du^\nu}{ds} \tag{A2}$$

and

$$u^\mu = \begin{pmatrix} \gamma c \\ \gamma \vec{v} \end{pmatrix} \text{ with } \gamma = \frac{1}{\sqrt{1 - \vec{V}^2/c^2}} \tag{A3}$$

Here, $\vec{V} = d\vec{X}/dt$ is the velocity of the accelerated frame **S** with respect to the laboratory frame, $ds^2 = c^2 d\tau^2 = c^2 dt^2 - d\vec{x}^2$ is the invariant measure in a Minkowski metric [i.e., $\eta_{\mu\nu} = (1,-1,-1,-1)$, where τ is the proper time], and $(t, \vec{x})$ are the coordinates of the particle in the laboratory frame.

The motion of the accelerated frame **S**, as seen from the laboratory frame, is:

$$V = \frac{dX}{dt} \rightarrow \frac{d}{dt}\left[\frac{V}{\sqrt{1-V^2/c^2}}\right] = a \rightarrow V = \frac{at}{\sqrt{1+(at/c)^2}} \rightarrow \gamma = \sqrt{1+(at/c)^2} \quad (A4)$$

where a is the constant acceleration. From the definition of the invariant measure, we have:

$$\tau = \int \frac{dt}{\sqrt{1+(at/c)^2}} \rightarrow t = \frac{c}{a}\sinh\left(\frac{a\tau}{c}\right) \quad (A5)$$

which leads to:

$$\gamma = \cosh\left(\frac{a\tau}{c}\right) \text{ and } \beta = \frac{V}{c} = \tanh\left(\frac{a\tau}{c}\right) \quad (A6)$$

The motion of the accelerated frame, after transformation to the proper time variable, is:

$$X(\tau) = \frac{c^2}{a}\left[\cosh\left(\frac{a\tau}{c}\right) - 1\right] \quad (A7)$$

which in the nonrelativistic limit ($a\tau \ll c$) yields $X \approx a\tau^2/2$ as expected. Note that HRP used a different expression [1, Eq. (9c)] for the distance $X(\tau) = (c^2/a)\cosh(a\tau/c)$. This leads to a distance (and a phase in the Fourier transform of the vacuum fields) that becomes infinite as the acceleration becomes very small, and does not make the two frames coincide at $\tau = 0$. By contrast, Boyer obtains the expression in Eq. (A7) but incorrectly applies it when computing the phase shift in the Fourier transforms. In both cases, this error does not change the results because it is equivalent to a constant at finite acceleration.

Let us now compute the reaction force using the following definition of the four-velocity, for an oscillator motion in the $\hat{y}$ direction only:

$$u^\mu = \begin{pmatrix} \gamma c \\ \gamma V \\ \gamma \cdot dy/dt \\ 0 \end{pmatrix} = \begin{pmatrix} \gamma c \\ \gamma V \\ \dot{r} \\ 0 \end{pmatrix} \quad (A8)$$

where we have now identified $\dot{r} = dy/d\tau$. The proper time derivative is:

$$\frac{du^\mu}{d\tau} = \begin{pmatrix} a\sinh(a\tau/c) \\ a\cosh(a\tau/c) \\ \ddot{r} \\ 0 \end{pmatrix} \tag{A9}$$

We have then

$$\frac{du_\nu}{ds}\frac{du^\nu}{ds} = -\frac{a^2 + \dot{r}^2}{c^2} \tag{A10}$$

and the *transverse* reaction force is

$$F_y = \frac{2e^2}{3c^3}\left[\dddot{r} - \frac{a^2 + \dot{r}^2}{c^2}\dot{r}\right] \tag{A11}$$

There is a significant difference between the formulation of Boyer [14] and HRP. Boyer dismisses the $\dot{r}^2\dot{r}$ term when compared with $\dddot{r}$ because the former is of the first order of the oscillator amplitude while the latter is of second order. In the limit of small oscillator, the $\dot{r}^2\dot{r}$ term can therefore be neglected when compared with $\dddot{r}$. This would be correct, but not entirely prudent, because Boyer also keeps the term proportional to a^2 to yield an equation of motion [see Eq. (7) in text]:

$$m_0\ddot{r} = -m_0\omega_0^2 r + \frac{2e^2}{3c^3}\left[\dddot{r} - \frac{a^2}{c^2}\dot{r}\right] + eE_0 \tag{A12}$$

This is valid as long as $\ddot{r} \ll a^2$. However, the assumption of small oscillator size can easily be compensated by the high frequencies of oscillations. Boyer [14] finds a solution for the average amplitude of oscillation (without considering the $\dot{r}^2\dot{r}$ term) to be:

$$<r^2> \approx \frac{1}{2}(\hbar/m\omega_0)\coth(\pi c\omega_0/a) \tag{A13a}$$

from which one can easily derive

$$<\ddot{r}^2> \approx \frac{1}{2}(\hbar\omega_0^3/m)\coth(\pi c\omega_0/a) \tag{A13b}$$

Approximation of small oscillating amplitude made by Boyer corresponds to:

$$<r^2> \ll c^2/\omega_0^2 \tag{A14}$$

However, to neglect $\ddot{r}$ against a^2 implies that one must also satisfy:

$$\omega_0^2 c^2 \ll a^2 \tag{A15}$$

In the limit of small acceleration, this implies very low frequencies. Because in the limit in Eq. (A15) one can approximate $\coth z \approx 1/z$, the average oscillating velocity is given by:

$$<\dot{r}^2> \leq \frac{kT_U}{4m_0} \ll c^2 \tag{A16}$$

where $kT_U = \hbar a/(2\pi c)$ is the Unruh–Davies temperature (here a is the magnitude of the acceleration, i.e., does not depend on the sign). Because the Boyer oscillator model is for a specific direction, the total energy is given by: $E = mc^2 = m_0 c^2 + 3kT_U/2$.

Appendix B: Estimate of Rest Mass from Classical Interaction with ZPF

One can estimate the impact of vacuum field fluctuations on the average energy of a dipole by very simple arguments. We consider a simple oscillator:[§§§§§]

$$m_0 \ddot{r} = eE \rightarrow m_0 |v_\omega| \approx \frac{e}{\omega}|E_\omega| \rightarrow <v^2> = \frac{e^2}{m_0^2} \int_0^\infty \frac{d\omega}{\omega^2} <E_\omega^2> \tag{B1}$$

The average energy density in the vacuum fluctuations can be obtained as:

$$\frac{1}{2} <\varepsilon_0 E^2 + B^2/\mu_0> = (2) \cdot \int d^3 k \left(\frac{\hbar \omega_k}{2}\right) = \int d\omega \frac{\hbar \omega^3}{2\pi^2 c^3} \tag{B2}$$

Using equipartition of energy, this leads to $<E_\omega^2> = \hbar \omega^3/(2\pi^2 \varepsilon_0 c^3)$, and therefore the energy density of a single mode of the vacuum field scales as ω^3, as previously explained. The kinetic energy of oscillation, in the nonrelativistic approximation, is:

$$<v^2> = \frac{2}{\pi} \alpha \cdot \left(\frac{\hbar}{mc}\right)^2 \int_0^\Lambda \omega\, d\omega \rightarrow \frac{1}{2} m_0 <v^2> = \frac{\alpha}{\pi} c^2 \frac{m_\Lambda^2}{m_0} \tag{B3}$$

and we have recovered the same mass scaling as in the HRP result [Eq. (15) in the text]. On time scales that are large compared to the

---

[§§§§§] This model is essentially the same as the one used by Welton [35] for the Lamb shift, where the average oscillator amplitude, with a logarithmic divergence, is used as a perturbation of the electrostatic potential of the electron-nucleus system; see also Ref. 30, p. 90.

fluctuations,¶¶¶¶¶ this average kinetic energy is part of the total energy of the oscillator, that is $E = mc^2 \approx m_0 c^2 (1 + \alpha \cdot m_\Lambda^2/m_0^2)$. The nonrelativistic treatment is valid as long as $<v^2> \ll c^2$, which implies that the upper cutoff is at most $m_\Lambda \approx m_0$. One can generalize to the relativistic case by taking into account the Lorentz dilation factor, such that $\gamma^2 \beta^2 \approx \alpha m_\Lambda^2/m_0^2$, from which the rest mass becomes: $m = \gamma m_0 \approx m_0 \sqrt{1 + \alpha m_\Lambda^2/m_0^2} \approx \alpha^{1/2} m_\Lambda$. Thus, in the relativistic case, the quadratic divergence Λ^2 is replaced by a linear divergence (Λ), a common occurrence in classical models of self-energy.

Appendix C: Boyer's Correlation Function and Poynting Vector

The trajectory of the center of mass of the oscillator under uniform acceleration with respect to the proper time (so-called "hyperbolic motion") is described by Eqs. (A6) and (A7). Thus, the Lorentz transformation that relates the inertial frame **L'** that is instantaneously at rest with **S** is determined by the time-dependent parameters $\beta(\tau)$, $\gamma(\tau)$ given by Eq. (A6):

$$L_\tau = \begin{pmatrix} \gamma & +\beta\gamma \\ \beta\gamma & \gamma \end{pmatrix} \text{ such that: } \begin{pmatrix} ct' \\ x' \end{pmatrix} = L_\tau \begin{pmatrix} ct \\ x \end{pmatrix} \text{ (only } t, x \text{ coordinates)} \quad \text{(C1)}$$

A plane wave "seen" by the moving observer will have its wave vector and frequency shifted, while the phase of the wave remains invariant, that is, $\phi = k_x x - \omega t = k_x' x' - \omega' t'$. Therefore,

$$\begin{pmatrix} \omega' \\ ck_x' \end{pmatrix} = L_\tau^{-1} \begin{pmatrix} \omega \\ ck_x \end{pmatrix} = \begin{pmatrix} \gamma & -\beta\gamma \\ -\beta\gamma & \gamma \end{pmatrix} \begin{pmatrix} \omega \\ ck_x \end{pmatrix} \quad \text{(C2)}$$

The electric and magnetic fields of the vacuum can be written by a Fourier transform:

$$\vec{E}(\vec{x}, t) = \sum_\lambda \int d^3 k f(\omega) \hat{\varepsilon}_{k\lambda} \cos(\vec{k}\vec{x} - \omega t - \theta_{k\lambda}) \quad \text{(C3a)}$$

$$\vec{B}(\vec{x}, t) = \sum_\lambda \int d^3 k f(\omega)(\hat{k} \times \hat{\varepsilon}_{k\lambda}) \cos(\vec{k}\vec{x} - \omega t - \theta_{k\lambda}) \quad \text{(C3b)}$$

¶¶¶¶¶An analogous situation occurs for electrons oscillating in a classical, externally imposed electromagnetic field, such as a microwave or laser beam; although the electron oscillates in the field, there is no net energy transfer between the field and electron. However, during a collision, this kinetic energy becomes available and participates in the collision process, leading to a net absorption of electromagnetic energy. The "jitter" motion of the classical electron is also unobservable until interaction with another field, which occurs on longer time scales than the high-frequency jitter motion.

Here, $\theta_{k\lambda}$ is a random phase, $\hat{\varepsilon}$ is the unit vector along the electric field of a given mode, and λ is the polarization ($\lambda = \pm 1$). Averaging over the random phases yields the following relations:

$$\langle\cos \theta_{k\lambda} \cos \theta_{k'\lambda'}\rangle = \langle\sin \theta_{k\lambda} \sin \theta_{k'\lambda'}\rangle = \tfrac{1}{2}\delta_{\lambda\lambda'}\delta(\vec{k} - \vec{k}') \quad \text{(C4a)}$$

$$\langle\cos \theta_{k\lambda} \sin \theta_{k'\lambda'}\rangle = \langle\sin \theta_{k\lambda}\rangle = \langle\cos \theta_{k\lambda}\rangle = 0 \quad \text{(C4b)}$$

The factor $f(\omega)$ in Eqs. (C3a) and (C3b) is the spectral density which, for Lorentz invariance of the vacuum field, must be proportional to $\omega^{1/2}$. From the Lorentz transformation laws of arbitrary electric and magnetic fields [12], we can write the fields observed in the L' frame (ignoring multiplicative constants) as:

$$\vec{E}'(0, \tau) \approx \sum_\lambda \int d^3 k \cdot \sqrt{\omega} \cdot \begin{pmatrix} \hat{\varepsilon}_x \\ \gamma\hat{\varepsilon}_y - \beta\gamma(\hat{k} \times \hat{\varepsilon})_z \\ \gamma\hat{\varepsilon}_z + \beta\gamma(\hat{k} \times \hat{\varepsilon})_y \end{pmatrix} \cos(k_x x - \omega t - \theta_{k\lambda}) \quad \text{(C5a)}$$

$$\vec{B}'(0, \tau) \approx \sum_\lambda \int d^3 k \cdot \sqrt{\omega} \cdot \begin{pmatrix} (\hat{k} \times \hat{\varepsilon})_x \\ \gamma(\hat{k} \times \hat{\varepsilon})_y + \beta\gamma\hat{\varepsilon}_z \\ \gamma(\hat{k} \times \hat{\varepsilon})_z - \beta\gamma\hat{\varepsilon}_y \end{pmatrix} \cos(k_x x - \omega t - \theta_{k\lambda}) \quad \text{(C5b)}$$

This description follows exactly Boyer's [13] and Haisch and Rueda [3]. We can now compute the product of such fields and average over the random phases, using Eqs. (C4a) and (C4b). Let us compute the correlation function between electric and magnetic fields at two proper times $\tau^\pm = \tau \pm \delta\tau$:

$$\langle E'_y(\tau^-)B'_z(\tau^+)\rangle = \sum_\lambda \int d^3 k \cdot \omega \cdot \gamma^- \lfloor \hat{\varepsilon}_y - (\hat{k} \times \hat{\varepsilon})_z \beta^- \rfloor$$
$$\times \gamma^+ \left[\hat{\varepsilon}_y - (\hat{k} \times \hat{\varepsilon})_z \beta^+\right] \cdot \cos\left[k_x(x^- - x^+) - \omega(t^- - t^+)\right] \quad \text{(C6)}$$

where $\gamma^\pm = \cosh(a\tau^\pm/c)$ and $\beta^\pm = \tanh(a\tau^\pm/c)$. Using also the notation $\zeta^\pm = a\tau^\pm/c$, this leads to:

$$\langle E_y(\tau^-)B_z(\tau^+)\rangle = \sum_\lambda \int d^3 k \cdot \omega \left[\varepsilon_y \cosh \zeta^- - (k \times \varepsilon)_z \sinh \zeta^-\right]$$
$$\times \left[\varepsilon_y \cosh \zeta^+ - (k \times \varepsilon)_z \sinh \zeta^+\right] \cos\left[(k_x c^2/a)\right.$$
$$\left.(\cosh \zeta^- - \cosh \zeta^+) - (\omega c/a)(\sinh \zeta^- - \sinh \zeta^+)\right] \quad \text{(C7)}$$

Let us examine the argument of the cosine function first which, using the hyperbolic identities, can be written as:

$$[\ldots] = \frac{k_x c^2}{a}\left[\cosh(\zeta)\cosh\left(\tfrac{1}{2}\delta\zeta\right) - \sinh(\zeta)\sinh\left(\tfrac{1}{2}\delta\zeta\right) - \cosh(\zeta)\cosh\left(\tfrac{1}{2}\delta\zeta\right)\right.$$
$$\left. - \sinh(\zeta)\sinh\left(\tfrac{1}{2}\delta\zeta\right)\right]$$
$$- \frac{\omega c}{a}\left[\cosh(\zeta)\sinh\left(\tfrac{1}{2}\delta\zeta\right) + \sinh(\zeta)\cosh\left(\tfrac{1}{2}\delta\zeta\right) - \cosh(\zeta)\sinh\left(\tfrac{1}{2}\delta\zeta\right)\right.$$
$$\left. - \sinh(\zeta)\cosh\left(\tfrac{1}{2}\delta\zeta\right)\right] \tag{C8}$$

where we have also defined $\zeta = a\tau/c$ and $\delta\zeta = a\delta\tau/c$. Thus,

$$[\ldots] = 2\frac{\omega c}{a}\cosh(\zeta)\sinh\left(\tfrac{1}{2}\delta\zeta\right) - 2\frac{k_x c^2}{a}\sinh(\zeta)\sinh\left(\tfrac{1}{2}\delta\zeta\right)$$
$$= 2\frac{c}{a}\sinh\left(\frac{a\delta\tau}{2c}\right)\cdot\left[\omega\cosh(a\tau/c) - ck_x\sinh(a\tau/c)\right] \tag{C9}$$

The last term in brackets can be identified from Eq. (C2) as the transformed frequency ω'. This finally leads to a compact expression for the trigonometric function:

$$\cos\left[2\omega'\frac{c}{a}\sinh\left(\frac{a\delta\tau}{2c}\right)\right] \tag{C10}$$

Let us now turn our attention to the product of amplitudes. Expanding again the hyperbolic functions, we have:

$$[\ldots]\cdot[\ldots] = \hat{\varepsilon}_y(\hat{k}\times\hat{\varepsilon})_z\left[\cosh^2(\zeta) + \sinh^2(\zeta)\right]$$
$$- \hat{\varepsilon}_y^2\left[\sinh(\zeta)\cosh(\zeta) + \sinh\left(\tfrac{1}{2}\delta\zeta\right)\cosh\left(\tfrac{1}{2}\delta\zeta\right)\right]$$
$$- (\hat{k}\times\hat{\varepsilon})_z^2\left[\sinh(\zeta)\cosh(\zeta) - \sinh\left(\tfrac{1}{2}\delta\zeta\right)\cosh\left(\tfrac{1}{2}\delta\zeta\right)\right] \tag{C11}$$

The sum over polarizations yields the following relations:

$$\sum_\lambda \hat{\varepsilon}_y^2 = 1 - k_y^2/k^2, \quad \sum_\lambda (\hat{k}\times\hat{\varepsilon})_z^2 = 1 - k_z^2/k^2$$

and $\sum_\lambda \hat{\varepsilon}_y(\hat{k}\times\hat{\varepsilon})_z = k_x/k$ \hfill (C12) while the hyperbolic functions can be replaced by the Lorentz transformation parameters, that is, $\gamma = \cosh \zeta$ and $\beta\gamma = \sinh \zeta$. The final result is:

$$[\ldots]\cdot[\ldots] = \underbrace{\gamma^2(1+\beta^2)\frac{k_x}{k}}_{(1)} - \underbrace{\beta\gamma^2\left[\left(1-\frac{k_y^2}{k^2}\right)+\left(1-\frac{k_z^2}{k^2}\right)\right]}_{(2)}$$

$$-\underbrace{\left[\left(1-\frac{k_y^2}{k^2}\right)-\left(1-\frac{k_z^2}{k^2}\right)\right]}_{(3)}\tfrac{1}{2}\sinh\left(\frac{a\delta\tau}{c}\right) \quad (C13)$$

To transform the complete expression in Eq. (C7) into an integral on the transformed modes (those seen by the accelerated observer), we need to make use of the transformation properties:

$$\begin{aligned}\omega' &= \gamma(\omega - \beta c k_x)\\ k'_x &= \gamma(k_x - \beta\omega/c)\\ k'_y &= k_y, \; k'_z = k_z\end{aligned} \quad (C14)$$

from which we derive:

$$\omega' c k'_x = \gamma^2(1+\beta^2)\omega c k_x - \beta\gamma^2(\omega^2 + c^2 k_x^2) \quad (C15a)$$

and the Lorentz-invariant measure:

$$\int \frac{d^3 k}{\omega} \equiv \int \frac{d^3 k'}{\omega'} \quad (C15b)$$

Consider the first term: using $\omega^2 \equiv c^2 k^2$ and Eq. (C15a),

$$\gamma^2(1+\beta^2)\int d^3 k \cdot \omega\frac{k_x}{k} = \gamma^2(1+\beta^2)\int \frac{d^3 k'}{\omega'} \cdot \omega c k_x$$

$$= \int \frac{d^3 k'}{\omega'}\cdot[\omega' c k'_x + \beta\gamma^2(\omega^2 + c^2 k_x^2)] \quad (C16)$$

The second term of Eq. (C13) yields:

$$\int \frac{d^3 k}{\omega}(-2\beta\gamma^2\omega^2) + \int \frac{d^3 k}{\omega}(c^2\beta\gamma^2)(k_y^2 + k_z^2) \quad (C17)$$

Combining Eqs. (C16) and (C17), we obtain:

$$\int \frac{d^3 k'}{\omega'} [\omega' c k'_x + \beta \gamma^2 (c^2 k^2 - \omega^2)] = \int d^3 k' \omega' \frac{k'_x}{k'} \quad \text{(C18)}$$

Consider now the last term (3) in Eq. (C13). Using $\omega^2 \equiv c^2 k^2$,

$$\int d^3 k \cdot \omega \cdot \left(\frac{k_y^2 - k_z^2}{k^2}\right) = c^2 \int \frac{d^3 k}{\omega} (k_y^2 - k_z^2) = c^2 \int \frac{d^3 k'}{\omega'} (k_y'^2 - k_z'^2)$$

$$= \int d^3 k' \cdot \omega' \cdot \left[-\left(1 - \frac{k_y'^2}{k^2}\right) + \left(1 - \frac{k_z'^2}{k^2}\right)\right] \quad \text{(C19)}$$

The complete result reads:

$$<E_y(\tau^-)B_z(\tau^+)> = \int d^3 k' \cdot \omega' \cdot \left\{\left[-\left(1 - \frac{k_y'^2}{k^2}\right) + \left(1 - \frac{k_z'^2}{k^2}\right)\right]\frac{1}{2}\sinh\left(\frac{a\delta\tau}{c}\right) + \frac{k'_x}{k'}\right\}$$

$$\cdot \cos\left[\frac{2\omega' c}{a}\sinh\left(\frac{a\delta\tau}{2c}\right)\right] \quad \text{(C20)}$$

This is exactly the result obtained by Boyer.[******] The integration over the wave vectors identically yields zero by symmetry arguments. Furthermore, if we evaluate Eq. (C20) at the *same* proper time τ (i.e., $\delta\tau \equiv 0$), the result is identically null irrespective of the integration over the wave vectors. Therefore, it is clear that the x-component of the Poynting vector should be zero. This result is very general; by symmetry arguments (time reversal), one could also evaluate the phase-averaged Poynting vector in the coincident $\mathbf{K}'$ frame and transform back into the laboratory frame and obtain similar results. The fact that the vacuum fluctuation spectrum is Lorentz-invariant leads to that conclusion.

Haisch and Rueda compute a similar integral function,[††††††] which is essentially the sum of contributions (1) and (2) of eq. (C13), leading again to Eq. (C18) which, after integration over angular variables, is identically null. This result is identified Haisch and Rueda [3][‡‡‡‡‡‡] then attempt to obtain an expression in the laboratory frame simply by replacing the variables of integration (wave vector). This is incorrect; the Poynting vector (or equivalently, the radiation momentum density) must be transformed back into the laboratory frame through a Lorentz transformation, thus acting on a *quadrivector* of energy-momentum density. The only net momentum density that can be obtained in the laboratory frame would then be the transform of the time-like component

---

[******]In Eq. (67) of Ref. 13, the cosine argument should be ω' instead of ω; this is a typographical error, because the computation of the argument is the same for all two-point correlations.
[††††††]See Ref. 3, eq. (A30).
[‡‡‡‡‡‡]See Ref. 3, eq. (C19).

in the accelerated frame, that is, the energy density. Because the acceleration leads to a thermalization of the vacuum fluctuations, the net momentum should then be equivalent to the pressure of thermal photons promoted from the vacuum by the acceleration; in other words, the Unruh–Davies effect should be recovered and the effect would be excessively small.

Appendix D: Unruh–Davies and Classical Oscillator

Here we offer a very simple (trivial) alternative approach to the coupling between a classical oscillator and ZPF in an accelerated frame. Consider first the complete radiation spectrum, including both thermal and zero-point components:

$$\rho(\omega) = \frac{\omega^3}{2\pi^2 c^2}\left[\frac{1}{e^{\hbar\omega/kT}-1} + \frac{1}{2}\right] \quad (D1)$$

which can be rewritten as:

$$\rho(\omega) = \frac{\hbar\omega^3}{4\pi^2 c^2}\coth\left(\frac{\hbar\omega}{2kT}\right) \quad (D2)$$

Consider now the oscillator model of Einstein and Hopf, who showed that the reaction force exerted by a background field onto an oscillator moving at a velocity v with respect to the field was:

$$dF = -\frac{4\pi^2 e^2}{5mc^2}\left[\rho - \frac{\omega}{3}\frac{d\rho}{d\omega}\right]d\omega \cdot v \quad (D3)$$

Of course the zero-point component does not contribute to the reaction force because it is Lorentz-invariant, but the thermal component does (a finite temperature breaks down Lorentz invariance). Using Eq. (D2), the derivative of the energy density is:

$$\frac{d\rho}{d\omega} = \frac{3}{\omega}\rho + \frac{\hbar}{2kT}\frac{\rho}{\sinh\varsigma \cdot \cosh\varsigma} \quad (D4)$$

with $\varsigma = \hbar\omega/2kT$. Therefore the force element is (barring the physical constants) of the form:

$$dF \propto -\frac{\hbar\omega^3}{2kT}\frac{d\omega}{\sinh^2(\varsigma)} \cdot v \quad (D5)$$

For a uniformly accelerated frame, we have, of course:

$$v = c\tanh(a\tau/c) \quad (D6)$$

and because the frame is accelerated, the effective temperature of the vacuum electromagnetic field is given by the Unruh–Davies temperature:

$kT_U = \hbar\, a/2\pi c$. In the limit of moderate acceleration (i.e., small temperature), we have approximately:

$$dF \propto -\frac{\hbar\omega^3}{2kT}\left(\frac{2kT}{\hbar\omega}\right)^2 d\omega \cdot \tanh(a\tau/c) \approx -|a| \cdot (\omega\, d\omega) \cdot (a\tau/c) \quad \text{(D7)}$$

This leads to an inertial mass that is:

$$\delta m_i(\tau) \propto |a|\tau \cdot \Lambda^2 \quad \text{(D8)}$$

Note that this is a *correction* to the inertial (or rest) mass of the particle, that is, this term is added to the existing rest mass. The term is quadratically divergent with respect to the cutoff energy, as expected from classical model in the nonrelativistic approximation. The correction grows in time until saturation occurs (when $a\tau/c \approx 1$), at which time the reaction force becomes constant and the mass correction changes behavior. Of course, we do not expect this result to be correct; instead, a QED calculation using finite-temperature formulation leads to a mass correction that is proportional to a^2/m, still very small for almost all cases of acceleration. The point of this exercise is to show that simple classical arguments can be used to yield acceleration effects, but leading to small corrections to the mass, rather than attempting to explain away the complete mass as an interaction with the ZPF.

References

[1] Haisch, B., Rueda, A., and Puthoff, H. E., "Inertia as a Zero-Point-Field Lorentz Force," *Physical Review A*, Vol. 49, 1994, pp. 678–694.

[2] Haisch, B., and Rueda, A. "Reply to Michel's 'Comment on Zero-Point Fluctuations and the Cosmological Constant'," *The Astrophysical Journal*, Vol. 488, 1997, pp. 563–565.

[3] Rueda, A., and Haisch, B., "Contribution to Inertial Mass by Reaction of the Vacuum to Accelerated Motion," *Foundations of Physics*, Vol. 28, 1998, pp. 1057–1108.

[4] Sunahata, H., "Interaction of the Quantum Vaccum with an Accelerated Object and its Contribution to Inertia Reaction Force," PhD Thesis, California State University, 2006.

[5] Haisch, B., and Rueda, A., "Reply to Michel's 'Comment on Zero-Point Fluctuations and the Cosmological Constant'," *The Astrophysical Journal*, Vol. 488, 1997, pp. 563–565.

[6] Dobyns, Y., Haisch, B., and Rueda, A., *Foundations of Physics*, Vol. 30, arXiv:gr-gc/002069, 2000, pp. 59–80.

[7] Rueda, A., and Haisch, B., "Gravity and the Quantum Vacuum Inertia Hypothesis," *Annalen der Physik*, Vol. 14, 2005, pp. 479–498.

[8] Thaller, B., *The Dirac Equation* (Text and Monographs in Physics), Springer-Verlag, Berlin, 1992.

[9] Einstein, A., and Hopf, L., "Statistische Untersuchung der Bewegung eines Resonators in einem Strahlungsfeld," *Annalen der Physik*, Vol. 338, 1910, pp. 1105–1115.

[10] Marshall, T. W., "Statistical Electrodynamics," *Proceedings of the Cambridge Philosophical Society*, Vol. 61, 1965, pp. 537–546.
[11] Boyer, T. H., "Derivation of the Blackbody Radiation Spectrum without Quantum Assumptions," *Physical Review*, Vol. 182, 1969, pp. 1374–1383.
[12] Jackson, J. D., *Classical Electrodynamics*, 2nd ed., Wiley, New York, 1982.
[13] Boyer, T. H., "Thermal Effects of Acceleration Through Random Classical Radiation," *Physical Review D*, Vol. 21, 1980, pp. 2137–2148.
[14] Boyer, T. H., "Thermal Effects of Acceleration for a Classical Dipole Oscillator in Classical Electromagnetic Zero-Point Radiation," *Physical Review D*, Vol. 29, 1984, pp. 1089–1095.
[15] Boyer, T. H., "Derivation of the Blackbody Radiation Spectrum from the Equivalence Principle in Classical Physics with Classical Electromagnetic Zero-Point Radiation," *Physical Review D*, Vol. 29, 1984, pp. 1096–1098.
[16] Miyamoto, U., and Maeda, H., "On the Resolution of Negative Mass Naked Singularities," *Journal of Physics: Conference Series*, Vol. 31, 2006, pp. 157–158.
[17] Lawrie, I. D., *A Unified Grand Tour of Theoretical Physics*, Institute of Physics Publishing, Bristol, U.K., 2002.
[18] Cebeci, H., Sarioglu, Ö., and Tekin, B., "Negative Mass Solitons in Gravity," *Physical Review D*, Vol. 73, 2006, 064020.
[19] Veltman, M., *Facts and Mysteries in Elementary Particle Physics*, World Scientific, New York, 2003.
[20] Itzykson, C., and Zuber, J.-B., *Quantum Field Theory*, McGraw-Hill, New York, 1980.
[21] Kaku, M., *Quantum Field Theory: A Modern Introduction*, Oxford University Press, New York, 1993.
[22] Coleman, S., in *Electromagnetism: Paths to Research*, Tiplitz, D., ed., Plenum, New York, 1982, pp. 183–210.
[23] Amit, D. J., *Field Theory, The Renormalization Group and Critical Phenomena*, McGraw-Hill, New York, 1978.
[24] Dolan, L., and Jakiw, R., "Symmetry Behavior at Finite Temperature," *Physical Review D*, Vol. 9, 1974, pp. 3320–3341.
[25] Cambier, J.-L., Primack, J. R., and Sher, M., "Finite Temperature Radiative Corrections to Neutron Decay and Related Processes," *Nuclear Physics B*, Vol. B209, 1982, pp. 372–388.
[26] Peresutti, G., and Skagerstam, B.-S., "Finite Temperature Effects in Quantum Field Theory," *Physics Letters B*, Vol. 110, 1982, pp. 406–410.
[27] Qader, M., Masood, S., and Ahmed, K., "Second-Order Electron Mass Dispersion Relation at Finite Temperature," *Physical Review D*, Vol. 44, 1991, 3322–3327.
[28] Davies, P. C. W., "Scalar Production in Schwarzschild and Rindler Metrics," *Journal of Physics A*, Vol. 8, 1975, pp. 609–616.
[29] Unruh, W. G., "Notes on Black-hole Evaporation," *Physical Review D*, Vol. 14, 1976, pp. 870–892.
[30] Chen, P., and Tajima, T., "Testing Unruh Radiation with Ultraintense Lasers," *Physical Review Letters*, Vol. 83, 1999, pp. 256–259.
[31] Leinass, J. M., "Unruh Effect in Storage Rings," arXiv:hep-th/0101054.
[32] Chen, P., "Press Release: Violent Acceleration and the Event Horizon," June 2006, http://home.slac.stanford.edu/pressre leases/2000/20000606.htm. [Cited 19 Nov. 2008].

[33] Milonni, P., *The Quantum Vacuum*, Academic Press, New York, 1994.
[34] Novoselov, K. S., Geim, A. K., Morozov, S. V., Jiang, D., Katsnelson, M. I., Grigorieva, I. V., Dubonos, S. V., and Firsov, A. A., "Two-Dimensional Gas of Massless Dirac Fermions in Graphene," *Nature*, Vol. 438, 2005, 04233.
[35] Welton, T. A., "Some Observable Effects of the Quantum-Mechanical Fluctuations of the Electromagnetic Field," *Physical Review*, Vol. 74, 1948, 1157–1167.

Chapter 14

Relativistic Limits of Spaceflight

Brice N. Cassenti*
Rensselaer Polytechnic Institute, Hartford, Connecticut

I. Introduction

THE Special Theory of Relativity is one of the foundations of modern physics. It provides a conceptual starting point for the General Theory of Relativity and is a key ingredient of Quantum Field Theory. Yet many of its predictions are counterintuitive. These counterintuitive predictions produce paradoxes that are not readily explained; yet experimental results have verified the predictions of Special Relativity to an extraordinary degree. Although the Special Theory of Relativity is formulated for inertial systems, it can also be applied to accelerating interstellar spacecraft and is useful for sizing interstellar spacecraft.

The Special Theory of Relativity was proposed by Einstein in 1905 [1] to reconcile experimental results that did not agree with classical Newtonian mechanics. Special Relativity assumes that Maxwell's equations for electromagnetism are exactly correct by hypothesizing that the speed of light in a vacuum takes on the same value in all inertial (constant velocity) reference frames. As a consequence of this assumption, time cannot flow at the same rate in all inertial reference systems, and this is usually at the heart of the paradoxes. An excellent introduction to both Special and General Relativity can be found in Einstein's own writings. A detailed mathematical treatment is presented in Ref. 2, while Ref. 3 gives a more popular presentation. Taylor and Wheeler have written an excellent introduction to Special Relativity [4].

Paradoxes are a common feature of the Special Theory of Relativity and are very useful at contrasting our common notions of space and time and the

---

Copyright © 2008 by the American Institute of Aeronautics and Astronautics, Inc. All rights reserved.
Based on "Faster-than-Light Paradoxes in Special Relativity," AIAA 2006-4607, presented at the 41st AIAA/ASME/SAE/ASEE Joint Propulsion Conference and Exhibit, Sacramento, CA, 10–12 July 2006.
*Associate Professor, Department of Engineering and Science.

results predicted from the foundations of Special Relativity. A detailed discussion of these paradoxes can be found in Ref. 5. The paradoxes are even more profound if speeds greater than the speed of light are considered. For example, in cases where objects travel faster than the speed of light, there are reference frames where an effect can precede its cause. Of course, this implies time travel. Modern physics sometimes removes these effects by postulating that nothing can move faster than light. But the General Theory of Relativity does not exclude the possibility, nor does the Special Theory, and quantum mechanics may actually require it. Of course, it may be that it is just our poor understanding of space and time that causes the confusion and that new concepts will help clarify.

Paradoxes, though, do not mean that the theory is inaccurate or in error. Empirical evidence strongly favors the Special Theory of Relativity, and has verified it to an extraordinary accuracy and, hence, we can use Special Relativity as a basis for sizing interstellar vehicles.

The sections that follow will introduce the Special Theory of Relativity and then cover some of the well-known paradoxes. Closing sections give some empirical results and a basis for the description of the motion of relativistic rockets.

II. Principle of Special Relativity

Special Relativity is based on the experimental fact that the speed of light in a vacuum remains the same in all inertial frames regardless of the observer's speed with respect to the light source. Many experiments were performed in the late 1800s and early 1900s that verified that the speed of light did not vary with the relative speed of the source, but a simple Galilean transformation of Maxwell's equations indicated that the speed of light should vary. Ad hoc theories were proposed to satisfy the observations. The most complete was that of Lorentz [1], who proposed a transformation where both space and time were connected to the relative speed and, although he modified Maxwell's equations to account for motion with respect to the stationary ether, the effects were forced to cancel exactly in order to agree with the experimental observations. Einstein, though, assumed that Maxwell's equations were exactly correct, and that the fault was in the Galilean transformation.

A. Postulates and Lorentz Transformation

The two postulates of Einstein's Special Theory of Relativity are that: 1) physical laws hold in all inertial reference systems, and 2) the speed of light (in a vacuum) is the same in all inertial reference frames. These two postulates are all that is required to find the transformations that Lorentz found in his ad hoc manner.

Consider two inertial space-time reference systems (primed and unprimed) as shown in Fig. 1. The coordinates (t, x) are for the stationary system, and the coordinates (t', x') are in the moving system. The origin of the primed system is moving at the constant speed v along the x-axis. Because Maxwell's equations must remain linear in both systems, the coordinates (t', x') should be a linear combination of the coordinates (t, x). Taking the origins of both systems to coincide,

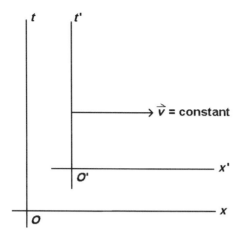

Fig. 1 Inertial reference frames.

the coordinates must be related by

$$ct' = Act + Bx$$
$$x' = Cct + Dx \qquad (1)$$

where c is the speed of light and A, B, C, D are constants that depend only on the relative speed v. The quantities ct and ct' are used because they are distances and hence have the same units as x and x'.

The speed of light, c, must be exactly the same in both coordinate systems. Hence, light moving through dx in time dt in the unprimed coordinate system at speed c must also move through dx' and dt' in the primed system at speed c. Then

$$c^2 dt^2 - dx^2 = c^2 dt'^2 - dx'^2 = c^2 d\tau^2 \qquad (2)$$

The quantity $d\tau$ is referred to as the proper time and represents time intervals for an object that is stationary in its own frame.

Substituting Eq. (1) into Eq. (2) and equating the coefficients of dx^2, dt^2, and $dx\,dt$ on both sides of the resulting equation gives

$$A = D = \cosh\theta$$
$$B = C = -\sinh\theta \qquad (3)$$

where $\theta = \theta(v)$. The function $\theta(v)$ can be found by noting that a particle stationary in the primed system is moving at v in the unprimed system. Using Eqs. (1) and (3)

$$\frac{dx'}{cdt'} = 0 = \frac{Cc + Dv}{Ac + Bv} \qquad (4)$$

Then

$$\frac{v}{c} = \tanh \theta \tag{5}$$

Equation (1) can now be written as

$$ct' = \gamma (ct - \beta x)$$
$$x' = \gamma (x - \beta ct) \tag{6}$$

where $\beta = \tanh \theta$ and $\gamma = \cosh \theta = 1/\sqrt{1 - \beta^2}$. Taylor and Wheeler [4] refer to θ as the velocity parameter, but in the physics community, θ is usually referred to as the rapidity. The transformation in Eq. (6) is commonly known as the Lorentz transformation [1].

We can now compare the coordinate systems by considering lines of constant ct' and x' (and constant ct and x) as shown in Fig. 2. Note that light emitted from the origin will travel along the diagonal in both coordinate systems. Also note that the primed coordinate system is collapsing around the diagonal of the unprimed system and, at a relative speed equal to the speed of light, the primed coordinates will have completely collapsed. Equation (5) indicates that v/c can only approach the speed of light if θ is finite and real. For $\beta > 1$, θ is imaginary, as well as γ. The proper time also becomes imaginary as indicated by Eq. (2). Another way to look at faster-than-light relative motion is that the spatial and temporal coordinates switch. This clearly means that objects can travel both ways in time for an observation of faster-than-light particles. Yet another way to look at this is to consider what Fig. 2 would look like for the primed system moving at greater than the speed of light. The ct' axis would be on the x-coordinate side,

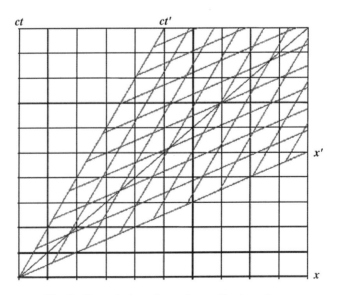

Fig. 2 Lorentz transformed coordinate systems.

and the x' axis would be on the ct-coordinate side. Time in the primed system would be a space-like coordinate to the observer in the primed system, and the space coordinate in the direction would be time-like in the unprimed system. In Special Relativity, physicists sometimes assume that nothing can go faster than light and this completely removes those cases from consideration.

B. Lorentz Contraction and Time Dilation

Two significant consequences of Special Relativity are the apparent contraction and the apparent slowing of time for moving objects. The contraction of moving objects is easiest to predict once a method of measuring length is defined. We will define length as the Euclidean difference in coordinates at the same value for the time coordinate. In our case this means the difference in the x-coordinate locations for a specific value of ct. For a measurement of a rod stationary along the x-axis in the unprimed system, we can set $ct' = 0$ in Eq. (6). Then in the primed system $ct = \beta x$, and

$$x' = \sqrt{1 - \beta^2}\, x \tag{7}$$

and stationary lengths in the unprimed system appear to be contracted in the primed system. Equation (7) is the Lorentz contraction. Of course a similar analysis will show that stationary lengths in the primed system appear contracted in the unprimed system.

Time dilation is the analogous situation for relative time measurements. Consider a stationary particle at the origin of the unprimed system (i.e., $x = 0$). Then Eq. (6) yields $ct = \beta x$ and

$$ct = \sqrt{1 - \beta^2}\, ct' \tag{8}$$

Hence, time in the moving coordinate system appears to move slower. However, for an observer in the primed system watching a clock stationary at the origin of the unprimed system stationary (i.e., $x' = 0$), we find from Eq. (6) that

$$ct' = \sqrt{1 - \beta^2}\, ct \tag{9}$$

Hence observers in both frames will predict the other's clocks are moving more slowly. If they continue to move at a constant relative velocity (i.e., both systems are inertial), then they can only meet once and there will be no contradictions.

III. Paradoxes in Special Relativity

A. Pole Vaulter Paradox

A popular paradox in Special Relativity is the "pole vaulter paradox." In this scenario a pole vaulter carries his pole horizontally at high speed. In our version of the paradox, the pole is 15 feet long according to the pole vaulter, but he is moving so fast that an observer standing at a stationary barn that is 10 feet long sees the pole contracted to half its length, which is now less than the length of the barn. The front door of the barn is open as the pole enters. When the back end of the pole enters, the front end has not reached the back door, and the front door closes. Hence to the observer the pole is entirely enclosed by the barn. When the front of the pole meets the back door, the back door is opened and the pole leaves the barn. Of course to the pole vaulter, the barn appears half as long (i.e., 5 feet) and the pole vaulter will note that the pole was never completely in the barn. The pole vaulter will also note that the back door opened before the front door closed. Obviously, the pole cannot both be longer and shorter than the barn. The paradox is solved by tracking the front and back end of the pole and barn in both the pole vaulter's and the observer's systems. The transformations will show that the order in which the doors open and close is reversed in each system, agreeing with the observations. The results will also show that the time interval when the doors open and close is shorter than the time it would take light to traverse the distance between the doors of the barn in the observer's coordinate system, and is also shorter than the time it would take light to traverse the length of the pole in the pole vaulter's coordinate system. Hence, closing the front door cannot influence opening the back door if the speed of light is the maximum speed in the universe.

This leads to the primary contradiction of Special Relativity with our common low-speed experience. Space and especially time are not what we picture. Newtonian physics exists in a universe where time is the same for all observers, and Special Relativity says it cannot be if the speed of light is the same in all inertial (nonaccelerating) reference systems.

B. Twin Paradox

The best-known paradox in Special Relativity is the twin paradox [4,6]. The problem presented by this paradox involves time directly. Consider two fraternal twins, Alice and Bob, preparing for a long space voyage. Bob is in mission control while Alice will do the traveling. Alice departs at a very high speed so that Bob sees Alice's clocks running at half speed, but, of course, Alice sees Bob's clock running at half speed also. At the half way point Alice reverses speed, so that on the return each sees the other's clock as running at half speed. When Alice stops to meet Bob, they both cannot think that each has aged half as much, thus the paradox. Certainly the Lorentz transformations apply to Bob, because at no time did he experience any acceleration, and hence, Alice will have aged half as much as Bob. Obviously the problem is with the acceleration Alice experiences. It can be shown that while Alice changes her speed she will see Bob's clock run faster. All three of her accelerations will exactly compensate for the slowing during the constant velocity portions of the trip, and both Bob and Alice will agree that each of their clocks is correct when they meet. Actually, there are many ways to resolve the paradox. Weiss [6] gives an excellent discussion. The section to follow on constant acceleration rockets can be used to show that each will agree with the clock readings on Alice's arrival.

C. Faster-than-Light Travel

Objects that move faster than the speed of light can reverse the order of events in some inertial systems. Figure 3 illustrates the appearance of two objects in two inertial systems. The object moving from A to B is moving less than the speed of light in the unprimed system. The projection of the two points A and B onto the ct-axis clearly shows that A precedes B. We can also project the points A and B onto the ct'-axis, which again clearly shows that A precedes B.

If we now consider the path from C to D, the slope of the line shows that in the unprimed system the speed of an object going from C to D is greater than the speed of light. Again projecting onto the ct-axis shows that C precedes D in the unprimed system, but in the primed system a projection onto the ct'-axis shows that D now precedes C. Hence, events that are observed for an object moving faster than the speed of light can have effects that precede their cause. For example, a dish that breaks going from C to D in the unprimed system will magically reassemble itself when observed in the primed system. The simplest resolution of the paradox is to again assume that physical objects cannot move faster-than-light but, of course, this may not be the actual explanation.

D. Instant Messaging

Now consider the case illustrated in Fig. 4. Two observers, Art (A) and Don (D), are stationary in the unprimed coordinate system. There are two other observers, Brenda (B) and Cathy (C), in the primed system, moving at a constant velocity $\bar{v}$ with respect to the unprimed system. Art hands off a message to Brenda as she goes by. Brenda then instantaneously transmits the message to Cathy. Cathy hands the message to Don, who instantaneously sends the message to Art. The

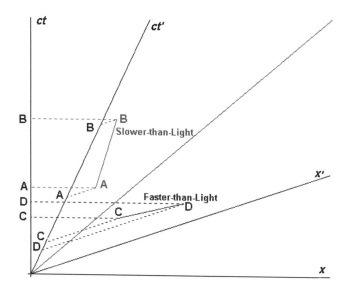

Fig. 3 Lorentz transformation for faster-than-light-travel.

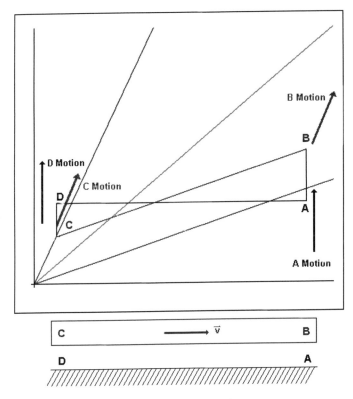

Fig. 4 Instant messaging.

loop in Fig. 4 shows that Art receives the message before he hands it off to Brenda. Art "now" has two messages. "After" completing the loop again Art has four, and the process continues to double infinitely many times, creating quite a puzzling paradox. The problem again lies with speeds greater than the speed of light. The objects that are moving faster than light are messages, which need not consist of a physical mass. Nevertheless, the information in the messages exceeds the speed of light.

Again, the simplest way to remove the paradox is to assume that nature does not allow any cause to propagate faster then the speed of light, but there is another explanation. The infinite number of messages arriving at A is clearly a nonlinear process and will not satisfy the linear relationships assumed in Special Relativity. This nonlinear process may be related to the collapse of the wave function in quantum mechanics, but a nonlinear process would also violate the postulates of quantum mechanics. The nonlinear theory could result in one of the infinite number of messages being chosen and resolve the paradox. That is, the collapse of the wave function in quantum mechanics could be the mechanism that resolves the paradox and may clear up the conceptual problems with entanglement [7]. Chapter 16 explores quantum entanglement in more detail.

A better illustration of the collapse of the wave function is the case where Art can randomly choose between two messages, one labeled Up and the other Down. Art randomly chooses one of the two. For example, Art chooses Up and hands it off to Brenda. Instantaneously Art will have three messages. Two of these are Up and the third is Down. A random choice now is more likely to be Up. Of course instantly there will be an infinite number of messages consisting of the actual choice for the collapse of the wave function, and the other choice will effectively not appear at all. Hence, the collapse of the wave function could be explained as a closed self-referential spacetime loop [6].

IV. Empirical Foundations

Paradoxes in the Special Theory of Relativity seem to imply that the theory may not be physically consistent, but the Special Theory of Relativity has been demonstrated to be accurate to an extraordinary precision. The text by Bergmann [8] contains an entire chapter on the experimental verification of the Special Theory of Relativity. Born [9] has a more popular description of the Special Theory of Relativity but it does contain references to experimental verification scattered throughout his chapter on Special Relativity. A readily accessible summary by Roberts [10] contains numerous references to experiments and their results.

All of the experiments so far performed indicate that the Special Theory of Relativity is an extremely accurate representation of the nature of the universe. Of course, with over a century of experimental verification it will be difficult to publish any experimental violations, nevertheless searches for violation do continue (e.g., see the article by Kostelecky [11], which contains a popular description for the continuing search for experimental violations). Several of these are summarized below.

A. Michelson–Morley Experiment

The Michelson–Morley experiment is the classic experiment referenced for the verification of the fundamental tenets of the Special Theory of Relativity. It consists of a Michelson interferometer where a single light beam is split into two beams moving at right angles. When the beams are brought back together, interference fringes appear according to the path lengths of the split beams and the speed of light in each direction. Variations in the travel times can then be measured along different directions. Because the Earth has a preferred direction for its orbital velocity about the sun, any change in the speed of light due to the motion of the Earth will be recorded. The most recent experiment shows that there is no change with direction for the velocity of light [10] to within 2 parts in 10,000.

B. Half-Lives of Relativistic Particles

One of the predictions of the Special Theory of Relativity is changes in the time observed by individuals moving at relativistic velocities. Cosmic rays are a natural source of high energy, (relativistic) particles and, in fact, the muon, which is produced when high-energy cosmic rays collide with atoms in the upper atmosphere, has been observed at ground level indicating that its life has been significantly extended beyond its nominal 2µs lifetime. Experiments done in particle accelerators are much more accurate. Experiments have shown accuracies in time dilation predictions to better than 0.1% [12].

C. Doppler Shifts

The Doppler shift on the spectra of quasars, which are moving away at relativistic velocities, is a very accurate test of the theory of Special Relativity for radial motion but the more important tests are those that involve transverse motion. Here laboratory tests have verified the theory with accuracies of better than 3×10^{-6} relative to the predicted value [9]. These tests are particularly important because the predictions of Special Relativity are at odds with all nonrelativistic (Galilean) theories.

D. Cerenkov Radiation

Cerenkov radiation is produced when charged particles move through a material at speeds faster than the speed of light in the material. This is analogous to the motion of an aircraft through the air at faster than the local speed of sound in the air. When these particles exceed the local speed of light in a material a "light boom"—instead of a sonic boom—is created from radiation emitted by the particle as it slows down. The emitted light propagates perpendicular to the wave front of the light boom. Also, the intensity of the light increases with the emitted frequency and therefore is preferentially blue. In a material the local speed of light is reduced by a factor equal to the index of refraction. For example, if particles move through water, which has an index of refraction of 4/3, then Cerenkov radiation will be observed with more than 3/4 of the speed of light. Such speeds are easily exceeded in cosmic ray and elementary particle physics experiments. In fact, Cerenkov radiation is widely used in elementary particle physics experiments to differentiate relativistic particles of the same momentum but different masses [11] and the measurements are completely consistent with the Special Theory of Relativity.

E. Tachyon Searches

Tachyons are hypothetical particles that move faster than the speed of light and, if they existed, could produce causality paradoxes. A brief popular description on tachyons can be found in Ref. 13. For a particle with a velocity exceeding the speed of light, Eq. (11) implies that both the energy and the momentum become imaginary (i.e., a multiple of $\sqrt{-1}$). This is consistent with a rest mass that has an imaginary value. There is nothing in the Special Theory of Relativity that precludes such particles from existing, but reported searches have been consistent with the observation that tachyons do not exist. The experimental limits on the imaginary mass-energy of the tachyon particles [14] have been reported as less than 10^4 eV.

Even if tachyons do exist, the Feinberg reinterpretation principle [15] could be used to show that tachyons will not violate causality. Recall that the order for cause and effect is dependent on the relative velocity of the observers. Causality would be violated if a tachyon could send information into its own past. The

Feinberg reinterpretation principle states that sending a negative energy tachyon backward in time (which could violate causality) would appear as positive energy tachyons moving forward in time. According to the Feinberg reinterpretation principle, the creation and absorption of tachyons cannot be differentiated by observers undergoing relative motion. Hence, if the Feinberg reinterpretation principle holds, it would be possible to use tachyons to send energy forward in time, but it would not be possible to communicate backward in time [13].

V. Relativistic Rockets

Although Special Relativity applies to constant velocity processes, it can be extended to accelerating objects for applications to interstellar spaceflight. The extension only requires that we constantly update the reference velocity to an accelerating system. Before proceeding we must find the relativistic form of the momentum and energy so that we may apply the conservation of momentum to relativistic rockets. If we multiply Eq. (2) by $m_0^2 c^2$, where m_0 is the mass in the frame at rest, then

$$m_0^2 c^4 = m_0^2 c^4 \left(\frac{dt}{d\tau}\right)^2 - m_0^2 c^2 \left(\frac{dx}{d\tau}\right)^2 \tag{10}$$

The first quantity on the right-hand side is the relativistic energy, E, and the second is the relativistic momentum, p [1,4]. Using the coefficients in Eq. (6), the relativistic energy and the three-dimensional momentum form a four-dimensional vector with components

$$E = \gamma m_0 c^2 \tag{11a}$$

and a total momentum

$$pc = \gamma \beta m_0 c^2 \tag{11b}$$

For small speeds ($\beta \ll 1$ and $\cosh\theta \approx 1 + \theta^2/2$), Eq. (5) yields

$$E \approx m_0 c^2 \left(1 + \frac{1}{2}\theta^2\right) \approx m_0 c^2 + \frac{1}{2} m_0 v^2 \tag{12}$$

The second term is the correct form for the kinetic energy and the first is the rest mass energy.

A. Constant Acceleration Rocket

Many authors have considered relativistic rockets (e.g., see Refs. 16 and 17), and all are based on the conservation of momentum and energy. Using the velocity parameter, the momentum is given by

$$p = m_0 c \sinh\theta \tag{13}$$

Then

$$\frac{dp}{d\tau} = m_0 c \cosh\theta \frac{d\theta}{d\tau} \tag{14}$$

In the accelerating frame, $\theta = 0$ and $\cosh\theta = 1$. Using Newton's Second Law

$$\frac{dp}{d\tau} = m_0 c \frac{d\theta}{d\tau} = m_0 a \tag{15}$$

where a is the acceleration felt in the accelerating (rocket) frame of reference. Then the acceleration is simply given by

$$a = c \frac{d\theta}{d\tau} \tag{16}$$

Using Eq. (2), it follows that for all inertial reference frames

$$\begin{aligned} \frac{dt}{d\tau} &= \cosh\theta \\ \frac{dx}{d\tau} &= c \sinh\theta \end{aligned} \tag{17}$$

For constant acceleration a, Eq. (16) can be integrated to yield

$$\theta = \frac{a\tau}{c} \tag{18}$$

where $\theta = 0$ at $\tau = 0$. Substituting Eq. (18) into Eq. (17) and integrating,

$$\begin{aligned} t &= \frac{c}{a} \sinh\left(\frac{a\tau}{c}\right) \\ x &= \frac{c^2}{a} \left[\cosh\left(\frac{a\tau}{c}\right) - 1\right] \end{aligned} \tag{19}$$

where $t = 0$, and $x = 0$ at $\tau = 0$. Dividing the last of Eq. (17) by the first yields

$$\frac{dx}{dt} = c \tanh\theta \tag{20}$$

which reproduces Eq. (5). For small proper times, τ, Eq. (19) and (20) become

$$t \approx \tau$$
$$x \approx \frac{a\tau^2}{2} \qquad (21)$$
$$\frac{dx}{d\tau} \approx a\tau$$

Equation (21) in the correct nonrelativistic expressions for constant acceleration.

Equations (18) through (20) can now be used to obtain results for a spacecraft accelerating to relativistic speeds (see Chapter 2 for more applications.) Note that Eq. (17) indicates that the rocket will cover distance dx in onboard time $d\tau$. Hence, when the hyperbolic sine of the velocity parameter is greater than one, the crew will experience travel at speeds greater than the speed of light. Although, the stationary observer will note that the rocket is always moving less than the speed of light according to Eq. (20), there is no contradiction. Time intervals in the stationary observer's frame, from Eq. (17), are always larger than the crew's changes in time. The speeds can become so large in the crew's frame of reference that they could circumnavigate our expanding universe in less than their working lifetime if they accelerate continuously at one Earth gravity. The conclusion is that relativity does not limit the speeds that can be achieved. Of course the observer on the ground will be long gone, along with the reference stellar system, before the circumnavigating crew returns.

We now have enough information to resolve the twin paradox. Time in the accelerating twin's reference frame can now be taken as the proper time, τ, and the readings of the clocks on return can be predicted for each twin. The analysis shows no paradox.

There is more in the consideration of accelerations than just a resolution of the twin paradox. The case of constant acceleration also points to an approach that will allow the inclusion of gravitational accelerations. Because, in both cases, the accelerations experienced do not depend on the mass, this fact can be used to lead directly to the fundamental principle of the General Theory of Relativity.

B. Photon Rocket

We can estimate the amount of mass, or energy, required by considering a photon rocket. Because a photon is a particle of light it has no rest mass and, hence, $m_0 = 0$ in Eq. (11). Then Eq. (10) in terms of energy and momentum becomes

$$m_0^2 c^4 = E^2 - p^2 c^2 = 0 \qquad (22)$$

and, taking the positive root of Eq. (22),

$$E = pc \qquad (23)$$

for a photon. Consider the conservation of momentum in the rocket's frame of reference. If dE_γ is the energy of the photons released in proper time $d\tau$, then

$$\frac{dp}{d\tau} = \frac{1}{c}\frac{dE_\gamma}{d\tau} = mc\frac{d\theta}{d\tau} \qquad (24)$$

where m is the current mass of the rocket in the reference frame moving with the rocket, but the conservation of energy yields, in the accelerating rocket frame

$$\frac{d(E_\gamma + mc^2)}{d\tau} = \frac{dE_\gamma}{d\tau} + c^2\frac{dm}{d\tau} = 0 \qquad (25)$$

Substituting for the energy in Eq. (25) using Eq. (24) and integrating, yields

$$m = m_i e^{-\theta} \qquad (26)$$

where m_i, is the mass of the rocket at $\theta = 0$. If the final mass occurs at $\theta = \theta_f$, then the mass ratio is

$$MR = \frac{m_i}{m_f} = e^{\theta_f} \qquad (27)$$

Although, for reasonable final masses the propellant mass required to circumnavigate the universe would be less than the mass of the universe, it would nevertheless be enormous.

It should be noted that the above could be readily extended to mass annihilation rockets [17,18] (i.e., antimatter rockets) and gives reasonable estimates for the performance requirements of interstellar antimatter rockets. Antimatter rockets require not only the application of energy and momentum conservation, but also the conservation of the baryon (nucleon) number must be applied. A baryon number of plus one is assigned to protons and neutrons, while antiprotons and antineutrons have a baryon number of minus one. The conservation laws can also be extended to the Bussard interstellar ramjet [19], where interstellar matter is collected by a moving spacecraft and used as propellant. In Refs. 20 and 21 all three conservation laws are applied to the interstellar ramjet.

VI. Conclusions

The Special Theory of Relativity is based on sound empirical evidence and demonstrates to an extraordinary degree of accuracy that the speed of light is constant in inertial reference frames. It is our concept of time that has had the most dramatic change as a result of the Special Theory of Relativity and, again, it is our concept of time that produces the most paradoxical results. This is especially true when objects are postulated to travel faster than the speed of light. But these faster-than-light paradoxes in the Special Theory of Relativity are more than curiosities based on our concepts of time and space; they can point the way to a more complete understanding of space and time. They can even add insight into other paradoxes such as the collapse of the wave function in quantum mechanics, and may even lead to new physical theories that may allow unlimited access to the universe. If our concept of time can be clearly defined, then it may be possible to resolve all of these paradoxes.

Although the Special Theory of Relativity is applicable only in inertial reference frames, it can be extended to accelerating spacecraft for sizing interstellar missions. Only the conservation of momentum, energy, and baryon (nucleon) number are required. Sizing, using a photon rocket as an example, will show that although practical interstellar flight will be difficult and expensive, it is not impossible.

Within the constraints of the Special Theory of Relativity it is clearly impossible to exceed the speed of light, and interstellar flight will always be handicapped by long travel times to outside observers. Our only hope of reducing the long travel times may be through the General Theory of Relativity (e.g., warp drives) and/or quantum mechanics.

References

[1] Loventz, H. A., Einstein, A., Minkowski, H., Weyl, H., Perret, W., Sommerfeld, A., and Jeffery, G. B., *The Principle of Relativity: A Collection of Original Memoirs on the Special and General Theory of Relativity*, Courier Dover Publications, New York, 1952.

[2] Einstein, A., *Relativity—The Special and General Theory*, Crown Publishers, New York, 1961.

[3] Einstein, A., *The Meaning of Relativity*, Princeton University Press, Princeton, NJ, 1956.

[4] Taylor, E. F., and Wheeler, J. A., *Spacetime Physics*, W. H. Freeman and Co., San Francisco, 1966.

[5] Yakov P., and Terletskii, Y., *Paradoxes in the Theory of Relativity* (translated from Russian), Plenum Press, NY, 1968.

[6] Weiss, M., "The Twin Paradox." URL: http://math.ucr.edu/home/baez/physics/Relativity/SR/TwinParadox/twin_vase.html.

[7] Goff, A., "Quantum Tic-Tac-Toe: A Teaching Metaphor for Superposition in Quantum Mechanics," *American Journal of Physics*, Vol. 74, No. 11, 2006, pp. 962–973.

[8] Bergmann, P. G., *Introduction to the Theory of Relativity*, Dover Publications, New York, 1976.

[9] Born, M., *Einstein's Theory of Relativity*, Dover Publications, New York, 1965.

[10] Roberts, T. "What Is the Experimental Basis of Special Relativity?" URL: http://math.ucr.edu/home/baez/physics/Relativity/SR/experiments.html#5.%20Twin%20paradox.

[11] Kostelecky, A., "The Search for Relativity Violations," *Scientific American*, Vol. 291, No. 3, Sept. 2004, pp. 93–101.

[12] Perkins, D. H., *Introduction to High Energy Physics*, 2nd ed., Addison-Wesley, Reading, MA, 1982.

[13] Weisstein, E., *World of Physics*. URL: http://en.wikipedia.org/wiki/Tachyon#_note-feinberg67.

[14] Herbert, N., *Faster Than Light—Superluminal Loopholes in Physics*, New American Library, Markham, Canada, 1988.

[15] Ramana Murthy, P. V., "Stability of Protons and Electrons and Limits on Tachyon Masses," *Physion Review D*, Vol. 7, No. 7, 1973, pp. 2252–2253.

[16] Forward, R. L., "A Transparent Derivation of the Relativistic Rocket Equation," AIAA/ASME/SAE/ASEE 31st Joint Propulsion Conference and Exhibit, AIAA Paper 95-3060, San Diego, CA, 10–12 July 1995.

[17] Cassenti, B. N., "Antimatter Rockets and Interstellar Propulsion," AIAA/ASME/SAE/ASEE 29th Joint Propulsion Conference and Exhibit, AIAA Paper 93-2007, San Diego, CA, 28–30 June 1993.

[18] Cassenti, B. N., "High Specific Impulse Antimatter Rockets," AIAA/ASME/SAE/ASEE 27th Joint Propulsion Conference and Exhibit, AIAA Paper 91-2548, Sacramento, CA, 24–26 June 1991.

[19] Bussard, R. W., "Galactic Matter and Interstellar Flight," *Astronatica Acta*, Vol. VI, 1960, pp. 179–195.

[20] Cassenti, B. N., and Coreano, L., "The Interstellar Ramjet," AIAA/ASME/SAE/ASEE 40th Joint Propulsion Conference and Exhibit, AIAA Paper 2004-3568, Fort Lauderdale, FL. 11–14 July 2004.

[21] Cassenti, B. N., "Design Concepts for the Interstellar Ramjet," *Journal of the British Interplanetary Society*, Vol. 46, 1993, pp. 151–160.

Chapter 15

Faster-than-Light Approaches in General Relativity

Eric W. Davis*
Institute for Advanced Studies at Austin, Austin, Texas

I. Introduction

IT was nearly two decades ago when science fiction media (TV, film, and novels) began to adopt traversable wormholes, and more recently "stargates," for interstellar travel schemes that allowed their heroes to travel throughout our galaxy. In 1985 physicists M. Morris and K. Thorne at CalTech discovered the principle of traversable wormholes based on Einstein's General Theory of Relativity published in 1915. Morris and Thorne [1] and Morris et al. [2] did this as an academic exercise and in the form of problems for a physics final exam, at the request of Carl Sagan who had then completed the draft of his novel *Contact*. This little exercise led to a continuous line of insights and publications in general relativity research, i.e., the study of traversable wormholes and time machines. Wormholes are hyperspace tunnels through spacetime connecting either remote regions within our universe or two different universes; they even connect different dimensions and different times. Space travelers would enter one side of the tunnel and exit out the other, passing through the throat along the way. The travelers would move through the wormhole at $\leq c$, where c is the speed of light (3×10^8 m/s) and therefore not violate Special Relativity, but external observers would view the travelers as having traversed multi-light-year distances through space at faster-than-light (FTL) speed. A "stargate" was shown to be a very simple special class of traversable wormhole solutions to Einstein's general relativistic field equation [3,4].

Copyright © 2008 by the American Institute of Aeronautics and Astronautics, Inc. All rights reserved.
*Senior Research Physicist.

This development was later followed by M. Alcubierre's formulation in 1994 of the "warp drive" spacetime metric,[†] which was another solution to the general relativistic field equation [5]. Alcubierre derived a metric motivated by cosmological inflation that would allow arbitrarily short travel times between two distant points in space. The behavior of the warp drive metric provides for the simultaneous expansion of space behind the spacecraft and a corresponding contraction of space in front of the spacecraft. The warp drive spacecraft would appear to be "surfing on a wave" of spacetime geometry. A spacecraft can be made to exhibit an arbitrarily large apparent FTL speed ($\gg c$) as viewed by external observers, but its moving local rest frame never travels outside of its local comoving light cone and thus does not violate Special Relativity. A variety of warp drive research ensued and papers were published during the next 14 years.

The implementation of FTL interstellar travel via traversable wormholes, warp drives, or other FTL spacetime modification schemes generally requires the engineering of spacetime into very specialized local geometries. The analysis of these via the general relativistic field equation plus the resultant source matter equations of state demonstrates that such geometries require the use of "exotic" matter in order to produce the requisite FTL spacetime modification. Exotic matter is generally defined by general relativity physics to be matter that possesses (renormalized) negative energy density (sometimes negative stress-tension = positive outward pressure, a.k.a. gravitational repulsion or antigravity), and this is a very misunderstood and misapplied term by the non-general relativity community. We clear up this misconception by defining what negative energy is and where it can be found in nature, as well as reviewing the experimental concepts that have been proposed to generate negative energy in the laboratory. Also, it has been claimed that FTL spacetimes are not plausible because exotic matter violates the general relativistic energy conditions. However, it has been shown that this is a spurious issue. The identification, magnitude, and production of exotic matter is seen to be a key technical challenge, however. FTL spacetimes also possess features that challenge the notions of causality and there are alleged constraints placed upon them by quantum effects. These issues are reviewed and summarized, and an assessment on the present state of their resolution is provided.

II. General Relativistic Definition of Exotic Matter and the Energy Conditions

In classical physics the energy density of all observed forms of matter (fields) is non-negative. What is exotic about the matter that must be used to generate FTL spacetimes is that it must have negative energy density and/or negative flux [6]. The energy density is "negative" in the sense that the configuration of matter fields that must be deployed to generate and thread a traversable wormhole

---

[†]A spacetime metric (ds^2) is a Lorentz-invariant distance function between any two points in spacetime that is defined by $ds^2 = g_{\mu\nu} dx^\mu dx^\nu$, where $g_{\mu\nu}$ is the metric tensor, which is a 4×4 matrix that encodes the geometry of spacetime and dx^μ is the infinitesimal coordinate separation between two points. The Greek indices ($\mu, \nu = 0 \ldots 3$) denote spacetime coordinates, $x^0 \ldots x^3$, such that $x^1 \ldots x^3 \equiv$ space coordinates and $x^0 \equiv$ time coordinate.

throat or a warp drive bubble must have an energy density, ρ_E ($=\rho c^2$, where ρ is the rest-mass density), that is less than or equal to its pressures, p_i [1,3].‡ In many cases, these equations of state are also known to possess an energy density that is algebraically negative, i.e., the energy density and flux are less than zero. It is on the basis of these conditions that we call this material property "exotic." The condition for ordinary, classical (nonexotic) forms of matter that we are all familiar with in nature is that $\rho_E > p_i$ and/or $\rho_E \geq 0$. These conditions represent two examples of what are variously called the "standard" energy conditions: weak energy condition (WEC: $\rho_E \geq 0$, $\rho_E + p_i \geq 0$), null energy condition (NEC: $\rho_E + p_i \geq 0$), dominant energy condition (DEC), and strong energy condition (SEC). These energy conditions forbid negative energy density between material objects to occur in nature, but they are mere hypotheses. Hawking and Ellis [7] formulated the energy conditions in order to establish a series of mathematical hypotheses governing the behavior of collapsed-matter singularities in their study of cosmology and black hole physics. More specifically, classical general relativity allows one to prove lots of general theorems about the behavior of matter in gravitational fields. Some of the most significant of these general theorems are:

1) Focusing theorems (for gravitational lensing)
2) Singularity theorems (for gravity-dominated collapse leading to black holes possessing a singularity)
3) Positive mass theorem (for positive gravitational mass)
4) Topological censorship (i.e., you cannot build a wormhole)

Each of these requires the existence of "reasonable" types of matter for their formulation, i.e., matter that satisfies the energy conditions. In what follows, we will not consider the impact or implications of the DEC or SEC because they add no new information beyond the WEC and NEC.

The bad news is that real physical matter is not "reasonable" because the energy conditions are, in general, violated by semiclassical quantum effects (occurring at order $\hbar$, which is Planck's reduced constant, 1.055×10^{-34} J·s) [3]. More specifically, quantum effects generically violate the averaged NEC (ANEC). Furthermore, it was discovered in 1965 that quantum field theory has the remarkable property of allowing states of matter containing local regions of negative energy density or negative fluxes [8]. This violates the WEC, which postulates that the local energy density is non-negative for all observers. And there are also general theorems of differential geometry that guarantee that there must be a violation of one, some, or all of the energy conditions (meaning exotic matter is present) for all FTL spacetimes. With respect to creating FTL spacetimes, "negative energy" has the unfortunate reputation of alarming physicists. This is unfounded because all the energy condition hypotheses have been tested in the laboratory and experimentally shown to be false—25 years before their formulation [9].

---

‡Latin indices (e.g., i, j, $k = 1 \ldots 3$) affixed to physical quantities denote the usual three-dimensional space coordinates, $x^1 \ldots x^3$, indicating the spatial components of vector or tensor quantities.

Further investigation into this technical issue showed that violations of the energy conditions are widespread for all forms of both "reasonable" classical and quantum matter [10–14]. Furthermore, Visser [3] showed that all (generic) spacetime geometries violate all the energy conditions. So the condition that $\rho_E > p_i$ and/or $\rho_E \geq 0$ must be obeyed by all forms of matter in nature is spurious. Violating the energy conditions commits no offense against nature. Negative energy has been produced in the laboratory and this will be discussed in the following sections.

A. Examples of Exotic or "Negative" Energy Found in Nature

The exotic (energy condition-violating) fields that are known to occur in nature are:

1) Static, radially dependent electric or magnetic fields. These are borderline exotic, if their tension were infinitesimally larger, for a given energy density [7,15].

2) Squeezed quantum vacuum states: electromagnetic and other (non-Maxwellian) quantum fields [1,16].

3) Gravitationally squeezed vacuum electromagnetic zero-point fluctuations [17].

4) Casimir effect, i.e., the Casimir vacuum in flat, curved, and topological spaces [18–24].

5) Other quantum fields, states, or effects. In general, the local energy density in quantum field theory can be negative due to quantum coherence effects [8]. Other examples that have been studied are Dirac field states: the superposition of two single particle electron states and the superposition of two multi-electron–positron states [25,26]. In the former/latter, the energy densities can be negative when two single/multi-particle states have the same number of electrons/electrons and positrons or when one state has one more electron/electron–positron pair than the other.

Cosmological inflation [3], cosmological particle production [3], classical scalar fields [3], the conformal anomaly [3], and gravitational vacuum polarization [10–13] are among many other examples that also violate the energy conditions. Because the laws of quantum field theory place no strong restrictions on negative energies and fluxes, then it might be possible to produce exotic phenomena such as FTL travel [5,27,28], traversable wormholes [1–3], violations of the second law of thermodynamics [29,30], and time machines [2,3,31]. There are several other exotic phenomena made possible by the effects of negative energy, but they lie outside the scope of the present study. In the following section, we review the previously listed items 1 through 4 and examine their applicability and technical maturity. Dirac field states are currently under study by investigators. Also, we do not consider the issue of capturing and storing negative energy in what follows because free-space negative energy sources appear to be a more desirable option for inducing FTL spacetimes than stored negative energy, and because there is very little technical literature that addresses how to capture and store negative energy (see Ref. 6). The issue of capturing and storing negative energy will be left for future investigations.

B. Generating Negative Energy in the Lab

1. Static Radial Electric and Magnetic Fields

It is beyond the scope of this study to include all the technical configurations by which one can generate static, radially dependent electric or magnetic fields. Suffice it to say that ultrahigh-intensity tabletop lasers have been used to generate extreme electric and magnetic field strengths in the lab. Ultrahigh-intensity lasers use the chirped-pulse amplification (CPA) technique to boost the total output beam power. All laser systems simply repackage energy as a coherent package of optical power, but CPA lasers repackage the laser pulse itself during the amplification process. In typical high-power short-pulse laser systems, it is the peak intensity, not the energy or the fluence, that causes pulse distortion or laser damage. However, the CPA laser dissects a laser pulse according to its frequency components and reorders it into a time-stretched lower-peak-intensity pulse of the same energy [32–34]. This benign pulse can be then amplified safely to high energy, and then only afterwards reconstituted as a very short pulse of enormous peak power, a pulse that could never itself have passed safely through the laser system. Made more tractable in this way, the pulse can be amplified to substantial energies (with orders of magnitude greater peak power) without encountering intensity-related problems.

The extreme output beam power, fields, and physical conditions that have been achieved by ultrahigh-intensity tabletop lasers are [34]:

1) Power intensity $\approx 10^{19}$ to 10^{30} W/m^2 (10^{34} W/m^2 using SLAC as a booster).
2) Peak power pulse $\leq 10^3$ fs.
3) Electric field, $E \approx 10^{14}$ to 10^{18} V/m. [Note: Compare this with the critical quantum electrodynamic (QED) vacuum breakdown E-field intensity, $E_c = 2m_e^2 c^3/\hbar e \approx 10^{18}$ V/m, defined by the total rest-energy of an electron–positron pair created from the vacuum divided by the electron's Compton wavelength, where m_e is the electron mass (9.11×10^{-31} kg) and e is the electron charge (1.602×10^{-19} C)].
4) Magnetic field, $B \approx$ several $\times 10^6$ Tesla (note: The critical QED vacuum breakdown B-field intensity is $B_c = E_c/c \approx 10^{10}$ Tesla).
5) Ponderomotive acceleration of electrons $\approx 10^{17}$ to $10^{30} g_\oplus$ ($g_\oplus$ is the acceleration of gravity near the Earth's surface, 9.81 m/s^2).
6) Light pressure $\approx 10^9$ to 10^{15} bars.
7) Plasma temperatures $> 10^{10}$ K.

The vigilant reader might assert that the electric and magnetic fields generated by ultrahigh-intensity lasers are not static. But in fact, these fields are static over the duration of the pulsewidth while at peak intensity. We find from the above data that ultrahigh-intensity lasers can generate an electric field energy density $\sim 10^{16}$ to 10^{28} J/m^3 and a magnetic field energy density $\sim 10^{19}$ J/m^3. However, there remains the problem of engineering this type of experiment because classical electromagnetic theory states that every observer associated with the experiment will see a non-negative energy density that is $\propto E^2 + B^2$, where E and B are measured in an observer's reference frame. It is not known how to increase the tension in these fields using current physics, but some new physics may provide an answer. This technical problem must be left for future investigation.

2. Squeezed Quantum Vacuum

Substantial theoretical and experimental work has shown that in many quantum systems the limits to measurement precision imposed by the quantum vacuum zero-point fluctuations (ZPF) can be breached by decreasing the noise in one observable (or measurable quantity) at the expense of increasing the noise in the conjugate observable; at the same time the variations in the first observable, say the energy, are reduced below the ZPF such that the energy becomes "negative." "Squeezing" is thus the control of quantum fluctuations and corresponding uncertainties, whereby one can squeeze the variance of one (physically important) observable quantity provided the variance in the (physically unimportant) conjugate observable is stretched/increased. The squeezed quantity possesses an unusually low variance, meaning less variance than would be expected on the basis of the equipartition theorem. One can in principle exploit quantum squeezing to extract energy from one place in the ordinary vacuum at the expense of accumulating excess energy elsewhere [1].

The squeezed state of the electromagnetic field is a primary example of a quantum field that has negative energy density and negative energy flux. Such a state became a physical reality in the laboratory as a result of the nonlinear-optics technique of "squeezing," i.e., of moving some of the quantum fluctuations of laser light out of the $\cos[\omega(t - z/c)]$ part of the beam and into the $\sin[\omega(t - z/c)]$ part (ω is the angular frequency of light, t is time, and z denotes the z-axis direction of beam propagation) [16,35–39]. The observable that gets squeezed will have its fluctuations reduced below the vacuum ZPF. The act of squeezing transforms the phase space circular noise profile characteristic of the vacuum into an ellipse, whose semimajor and semiminor axes are given by unequal quadrature uncertainties (of the quantized electromagnetic field harmonic oscillator operators). This applies to coherent states in general, and the usual vacuum is also a coherent state with eigenvalue zero. As this ellipse rotates about the origin with angular frequency ω, these unequal quadrature uncertainties manifest themselves in the electromagnetic field oscillator energy by periodic occurrences, which are separated by one quarter cycle of both smaller and larger fluctuations compared to the unsqueezed vacuum.

Morris and Thorne [1] and Caves [40] point out that if one squeezes the vacuum (i.e., if one puts vacuum rather than laser light into the input port of a squeezing device), then one gets at the output an electromagnetic field with weaker fluctuations and thus less energy density than the vacuum at locations where $\cos^2[\omega(t - z/c)] \cong 1$ and $\sin^2[\omega(t - z/c)] \ll 1$; but with greater fluctuations and thus greater energy density than the vacuum at locations where $\cos^2[\omega(t - z/c)] \ll 1$ and $\sin^2[\omega(t - z/c)] \cong 1$. Because the vacuum is defined to have vanishing energy density, any region with less energy density than the vacuum actually has a negative (renormalized) expectation value for the energy density. Therefore, a squeezed vacuum state consists of a traveling electromagnetic wave that oscillates back and forth between negative energy density and positive energy density, but has positive time-averaged energy density.

For the squeezed electromagnetic vacuum state, the energy density $\rho_{\text{E-sqvac}}$ is given by [41]:

$$\rho_{\text{E-sqvac}} = \left(\frac{2\hbar\omega}{L^3}\right) \sinh\xi \, [\sinh\xi + \cosh\xi \cos(2\omega(t - z/c) + \delta)] \quad (J/m^3) \quad (1)$$

where L^3 is the volume of a large box with sides of length L (i.e., we put the quantum field inside a box with periodic boundary conditions), ξ is the squeezed state amplitude (giving a measure of the mean photon number in a squeezed state), and δ is the phase of squeezing. Equation (1) shows that $\rho_{\text{E-sqvac}}$ falls below zero once every cycle when the condition $\cosh\xi > \sinh\xi$ is met. It turns out that this is always true for every nonzero value of ξ, so $\rho_{\text{E-sqvac}}$ becomes negative at some point in the cycle for a general squeezed vacuum state. On another note, when a quantum state is close to a squeezed vacuum state, there will almost always be some negative energy densities present.

Negative energy can be generated by an array of ultrahigh-intensity lasers using an ultrafast rotating mirror system [42]. In this scheme a laser beam is passed through an optical cavity resonator made of a lithium niobate ($LiNbO_3$) crystal that is shaped like a cylinder with rounded silvered ends to reflect light. The resonator will act to produce a secondary lower frequency light beam in which the pattern of photons is rearranged into pairs. The squeezed light beam emerging from the resonator will contain pulses of negative energy interspersed with pulses of positive energy.

In this concept both the negative and positive energy pulses are $\sim 10^{-15}$ second in duration. We could, in principle, arrange a set of rapidly rotating mirrors to separate the positive and negative energy pulses from each other. The light beam would be set to strike each mirror surface at a very shallow angle while the rotation would ensure that the negative energy pulses would be reflected at a slightly different angle from the positive energy pulses. A small spatial separation of the two different energy pulses would occur at some distance from the rotating mirror. Another system of mirrors would be needed to redirect the negative energy pulses to an isolated location and concentrate them there. Figure 1 illustrates this concept.

The rotating mirror system can actually be implemented via nonmechanical means. A chamber of sodium gas is placed within the squeezing cavity and a laser beam is directed through the gas. The beam is reflected back on itself by a mirror to form a standing wave within the sodium chamber. This wave causes rapid variations in the optical properties of the sodium, thus causing rapid variations in the squeezed light so that we can induce rapid reflections of pulses by careful design [36]. An illustration of this is shown in Fig. 2.

Another way to generate negative energy via squeezed light would be to manufacture extremely reliable light pulses containing precisely one, two, three, etc., photons each and combine them together to create squeezed states to order [42]. Superimposing many such states could theoretically produce bursts of intense negative energy. Figure 3 provides a conceptual diagram of this concept. Photonic crystal research has already demonstrated the feasibility of using photonic crystal waveguides (mixing together the classical and quantum properties of

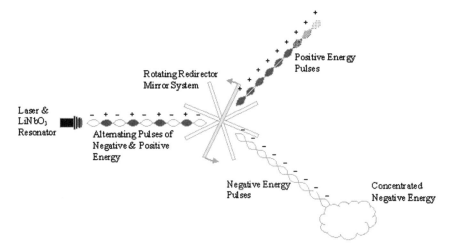

Fig. 1 Proposed squeezed light negative energy generator.

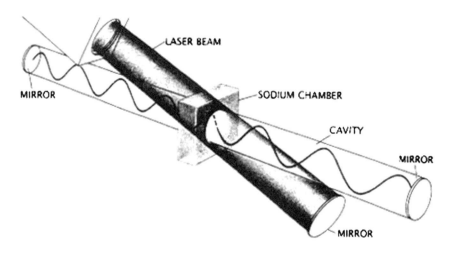

Fig. 2 Sodium chamber negative energy separator [36].

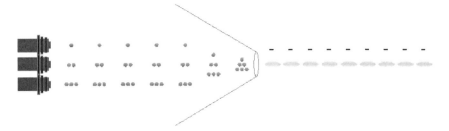

Fig. 3 Alternative squeezed light negative energy generator.

optical materials) to engineer light sources that produce beams containing precisely one, two, three, etc., photons. For example, researchers at Melbourne University used a microwave oven to fuse a tiny diamond, just 1/1000 of a millimeter long, onto an optical fiber, which could be used to create a single photon beam of light [43,44]. The combining of different beams containing different (finite integer) numbers of photons is already state-of-the-art practice via numerous optical beam combining methods that can readily be extended to our application.

Finally, Ries et al. [45] experimentally demonstrated the first simple, scalable squeezed vacuum source in the laboratory that consisted of a continuous-wave diode laser and an atomic rubidium vapor cell. The experimental tools we need to begin exploring the generation of negative energy for the purpose of creating FTL spacetimes are just now becoming available.

3. Gravitationally Squeezed Electromagnetic Zero-Point Fluctuations

A natural source of negative energy comes from the effect product by gravitational fields (of astronomical bodies) in space on the surrounding vacuum. For example, the gravitational field of the Earth produces a zone of negative energy around it by dragging some of the virtual quanta (a.k.a. vacuum ZPF) downward. This concept was initially developed in the 1970s as a byproduct of studies on quantum field theory in curved space [21]. However, Hochberg and Kephart [17] derived an important application of this concept to the problem of creating and stabilizing traversable wormholes. They showed that one can utilize the negative energy densities, which arise from distortion of the vacuum ZPF due to the interaction with a prescribed gravitational background, to provide a violation of the energy conditions. The squeezed quantum states of quantum optics provide a natural form of matter having negative energy density.

The analysis, via quantum optics, showed that gravitation itself provides the mechanism for generating the squeezed vacuum states needed to support stable traversable wormholes. The production of negative energy densities via a squeezed vacuum is a necessary and unavoidable consequence of the interaction or coupling between ordinary matter and gravity, and this defines what is meant by gravitationally squeezed vacuum states. The magnitude of the gravitational squeezing of the vacuum can be estimated from the quantum optics squeezing condition for given transverse momentum and (equivalent) energy eigenvalues, j, of two electromagnetic ZPF field modes, such that this condition is subject to $j \to 0$, and it is defined as [17]:

$$j \equiv \frac{4\pi c^2}{\lambda g_\oplus} \left(\frac{r}{R_\oplus}\right)^2 \frac{M_\oplus}{M}$$
$$= \frac{8\pi r_S}{\lambda} \quad (2)$$

where λ is the ZPF mode wavelength, r is the radial distance from the center of the astronomical body in question, $R_\oplus$ is the radius of the Earth (6.378×10^6 m), $M_\oplus$ is the mass of the Earth (5.972×10^{24} kg), M is the mass of the astronomical

Table 1 Substantial gravitational squeezing occurs for vacuum ZPF when $\lambda \geq 8\pi r_S$

Mass of body (kg)	r_S (m)	λ (m)	$\rho_{\text{E-gsvac}}$ (J/m^3)
Sun = 2.00×10^{30}	2.95×10^3	$\geq 78.0 \times 10^3$	-1.69×10^{-44}
Jupiter = 1.90×10^{27}	2.82	≥ 74	-2.08×10^{-32}
Earth = 5.98×10^{24}	8.87×10^{-3}	≥ 0.23	-2.23×10^{-22}
Typical mountain $\approx 10^{11}$	$\approx 10^{-16}$	$\geq 10^{-15}$	-6.25×10^{35}
Planck mass = 2.18×10^{-8}	3.23×10^{-35}	$\geq 8.50 \times 10^{-34}$	-1.20×10^{108}
Proton = 1.67×10^{-27}	2.48×10^{-54}	$\geq 6.50 \times 10^{-53}$	-3.50×10^{184}

body, and r_S is the Schwarzschild radius of the astronomical body.[§] Note that r_S is only a convenient radial distance parameter for any object under examination and so there is no black hole collapse involved in this analysis. We can actually choose any radial distance from the body in question to perform this analysis, but using r_S makes the equation simpler in form. Also note that Eq. (2) contains an extra factor of two (compared to the j derived in Ref. 17) in order to account for the photon spin. The squeezing condition plus Eq. (2) simply states that substantial gravitational squeezing of the vacuum occurs for those ZPF field modes with $\lambda \geq 8\pi r_S$ of the mass in question (whose gravitational field is squeezing the vacuum). The corresponding local vacuum state energy density will be $\rho_{\text{E-gsvac}} = -2\pi^2 \hbar c / \lambda^4$.

The general result of the gravitational squeezing effect is that as the gravitational field strength increases, the negative energy zone (surrounding the mass) also increases in strength. Table 1 shows when gravitational squeezing becomes important for sample masses and their corresponding $\rho_{\text{E-gsvac}}$. It shows that in the case of the Earth, Jupiter, and the Sun, the squeezing effect is extremely feeble because only ZPF mode wavelengths above 0.2 m to 78 km are affected, each having very minute $\rho_{\text{E-gsvac}}$. For a solar mass black hole (2.95 km radius), the effect is still feeble because only ZPF mode wavelengths above 78 km are affected. But note that Planck mass objects will have an enormously strong negative energy zone surrounding them because all ZPF mode wavelengths above 8.50×10^{-34} m will be squeezed, in other words, all wavelengths of interest for vacuum fluctuations. Protons will have the strongest negative energy zone in comparison because the squeezing effect includes all ZPF mode wavelengths above 6.50×10^{-53} m. Furthermore, a body smaller than a nuclear diameter ($\approx 10^{-16}$ m) and containing the mass of a mountain ($\approx 10^{11}$ kg) has a fairly strong negative energy zone because all ZPF mode wavelengths above 10^{-15} m will be squeezed. In each of these cases, the magnitude of the corresponding $\rho_{\text{E-gsvac}}$ is very large.

However, the estimates for the wavelengths in Table 1 might be too small. L. H. Ford (private communication, 2007) argues that the analysis in Ref. 17 is

---

[§]$r_S = 2GM/c^2$, where G is Newton's universal gravitation constant (6.673×10^{-11} Nm^2/kg^2). According to General Relativity Theory, this is the critical radius at which a spherically symmetric massive body becomes a black hole, that is, at which light is unable to escape from the body's surface.

in error because spacetime is flat on scales smaller than the local radius of curvature, which is defined by the inverse square root of the typical Riemann curvature tensor component in a local orthonormal frame, or $\ell_C \approx (r^3 c^2/GM)^{1/2}$. According to Ford, only ZPF modes with $\lambda \geq \ell_C$ will be squeezed by the gravitational field. This leads to a different local vacuum state energy density (for $r \gg r_S$) [11]:

$$\rho_{\text{E-gsvac}} = -\frac{2\pi^2 \hbar c}{\lambda^4}$$

$$\approx -\frac{2\pi^2 \hbar c}{\ell_C^4} \qquad (3)$$

$$\approx -\frac{2\pi^2 \hbar G^2 M^2}{c^3 r^6} \quad (J/m^3)$$

For example, near the surface of the Earth ($r \approx R_\oplus, M = M_\oplus$), $\ell_C \approx 2.42 \times 10^{11}$ m and hence, Eq. (3) gives $\rho_{\text{E-gsvac}} \approx -1.82 \times 10^{-70}$ J/m^3. Compare these values with $\lambda \geq 0.23$ m and $\rho_{\text{E-gsvac}} \approx -2.23 \times 10^{-22}$ J/m^3 in Table 1. The resolution of this disagreement remains an open question.

We are presently unaware of any way to artificially generate gravitational squeezing of the vacuum in the laboratory. This will be left for future investigation. However, it is predicted to occur in the vicinity of astronomical matter. Naturally occurring traversable wormholes in the vicinity of astronomical matter would therefore become possible.

4. Vacuum Field Stress: Negative Energy from the Casimir Effect

The Casimir effect is by far the easiest and most well-known way to generate negative energy in the lab. The Casimir effect that is familiar to most people is the force associated with the electromagnetic quantum vacuum [46]. This is an attractive force that must exist between any two neutral (uncharged), parallel, flat, conducting surfaces (e.g., metallic plates) in a vacuum. This force has been well measured and it can be attributed to a minute imbalance in the electromagnetic zero-point energy density inside the cavity between the conducting surfaces versus the region outside of the cavity [47–49].

It turns out that there are many different types of Casimir effects found in quantum field theory [18–20,24,50]. For example, if one introduces a single infinite plane conductor into the Minkowski (flat spacetime) vacuum by bringing it adiabatically from infinity so that whatever quantum fields are present suffer no excitation but remain in their ground states, then the vacuum (electromagnetic) stresses induced by the presence of the infinite plane conductor produces a Casimir effect. This result holds equally well when two parallel plane conductors (with separation distance d) are present, which gives rise to the familiar Casimir effect inside a cavity. Note that in both cases, the spacetime manifold is made incomplete by the introduction of the plane conductor boundary condition(s). The vacuum region put under stress by the presence of the plane conductor(s) is called the Casimir vacuum. The generic expression for the energy density of the Casimir effect is $\rho_{CE} = -A(\hbar c)d^{-4}$, where $A = \zeta(D)/8\pi^2$ in spacetimes of arbitrary dimension D [18–20]. The appearance of the zeta-function $\zeta(D)$ is characteristic of expressions for vacuum stress-energy tensors, $\langle T^{\mu\nu}\rangle_{\text{vac}}$.¶ In our familiar four-dimensional spacetime ($D = 4$) we have that $A = \pi^2/720$. To calculate $\langle T^{\mu\nu}\rangle_{\text{vac}}$ for a given quantum field is to calculate its associated Casimir effect.

Analogs of the Casimir effect also exist for fields other than the electromagnetic field. When considering the vacuum state of other fields, one must consider boundary conditions that are analogous to the perfect-conductor boundary conditions for the electromagnetic field at the surfaces of the plates [18–20, 24]. Other fields are not electromagnetic in nature, that is to say they are non-Maxwellian, and so the perfect-conductor boundary conditions do not apply to them. It turns out that complete manifolds exhibit what is called the topological Casimir effect for any non-Maxwellian fields. In order to define boundary conditions for other fields we replace the conductor boundary conditions and Minkowski spacetime by a manifold of the form $\Re \times \Sigma$ (i.e., a product space), where $\Re$ is the real line defining the time dimension for this particular product space and Σ is a flat three-dimensional manifold (three-manifold) having any one of the following topologies: $\Re^2 \times S^1$, $\Re \times T^2$, T^3, $\Re \times K^2$, etc., $\Re$ being the real line that defines any linear space dimension (e.g., $\Re =$ line, $\Re^2 =$ two-dimensional plane, etc.), T^n being the n-torus, K^2 the two-dimensional Klein bottle, S^1 the circle, etc.

The case $\Sigma = \Re^2 \times S^1$ has the closest resemblance to the electromagnetic Casimir effect, the difference being that instead of imposing conductor boundary conditions, one imposes periodic boundary conditions on some of the space coordinates in the three-manifold. When imposing this topological constraint on the field theoretic calculation of the topological Casimir effect (for linear massless fields), one finds that the generic expression for the energy density is also $\rho_{\text{CE}} = -A(\hbar c)d^{-4}$, where $A = \pm d_f(\pi^2/90)$, d_f is the number of degrees of freedom (e.g., helicity states) per spatial point, the plus sign holds for boson fields (giving a negative energy density), and the negative sign for fermion fields (giving a positive energy density).

If one were to admit spin structure in the manifolds described above and the field is spinorial, then there is another important subtlety that must be taken into account when evaluating $\langle T^{\mu\nu}\rangle_{\text{vac}}$. However, this introduces an additional complexity involving the relationship between the spin structure and the global structure (i.e., the configuration space or fibre bundle) of the field in question whereby the topology not only of the base manifold but of the fibre bundle itself has an effect on $\langle T^{\mu\nu}\rangle_{\text{vac}}$. In addition to this, there are (compactified) extra-space dimensional quantum field (i.e., D-Brane or "brane world") analogs of the Casimir effect yet to be explored. But a detailed consideration of these for producing FTL spacetimes is beyond the scope of this chapter and will be left for future investigation. One proposal that exploits a brane world analog of the Casimir effect to create a warp drive is described in Sec. III.B.

As a final note, we point out that the methods used to obtain the electromagnetic $\langle T^{\mu\nu}\rangle_{\text{vac}}$ between parallel plane conductors can be used also when the

---

¶The angular brackets denote the quantum (vacuum state) expectation value of the stress-energy tensor $T^{\mu\nu}$. Also note that stress-energy is synonymous with energy-momentum.

conductors are not parallel but are joined together along a line of intersection. If the conductors have curved surfaces instead, then one obtains results that are similar to the case of intersecting conductors. These geometries have also been evaluated for the case of dielectric media. These particular cases will not be considered further because there are technical subtleties involved that complicate the calculations and application of the different approaches. This topic will also be left for future investigation.

5. Casimir Effect: Moving Mirrors

Negative energy can be created by a single moving reflecting (conducting) surface (a.k.a. a moving mirror). A mirror moving with increasing acceleration generates a flux of negative energy that emanates from its surface and flows out into the space ahead of the mirror [21,51]. This is essentially the simple case of an infinite plane conductor undergoing acceleration perpendicular to its surface. If the acceleration varies with time, the conductor will generally emit or absorb photons (i.e., exchange energy with the vacuum), even though it is neutral. This is an example of the well-known quantum phenomenon of parametric excitation. The parameters of the electromagnetic field oscillators (e.g., their frequency distribution function) change with time owing to the acceleration of the mirror [52]. However, this effect is known to be exceedingly small, and it is not the most effective way to produce negative energy for our purposes. We will not consider this scheme any further.

6. Casimir Effect: Negative Energy for Traversable Wormholes

The electromagnetic Casimir effect can, in principle, be used to create a traversable wormhole. The energy density $\rho_{CE} = -(\pi^2 \hbar c/720)d^{-4}$ within a Casimir cavity is negative and manifests itself by producing a force of attraction between the cavity walls. But cavity dimensions must be made exceedingly small in order to generate a significant amount of negative energy for our purposes. In order to use the Casimir effect to generate a spherically symmetric traversable wormhole throat of radius r_{throat}, we will need to design a cavity made of perfectly conducting spherically concentric thin plates with a plate separation d of [2]:

$$d = \left(\frac{\pi^3}{30}\right)^{\frac{1}{4}} \left(r_{\text{throat}} \sqrt{\frac{\hbar G}{c^3}}\right)^{\frac{1}{2}} \quad (4)$$

$$= \left(4.05 \times 10^{-18}\right) \sqrt{r_{\text{throat}}} \quad (m)$$

To counteract the collapse of the cavity due to the Casimir force acting between the plates, the plates will have equal electric charges placed upon them to establish adequate Coulomb repulsion. (In a detailed analysis, the electrostatic energy required to support the Coulomb repulsion between the plates would be considered separately.) Equation (4) shows that a 1-km radius throat will require a cavity plate separation of 1.28×10^{-16} m (smaller than a nuclear diameter), which gives $\rho_{EC} = -1.62 \times 10^{36}$ J/m^3

for this configuration. In contrast, a wormhole with a throat radius of 1 AU (mean Earth–Sun distance, 1.50×10^{11} m) will require a plate separation of 1.57×10^{-12} m (or 35% smaller than the electron's Compton wavelength), which results in an energy density of -7.14×10^{19} J/m^3. There is no technology known today that can engineer a cavity with such minuscule plate separations. In addition, such minuscule plate separations are unrealistic because the Casimir effect switches over to the nonretarded field behavior ($\sim d^{-3}$) of van der Waals forces when plate separations go below the wavelength (≈ 10 nm) where they are no longer perfectly conducting [53]. We will not consider this scheme further. However, future work will be necessary to elucidate whether the various quantum field analogs of the Casimir effect can provide a more reasonable technical solution.

III. Brief Review of Faster-than-Light Spacetimes

In Section I we briefly described the two primary classes of FTL spacetimes found in General Relativity Theory: traversable wormholes and warp drives. There are as yet no other (credible) classes of FTL spacetimes having to do with General Relativity Theory, higher dimensional D-Brane theory, or (credible) alternative theories of gravity (e.g., Puthoff's Polarizable-Vacuum Representation of General Relativity, Yilmaz theory, etc.). Investigators have discovered that General Relativity Theory and all of the (credible) alternative theories of gravity are infested with traversable wormholes, time machines, and some form of a warp drive. The small number of proposed alternative FTL spacetimes is largely speculative and not yet on the same level of rigor as the warp drive and traversable wormhole spacetimes.

How does one study the physics of FTL spacetimes within the framework of General Relativity Theory? When studying spacetime physics, the normal philosophy is to take the general relativistic field equation, add some form of matter, make simplifying assumptions, and then solve to deduce what the geometry of spacetime will be. This is very difficult to do because there are 10 nonlinear, second-order partial differential equations with four redundancies (arbitrary choice of spacetime coordinates) and four constraints (energy-momentum conservation). There is a tremendous body of research that takes exactly this approach, either analytically or numerically. However, this is not the best strategy for understanding FTL spacetimes. The appropriate strategy is to decide beforehand on the definition of the desired traversable wormhole or warp drive and decide what the spacetime geometry should look like. Given the desired geometry, use the general relativistic field equation to calculate the distribution of matter required to set up this geometry. Then one needs to assess whether the required distribution of matter is physically reasonable and whether it violates any basic rules of physics, etc. In the following sections we briefly outline the key results for traversable wormhole and warp drive spacetimes.

A. Traversable Wormholes

Traversable wormholes represent a class of exact metric solutions of the general relativistic field equation. The solutions are "exact" in the sense that no approximations requiring a plethora of physical assumptions have to be made to derive the appropriate spacetime geometry. To define a stable traversable wormhole one needs to define the desirable physical requirements it is to have in order to achieve the desired FTL travel benefit. The desired requirements are the following [1,3]:

1) Travel time through the wormhole tunnel or throat should be ≤1 year as seen by both the travelers and outside static observers.

2) Proper time as measured by travelers should not be dilated by relativistic effects.

3) The gravitational acceleration and tidal-gravity accelerations between different parts of the travelers' body should be $\leq 1g_\oplus$ when going through the wormhole.

4) Travel speed through the tunnel/throat should be $<c$.

5) Travelers (made of ordinary matter) must not couple strongly to the material that generates the wormhole curvature; the wormhole must be threaded by a vacuum tube through which the travelers can move.

6) There is no event horizon at the wormhole throat.

7) There is no singularity of infinitely collapsed matter residing at the wormhole throat.

See item 5b in Box 3 of Ref. 1 for the technical description of a trip through a wormhole. These requirements then lead us to define a spherically symmetric Lorentzian spacetime metric, ds^2, that prescribes the required traversable wormhole geometry [1,3]:

$$ds^2 = -e^{2\phi(r)}c^2dt^2 + [1 - b(r)/r]^{-1}dr^2 + r^2 d\Omega^2 \qquad (5)$$

where standard spherical coordinates are used (r: $2\pi r =$ circumference; $0 \leq \theta \leq \pi$; $0 \leq \varphi \leq 2\pi$), t is time ($-\infty < t < \infty$), $d\Omega^2 = d\theta^2 + \sin^2\theta d\varphi^2$, $\phi(r)$ is the freely specifiable redshift function that defines the proper time lapse through the wormhole throat, and $b(r)$ is the freely specifiable shape function that defines the wormhole throat's spatial (hypersurface) geometry. The throat is spherically shaped. There are several variations of Eq. (5), which define traversable wormholes having different properties. The reader should consult Ref. 3 for further details.

Figure 4 shows two diagrams representing the embedded space (Flamm) representation of Eq. (5), which depicts the geometry of an equatorial ($\theta = \pi/2$) slice through space at a specific moment of time ($t =$ constant). The top of Fig. 4 shows the embedding diagram for a traversable wormhole that connects two different universes (i.e., an inter-universe wormhole). The bottom diagram is an intra-universe wormhole with a throat that connects two distant regions of our own universe. These diagrams aid in visualizing traversable wormhole geometry and are merely a geometrical exaggeration.

There was originally one other criterion for defining a traversable wormhole, which was that it must be embedded within the surrounding (asymptotically) flat spacetime. However, Hochberg and Visser [54] proved that it is only the behavior near the wormhole throat that is critical to understanding the physics, and that

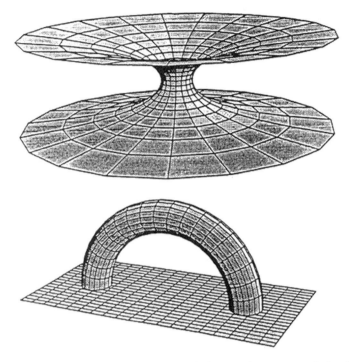

Fig. 4 Inter-universe wormhole (*top*) and intra-universe wormhole (*bottom*). (Courtesy of Ron Koczor, NASA-Marshall.)

a generic throat can be defined without having to make all the symmetry assumptions and without assuming the existence of an asymptotically flat spacetime in which to embed the wormhole. Therefore, one only needs to know the generic features of the geometry near the throat in order to guarantee violations of the NEC for certain open regions near the throat. So we are free to place our wormhole anywhere in spacetime we want because it is only the geometry and physics near the throat that matters for any analysis. This fact led to the development of a number of different traversable wormhole throat designs that are cubic shaped, polyhedral shaped, flat-face shaped, generic shaped, etc. The reader should consult Ref. 3 for a complete technical review of the various types (and shapes) of traversable wormhole solutions found in General Relativity Theory.

We know that we need exotic or negative energy to create and thread open a traversable wormhole. In this regard, we ask what kind of wormhole can one make with less effort. To answer this question we can relate the local wormhole geometry to the global topological invariant of the spacetime via the Gauss–Bonnet theorem [55]. In the Gauss–Bonnet theorem the local wormhole geometry is quantified by the energy density, U (in geometrodynamic units, $\hbar = G = c = 1$), threading the wormhole throat plus a spatial curvature constant (for the throat). The global topological invariant of spacetime is quantified by the Euler number, χ_e, which is itself defined in terms of the genus, g, representing the number of handles (or throats or tunnels) a wormhole can be assigned. These two topological quantities are related via $\chi_e = 2(1 - g)$. Therefore, the (static) wormhole Gauss–Bonnet relation is given by $U \le \chi_e/4$ or $U \le (1 - g)/2$ [55]. (The case for dynamic traversable wormholes has results that are similar to the static case.) This relation will help us decide if we want to build a traversable wormhole having one throat, or two or more throats, and what energy cost this will incur.

The following is the result of our analysis for traversable wormholes having:

1) One-handle/throat (i.e., flat torus or spherical wormhole topology) giving $g = 1$, thus $\chi_e = 0$, and so $U \le 0$.
2) Two-handles/throats giving $g = 2$, thus $\chi_e = -2$, and so $U \le -1/2$.
3) Three-handles/throats giving $g = 3$, thus $\chi_e = -4$, and so $U \le -1$; and so on.

It is clear from this that, as the number of wormhole handles/throats increases, the amount of negative energy required to create the wormhole will grow larger in magnitude. This is an undesirable demand on any putative negative energy generator. It is clear then that item 1 defines the most desirable engineering solution we can hope for: a one-handle/throat traversable wormhole that will require zero or (arbitrarily) little negative energy to create. The magnitude of energy condition violations and the amount of negative energy required to build a traversable wormhole will be addressed in Sec. IV.

B. Warp Drives

Alcubierre [5] derived a spacetime metric motivated by cosmological inflation that would allow arbitrarily short travel times between two distant points in space. The "warp drive" metric uses coordinates (t, x, y, z) and curve (or worldline) $x = x_{sh}(t)$, $y = 0$, $z = 0$, lying in the $t - x$ plane passing through the origin. Note that x_{sh} is the x-axis coordinate position of the moving spaceship (or warp bubble) frame. The metric specifying this spacetime is (also an exact solution to the general relativistic field equation) [5]:

$$ds^2 = -c^2 dt^2 + [dx - v_{sh}(t)f(r_{sh}(t))dt]^2 + dy^2 + dz^2 \tag{6}$$

where $v_{sh}(t) \equiv dx_{sh}(t)/dt$ is the speed associated with the curve (or warp bubble speed) and $r_{sh}(t) \equiv [(x - x_{sh}(t))^2 + y^2 + z^2]^{1/2}$ is the Euclidean distance from the curve. The warp bubble shape function $f(r_{sh})$ is any smooth positive function that satisfies $f(0) = 1$ and decreases away from the origin to vanish when $r_{sh} > R$ for some distance R. The geometry of each spatial slice is flat, and spacetime is flat where $f(r_{sh})$ vanishes but is curved where it does not vanish.

The driving mechanism of Eq. (6) is the York extrinsic time, ϑ. This quantity is defined as [5]:

$$\vartheta = \frac{v_{sh} x_{sh}}{c} \frac{df}{dr_{sh}} \tag{7}$$

The ϑ behavior of the warp drive bubble provides for the simultaneous expansion of space behind the spacecraft and a corresponding contraction of space in front

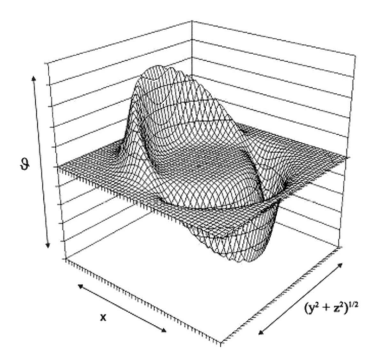

Fig. 5 York extrinsic time (ϑ) plot.

of the spacecraft. Figure 5 illustrates the ϑ behavior of the warp drive bubble geometry. Thus the spacecraft is enveloped within a warp bubble and can be made to exhibit an arbitrarily large FTL speed ($v_{sh} \gg c$) as viewed by external coordinate observers. Even though the worldlines inside the warp bubble region are spacelike for all external observers, the moving spaceship (warp bubble) frame itself never travels outside of its local comoving light cone and thus does not violate Special Relativity. However, Alcubierre's warp drive suffers from the problem that at superluminal speed, the interior of the warp bubble is causally disconnected from its surface and the exterior region, which means that photons cannot pass from the inside to the outside of the bubble and so there is no way of controlling the space warp in order to start, stop, or steer it. And the starship passengers will have no way of seeing where they are going.

Other investigators have designed warp drive metrics similar to Eq. (6) but with some modifications to the space warp geometry and corresponding changes to the negative energy requirement [56–63]. A well-known example is the proposal by Krasnikov [56] to fix the warp bubble causal disconnection problem by specifying a "warp tube" spacetime in which a starship traveling one way at ultra-relativistic speed creates a tube-shaped space warp behind itself, and then it would return by traveling back through the warp tube at FTL speed. Thus a starship can return from its interstellar journey shortly after it left no matter how far away it traveled. White and Davis [64] give another example in which Alcubierre's warp drive can be reinterpreted in extra-space dimensional brane-world theory as a spacetime expansion boost (i.e., like a scalar multiplier acting) on the initial spacecraft speed. This mechanism recasts the warp drive energy requirement into the equation of state for dark energy (a.k.a. the vacuum energy of Einstein's cosmological constant; see Section III.D in Chapter 4 for more details), whereby there is no negative energy density but only a negative pressure that is required to construct the warp drive effect. An alternative model is the proposal by Obousy [65] and Obousy and Cleaver [66] to engineer a version of Alcubierre's warp drive via exploiting the Casimir energy of the compactified extra-space dimensions in brane-world theories. They propose that the vacuum energy of Einstein's cosmological constant is a function of the size of the extra-space dimensions and that if one could locally manipulate the size of an extra-space dimension and thus locally manipulate the cosmological constant, then one could facilitate the expansion or contraction of the spacetime surrounding a spacecraft at will. Their model utilizes theoretical results from brane-world theories, which show that the Casimir energy of extra-space dimensions is related to the cosmological constant. But their warp drive model differs entirely in that the approach they use is based on the physics of higher-dimensional quantum field theory and not General Relativity Theory. Finally, Puthoff [67] showed that Alcubierre's warp drive is a particular case of a broad, general approach that is called "metric engineering," the details of which provide support for the concept that reduced-time interstellar travel is not fundamentally constrained by physical principles. The magnitude of energy condition violations and the amount of negative energy required to build an Alcubierre-type warp drive will be addressed in the next section.

IV. Make or Break Issues

A. Negative Energy Requirements and Energy Condition Violations

We know how to make small quantities of negative energy in the lab, but we do not know if it is possible to make large quantities of negative energy. It was pointed out in Section II that one, some, or all of the energy conditions must be violated in order to build a FTL spacetime. And we also cautioned that this was not a showstopper because the energy conditions have all been violated by nature or by lab experiment prior to their formulation. However, the reader should be forewarned that there are a number of published claims that the energy condition violations can be avoided. M. Visser (unpublished lecture, 1998) points out that these claims are just semantic games whereby investigators universally invoke the following scenario: Divide the total stress-energy into weird matter plus normal matter, push all the energy condition violations into the weird matter so that the normal matter does not violate the energy conditions. Given that the energy conditions are not absolute, such rearranging approaches are not necessary.

1. Traversable Wormhole Requirements

Traversable wormhole throats violate the NEC (or ANEC). So we ask how big a violation is required. The answer is that we only need to calculate the amount of negative energy that will be needed to generate and hold open a wormhole throat. A simple formula for short-throat wormholes using the thin shell formalism gives this quantity in terms of the equivalent mass (note: the energy density derived from the general relativistic field equation is too complex to use for this mass comparison) [3]:

$$M_{wh} = -\frac{\ell_{throat} c^2}{G}$$

$$= -(1.35 \times 10^{27} \text{ kg}) \frac{\ell_{throat}}{1 \text{ m}} \quad (8)$$

$$= -(0.71 \, M_J) \frac{\ell_{throat}}{1 \text{ m}}$$

where M_{wh} is the equivalent mass required to build the wormhole, ℓ_{throat} is a suitable measure of the linear dimension (width or diameter) of the throat, and M_J is the mass of the planet Jupiter. One can also obtain the required energy, $\mathcal{E}_{wh}$, by multiplying both sides of Eq. (8) by c^2. Equation (8) shows that a mass of $-0.71 \, M_J$ will be required to build a wormhole 1 m in size. As the wormhole size increases, the mass requirement grows negative-large. Table 2 presents a tabulation of the required negative (equivalent) mass as a function of sample wormhole throat sizes. After being alarmed by the magnitude of the results, one should note that M_{wh} is not the total mass of the wormhole as seen by remote observers. The nonlinearity of the general relativistic field equation dictates that the total mass is zero (actually, the total net mass being positive, negative, or zero in the Newtonian approximation depending on the details of the negative energy configuration constituting the wormhole system). Finally, Visser et al. [68] demonstrated the existence of spacetime geometries containing traversable wormholes that are supported by arbitrarily small quantities of negative energy, and this was proved to be a general result. We will expand on this further in Sec. IV.B.

2. Warp Drive Requirements

Warp drive spacetimes violate the WEC and NEC (also the DEC and SEC). The amount of negative energy required to create a warp drive gives a

Table 2 Negative equivalent mass required for traversable wormhole[a]

ℓ_{throat} (m)	M_{wh}
1000	$-709.9 \, M_J$
100	$-71 \, M_J$
10	$-7.1 \, M_J$
1	$-0.71 \, M_J$
0.1	$-22.6 \, M_\oplus$
0.01	$-2.3 \, M_\oplus$

[a]$M_J = 1.90 \times 10^{27}$ kg, $M_\oplus = 5.98 \times 10^{24}$ kg.

measure of the WEC/NEC violations. Because the energy density for the Alcubierre [5]—and Natário [69]—warp drive that is derived from the general relativistic field equation is too complex to use for these comparative purposes, we instead use a more simple formula to express the net energy required, $\mathcal{E}_{\text{warp}}$, to build a warp bubble around a spaceship [70]:

$$\mathcal{E}_{\text{warp}} = -\frac{v_{\text{warp}}^2 c^4 R^2 \sigma}{G} \qquad (9)$$
$$= -(1.21 \times 10^{44}) v_{\text{warp}}^2 R^2 \sigma$$

where v_{warp} [v_{warp}: $(0, \infty)$] is the dimensionless speed of the warp bubble, R (> 0) is the radius of the warp bubble, and σ (> 0) is proportional to the inverse of the warp bubble wall thickness Δ (i.e., $\sigma \sim 1/\Delta$). One can also obtain the equivalent mass, M_{warp}, by dividing both sides of Eq. (9) by c^2. Equation (9) characterizes the amount of negative energy that one needs to localize in the walls of the warp bubble. Table 3 presents a tabulation of the required negative energy as a function of the "warp factor," v_{warp}. One can compare the values of $\mathcal{E}_{\text{warp}}$ in the table with the positive rest-energy contained in the Sun (1.79×10^{47} J). The consequence of Eq. (9) and Table 3 is that if one wants to travel at hyperlight speeds, then the warp bubble energy requirement will be an enormous negative number. And this remains true even if one engineers an arbitrarily low sublight-speed warp bubble. Engineering a warp drive bubble is quite daunting given these results.

One further complication arises from the serious flaw that was discovered in the Alcubierre and Natário warp drives by Lobo and Visser [70]. They point out that in the original version of the warp drive, the point at the center of the warp bubble moves on a geodesic and is massless. The spaceship is always treated as a test particle in Alcubierre's warp drive and in all other incarnations of it. Lobo and Visser corrected this flaw by constructing a more realistic model of the warp drive spacetime by applying linearized gravity to the weak-field warp drive case and testing the energy conditions to first and second orders of v_{warp}. The fundamental basis of their model is that it specifically includes a finite mass spaceship that interacts with the warp bubble. Their results verified that

Table 3 Negative energy required for warp bubble[a]

v_{warp}	$\mathcal{E}_{\text{warp}}$ (J)
10^{-5} (= 3 km/s)	-3.03×10^{40}
10^{-4} (= 30 km/s)	-3.03×10^{42}
0.01 (= 3000 km/s)	-3.03×10^{46}
0.5 (= 150,000 km/s)	-7.59×10^{49}
1 (= light speed)	-3.03×10^{50}
2 (= 600,000 km/s)	-1.21×10^{51}
10 (= 3.0×10^6 km/s)	-3.03×10^{52}
100 (= 3.0×10^7 km/s)	-3.03×10^{54}

[a]Assume: $R = 50$ m, $\sigma = 10^3$ m^{-1}.

all warp drive spacetimes violate the energy conditions and will continue to do so for arbitrarily low warp bubble speed. They also found that the energy condition violations in this class of spacetimes is generic to the form of the geometry under consideration and is not a side effect of the superluminal properties. Based on these facts plus Eq. (9) and Table 3, it appears that for all conceivable laboratory experiments in which negative energy can be created in very small amounts, the warp bubble speed will be absurdly low. It therefore appears unlikely that warp drives will ever prove to be technologically practical unless new warp bubble geometries or ways to generate astronomical amounts of negative energy are found.

B. Quantum Inequalities and Spatial Distributions of Negative Energy

The quantum inequalities (QI) conjecture is an extension of the Heisenberg uncertainty principle to curved spacetimes. Much research has been conducted around this one topic alone. The literature is too numerous to cite here but the reader should consult Refs. 6 and 41 for detailed information. The QI conjecture relates (via model-dependent time integrals of the energy density along geodesics) the energy density of a free quantum field and the time during which this energy density is observed. This conjecture was devised as an attempt to quantify the amount of negative energy or energy condition violations required to build a FTL spacetime. Investigators have invoked the QI to rule out many of the macroscopic wormhole and warp drive spacetimes. When generating negative energy the QI postulate that: 1) the longer the pulse of negative energy lasts, the weaker it must be; 2) a pulse of positive energy must follow and the magnitude of the positive pulse must exceed that of the initial negative pulse; and 3) the longer the time interval between the two pulses, the larger the positive pulse must be. This actually sounds quite reasonable on energy conservation grounds until one discovers that the Casimir effect and its non-Maxwellian quantum field analogs violate all three conditions. There are also a number of squeezed vacuum sources and Dirac field states that manifestly violate all three conditions. Cosmological inflation, cosmological particle production, classical scalar fields, the conformal anomaly, and gravitational vacuum polarization are among the many other examples that also violate the QI. Visser [71] also points out that observational data indicate that large amounts of "exotic matter" are required to exist in the universe in order to account for the observed cosmological evolution parameters. The QI have also not been verified by laboratory experiments. The assumptions used to derive the QI and the efficacy of their derivation for various cases has been called into question by numerous investigators. Krasnikov [72] constructed an explicit counterexample for generalized FTL spacetimes showing that the relevant QI breaks down even in the simplest FTL cases. And he also addressed Fewster's [73] technical arguments on this issue. It is important to point out that the QIs have been mainly proven for free massless scalar fields in flat two-dimensional Minkowski spacetime, so there remains the unanswered questions of extending the QI into a four-dimensional curved spacetime model (with or without boundaries) and how much negative energy density can arise for interacting fields.

It turns out that Visser and coworkers [60,68,70,74] developed a superior way to properly quantify the amount of negative energy or energy condition violations required to build a FTL spacetime. They propose a quantifier in terms of a spatial volume integral, which amounts to calculating the following definite integrals [68,70,74]:

$$\int \rho_E dV \leq 0; \quad \int (\rho_E + p_i) dV \leq 0 \tag{10}$$

with an appropriate choice of the integration measure $dV [= 4\pi r^2 dr$ or $g^{1/2} dr\, d\theta\, d\varphi$, where $g \equiv \det(g_{\mu\nu})$ is the matrix determinant of $g_{\mu\nu}$]. The amount of energy condition violation is defined as the extent to which Eq. (10) can become negative. The value of Eq. (10) provides information about the total amount of energy condition violating matter that must exist for any given FTL spacetime under study. It was further shown that Eq. (10) can be adjusted to become vanishingly small by appropriate choice of parameters; therefore, examples can be constructed whereby the energy condition violation can be made arbitrarily small. But the violation cannot be made to vanish entirely.

Equation (9) is also a consequence of Eq. (10). Coupling of the finite spaceship mass with the warp bubble leads to the quite reasonable condition that the net total energy stored in the warp bubble be less than the total rest-energy of the spaceship itself, which places a strong constraint upon the dimensionless speed of the warp bubble [70]:

$$v_{\text{warp}} \leq \left[\frac{G}{c^2} \left(\frac{M_{\text{ship}}}{R_{\text{ship}}} \frac{R_{\text{ship}} \Delta}{R^2} \right) \right]^{\frac{1}{2}}$$

$$\leq \left[(7.41 \times 10^{-28}) \left(\frac{M_{\text{ship}} \Delta}{R^2} \right) \right]^{\frac{1}{2}} \tag{11}$$

where M_{ship} and R_{ship} are the mass and size of the spaceship, respectively, and R is the radius of the warp bubble. [One can multiply both sides of Eq. (11) by c to convert to the MKS unit of speed.] Equation (11) indicates that for any reasonable values of the engineering parameters inside the brackets, v_{warp} will be absurdly low. This result is due to the intrinsic nonlinearity of the general relativistic field equation. To illustrate this point, we insert the example starship parameters from Table 3 ($R = 50$ m, $\Delta \sim 1/\sigma = 10^{-3}$ m) into Eq. (11) and assume $M_{\text{ship}} = 10^6$ kg to find that $v_{\text{warp}} \leq 1.72 \times 10^{-14}$ (or 5.16×10^{-6} m/s). Garden snails can crawl faster than this. And if R and M_{ship} are kept constant, then $\Delta = 3.37 \times 10^{24}$ m (or 3.57×10^8 light-years) in order for $v_{\text{warp}} \leq 1$, which is an unrealistic requirement on the warp bubble design.

Equation (10) also gives the result that traversable wormholes require arbitrarily small amounts of negative energy to build [whereby Eq. (8) serves only as a gross upper limit] such that within a wormhole spacetime we must have that [68]:

$$\rho_E = 0; \quad \int_C p_r dV \to 0 \qquad (12)$$

where p_r is the outward radial pressure required to hold the wormhole throat open. We should point out that the Gauss–Bonnet theorem (discussed in Sec. III.A) predicted this result beforehand. Equation (12) is a result that is also due to the intrinsic nonlinearity of the general relativistic field equation. This nonlinearity also impacts the coupling of a finite spaceship mass with each side of a wormhole's throat (or the mouth on each side of the throat) leading to a specialized mass conservation law for the combined system of spacecraft and wormhole: When finite mass spaceships traverse a wormhole they alter the equivalent mass of the wormhole mouths they pass through [3]. The entrance mouth absorbing the spacecraft gains equivalent mass while the exit mouth emitting it loses equivalent mass.** (This mass coupling and conservation law takes into account the possibility that spaceships traversing the wormhole may lose or gain some momentum and kinetic energy in the process, and it is assumed that the two mouths are sufficiently far apart that their mutual gravitational interaction is negligible.) This unusual result suggests, but does not prove, the possibility of a fundamental limit on the total mass that can traverse a wormhole. The coupled mass conservation law shows that for a sufficiently large net transfer of mass the final equivalent mass of the exit mouth becomes negative. This is actually a beneficial result because ANEC violations are required just to hold the wormhole throat open in the first place. If it appears that a runaway reaction might occur, then it would be prudent for wormhole engineers to simply "turn off" the wormhole for a brief moment and then "turn it back on" (i.e., "reset" the wormhole) to restart space transportation operations.

It is on the basis of the foregoing discussion that we can say with confidence that traversable wormholes appear to be the most viable form of FTL transport. However, we still do not know how to construct a traversable wormhole because General Relativity Theory only provides a recipe for the essential geometric and material ingredients required to open and maintain one, but not the required assembly instructions. Will we need to pull a traversable wormhole out of the quantum spacetime foam and enlarge it to macroscopic scale, or will we need to use extremely large spacetime curvatures to "punch a hole" through space? Or are there construction techniques yet to be identified? The author is convinced that the answer can only be found through empirical studies designed to decide whether the present general relativistic recipe is enough to work with or an additional construction mechanism is required.

On physical grounds Eq. (10) appears to be the correct negative energy/ energy condition violation quantifier. However, further work is needed to establish whether Eq. (10) is the correct quantifier to use overall and whether all (averaged) energy condition theorems can be extended to include it.

**Similar coupling and conservation law holds for the case of electrically charged matter that traverse a (charged or uncharged) wormhole.

On another note, Borde et al. [75] have recast the QI conjecture into a new program that seeks to study the allowed spatial distributions of negative energy density in quantum field theory. Their study models free massless scalar fields in flat two-dimensional Minkowski spacetime. Several explicit examples of spacetime averaged QI were studied to allow or rule out some particular model spatial distributions of negative energy. Their analysis showed that some geometric configurations of negative energy can either be ruled out or else constrained by the QI restrictions placed upon the allowable spatial distributions of negative energy. And there were found to be allowable negative energy distributions in which observers would never encounter the accompanying positive energy distribution so long as the QI restrictions and corresponding energy conditions are violated. The extent to which the results of Borde et al.'s analysis can be generalized to a four-dimensional curved spacetime (with or without boundaries) and interacting fields remains unsolved.

C. Observing Negative Energy in the Lab

Negative energy should be observable in lab experiments. The presence of naturally occurring negative energy regions in space is predicted to produce a unique signature corresponding to lensing, chromaticity, and intensity effects in micro- and macro-lensing events on galactic and extragalactic/cosmological scales [76–81]. It has been shown that these effects provide a specific signature that allows for discrimination between ordinary (positive energy) and negative energy lenses via the spectral analysis of astronomical lensing events. Theoretical modeling of negative energy lensing effects has led to intense astronomical searches for naturally occurring traversable wormholes in the universe. Computer model simulations and comparison of their results with recent satellite observations of gamma ray bursts (GRBs) have shown that putative negative energy (i.e., traversable wormhole) lensing events very closely resemble the main features of some GRBs. Other research has found that current observational data suggest that large amounts of naturally occurring "exotic matter" must have existed sometime between the epoch of galaxy formation and the present in order to properly quantitatively account for the "age-of-the-oldest-stars-in-the-galactic halo" problem and the cosmological evolution parameters [71].

When background light rays strike a negative energy lensing region, they are swept out of the central region thus creating an umbra region of zero intensity. At the edges of the umbra the rays accumulate and create a rainbow-like caustic with enhanced light intensity. The lensing of a negative energy region is not analogous to a diverging lens because in certain circumstances it can produce more light enhancement than does the lensing of an equivalent positive energy region. Real background sources in lensing events can have non-uniform brightness distributions on their surfaces and a dependency of their emission with the observing frequency. These complications can result in chromaticity effects, that is, in spectral changes induced by differential lensing during the event. The quantification of such effects is quite lengthy, somewhat model dependent, and with recent application only to astronomical lensing events. Suffice it to say that future work is necessary to scale down the predicted lensing parameters and characterize their effects for lab experiments in which the negative energy will not be of astronomical magnitude. Present ultrahigh-speed optics and optical cavities, lasers, photonic crystal (and related switching) technology, sensitive nano-sensor technology, and other techniques are very likely capable of detecting the very small magnitude lensing effects expected in lab experiments.

A nonoptical scheme for detecting negative energy in experiments was recently reported by Davies and Ottewill [82] who studied the response of switched particle detectors to static negative energy densities and negative energy fluxes. Their model is based on a free massless scalar field in flat four-dimensional Minkowski spacetime and utilized a simple generalization of the standard monopole detector, which is switched on and off to concentrate the measurements on periods of isolated negative energy density or negative energy flux. The detector model includes an explicit switching factor whereby five different switching functions (based on data windowing theory) are defined and evaluated. In order to isolate the effects of negative energy a comparison is made for the response of a detector switched on and off during a period of negative energy density (or negative energy flux) and that switched on and off in the vacuum. The results shed light on the response of matter detectors to pulses of negative energy of finite duration, and they showed that negative energy should have the effect of enhancing de-excitation (i.e., induce cooling) of the detector. This is the opposite of our experience with detectors that undergo excitation when encountering "normal" matter or energy, and isolated detectors placed in a vacuum naturally cool due to the usual thermodynamic reasons. But Davies and Ottewill point out that the enhanced cooling effect they discovered cannot be used to draw a thermodynamic conclusion because their modeling was restricted to first order in perturbation theory. It is not possible at first order to determine whether the enhanced cooling effects are due to the small violation of energy conservation expected in any process in which a general quantum state collapses to an energy eigenstate, or whether they predict a systematic reduction in the energy of the detector which has serious thermodynamic implications. However, Davies and Ottewill point out that their results are model dependent and they found for their standard monopole detector model that there is not always a simple relationship between the strength of the negative energy density/flux and the behavior of the detector. Further research will be necessary to resolve these issues.

D. Time Machines and Faster-than-Light Causality Paradox Issues

Simply put, any closed timelike curve (CTC), not necessarily a geodesic, in spacetime defines a time machine. Traversable wormholes imply time machines, and this discovery spawned numerous follow-on research on time machines [2,3,42,83–91]. Given a traversable wormhole, it appears to be very easy to build a time machine, although it will require a Herculean effort to induce a time shift between the two wormhole mouths (via special relativistic or general relativistic time dilation techniques). Comprehensive theoretical studies show that the creation of a time machine might be the generic fate of a traversable wormhole [3]. One should not become alarmed by this revelation because it is a well-known and widely accepted fact that classical (and semi-classical) General Relativity is infested with time machines whereby there are numerous spacetime geometry solutions that exhibit time travel and/or have the properties of time machines [3]. It should be pointed out that various alternative theories of gravity, and quantum gravity models, are also infested with time machines, traversable wormholes, and warp drives. Because we are only interested in exploiting (static or dynamic) traversable wormholes to facilitate FTL travel between planets or stars, we will not consider their time machine properties any further. It might be worth revisiting this topic when a traversable wormhole can be successfully created in the lab (or in space).

Not surprisingly, warp drives also engender the appearance of CTCs [31,92]. Technical fixes have been proposed to mitigate against the formation of CTCs in and around the warp bubble by performing some minor "surgery(ies)" to the warp drive geometry. However, because we previously showed that warp drives may not be technologically practical to implement, we will not consider this issue any further.

There is ongoing, vibrant debate among some physicists over the effects of FTL (i.e., superluminal) motion on causality. Because a comprehensive discussion of this topic is beyond the scope of this chapter, we make the following points in this regard:

1) Einstein's general relativistic field equation admits solutions that violate causality (e.g., the Kerr solution, Gödel's universe, etc.).

2) There are no grounds for microcausality violations in accordance with Ref. 93.

3) A new definition of chronology and/or causality is in order for FTL motion.

4) Investigators have found that CTCs/time machines do not affect Gauss's theorem, and thus do not affect the derivation of global conservation laws from differential ones [94]. The standard conservation laws remain globally valid while retaining a natural quasi-local interpretation for spacetimes possessing time machines (e.g., asymptotically flat wormhole spacetimes).

Thorne [95] states that it may turn out that causality is violated at the macroscopic scale. Even if causality is obeyed macroscopically, then quantum gravity might offer finite probability amplitudes for microscopic spacetime histories possessing time machines. Li and Gott [96] found a self-consistent vacuum for quantum fields in Misner space (a simple flat space with CTCs) for which the renormalized stress-energy tensor is regular (in fact zero) everywhere. This implies that CTCs could exist at least at the level of semi-classical quantum gravity theory.

As for items 1 and 3 in the list above, we point out that the fact that the spacetime metric defines a global chronology in General Relativity cannot be proven. That is because local physics does not determine the topology of spacetime, which could prevent the existence of a global chronology. In fact, Bruneton [97] emphasizes the fact that relativistic field theories (including Einstein's General Relativity) do not involve a pre-existing notion of time and chronology. The basis for this being that any relativistic field defines its own chronology on the manifold by means of the Lorentzian metric along which it propagates. Thus any Lorentzian metric induces a local chronology in the tangent space through the usual special relativistic notions of absolute Lorentz-invariant future and past. Because spacetime may be curved and has a nontrivial topology, the existence of a local chronology in the tangent space (of Special Relativity) does not imply that a chronology over the whole manifold exists. As a result of this consideration, Bruneton showed that if a signal made up of waves of some field propagates between two spacetime points, then these points can be time ordered with the help of the field's metric, and causality may be preserved even if the field propagates superluminally. It is standard practice for physicists to use the gravitational metric field to define a chronology on the manifold when establishing equations of motion for problems of interest. However, it is just one particular field that exists on the manifold, and so there are no clear reasons why it should be favored over the other fields in order to define a particular chronology. Because the global properties of spacetime break causality in General Relativity, this cannot be related to some intrinsic disease of superluminal motion.

Bruneton [97] further addresses item 3 by showing that causality requires the existence of a global mixed chronology in spacetime in which classical fields can propagate superluminally without threatening causality provided that we do not refer to any preferred chronology in spacetime. To prove this Bruneton defined a minimal formulation of causality in which no prior chronology is assumed. This was done by considering a collection $\mathcal{I}$ of relativistic fields $\Psi_\mathcal{I}$ (gravity included) that propagate along some metrics $g_\mathcal{I}$ (tensor indices suppressed), all of which together establishes a finite set of causal cones at each point of spacetime. The cones defined by $g_\mathcal{I}$ may be in any relative position with respect to one another and they may even tip each other over depending on the location in spacetime. The fields may also interact with each other, which will allow a physical signal to propagate in spacetime along different metrics. One, therefore, does not want to prefer one metric with respect to the others because there is no reason why some sets of local coordinates, or some rods and clocks, should be preferred because General Relativity is covariant (i.e., coordinates are meaningless). This framework then leads to the definition of an "extended" notion of future-directed and timelike curves that requires that at each point on the curve, the tangent vector is future-directed and timelike with respect to at least one of the $g_\mathcal{I}$. From this construction, all of the extended notions of future, past, initial conditions (for equations of motion), and other causal properties of spacetime then follow.

This new causal framework then leads to an immediate resolution of the so-called "grandfather paradox." This paradox traditionally assumes that causality (or determinism) holds although it cannot be defined because there is no available notion of time ordering along a CTC. Because of this important key fact, the claim that CTCs have bad causal behavior (in the form of grandfather-type paradoxes) is unjustified. Bruneton [97] emphasizes that the paradox only arises if one asks "the wrong question," and so temporal paradoxes do not exist because they are not related to causality but only to logic. He proved that there are CTCs that cannot exist because an event along their curve will influence itself (i.e., the event is not freely specifiable), while other types of CTCs can exist whereby an event along their curve is freely specifiable (up to possible gauge invariance constraints, if any) [97]. It is important to stress that the latter type of CTCs already occur in QED. When nontrivial QED vacua involving electromagnetic or gravitational backgrounds, or finite temperature effects, and boundaries (e.g., the Casimir effect) are taken into account, local Lorentz invariance is broken at the loop level through vacuum polarization. Some of these cases lead to superluminal light propagation [93,98–103].

E. Conservation of Momentum in Faster-than-Light Spacetimes

Are FTL spacetimes in the form of traversable wormholes or warp drives a representative form of propellantless propulsion? If so, then what about the conservation of momentum in such spacetimes? When studying FTL spacetimes, one finds a remarkable absence of any published technical papers that address the topic of momentum conservation. What is not remarkable is that the conservation laws are already incorporated into General Relativity Theory. Conservation laws are based on uniform time and isotropic space. Once there is a condition in the universe where time is not uniform and space is not isotropic, then those conservation laws break down. For example, conservation of energy and momentum in flat spacetime is rigorously given by the following divergence condition on a source of matter: $\partial_\nu T^{\mu\nu} = 0$, where $T^{\mu\nu}$ (in combination with the coefficient $8\pi G c^{-4}$) is the stress-energy tensor that encodes the density and flux of the source's energy and momentum (i.e., $8\pi G c^{-4} T^{\mu 0}$ is the density and $8\pi G c^{-4} T^{\mu i}$ is the flux), and ∂_ν denotes the usual flat-spacetime partial derivative with respect to the spacetime coordinates. However, in curved spacetime this condition becomes $\nabla_\nu T^{\mu\nu} = 0$, where ∇_ν is the covariant derivative or spacetime curvature gradient, and the conservation of energy and momentum is now only approximate such that the error is attributed to the gravitational field acting on the matter and itself having some energy and momentum.

It appears that many physicists assert that the conservation laws are already incorporated into General Relativity and so a discussion of momentum conservation in their published papers is unnecessary. We disagree because momentum conservation for any FTL spacetime model must be clearly demonstrated in order to gain confidence that a given model is correct, plus exploring such issues is another venue for further discoveries.

The problem that needs to be addressed can be established via the following scenarios:

1) *Warp drives:* Imagine there is a spacecraft that is surrounded by a warp bubble. Place the coupled spacecraft warp bubble system inside an imaginary box. The specific warp drive spacetime engineering and energy used to implement the warp bubble is not important. Because space for all warp drive geometries becomes asymptotically flat at large distances away from the coupled system inside the box, we can place observers in remote, asymptotically flat spacetime. These observers see the center-of-mass (COM) of the box translate at FTL speeds across space.

2) *Traversable wormholes:* Imagine there is a spacecraft that passes through the throat of a traversable wormhole. Place the coupled spacecraft traversable wormhole system inside an imaginary box. The specific traversable wormhole spacetime engineering and energy used to implement the traversable wormhole is not important. Because space for all traversable wormhole geometries becomes asymptotically flat at large distances away from the coupled system

inside the box, we can place observers in remote, asymptotically flat spacetime. These observers see the spacecraft disappear as it enters one side of the wormhole throat and reappear as it exits the other side such that the spacecraft appears to translate at FTL speeds across space. In this case, the COM of this coupled system moves as well.

3) *Given these two scenarios, the problem we are interested in is the following:* Begin with a spacecraft of mass M_{ship} in flat spacetime (using a flat spacetime coordinate system) at fixed location $\mathcal{A}$ before the warp bubble or traversable wormhole is switched on. Switch on the warp bubble or traversable wormhole. Assume that the spacecraft moves on a geodesic. Switch off the warp bubble or traversable wormhole. The spacecraft is now in flat spacetime at fixed location $\mathcal{B}$ in the original flat spacetime.

Note that both scenarios appear to be classifiable as a form of propellantless propulsion as seen within the reference frame of the spacecraft: in both cases the spacecraft does not require rocket engines (to exert thrust) in order to move FTL across space because it is the FTL spacetime effect that does all the work. But there is a slight difference for the case of a traversable wormhole. In that case, the spacecraft must be self-propelled (under the action of a rocket engine) in order to move through the throat. But it can move through the throat at any speed $\leq c$ because a traversable wormhole, which is a hyperspace shortcut, allows the spacecraft to undergo apparent FTL motion as seen by remote observers.

In both FTL scenarios remote observers in flat spacetime view the coupled spacecraft warp bubble/wormhole system inside the imaginary box as a "particle" that is embedded in flat spacetime. And this "particle" appears to them as spontaneously moving from one place to another. But such a "particle" cannot spontaneously move from one place to another without emitting a propellant along with a corresponding loss of mass from the coupled system. The particle's acceleration must be accompanied by a reciprocal momentum ("propellant"). In other words, propellant here means something the remote observers see as compensating for the momentum of the coupled spacecraft warp bubble/wormhole system in the observers' flat spacetime coordinate system.

We conjecture that some of the mass that must have been expended as propellant is some form of reaction mass, which we provisionally assume to be gravitational radiation. Now the passengers inside the spacecraft would be reluctant to call this gravitational radiation "propellant" because they do not feel themselves being accelerated because they move along a geodesic. By contrast, sufficiently remote observers can only follow the COM of the spontaneously accelerated ponderable matter and spacetime is flat elsewhere. The spacecraft's spontaneous acceleration is explained to them as being compensated for by the gravitational radiation emitted in the other direction. That is how they see momentum to be conserved.

This issue needs to be studied in detail with quantitative results stated. This is an excellent subject for future research.

V. Conclusions

FTL spacetimes hold the promise of allowing mankind to reach the far planets and stars in our galaxy within a human lifetime to fulfill "the compelling urge of man to explore and to discover," which is guided by "the thrust of curiosity that leads men to try to go where no one has gone before" [104]. In this assessment of FTL spacetimes, we delineated a substantial number of crucial technical details along with their accompanying "make or break" issues having either tentative or firm results. The tentative results suggest a path for future research that is needed to put certain crucial technical issues on a firmer footing.

We identified the two primary forms of FTL spacetimes found in General Relativity Theory that can be created in principle: traversable wormholes and warp drives. These specialized spacetimes require the introduction of negative energy densities or fluxes in order to implement their geometries and FTL effects. Our assessment concludes that we already make small amounts of negative energy in the lab, but we do not yet know if we can access larger amounts for extended periods of time over extended spatial distributions for the purpose of engineering a particular FTL spacetime. We found that there are proposals for observing negative energy in outer space and in the lab, but further work is needed to downscale astronomical techniques for use at the lab scale, and we need to firm up our understanding of how lab detectors will respond to negative energy in situ.

A detailed energy analysis showed that warp drives are not technologically practical to implement due to their requirement for extremely large amounts of negative energy (in order to achieve absurdly low "warp speeds"), while traversable wormholes appear to be the most practical to implement because of their very minimal negative energy requirement. The appearance of time machines via warp drives or traversable wormholes, which can induce grandfather-type causality paradoxes, might not be an issue because investigators have shown that causality can be preserved in relativistic field theories even when there is superluminal motion. A resolution to the causality paradoxes is found by using an extended definition of causality which accounts for the fact that all relativistic field theories (including Einstein's General Relativity) do not involve a pre-existing notion of time and chronology.

The issue of momentum conservation in FTL spacetimes has not been addressed in the published literature. Here we introduce the nature of the problem and propose a solution. We suggest that the coupled spacecraft warp drive/wormhole system must eject some form of reaction mass and subsequently lose mass in order to conserve momentum as seen by remote observers in asymptotically flat spacetime, whereas the passengers on board the spacecraft would not see the loss of mass because they do not feel themselves being accelerated as they move FTL across space. This proposed solution has a direct consequence on whether one can consider FTL spacetimes to be a true form of propellantless propulsion. This appears to depend upon the frame of reference one is situated.

Limitations on length prevented us from giving a detailed discussion and analysis of FTL spacetimes within the framework of alternative theories of gravity or quantum gravity theories. Such theories remain speculative at present due to a lack of experimental validation, but we encourage the interested investigator to pursue this line of research to find any comparisons or contrasts with the results obtained from Einstein's General Relativity Theory. We hope that this work will serve as a useful guide for the interested investigator who wishes to pursue this topic on their own.

Acknowledgments

The author would like to thank the Institute for Advanced Studies at Austin and H. E. Puthoff for supporting this work. Parts of this work previously originated under Air Force Research Laboratory (AFMC) contract F04611-99-C-0025. The author extends thanks and appreciation to F. B. Mead, Jr. (AFRL/PRSP, Edwards AFB, CA) and members of the Air Staff (Air Force HQ, Pentagon) for encouraging our exploration into this topic. He also thanks V. Teofilo (Lockheed Martin), Lt. Col. P. A. Garretson (Air Force HQ, Pentagon), M. G. Millis (NASA-Glenn), M. Ibison (Inst. for Advanced Studies at Austin), S. Little (EarthTech Int'l), R. M. Wald (Univ. of Chicago), T. W. Kephart (Vanderbilt Univ.), P. C. W. Davies (Arizona State Univ.), L. H. Ford (Tufts Univ.), and H. G. White (NASA-Johnson) for many useful discussions or their contributions to this chapter.

References

[1] Morris, M. S., and Thorne, K. S., "Wormholes in Spacetime and Their Use for Interstellar Travel: A Tool for Teaching General Relativity," *American Journal of Physics*, Vol. 56, 1988, pp. 395–412.

[2] Morris, M. S., Thorne, K. S., and Yurtsever, U., "Wormholes, Time Machines, and the Weak Energy Conditions," *Physical Review Letters*, Vol. 61, 1988, pp. 1446–1449.

[3] Visser, M., *Lorentzian Wormholes: From Einstein to Hawking*, AIP Press, New York, 1995.

[4] Davis, E. W., "Teleportation Physics Study," Air Force Research Laboratory, Final Report AFRL-PR-ED-TR-2003-0034, Air Force Materiel Command, Edwards AFB, CA, 2004, pp. 3–11.

[5] Alcubierre, M., "The Warp Drive: Hyper-fast Travel Within General Relativity," *Classical and Quantum Gravity*, Vol. 11, 1994, pp. L73–L77.

[6] Ford, L. H., and Roman, T. A., "Negative Energy, Wormholes and Warp Drive," *Scientific American*, Vol. 13, 2003, pp. 84–91.

[7] Hawking, S. W., and Ellis, G. F. R., *The Large-Scale Structure of Space-Time*, Cambridge Univ. Press, Cambridge, U.K., 1973, pp. 88–91, 95–96.

[8] Epstein, H., Glaser, V., and Jaffe, A., "Nonpositivity of the Energy Density in Quantized Field Theories," *Nuovo Cimento*, Vol. 36, 1965, pp. 1016–1022.

[9] Visser, M., "Wormholes, Baby Universes, and Causality," *Physical Review D*, Vol. 41, 1990, pp. 1116–1124.

[10] Visser, M., "Gravitational Vacuum Polarization. I. Energy Conditions in the Hartle–Hawking Vacuum," *Physical Review D*, Vol. 54, 1996, pp. 5103–5115.

[11] Visser, M., "Gravitational Vacuum Polarization. II. Energy Conditions in the Boulware Vacuum," *Physical Review D*, Vol. 54, 1996, pp. 5116–5122.

[12] Visser, M., "Gravitational Vacuum Polarization. III. Energy Conditions in the $(1+1)$-Dimensional Schwarzschild Spacetime," *Physical Review D*, Vol. 54, 1996, pp. 5123–5128.

[13] Visser, M., "Gravitational Vacuum Polarization. IV. Energy Conditions in the Unruh Vacuum," *Physical Review D*, Vol. 56, 1997, pp. 936–952.

[14] Barcelo, C., and Visser, M., "Twilight for the Energy Conditions?," *International Journal of Modern Physics D*, Vol. 11, 2002, pp. 1553–1560.

[15] Herrmann, F., "Energy Density and Stress: A New Approach to Teaching Electromagnetism," *American Journal of Physics*, Vol. 57, 1989, pp. 707–714.

[16] Drummond, P. D., and Ficek, Z. (eds.), *Quantum Squeezing*, Springer-Verlag, Berlin, 2004.
[17] Hochberg, D., and Kephart, T. W., "Lorentzian Wormholes from the Gravitationally Squeezed Vacuum," *Physics Letters B*, Vol. 268, 1991, pp. 377–383.
[18] DeWitt, B. S., "Quantum Field Theory in Curved Spacetime," *Physics Reports*, Vol. 19C, 1975, pp. 295–357.
[19] DeWitt, B. S., "Quantum Gravity: The New Synthesis," *General Relativity: An Einstein Centenary Survey*, Hawking, S. W., and Israel W. (eds.), Cambridge Univ. Press, Cambridge, MA, 1979, pp. 680–745.
[20] DeWitt, B. S., "The Casimir Effect in Field Theory," *Physics in the Making, Essays on Developments in 20th Century Physics, In Honour of H. B. G. Casimir*, Sarlemijn, A., and Sparnaay, J. (eds.), North-Holland Elsevier Science Publ., New York, 1989, pp. 247–272.
[21] Birrell, N. D., and Davies, P. C. W., *Quantum Fields in Curved Space*, Cambridge Univ. Press, Cambridge, U.K., 1984.
[22] Saunders, S., and Brown, H. R. (eds.), *The Philosophy of Vacuum*, Clarendon Press, Oxford, 1991.
[23] Milonni, P. W., *The Quantum Vacuum: An Introduction to Quantum Electrodynamics*, Academic Press, San Diego, CA, 1994.
[24] Milton, K. A., *The Casimir Effect: Physical Manifestations of Zero-Point Energy*, World Scientific, NJ, 2001.
[25] Vollick, D. N., "Negative Energy Density States for the Dirac Field in Flat Spacetime," *Physical Review D*, Vol. 57, 1998, pp. 3484–3488.
[26] Yu, H., and Shu, W., "Quantum States with Negative Energy Density in the Dirac Field and Quantum Inequalities," *Physics Letters B*, Vol. 570, 2003, pp. 123–128.
[27] Olum, K. D., "Superluminal Travel Requires Negative Energies," *Physical Review Letters*, Vol. 81, 1998, pp. 3567–3570.
[28] Gao, S., and Wald, R. M., "Theorems on Gravitational Time Delay and Related Issues," *Classical and Quantum Gravity*, Vol. 17, 2000, pp. 4999–5008.
[29] Ford, L. H., "Quantum Coherence Effects and the Second Law of Thermodynamics," *Proceedings of the Royal Society of London, Series A, Mathematical and Physical Sciences*, Vol. 364, 1978, pp. 227–236.
[30] Davies, P. C. W., "Can Moving Mirrors Violate the Second Law of Thermodynamics?," *Physics Letters B*, Vol. 11, 1982, p. 215.
[31] Everett, A. E., "Warp Drive and Causality," *Physical Review D*, Vol. 53, 1996, pp. 7365–7368.
[32] Perry, M. D., "Crossing the Petawatt Threshold," *Science & Technology Review*, Lawrence-Livermore National Laboratory, Livermore, CA, Dec. 1996, pp. 4–11.
[33] Perry, M. D., "The Amazing Power of the Petawatt," *Science & Technology Review*, Lawrence-Livermore National Laboratory, Livermore, CA, March 2000, pp. 4–12.
[34] Mourou, G. A., Barty, C. P. J., and Perry, M. D., "Ultrahigh-Intensity Lasers: Physics of the Extreme on a Tabletop," *Physics Today*, Vol. 51, 1998, pp. 22–28.
[35] Slusher, R. E., et al., "Observation of Squeezed States Generated by Four-Wave Mixing in an Optical Cavity," *Physical Review Letters*, Vol. 55, 1985, pp. 2409–2412.
[36] Slusher, R. E., and Yurke, B., "Squeezed Light," *Scientific American*, Vol. 254, 1986, pp. 50–56.
[37] Robinson, A. L., "Bell Labs Generates Squeezed Light," *Science*, Vol. 230, 1985, pp. 927–929.

[38] Robinson, A. L., "Now Four Laboratories Have Squeezed Light," *Science*, Vol. 233, 1986, pp. 280–281.
[39] Saleh, B. E. A., and Teich, M. C., *Fundamentals of Photonics*, Wiley, New York, 1991, pp. 414–416.
[40] Caves, C. M., "Quantum-Mechanical Noise in an Interferometer," *Physical Review D*, Vol. 23, 1981, pp. 1693–1708.
[41] Pfenning, M. J., "Quantum Inequality Restrictions on Negative Energy Densities in Curved Spacetimes," Ph.D. Dissertation, Dept. of Physics and Astronomy, Tufts Univ., Medford, MA, 1998.
[42] Davies, P. C. W., *How to Build a Time Machine*, Penguin Books, New York, 2001.
[43] Rabeau, J. R., Chin, Y. L., Prawer, S., Jelezko, F., Gaebel, T., and Wrachtrup, J., "Fabrication of Single Nickel-Nitrogen Defects in Diamond by Chemical Vapor Deposition," *Cornell Univ. Library arXiv.org e-Print Archive*. URL: http://arxiv.org/cond-mat/0411245.pdf [cited 11 April 2004].
[44] Rabeau, J. R., Huntington, S. T., Greentree, A. D., and Prawer, S., "Diamond Chemical Vapor Deposition on Optical Fibers for Fluorescence Waveguiding," *Cornell Univ. Library arXiv.org e-Print Archive*. URL: http://arxiv.org/cond-mat/0411249.pdf [cited 11 April 2004].
[45] Ries, J., Brezger, B., and Lvovsky, A. I., "Experimental Vacuum Squeezing in Rubidium Vapor via Self-rotation," *Physical Review A*, Vol. 68, 2003, 025801.
[46] Casimir, H. B. G., "On the Attraction Between Two Perfectly Conducting Plates," *Proceedings of the Koninklijke Nederlandse Akademie van Wetenschappen*, Vol. 51, 1948, pp. 793–796.
[47] Lamoreaux, S. K., "Demonstration of the Casimir Force in the 0.6 to 6 µm Range," *Physical Review Letters*, Vol. 78, 1997, pp. 5–8.
[48] Mohideen, U., "Precision Measurement of the Casimir Force from 0.1 to 0.9 µm," *Physical Review Letters*, Vol. 81, 1998, pp. 4549–4552.
[49] Chen, F., Mohideen, U., Klimchitskaya, G. L., and Mostepaneko, V. M., "Theory Confronts Experiment in the Casimir Force Measurements: Quantification of Errors and Precision," *Physical Review A*, Vol. 69, 2004, 022117.
[50] Brown, L. S., and Maclay, G. J., "Vacuum Stress Between Conducting Plates: An Image Solution," *Physical Review*, Vol. 184, 1969, pp. 1272–1279.
[51] Walker, W. R., "Negative Energy Fluxes and Moving Mirrors in Curved Space," *Classical and Quantum Gravity*, Vol. 2, 1985, pp. L37–L40.
[52] Moore, G. T., "Quantum Theory of the Electromagnetic Field in a Variable-Length One-Dimensional Cavity," *Journal of Mathematical Physics*, Vol. 11, 1970, pp. 2679–2691.
[53] Forward, R. L., "Alternate Propulsion Energy Sources," Air Force Rocket Propulsion Laboratory, Air Force Space Tech. Ctr. Space Div., Air Force Systems Command, Final Report AFRPL TR-83-067, Edwards AFB, CA, 1983, pp. A1–A14.
[54] Hochberg, D., and Visser, M., "Geometric Structure of the Generic Static Traversable Wormhole Throat," *Physical Review D*, Vol. 56, 1997, pp. 4745–4755.
[55] Ida, D., and Hayward, S. A., "How Much Negative Energy Does a Wormhole Need?," *Physics Letters A*, Vol. 260, 1999, pp. 175–181.
[56] Krasnikov, S. V., "Hyperfast Travel in General Relativity," *Physical Review D*, Vol. 57, 1998, pp. 4760–4766.
[57] Everett, A. E., and Roman, T. A., "Superluminal Subway: The Krasnikov Tube," *Physical Review D*, Vol. 56, 1997, pp. 2100–2108.

[58] Broeck, C. V. D., "A 'Warp Drive' with More Reasonable Total Energy Requirements," *Classical and Quantum Gravity*, Vol. 16, 1999, pp. 3973–3979.
[59] Natário, J., "Newtonian Limits of Warp Drive Spacetimes," *General Relativity and Gravitation*, Vol. 38, 2006, pp. 475–484.
[60] Lobo, F., and Crawford, P., "Weak Energy Condition Violation and Superluminal Travel," *Lecture Notes in Physics*, Vol. 617, Springer, Berlin, 2003, pp. 277–291.
[61] Pfenning, M. J., and Ford, L. H., "The Unphysical Nature of Warp Drive," *Classical and Quantum Gravity*, Vol. 14, 1997, pp. 1743–1751.
[62] Coule, D. H., "No Warp Drive," *Classical and Quantum Gravity*, Vol. 15, 1998, pp. 2523–2527.
[63] Hiscock, W. A., "Quantum Effects in the Alcubierre Warp-Drive Spacetime," *Classical and Quantum Gravity*, Vol. 14, 1997, pp. L183–L188.
[64] White, H. G., and Davis, E. W., "The Alcubierre Warp Drive in Higher Dimensional Spacetime," *Proceedings of the STAIF-2006: 3rd Symposium on New Frontiers and Future Concepts*, El-Genk, M. S. (ed.), AIP Conference Proceedings, Vol. 813, AIP Press, New York, 2006, pp. 1382–1389.
[65] Obousy, R., "Supersymmetry Breaking Casimir Warp Drive," *Cornell Univ. Library arXiv.org e-Print Archive*. URL: http://arxiv.org/gr-qc/0512152.pdf [cited 27 Dec. 2005].
[66] Obousy, R., and Cleaver, G., "Supersymmetry Breaking Casimir Warp Drive," *Proceedings of the STAIF-2007: 4th Symposium on New Frontiers and Future Concepts*, El-Genk, M. S. (ed.), AIP Conference Proceedings, Vol. 880, AIP Press, New York, 2007, pp. 1163–1169.
[67] Puthoff, H. E., "SETI, the Velocity-of-Light Limitation, and the Alcubierre Warp Drive: An Integrating Overview," *Physics Essays*, Vol. 9, 1996, pp. 156–158.
[68] Visser, M., Kar, S., and Dadhich, N., "Traversable Wormholes with Arbitrarily Small Energy Condition Violations," *Physical Review Letters*, Vol. 90, 2003, 201102.
[69] Natário, J., "Warp Drive with Zero Expansion," *Classical and Quantum Gravity*, Vol. 19, 2002, pp. 1157–1165.
[70] Lobo, F. S. N., and Visser, M., "Fundamental Limitations on 'Warp Drive' Spacetimes," *Classical and Quantum Gravity*, Vol. 21, 2004, pp. 5871–5892.
[71] Visser, M., "Energy Conditions in the Epoch of Galaxy Formation," *Science*, Vol. 276, 1997, pp. 88–90.
[72] Krasnikov, S., "Counter Example to a Quantum Inequality," *Cornell Univ. Library arXiv.org e-Print Archive*. URL: http://arxiv.org/gr-qc/0409007.pdf [cited 25 May 2005].
[73] Fewster, C. J., "Comments on 'Counter Example to the Quantum Inequality,'" *Cornell Univ. Library arXiv.org e-Print Archive*. URL: http://arxiv.org/gr-qc/0409043.pdf [cited 10 Sept. 2004].
[74] Kar, S., Dadhich, N., and Visser, M., "Quantifying Energy Condition Violations in Traversable Wormholes," *Pramana*, Vol. 63, 2004, pp. 859–864.
[75] Borde, A., Ford, L. H., and Roman, T. A., "Constraints on Spatial Distributions of Negative Energy," *Physical Review D*, Vol. 65, 2002, 084002.
[76] Cramer, J. G., Foward, R. L., Morris, M., Visser, M., Benford, G., and Landis, G.L., "Natural Wormholes as Gravitational Lenses," *Physical Review D*, Vol. 51, 1995, pp. 3117–3120.

[77] Torres, D. F., Anchordoqui, L. A., and Romero, G. E., "Wormholes, Gamma Ray Bursts and the Amount of Negative Mass in the Universe," *Modern Physics Letters A*, Vol. 13, 1998, pp. 1575–1581.

[78] Torres, D. F., Romero, G. E., and Anchordoqui, L. A., "Might Some Gamma Ray Bursts be an Observable Signature of Natural Wormholes?," *Physical Review D*, Vol. 58, 1998, 123001.

[79] Anchordoqui, L. A., Romero, G., Torres, D. F., and Andruchow, I., "In Search for Natural Wormholes," *Modern Physics Letters A*, Vol. 14, 1999, pp. 791–797.

[80] Safonova, M., Torres, D. F., and Romero, G. E., "Macrolensing Signatures of Large-Scale Violations of the Weak Energy Condition," *Modern Physics Letters A*, Vol. 16, 2001, pp. 153–162.

[81] Eiroa, E., Romero, G. E., and Torres, D. F., "Chromaticity Effects in Microlensing by Wormholes," *Modern Physics Letters A*, Vol. 16, 2001, pp. 973–983.

[82] Davies, P. C. W., and Ottewill, A. C., "Detection of Negative Energy: 4-Dimensional Examples," *Physical Review D*, Vol. 65, 2002, 104014.

[83] Nahin, P. J., *Time Machines: Time Travel in Physics, Metaphysics, and Science Fiction*, 2nd ed., AIP Press, New York, 1999.

[84] Davies, P. C. W., *The Physics of Time Asymmetry*, Univ. of California Press, Berkeley and Los Angeles, 1974.

[85] Davies, P. C. W., *About Time: Einstein's Unfinished Revolution*, Simon and Schuster, New York, 1995.

[86] Gott, J. R., *Time Travel in Einstein's Universe: The Physical Possibilities of Travel Through Time*, Houghton Mifflin Co., Boston, 2001.

[87] Thorne, K. S., *Black Holes and Time Warps: Einstein's Outrageous Legacy*, W. W. Norton & Co., New York, 1994.

[88] Pickover, C., *Time: A Traveler's Guide*, Oxford Univ. Press, Oxford, 1999.

[89] Novikov, I. D., *The River of Time*, Cambridge Univ. Press, Cambridge, U.K., 1998.

[90] Deutsch, D., *The Fabric of Reality*, Penguin Press, London, 1997.

[91] Al-Khalili, J., *Black Holes, Wormholes and Time Machines*, Inst. of Physics Publishing, Bristol, UK, 1999.

[92] González-Díaz, P. F., "Warp Drive Space-Time," *Physical Review D*, Vol. 62, 2000, 044005.

[93] Drummond, I. T., and Hathrell, S. J., "QED Vacuum Polarization in a Background Gravitational Field and Its Effect on the Velocity of Photons," *Physical Review D*, Vol. 22, 1980, pp. 343–355.

[94] Friedman, J., et al., "Cauchy Problem in Spacetimes with Closed Timelike Curves," *Physical Review D*, Vol. 42, 1990, pp. 1915–1930.

[95] Thorne, K. S., "Closed Timelike Curves," GRP-340, CalTech, Pasadena, CA, 1983.

[96] Li, L.-X., and Gott, J. R., "Self-Consistent Vacuum for Misner Space and the Chronology Protection Conjecture," *Physical Review Letters*, Vol. 80, 1998, pp. 2980–2983.

[97] Bruneton, J.-P., "Causality and Superluminal Behavior in Classical Field Theories: Applications to k-Essence Theories and Modified-Newtonian-Dynamics-like Theories of Gravity," *Physical Review D*, Vol. 75, 2007, 085013.

[98] Liberati, S., Sonego, S., and Visser, M., "Faster-Than-c Signals, Special Relativity, and Causality," *Annals of Physics*, Vol. 298, 2002, pp. 167–185.

[99] Scharnhorst, K., "On Propagation of Light in the Vacuum Between Plates," *Physics Letters B*, Vol. 236, 1990, pp. 354–359.

[100] Scharnhorst, K., "The Velocities of Light in Modified QED Vacua," *Annalen der Physik (Leipzig), 8*, Vol. 7, 1998, pp. 700–709.
[101] Latorre, J. I., Pascual, P., and Tarrach, R., "Speed of Light in Non-trivial Vacua," *Nuclear Physics B*, Vol. 437, 1995, pp. 60–82.
[102] Dittrich, W., and Gies, H., "Light Propagation in Nontrivial QED Vacua," *Physical Review D*, Vol. 58, 1998, 025004.
[103] Shore, G. M., "Causality and Superluminal Light," *Cornell Univ. Library arXiv.org e-Print Archive*. URL: http://arxiv.org/gr-qc/0302116.pdf [cited 27 Feb. 2003].
[104] Killian, J. R., Jr., et al., "Introduction to Outer Space: An Explanatory Statement Prepared by the President's Science Advisory Committee," The White House, U.S. Government Printing Office, Washington, DC, March 1958.

Chapter 16

Faster-than-Light Implications of Quantum Entanglement and Nonlocality

John G. Cramer*
University of Washington, Seattle, Washington

I. Introduction

WHEN two photons emerge from a single quantum event, the state of one may be subtly connected to that of the other. The classical view is that, once separated, such states are fixed according to mechanics and conservation relations that act at the point of their origin, so that modifying one state later will not affect the other. In quantum physics, however, as borne out by experiment [1,2], the outcome of a measurement of the state of one of the photons, even well after the point of joint creation, does affect the state of the other photon. This connection is referred to as *quantum entanglement*, a phrase first coined by Erwin Schrödinger [3]. The questions at issue in such experiments are: 1) what is the *causal* connection between states acting in such phenomena, and 2) can it possibly be used for sending state-to-state signals? The chapter takes a close look at quantum entanglement, quantum nonlocality, the experiments to explore them, and proposed experiments to test the causal and faster-than-light (FTL) communication issues evoked by such physics.

Quantum entanglement describes the condition of separated parts of the same quantum system in which each of the parts can only be described by referencing the state of other part. This is one of the most counterintuitive aspects of quantum mechanics, because classically one would expect system parts out of "local" contact to be completely independent. Thus, entanglement represents a kind of quantum "connectedness" in which measurements on one isolated part of an entangled quantum system have nonclassical consequences for the outcome of measurements performed on the other (possibly very distant) part of the same system. This quantum connectedness that enforces the measurement correlation and state-matching in entangled quantum systems is called *quantum nonlocality*.

Copyright © 2008 by John G. Cramer. Published by the American Institute of Aeronautics and Astronautics, Inc., with permission.
*Professor, Department of Physics.

Nonlocality was first highlighted by Albert Einstein and his coworkers Boris Podolsky and Nathan Rosen in their famous EPR paper [4]. They argued that the nonlocal connectedness of quantum systems requires a FTL connection that appears to be in conflict with Special Relativity. Despite this objection, quantum nonlocality has been demonstrated (see Sec. II) in many quantum systems [1,2]. In the physics community it is now generally acknowledged to be implicit in the quantum formalism as applied to entangled systems, although there are a few "holdouts" who would require an explicit demonstration of nonlocal signaling before admitting that nonlocality can be considered a real quantum phenomenon.

The question that will be investigated here is whether quantum nonlocality is the private domain of nature, or whether it can be used in experimental situations to send signals from one observer to another. As we will see, there is presently no compelling answer to this question. However, it is clear that if such nonlocal communication were possible, it would have far reaching implications. In particular, it would represent an enabling technology for superluminal and retrocausal signaling and communications.

II. Quantum Entanglement, Nonlocality, and EPR Experiments

In a quantum mechanical description of elementary entities like photons there is a duality between the description as a particle and as a wave. Such objects can be thought of as traveling through space as waves but delivering energy (and other conserved quantities) at detection as particles. By choosing the kinds of measurements made on such objects, one can force wave-like or particle-like behavior to be exhibited in the measurements results. Between the entangled parts of a quantum system (e.g., the emission of a pair of entangled photons), this wave-like or particle-like behavior in a measurement on one part of the system may force similar behavior in the other part. This will be considered further in Sec. V.

The quantum entanglement condition is usually a consequence of some conservation law acting within the system, so that the subsystems are connected by the conservation law. For example, if two photons are emitted back to back in a joint state that has zero angular momentum and positive parity, then whatever linear or circular polarization state one photon is measured to have, the other photon must have an identical polarization if measured in the same basis (linear or circular). This condition must exist to ensure that the net angular momentum of the two photon state is zero. In this situation, if the photons are measured for circular polarization they must both be in states of right circular polarization or of left circular polarization. Because linear polarization is a coherent superposition of circular polarization states, if measured in the vertical/horizontal linear polarization basis, they must be in the same vertical or horizontal polarization state; if measured in the 45° left/right linear polarization basis, they must be in the same 45° left or right polarization state.

Classically, such a polarization correlation condition could, in principle, exist in some *particular* polarization basis but not in all of the several polarization bases simultaneously. This is the underlying physics of the Bell inequalities [5], which deal with the falloff rate of the correlations as the polarization basis of one of the measurements is rotated in angle. The Bell inequalities demonstrate mathematically that the predictions of semiclassical, local, hidden-variable theories are inconsistent with those of standard quantum mechanics. Tests of such polarization correlations have been the basis for a number of Bell inequality tests (or so-called EPR experiments), in which the validity of the predictions of quantum mechanics and the inadequacies of semi-classical local hidden-variable theories have been demonstrated to high statistical precision [1,2].

It was later demonstrated [6,7] that the issues surrounding a violation of the Bell inequalities could be separated into violations of either *parameter independence* (i.e., the outcome probability of a measurement on one of a pair of entangled particles is independent of the choice of *parameters* of a measurement performed on the other member of the entangled pair) and violations of *outcome independence* (i.e., the outcome probability of a measurement on one of a pair of entangled particles is independent of the *outcome* of a measurement performed on the other member of the entangled pair). The observation of a violation of the Bell inequalities indicates a violation of either *parameter independence* or *outcome independence* (or both). Outcome independence is fairly evident in the quantum formalism, while parameter independence is more elusive and depends on specific assumptions. We will consider the implications of this dichotomy in the context of the "no-signal" theorems.

We note that there is some misinformation in the literature concerning the chronology of successful EPR polarization correlation experiments, and here we wish to set the record at least somewhat straighter. The experimental measurement that first demonstrated a polarization correlation related to EPR nonlocality was performed by C. S. Wu and I. Shanknov in 1949 [8], well before Bell's work and the subsequent interest in testing Bell's inequality. Wu and Shanknov showed that the linear polarizations of back-to-back entangled gamma rays from electron–positron annihilation (an $L = 0$ *negative* parity state) were anticorrelated, for example, if one photon was polarized vertically then the other was polarized horizontally. They did not, however, investigate the falloff of the correlation with angle, which is the basis of Bell inequality tests, nor did they depict their results as the action of quantum nonlocality.

It required two more decades for the publication of John Bell's pivotal work in 1966 [5]. In 1972 Freedman and Clauser [1] performed the first definitive Bell inequality test by measuring the polarization correlation of entangled photons from a positive parity $L = 0$ atomic cascade in calcium (Fig. 1). Their results were in agreement with the predictions of quantum mechanics and were inconsistent with local hidden-variable theories by 6.7 standard deviations. A decade later in 1982, EPR measurements of the Aspect group [2] eliminated several "loophole" scenarios that might constitute unlikely ways of preserving classical locality and again demonstrated agreement with quantum mechanics and inconsistency with local hidden-variable theories, this time by 46 standard deviations. A more recent example of an EPR experiment is the work of the Gisin group [9]. They used the fiber-optics cables owned by the Swiss Telephone System to demonstrate the nonlocal connection between EPR measurements made at locations in Geneva and Bern, Swiss cities with a line-of-sight separation of 156 km, a direct demonstration, if one was required, that quantum nonlocality can operate over quite large distances.

Do these EPR experiments constitute a demonstration of the existence of quantum nonlocality? There is more than one way of interpreting the implications of the experimental results [1,2]. One can find much discussion in the literature as

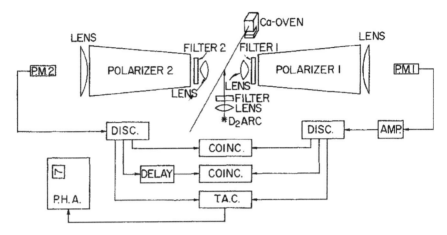

Fig. 1 Schematic of the 1972 Freedman–Clauser experiment [5].

to whether it is locality or "realism" (the objective observer–independent reality of external events) that has been refuted by these EPR measurements. We here adopt the view that reality should be a given, and so we regard these experiments as direct demonstrations of the intrinsic nonlocality of standard quantum mechanics.

We note that the several polarization bases used in these kinds of polarization EPR experiments make it straightforward to demonstrate the quantum nonlocal connections but also make it effectively impossible to use those connections for observer-to-observer signaling [10] because one would need to deduce from the arriving photons the polarization basis that was being used in the distant measurements. This is an aspect of the parameter independence mentioned earlier. Although each observer is free to choose the polarization basis (e.g., circular right/left, linear horizontal/vertical, linear 45° left/right) for the measurement, the observer is not free to force the photon into a particular state of that basis, as would be required for nonlocal communication. However, measuring *polarization correlations* in a system with angular momentum constraints is not the only way of demonstrating the nonlocal connection between the entangled separated parts of a quantum system. We will discuss EPR experiments that use *momentum entanglement* and will explore the question of whether such quantum systems might provide a better vehicle for observer-to-observer nonlocal communication, because by using momentum entanglement an observer is able to force the photon into particle-like or wave-like behavior.

III. Quantum No-Signal Theorems

As Einstein asserted with his well-known "spooky actions at a distance" comment, the enforcement of quantum correlations across space-like and negative time-like intervals by nonlocality is very counterintuitive. It appears to imply the twin possibilities of superluminal communication between observers and of reverse causation through back-in-time communication between observers. However, over the years, a number of authors [11] have presented "proofs" that such nonlocal observer-to-observer communication is impossible within the formalism of standard quantum mechanics. These theorems assert that, in separated measurements involving entangled quantum systems, the quantum correlations will be preserved but there will be no effect apparent to an observer in one sub-system if the character of the measurement is changed in the other sub-system. Thus, it is asserted, nonlocal signaling is impossible.

As mentioned above, EPR experiments can be viewed [5,6] as demonstrating violations of outcome independence or parameter independence or both. Outcome independence cannot be used for nonlocal signaling, while parameter independence could be used for such signaling. Thus, any test of nonlocal signaling is, in effect, a test of the parameter independence of quantum phenomena and the no-signal theorems are "proofs" of parameter independence.

Do these no-signal "proofs" really have the status of mathematical theorems? Perhaps not. Recently it has been pointed out [12] that at least some of these "proofs" ruling out nonlocal signaling are tautological, assuming that the measurement process and its associated Hamiltonian are local, thereby building the final conclusion of no signaling into their starting assumptions. Standard quantum mechanical Bose–Einstein symmetrization has been raised as a counter-example, shown to be inconsistent with the initial assumptions of some of these "proofs." Therefore, at least from some perspectives, the possibility of nonlocal communication in the context of standard quantum mechanics remains open and appropriate for experimental testing.

IV. Nonlocality vs Special Relativity?

If nonlocal communication is possible, would it be in conflict with Special Relativity, with its well-known prohibition against FTL signals? The answer to this question is *no*.

The prohibition of signals with superluminal speeds by Einstein's Theory of Special Relativity is related to the fact that the definite simultaneity of two separated spacetime points is not Lorentz invariant. Because some hypothetical superluminal signal could be used to establish a fixed simultaneity relation between two such points, this would imply a preferred inertial frame and would be inconsistent with Lorentz invariance and Special Relativity. In other words, it would be inconsistent with the even-handed treatment of all inertial reference frames in Special Relativity.

However, if a nonlocal signal could be transmitted through measurements at separated locations performed on two entangled photons, the signal would be "sent" at the time of the arrival of the photon in one location and "received" at the time of arrival of the other photon. By varying path lengths to the two locations, these events could be made to occur in any order and time separation in any reference frame. Therefore, nonlocal signals (even superluminal and retrocausal ones) could *not* be used to establish a fixed simultaneity relation between two separated spacetime points, because the sending and receiving of such signals do not have fixed time relations. The transmission and arrival instants of a nonlocal signal cannot be used for synchronization because the transmission and reception instants are path and delay-dependent variables.

To put it another way, the nonlocal connections of entangled photons lie along segmented, light-like world lines that transform properly under Lorentz transformations. Therefore, there is no conflict between nonlocal signaling and the Lorentz invariance of special relativity. On the other hand, the principle of causality (i.e., cause must precede effect in all reference frames) appears very likely to be violated (or at least violate-able) if nonlocal signaling is possible.

Is it possible that the universe *does* have some preferred reference frame, perhaps that laid down by the cosmic microwave background or implied by Mach's principle? Perhaps, but if such a preferred frame existed, its existence could not be established by nonlocal communication.

V. Momentum Domain Entanglement and EPR Experiments

Einstein's original objection [4] that quantum mechanics appeared to be nonlocal was made with arguments based on a *gedanken experiment* in the momentum domain. However, almost all of the modern ERP experiments testing the Bell inequality and demonstrating quantum nonlocality have been performed in the polarization (i.e., angular momentum) domain, usually with linearly polarized photons. Interestingly, it appears that if nonlocal quantum communication is possible at all, it may be more easily achieved in the momentum domain of Einstein's original focus.

The optical process of spontaneous parametric down-conversion [11] turns out to be a very useful way of generating photon pairs entangled in either the polarization or the momentum domains. In this process a photon from a "pump laser" interacts with a nonlinear crystal and is transformed into two photons with energies and vector momenta that add up to those of the original pump photon. Depending on the type of down-conversion process, there are well-defined polarization correlations between the entangled photons. The down-converted photons may also be easily prepared in momentum-entangled states, because within the nonlinear medium the vector momenta of the down-converted pair of photons must add to give that of the pump photon.

The first measurement using momentum-entangled down-conversion photons that might be related to nonlocal communication is the "ghost interference" experiment reported in 1995 by Shih group [13]. It is shown schematically in Fig. 2. They used degenerate collinear type-II down-conversion of 351 nm UV pump radiation from an argon-ion laser passed through a 3.0 mm long BBO (β-BaB$_2$O$_4$) crystal that had been cut with the optic axis at a phase-matching angle of 42.2° to the pump beam to produce a pair of collinear momentum-entangled 702-nm photons with opposite polarizations. The entangled photons emerge from the crystal very nearly parallel with the pump beam. The pump beam is then split off from the pair using refraction in a quartz prism (UV prism), and the entangled photons are separated with a polarization-selecting beam splitter (BS) that reflects the "extraordinary" vertically polarized photon (e), and transmits the "ordinary" horizontally polarized photon (o). Both photons are passed through 702 ± 10 nm wavelength selective filters ($f_{1,2}$) and then detected ($D_{1,2}$).

The experimenters demonstrated that passing the vertically polarized photon (e) through a double or a single slit system before detection at D_1 produced a

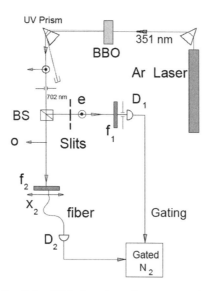

Fig. 2 Schematic of the 1995 ghost interference experiment of the Shih group.

"comb" interference distribution or a "bump" diffraction distribution, respectively in the position X_2 of the horizontally polarized photon (o) detected at D_2, when the pair of photons is examined in coincidence. In other words, the position distribution of the straight-through photon shows patterns characteristic of the single or double slit system through which its twin entangled photon passed. Figure 3 shows the observed position distributions for the two cases.

From the viewpoint of nonlocal communication, we note that modifying the slit system before D_1 through which the reflected photon passes, which can be thought of as the action of a "sending" observer, nonlocally causes an observable change in the X_2-position distribution of the undeflected photon, as detected by a "receiving" observer at D_2. This is a nonclassical effect that demonstrates the nonlocal connection between the entangled pair that might form the basis for transmission of a nonlocal signal between the two observers. However, the ghost interference experiment does *not*, in the form reported, demonstrate nonlocal communication because of its use of a classical communication link in imposing the coincidence requirement between the detected photons.

In their paper, the authors comment that with the two-slit system in place, in the absence of coincidences there is no observable two-slit interference pattern distributions at either D_1 or D_2. They attribute this lack of an interference "signal" to the horizontal variation in the creation position of the down-converted photons. The variation is enough to cause the e photons to arrive at the two slits with relative path lengths that may differ by more than a wavelength, thereby randomly shifting and washing out any interference pattern.

The authors point out that there is a simple way of thinking about momentum-entanglement measurements involving entangled photons. It can be shown from Snell's Law and conservation of momentum in the crystal that if one photon has a

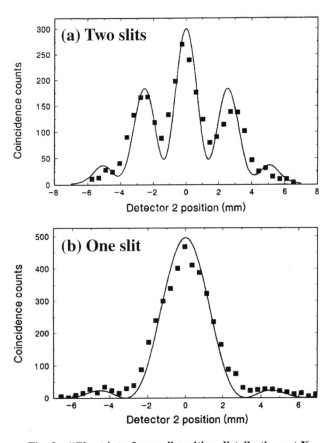

Fig. 3 "Ghost interference" position distributions at X_2.

small momentum that causes it to be slightly deflected to the right of the pump beam by an angle θ, then the twin entangled photon will be deflected to the left by the same angle θ, a situation reminiscent of reflection from a mirror. This allows the experiment to be "unfolded" by replacing the effective reflection by a straight-through path, as shown in Fig. 4. The point of the unfolding is that the entangled photons behave exactly as they would if the direction of the deflected photon was reversed, so that it originated at the detection point D_1, passed through one or two slits at C, D, and produced a one- or two-slit interference pattern at D_2.

Why is the coincidence needed? First, it should be clear from Fig. 2 that D_2 detects not only the entangled twins of the photons that pass through the slit openings, but also the entangled twins of the much larger number of photons that are stopped by the opaque parts of the slits. Therefore, without coincidences no interference pattern could be observed. Moreover, one can see from Fig. 4b that detector D_1 behind the slits receives light in a very localized region, and if it were moved vertically in the diagram, the interference pattern at D_2 would

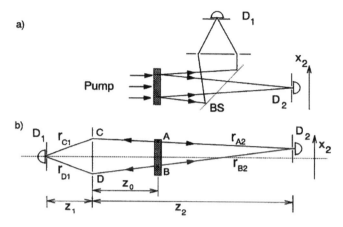

Fig. 4 "Unfolding" the ghost interference experiment.

be shifted, with maxima becoming minima, and vice versa. Without coincidences requiring a particular location for the detection at D_1, the D_2 distribution would have to average over all possible D_1 positions, washing out the two-slit interference pattern. Therefore, because of the geometry used, the ghost interference experiment required a coincidence to observe a two-slit interference pattern like the one shown in Fig. 3a.

Another momentum-entangled EPR experiment was performed in 1998 at the University of Innsbruck by Birgit Dopfer [14] and shown schematically in Fig. 5. In the Dopfer experiment, moving a detector in one arm nonlocally changes the observed interference pattern in the other arm. Dopfer used 351-nm UV pump radiation from an argon-ion laser with type I down-conversion in a nonlinear $LiIO_3$ crystal cut with the optic axis at 90° to the pump beam. This produced a pair of 702-nm momentum-entangled photons that emerged from the crystal at angles of 28.2° to the right and left of the pump axis, as shown in Fig. 5.

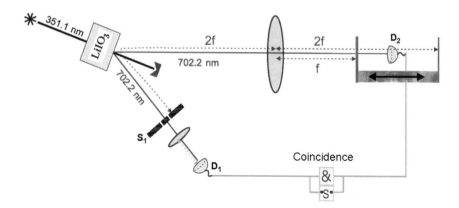

Fig. 5 Schematic of the 1998 Dopfer experiment [5].

The lower entangled photon passed through a pair of slits and into a detector, while the upper photon passed through a lens that could image the two slits to perform a "which-way" measurement if detector D_2 was placed two focal lengths behind the lens (2f). However, if detector D_2 was placed in a position one focal length behind the lens (f), the slits were not imaged and light on the reflected line passing through either slit could reach the detector at the same points. This produced a result similar to that of the ghost interference experiment. A structured two-slit interference pattern could be switched on and off by moving a detector in the other arm of the experiment between the f and 2f positions.

Again, from the viewpoint of nonlocal communication, we note that moving detector D_2, which can be thought of as the action of a "sending" observer, nonlocally causes an observable change in the position distribution of the second photon, as detected at "receiver" position D_1. However, the Dopfer experiment does not demonstrate nonlocal communication because, like the ghost interference experiment, it requires a classical communication link to impose the coincidence requirement between the detected photons because of the geometry of the experiment.

Examination of these two experiments raises a very interesting question: Can the coincidence requirement be removed? The answer to this question is not clear. In principle, the two entangled photons are connected by nonlocality whether or not they are detected in coincidence. The coincidence should therefore be removable. However, in both experiments the authors report that *no* two-slit interference distribution is observed when the coincidence requirement is relaxed. This may be explained by the action of coherence-entanglement complementarity, as discussed in the next section.

VI. Coherence-Entanglement Complementarity

As discussed, the finite extent of the source is expected to limit the possibility of observing a 2-slit interference pattern, which would be the "signal" if nonlocal communication was possible. Figure 6 shows schematically this "thick source" effect. The source volume on the left is the region of the nonlinear crystal that is illuminated by the UV pump-laser beam directed along the u axis. The source volume is a cylinder a few mm thick and a mm or so in radius with a center point C. The source cylinder is assumed to be tilted at an angle θ with respect to the horizontal z axis on which the slit system and detector plane are symmetrically centered. We note that $\theta = 0°$ in the ghost interference experiment and $\theta = 28.2°$ in the Dopfer experiment. A horizontal distance L_{xs} away from the source is a two-slit system, a pair of apertures a with center-to-center separation d. Light passing through the slit system travels a horizontal distance L_{sd} and is detected at detector plane at position x_1.

If the point of photon production is off the z-axis, there will be a path length difference between waves as they arrive at the two slits. In Fig. 6, waves from points A and B could have path length differences greater that half a wavelength and phase differences of greater than 180°. Roughly speaking, this shifts the interference pattern relative to waves created at central point C, so that maxima become minima and vice versa. The net effect of averaging over all points in the source volume therefore would be to wash out the two-slit interference

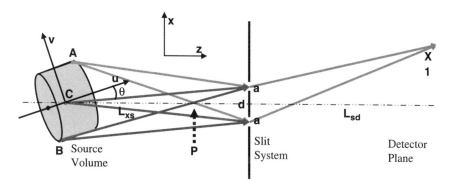

Fig. 6 Thick-source effect (not to scale). Waves arriving at the two slits from points A and B at the extrema of the source volume may have significant path length and phase differences, while waves from the central point C are in phase at the slits.

pattern. That two-slit interference pattern must be observed unambiguously, because it is the "signal" that would be used in any nonlocal communication.

The constancy of the relative phase at the two slits for photons arriving from various parts of the source is called "coherence." It should be clear that a point-like source has perfect coherence and a source with a large solid angle as viewed from the slits will have reduced coherence. The path length difference at the slits is inversely proportional to L_{xs} to a good approximation, so making L_{xs} large (i.e., placing the slits some distance from the source volume), can reduce the path length differences to a value that increases coherence and allows observation of a sharp two-slit interference pattern signal at detector D_1. Alternatively, placing a thin double-concave diverging lens at point P of Fig. 6 can have the same effect by causing the shorter path lengths to pass through a greater thickness of lens glass. Such a lens would also demagnify the source, producing the equivalent of a longer path length and smaller source solid angle.

However, increasing source coherence has another consequence. The momentum entanglement of photons from the source arises from momentum conservation. Restricting the solid angle of the source as viewed from the slits means that fewer photon pairs can be entangled and still satisfy the geometrical constraints of the experimental configuration. The Saleh group at Boston University has studied the complementarity between source coherence and two photon entanglement [15] and has shown that there is a complementary relation between these two characteristics. As the source-slit distance L_{xs} is increased, there is a smooth transition from one-slit to two-slit interference patterns and a smooth transition from a highly entangled source to a highly coherent source.

Nonlocal communication using momentum entanglement requires source coherence. Source coherence is needed in order to observe the "signal" of a two-slit interference pattern and two photon entanglement so that a measurement of one of the photons "connects" with the interference pattern produced by the other photon. Where there is coherence without entanglement or entanglement without coherence, nonlocal communication with momentum-entangled photons is not possible. The open question is whether there is a "sweet spot" in the experimental design that embraces both partial coherence and partial entanglement and that permits the transmission of nonlocal signals. This is an unresolved issue that requires further theoretical consideration and experimental testing.

VII. Nonlocal Communication vs Signaling

As we have pointed out above, the possibility of nonlocal communication is an unresolved issue. It is perhaps likely that the coherence/entanglement trade-off is nature's way of preventing nonlocal signaling, but that has not been demonstrated. In this section, we will assume than nonlocal signaling is possible and will examine its implications. As will be seen, they are so far-reaching that they could be taken as a syllogism that nature would not allow such things and therefore nonlocal signaling must be impossible.

Figure 7 shows a variation of the ghost interference experiment [13] in which the slit imaging procedure of the Dopfer experiment [14] is used to ensure that entangled photon pairs passing through slits reach both detectors, and those intercepted by the opaque regions of the slits reach neither detector. In particular, a lens of focal length f is placed in the path after the BBO crystal, and before the polarization splitter, so that both entangled photons pass through the lens. A pair of slits S_1 is placed at a path distance f beyond the lens in the path of the "o" photons, which are linearly polarized horizontally (HLP) and are transmitted by the splitter. As Dopfer has shown, because of momentum entanglement, an image of slit system S_1 will be formed by the "e" photons linearly polarized vertically (VLP) at a path length f beyond the lens on the deflected path at position S_2, where a pair of "cleanup" slits are located that pass only those photons whose entangled twins passed through S_1. We note that because of the optical geometry, this imaging occurs even for waves that pass through the image points and ultimately interfere.

At the image position of each slit at S_2 we place an optical fiber, as shown. The fibers conduct the light to an optical switch, at which the light is either sent directly to two avalanche photodiode detectors D (providing which-way information about which S_1 slit the photon entered), or alternatively is routed to an optical combiner C and then detected, so that waves passing through both slits can contribute constructively to the detection event. We note that this fiber switching system is the fiber-optics equivalent of a Mach–Zehnder interferometer [16] in which one can activate and deactivate the last half-silvered mirror by switching, so that which-way information can be switched on and off.

A quantum sensitive cooled CCD camera is substituted for detector D_1 of the ghost interference experiment [14] and is set to measure distributions like those shown in Fig. 3. In the arrangement in Fig. 7, switching the optical fiber routing can be considered to be an act of transmitting a binary 0 or 1 signal. If the switch is in the position leading to the outer detectors, then which-way information is available and the pattern detected by the camera should be a single-slit diffraction pattern labeled "1" in Fig. 7. If the switch is in the position leading to the combiner and middle detector, waves from both slits contribute to the detection, no

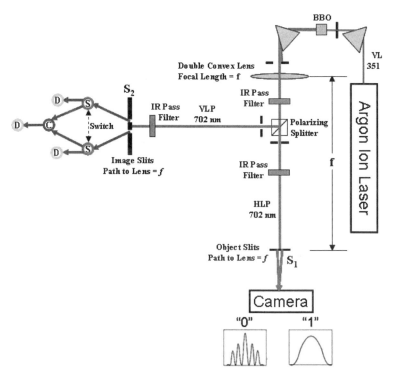

Fig. 7 Slit-imaging coincidence-free version of the ghost interference experiment to demonstrate nonlocal communication.

which-way information is available, and the pattern detected by the camera should be the two-slit interference pattern labeled "0" in Fig. 7.

If the pattern observed by the camera can indeed be changed by switching the optical fiber routing, then this would constitute a direct demonstration on nonlocal communication. Such an observation would falsify the no-signal theorems mentioned above, which require that in a non-coincidence scenario, no action on one entangled photon can produce a "signal-capable" observable result at the detection of the other entangled photon of the pair.

We should emphasize that demonstrating nonlocal communication with momentum-entangled photons, as described above, is not the same as actually sending a signal. It should be clear that no real signal can be communicated with a single photon pair. It is only when multiple photons are detected that the underlying distribution function can become apparent. We estimate that if the distribution functions to be distinguished are a "pure" two-slit interference pattern modulated by a diffraction envelope and a "pure" one-slit diffraction pattern, then about 10 photon detections would be required for a 3σ decision between these two possibilities.

However, as mentioned, it is likely that if nonlocal communication is possible at all, it would have to be accomplished in a situation where some compromise between entanglement and coherence has been achieved, and such a compromise would inevitably cause the two patterns to be distinguished to be more similar and more difficult to separate. Therefore, the 10 photon detections cited above must be taken as a rather optimistic lower limit, and it is likely that a significantly larger number of detections (perhaps ~ 100) would be required. The time required to send a single bit of information would then be the product of the photon detection rate in the two arms of the experiment times the number of photons that must be detected to receive the signal. In principle, such a transmission rate might be improved by pulsing the pump laser, so that "clusters" of entangled photons would be received with each such pulse.

VIII. Superluminal and Retrocausal Nonlocal Communication

As mentioned, we will assume, for the sake of discussion, that nonlocal signaling is possible and will consider its implications for the speed of transmission of signals. For definiteness, schemes for doing this are based on the slit-imaging, coincidence-free version of the ghost interference experiment already described and shown in Fig. 7. In that system, the instant at which a nonlocal signal is sent is the arrival of the VLP photon at the fiber-optics system on the left, and the instant at which the signal is received is the arrival of the HLP photon at the camera at the bottom of the diagram. Assuming the workability of this scheme, both the instants of sending and receiving can be delayed, in principle, by the introduction of delay paths (e.g., runs of fiberoptic cables) in the system. In particular, the send and receive instants, occurring at widely separated locations connected by runs of fiber-optics, could be tuned to occur simultaneously in any desired reference frame. This would constitute a direct demonstration of superluminal signaling.

Additionally, the "send" instant could be made to occur well *after* the "receive" instant in the system, constituting a direct demonstration of retrocausal signaling. This is shown in Fig. 8. Here the cleanup two-slit system S_2 becomes the entrance for two 10-km long runs of fiber-optics that are carefully matched to have identical exit phases at S_3, the end of the fiber runs where the light enters the optical switching arrangement described above. If the index of refraction of the fiber is 1.5, light transiting the 10-km path requires about 50 μs. In the presence of detection noise or the degradation of pattern visibility because of compromises between entanglement and coherence, considerably more photon detection events, say 100, might be required.

Assume that the fiber coils of Fig. 8 remain rolled up and are stored in a corner of the laboratory and that the source can be made strong enough so that the average rate at which the entangled photon pairs are detected is 10 MHz (which would correspond to the efficient detection of about 3.0 nW of 702 nm photons). Then the "sending" detector system and the "receiving" camera are in the same room and separated by a distance of 1 m or less. If the switch is set on the 0 or 1 position, the "message" that it is in that position begins to arrive at the camera 50 μs *before* the switch position is moved. If 100 photon counts constitute a signal, then the message could be received 40 μs *before* it was sent. This would be a direct demonstration of retrocausal signaling using nonlocal communication, and would constitute a direct violation of the principle of causality.

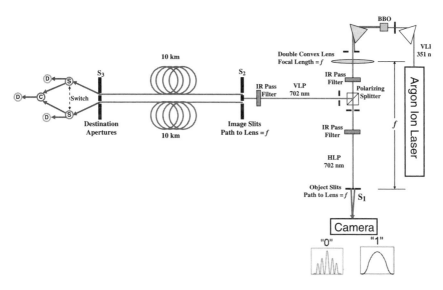

Fig. 8 Slit-imaging coincidence-free version of the ghost interference experiment to demonstrate superluminal and retrocausal signaling.

IX. Paradoxes and Nonlocal Communication

The setup described above, with its retrocausal communication link, raises some time-communication paradoxes. First, let us consider the issue of "bilking." Suppose that we construct one million linked systems of the type shown in Fig. 8. Then the transmitted message would be received 40 sec before it was sent. Now suppose that a tricky observer receives a message from himself 40 sec in the future, and then decides not to send it. This produces an inconsistent timelike loop, which has come to be known as a "bilking paradox." Could this happen? If not, what would prevent it?

There are discussions of such bilking paradoxes in the physics literature by Wheeler and Feynman [17] who were considering the retrocausal aspects of the advanced waves of absorber theory, and by Thorne and colleagues [18] who were considering the paradoxes that might arise from timelike wormholes. The general consensus of this work is that nature will forbid it and require a consistent set of conditions. Thorne and co-workers showed that "nearby" to any inconsistent paradoxical situation involving timelike wormholes there is always a self-consistent situation that does not involve a paradox. As Sherlock Holmes said, "When the impossible is eliminated, whatever remains, however improbable, must be the truth." These speculations assert that equipment failure producing a consistent sequence of events is more likely than producing an inconsistency between the send and receive events. The implications of this are that bilking itself is impossible, but very improbable events could be produced in avoiding it.

The other issue raised by retrocausal signaling might be called the "immaculate conception" paradox. Suppose that you are using the setup described above, and you receive from yourself in the future the manuscript of a best-selling novel with your name listed as the author. You sell it to a publisher and become rich and famous. And when the time subsequently comes for transmission, you duly send the manuscript back to yourself, thereby closing the timelike loop and producing a completely consistent set of events. But the question is, just who wrote the novel? Clearly, you did not; you merely passed it along to yourself. Yet highly structured information (the novel) has been created out of nothing. And in this case, nature should not object, because there was no bilking and you produced no inconsistent timelike loops.

It is not known how to resolve either of these paradoxes. All that can be said is the following:

1) If nonlocal signaling is impossible, then it needs no resolution, but better, more "air-tight" proofs of the impossibility of nonlocal signaling would be needed.

2) If nonlocal signaling is possible and can be used to form timelike loops, then paradoxes become important subjects for further experimental testing, study, and theoretical treatment.

X. Nonlinear Quantum Mechanics and Nonlocal Communication

So far, we have focused on the possibility that nonlocal communication may be possible within the framework of standard quantum mechanics. However, even if nonlocal communication is impossible in standard quantum mechanics, there could also be another path to nonlocal communication.

The no-signal theorems described in Section III are based on the formalism of standard quantum mechanics. Such "proofs" become invalid if quantum mechanics is allowed to be slightly "nonlinear," a technical term meaning that when quantum waves are superimposed they may generate a small cross-term not present in the standard formalism. Steven Weinberg, Nobel laureate for his theoretical work in unifying the electromagnetic and weak interactions, investigated a theory that introduces small nonlinear corrections to standard quantum mechanics [19]. The onset of nonlinear behavior is seen in other areas of physics (e.g., laser light in certain media) and, he suggested, might also be present but unnoticed in quantum mechanics itself. Weinberg's nonlinear quantum mechanics subtly alters certain properties of the standard theory, producing new physical effects that can be detected through precise measurements.

Two years after Weinberg's nonlinear quantum mechanics theory was published, Joseph Polchinski published a paper demonstrating that Weinberg's nonlinear corrections upset the balance in quantum mechanics that prevents superluminal communication using EPR experiments [20]. Through the new nonlinear effects, separated measurements on the same quantum system begin to "talk" to each other and FTL and/or backward-in-time signaling becomes possible. Polchinski describes such an arrangement as an "EPR telephone."

The work by both Weinberg and Polchinski had implications that are devastating for the Copenhagen representation of the wave function as "observer knowledge." Polchinski has shown that a tiny nonlinear modification transforms the "hidden" nonlocality of the standard quantum mechanics formalism into a manifest property that can be used for nonlocal observer-to-observer communication. This is completely inconsistent with the Copenhagen "knowledge" interpretation.

Finally there is also the possibility of a nonlinear form of quantum mechanics that reduces to linear quantum mechanics in the limiting case of weak gravity or no gravity. Using the wave picture, it is possible to formulate a Laplace–Beltrami wave equation for curved space. The Laplace–Beltrami operator on the left-hand side contains information about the spacetime geometry (the metric tensor) and operates on the wave function. On the right-hand side, is the same term found in the flat space Klein–Gordon wave equation. In flat space (no gravity), this "curved-space" wave equation reduces to the Klein–Gordon wave equation. In addition, some theoretical physicists have studied Dirac-like wave operators on curved space to handle spin.

There are many possibilities for this type of nonlinear curved space quantum mechanics that can be explored. The Laplace–Beltrami operator can be expanded for any type of space-time geometry. Schwartzschild, Yilmaz, Friedmann–Walker, or even Morris–Thorne traversable wormhole geometries can all be explored. Nonlinear differential equations will result. But do physically meaningful solutions exist? Most likely, they will not obey the principle of superposition. However, there could be some intriguing possibilities. Analogous to tunneling through a potential barrier, elementary particles might be able to tunnel through spacetime geometries at superluminal speed.

But in the everyday world of weak gravity and fairly flat space, is quantum mechanics indeed nonlinear? Atomic physics experiments have been used by a number of experimental groups to test Weinberg's nonlinear theory. So far, these tests have all been negative, indicating that any nonlinearities in the quantum formalism must be extremely small, if they exist at all. These negative results are not surprising, however, because the atomic transitions used involve only a few electron-volts of energy. If quantum mechanics does have nonlinear properties, they would be expected to depend on energy and to appear only at a very high energy scale and particularly at the highest energy densities. Nonlocal communication tests should be made, if possible, with the highest energy particle accelerators.

XI. Issues and Summary

A summary of key unresolved issues follows:

1) Can the intrinsic nonlocality of quantum mechanics be used for observer-to-observer communication?

2) If nonlocal communication is possible, can it be used to send messages faster than the speed of light (i.e., across spacelike intervals)?

3) If nonlocal communication is possible, can it be used to send messages backward in time (i.e., across negative timelike intervals)?

4) If nonlocal back-in-time communication is possible, how can the paradoxes that result from this capability be resolved?

5) Is quantum mechanics perfectly linear, or are there small nonlinearities, perhaps consequences of quantum gravity, that could be exploited for faster-than-light or backward-in-time communication?

XII. Conclusions

Ultimately, the question of whether nonlocal communication is possible is an experimental one. The issue should be resolvable by testing for nonlocal communication and observing what experimental limits appear. In particular, are the limits of coherence/entanglement complementarity so severe that signaling is precluded? Currently there is at least one experiment in progress that aims at producing a coincidence free version of the ghost interference experiment. We eagerly await the outcome of such tests.

Appendix: Glossary and Description of Key Concepts

Basis: In quantum mechanics, a choice of an observable quantity that may be complimentary to another variable, so that both cannot be measured at the same time. An example is the choice of measuring position, which prevents the simultaneous measurement of momentum. In EPR experiments, one must choose a polarization basis; for example, linear polarization that may be either vertical or horizontal. Because both circular polarization and 45° left/right polarization are linear superpositions of vertical/horizontal polarization, they may not be measured simultaneously. In quantum mechanics, the measurement causes the wave function to collapse to a particular basis value, excluding other possible values.

Bell's Theorem: A mathematical proof by John S. Bell [5] demonstrating that in a polarization-based EPR experiment, the falloff of correlations as the basis angle of a polarization measurement is changed is qualitatively different, as predicted by local hidden-variable theories and by standard quantum mechanics. In particular, local hidden-variable theories predict a linear falloff while quantum mechanics predicts a quadratic falloff. This difference in predictions is represented as an inequality in measurement intensity ratios that all local hidden variable theories must satisfy while quantum mechanics does not. Tests of these predictions have been found to agree with quantum mechanics and to falsify local hidden-variable theories.

Bilking Paradox: A type of back-in-time communication paradox in which an inconsistent causal loop is created. A well-known example is the grandmother paradox, a time travel scenario from science fiction in which a time traveler travels to the past and kills his grandmother before she had children. The question then arises, how he could have been born if his grandmother had no children? Several works in the physics literature [17,18] have concluded that such transtemporal bilking is impossible, that nature will not permit inconsistent timelike loops, and that it is more likely that some apparatus will fail than that a "bilk" of nature could be achieved.

Causality: The observation, which is regarded as a law of physics, that a cause must precede its effects as viewed in any and all reference frames. Sometimes referred to as "cause and effect" or "the law of cause and effect."

Correlations: The mathematical connection between two variables or two measured quantities. As an example, in an EPR measurement, the basis polarization of one photon is selected, the basis polarization of the twin entangled photon is varied, and the coincidence counting rate versus varied angle is measured to establish the correlation between the two polarizations.

Coherence: Describes whether two waves (e.g., those arriving at a pair of slits or at a detector) have a definite phase relation (in which case they are completely coherent) or a random phase relation (in which case they are completely incoherent) or something in between.

Coherence-Entanglement Complementarity: The theoretical expectation and experimental observation [15] that perfect coherence and perfect entanglement cannot be achieved for an entangled pair of photons at the same time.

Coherent Superposition: The formation of a quantum mechanical state (e.g., right circular polarization) by adding components of other states (e.g., left and right polarization) with a definite complex phase between the added states.

Collapse: A quantum mechanical wave function is said to collapse to a particular basis value when a measurement is made in that basis. For example, if a photon is emitted isotropically (with equal probability in all directions), its wave function is distributed uniformly over a sphere with a radius that grows at the speed of light until it is detected. On detection, the photon's wave function is localized at the detection point and disappears everywhere else.

Entangled: The separated parts of the same quantum system are said to be entangled when each of the parts can only be described by referencing the state of other part. This is one of the most counterintuitive aspects of quantum mechanics, because classically system parts out of "local" contact should be completely independent. Thus, entanglement represents a kind of quantum "connectedness" in which measurements on one isolated part of an entangled quantum system have nonclassical consequences for the outcome of measurements performed on the other (possibly very distant) part of the same system.

EPR Experiment: A class of experiments with entangled particles, usually photons, that demonstrate quantum nonlocality. A gedanken experiment of this kind was first suggested in the famous 1935 paper by Einstein, Podolsky, and Rosen [4], in which a set of criticisms of quantum mechanics were presented.

Hidden-Variable Theories: A set of alternatives to quantum mechanics intended to satisfy the objections of the EPR paper, in which the uncertainty principle does not apply and a quantum system can simultaneously have definite values of complementary variables like position and momentum, provided one of these values is somehow "hidden." Hidden-variable theories are usually also "local" (see *Locality*) to deal with Einstein's objection to the nonlocality of quantum mechanics.

Immaculate Conception Paradox: A type of back-in-time communication paradox in which a completely consistent causal loop produces information with no known origin. An example is the book paradox, in which an author receives a book in a message from the future. He publishes it, and when the time comes, he transmits the manuscript to himself in the past. The question then arises, who wrote the book? In this case, no inconsistent timelike loops are involved, the arguments against bilking do not apply in this case.

Locality: The assumption that the correlations between parts of a system can only be established while the subsystems are in contact (or speed-of-light communication) and that, once out of such contact, no changes in such correlations are possible.

Nonlocality: The situation, apparently present in quantum mechanics, that correlations between parts of a system can be established independent of the separation of the parts in time and space.

Retro-Causal: Situations in theory or in the real world in which the effect precedes the cause, in violation of the principle of causality.

References

[1] Freedman, S. J., and Clauser, J. F., "Experimental Test of Nonlocal Hidden-Variable Theories," *Physical Review Letters*, Vol. 28, 1972, pp. 938–942.

[2] Aspect, A., Dalibard, J., and Roger, G., "Experimental Realization of Einstein-Podolsky-Rosen-Bohm Gedankenexperiment: A New Violation of Bell's Inequalities," *Physical Review Letters*, Vol. 49, 1982, pp. 91–95; Aspect, A., Dalibard, J., and Roger, G., "Experimental Test of Bell's Inequalities Using Time-Varying Analyzers," *Physical Review Letters*, Vol. 49, 1982, p.1804.

[3] Schrödinger, E., "Discussions of Probability Relations between Separated Systems," *Proceedings of Cambridge Philosophical Society*, Vol. 31, 1935, pp. 555–563; Vol. 32, 1936, pp. 446–451.

[4] Einstein, A., Podolsky, B., and Rosen, N., "Can Quantum-Mechanical Description of Physical Reality Be Considered Complete?," *Physical Review*, Vol. 47, 1935, pp. 777–785.

[5] Bell, J. S., *Physics*, Vol. 1, 1964, p. 195; "On the Problem of Hidden Variables in Quantum Mechanics," *Review of Modern Physics*, Vol. 38, 1966, p. 447.

[6] Jarrett, J. P., "On the Physical Significance of the Locality Condition in the Bell Argument," *Noûs*, Vol. 18, 1984, p. 569.

[7] Shimony, A., "Events and Processes in the Quantum World," *Quantum Concepts in Space and Time*, Penrose, R., and Isham, C. J. (eds.), Clarendon Press, Oxford, 1986, pp. 182–203.

[8] Wu, C. S., and Shanknov, I., "The Angular Correlation of Scattered Annihilation Radiation," *Physical Review*, Vol. 77, 1950, p. 136.

[9] Tittel, W., Brendel, J., Zbinden, H., and Gisin, N., "Violation of Bell Inequalities by Photons More Than 10 km Apart," *Physical Review Letters*, Vol. 81, 1998, pp. 3563–3566.

[10] Pagels, H., *The Cosmic Code*, Simon & Schuster, New York, 1982.

[11] Eberhard, P. H., "Bell's Theorem Without Hidden Variables," *Nuovo Cimento*, Vol. B38, 1977, p. 75; "Bell's Theorem and the Different Concepts of Locality," *Nuovo Cimento*, Vol. B46, 1978, p. 392; Ghirardi, G. C., Rimini, A., and Weber, T., "A General Argument Against Superluminal Transmission Through the Quantum-Mechanical Measurement Process," *Lett. Nuovo Cimento*, Vol. 27, 1980, pp. 293–298; Yurtsever, U., and Hockney, G., "Signaling, Entanglement, and Quantum Evolution Beyond Cauchy Horizons," *Classical and Quantum Gravity*, Vol. 22, 2005, pp. 295–312.

[12] Peacock, K. A., and Hepburn, B., "Begging the Signaling Question: Quantum Signaling and the Dynamics of Multiparticle Systems," *Proceedings of the Meeting of the Society of Exact Philosophy*, 1999, quant-ph/9906036.

[13] Strekalov, D. V., Sergienko, A. V., Klyshko, D. N., and Shih, Y. H., "Observation of Two-Photon 'Ghost' Interference and Diffraction," *Physical Review Letters*, Vol. 74, No. 17, 1995, pp. 3600–3603.

[14] Dopfer, B., "Zwei Experimente zur Interferenz von Zwei-Photonen Zuständen, Ein Heisenbergmikroskop und Pendellösung," PhD Thesis, University of Innsbruck, 1998; Zeilinger, A., "Experiment and the Foundations of Quantum Physics," *Review Modern Physics*, Vol. 71, 1999, S288-S297.

[15] Abouraddy, A. F., Nasr, M. B., Saleh, B. E. A., Sergienko, A. V., and Teich, M. C., "Demonstration of the Complementarity of One- and Two-Photon Interference," *Physical Review A*, Vol. 63, 2001, 063803.

[16] Zehnder, L., "Ein neuer Interferenzrefraktor," *Z. Instrumentenkunde*, Vol. 11, 1891, p. 275; L. Mach, "Über einen Interferenzrefraktor," *Z. Instrumentenkunde*, Vol. 12, 1892, p. 89.

[17] Wheeler, J. A., and Feynman, R. P., "Classical Electrodynamics in Terms of Direct Interparticle Action," *Review of Modern Physics*, Vol. 21, 1949, pp. 425–433.

[18] Echeverria, F., Klinkhammer, G., and Thorne, K. S., "Billiard Balls in Wormhole Spacetimes with Closed Timelike Curves: Classical Theory," *Physical Review D*, Vol. 44, 1991, pp. 1077–1099.

[19] Weinberg, S., "Precision Tests of Quantum Mechanics," *Physical Review Letters*, Vol. 62, 1989, pp. 485–490.

[20] Polchinski, J., "Weinberg's Nonlinear Quantum Mechanics and the Einstein-Podolsky-Rosen Paradox," *Physical Review Letters*, Vol. 66, 1991, pp. 397–401.

Chapter 17

Comparative Space Power Baselines*

Gary L. Bennett[†]
Metaspace Enterprises, Emmett, Idaho

I. Introduction

SPACE power is an integral part of any space mission—without a source of electrical power nothing else works, including propulsion. Moreover, in a number of advanced space propulsion concepts (e.g., electric propulsion, fusion, etc.) power is essential for the advanced space propulsion concept to function. This chapter examines space power technology to convey its existing and projected performance limits that are based on established science. This information is provided both as a baseline for contemplating powering new propulsion science techniques with known power methods, as well as to define the limits that any new power methods would have to exceed to become competitive. Also, to help convey a sense of progress, much of this information is presented in its historical context. Because missions to the outer solar system and beyond will require power sources independent of the sun, the focus of this chapter is on nuclear power sources.

Figure 1 illustrates the basic subsystems of a typical Earth-orbiting spacecraft. Within the electric power system (EPS) of a typical spacecraft, one can further subdivide into the basic configuration shown in Fig. 2. In very simple terms, an EPS can be thought of as consisting of three basic elements: 1) power source; 2) power management and distribution (PMAD); and 3) energy storage

---

Copyright © 2008 by Gary L. Bennett. Published by the American Institute of Aeronautics and Astronautics, Inc., with permission.

Unless otherwise indicated, mission information in this chapter was taken from Web sites maintained by the mission organizations or by the National Space Science Data Center <http://nssdc.gsfc.nasa.gov/>.

*First presented as G. L. Bennett, "Space Nuclear Power: Opening the Final Frontier," AIAA Paper 2006-4191, 4th International Energy Conversion Engineering Conference, 26–29 June 2006, San Diego, California.

[†]Director.

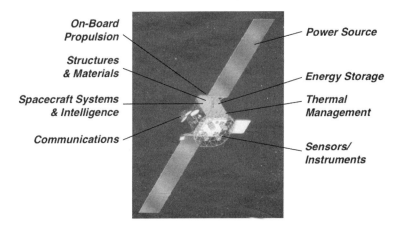

Fig. 1 Typical subsystems on an Earth-orbiting spacecraft. (Artist's illustration of the U.S. Navy ultra-high-frequency follow-on satellites, courtesy of Hughes Space and Communications.)

[1]. This chapter will focus on power sources and energy storage; however, the spacecraft designer must not forget the ever-important PMAD element.

In this chapter we shall review the basic space power concepts and list their current attributes. Historically, space power sources have been divided into nuclear and non-nuclear where the former has included radioisotope power sources and fission power sources, and the latter has generally included solar power (e.g., photovoltaic systems) and energy storage (e.g., batteries or fuel cells). Looking to advanced "classical" systems one can include fusion and antimatter. For the purposes of this chapter, both fusion and antimatter power sources will be included under the umbrella term "nuclear."

Figure 3, which is based on studies conducted at the Jet Propulsion Laboratory, shows the attributes of the major classical power sources in terms of the theoretical

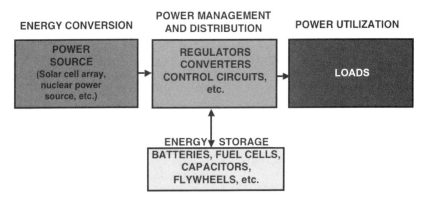

Fig. 2 Basic configuration of a spacecraft electric power system. (Adapted from Ref. 1.)

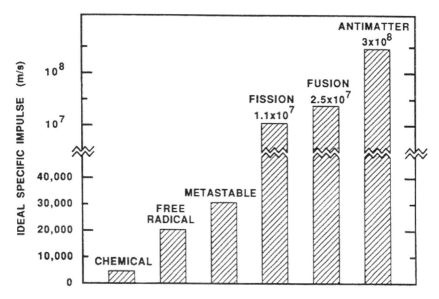

Fig. 3 Theoretical performances of advanced non-nuclear and nuclear power sources. (From Ref. 2.)

highest specific impulse they can produce. (Here specific impulse is expressed in its natural unit of velocity, i.e., meters per second.) *Free radicals* are neutral monatomic or polyatomic fragments produced by the dissociation of molecules. An electronically excited state of an atom or molecule is said to be *metastable* if its radiative lifetime is greater than a microsecond.

In terms of performance, whether energy release or specific impulse, the classical nuclear systems (fission, fusion, antimatter) have the clear advantage as shown in Fig. 4 where candidate near-term missions such as a Comet Nuclear Sample Return (CNSR) and Mars Rover Sample Return (MRSR) are compared with more challenging missions such as the proposed Thousand Astronomical Unit (TAU) mission beyond the solar system and a mission to Alpha Centauri. However, there are occasions when non-nuclear power sources can help in advancing human exploration of the solar system and beyond, so it is important to consider them as well. For example, a mission planner might want to use a solar electric propulsion (SEP) stage to send a spacecraft beyond Earth, dropping that SEP stage when the sunlight becomes too dim, and continuing with gravity assists and some type of nuclear power. Some have proposed using the Sun for a gravity assist to send a spacecraft beyond the solar system.[‡]

[‡]The use of another planetary body's gravity to accomplish a mission has been successfully employed on a range of NASA missions, e.g., Pioneers 10 and 11; Voyagers 1 and 2; Mariner 10; Galileo; Ulysses; Cassini; MESSENGER; and New Horizons. Because energy is conserved in such encounters the increase in velocity and/or change in direction for the spacecraft comes at the expense of a minuscule decrease in the velocity of the planetary body.

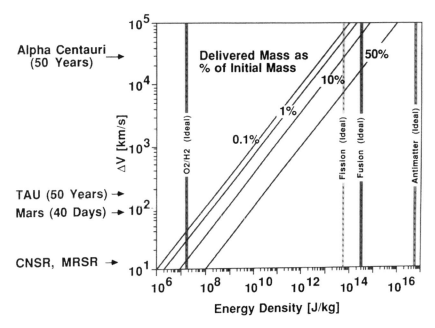

Fig. 4 Mission velocity changes (Δv) as a function of specific energy. (Courtesy of the Jet Propulsion Laboratory.)

The performance characteristics are also fundamentally clear when one considers that the energies of chemical reactions are typically measured in electron volts (eV) while nuclear reactions are orders of magnitude greater. For example, the alpha particle (^{4}He) decay of plutonium-238 (^{238}Pu or Pu-238) carries almost 5.5 MeV of energy. The fission of a uranium-235 nucleus (^{235}U or U-235) is about 200 MeV. The fusion reaction of a deuteron (^{2}H) and a tritium nucleus ("triton" or ^{3}H)—the so-called "d-t reaction"—yields 17.6 MeV. While this may not seem like much compared to fission reactions, the energy per nucleon is much greater (3.5 MeV/nucleon for d-t fusion versus about 0.8 MeV/nucleon for ^{235}U fission). In contrast, the annihilation of a nucleon via a matter–antimatter reaction would release over 938 MeV/nucleon. Radioactive decay rates are fixed; however, in principle, the reaction rates for fission, fusion, and antimatter can be increased as needed for the application.

Mission planners and space power designers have several top-level figures of merit that they use. For power sources, the interest is in having a high specific power (We/kg) and usually a high power density (We/m^3). In the electric propulsion world, there is often a focus on the inverse, having a low specific mass (kg/kWe). In the solar power community, there is also a desire to reduce the area of the solar arrays; hence, the occasional requirement on areal power density (We/m^2). For energy storage systems, the focus is usually on having a high specific energy (We-h/kg) and usually a high energy density (We-h/m^3). In addition, the mission planner needs to know the values at beginning-of-life (BOL), beginning-of-mission (BOM), and end-of-mission (EOM).

II. Non-nuclear Space Power Sources

In the space power community, non-nuclear power sources typically include solar and chemical. Solar can be divided into photovoltaic power sources and solar dynamic power sources. In the former, sunlight is directly converted into electrical power by means of a solar cell. Solar dynamic power sources utilize solar heat to drive a working fluid in a "dynamic" conversion system such as Brayton (gas-operated turbine-alternator-compressor), Rankine (two-phase fluid-operated turbine-generator), and Stirling (gas-operated linear alternator). Because dynamic conversion systems will be covered in Section III, this section will focus on photovoltaic power sources.

Chemically produced power can come from batteries and fuel cells. Such devices are usually referred to as energy storage systems (along with capacitors and flywheels). Energy storage systems tend to be power rich but energy limited (e.g., once the battery is discharged that is the end of the power until it can be recharged if rechargeable). Power sources tend to be energy rich but power limited (e.g., a radioisotope power source can operate for decades, which translates into a large energy release, but it is fixed in power).

Figure 3 shows the potential benefits of two other non-nuclear power sources: free radicals and metastable propellants. In principle, these concepts offer the potential of high specific impulses and high thrusts without the nuclear radiation associated with "classical" nuclear propulsion concepts. However, storage and control are issues yet to be satisfactorily addressed with some studies indicating that when these factors are included, the performance of free radicals and metastable propellants drops to that of nuclear thermal propulsion systems which do not have these same storage and control issues [2]. With the ongoing research on Bose–Einstein condensates (BECs) and related laser and magnetic field studies, there may yet be a breakthrough in the use of free radicals and metastable propellants.

A. Photovoltaic Space Power Sources

In the early days of the space program, solar arrays produced about 15 We/kg using solar cells that were about 10% efficient. (Unless otherwise specified solar array and solar cell performance values are presented for 1 AU from the sun with no intervening atmosphere, i.e., "air mass zero" or AM0.) The early space solar cells were usually based on a single-crystal silicon material. The early Earth-orbiting spacecraft received a few tens of watts from those early photovoltaic power sources. Gradually the power production increased into the hundreds of watts to today's tens of kilowatts. The solar arrays on the International Space Station (ISS) produce 110 kWe [3].

Since those early missions, the U.S. space community has focused on producing more power in lighter structures. One approach is to develop more efficient solar cells. Efficiencies have been raised from the early ~10% to over 27% in production by using several layers of different materials (e.g., GaAs, GaInP, Ge), which are typically referred to as dual-junction and triple-junction solar cells. Efficiencies in laboratory cells beyond 35% are being measured [3].

It is possible to increase the specific power of solar arrays without using advanced solar cells by employing low-mass structures. This approach was used on NASA's Advanced Photovoltaic Solar Array (APSA) program where specific powers in the range of 130 We/kg for 12-kWe applications were demonstrated with thin silicon solar cells for geosynchronous Earth orbit (GEO) conditions [4,5].

Beginning in the late 1980s, NASA had two long-range goals for space solar power: 1) to develop solar arrays that can provide 300 We/kg for deployable planar arrays; and 2) to develop solar arrays that can provide 300 We/m^2 at 100 We/kg in concentrator arrays (both at 1 AU from the sun) [5]. Because the dual-junction and triple-junction solar cells are heavier than thin silicon cells and thin-film cells, it may not be possible to have both attributes in the same array. Already there are reports of arrays with specific powers of 300 We/kg with stowed packing densities on the order of 50 kWe/m^3 [3].

Occasionally one will find studies in the literature of using extremely large solar arrays to capture starlight for "generation ships" or "world ships" moving between the stars. A quick calculation based on the inverse-square reduction in insolation with distance from a star will easily show how unrealistically large (with today's technologies) such arrays would have to be. (A solar array that would produce 1000 We at Jupiter would have to be over 25 times the size of a 1000-We array operating in Earth orbit. At Pluto the array would have to be over 1500 times as large!)

In choosing a solar array design, the mission planner and the spacecraft power engineer need to consider such factors as [3]:

1) The range of insolation (solar intensity) of the mission (e.g., a solar probe mission might travel between Jupiter and to within a few solar radii of the sun)
2) High intensity/high temperatures (e.g., Mercury and solar missions)
3) Low intensity/low temperatures (e.g., missions beyond Mars)
4) Very high power for solar electric propulsion (SEP)
5) High radiation fields (e.g., Europa, Jupiter)
6) Electrostatically clean arrays for fine magnetic measurements
7) Solar power in dusty environments (e.g., surface of Mars)

Of particular concern for mission planners considering sending a spacecraft to the outer planets and beyond is the "LILT" (low-intensity, low-temperature) effect. Solar arrays operating in Earth's orbit typically have steady-state temperatures on the order of 313 K to 343 K. Some benefit can be obtained by reducing the temperature to about 223 K (at a solar distance of about 3 AU). Beyond about 3 AU, the combined effects of the low temperature and the low intensity degrade the conversion efficiency of the solar cell [3].

Large solar arrays obviously have impacts on the attitude control of a spacecraft, which in turn affects the amount of attitude control propellant that must be carried. In addition, concentrator arrays can have an even larger attitude control effect than planar arrays because of their requirement to be more precisely pointed at the sun.

B. Chemical Space Energy Storage Systems

Energy storage systems can provide backup power in case the primary power source shuts down. Energy storage systems can also provide extra power for short periods of time (e.g., "burst power") when the power source is limited. The two principal types of chemical energy storage systems are batteries and fuel cells. Capacitors and flywheels are also candidates for energy storage in certain applications.

Batteries have been the energy storage system of choice for most space missions. Batteries come in two types: 1) primary (single discharge) and 2) secondary (rechargeable). Primary batteries are typically used on missions that require a one-time use of electrical power for a few minutes to several hours. The state-of-practice on primary batteries is 250 We-h/kg with the goal of achieving 600 We-h/kg in 10 years [6].

Rechargeable batteries have been used principally for load leveling and to provide power during eclipse periods. Historically, nickel-cadmium batteries have been used; however, more recent space missions have used nickel-hydrogen and lithium-based batteries. The state-of-practice on rechargeable, low-temperature batteries is about 100 We-h/kg with the goal of reaching 200 We-h/kg in 10 years [6].

Fuel cells have been used on human missions (e.g., Gemini, Apollo, Space Shuttle) because they can provide kilowatts of power. Fuel cells that can be recharged are referred to as regenerative fuel cells (RFCs). NASA has had a goal of developing the technology for a 1000 We-h/kg RFC with high efficiency and greater than 5000-h reliable operation [5].

III. Nuclear Power Sources

Four types of classical nuclear power sources will be considered: 1) radioisotope power; 2) fission power; 3) fusion power; and 4) antimatter power.

A. Radioisotope Power Sources

The development and use of nuclear power in space has enabled the human race to extend its vision into regions that would not have been possible with non-nuclear power sources. For example, in the bitterly cold, radiation-rich, and poorly lit environments of the outer planets, only a rugged, solar-independent power source has the capability to survive and function for long periods of time. Even closer to the sun, environments can be too harsh or otherwise inhospitable for more conventional power sources. For example, the long lunar nights ($\sim$14 Earth days) create a major penalty for solar-powered systems. The frigid, dimly lit polar regions of Mars are another potentially difficult location in which to operate non-nuclear power sources for long periods of time.

Serious studies (such as Project Feedback) of using the immense potential energy of the nucleus for power and propulsion in space began in the United States shortly after the end of World War II. At a time when the full potential of more conventional power and propulsion systems was not yet fully realized, nuclear systems promised advantages in opening up the final frontier of space. These early studies identified two types of nuclear power sources of potential interest: radioisotope power sources and nuclear reactors [7].

Figure 5 illustrates the basic features of a radioisotope thermoelectric generator (RTG). The heart of the RTG is the radioisotope heat source that contains the

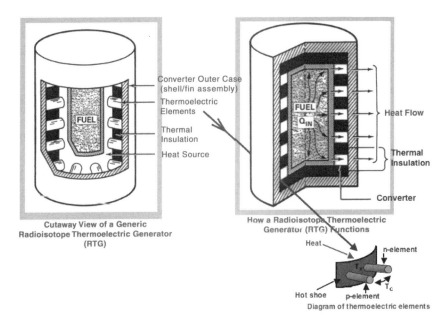

Fig. 5 Diagram of how a radioisotope thermoelectric generator works. (From [7,8].)

heat-producing radioisotope (plutonium-238 is used for U.S. space RTGs). The thermal power from the heat source is transmitted to the thermoelectric elements housed in the converter. The thermoelectric elements convert the thermal power, into electrical power, which is sent to the spacecraft.

With advances in the more conventional technologies such as solar cells and batteries it was more advantageous to use those technologies for the first space flights, particularly because those early missions were in Earth orbit and of short duration.

However, NASA and its predecessor, the National Advisory Committee on Aeronautics (NACA), realized early on the need for nuclear systems in space if the solar system was to be explored. During the Cold War, the U.S. Department of Defense also realized the inherent advantages of nuclear systems, particularly their hardness against certain types of countermeasures. ("Hardness" refers to the ability to resist certain threats such as radiation or electromagnetic fields. Because, by their nature, nuclear power sources operate in their own self-generated radiation environment, their designs incorporate features that make them resistant or "hard" to radiation effects. Reference 9 discusses some aspects of survivability requirements for DoD spacecraft.)

Following on these early mission studies, the U.S. Atomic Energy Commission (USAEC), beginning in 1951, requested several studies on using radioisotopes and nuclear reactors to power spacecraft. As these studies were being completed, K. C. Jordan and J. H. Birden, two researchers working at what was then the USAEC's Mound Laboratory (then operated by Monsanto Research

Corporation), built the first RTG in 1954. Even though this first RTG produced only 1.8 mWe of power, it demonstrated the feasibility of coupling radioisotopes with thermocouple-type (thermoelectric) conversion systems [7].

Table 1 summarizes the U.S. space missions that have utilized or are utilizing radioisotope and fission nuclear power (no fusion or antimatter missions have been flown). The following sections discuss the various missions shown in Table 1. The interested reader is referred to Refs. 10, 11, and 12 for more details. Table 1 does not include those missions that used only radioisotope heater units (RHUs).

All of the U.S. space nuclear power systems used thermoelectric technology to convert the heat (thermal power) from the radioisotope heat source or nuclear reactor into electrical power. The United States did, however, support research into more advanced conversion technologies, including dynamic conversion (Brayton, Rankine, and Stirling). The early RTGs used telluride-based materials in their thermoelectric elements. The Systems for Space Nuclear Power (SNAP) 10A space nuclear reactor power system and the RTGs on all missions beginning with Lincoln Experimental Satellites (LES) 8 and 9 (1976) used silicon-germanium alloys.

Table 1 Uses of space nuclear power by the United States[a]

Transit Navy Navigational Satellites
- Transits 4A and 4B (1961) SNAP-3B (2.7 We)
- Transits 5BN-1 and 5BN-2 (1963) SNAP-9A (≥ 25 We)
- Transit TRIAD (1972) Transit-RTG (35 We)

SNAPSHOT Space Reactor Experiment
- SNAP-10A nuclear reactor (1965) (≥ 500 We)

Nimbus-3 Meteorological Satellite
- SNAP-19B RTGs (1969) (2 at 28 We each)

Apollo Lunar Surface Experiments Packages
- Apollos 12 (1969), 14 (1971), 15 (1971), 16 (1972), 17 (1972) SNAP-27 (≥ 70 We each)

Lincoln Experimental Satellites (Communications)
- LES 8 and LES 9 (1976) MHW-RTG (2 per spacecraft at ~ 154 We each)

Interplanetary Missions
- Pioneer 10 (1972) and Pioneer 11 (1973) SNAP-19 (4 per spacecraft at ~ 40 We each)
- Viking Mars Landers 1 and 2 (1975) SNAP-19 (2 per Lander at ~ 42 We each)
- Voyager 1 and Voyager 2 (1977) MHW-RTG (3 per spacecraft at ~ 158 We each)
- Galileo (1989) GPHS-RTG (2 at ~ 288 We each) (>3-year delay)
- Ulysses (1990) GPHS-RTG (289 We) (>4-year delay)
- Cassini (1997) GPHS-RTG (3 at ~ 295 We each)
- New Horizons (2006) GPHS-RTG (1 at 245.7 We) (most of fuel >21 years old)

[a]Data are presented in the following reference: spacecraft/year launched, type of nuclear power source, beginning-of-mission power. The odd-numbered SNAP power sources used radioisotope heating; the even-numbered SNAP power sources used nuclear reactors to produce heat.

Abbreviations: GPHS-RTG = General-purpose heat source radioisotope thermoelectric generator; MHW-RTG = multi-hundred watt radioisotope thermoelectric generator; SNAP = systems for nuclear auxiliary power.

1. Transit Navy Navigational Satellites

In 1958, researchers at the Johns Hopkins University Applied Physics Laboratory (JHU/APL) conceived the idea of a navigational satellite based on Doppler technology as part of the APL-proposed Navy Navigation Satellite System. The navigation satellite technologies developed under the Transit program are now in use in the Global Positioning System (GPS). While solar cells and batteries powered these early "Transit" satellites, JHU/APL accepted an offer from the USAEC to have an auxiliary radioisotope power source, denoted by the acronym SNAP-3B, for its Transit 4A and Transit 4B satellites [13]. And with that act the era of space nuclear power was literally launched. The following sections discuss the Transit satellites in the order of launching.

a. Transits 4A and 4B. Transit 4A was launched on 29 June 1961 and was quickly followed by the launch of Transit 4B on 15 November 1961. The objectives of both missions were 1) to conduct navigation trials and demonstrations; 2) to improve the understanding of the effects of ionospheric refraction on radio waves; and 3) to increase knowledge of Earth's shape and gravitational field. Both spacecraft met all launch objectives [14].

Each 2.1-kg SNAP-3B RTG (one per spacecraft) produced about 2.7 We at BOM from a radioisotope heat source that provided a thermal power of about 52.5 Wt. While these power levels seem low, both RTGs served to fill a critical niche by powering the crystal oscillator that was the heart of the electronic system used for Doppler-shift tracking. In addition, the RTGs powered the buffer-divider-multiplier, phase modulators, and 54- and 324-MHz power amplifiers. While various component failures interfered with a full analysis of the RTG performance, there were sufficient data from other measurements to show that both RTGs operated well beyond their design life of five years [14,15].

Figure 6 is an artist's illustration of Transit 4A in space and Fig. 7 is a cutaway of the SNAP-3B RTG.

b. Transits 5BN-1 and 5BN-2. Based on the successful performance of the SNAP-3B RTGs on the Transit 4A and Transit 4B satellites, JHU/APL had the confidence to design and fly two totally nuclear-powered navigational satellites (Transit 5BN-1 and Transit 5BN-2). Each satellite used a new, higher power RTG designated SNAP-9A. Each 12.3-kg SNAP-9A was designed to provide 25 We at a nominal 6 V for five years in space after one year of storage on Earth [14,16].

Transit 5BN-1 was launched on 28 September 1963 from Vandenberg Air Force Base (VAFB) in California. Transit 5BN-2 followed a little over two months later on 6 December 1963 [14].

One of the objectives of the Transit 5BN series was to provide "a means by which U.S. Navy ships may navigate anywhere in the world" [14]. Because of some electronic problems, Transit 5BN-1 partially achieved this objective; nevertheless, JHU/APL reported that: "All APL satellite navigational concepts were validated using Satellite 5BN-1" [14].

In a summary report, JHU/APL stated that: "Satellite 5BN-1 was the first artificial earth satellite to employ nuclear energy as its primary power source. . . . In her role as a pioneer nuclear satellite, 5BN-1 demonstrated the extreme

Fig. 6 Artist's concept of the Transit-4A satellite in orbit. The SNAP-3B RTG is shown on "top" of the satellite. (Courtesy of Martin Nuclear.)

simplicity with which thermoelectric generators may be integrated into the design, not only to provide the electrical power but also to aid in thermal control" [14].

Transit 5BN-2 reportedly met all launch objectives and was described by JHU/APL as "the first truly operational navigation satellite" [14].

Figure 8 is an artist's conception of the Transit 5BN-1 satellite showing it in the gravity gradient stabilization mode with the SNAP-9A RTG mounted at the aft end. Figure 9 is a photograph of a SNAP-9A RTG.

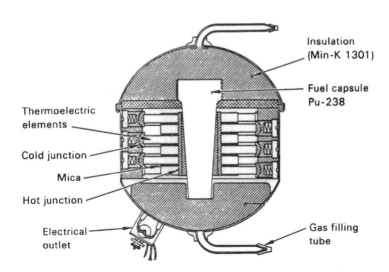

Fig. 7 Cutaway of the SNAP-3B RTG. SNAP-3B was 12.1 cm in diameter and 14-cm high. The power was 2.7 We and the mass was 2.1 kg. (Courtesy of Martin Nuclear.)

542   G. L. BENNETT

Fig. 8   Artist's illustration of the Transit 5BN-1 navigational satellite. (From Ref. 14.)

   c. Transit TRIAD.   The TRIAD satellite (Fig. 10), launched on 2 September 1972 from VAFB, "was the first in a series of three experimental/operational spacecraft designed to flight test improvements to the Navy Navigation Satellite System" [14]. The principal power source was the Transit RTG (Fig. 11), which was to provide a minimum EOM power of 30 We after five years at a minimum of 3 V. Four solar-cell panels and a 6-Ah nickel-cadmium battery provided auxiliary power [17].
   The 13.6-kg Transit RTG used a SNAP-19 radioisotope heat source (Section III.A.6) coupled with light-weight thermoelectric panels using lead-telluride

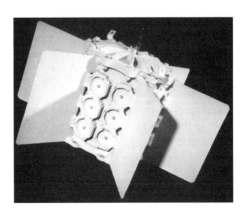

Fig. 9   SNAP-9A RTG. The height was 26.7 cm and the fin span was 50.8 cm. The power was 26.8 We and the mass was 12.3 kg. (Courtesy of Martin Nuclear.)

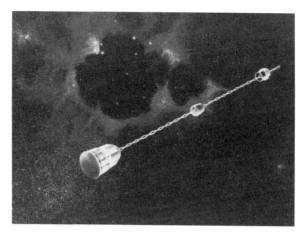

Fig. 10 Transit TRIAD satellite in orbit (artist's conception). The TRANSIT RTG is at the lower left.

technology. In effect, the heat source radiated to the panels so the RTG did not have to be sealed [17].

JHU/APL reported that: "All TRIAD satellite and space technology experiments were exercised and the TRIAD short-term objectives were

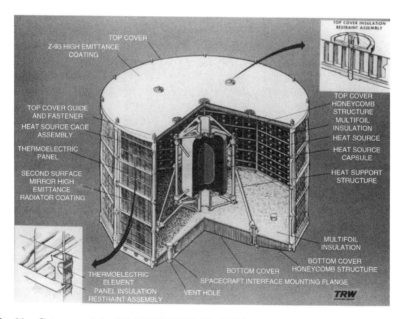

Fig. 11 Cutaway of the TRANSIT RTG. The RTG was approximately 36.3-cm high by 61-cm across the flats. The TRANSIT RTG produced 35.6 We BOM with a mass of 13.6 kg. (Courtesy of TRW/GA/AEC.)

demonstrated" [14]. Despite a telemetry loss about one month into the mission that precluded measuring the Transit RTG power, the functioning of various TRIAD experiments showed that the Transit RTG more than met its objective [14].

2. Nimbus-3 Meteorological Satellite

The SNAP-9A development program moved on to the SNAP-19 technology improvement program, which became the basis for NASA's first outerplanetary missions. First, though, came the checkout on NASA's Nimbus-3 meteorological satellite launched on 14 April 1969 from VAFB. Nimbus-3 was the first U.S. weather satellite to make day and night global measurements from space of temperatures at varying levels in the atmosphere. Figure 12 is an artist's conception of the Nimbus-3 meteorological satellite showing two SNAP-19 RTGs mounted to the base platform. In addition to the two SNAP-19 RTGs, Nimbus-3 carried solar arrays.

The requirement on the two SNAP-19 RTGs was to deliver 50 We to the regulated-power bus after one year in orbit. The two 13.4-kg RTGs produced a combined 56.4 We (49.4 We usable) at launch and 47 We one year later. This nuclear power comprised about 20% of the total power delivered to the regulated-power bus during that time, allowing a number of extremely important atmospheric-sounder experiments to operate in a full-time duty cycle. Without the RTGs, the total delivered power would have fallen below the load line about two weeks into the mission [18,19].

Fig. 12 Nimbus-3 in orbit (artist's conception). The two SNAP-19 RTGs are shown mounted on the left. The RTGs had a height of 26.7 cm and a fin span of 53.8 cm. The average power per RTG was 28.2 We with a mass of 13.4 kg. (Courtesy of NASA.)

3. Apollo Lunar Surface Experiments Packages

Nuclear power on the moon provides a number of important advantages, chiefly the ability to deliver full power throughout the long (14 Earth day) lunar night. This was one of the factors that led NASA to select RTG power for its Apollo Lunar Surface Experiments Packages (ALSEPs). The program objectives included acquiring scientific data to aid in determining the internal structure and composition of the moon and the composition of the lunar atmosphere [20].

Five ALSEPs were placed on the moon, beginning with the Apollo 12 mission (Fig. 13) whose ALSEP downlink acquisition started on 19 November 1969. The Apollo 11 crew deployed a forerunner of ALSEP, known as the Early Apollo Scientific Experiment Package (EASEP). The EASEP used solar cells for power and two 15-Wt radioisotope heater units (RHUs) for warmth [20].

Power for each ALSEP was provided by a new RTG designated SNAP-27 (Fig. 14). The power requirement for the 19.6-kg RTG was to provide at least 63.5 We at 16-V DC for one year after lunar emplacement. (For Apollo 17, the requirement was 69 We two years after emplacement [21].) All five ALSEP SNAP-27 RTGs (Apollo 12, 14, 15, 16, and 17) exceeded their mission requirements in both power and lifetime, which enabled the ALSEP stations to gather long-term scientific data on the moon [20].

4. Lincoln Experimental Satellites 8 and 9

Lincoln Experimental Satellites 8 and 9 (LES-8/9), which were built for the U.S. Air Force, "were developed with the goal of demonstrating, in full-scale operation (terminals as well as satellites), advanced technologies for strategic communications links" (Fig. 15) [22]. Each spacecraft was powered by two RTGs of a completely new design (Fig. 16). Termed "Multi-Hundred Watt

Fig. 13 Alan Bean removing the SNAP-27 on the Apollo 12 mission in November 1969. (Courtesy of NASA.)

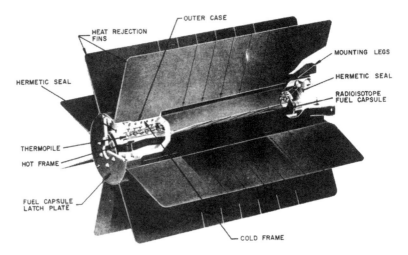

Fig. 14 Cutaway of the SNAP-27 RTG. The converter was 46-cm high and 40.0-cm across the fins. The RTG power was ≥63.5 We with a mass of 19.6 kg. (Courtesy of GE.)

Radioisotope Thermoelectric Generator" (MHW-RTG), these RTGs were being developed for multiple missions; however, the initial focus was on NASA's Grand Tour (later Voyager) mission. With minimal modifications, the MHW-RTGs were used on LES-8/9 because they "offer impressive advantages of physical survivability by comparison with solar-cell arrays" [22].

For LES-8/9, each 39.69-kg MHW-RTG was to produce 125 We at 26 V at the end of mission—an operational life of at least five years after launch (which occurred on 14 March 1976) [23]. BOM powers averaged about 154 We per RTG. The MHW-RTGs more than met this goal. LES-8 was turned off on 2 June 2004 because of control difficulties (its two MHW-RTGs were still providing usable electrical power). *LES-9 continues to operate over 30 years after launch.*

Writing 13 years after the launch, two of the LES-8/9 managers concluded, "The RTGs were well worth the effort ... they have performed superbly. They provide continuous electrical power through the 70-min eclipses of the sun by the Earth that LES-8 and LES-9 experience every day. The compatibility of these rugged power sources with complex signal-processing circuitry has been well established ... the measures that were taken to assure the success of the RTGs in LES-8 and 9 contributed directly to the success of the Voyager missions" [24].

5. Interplanetary Missions

From a public standpoint, the most spectacular uses of nuclear power sources in space have come from the interplanetary missions, beginning with the launches of the Pioneer 10 and Pioneer 11 spacecraft in 1972 and 1973, respectively. When a spacecraft is sent where the sunlight is low (at Jupiter it is 25 times less than at Earth; at Pluto, >900 times less than at Earth), where the temperatures are quite

Fig. 15 LES-8/9 in orbit. The two MHW-RTGs are shown mounted on top of one of the satellites.

low (~130 K at Jupiter), and where the radiation belts are very severe, the only option is nuclear power. This section summarizes the RTG performance on U.S. interplanetary missions from the Pioneers (Jupiter, Saturn) to the Vikings (Mars), Voyagers (Jupiter, Saturn, Uranus, Neptune), Galileo (Jupiter), Ulysses (solar polar), and Cassini (Saturn) to the most recent, the New Horizons mission to Pluto.

a. Pioneers 10 and 11. Pioneers 10 and 11 were built to explore the environment of Jupiter and the interplanetary medium beyond the orbit of Mars and the asteroid belt. Pioneer 10 was launched on 2 March 1972 on a fast trajectory to Jupiter (closest approach on 3 December 1973). The twin spacecraft Pioneer 11 was launched on 5 April 1973 to follow Pioneer 10 to Jupiter (closest approach on 4 December 1974).

Each Pioneer spacecraft carried four 13.6-kg SNAP-19 RTGs, producing an average of 40.3 We per RTG at BOM. The mission requirement was that the four SNAP-19 RTGs on each Pioneer spacecraft had to produce 120 We total at the Jupiter flyby. The RTGs more than met this requirement, which enabled NASA to direct Pioneer 11 on for a first encounter with Saturn (closest approach on 1 September 1979). The Pioneer 11 SNAP-19 RTGs exceeded the 90 We required at Saturn by producing 119.3 We. By this time the RTGs on both spacecraft had operated well beyond their original 21-month requirement [25]. Both Pioneer spacecraft also carried a number of small radioisotope heater units

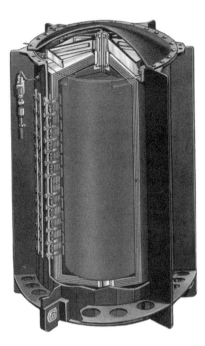

Fig. 16 Cutaway of the MHW-RTG. The length was 58.31 cm and the overall diameter was 39.73 cm. For LES-8/9 the average BOM power was ∼154 We per RTG with an average mass of 39.69 kg. For Voyager 1/2, the average BOM power was ∼158 We/RTG with an average mass of 37.69 kg. (Courtesy of GE.)

(RHUs) to keep equipment warm without the potential electromagnetic interference of electrical heaters.

Both spacecraft continued to operate for decades after their original mission objectives had been met and so gave scientists a wealth of information on the interplanetary medium in the outer solar system. The last signal from Pioneer 10 was received on 22 January 2003, more than 30 years after launch. The last signal from Pioneer 11 was received on 30 September 1995. Figure 17 is an artist's illustration of Pioneer 10, and Fig. 18 is a cutaway of the Pioneer SNAP-19 RTG.

  b. Viking Landers 1 and 2. In 1975, the United States embarked on the first surface exploration of Mars with the launches of the two Viking missions. Each Viking spacecraft consisted of an orbiter and a lander. In view of the hostile environment of Mars (including dust storms and Antarctic-style temperatures), NASA selected a modified SNAP-19 RTG to power the Viking landers (two RTGs per lander). The modifications included the addition of a dome for gas exchange with the main body of the RTG. The average BOM power of these 15.2-kg RTGs was 42.7 We per RTG. The requirement was to produce a minimum of 35 We during the 90-day primary mission [25,26]. Figure 19 shows how the SNAP-19 RTGs were placed on a Viking lander. Viking 1 was

Fig. 17 Pioneer 10 leaving the solar system (artist's conception). (Courtesy of NASA.)

launched on 20 August 1975 and its lander reached the Martian surface on 20 July 1976. Viking 2 was launched on 9 September 1975 and its lander reached the Martian surface on 3 September 1976.

All four SNAP-19 RTGs easily met the 90-day requirement allowing the Viking landers to operate for years until other system failures led to a loss of

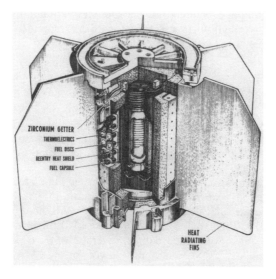

Fig. 18 Cutaway of the Pioneer SNAP-19 RTG. The height was 28.2 cm and the fin span 50.8 cm. The average BOM power was 40.3 We per RTG. (Courtesy of TES.)

Fig. 19 Model of the Viking Mars lander showing the locations of the two Viking SNAP-19 RTGs. The RTGs had a height of 40.4 cm and a fin span of 58.7 cm. The average BOM power per RTG was 42.7 We and the average mass was 15.2 kg per RTG. (Courtesy of NASA/TES.)

data. When the last data were received from Viking Lander 1 in November 1982, it had been estimated that the RTGs were capable of providing sufficient power for operation until 1994—*18 years beyond the original mission requirement.* Had an accidental shutdown not occurred, it is fascinating to speculate if Viking Lander 1 might have been ready to greet the Mars Sojourner when it landed on 4 July 1997.

It is worth noting that all three U.S. Mars Rovers (Sojourner, Opportunity, and Spirit) carried 1-Wt light-weight radioisotope heater units (LWRHUs) to keep them warm during the cold Martian nights.

  c. Voyager 1 and Voyager 2. Following on the paths of Pioneers 10 and 11, the more advanced Voyager 1 and Voyager 2 spacecraft were launched in 1977 to conduct more detailed studies of Jupiter and Saturn, their satellites and magnetospheres as well as studies of the interplanetary medium. The successful completion of the Saturn flybys allowed NASA to send Voyager 2 to the first flybys of Uranus and Neptune, completing most of the objectives for the originally planned Grand Tour mission.

To power these two spacecraft so far from the sun, the Voyagers used multi-hundred watt radioisotope thermoelectric generators (MHW-RTGs) similar in design to those flown the year before on the U.S. Air Force communications satellites LES-8/9 (Fig. 16 for LES-8/9 version). With some modifications, the MHW-RTG mass was reduced to an average of 37.69 kg. The Voyager mission required an EOM minimum power of 128 We per RTG four years after launch. BOM powers averaged about 158 We per RTG [10,11]. (Both Voyager spacecraft also carried a number of small RHUs to keep equipment warm.)

Fig. 20 Artist's concept of the path of a Voyager spacecraft past Jupiter and Saturn. The three MHW-RTGs are shown mounted below the spacecraft on a boom.

The two Voyager spacecraft (Fig. 20) have probably explored more territory than any other spacecraft in human history. Discoveries ranged from finding 22 new satellites (3 at Jupiter; 3 at Saturn; 10 at Uranus; and 6 at Neptune) to witnessing the first volcanic eruption on another solar system body (Io) and geysers on another satellite (Triton). The two Voyager spacecraft are now operating in an interstellar mode where they continue to provide new information beyond the orbit of Pluto.

d. Galileo. With the completion of the successful Pioneer and Voyager flybys scientists turned to an in-depth study of the largest planet in the solar system. Conceived at the time of the Voyager launches, the Galileo spacecraft was launched on 18 October 1989 on a series of gravity assist maneuvers to take this technologically sophisticated orbiter and atmospheric probe to Jupiter. There Galileo investigated the Jovian atmosphere; the Jovian satellites; the Jovian magnetosphere; and energetic particles and plasma.

Powering Galileo in this hostile environment were two new RTGs designated by the acronym GPHS-RTG (general-purpose heat source radioisotope thermoelectric generator). At the time of fueling (BOL), a GPHS-RTG is capable of producing about 300 We in a mass envelope of about 55.9 kg making this the highest specific power nuclear power source ever flown by the United States. Because of launch delays caused by the Challenger accident, the power requirement for the Galileo GPHS-RTGs was reduced to 470 We (235 We per RTG) at EOM

Fig. 21 Galileo orbiter communicating with the Galileo probe. One of the two GPHS-RTGs is shown on a boom "above" Galileo. (Courtesy of NASA/JPL.)

(71,000 h after BOM) [27]. The combined BOM power of the two GPHS-RTGs was 577.2 We [28]. Galileo went into orbit on 7 December 1995 as the probe entered the atmosphere of Jupiter. Figure 21 is an artist's conception of the Galileo spacecraft at Jupiter and Fig. 22 illustrates the features of the GPHS-RTG. Both the Galileo orbiter and probe carried new, lower-mass 1-Wt RHUs, which were termed the light-weight radioisotope heater unit or LWRHU.

Both GPHS-RTGs met their EOM power requirements allowing NASA and JPL to extend the Galileo mission three times. Finally, on 21 September 2003, after 35 orbits of Jupiter and with its propellant running low, Galileo was sent into the atmosphere of Jupiter so that it would not collide with the

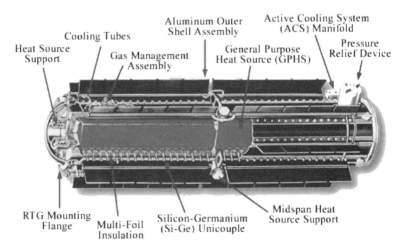

Fig. 22 Cutaway of the GPHS-RTG. The length is 114 cm and the fin span is 42.2 cm. At time of fueling, the GPHS-RTG can produce 300 We. For Galileo and Ulysses, the average RTG mass was 55.9 kg. (Courtesy of LMA/DOE.)

oceanic satellite Europa. In its 14-year life, Galileo sent back data on Earth, Venus, two asteroids (Gaspra and Ida and its moon Dactyl) and the Jovian system plus views of comet Shoemaker-Levy 9 crashing into Jupiter. Galileo also sent the first atmospheric probe into the atmosphere of a giant planet, a probe kept warm with LWRHUs.

e. Ulysses. Originally planned as part of a two-spacecraft mission, the Ulysses spacecraft was launched on 6 October 1990 "to investigate for the first time as a function of heliographic latitude, the properties of the solar wind, the structure of the sun/wind interface, the heliospheric magnetic field, solar radio bursts and plasma waves, solar x-rays, solar and galactic cosmic rays and both interstellar and interplanetary neutral gas and dust" [29]. Even though Ulysses is essentially a solar polar mission, it needed the immense gravity of Jupiter to bend its orbit out of the plane of the ecliptic. These regular trips out to 5 AU (along with the Jovian radiation belts) dictated a nuclear power source (GPHS-RTG) for Ulysses.

The Ulysses GPHS-RTG power requirements were modified because of the more than four-year delay following the Challenger accident: provide a BOM power of 277 We and an EOM power (42,000 h) of 245 We [27]. Telemetry measurements indicated a BOM power of 284 We at the bus and 289 We at the RTG connector, both in excess of the minimum required. The August 1995 'EOM' was reported to be 248 We, 3 watts above the minimum [28]. Because of the excellent performance of the Ulysses GPHS-RTG, the mission has been extended several times. In November 2006, Ulysses started its third polar orbit of the sun and, beginning in November 2007, the spacecraft moved over the sun's north polar latitudes. The Ulysses mission ended in June 2008, having operated more than three times longer than planned. In this time Ulysses has returned a wealth of information on the latitudinal distribution of particles and fields from our closest star.

Figure 23 is an artist's illustration of the Ulysses mission showing its GPHS-RTG mounted on the side of the main body of the spacecraft.

f. Cassini. Following on the successful Pioneer and Voyager flybys of Saturn, scientists turned next to an orbiter/probe mission to provide a more in-depth study of the ringed world. On 15 October 1997, the Cassini spacecraft was launched to do, in a sense, what Galileo had done at Jupiter: carry out a multiyear scientific study of the second largest gas giant in the solar system. Mission objectives included studying the rings, the satellites, the magnetosphere, and, in particular, Titan, the cloud-covered largest satellite of Saturn. Like Galileo, Cassini carried a probe (Huygens), this one built by the European Space Agency (ESA) and designed to land on Titan.

Cassini successfully went into orbit around Saturn in July 2004 after traveling past Venus, Earth, and Jupiter. Huygens, which carried Galileo-style lightweight radioisotope heater units (LWRHUs) to keep it warm, successfully landed on Titan and provided dramatic views of that shrouded world. Cassini is now well into its 4-year main mission of 74 orbits around Saturn with 44 flybys of Titan.

Because of mission complexity, Cassini needed more power than used on previous missions of this type. Three modified GPHS-RTGs provide that power. The

Fig. 23 Ulysses spacecraft shown in its second Jupiter encounter (November 2003 to April 2004). The GPHS-RTG is shown mounted on "top" of the spacecraft. (Courtesy of NASA/ESA/JPL.)

specification requirement for BOM was 826 We. The three GPHS-RTGs provided 887 We total at BOM. The average mass per RTG is 56.4 kg. The projected power at 16 years after BOM is 640 We, which will exceed the specification requirement of 596 We [28]. The prognosis is excellent that the Cassini RTGs will provide sufficient power for an extended mission beyond the main mission. Figure 24 is an artist's illustration of the Cassini spacecraft showing two of the three GPHS-RTGs. Cassini also carried LWRHUs to keep equipment warm.

g. New Horizons Pluto Kuiper Belt Flyby. One planet remained unexplored after the Pioneer and Voyager flyby missions: Pluto, once thought to be the "last planet." On 19 January 2006, the New Horizons spacecraft was launched to fly by Pluto and its moon Charon. New Horizons will then continue on into the Kuiper Belt where it will fly by one or more Kuiper belt objects (KBOs). On its way to the Pluto/Charon system, New Horizons flew by Jupiter in late February 2007 for a gravity assist. During that four-month encounter period, New Horizons studied Jupiter, adding to our knowledge of the solar system's largest planet.

New Horizons is expected to fly by Pluto on 14 July 2015 (Fig. 25). From about 2016 to 2020, following the Pluto/Charon flyby, New Horizons will enter the KBO study phase of the mission.

In a region of space where the sunlight is less than 0.001 of what it is at Earth, nuclear power is the only viable option. Thus, New Horizons carries a single GPHS-RTG to provide power in the far reaches of the solar system. With various modifications (including changes to the heat source modules), the

Fig. 24 Cassini shown over Titan as the Huygens probe descends toward Titan (artist's conception). Two of the three GPHS-RTGs are shown mounted at the "upper end" in this figure. (Courtesy of NASA/JPL.)

flight mass is 57.8 kg. The BOM power from this GPHS-RTG was 245.7 We versus a specification requirement of 237 We. This power was lower than what has been produced in the GPHS-RTGs flown on earlier missions, primarily because the New Horizons GPHS-RTG used some older fuel that had decayed for over 21 years [30,31]. Still, it is expected that the GPHS-RTG will produce about 200 We (versus the specification minimum of 191 We) at the time of Pluto/Charon flyby, enough for New Horizons to conduct exciting scientific studies of what was once thought to be our farthest planet [31].

6. Future Uses of Radioisotope Power Sources

In September 2009, NASA plans to launch the Mars Science Laboratory (MSL), which will use a modified SNAP-19 RTG known as the Multi-Mission Radioisotope Thermoelectric Generator (MMRTG). The MMRTG, like the Viking SNAP-19, is designed to operate on the surface of Mars. The projected power of the MMRTG at BOM is 121 We at a thermal power of 1982 Wt for Mars noon and 123 We in deep space. After 14 years, these powers are projected to be 96 We and 97 We, respectively. With a mass of about 45 kg, the specific power works out to be about 2.7 We/kg.

Separately, NASA and the U.S. Department of Energy are sponsoring the development of an advanced Stirling radioisotope generator (ASRG) to produce 140 We BOM. With a mass of 20.2 kg, the specific power will be

Fig. 25 New Horizons flying over Pluto with Charon in the background. The GPHS-RTG is shown mounted on the left side in this illustration. (Courtesy of JHU/APL; SwRI.)

almost 7 We/kg. The additional advantage of the ASRG is that the higher efficiency ($\sim 29\%$) of the Stirling conversion system will translate into much less ^{238}Pu being required; a definite advantage at a time when the United States is no longer producing ^{238}Pu.

Radioisotope power sources have been considered to power electric propulsion thrusters, a concept designated radioisotope electric propulsion (REP). After being sent on an Earth escape trajectory by a chemical upper stage, REP used in conjunction with gravity assists can enable some challenging science missions such as orbiting the satellites of the outer planets. One can envision a combination of SEP, REP, and gravity assists to send an interstellar explorer on its way.

B. Fission Space Power

Both the United States and the former Soviet Union have had active technology and flight programs in fission space power. In addition to the fission space power programs, both nations have had technology programs in fission propulsion (both nuclear thermal propulsion, NTP, and nuclear electric propulsion, NEP).

1. U.S. Fission Space Power

The first nuclear reactor to be employed in space was launched by the U.S. Air Force from VAFB on 3 April 1965. The reactor, which was designated SNAP-10A,

Fig. 26 SNAP-10 in orbit attached to the Agena bus (artist's conception). The overall length was 3.48 m and the mounting base diameter was 1.27 m. The total system mass of the final flight unit was 435 kg. SNAP-10A was designed to produce at least 500 We for 1 year. (Courtesy of Atomics International/AEC.)

built upon the heritage of the earlier SNAP reactor programs, in particular, SNAP-2 and SNAP-10. SNAPSHOT, as the experiment was named, was a test of the operability of an automated space reactor. The power requirement for the 435-kg SNAP-10A was to produce at least 500 We for 1 year [32]. Figure 26 is an artist's conception of the SNAP-10A reactor coupled to the Agena "spacecraft."

Once in its 1288-km by 1307-km nuclear safe orbit, the reactor was started up and operated. All went well for 43 days until a failure of a voltage regulator in the Agena "spacecraft" caused a termination of the power operation. Nevertheless, SNAP-10A showed that it was feasible to remotely operate a liquid metal-cooled nuclear reactor in space. The capability of SNAP-10A to operate unattended for one year was demonstrated in a ground-test twin to the flight reactor [32].

In addition to the SNAP-10A program, during the late 1950s to the early 1970s the United States sponsored a number of other space reactor programs (e.g., SNAP-10, SNAP-50/SPUR) that would use advanced power conversion systems (e.g., Brayton and Rankine) to reach even higher powers (e.g., up to 1 MWe for SNAP-50). In particular, the SNAP-50/SPUR (Space Power Unit Reactor) program had nuclear electric propulsion (NEP) as one of its proposed uses.

From 1983 to 1994, the United States sponsored the SP-100 space nuclear reactor power system technology program (Fig. 27). The goal was to develop a space nuclear reactor technology that could support a range of projected future missions including planetary surface operations and NEP for science missions. The generic flight system (GFS) was established to support operational missions requiring relatively high power (100-kWe class) for 10-year mission durations, but scalable from about 10 kWe to 1000 kWe and with high specific power.

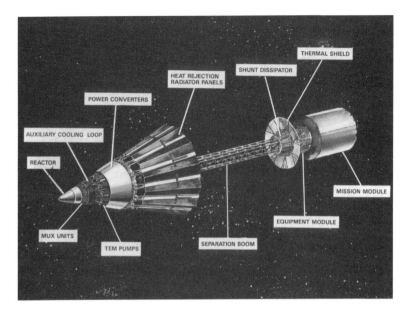

Fig. 27 SP-100 space nuclear reactor power system module showing the key subsystems. The diameter and length of the main body (less the radiator panels) are 3.5 m and 6 m, respectively, and the mass is 4575 kg for the 100-kWe GFS. A deployable boom is used to maintain a separation distance of 22.5 m between the reactor and the payload plane to keep the neutron and gamma doses within specified values. (Courtesy of GE.)

The 100-kWe version with mature thermoelectric conversion was estimated to have a specific power of 26 We/kg [33,34].

One study of coupling the SP-100 nuclear reactor with either Brayton or Stirling dynamic power conversion systems at various temperatures for a lunar base showed that for a 550-kWe power output, all of them were within 30% of 12,000 kg, corresponding to a specific mass of 21.8 kg/kWe or a specific power of 45.8 We/kg [35].

Nuclear reactors have the advantage that they scale up well in terms of reduced specific mass as the power increases. A scaling study completed in 1991 showed that coupling a nuclear reactor based on SP-100 technology to a potassium Rankine power conversion system could produce 1000 kWe with a mass of about 7600 kg, yielding a specific power of over 131 We/kg or a specific mass of 7.6 kg/kWe. The corresponding masses for the Brayton and Stirling systems were about 13,600 kg (Brayton) and over 15,000 kg (Stirling) [36].

2. U.S. Fission Propulsion

The classic nuclear propulsion system involves using a nuclear reactor as a kind of heat exchanger to heat the propellant (hydrogen) and to expel it through a nozzle as shown in Fig. 28. From about 1955 to about 1972, the

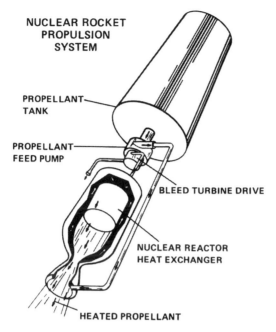

Fig. 28 Schematic of a generic nuclear rocket. (Courtesy of AEC/NASA.)

United States conducted research on a series of such reactors under the Rover and NERVA (Nuclear Engine for Rocket Vehicle Applications) programs. Under the NERVA program, the goal was to develop a nuclear rocket capable of producing a thrust of 334 kN at a specific impulse of 8.1 km/s, which is almost double that of the best classical chemical propulsion system [37,38].

Seventeen reactors, one nuclear safety reactor, and two ground experimental engines were tested with the summary record performances listed in Table 2.

Table 2 Summary of U.S. nuclear rocket performance[a]

Power (Phoebus 2A)	4100 MWt
Thrust (Phoebus 2A)	~930 kN
Hydrogen flow rate (Phoebus 2A)	120 kg/s
Equivalent specific impulse (Pewee)	~8.3 km/s
Minimum reactor specific mass (Phoebus 2A)	2.3 kg/MW
Average coolant exit temperature (Pewee)	2550 K
Peak fuel temperature (Pewee)	2750 K
Core average power density (Pewee)	2340 MW/m^3
Peak fuel power density (Pewee)	5200 MW/m^3
Accumulated time at full power (NF-1)	109 minutes
Greatest number of restarts	28

[a]From Refs. 37 and 38.

In theory, a nuclear rocket should have a specific impulse about three times that of the best state-of-practice chemical rocket (liquid hydrogen–liquid oxygen, LH_2–LO_2). In practice, materials have limited the nuclear rocket's performance to about twice that of the chemical rocket. In an effort to overcome the material limits, researchers have investigated nuclear rockets that are not so dependent on materials, that is, the particle bed reactor (PBF), the liquid-core reactor, the gas-core nuclear rocket (GCNR), and the closed-cycle gas-core nuclear rocket or "nuclear light bulb" reactor (essentially a reacting ^{235}U plasma radiating through a transparent medium to the hydrogen propellant). In theory, the GCNR offers the highest specific impulse ($\sim 10^5$ m/s) although the engine mass would be on the order of 200 MT [39,40].

In the early days of the space program, a proposal was made to use fission bombs as a propulsion system (Project Orion). In this concept, small fission bombs would be released at the rate of 1 bomb every 1 to 10 s and exploded at a distance on the order of 30 m to 300 m from the vehicle. The blast would interact with a pusher plate that in turn would transmit the impulse to the vehicle through a shock attenuation system. The specific impulse was estimated to be in the range of 18 km/s to 25 km/s with a thrust-to-weight (T/W) of about 4. Approximately 2000 fission bombs were estimated to be required for a 250-day roundtrip mission to Mars. The program was ended in 1965 after some sub-scale tests with chemical explosives were conducted [40].

Another approach to realize the benefits of nuclear fission is the fission fragment rocket in which the fission fragments themselves become the propellant through the use of magnetic fields to guide them. Specific impulses on the order of 10^6 m/s would be possible if this concept could be developed [41]. A number of technology hurdles would have to be overcome including the control of the fission fragments.

The high power densities achieved in the Rover/NERVA program indicate what is possible with multi-megawatt nuclear reactors. If these high power densities were coupled to advanced electric propulsion thrusters the solar system would be open to human exploration. Comparisons of chemical, SEP, NEP, and NTP for human missions to Mars have shown that NEP is a competitor in terms of mass that can be delivered [42]. With reactors providing 200 MWe and specific masses on the order of 1 kg/kWe to 2 kg/kWe, fast trip times could be achieved. An artist's illustration of one such concept, Odyssey, is shown in Fig. 29. The reactor concept would produce 200 MWe and deliver 100 MT of payload. The length of the NEP spacecraft would be over 100 m [43].

A JPL top-level study indicated that the Odyssey spacecraft could make 230-day roundtrip times to Mars and 2-to-3-year roundtrip times to Jupiter.

3. Former Soviet Union Fission Program

The former Soviet Union also had an active space nuclear program, launching about 31 (and perhaps 33) BUK space nuclear reactors to power spacecraft used for marine radar observations. The BUK liquid metal-cooled fast reactor produced about 3 kWe from 100 kWt of thermal power using two-stage thermoelectric elements. Launches of BUK may have begun as early as 1967 with the last publicly known launch occurring in 1988 [44,45].

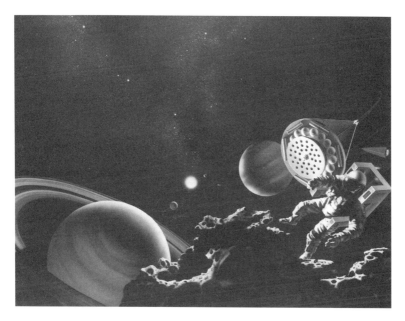

Fig. 29 Project Odyssey: a 200 MWe NEP concept for facilitating human exploration of the solar system. (Courtesy of JPL.)

In 1987, the former Soviet Union launched two experimental thermal reactors designated TOPAZ. These experimental reactors, which used an in-core thermionic conversion system, reportedly produced about 5 kWe from a reactor thermal power of 150 kWt [44,45]. Both the BUK and TOPAZ space reactor power systems had masses on the order of 1 T. In the late 1980s and early 1990s, certain elements of the U.S. Department of Defense showed some interest in testing and flying another Russian thermionic reactor (ENISEY) that had not been flown by the Russians [44]. In the United States, this reactor was erroneously dubbed "TOPAZ 2" (but it was not built by the Russian group that had built TOPAZ).

In addition to nuclear reactors, the former Soviet Union has used RHUs and small RTGs on some missions [45], and has also conducted tests of nuclear fuel elements for a nuclear rocket.

C. Fusion Power and Propulsion

Nuclear fusion powers the stars and it offers the potential of powering interplanetary and, perhaps, interstellar spacecraft. Very light nuclides such as deuterons and tritons have a relatively small binding energy per nucleon compared to heavy nuclei such as uranium-235 (^{235}U) and plutonium-239 (^{239}Pu). Therefore, the fusion of light nuclei generally results in the release of a relatively large amount of energy per nucleon. Table 3 lists some of the more important fusion reactions together with uranium (^{235}U) fission for comparison.

Table 3 Fusion reactions

Reaction	Energy release	
	Total (MeV)	E/m (MeV/amu)
$^1H + {}^1H \rightarrow {}^2H + \beta^+$	1.4	0.7
$^1H + {}^2H \rightarrow {}^3He + \gamma$	5.5	1.8
$^1H + {}^7Li \rightarrow {}^4He + {}^4He$	17.4	2.2
$^3He + n$ (~50%)	3.3	0.8
$^2H + {}^2H \rightarrow {}^3H + {}^1H$ (~50%)	4.0	1.0
$^2H + {}^3H \rightarrow {}^4He + n$	17.6	3.5
$^2H + {}^3He \rightarrow {}^4He + {}^1H$	18.4	3.8
$^2H + {}^6Li \rightarrow {}^4He + {}^4He$	22.4	2.8
$^3H + {}^3H \rightarrow {}^4He + n + n$	11.3	1.9
$^3He + {}^3He \rightarrow {}^4He + {}^1H + {}^1H$	12.9	2.1
$^{235}U + n \rightarrow$ (fission)	~200	0.8

Note: ^1H = proton; ^2H = deuteron; ^3H = triton; He = Helium; Li = Lithium; U = Uranium β^+ = positron; γ = gamma ray; n = neutron.
From Ref. 46.

The principal issues in making fusion work include using enough energy ("temperature") to overcome the electrostatic repulsion of the positive charges in order to make the reaction happen and to confine the plasma long enough for the reactions to occur. However, if fusion propulsion is achieved, the benefits of fusion propulsion would greatly outweigh those of chemical and fission propulsion.

Many nations have actively pursued fusion power for terrestrial applications. For space propulsion, there is a very close coupling of the fusion power source design and the propulsion technology. Classically, two different approaches have been proposed to confine the plasma long enough to sustain a fusion reaction: 1) inertial confinement fusion (ICF) and 2) magnetic confinement fusion (MCF).

ICF utilizes a high-power source (e.g., lasers) to compress and heat a pellet of fusion fuel to fusion ignition conditions. Figure 30 illustrates the ICF propulsion concept. In the 1970s, the British Interplanetary Society investigated an interstellar ICF concept under their Project Daedalus study. In the early 1990s, the JPL, the Lawrence Livermore National Laboratory (LLNL), the Energy Technology Engineering Center (ETEC), and NASA's Johnson Space Center (JSC) studied an ICF propulsion concept termed VISTA (vehicle for interplanetary space transport applications). The VISTA vehicle was sized for a fast (100-day roundtrip) human mission to Mars. The payload was to be 100 MT for a total mass of 5800 MT, of which 4100 MT would be hydrogen expellant and 40 MT would be deuterium-tritium (d-t) fuel. A jet power of 20,000 MW was calculated for 30 Hz operation (30 d-t pellets ignited per second) with a specific impulse of 170 km/s [40].

MCF (shown diagrammatically in Fig. 31) uses strong magnetic fields to confine the fusion plasma. In principle, MCF should provide similar performance

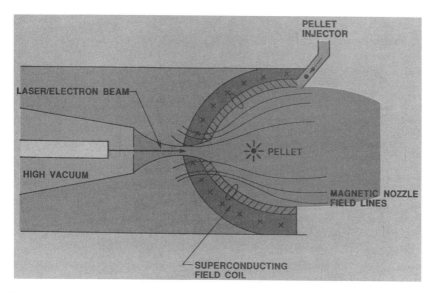

Fig. 30 Basic features of a fusion propulsion system using inertial confinement fusion. (Courtesy of JPL.)

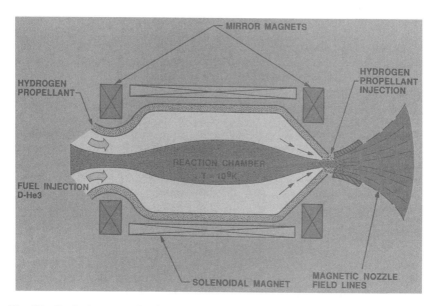

Fig. 31 Basic features of a fusion propulsion system using magnetic confinement fusion. (Courtesy of JPL.)

to the ICF propulsion concepts, although the use of large magnets may make MCF devices heavier than ICF devices.

D. Antimatter Power and Propulsion

As noted in Section I and Figs. 3 and 4, matter–antimatter annihilation offers the highest possible classical physical specific energy of any known reaction substance. The matter–antimatter annihilation specific energy of 9×10^{16} J/kg is many orders of magnitude greater than chemical (1×10^7 J/kg), fission (8×10^{13} J/kg), or even fusion (3×10^{14} J/kg) reaction energies. Moreover, matter–antimatter annihilation can be induced spontaneously without the use of complex reactor systems.

While no antimatter power sources have been built, studies have shown that the design of the antimatter "power source" is strongly coupled to how it would be used in propulsion. Historically, four basic antimatter thruster concepts have been considered (Fig. 32).

Most antimatter power and propulsion studies have assumed the use of antiprotons or neutral anti-hydrogen that would be produced in a terrestrial accelerator. Because antimatter annihilates spontaneously when it comes in contact with matter, the antimatter would have to be contained in an electromagnetic field and stored separately in a high vacuum.

One Monte Carlo analysis of the four concepts shown in Fig. 32 found the efficiencies for conversion of annihilation energy to energy available for propulsion "to be approximately 75% for the solid core concept, 13% for the gas core concept, 3% or less for the plasma core concept and 20% or less for the beam core concept." This study also observed that "the gas core, plasma core and

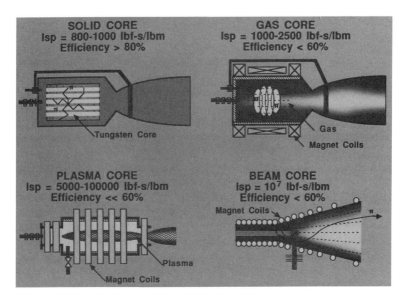

Fig. 32 Four basic antimatter propulsion concepts. (Courtesy of JPL.)

beam core concepts all suffered from additional plasma radiation losses which are severe. The solid core concept, because of its higher density core, experiences a substantial neutron production problem from nuclear charge-exchange" [47].

A separate analysis of a pulsed, antiproton-powered magnetically confined hydrogen plasma engine generally found energy transfer efficiencies below 1% [48].

As a result of such studies, advocates have focused on alternative concepts to raise the efficiency of conversion.

IV. Conclusions

In this chapter we briefly covered a number of non-nuclear and nuclear space power sources that either are available now (solar, batteries, fuel cells, radioisotopes) or could be available in the future (fission, fusion, antimatter). Particular attention was paid to radioisotope and fission power sources because these have been used in a number of space missions and are most directly applicable to missions to the outer solar system and beyond. The development of fusion and antimatter power and propulsion systems would greatly facilitate these explorations. From a "classical" standpoint, antimatter represents the "ultimate" source of power for space exploration.

In terms of classical power sources for interplanetary and early interstellar travel, one can summarize the foregoing sections with the following projected or "ultimate" performance metrics as described in the parenthetical comments):

Photovoltaic: 1000 We/kg (future achievable at 1 AU with thin films)
Energy storage: 1000 We-h/kg (future goal)
Radioisotope: ~ 10 We/kg (future)
Fission: ~ 1000 We/kg (future at 100 s of MWe)
Fusion: ~ 1000 We/kg (future)
Antimatter: 9×10^{16} J/kg (ultimate)

References

[1] Brandhorst, H. W., Chetty, P. R. K., Doherty, M. J., and Bennett, G. L., "Technologies for Spacecraft Electric Power Systems," *Journal of Propulsion and Power*, Vol. 12, No. 5, Sept.–Oct. 1996, pp. 819–827.

[2] Bennett, G. L., and Stone, J. R., "The NASA Advanced Propulsion Concepts Program," *1989 JANNAF (Joint Army-Navy-NASA-Air Force) Propulsion Meeting*, Cleveland, OH, 23–25 May 1989.

[3] Surampudi, R., Hamilton, T., Rapp, D., Stella, P., Mardesich, N., Mondt, J., Nesmith, B., Bunker, R. L., Cutts, J., Bailey, S. G., Curtis, H. B., Piszczor, M., Gaddy, E., Marvin, D., and Kazmerzki, L., *Solar Cell and Array Technology for Future Space Science Missions*, JPL D-24454, Jet Propulsion Laboratory, Pasadena, CA, Aug. 2002.

[4] Kurland, R. M., and Stella, P. M., "The Advanced Photovoltaic Solar Array Program Update," *Proceedings of the 3rd European Space Power Conference* (Graz, Austria), European Space Agency Publication ESA WPP-054, Paris, France, Aug. 1993.

[5] Bennett, G. L., "The OAST Space Power Program," *Space Photovoltaic Research and Technology—1989*, NASA Conf. Publication 3107, NASA Lewis Research Center, Cleveland, OH, 7–9 Nov. 1989, published 1991.

[6] Mondt, J., Halpert, G., Rapp, D., Surampudi, S., Peterson, C., Frank, H., Burke, K., Manzo, M., Wolff, F., Bragg, B., Rao, G., Vukson, S., Marsh, R., Plichta, E., Sutton, R., Browning, V., Methlie, G., Hwang, W., and Savinall, R., *Energy Storage Technology for Future Space Science Missions*, JPL D-30268, Jet Propulsion Laboratory, Pasadena CA, Nov. 2004.

[7] Corliss, W. R., and Harvey, D. G., *Radioisotopic Power Generation*, Prentice-Hall, Englewood Cliffs, NJ, 1964.

[8] Corliss, W. R., and Mead, R. L., *Power from Radioisotopes*, U.S. Atomic Energy Commission, No. 74-169081, 1971 (rev.).

[9] Bennett, G. L., "Survivability Considerations in the Design of Space Power Systems," *Proceedings of the 23rd Intersociety Energy Conversion Engineering Conference*, Paper No. 889159, Denver, CO, 31 July–5 Aug. 1988.

[10] Bennett, G. L., Lombardo, J. J., and Rock, B. J., "U.S. Radioisotope Thermoelectric Generator Space Operating Experience" (June 1961–Dec. 1982), *Proceedings of the 18th Intersociety Energy Conversion Engineering Conference*, Orlando, FL, 21–26 Aug. 1983; reprinted in *The Nuclear Engineer*, Vol. 25, No. 2, March/April 1984, pp. 49–58.

[11] Bennett, G. L., and Skrabek, E. A., "Power Performance of U.S. Radioisotope Thermoelectric Generators," *Proceedings ICT '96, 15th International Conference on Thermoelectrics*, Pasadena, CA, 26–29 March 1996, IEEE No. 96TH8169, Library of Congress No. 96-75531, pp. 357–372.

[12] Bennett, G. L., "Space Nuclear Power," *Encyclopedia of Physical Science and Technology*, 3rd ed., Vol. 15, Academic Press, New York, 2002, pp. 537–553.

[13] Wyatt, T., "The Gestation of Transit as Perceived by One Participant," *Johns Hopkins APL Technical Digest*, Vol. 2, No. 1, Jan.–March 1981, pp. 32–38.

[14] Johns Hopkins University Applied Physics Laboratory, *Artificial Earth Satellite Designed and Fabricated by the Johns Hopkins University Applied Physics Laboratory*, JHU/APL Report SDO-1600 (rev.), Aug. 1980.

[15] Dick, P. J., and Davis, R. E., "Radioisotope Power System Operation in the Transit Satellite," *AIEE Summer General Meeting*, No. CP 62-1173, Denver, CO, 1962, pp. 17–22.

[16] Dick, P. J., *SNAP-9A Quarterly Progress Report No. 2, Radioisotope-Fueled Thermoelectric Power Conversion System Development, 1 Dec. 1961–Feb. 28 1962*, MND-P-2700-2.

[17] Bradshaw, G. B., and Postula, F. D., "Beginning of Mission Flight Data on the TRANSIT RTG," *Proceedings of the 8th Intersociety Energy Conversion Engineering Conference*, Paper 739091, Philadelphia, PA, 13–16 Aug. 1973.

[18] Fihelly, A. W., and Baxter, C. F., "Orbital Performance of the SNAP-19 Radioisotopic Thermoelectric Generator Experiment," *Proceedings of the 6th Intersociety Energy Conversion Engineering Conference*, Paper 719152, Boston, MA, 3–5 Aug. 1971.

[19] Jaffe, H., and O'Riordan, P., "Isotope Power Systems for Unmanned Spacecraft Applications," *Proceedings of the 7th Intersociety Energy Conversion Engineering Conference*, Paper 729088, San Diego, CA, 25–29 Sept. 1972.

[20] Bates, J. R., Lauderdale, W. W., and Kernaghan, H., *ALSEP Termination Report*, NASA Ref. Publication 1036, 1979.

[21] Pitrolo, A. A., Rock, B. J., Remini, W. C., and Leonard, J. A., "SNAP-27 Program Review," *Proceedings of the 4th Intersociety Energy Conversion Engineering Conference*, Washington, DC, 22–26 Sept. 1969.

[22] Ward, W. W., "Developing, Testing, and Operating Lincoln Experimental Satellites 8 and 9 (LES-8/9)," M.I.T. Lincoln Laboratory Tech. Note 1979-3, 16 Jan. 1979.

[23] Kelly, C. E., "The MHW Converter (RTG)," *Record of the 10th Intersociety Energy Conversion Engineering Conference*, Paper 759132, Newark, DE, 18–22 Aug. 1975.

[24] Ward, W. W., and Floyd, F. W., "Thirty Years of Research and Development in Space Communications at Lincoln Laboratory," *The Lincoln Laboratory Journal*, Vol. 2, No. 1, Spring 1989.

[25] Brittain, W. M., and Skrabek, E. A., "SNAP 19 RTG Performance Update for the Pioneer and Viking Missions," *Proceedings of the 18th Intersociety Energy Conversion Engineering Conference*, Orlando, FL, 21–26 Aug. 1983.

[26] Brittain, W. M., "SNAP-19 Viking RTG Mission Performance," *Proceedings of the 11th Intersociety Energy Conversion Engineering Conference*, Paper 769255, State Line, NV, 12–17 Sept. 1976.

[27] Bennett, G. L., Hemler, R. J., and Schock, A., "Development and Use of the Galileo and Ulysses Power Sources," *Space Technology*, Vol. 15, No. 3, 1995, pp. 157–174; originally presented as Paper IAF-94-R-1-362, *45th Congress of the International Astronautical Federation*, Jerusalem, Israel, 9–14 Oct. 1994.

[28] Lockheed Martin, *GPHS-RTGs in Support of the Cassini RTG Program, Final Technical Report, 11 Jan. 1991 30 April 1998*, Lockheed Martin Astronautics, Doc. No. RR18, Valley Forge Operations, Aug. 1998.

[29] Ulysses objectives. URL: http://sci.esa.int/science-e/www/object/index.cfm?fobjectid=30990 [cited on 20 Jan. 2008].

[30] Ottman, G. K., and Hersman, C. B., "The Pluto–New Horizons RTG and Power System Early Mission Performance," *4th International Energy Conversion Engineering Conference*, AIAA Paper-2006-4029, San Diego, CA, 26–29 June 2006.

[31] Cockfield, R. D., "Preparation of RTG F8 for the Pluto New Horizons Mission," *4th International Energy Conversion Engineering Conference*, AIAA Paper-2006-4031, San Diego, CA, 26–29 June 2006.

[32] Staub, D. W., *SNAP 10A Summary Report*, NAA-SR-12073, Atomics International, Canoga Park, CA, 25 March 1967.

[33] Josloff, A. T., Shepard, N. F., Chan, T. S., Greenwood, F. C., Deane, N. A., Stephen, J. D., and Murata, R. E., "SP-100 Generic Flight System Design and Early Flight Options," *Proceedings of the 11th Symposium on Space Nuclear Power and Propulsion*, American Institute of Physics, CP 301, Pt. 2, Woodbury, NY, 1994, pp. 533–538.

[34] Josloff, A. T., and Mondt, J. F., "SP-100 Space Reactor Power System Readiness to Support Emerging Missions," *Proceedings of the 11th Symposium on Space Nuclear Power and Propulsion*, American Institute of Physics, CP 301, Pt. 2, Woodbury, NY, 1994, pp. 539–545.

[35] Nainiger, J. J., and Mason, L. S., "Nuclear Reactor Power Systems for Lunar Base Applications," *NTSE-92, Nuclear Technologies for Space Exploration*, Proceedings of the American Nuclear Society Topical Meeting on Nuclear Technologies for Space Exploration, 16–19 Aug. 1991, Jackson, WY, American Nuclear Society Order No. 700177, La Grange, IL, pp. 247–256.

[36] Cropp, L. O., Gallup, D. R., and Marshall, A. C., "Mass and Performance Estimates for 5 to 1000 kW(e) Nuclear Reactor Power Systems for Space Applications," *Space Power*, Vol. 10, No. 1, 1991, pp. 43–78.

[37] Black, D. L., and Gunn, S. V., "Space Nuclear Propulsion," *Encyclopedia of Physical Science and Technology*, 3rd ed., Vol. 15, Academic Press, New York, 2002, pp. 555–575.

[38] Bennett, G. L., Finger, H. B., Miller, T. J., Robbins, W. H., and Klein, M., "Prelude to the Future: A Brief History of Nuclear Thermal Propulsion in the United States," *A Critical Review of Space Nuclear Power and Propulsion, 1984–1993*, El-Genk, M. S. (ed.), American Institute of Physics, New York, 1994, pp. 221–267.

[39] Ragsdale, R., "Open Cycle Gas Core Nuclear Rockets," *Nuclear Thermal Propulsion, A Joint NASA/DOE/DOE Workshop*, Proceedings of the Nuclear Thermal Propulsion Workshop, Strongsville, OH, NASA Lewis Research Center, Cleveland, OH, NASA Conference Publication 10079, 10–12 July 1990, pp. 343–357.

[40] Burke, K. A. (ed.), *Advanced Propulsion Concepts*, Jet Propulsion Laboratory, Jan. 1989.

[41] Schnitzler, B. G., Jones, J. L., and Chapline, G. F., *Fission Fragment Rocket Scientific Feasibility Assessment*, Idaho National Engineering Laboratory, EGG-NERD-8585, June 1989.

[42] Braun, R. D., and Blersch, D. J., "Propulsive Options for a Manned Mars Transportation System," *Journal of Spacecraft*, Vol. 28, No. 1, Jan.–Feb. 1991, pp. 85–92.

[43] Sercel, J., and Sargent, M., "Odyssey: A Vision of Human Exploration," NASA Headquarters presentation, Jet Propulsion Laboratory, Pasadena, CA, 2 Feb. 1989.

[44] Ponomarev-Stepnoi, N. N., Talyzin, V. M., and Usov, V. A., "Russian Space Nuclear Power and Nuclear Thermal Propulsion Systems," *Nuclear News*, Vol. 43, No. 13, pp. 33–46, Dec. 2000.

[45] Bennett, G. L., "A Look at the Soviet Space Nuclear Power Program," *Proceedings of the 24th Intersociety Energy Conversion Engineering Conference*, Paper 899009, Crystal City, VA, 7–11 Aug. 1989.

[46] Hoisington, D. B., *Nucleonics Fundamentals*, McGraw-Hill, New York, 1959.

[47] Callas, J. L., *The Application of Monte Carlo Modeling to Matter-Antimatter Annihilation Propulsion Concepts*, Jet Propulsion Laboratory, JPL D-6830, CA, 1 Oct., 1989.

[48] LaPointe, M. R., "Antiproton Powered Propulsion with Magnetically Confined Plasma Engines," *Journal of Power and Propulsion*, Vol. 7, No. 5, 1991, pp. 749–759; based on LaPointe, M. R., "Antiproton Annihilation Propulsion with Magnetically Confined Plasma Engines," Ph.D. Dissertation, Department of Chemical and Nuclear Engineering, The University of New Mexico, Albuquerque, NM, May 1989.

Chapter 18

On Extracting Energy from the Quantum Vacuum

Eric W. Davis* and H. E. Puthoff[†]
Institute for Advanced Studies at Austin, Austin, Texas

I. Introduction

QUANTUM theory predicts that the vacuum of space throughout the universe is filled with electromagnetic waves, random in phase and amplitude, propagating in all possible directions, and with a cubic frequency distribution. This differs from the cosmic microwave background radiation, and is referred to as the electromagnetic quantum vacuum, which is the lowest energy state of otherwise empty space. When integrated over all frequency modes up to the Planck frequency, ν_P ($\sim 10^{43}$ Hz), it represents an energy density of as much as 10^{113} J/m^3, which is far in excess of any other known energy source, although only an infinitesimal fraction of it is accessible. Even if we are constrained to integrate over all frequency modes only up to the nucleon Compton frequency ($\sim 10^{23}$ Hz, the characteristic frequency associated with the size of nucleons), this energy density is still enormous ($\sim 10^{35}$ J/m^3). In addition, the electromagnetic quantum vacuum is not alone; it intimately couples to the charged particles in the Dirac sea of virtual particle–antiparticle pairs, and thereby couples to the other interactions inherent in the Standard Model (weak and strong force vacua). Therefore, all the numbers just mentioned are subject to further refinement. However, it should be noted that we can safely ignore any coupling of the quantum electromagnetic vacuum to the quantum chromodynamic (QCD) vacuum in the context of this chapter because the latter coexists in two phases: 1) the ordinary vacuum exterior to the hadron, which is impenetrable to quark color, and 2) the vacuum interior of the hadron in which the Yang–Mills fields that carry color (gluons) propagate freely. Both vacuum phases are separated

---

Copyright © 2008 by the American Institute of Aeronautics and Astronautics, Inc. All rights reserved.
*Senior Research Physicist.
[†]Director.

by a boundary at the surface of the hadron on which the Yang–Mills and quark fields satisfy boundary conditions.

Even though this zero-point field (ZPF) energy seems to be an inescapable consequence of quantum field theory, its energy density is so enormous as to make it difficult to reconcile. Instead, many quantum calculations subtract away the ZPF energy by ad hoc means (e.g., renormalization). However, we observe the effects of the quantum vacuum ZPF that are responsible for a variety of well-known physical effects, such as:

1) Lamb shift
2) Spontaneous atomic emission
3) Low-temperature van der Waals forces
4) Casimir effect
5) Source of photon shot and fluctuating radiation-pressure noise in lasers
6) Astronomically observed cosmological constant (a.k.a. dark energy, a form of Casimir energy according to the Schwinger–DeWitt quantum ether prescription [1–4])

Rather than eliminate the ZPF energy (a.k.a. ZPE) from the equations, there is much left to be learned by exploring the possibility that it is a real energy. From this perspective, the ordinary world of matter and energy is like foam atop the quantum vacuum sea. If the ZPF is real, then there is the possibility that it can be tapped as a source of power or harnessed to generate a propulsive force for space travel. This notion, of exchanging energy with the quantum vacuum, is the focus of this chapter.

The propeller or the jet engine of an aircraft can push air backward to propel the aircraft forward. A ship or boat propeller does the same thing in water. On Earth there is air or water to push against. But a rocket in space has no material medium to push against, and so it needs to carry propellant to eject in order to provide momentum. A deep space rocket must start out with all the propellant it will ever require, and this quickly results in the need to carry additional propellant just to propel the propellant. The breakthrough one wishes to achieve in deep space travel is to eliminate the need to carry propellant at all. How can one generate a propulsive force without carrying and ejecting propellant?

II. Early Concepts for Extracting Energy and Thermodynamic Considerations

The Casimir force is a force associated with the electromagnetic quantum vacuum [5]. This force is an attraction between parallel uncharged metallic plates that has now been well measured and can be attributed to a minute imbalance in the ZPE density inside the cavity between the plates versus the region outside the plates (Fig. 1) [6–8]. As shown in the figure, the vacuum is full of virtual photons, but photons with wavelengths, λ, more than twice the plate separation, d, are excluded from the space between them, which causes the imbalance that pushes the plates together. However, this is not useful for propulsion because it symmetrically pulls on the plates. If some asymmetric variation of the Casimir force could be identified, then one could in effect sail through space as if propelled by a kind of quantum fluctuation wind. This specific notion is explored in Chapter 12.

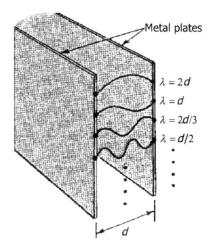

Fig. 1 Schematic of Casimir effect cavity.

The other requirement for space travel is energy. It is sometimes assumed that attempting to extract energy from the vacuum ZPF would somehow violate the laws of thermodynamics. Fortunately, it turns out that this is not the case. A thought experiment published by Forward [9,10] demonstrated how the Casimir force could, in principle, be used to extract energy from the vacuum ZPF. Forward showed that any pair of conducting plates at close distance experiences an attractive Casimir force that is due to the electromagnetic ZPF of the vacuum. A "vacuum-fluctuation battery" can be constructed by using the Casimir force to do work on a stack of charged conducting plates, as shown in Fig. 2. By applying a charge of the same polarity to each conducting plate, a repulsive electrostatic force will be produced that opposes the Casimir force. If the applied electrostatic force is adjusted to be always slightly less than the

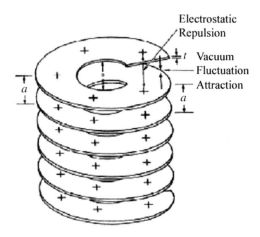

Fig. 2 Vacuum-fluctuation battery. (From Ref. 9.)

Casimir force, the plates will move toward each other and the Casimir force will add energy to the electric field between the plates. The battery can be recharged by making the electrical force slightly stronger than the Casimir force to re-expand the foliated conductor.

Cole and Puthoff [11] verified that (generic) energy extraction schemes are not contradictory to the laws of thermodynamics. For thermodynamically reversible processes, no heat will flow at temperature $T = 0$. However, for thermodynamically irreversible processes, heat can be produced and made to flow, either at $T = 0$ or at any other $T > 0$ situation, such as by taking a system out of mechanical equilibrium. Moreover, work can be done by or on physical systems, either at $T = 0$ or $T > 0$ situations, whether for a reversible or irreversible process. However, if one is considering a net cyclical process on the basis of, say, the Casimir effect, then energy would not be able to be continually extracted without a violation of the Second Law of Thermodynamics. Thus, Forward's process cannot be cycled to yield a continuous extraction of energy. Here, the recharging of the battery would, owing to frictional and other losses, require more energy than is gained from the ZPF. There is no useful engine cycle in this process; nonetheless, the plate-contraction phase of the cycle does demonstrate the ability to cause "extraction" of energy from the ZPF. It does reflect work done by the ZPF on matter.

Another illustrative example of an early scheme for extracting energy from the ZPF is described in a patent by Mead and Nachamkin [12]. They propose that a set of resonant dielectric spheres be used to extract energy from the ZPF and convert it into electrical power. They consider the use of resonant dielectric spheres, slightly detuned from each other, to provide a beat-frequency downshift of the more energetic high-frequency components of the ZPF to a more easily captured form. Figure 3 shows two embodiments of the invention. The device includes a pair of dielectric structures (items 12, 14, 112, and 114 in the figure) that are positioned proximal to each other and which intercept incident ZPE radiation (items 16 and 116). The volumetric sizes of the structures are selected so that they resonate at a particular frequency of the incident radiation. But the volumetric sizes of the structures are chosen to be slightly different, so that the secondary radiations emitted from them (items 18, 20, 24, 118, 120, and 124) at resonance interfere with each other, thus producing a beat-frequency radiation that is at a much lower frequency than that of the incident radiation, and can be converted into electrical energy. A conventional metallic antenna (loop or dipole type, or a RF cavity structure; items 22 and 122 in the figure) can then be used to collect the beat-frequency radiation. This radiation is next transmitted from the antenna to a converter via an electrical conductor or waveguide (items 26 and 126) and converted to electrical energy. The converter must include: 1) a tuning circuit or comparable device so that it can effectively receive the beat frequency radiation, 2) a transformer to convert the energy to electrical current having a desired voltage, and 3) a rectifier to convert the energy to electrical current having a desired waveform (items 28, 30, 32, 34, 128, 130, and 132 in the figure).

The receiving structures are composed of dielectric material in order to diffract and scatter the incident ZPE radiation. The volumetric sizing requirements for the receiving structures are selected to enable them to resonate at a high frequency

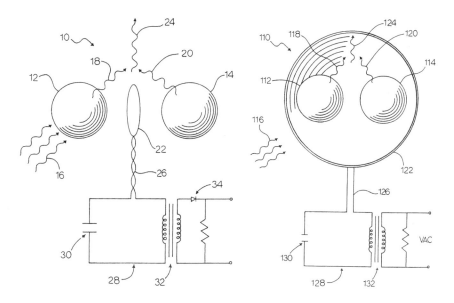

Fig. 3 ZPE resonant dielectric spheres electrical power generator. (From Ref. 12.)

corresponding to the incident ZPE radiation, based on the parameters of frequency of the incident ZPE radiation, and the propagation characteristics of the medium (vacuum or otherwise) and the receiving structures. Because the ZPE radiation energy density increases with increasing frequency, greater amounts of electromagnetic energy are potentially available at higher frequencies. Consequently, the size of the receiving structures must be miniaturized in order to produce greater amounts of energy from a system located within a space or volume of a given size. Therefore, the smaller the size of the receiving structures, the greater the amount of energy that can in principle be produced by the system.

Although a computer model study performed at the Air Force Research Laboratory (Edwards Air Force Base, California) indicates that the invention could work, no experimental study has been performed to validate this in the lab (F. B. Mead, private communication, 2002). Regarding critiques, it is not clear how the beat frequency can be picked up by the receiving loop antenna. There is no nonlinear method in the invention showing that an electromagnetic beat frequency can be generated and coupled to the loop. Without a nonlinear coupling method there will be no sidebands, one of which would be frequency downshifted and called the beat frequency. The coupling method requires the generation of sidebands in the mixing of two different frequencies via a nonlinear technique. However, an easy resolution to this potential deficiency is that the resonant dielectric spheres could be constructed of a nonlinear dielectric material.

Although several novel ZPF energy extraction mechanisms have been proposed in the popular and technical literature, no practicable technique has been successfully demonstrated in the laboratory. To better understand how ZPE extraction methods might work, it is necessary to characterize the physics of the ZPF and proposed energy extraction techniques, and to evaluate their feasibility for application to space power and propulsion systems. In what follows, we summarize the physics of the ZPF and the experimental investigations being pursued to address the question of extracting energy from the quantum vacuum.

III. Origin of Zero-Point Field Energy

A. Quantum Electrodynamics Theory

The basis of the ZPF is typically attributed to the Heisenberg uncertainty principle. According to this principle, A and B are any two conjugate observables that we are interested in measuring, and they obey the commutation relation $[A, B] = i\hbar$ (i is the unit complex number and $\hbar$ is Planck's reduced constant, 1.055×10^{-34} J · s). Their corresponding uncertainty relation is $\Delta A \Delta B \geq \hbar/2$, where ΔA is the variance (a.k.a. uncertainty) of observable A and ΔB is that of the conjugate observable B. This relation states that if one measures observable A with very high precision (i.e., its uncertainty ΔA is very small), then a simultaneous measurement of observable B will be less precise (i.e., its uncertainty ΔB is very large), and vice versa. In other words, it is not possible to simultaneously measure two conjugate observable quantities with infinite precision. This minimum uncertainty is not due to any correctable flaws in measurement, but rather reflects the intrinsic fuzziness in the quantum nature of energy and matter. Substantial theoretical and experimental work has shown that in many quantum systems the limits to measurement precision is imposed by the quantum vacuum ZPF embodied within the uncertainty principle. Nowadays we rather see the Heisenberg uncertainty principle as a necessary consequence, and therefore a derived result of the wave nature of quantum phenomena. The uncertainties are just a consequence of the Fourier nature of conjugate pairs of quantities (observables). For example, the two Fourier wave conjugates, time and frequency, become the pair of quantum-particle conjugates, time and energy; the two Fourier wave conjugates, displacement and wavenumber, become the pair of quantum-particle conjugates, position and momentum. (For more on this see, e.g., Ref. 13.)

Radio and microwaves, infrared light, visible light, ultraviolet light, X-rays, and gamma rays are all forms of electromagnetic radiation. Classically, electromagnetic radiation can be pictured as waves flowing through space at the speed of light. The waves are not waves of anything substantive, but are in fact ripples in the state of a field. These waves carry energy, and each wave has a specific direction, frequency, and polarization state. This is called a "propagating mode of the electromagnetic field." A useful tool for modeling the propagating mode of the electromagnetic field in quantum mechanics is the ideal quantum mechanical harmonic oscillator, a hypothetical charged mass on a perfect spring oscillating back and forth under the action of the spring's restoring force. The Heisenberg uncertainty principle dictates that a quantized harmonic oscillator (a.k.a. a photon state) can never come entirely to rest, because that would be a state of exactly zero energy, which is forbidden by the commutation relation outlined above. Instead, every mode of the field has $\hbar\omega/2$ (ω is the mode or photon frequency and $\hbar\omega$ is the energy of a single mode or photon) as its average minimum energy in the vacuum. (This is a small amount of energy, but the number of modes is enormous, and indeed increases as the square of the frequency. The product of this minuscule energy per mode, multiplied by the huge spatial density of modes, yields a very high theoretical energy density per unit volume.) This ZPE term is added to the classical blackbody spectral radiation energy density $\rho(\omega)d\omega$ (i.e., the energy per unit volume of radiation in the frequency interval $(\omega, \omega + d\omega)$) [14]:

$$\rho(\omega)d\omega = \frac{\omega^2}{\pi^2 c^3}\left[\frac{\hbar\omega}{\exp(\hbar\omega/kT) - 1} + \frac{\hbar\omega}{2}\right]d\omega$$

$$= \frac{\hbar\omega^3}{2\pi^2 c^3}\coth\left(\frac{\hbar\omega}{2kT}\right)d\omega \quad (1)$$

where c is the speed of light (3.0×10^8 m/s), k is Boltzmann's constant (1.3807×10^{-23} J/K), T is the absolute temperature, and $\omega = 2\pi\nu$ is the angular frequency. The factor outside the square brackets in the first line of Eq. (1) is the density of mode (or photon) states (i.e., the number of states per unit frequency interval per unit volume); the first term inside the square brackets is the standard Planck blackbody radiation energy per mode; and the second term inside the square brackets is the quantum zero-point energy per mode. Equation (1) is called the Zero–Point Planck (ZPP) spectral radiation energy density. Planck first added the ZPE term to the classical blackbody spectral radiation energy density in 1912, although it was Einstein, Hopf, and Stern who actually recognized the physical significance of this term in 1913 [14]. Direct spectroscopic evidence for the reality of ZPE was provided by Mulliken's boron monoxide spectral band experiments in 1924, several months before Heisenberg first derived the ZPE for a harmonic oscillator from his new quantum matrix mechanics theory [15].

Following this line of reasoning, quantum physics predicts that all of space must be filled with electromagnetic zero-point fluctuations (a.k.a. the zero-point field or ZPF) creating a universal sea of ZPE. The density of this energy depends critically on where the frequency of the zero-point fluctuations ceases. Because space itself is currently thought to break up into a kind of "quantum foam" at the Planck length, ℓ_P ($\sim 10^{-35}$ m), it is argued that the ZPF must cease at the corresponding ν_P. If true, then the ZPE density would be $\sim 10^{113}$ J/m^3, 108 orders of magnitude greater than the radiant energy at the center of the sun! Formally, in quantum electrodynamics (QED) theory, the ZPE energy density is taken as infinite; however, arguments based on quantum gravity considerations yield a finite cutoff at ν_P. Therefore, the spectral energy density is given by $\rho(\omega)d\omega = (\hbar\omega^3/2\pi^2 c^3)d\omega$, which integrates to an energy density, $\rho_E = \hbar\nu_P^4/8\pi^2 c^3 \approx 10^{113}$ J/m^3. As large as the ZPE is, interactions with it are typically cut off at lower frequencies depending on the particle coupling constants or their structure. Nevertheless, the potential ZPF energy density predicted by quantum physics is enormous.

B. Stochastic Electrodynamics Theory

An alternative to QED, stochastic electrodynamics (SED) identifies the origin of the ZPF as a direct consequence of a classical ZPF background. SED begins with the ordinary classical electrodynamics of Maxwell and Lorentz. However, instead of assuming the traditional homogeneous solution of the source-free differential wave equations for the electromagnetic potentials, it instead considers that due to multiple charged particles moving throughout the universe, there is always a random electromagnetic radiation background present that affects the particle(s) in any experiment. This new boundary condition (random radiation background) replaces the prior null background of traditional classical electrodynamics. Moreover, the principle of relativity dictates that identical experiments performed in different inertial frames must yield the same result, and that this random classical electromagnetic radiation must be isotropic in all inertial frames; it is invariant under scattering by a dipole oscillator, invariant under redshift (Doppler, cosmological, gravitational, no Einstein–Hopf drag force), and must therefore have a Lorentz-invariant energy density spectrum. The only energy density spectrum that obeys such conditions is one that is proportional to the cubic power of the frequency. Interestingly, this is exactly the same frequency dependence as that of the QED spectral ZPF energy density described above, when the temperature T is set to zero in Eq. (1). Thus in SED, the random radiation assumes the role of the ZPE of QED, and is termed the classical electromagnetic ZPE. Planck's constant appears then in SED as an adjustable parameter that sets the scale of the ZPE spectral density.

The formulation of the SED model has evolved over time, beginning with the work of Nernst in 1916 and the later foundational work of Marshall and Boyer in the 1960s [14]. The original standard SED model was based on random phases with fixed electric-field mode amplitudes. The more recent modified SED model employs random phases with random electric-field mode amplitudes and a full probability distribution for the ground state amplitude, in agreement with quantum theory [16]. A comparison of SED with quantum theory shows that the first and second moments of the spectral energy distribution are identical, but beyond that, the distributions diverge widely. Nevertheless, several quantum theory results have been reproduced by means of the SED approach, such as [14,17]:

1) Quantum mechanical harmonic oscillator
2) Lamb shift
3) Blackbody radiation
4) van der Waals forces
5) Casimir forces
6) Diamagnetism
7) Davies–Unruh effect

The strength of the SED model is that it is heuristically appealing, with transparent derivations, and it is applicable to linear systems. SED calculations have also been shown to be in one-to-one correspondence with the expectation values of the Heisenberg quantum equations of motion for linear systems. Both SED and QED will play a role in the discussions to follow.

IV. Review of Selected Experiments

In what follows, we outline each of the proposed experimental concepts that were selected for theoretical and laboratory investigation. A subset of our proposed concepts has undergone preliminary evaluation by Lockheed–Martin review panels involving both internal R&D personnel and outside experts on theory and experimentation (V. Teofilo, private communication, 2005).

A. Voltage Fluctuations in Coils Induced by ZPF at High Frequency

In a series of experiments, Koch et al. [18–20] measured voltage fluctuations in resistive wire circuits that are induced by the ZPF. Their result is striking corroboration of the reality of the ZPF and proves that the ZPF can do real work (cause measurable currents). Although the Koch et al. experiment detected minuscule amounts of ZPF energy, it shows the principle of ZPF energy circuitry to detect vacuum fluctuations and opens the door to consideration of means to extract useful amounts of energy. The secondary consequences on other phenomena, if energy can be successfully extracted, have not yet been investigated.

Blanco et al. [21] have proposed a method for enhancing the ZPF-induced voltage fluctuations in circuits. Theoretically treating a coil of wire as an antenna, they argue that the antenna-like radiation resistance of the coil should be included in the total resistance of the circuit, and suggest that this total resistance should be used in the theoretical computation of ZPF-induced voltage fluctuations. Because of the strong dependence of the radiation resistance on the number of coil turns (quadratic scaling), coil radius (quartic scaling), and frequency (quartic scaling), any enhanced ZPF-induced voltage fluctuations should be measurable in the laboratory at readily accessible frequencies (100 MHz compared to the 100 GHz range necessary in the Koch et al. experiments).

In the theory of Blanco et al., random voltage fluctuations are conveniently described by their frequency spectrum. That is, given a sufficient time interval of measured voltages, the measurements are Fourier transformed to the frequency domain to determine how the voltage fluctuations are distributed (e.g., quantity of low-frequency, long-duration fluctuations relative to high-frequency, short-duration fluctuations). Theoretically, the spectrum of voltage fluctuations, $S(\omega, T)$, of a resistive circuit is given by [21]:

$$S(\omega, T) = \frac{R(\omega, T) \hbar \omega}{\pi} \frac{1}{2} \coth\left(\frac{\hbar \omega}{2kT}\right) \quad (2)$$

where $R(\omega, T)$ is the total resistance (ohmic plus radiative), ω is the (angular) frequency, and T is the absolute temperature. The resistance $R(\omega, T)$ is temperature dependent through its ohmic contribution (the radiation resistance depends only on frequency). Note the similar hyperbolic cotangent functions appearing in Eq. (2) and in the second line of Eq. (1). The postulate of Blanco et al. is that the total resistance must include the radiation resistance of the circuit [21]:

$$R(\omega, T) = R_{\text{ohmic}}(\omega, T) + R_{\text{rad}}(\omega) \quad (3)$$

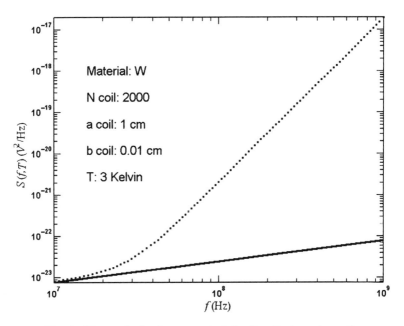

Fig. 4 Theoretical voltage spectral density of a tungsten coil.

Under the assumption that the wavelengths of the ZPF modes of interest are larger than the dimensions of the circuit, the radiation resistance of a coil is given by [21]:

$$R_{rad}(\omega) = \frac{2}{3}\frac{\pi^2 N^2}{c}\left(\frac{a\omega}{c}\right)^4 \qquad (4)$$

where N is the number of coil turns, and a is the radius of the coil winding.

According to Blanco et al., large enhancements in ZPF-induced voltage fluctuations are possible. By reducing the temperature to minimize ohmic resistance, making the coil of many turns and a large radius, and performing measurements at high frequency, it should be possible to investigate this amplification effect. The predicted coil-enhanced voltage spectrum can readily be computed. The result is shown in Fig. 4 for a 1-cm diameter coil of 2000 turns, made of 38 AWG tungsten wire, and kept at a temperature of 3K. In Fig. 4, the upper (dotted) curve represents the predicted voltage spectral density for the combined ohmic plus radiation resistance. The lower (solid) curve is the predicted result when radiation resistance is ignored. If the postulate of Blanco et al. is correct, the enhancement in voltage fluctuations due to the antenna-like nature of the coil should be easily measured at frequencies as low as 100 MHz (where the coil enhancement effect is $\sim$100-fold for tungsten).

To successfully measure the ZPF-induced voltage fluctuations, the requirements of low temperature, large coil, and high frequency must be met. The low-temperature requirement is met by performing the experiment in a cooled dewar. Existing high-quality cryogenic dewars (pumped down to 3K) and sensitive laboratory instruments are suitable for the measurements. The cold spot in one particular dewar under consideration is cylindrical, 2.5 cm in both diameter and height. The largest coil that can be installed will thus have a coil radius of approximately $a = 1$ cm. To keep the linear dimension of the coil small will require a small wire thicknesses, perhaps $b = 0.01$ cm (gauge 38 AWG). By winding the coil in a number of layers (10 or 12 layers), a large number of turns can be accommodated, perhaps $N = 2000$ turns. To minimize ohmic resistance, wire made of tungsten (W) is preferred; however, copper (Cu) is a suitable alternative.

Voltage fluctuations in the 100 MHz range are easily detected using commercially available laboratory equipment; hence this experiment could be performed using tungsten without resorting to the more sophisticated Josephson junction techniques required by Koch et al. for their higher frequency measurements. For a copper wire coil, the magnitude of the enhancement effect is reduced somewhat compared to the tungsten results shown in Fig. 4. But for frequencies approaching the GHz regime, the radiation resistance enhancement effect in copper wire is still predicted to be over four orders of magnitude larger. Commercial equipment readily allows measurements of the voltage spectrum in the GHz regime. Therefore, given a cost tradeoff of copper versus tungsten coil fabrication, the use of copper coils may be preferred. A second coil can be used in a control experiment constructed with the same parameters as the first coil, but with half of its turns wound in the reverse direction. This will make the coil non-inductive so that its voltage spectral density should correspond to the lower solid curve in Fig. 4.

B. Zero-Point Field Energy Extraction by Ground State Energy Reduction

As first analyzed by Boyer [22], and later refined by Puthoff [23], the following paradox was addressed: although atomic ground states involve electrons in accelerated motion, such states are nonetheless radiationless in nature, even though it is well known from classical electrodynamics that charged particles undergoing acceleration must always emit radiation. For the standard Bohr ground state orbit of the hydrogen atom, this was interpreted as an equilibrium process in which radiation by the electron in its ground state orbit was compensated by absorption of radiation from the background vacuum electromagnetic ZPE. This interpretation has recently been strengthened by the analyses of Cole and Zou [24,25] using a SED model for the vacuum ZPE. Because the balance between emitted orbital-acceleration radiation and absorbed ZPE radiation is modeled as taking place primarily at the ground state orbital frequency, one can consider the possibility of using this feature in some type of mechanism to extract energy from the ZPF. One fundamental difference between the SED interpretation and that of quantum mechanics is that in quantum mechanics, the $1s$ state of the electron is regarded as having zero angular momentum, whereas in the SED interpretation, the electron has an angular momentum of $m_e c r_e / 137$ (m_e = electron mass, r_e = electron radius, atomic fine structure constant $\alpha = 1/137$, and $c/137$ is the classical orbital velocity of the ground state electron).

The Bohr radius of the hydrogen atom in the SED view is 0.529 Å. This implies that the wavelength (λ) of zero-point radiation responsible for sustaining the orbit is $2\pi \cdot 0.529 \cdot 137 = 455$ Å (or 0.0455 µm). It has been conjectured by Puthoff and Haisch (private communication, 2004) that suppression of zero-point radiation at this wavelength (and at shorter wavelengths) inside a Casimir microcavity could result in the decay of the electron to a lower energy state determined by a new balance between classical emission of an accelerated charge and absorption of zero-point radiation at $\lambda < 455$ Å, where λ depends on the microcavity plate separation (d). Because the frequency of this orbit is 6.6×10^{15} Hz, no matter how quickly the atom were to be injected into a Casimir microcavity, one would assume that the decay process would be a slow one as experienced by the orbiting electron. Figure 5 shows a schematic representation of a hydrogenic atom in free space and inside a microcavity.

Consider the possibility that the decay to a new sub-Bohr ground state would involve gradual release of energy in the form of heat, rather than a sudden optical radiation signature. Because the binding energy of the electron is 13.6 eV, it is estimated that the amount of energy released in this process could be on the order of 1 to 10 eV for injection of the hydrogen atom into a Casimir cavity of $d = 250$ Å. Furthermore, consider the possibility that when the electron exits the cavity it would reabsorb energy from the ZPF and be re-excited to its normal state. If these conjectures were to be verified by experiment, then the energy extracted in the process comes at the expense of the ZPF, which in the SED interpretation propagates at the speed of light throughout the universe. In effect, the energy would be extracted locally and replenished globally. The secondary consequences on other phenomena, if this energy conversion were to succeed, have not yet been investigated. However, on a cautionary note, the conflicts between SED and QED theories (discussed in Sec. V) raise questions as to

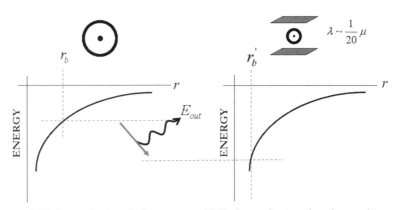

a) Hydrogenic atom in free space   b) Hydrogenic atom in microcavity

Fig. 5 Energy released from ground state suppression of hydrogenic atom in a microcavity. (r_b = free-space Bohr orbit radius, r_b' = suppressed Bohr orbit radius, λ = resonant wavelength of Bohr orbit, and E_{out} = released energy.)

whether the conjectured approach discussed here is viable. This issue is perhaps best addressed by experiment for its resolution.

In terms of an experimental test, consider using monatomic gases or liquids flowing in a block with Casimir tunnels, which has the following attributes: 1) no dissociation process is required for monatomic gases or liquids, 2) heavier element atoms are approximately two to four times larger than hydrogen and thus can utilize and be affected by a larger Casimir cavity, and 3) heavier elements have numerous outer shell electrons, several of which may be simultaneously affected by the reduction of zero-point radiation in a Casimir cavity.

All of the noble gas elements contain ns electrons. He ($Z = 2$, $r = 1.2$ Å) has two $1s$ electrons. Ne ($Z = 10$, $r = 1.3$ Å) has two each of $1s$ and $2s$ electrons. Ar ($Z = 18$, $r = 1.6$ Å) has two each of $1s$, $2s$, and $3s$ electrons. Kr ($Z = 36$, $r = 1.8$ Å) has two of each of $1s$, $2s$, $3s$, and $4s$ electrons. Xe ($Z = 54$, $r = 2.05$ Å) has two of each of $1s$, $2s$, $3s$, $4s$ and $5s$ electrons. Larger Casimir cavities would also be expected to have an effect on the energetics of the outer electron shells (at larger radii). One could therefore expect that a Casimir cavity having $d = 0.1$ μm could have an effect on reducing the energy levels of the outermost pair of s electrons, and possibly also p electrons and intermediate shell s electrons as well.

Continuing with this model, it is reasonable to expect that a 0.1 μm Casimir cavity could result in a release of 1 to 10 eV for each injection of a He, Ne, Ar, Kr, or Xe atom into such a cavity. According to Maclay [26], a long cylindrical Casimir cavity results in an inward force on the cavity walls due to the exclusion of interior ZPF modes. In the "exclusion of modes" interpretation of the Casimir force, this implies that a cylindrical cavity of diameter 0.1 μm could yield the desired decay of outer shell electrons and subsequent release of energy. If we let the length of the cylinder be 100 times the width, this results in $\ell = 10$ μm for the length of the Casimir tunnel. Taking advantage of this effect, Puthoff and Haisch (private communication, 2004) propose a segmented tunnel consisting of alternating conducting and nonconducting materials, each 10 μm in length. In a length of 1 cm, there could be 500 such pairs in segments, resulting in 500 energy releases (each yielding 1 to 10 eV) for each transit of an atom through the entire 1-cm long Casimir tunnel.

Now consider a 1-cm^3 block that is built up of 10 μm thick alternating layers as described above. Assume that tunnels of 0.1-μm diameter could be drilled through the cube perpendicular to the layers (this is not physically possible, of course; tunnel manufacture must be done differently). If 10% of the cross section comprises entrance to some 1.3 billion tunnels, then the amount of energy released would be proportional to the flow rate of the gas through the tunnels (for the number of entrances and exits through Casimir segments). A flow rate of 10 cm/s through a total cross-sectional area of 0.1 cm^2 yields 1 cm^3 of gas per second flowing through the tunnels, which at STP would be 2.7×10^{19} atoms. A very simple sealed, closed-loop pumping system could maintain such a continuous gas flow. Because each atom interacts 500 times during its passage, there would be 1.3×10^{22} transitions per second in the entire cube of 1 cm^3. An energy release of 1 to 10 eV per transition corresponds to 2150 to 21,500 W of power released for the entire Casimir cube of tunnels. However, again, all of this assumes that the chain of conjectures detailed above is correct. Fortunately, this can be experimentally tested.

Microcavity fabrication to match the atomic ground states is daunting because there will potentially be fabrication irregularities that cause edge and surface effects that act upon the particles as they enter or exit the Casimir region. And it is not possible to drill 1.3 billion tunnels having diameters of 0.1 μm. However, it should be feasible to use microchip technology to etch holes into the individual layers first and then assemble the stack. Extremely fine coregistration and alignment of stacks would be an issue, but a surmountable one. A much smaller number of layer pairs and tunnels would suffice for a measurable demonstration of release of ZPE by this process. If such a small-scale demonstration succeeds, larger versions that convert more energy could be built that also take advantage of more efficient thermal-to-electrical energy conversion methods. If successful, such apparatus could also be used to explore for secondary effects of converting quantum vacuum energy into thermal then electrical energy.

Further investigation by Puthoff et al. [27] was based on the premise that the above principle is broadly applicable to other than just atomic ground states. In their experiment, H_2 gas was passed through a 1-μm Casimir cavity to suppress the ZPE radiation at the vibrational ground state of the H_2 molecule. The anticipated signature for such a process would be an increase in the dissociation energy of the molecule. Initial experiments, shown in Fig. 6, were carried out at the Synchrotron Radiation Center at the University of Wisconsin at Madison, where an intense UV beam is available to disassociate gas molecules. Further experimentation to investigate this hypothesis has yet to be completed. (See Chapter 20 in this text for additional details.)

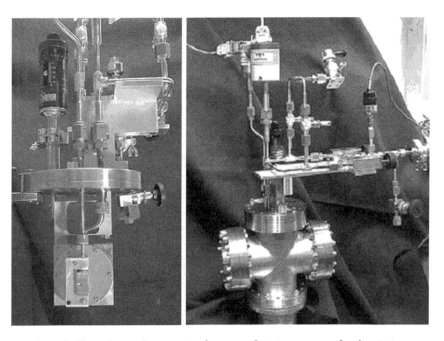

Fig. 6 **Experimental apparatus for ground state energy reduction tests.**

C. Tunable Casimir Effect

As previously discussed, the Casimir effect is a unique ZPF-driven quantum force that occurs between closely spaced conductive cavity walls (or plates). If left unfettered, the plates will collapse together and energy is converted from the ZPF into heat (or other forms of energy) in accordance with the expression $E/A = -\pi^2 \hbar c/720 d^3$, where E/A is the energy per unit area of the plates and d is the plate separation. Investigation of this mechanism by Cole and Puthoff [11] showed that this process fully obeys energy conservation and thermodynamic laws.

Although the Casimir force is conservative (thus the Casimir device might appear to be a one-shot device), the fact that the attractive Casimir force is weaker for dielectric plates compared to conductive plates raises the possibility of the use of thin-film switchable mirrors to obtain a recycling engine [28–30]. Figure 7 shows a comparison of the strength of the Casimir force in a conductive cavity with that in a dielectric cavity. In such an application the plates are drawn together by the stronger force associated with the conducting state and withdrawn after switching to the dielectric state. The engine cycle for this concept is shown in Fig. 8. Assuming optimistic conditions for practical devices (negligible energy required for switching; plate separation oscillations between 30 nm and 15 nm for 1 cm² plates; driving circuit ≈ 10 times the weight of the Casimir plates, etc.), an estimate of the achievable power might be obtained. Based on the described parameters, and assuming a switching from a purely conductive state to a dielectric constant of $K = 4$, yields a figure of merit of $\approx 35 \times f(\mathrm{MHz})$ W/kg (f = switching rate) for the power density [28]. This can be compared to the power density of ≈ 5 W/kg achieved by current radioisotope thermoelectric generators. The predicted output power per unit area for this experimental device is $\approx 10^{-6} f(\mathrm{MHz})/4[d(\mu m)]^3$ W/cm².

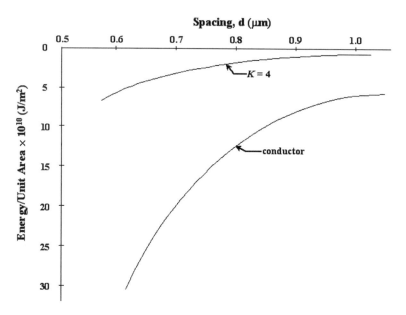

Fig. 7 Tunable Casimir effect: conductor versus dielectric.

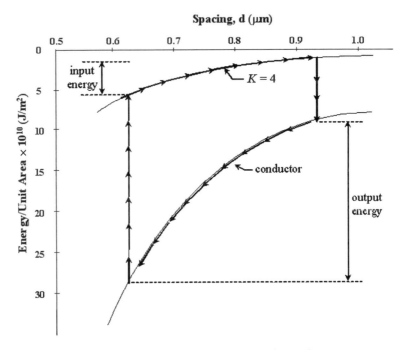

Fig. 8 Tunable Casimir effect: engine cycle.

Another "tunable" conductive-type plate experiment under consideration involves the use of plates consisting of three-dimensional photonic crystals, with the bandgap of the photons that can transmit through the structure being a "tunable" value. Using microelectro-mechanical processing methods, Sandia National Laboratories has produced such crystals and are researching methods of actively modifying the structures while in use [31]. The technology requirements for this concept are the nano-fabrication of microcavities with thin-film deposited surfaces, RF-driven piezoelectric mounts for cavity oscillation, mirror-switching modality (e.g., hydrogen pressure modulation), and calorimetric measurement of energy/heat production.

An initial experiment to explore this concept was recently performed by Iannuzzi et al. [32]. They investigated the effect of hydrogen switchable mirrors (HSMs) on the Casimir force. HSMs are shiny metals in their "as deposited" state. However, when they are exposed to a hydrogen-rich atmosphere, they become optically transparent. Because the electromagnetic ZPF depends on the optical properties of the surfaces, the Casimir force of attraction between two HSMs in air should be different than the attraction between the same HSMs immersed in a hydrogen-rich atmosphere. That is because one expects that the Casimir force will be much weaker when the HSM is in the transparent state rather than in the reflective state. The experiment tested this for plate separations of 70 to 400 nm.

Iannuzzi et al.'s experimental results showed that the Casimir force did not noticeably decrease after filling the experimental apparatus with hydrogen.

This may have occurred for two reasons. First, the dielectric properties of the HSMs used in the experiment are known only in a limited range of wavelengths spanning 0.3 to 2.5 μm, while the experiment measured the transparency of the HSMs over a wavelength range of 0.5 to 3 μm. This narrower wavelength span excludes the rest of the electromagnetic ZPF modes having wavelengths shorter than 0.5 μm and longer than 3 μm. The ZPF modes lying outside this narrow wavelength span were not affected by the hydrogenation-induced transparency of the HSMs, hence their contribution to the total Casimir force acting between the HSMs was not included. One would expect to see a significant decrease of the Casimir force if the hydrogenation-induced transparency of the HSMs had affected all of the ZPF mode wavelengths ranging from IR to UV (ZPF modes with $\lambda \gg 2.5$ μm will not give rise to large contributions to the force). Second, the experiment demonstrated a property of the Lifshitz theory (see Ref. 32 for more detail), that in order to significantly change the Casimir force between surfaces at separations on the order of 100 nm it is not sufficient just to change their optical (IR and visible) reflectivity, but it is necessary to modify their dielectric functions over a much wider spectral range. This comports with the first reason, and indicates that more theoretical and experimental work is needed to overcome the shortcomings of this experiment, and to allow for the design and testing of new experiments that can achieve Casimir plate transparency over a wider spectral range.

A notion similar to the tunable Casimir effect involves changing the dimensions of a rectangular "Casimir box." Forward [33] proposed a paradox in which energy could be extracted by altering the aspect ratio of a conductive rectangular Casimir cavity over a specific cycle of dimension changes (e.g., varying width while holding length constant). It was subsequently shown by Maclay [26,34], that the Casimir energy inside the box is not isotropic, varying in such a way that more work is expended in cycling the box dimensions than can be extracted. It appears that no net gain of energy is theoretically possible in this scheme. Whether such considerations apply to the tunable Casimir cavity concept remains to be assessed.

D. Electromagnetic Vortex Phenomenon

Shoulders [35] developed an experimental program to explore the physics of microscopic plasma vortices (a.k.a. force-free plasmoids), which are thought to be a form of ball lightning [36]. This study was motivated by the earlier experimental work of Wells at the Princeton University Plasma Physics Laboratory, Bostick and Nardi at the Stevens Institute of Technology, and their collaborators [37–44]. Shoulders became interested in the possibility of stable, quantized force-free structures that could be taken apart by some process to yield a net energy gain for power generation. The foundation for this speculation was Nardi et al.'s [44] observation of strange electron concentrations they called vortex filaments that formed in an electron beam made by plasma focus or relativistic electron beam machines, which exhibited electron concentrations that appeared to violate the space charge law. Furthermore, Nardi et al. observed that the vortex filaments were striking exposed materials (metals, dielectrics, ceramics, glass, etc.), boring smooth channels straight through them and sometimes exploding with such a large force that they created impact craters or holes in the materials. Piestrup et al. [45] performed more recent experiments to investigate this unusual phenomenon. This discovery inspired Shoulders to consider vortex filaments as a potential new source of energy, and hence he named them electromagnetic vortices or "EVs." However, given that he could not experimentally verify the vortex nature of the phenomenon, he later redefined EV to mean electrum validum (roughly translated as *strong electron*).

Bostick and Shoulders began collaborating and realized that EVs were much easier to generate and observe using micro-arc discharge devices because they are usually obscured in large high-power plasma machines by surrounding plasma. This led Shoulders to design a series of low-voltage, low-power micro-arc discharge (or condensed-charge emission) devices to produce EVs in the lab. Figure 9 shows a schematic diagram for one embodiment of an EV (pulse discharge source) device. The EVs are generated at the cathode tip and then follow the path (dashed line above the dielectric) to the impact site on the ground plane. The EVs generated by such devices were able to reproduce the material damage observed in Nardi et al.'s earlier experiments. Figure 10 shows a scanning electron microscope (SEM) photograph of the damage inflicted by a single EV burst fired along an aluminum-oxide ceramic plate. The EV bored through the ceramic, forming a smooth symmetrical channel along its path.

Shoulders' experimental studies claim that EVs have physical characteristics corresponding to the phenomenon observed by Nardi et al. His conclusions were that EVs are compact, spherically shaped balls (diameter ≈ 1 to 20 μm) of condensed high-density charge ($\sim 10^{30}$ electrons/m^3) with an internal electric field $> 10^8$ V/m, a charge-to-mass ratio of 1.7588×10^{11} Coulomb/kg ($\approx$ electron's charge-to-mass ratio), and a surface current density of 6×10^{15} Amps/m^2 [35]. Shoulders also reported that EVs are a source of (copious) X-rays; a single EV discharge gun can produce multiple EVs in which the coupling between adjacent EVs produces quasi-stable structures (chains); and EVs respond like an electron under deflection by external fields of known polarity.

Because electrons would not be expected to bind together due to their mutual Coulomb repulsion, a speculative model based on the vacuum electromagnetic ZPF was formed to explain the existence of EVs. The emerging laboratory

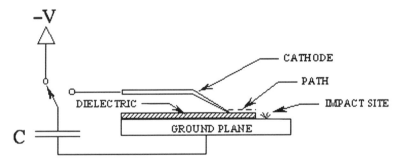

Fig. 9 Schematic of EV (pulse discharge source) device (C = capacitor; V = voltage). (From Ref. 46.)

ON EXTRACTING ENERGY FROM THE QUANTUM VACUUM 587

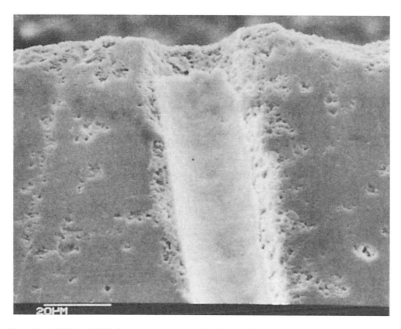

Fig. 10 SEM of EV damage to ceramic plate (20-μm scale). (From Ref. 35.)

evidence led them to consider the hypothesis that the Casimir effect may be a major contributing mechanism to the formation of EVs in micro-arc discharges. This conjecture is based on models by Casimir [47] and Puthoff and Piestrup [48] suggesting that the generation of a relatively cold, dense, non-neutral (charged) plasma results in charge-condensation effects that may be attributable to a Casimir-type pinch effect (i.e., ZPF-induced pressure forces) in which the inverse square-law Coulomb repulsion is overcome by an attractive inverse fourth-law Casimir force to yield a stable configuration of bound charges at small dimensions. This is a derivative of Casimir's semi-classical model of the electron in which a dense shell-like distribution of charge might suppress vacuum fields in the interior of the shell [47]. However, initial application of Casimir's model found that the vacuum field inside the modeled electron was found to augment rather than offset the divergent Coulomb field thus rendering the electron's self-energy divergent. Puthoff [49] later resolved this problem by developing a self-consistent vacuum fluctuation-based model in which the net contribution to the point-like electron's self-energy by its Coulomb and vacuum fields vanishes, thus rendering a stable finite-mass electron.

Shoulders and collaborators subsequently investigated different approaches to extracting useful energy from the vacuum ZPF by way of exploiting the EV phenomenon. Even though EVs can be easily produced in the lab, efforts to test this hypothesis have not met with success (see Chapter 20 for more details). However, this topic is suitable for future research.

V. Additional Considerations and Issues

A. Quantum Electrodynamic Vacuum Revisited

1. Quantum Electrodynamic Vacuum as a Plenum

Continued theoretical and experimental research has revealed that the vacuum constitutes an active agent that contributes to a host of phenomena ranging from microscopic level shifts of atomic states to possible connections to the cause of cosmological expansion [14,50]. As more of its attributes are explored, the vacuum has been found to exhibit phenomena characteristic of an optical medium, such as induced birefringence in the presence of an applied magnetic field [51], and breakdown (decay) in the presence of external electric fields [52–54]. The current view is that the vacuum has structure, and can be considered much like a medium of classical physics. However, the vacuum differs significantly from that of a classical medium due to the existence of quantum fluctuations. A primary attribute of quantum theory is the concept of matter and field fluctuations, rooted in Heisenberg's uncertainty principle.

In second-quantized QED theory, the theory that applies to the electromagnetic vacuum, the canonical approach to representing fluctuations of the free vacuum electromagnetic field is to express the field distribution in terms of standing- or traveling-wave normal modes. Section I suggested that the large value of the integrated ZPE density fuels the concept of potentially useful vacuum energy conversion to other forms, should even some small part of the spectral distribution be accessible for conversion by technological means.

2. Quantum Electrodynamic Vacuum as a Mathematical "Placeholder" for Fluctuating Matter Fields

The treatment of the QED vacuum as a fluctuating plenum with (formally) infinite energy density has caused some physicists to question the viability of the second-quantized QED formalism. Jaynes, for example, in considering the consequences for calculation of the Lamb shift of the 2s level of the hydrogen atom under the assumption of a much more modest electron Compton frequency cutoff ($\sim 10^{21}$ Hz), calculates a fluctuating power flow for the Poynting vector of 6×10^{20} MW/cm^2—comparable in every square centimeter to the total power output of the sun—and states that "real radiation of that intensity would do a little more than just shift the 2s level by 4 microvolts [55]." However, despite alternatives to the formalism of QED that have been suggested (more on this later), second-quantized QED cannot be lightly dismissed. This is so even though the infinities that must be dealt with by such procedures as renormalization caused even one of its founders, Paul Dirac, to remark: "This is just not sensible mathematics. Sensible mathematics involves neglecting a quantity when it turns out to be small—not neglecting it because it is infinitely great and you do not want it! [56]."

A second argument that can be raised against using the QED formalism to further explore vacuum fluctuation physics is that, despite the magnitude of the energy density potentially associated with vacuum electromagnetic fluctuations, observation of the cosmological constant—a measure of net vacuum energy density—has a value that is only on the order of the average energy density of matter in the universe $\approx 10^{-9}$ J/m^3 [57, 58]. This leads to what is often referred to as the 120 orders-of-magnitude problem, or "cosmological coincidence." Rather than the QED value being discounted, the resolution of this problem is thought to lie in the domain of infinity (or divergent integral) cancellations, requiring instead an accounting for the fine-tuning requirements of such cancellations [59,60].

Again, the root cause of the difficulties that accompany second quantization of the vacuum field is that an unbounded plenum possesses an infinite number of degrees of freedom, each with its assigned ground-state fluctuation energy. In an attempt to circumvent the difficulties associated with an unbounded, second-quantized plenum, alternative approaches to QED have been explored in the literature in some detail, a few of which are discussed in Sec. V.E. A number of these alternative viewpoints interpret the second-quantized QED vacuum with its infinite degrees of freedom as simply an overidealized mathematical placeholder for "real" fields that originate in matter fluctuations whose number of degrees of freedom is necessarily always limited. Nevertheless, though the alternative formalisms and associated interpretations differ significantly from the canonical approach, detailed calculations yield results identical to those generated by the second-quantized field formalism. As a result, even treated as a mathematical placeholder for matter fluctuation fields, at this point in our discussion the QED value must be taken seriously. The proposed corollary concerning the potentially significant conversion of QED vacuum energy to other forms is further evaluated in the sections that follow.

B. Casimir Effect Revisited

The most-quoted quintessential configuration for the conversion of vacuum energy to other forms of energy is the Casimir effect. As previously discussed, when parallel conducting plates are placed in a vacuum, they attract one another by a very weak force that varies inversely as the fourth power of the distance between them. First computed by Casimir in terms of van der Waals forces (a matter-fields approach; see below), he soon realized that, because the force turns out to be independent of the molecular details of the conductors, it could be computed as a problem in vacuum energy, and that is the way it is now generally presented in the literature (the "plenum approach") [5,61].

1. Casimir Effect in the Plenum Picture

One begins with the free quantum vacuum electromagnetic field fluctuations, and then determines their modification due to the insertion of two parallel plane conductors (i.e., plates) as additional boundary conditions, which constrain a discrete set of intra-cavity modes of integer half-wavelengths. Aside from an unobservable, high-frequency-cutoff-dependent, free-field term that remains from the mathematical regularization procedure, the resulting (renormalized) vacuum stress-energy tensor is given by $\langle T^{\mu\nu} \rangle_{\text{vac}} = (\pi^2 \hbar c / 720 d^4) \text{diag}(-1, 1, 1, -3)$, where the angular brackets denote the quantum (vacuum state) expectation value of the tensor $T^{\mu\nu}$, d is the plate separation, and $\text{diag}(-1, 1, 1, -3)$ denotes the diagonal elements of a 4×4 matrix [1–3,61]. (For this derivation, the vacuum fluctuations of other quantum fields are essentially undisturbed by the presence of the conductors or are affected only in the immediate vicinity of the atomic nuclei that they contain.) $\langle T^{\mu\nu}\rangle_{\text{vac}}$ represents the real physical stress carried by the vacuum field fluctuations in the presence of the parallel plane conductors, and it encodes the Casimir effect in terms of 1) an interaction energy per unit area, $E/A = -\pi^2 \hbar c/720d^3$, and 2) a corresponding force per unit area, $F/A = -\pi^2 \hbar c/240d^4$. If free to move in response to the attractive Casimir force, the motion of the plates toward each other is understood in the plenum approach to progressively eliminate intra-cavity modes, converting their associated ground-state energies first into kinetic energy, and then, upon collision of the plates, into heat. In Section II we described the Casimir force-driven collapse of Forward's charged slinky as a Casimir-type configuration for building up an electric field to charge a battery, and how such processes were shown not to violate either conservation of energy or thermodynamic constraints.

2. Casimir Effect in the Fluctuating Matter Fields Picture

Complementary to the vacuum mode description (plenum approach), the Casimir effect can be described, like van der Waals attraction, as arising from correlations in the state of electrons in the two plates through the intermediary of their coupled fields. From this standpoint (matter-fields approach), there is no requirement for the high energy density vacuum field of the plenum approach to reside throughout all space.

Unfortunately with regard to energy generation, though the Casimir forces involved can be of significance for microelectromechanical systems (MEMS) applications [62], the associated Casimir energies involved are too small to be considered of significance for energy applications, so if the possibility for vacuum energy conversion exists, one must look elsewhere to other types of matter-vacuum interactions.

C. Type I (Transient) and Type II (Continuous) Machines

A key feature of the Casimir process just described, regardless of viewpoint (plenum or matter-fields), is that it is a "one-shot," transient, energy-producing machine. That is, after delivering its energy, E, the matter that comprises the machine is in a "used" state (we commonly refer to this used matter as "ash") and cannot be restored to the original state without an input of energy that is greater than or equal to E. This "one-shot" feature can be generalized to define a category of machine we call a type I transient machine, with the Casimir machine constituting the prototypical representative. Should gravitation eventually be traced to a vacuum ZPF origin as proposed by Sakharov [63], then the fall of an object of mass m through a height h in a gravitational field (g = acceleration of gravity at Earth's surface), delivering its gravitational energy mgh upon impact with the ground, would constitute another example.

In contrast, one can envision a type II (continuous) machine in which vacuum ZPF energy is converted to a useful form on a recycling basis without net alteration to its own matter state. A hypothetical example is the tunable Casimir device that we reviewed in Sec. IV.C. The cycle of energy generation would consist of the collapse of conducting plates with delivery of energy, followed by separation of plates switched to insulating mode for which the attractive force is considerably weaker, only to be switched back to conducting mode for the next cycle, etc. Provided the input switching energy required per cycle is less than the output energy delivered per cycle, a continuous generation of energy without a net change in matter configuration would result. A second example would be a nonlinear oscillator that continuously, on a steady-state basis, downshifted high-frequency components of the vacuum ZPF spectrum to lower frequencies for convenient collection and application, without a net change in its own operation.

Clearly, a type II machine would be far more useful than a type I machine for energy extraction. Type II machines would constitute a fuelless energy source, with the ambient vacuum ZPF providing essentially unlimited energy. For this to be the case, however, another requirement needs to be satisfied, which we address in the next section.

D. Degradability of the Vacuum

The possibility of continuous conversion of vacuum ZPE to other forms (i.e., by a type II machine) requires that, in principle, vacuum energy must be degradable (i.e., continuously consumable), not just that there be a surfeit of energy in place to harvest. This perspective leads to a remarkable question for deeper explorations of QED. It turns out that the mathematical structure of QED is based on a formalism in which the vacuum mode structure and vacuum fluctuation energy per mode are quantized in what could be called a "hardwired" fashion, that is, they possess fixed immutable values. Therefore, at the end of a cycle of a hypothetical type II machine, in which both matter and vacuum mode structure have been returned to their original states, the vacuum modes must of necessity contain at a minimum the same "hardwired" energetic content as before the cycle. Therefore, assuming local detailed-balance energy conservation, continuous conversion of vacuum ZPE to other forms via a type II machine is, from the QED viewpoint, forbidden in principle because the vacuum as described by the QED formalism is nondegradable. (Globally, vacuum energy is not conserved during cosmological expansion, with work being done by the negative vacuum pressure to maintain positive constant vacuum energy density and therefore increasing the vacuum energy [64].) This outcome of second-quantized QED theory permits but two interpretations with regard to continuous vacuum energy conversion: 1) QED theory, despite criticisms that can be leveled against it, is correct in its description of vacuum fluctuation dynamics, and even though vacuum ZPE exists, it cannot be continuously converted to other forms, or 2) the axiomatic inconvertibility is an artifact of an over-idealized mathematical structure, and therefore the possibility of conversion remains an open question.[‡] What is not in question, however, is that QED, as an axiomatic, quantum formalism based on the concept of an immutable,

---

[‡]A number of publications by E. T. Jaynes, A. O. Barut, and their collaborators are based on the premise that second quantization is an unnecessary artifact of an overidealized formalism.

nondegradable vacuum, does not support the concept of continuous vacuum energy conversion.

E. Alternatives to Quantum Electrodynamics

As noted in Section V.A.1, despite its successes the second-quantized QED formalism with its infinite vacuum degrees of freedom and associated infinite energy density has been the subject of criticism and, as a result, alternatives have been proposed and investigated in the literature. The alternatives run the gamut from neoclassical theories in which matter is quantized but the fields are not (e.g., the voluminous work of E. T. Jaynes), through classical theories where both matter and fields are treated classically, with vacuum fluctuation fields taken to be real but of a classical nature (e.g., SED), to formalisms which eliminate the concept of vacuum fields altogether (e.g., direct-action approaches investigated by A. O. Barut and others; see the references cited below). We examine each of these briefly with regard to the possibility of useful "vacuum energy conversion."

1. Neoclassical Theories of Quantum Electrodynamic Vacuum Fluctuation Effects

A major proponent of the neoclassical approach has been E. T. Jaynes, who has questioned whether the quantized vacuum field is physically real or merely an artifice of the second-quantized QED formalism. Based on the fact that the QED formalism permits expression of effects in terms of quantized "self" or "source" fields as an alternative to expression in terms of quantized vacuum fluctuation fields, Jaynes advanced the hypothesis that QED effects can be attributed to the self-fields of quantized matter without considering independent quantization of the vacuum fields, expressions in terms of the latter just being a placeholder for the former. Pointedly, with regard to QED being "the jewel of physics because of its extremely accurate predictions," Jaynes's position is that "those accurate experimental confirmations of QED come from the local source fields, which are coherent with the local state of matter," and that "the quantized free field only tags along" [65]. Jaynes nonetheless arrived at a conclusion that we might call Jaynes' axiom, namely, "This complete interchangeability of source-field effects and vacuum-fluctuation effects ... shows that source-field effects are the same as if vacuum fluctuations were present." Applied to the case of a radiating atom, Jaynes provides a specific example of his conclusion with the statement "The radiating atom is indeed interacting with an electromagnetic field of the intensity predicted by the zero-point energy, but this is just the atom's own radiation reaction field" [66]. As a result, with the axiomatic second-quantized field formalism set aside, in the neoclassical approach any consideration of the conversion of vacuum ZPE for use must be displaced to consideration of the conversion and degradability of source or matter-fields fluctuation energy for use, issues yet to be addressed in the literature.

2. Stochastic Electrodynamics Model Revisited

SED is a classical (i.e., nonquantized) theory of particle-field interactions that assumes the existence of classical particles and a classical random background electromagnetic field distribution whose Lorentz-invariant spectral energy density is chosen to match that originally appearing in second-quantized QED. Given SED's heuristic value of classical-like modeling and ease of calculation, and its seeming ability to address many quantum mechanical problems with success (as outlined in Sec. III.B), the SED approach has been employed in the literature to explore vacuum energy conversion. In the absence of a formalism for vacuum field quantization, there are no fundamental immutability constraints that would mitigate against vacuum energy degradability, so that issue is not testable under this formalism.

Investigations to date have included the use of cavity-QED techniques to suppress atomic or molecular ground states [27], and evaluation of the use of a nonlinear oscillator to continuously downshift high-frequency components of the vacuum fluctuation spectrum to lower frequencies for convenient collection and use. With regard to the latter, the result of a nonrelativistic SED analysis is that the downshifting process acts to convert an initial hypothetical cubic-frequency vacuum fluctuation spectrum toward a Rayleigh–Jeans rather than a Planck heat spectrum (the former being a low-energy approximation of the latter) [67,68]. Extension of the analysis to the relativistic regime does not alter this conclusion [69,70]. Though further work remains, these considerations lead us to conclude that SED in its present form is incomplete and may not be useful for the assessment of the potential conversion of vacuum energy to other forms; its predictions concerning such must be treated with caution.

Additional shortcomings of the SED model include convoluted attempts to derive interference effects or Schrödinger's equation, and the difficulty in explaining sharply defined stationary states (i.e., sharp atomic spectra), though there have been many attempts [17]. QED and SED do not, in general, yield the same results for nonlinear systems, although they are in agreement for the range of linear systems examined. The apparent disagreements between SED and QED are quite serious, and occur in areas in which QED is highly successful. Perhaps the source of these difficulties lies in accurately dealing with the nonlinear stochastic differential equations in SED for these problems. Even still, it is likely that differences will remain, which should clearly be testable by experimental means [71]. For a very thorough, detailed and scholarly review of SED, see Ref. 17 and the corresponding review by Cole and Rueda [72].

Given the heuristic value of certain aspects of SED modeling, but with the shortcomings outlined above noted, SED theorists de la Peña and Cetto have proposed a modification to SED they call LSED (linear SED) [73]. The modification consists of the addition of three new constraining principles that result in a form of convergence with nonrelativistic quantum mechanics while retaining some of the appealing attributes of standard SED (e.g., quantum states being stable on the basis of a dynamic balance between absorption and emission of background vacuum fluctuation fields). The added constraints (e.g., an added constraint that invokes detailed energy balance for separate frequencies) result in correcting several known problems with standard SED. For example, now the equilibrium spectrum is Planck's, not Rayleigh–Jeans, and wavelike behavior of matter and nonlocality issues can be addressed, etc. The issue of continuous vacuum energy conversion has yet to be addressed in this new formalism, however, so that remains for the future.

3. Quantum Electrodynamics Without Second-Quantized Fields

As yet another alternative to canonical second-quantized QED, Barut [74] has proposed that effects attributable to vacuum ZPF can be derived with a theory in which there are source (matter) fluctuation fields but no vacuum fluctuation fields, and that even the former can be eliminated. Barut's approach is developed in considerable detail as an independent, self-consistent, formulation of QED in its own right. Barut argues that effects normally attributed to vacuum fluctuations in the second quantized, linear theory of the radiation field can be equally well computed within the framework of a non-second-quantized, nonlinear theory based entirely on matter wave functions alone. His program is to assess how far one can go in understanding radiative processes without second quantization or vacua that fluctuate. Barut and his collaborators have successfully applied the theory to the Lamb shift and spontaneous emission [75,76], problems of cavity QED [77], Casimir–Polder and van der Waals forces [78], calculations of the $g_e - 2$ (gyromagnetic ratio, g_e) factor of an electron [79–81], and the Davies–Unruh effect [82], among others.

Given that the formalism of second-quantized field operators are not used at all in the Barut approach, the seemingly quantized properties of fields are taken to simply reflect first quantization of the sources. Therefore, in the absence of the independent existence of second-quantized field fluctuations, the QED arguments concerning immutability and nondegradability of quantized vacuum fluctuation fields, and the corollary proscription against potential conversion of energy from such fields, do not apply. As in the neoclassical approach, the question of the conversion of quantum ZPE to other forms must be diverted to consideration of the global properties of matter fluctuation interactions in the as-yet incomplete development of the Barut approach.

F. Examples of a Degradable Vacuum

In closing, we review three examples of a degradable vacuum that are predicted by the quantum field theory of curved spacetime and the Standard Model of elementary particle physics.

1. Gravitational Squeezing of the Vacuum

In their study of traversable wormholes, Hochberg and Kephart [83] discovered that the gravitational field of any astronomical body produces a zone of negative energy around it by "dragging" some of the virtual quanta (a.k.a. vacuum ZPF) downward. They applied their discovery to the problem of creating and stabilizing traversable wormholes. Their quantum optics analysis showed that there is a distortion of the vacuum electromagnetic ZPF due to the interaction with a prescribed gravitational background, which results in "squeezed" vacuum

ON EXTRACTING ENERGY FROM THE QUANTUM VACUUM 595

states that possess a negative energy density. Squeezing of the vacuum is a quantum process that is roughly analogous to the compression of an ordinary fluid. (See Sec. II.B.2 in Chapter 15 for further details.) This means that, as the vacuum field is continuously being squeezed by the gravitational field of a body, its energy is continuously being degraded with respect to the undisturbed remote vacuum field.

The magnitude of the gravitational squeezing of the vacuum can be estimated from the quantum optics squeezing condition for given transverse (to direction of gravitational acceleration) momentum and (equivalent) energy eigenvalues, $j = 8\pi r_S/\lambda$, of two electromagnetic ZPF field modes, subject to $j \to 0$, where λ is the ZPF mode wavelength and r_S is the Schwarzschild radius of any astronomical body under study [83]. (Note that $r_S = 2GM/c^2$ is the critical radius at which a body of mass M collapses into a black hole, used here as a convenient radial distance parameter to simplify the inequality, but there is no actual black hole collapse involved in this mechanism; and G is Newton's universal gravitation constant, $6.673 \times 10^{-11}\,\mathrm{Nm^2/kg^2}$.) This condition simply states that substantial gravitational squeezing of the vacuum occurs for ZPF field modes with $\lambda \geq 8\pi r_S$ of the mass under study. (See Sec. II.B.3 in Chapter 15 for further details.)

It is not clear whether this mechanism can be exploited to extract energy from the vacuum. Conservation of energy suggests one of two possible outcomes: 1) the lost energy is injected into the gravitational energy of the body, or 2) the lost energy reappears as an accumulation of positive energy density ZPF modes elsewhere in the universe. Further research will be needed to address this question.

2. Redshifting the Vacuum

Calloni et al. [84,85] explored the possibility of verifying the equivalence principle for the ZPE of QED. They used semiclassical quantum gravity theory to evaluate the net force produced by the quantum vacuum ZPF acting on a rigid Casimir cavity in a weak gravitational field that is modeled using the standard Schwarzschild spacetime metric geometry.§ They evaluated the regularized (or renormalized) stress-energy tensor $\langle T^{\mu\nu}\rangle_\mathrm{vac}$ of the quantized vacuum electromagnetic field between two plane-parallel ideal metallic plates lying in a horizontal plane. $\langle T^{\mu\nu}\rangle_\mathrm{vac}$ encodes the Casimir effect, which has a negative energy density and a negative pressure along the vertical (gravitational acceleration) axis between the plates. Bimonte et al. [86,87] also studied this problem using Green-function techniques in the Schwinger–DeWitt quantum ether prescription for $\langle T^{\mu\nu}\rangle_\mathrm{vac}$ in a curved spacetime. The results from these studies agreed with the equivalence principle and showed that quantum vacuum ZPF does gravitate because the energy of each ZPF mode is redshifted by the factor

---

§A spacetime metric is a Lorentz-invariant distance function between any two points in spacetime, which is defined in terms of a metric tensor, $g_{\mu\nu}$, that encodes the geometry of spacetime (Greek indices $\mu,\nu = 0\ldots 3$ denote spacetime coordinates, $x^0\ldots x^3$, such that $x^1\ldots x^3 \equiv$ space coordinates and $x^0 \equiv$ time coordinate).

$(-g_{00})^{1/2} = [1 - (2GM/c^2 r)]^{1/2}$ even though the modes remain unchanged (here, M is the mass of a gravitating body, r is the radial distance from the body, and g_{00} is the time-time component of the Schwarzschild metric tensor).

These studies suggest that cavity electromagnetic vacuum states are continuously degrading inside a background gravitational field. But can we extract energy from this mechanism? The answer to this question is not known at present, but consideration of the conservation of energy suggests that the same two possible outcomes given in the previous section would seem to apply: 1) the lost energy is injected into the gravitational energy of the body, or 2) the lost energy reappears as positive energy density ZPF modes elsewhere in the universe. Further research will be needed to address this question as well.

3. Melting the Vacuum

In their study of the structured vacuum, Rafelski and Müller [53,54] analyze the nature of the strongly interacting (QCD) vacuum and elucidate its character from the Standard Model of particle physics and high energy particle accelerator data. They concluded that in addition to the electroweak vacuum (i.e., the unified electromagnetic and weak force vacua) there exists a dual QCD vacuum structure: one vacuum structure that is everywhere in space and consists of a complicated soup of interacting gluons which confine the quarks (called the ordinary or "frozen" vacuum); and another vacuum structure that is found inside elementary particles (e.g., hadrons) and behaves like the dielectric vacuum of electrodynamics. In this second vacuum structure, particles that have a strong charge (such as quarks or gluons) can move freely, but are confined by the frozen vacuum that is everywhere else. This is called the perturbative or gluon or "melted" vacuum, which can also be pictured as a quark–gluon plasma. They estimate that there is a "latent heat" of ~ 1 GeV/fm^3 (or 10^{35} J/m^3)¶ associated with the phase change of transforming from one vacuum structure to another when the gluonic structures of the perturbative vacuum are melted. It is important to point out here that this is a degradable vacuum structure.

This unusual dual vacuum structure led Rafelski and Müller to speculate on a mechanism for the "burning of matter" as the ultimate source of energy in which it might be possible that the energy contained within baryons could be converted into useful energy. Their idea is to remove or destroy the three quarks residing inside a baryon in order to gain energy, the latent heat, from the melted vacuum inside the baryon. This process also entails the decay of the quarks via lepton-quark interactions, which is a topic that is beyond the scope of this chapter. They suggest that it might be possible that producing a quark–gluon plasma in high energy nuclear collisions could be a very efficient source of energy. In this process atomic nuclei would be collided at high energy in order to form a compressed high density zone in the region where the two nuclei overlap. This would lead to the melting of the vacuum and the subsequent direct conversion of matter into radiation, thus releasing $\sim 10^{35}$ J/m^3 of energy density. This magnitude of energy density would be very useful as a source of energy for space propulsion applications.

¶1 eV = 1.602×10^{-19} J; 1 GeV = 10^9 eV; 1 fm = 10^{-15} m.

Rafelski and Müller point out that the commonly held view that the centers of neutron stars are dead and cold, due to their nuclear fuel having burnt out and the energy of gravitational collapse having been expended for the conversion of the collapsed star into a gigantic atomic nucleus, is not the complete story. They hold open the possibility that the entire rest-mass of all the baryons inside neutron stars might become available and converted into heat. In their scenario, the core of a neutron star is actually composed of condensed quark matter, and the rest-mass of baryons is burnt up into radiation inside the quark core. They also point out that supernovae explosions, gamma ray bursts, positron emission from the center of our galaxy, quasars, and galactic nuclei have been observed to emit extreme amounts of thermal energy, the mechanisms of which are still not understood today.

Theoretical and laboratory studies of the dual QCD vacuum have been underway for over 20 years. The progress in experimental particle physics is such that one gains an order of magnitude in the resolution (i.e., energy) of elementary particle structures roughly every decade. It is hoped that the commissioning of the Large Hadron Collider in 2008 will lead to higher resolution probing of the dual QCD vacuum structure, and help to determine whether there are deeper grand unified and/or Higgs vacuum structures residing within quarks. This topic is a rich area for future research, in which investigators can explore ways to exploit the energy residing within new vacuum structures for space power and propulsion applications.

4. *Zero-Point Fluctuation Modes and Vacuum Field Energy*

The above examples illustrate how the vacuum becomes degradable when perturbed under certain conditions. In each of the examples, the vacuum ZPF modes were perturbed in such a way as to drive the QED field's vacuum state energy below zero, or the vacuum undergoes a phase change as in the QCD case. The Casimir effect is another example in which certain ZPF modes are excluded by boundary conditions that perturb the free-space vacuum ZPF modes, thus driving the QED field energy below zero inside a Casimir cavity. In accordance with the discussion in Secs. III.A and V.D, the ZPF modes serve only as placeholders for a quantum field's vacuum state calculations. Therefore, the "hardwired" ZPF modes cannot be driven below the ground state. It is only a quantum field's overall vacuum state energy that can be driven below zero. So we conjecture that the key to exploring the possibility of extracting energy from the vacuum is to discover or invent additional mechanisms that perturb the ZPF modes of a quantum field, which could then be technologically implemented and tested in laboratory experiments.

VI. Conclusions

What are the conclusions that can be drawn from the considerations presented here regarding the concept of continuous conversion of energy from the vacuum electromagnetic ZPF?

First, we see that although the original inspiration for the concept of continuous vacuum energy extraction came from second-quantized QED theory, it must be acknowledged that QED, as an axiomatic, quantized formalism based on the concept of an immutable, non-degradable vacuum, does not support the concept of continuous vacuum energy conversion. Given that second-quantized QED is our most comprehensive quantum theory to date, its lack of support for continuous vacuum energy conversion must be given serious consideration.

Second, SED as an alternative theory, whose formalism has been taken to support the concept of continuous vacuum energy conversion, has enough shortcomings in its current state of development that one must conclude that it is not at present an adequate tool for the assessment of potential vacuum energy conversion. SED predictions must thus remain suspect in the absence of experimental confirmation. The purpose of the experimental program outlined in Sec. IV is meant to address this need.

Third, the concept of the conversion of energy from vacuum fluctuations is in principle not falsifiable, given the unknowns that present theory has yet to resolve (e.g., cosmological dark energy), and the numerous approaches currently being brought to bear in the development of quantum theory.

Finally, even though experimental efforts at energy extraction from the vacuum have been proposed or are already underway at various laboratories, definitive theoretical support underpinning the concept of useful extraction of energy from quantum fluctuations is not yet in place. Such support awaits theoretical developments that either posit a plenum that (unlike second-quantized QED) can be shown to be degradable, or posit conversion of energy associated with matter fluctuations, also in a degradable fashion. Because the quantum fluctuations of interest are associated with quantum ground states, what is minimally required are particle–vacuum or particle–particle interactions that result in the formation of alternate lower-energy, ground states of matter/field configurations. Suggested approaches to be explored are those which are known to yield results consistent with the existence of vacuum fluctuation fields, but without the formalism of independently postulated second-quantized vacuum fields. Whether useful conversion of energy from quantum fluctuations can be accomplished, and identifying the unequivocal conditions under which this can be achieved, are yet to be determined.

It has been argued that the QED vacuum is degradable under the action of gravitation-induced quantum squeezing or redshifting. However, it is not known whether these effects can be exploited for the extraction of energy from the vacuum. The concept of a dual, degradable vacuum structure in QCD can possibly lead to the generation of useful energy via the release of latent heat from melting the QCD vacuum. The new generation of high energy particle colliders coming online in the very near future may yield new information about the complex vacuum structure of the universe, allowing us to find ways to exploit its energy content for revolutionary space propulsion applications.

Acknowledgments

The authors wish to thank our colleagues, V. Teofilo (Lockheed Martin), B. Haisch (ManyOne Networks), L. J. Nickisch (Northwest Research Assoc.), A. Rueda (California State University–Long Beach), D. C. Cole (Boston University), M. Ibison (Institute for Advanced Studies at Austin), S. Little

(EarthTech International), and M. Little (EarthTech International), for their extensive technical contributions to this chapter. We also thank J. Newmeyer (Lockheed Martin), E. H. Allen (Lockheed Martin), T. W. Kephart (Vanderbilt University), P. C. W. Davies (Arizona State University), M. G. Millis (NASA-Glenn), and M. R. LaPointe (NASA-Marshall) for their very useful comments.

References

[1] DeWitt, B. S., "Quantum Field Theory in Curved Spacetime," *Physics Reports*, Vol. 19C, 1975, pp. 295–357.

[2] DeWitt, B. S., "Quantum Gravity: The New Synthesis," *General Relativity: An Einstein Centenary Survey*, Hawking, S. W., and Israel, W. (eds.), Cambridge University Press, Cambridge, MA, 1979, pp. 680–745.

[3] DeWitt, B. S., "The Casimir Effect in Field Theory," *Physics in the Making, Essays on Developments in 20th Century Physics, In Honour of H. B. G. Casimir*, Sarlemijn, A., and Sparnaay, J. (eds.), North-Holland Elsevier Science Publ., New York, 1989, pp. 247–272.

[4] Birrell, N. D., and Davies, P. C. W., *Quantum Fields in Curved Space*, Cambridge University Press, Cambridge, U.K., 1984.

[5] Casimir, H. B. G., "On the Attraction Between Two Perfectly Conducting Plates," *Proceedings of the Koninklijke Nederlandse Akademie van Wetenschappen*, Vol. 51, 1948, pp. 793–796.

[6] Lamoreaux, S. K., "Demonstration of the Casimir Force in the 0.6 to 6 μm Range," *Physical Review Letters*, Vol. 78, 1997, pp. 5–8.

[7] Mohideen, U., "Precision Measurement of the Casimir Force from 0.1 to 0.9 μm," *Physical Review Letters*, Vol. 81, 1998, pp. 4549–4552.

[8] Chen, F., Mohideen, U., Klimchitskaya, G. L., Mostepanenko, V. M., "Theory Confronts Experiment in the Casimir Force Measurements: Quantification of Errors and Precision," *Physical Review A*, Vol. 69, 2004, 022117.

[9] Forward, R. L., "Alternate Propulsion Energy Sources," Air Force Rocket Propulsion Laboratory, Air Force Space Tech. Ctr. Space Div., Air Force Systems Command, Final Report AFRPL TR-83-067, Edwards AFB, CA, 1983, pp. A1–A14.

[10] Forward, R. L., "Extracting Electrical Energy from the Vacuum by Cohesion of Charged Foliated Conductors," *Physical Review B*, Vol. 30, 1984, pp. 1700–1702.

[11] Cole, D. C., and Puthoff, H. E., "Extracting Energy and Heat from the Vacuum," *Physical Review E*, Vol. 48, 1993, pp. 1562–1565.

[12] Mead, Jr., F. B., and Nachamkin, J., "System Converting Electromagnetic Radiation Energy to Electrical Energy," U.S. Patent No. 5,590,031, 1996.

[13] Peres, A., *Quantum Theory: Concepts and Methods*, Kluwer, Dordrecht, The Netherlands, 1993.

[14] Milonni, P. W., *The Quantum Vacuum: An Introduction to Quantum Electrodynamics*, Academic Press, San Diego, CA, 1994, Chaps. 1, 2, 8.

[15] Mulliken, R. S., "The Band Spectrum of Boron Monoxide," *Nature*, Vol. 114, 1924, p. 349.

[16] Ibison, M., and Haisch, B., "Quantum and Classical Statistics of the Electromagnetic Zero-Point Field," *Physical Review A*, Vol. 54, 1996, pp. 2737–2744.

[17] de la Peña, L., and Cetto, A. M., *The Quantum Dice: An Introduction to Stochastic Electrodynamics*, Kluwer, Dordrecht, The Netherlands, 1996.

[18] Koch, R. H., Van Harlingen, D. J., and Clarke, J., "Quantum-Noise Theory for the Resistively Shunted Josephson Junction," *Physical Review Letters*, Vol. 45, 1980, pp. 2132–2135.

[19] Koch, R. H., Van Harlingen, D. J., and Clarke, J., "Observation of Zero-Point Fluctuations in a Resistively Shunted Josephson Tunnel Junction," *Physical Review Letters*, Vol. 47, 1981, pp. 1216–1219.

[20] Koch, R. H., Van Harlingen, D. J., and Clarke, J., "Measurements of Quantum Noise in Resistively Shunted Josephson Junctions," *Physical Review B*, Vol. 26, 1982, pp. 74–87.

[21] Blanco, R., França, H. M., Santos, E., and Sponchiado, R. C., "Radiative Noise in Circuits with Inductance," *Physics Letters A*, Vol. 282, 2001, pp. 349–356.

[22] Boyer, T. H., "Random Electrodynamics: The Theory of Classical Electrodynamics with Classical Electromagnetic Zero-Point Radiation," *Physical Review D*, Vol. 11, 1975, pp. 790–808.

[23] Puthoff, H. E., "Ground State of Hydrogen as a Zero-Point-Fluctuation-Determined State," *Physical Review D*, Vol. 35, 1987, pp. 3266–3269.

[24] Cole, D. C., and Zou, Y., "Quantum Mechanical Ground State of Hydrogen Obtained from Classical Electrodynamics," *Physics Letters A*, Vol. 317, 2003, pp. 14–20.

[25] Cole, D. C., and Zou, Y., "Analysis of Orbital Decay Time for the Classical Hydrogen Atom Interacting with Circularly Polarized Electromagnetic Radiation," *Physical Review E*, Vol. 69, 2004, 016601.

[26] Maclay, G. J., "Analysis of Zero-Point Electromagnetic Energy and Casimir Forces in Conducting Rectangular Cavities," *Physical Review A*, Vol. 61, 2000, 052110.

[27] Puthoff, H. E., Little, S. R., and Ibison, M., "Discussion of Experiment Involving Lowering of Ground State Energy for Oscillation in a Casimir Cavity," *International Conference on Squeezed States and Uncertainty Relations (ICSSUR 2001): Special Session on Topics of Physics Related to Stochastic Electrodynamics*, Boston University, 2001.

[28] Puthoff, H. E., "Vacuum Energy Extraction by Conductivity Switching of Casimir Force," Technical Memo, Inst. for Advanced Studies at Austin, Austin, TX, 1985, unpublished.

[29] Lipkin, R., "Thin-Film Mirror Changes into a Window," *Science News*, Vol. 149, 1996, p. 182.

[30] Pinto, F., "Engine Cycle of an Optically Controlled Vacuum Energy Transducer," *Physical Review B*, Vol. 60, 1999, pp. 14740–14755.

[31] Lin, S. Y., Moreno, J., and Fleming, J. G., "Three-Dimensional Photonic-Crystal Emitter for Thermal Photovoltaic Power Generation," *Applied Physics Letters*, Vol. 83, 2003, pp. 380–382.

[32] Iannuzzi, D., Lisanti, M., Munday, J. N., and Capasso, F., "The Design of Long Range Quantum Electrodynamical Forces and Torques Between Macroscopic Bodies," *Solid State Communications*, Vol. 135, 2005, pp. 618–626.

[33] Forward, R. L., "Apparent Endless Extraction of Energy from the Vacuum by Cyclic Manipulation of Casimir Cavity Dimensions," *NASA Breakthrough Propulsion Physics Workshop Proceedings*, NASA/CP-1999-208694, NASA Lewis Research Center, Cleveland, OH, 1999, pp. 51–54.

[34] Maclay, G. J., "Unusual Properties of Conductive Rectangular Cavities in the Zero Point Electromagnetic Field: Resolving Forward's Casimir Energy Extraction

Cycle Paradox," *Proceedings of the STAIF-1999: Conference on Applications of Thermophysics in Microgravity and Breakthrough Propulsion Physics*, El-Genk, M. S. (ed.), AIP Conf. Proc. Vol. 458, AIP Press, New York, 1999, pp. 968–973.

[35] Shoulders, K. R., *EV: A Tale of Discovery*, Jupiter Technologies, Austin, TX, 1987.

[36] Stenhoff, M., *Ball Lightning: An Unsolved Problem in Atmospheric Physics*, Kluwer Academic/Plenum Publ., New York, 1999.

[37] Bostick, W. H., "Experimental Study of Ionized Matter Projected Across Magnetic Field," *Physical Review*, Vol. 104, 1956, p. 292.

[38] Bostick, W. H., "Experimental Study of Plasmoids," *Physical Review*, Vol. 106, 1957, p. 404.

[39] Bostick, W. H., Prior, W., and Farber, E., "Plasma Vortices in the Coaxial Plasma Accelerator," *Physics of Fluids*, Vol. 8, 1965, p. 745.

[40] Wells, D. R., "Observation of Plasma Vortex Rings," *Physics of Fluids*, Vol. 5, 1962, p. 1016.

[41] Wells, D. R., "Axially Symmetric Force-Free Plasmoids," *Physics of Fluids*, Vol. 7, 1964, p. 826.

[42] Wells, D. R., "Injection and Trapping of Plasma Vortex Rings," *Physics of Fluids*, Vol. 9, 1966, p. 1010.

[43] Wells, D. R., and Schmidt, G., "Observation of Plasma Rotation Produced by an Electrodeless Plasma Gun," *Physics of Fluids*, Vol. 6, 1963, p. 418.

[44] Nardi, V., Bostick, W. H., Feugeas, J., and Prior, W., "Internal Structure of Electron-Beam Filaments," *Physical Review A*, Vol. 22, 1980, pp. 2211–2217.

[45] Piestrup, M. A., Puthoff, H. E., and Ebert, P. J., "Measurement of Multiple-Electron Emission in Single Field-Emission Events," *Journal of Applied Physics*, Vol. 82, 1997, pp. 5862–5864.

[46] Shoulders, K. R., "Method and Apparatus for Production and Manipulation of High Density Charge," U.S. Patent No. 5,054,046, 1991.

[47] Casimir, H. B. G., "Introductory Remarks on Quantum Electrodynamics," *Physica*, Vol. 19, 1953, pp. 846–849.

[48] Puthoff, H. E., and Piestrup, M. A., "On the Possibility of Charge Confinement by van der Waals/Casimir-type Forces," *Cornell University Library E-Print Physics Abstracts Archive arXiv.org Database* [online database]. URL: http://arxiv.org/physics/papers/0408/0408114.pdf [cited Aug. 2004, revised 2005].

[49] Puthoff, H. E., "Casimir Vacuum Energy and the Semiclassical Electron," *International Journal of Theoretical Physics*, Vol. 46, 2007, pp. 3005–3008.

[50] Volovik, G. E., *The Universe in a Droplet of Helium*, Clarendon Press, Oxford, 2003.

[51] Zavattini, E., Zavatti, G., Ruoso, G., Palacco, E., Milotti, E., Karuza, M., Di Domenico, G., Della Valle, F., Ciminio, R., Carusotto, S., Cantatore, G., and Bregant, M., "Experimental Observation of Optical Rotation Generated in Vacuum by a Magnetic Field," *Physical Review Letters*, Vol. 96, 2006, 110406.

[52] Fulcher, L. P., Rafelski, J., and Klein, A., "The Decay of the Vacuum," *Scientific American*, Vol. 241, 1979, pp. 150–159.

[53] Rafelski, J., and Müller, B., *The Structured Vacuum: Thinking About Nothing*, Verlag Harri Deutsch Publ., Frankfurt, Germany, 1985.

[54] Rafelski, J., "Vacuum Structure—An Essay," *Vacuum Structure in Intense Fields*, Fried, H. M., and Müller, B. (eds.), NATO ASI Series, Series B: Physics Vol. 255, Plenum Press, New York, 1991, pp. 1–28.

[55] Jaynes, E. T., "Survey of the Present Status of Neoclassical Radiation Theory," *Coherence and Quantum Optics*, Mandel, L., and Wolf, E. (eds.), Plenum Press, New York, 1973, pp. 35–81.

[56] Dirac, P. A. M., *Directions in Physics*, H. Hora and J. R. Shepanski (eds.), Wiley, New York, 1978, pp. 76–77.

[57] Schwarzschild, B., "High-Redshift Supernovae Indicate that Dark Energy Has Been Around for 10 Billion Years," *Physics Today*, Vol. 60, 2007, pp. 21–25.

[58] Particle Data Group, "Review of Particle Physics," *Journal of Physics G*, Vol. 33, 2006, pp. 210–232.

[59] Weinberg, S., "The Cosmological Constant Problem," *Reviews of Modern Physics*, Vol. 61, 1989, pp. 1–23; see also, "The Cosmological Constant Problems," *Cornell University Library E-Print Physics Abstracts Archive arXiv.org Database* [online database]. URL: http://arxiv.org/PS_cache/astro-ph/pdf/0005/0005265.pdf [cited 12 May 2000].

[60] Volovik, G. E., "Vacuum Energy: Myths and Reality," *International Journal of Modern Physics D*, Vol. 15, 2006, pp. 1987–2010.

[61] Milton, K. A., *The Casimir Effect: Physical Manifestations of Zero-Point Energy*, USA World Scientific, NJ, 2001, p. 13.

[62] Chan, H. B., Aksyuk, V., Kleiman, R., Bishop, D., and Capasso, F., "Quantum Mechanical Actuation of Microelectromechanical Systems by the Casimir Force," *Science*, Vol. 291, 2001, pp. 1941–1944.

[63] Sakharov, A. D., "Vacuum Quantum Fluctuations in Curved Space and the Theory of Gravitation," *Soviet Physics Doklady*, Vol. 12, 1968, pp. 1040–1041.

[64] Peacock, J. A., *Cosmological Physics*, Cambridge University Press, Cambridge, U.K., 1999, pp. 25–26.

[65] Jaynes, E. T., "Probability in Quantum Theory," *Complexity, Entropy, and the Physics of Information*, Zurek, W. (ed.), Addison-Wesley, Reading, MA, 1990, pp. 381–403.

[66] Jaynes, E. T., "Electrodynamics Today," *Coherence and Quantum Optics*, Mandel, L., and Wolf, E. (eds.), Plenum Press, New York, 1978, pp. 495–509.

[67] Boyer, T. H., "Equilibrium of Random Classical Electromagnetic Radiation in the Presence of a Nonrelativistic Nonlinear Electric Dipole Oscillator," *Physical Review D*, Vol. 13, 1976, pp. 2832–2845.

[68] Pesquera, L., and Claverie, P., "The Quartic Anharmonic Oscillator in Stochastic Electrodynamics," *Journal of Mathematical Physics*, Vol. 23, 1982, pp. 1315–1322.

[69] Blanco, R., Pesquera, L., and Santos, E., "Equilibrium between Radiation and Matter for Classical Relativistic Multiperiodic Systems. I. Derivation of Maxwell–Boltzmann Distribution from Rayleigh–Jeans Spectrum," *Physical Review D*, Vol. 27, 1983, pp. 1254–1287.

[70] Blanco, R., Pesquera, L., and Santos, E., "Equilibrium between Radiation and Matter for Classical Relativistic Multiperiodic Systems. II. Study of Radiative Equilibrium with Rayleigh–Jeans Radiation," *Physical Review D*, Vol. 29, 1984, pp. 2240–2254.

[71] Cole, D. C., "Simulation Results Related to Stochastic Electrodynamics," *Proceedings of the International Conference on Quantum Theory: Reconsideration of Foundations-3*, Adenier, G., Khrennikov, A., and Nieuwenhuizen, T. M. (eds.), AIP Conference Proceedings Vol. 810, No. 1, AIP Press, New York, 2005, pp. 99–113.

[72] Cole, D. C., and Rueda, A., "The Quantum Dice: An Introduction to Stochastic Electrodynamics," *Foundations of Physics*, Vol. 26, 1996, pp. 1559–1562.

[73] de la Peña, L., and Cetto, A. M., "Contribution from Stochastic Electrodynamics to the Understanding of Quantum Mechanics," *Cornell University Library E-Print Physics Abstracts Archive arXiv.org Database* [online database]. URL: http://arxiv.org/PS_cache/quant-ph/pdf/0501/0501011.pdf [cited 4 Jan. 2005].

[74] Barut, A. O., "Quantum Electrodynamics Based on Self-Energy versus Quantization of Fields: Illustration by a Simple Model," *Physical Review A*, Vol. 34, 1986, pp. 3502–3503.

[75] Barut, A. O., and van Huele, J. F., "Quantum Electrodynamics Based on Self-Energy: Lamb Shift and Spontaneous Emission without Field Quantization," *Physical Review A*, Vol. 32, 1985, pp. 3187–3195.

[76] Barut, A. O., Kraus, J., Salamin, Y., and Unal, N., "Relativistic Theory of the Lamb Shift in Self-Field Quantum Electrodynamics," *Physical Review A*, Vol. 45, 1992, pp. 7740–7745.

[77] Barut, A. O., and Dowling, J. P., "Quantum Electrodynamics Based on Self-Energy: Spontaneous Emission in Cavities," *Physical Review A*, Vol. 36, 1987, pp. 649–654.

[78] Barut, A. O., and Dowling, J. P., "Quantum Electrodynamics Based on Self-Energy, without Second Quantization: The Lamb Shift and Long-Range Casimir–Polder van der Waals Forces near Boundaries," *Physical Review A*, Vol. 36, 1987, pp. 2550–2556.

[79] Barut, A. O., Dowling, J. P., and van Huele, J. F., "Quantum Electrodynamics Based on Self-Fields, without Second Quantization: A Nonrelativistic Calculation of $g-2$," *Physical Review A*, Vol. 38, 1988, pp. 4405–4412.

[80] Barut, A. O., and Dowling, J. P., "Quantum Electrodynamics Based on Self-Fields, without Second Quantization: Apparatus Dependent Contributions to $g-2$," *Physical Review A*, Vol. 39, 1989, pp. 2796–2805.

[81] Barut, A. O., and Dowling, J. P., "QED Based on Self-Fields: A Relativistic Calculation of $g-2$," *Zeitschrift für Naturforschung*, Vol. A44, 1989, p. 1051.

[82] Barut, A. O., and Dowling, J. P., "Quantum Electrodynamics Based on Self-Fields: On the Origin of Thermal Radiation Detected by an Accelerated Observer," *Physical Review A*, Vol. 41, 1990, pp. 2277–2283.

[83] Hochberg, D., and Kephart, T. W., "Lorentzian Wormholes from the Gravitationally Squeezed Vacuum," *Physics Letters B*, Vol. 268, 1991, pp. 377–383.

[84] Calloni, E., Di Fiore, L., Esposito, G., Milano, L., and Rosa, L., "Gravitational Effects on a Rigid Casimir Cavity," *International Journal of Modern Physics A*, Vol. 17, 2002, pp. 804–807.

[85] Calloni, E., DiFiore, L., Esposito, G., Milano, L., and Rosa, L., "Vacuum Fluctuation Force on a Rigid Casimir Cavity in a Gravitational Field," *Physics Letters A*, Vol. 297, 2002, pp. 328–333.

[86] Bimonte, G., Calloni, E., Esposito, G., and Rosa, L., "Energy-Momentum Tensor for a Casimir Apparatus in a Weak Gravitational Field," *Physical Review D*, Vol. 74, 2006, 085011.

[87] Bimonte, G., Calloni, E., Esposito, G., and Rosa, L., "Erratum: Energy-Momentum Tensor for a Casimir Apparatus in a Weak Gravitational Field," *Physical Review D*, Vol. 75, 2007, 089901(E).

Chapter 19

Investigating Sonoluminescence as a Means of Energy Harvesting

John D. Wrbanek,* Gustave C. Fralick,[†] Susan Y. Wrbanek,[‡] and Nancy R. Hall[§]

NASA Glenn Research Center, Cleveland, Ohio

I. Introduction

SONOLUMINESCENCE has risen to a source of interest to those outside of the ultrasonic community over the last decade [1]. The processes of understanding the effect lead to the challenge of utilizing some of its more interesting properties in practical applications.

The sonoluminescence phenomenon is defined as the generation of light from sound waves, first discovered in the 1930s as a by-product of early work on sonar [2]. The report in 1992 of the ultrasonic trapping of a single glowing bubble in a flask of water generated a cascade of research [3]. The glow from the bubble was found to be generated in bubbles compressed to at least 150 kPa in an extremely short duration flash (<12 picoseconds), and had temperatures of at least 25,000 K for the single bubble [4,5]. Bubbles of noble gases were seen to flash brighter, but the nature of the liquid was seen as playing a large role in the flashes as well [6–8].

Shortly after experimental results on trapped single bubbles were published, models were developed to explain these measurements. Simple shock calculations showed that peak temperatures inside the sonoluminescent bubbles

---

This material is a work of the U.S. Government and is not subject to copyright protection in the United States.
Trade names and trademarks used in this report are for identification only. The usage does not constitute an official endorsement, either express or implied, by NASA.
*Sensor Electronics Engineer, Research and Technology Directorate.
[†]Senior Sensor Electronics Engineer, Research and Technology Directorate.
[‡]Optical Instrumentation Electronics Engineer, Research and Technology Directorate.
[§]Fluid Physics Aerospace Engineer, Research and Technology Directorate.

could reach 3×10^8 K based on the collapse of an ideal spherical gas bubble [9]. Assuming a nonspherical collapse, the high-speed jet striking the opposite side of the bubble gave rise to the possibility of the water being fractured on the molecular scale and generating light as fractoluminescence [10,11]. The extremely rapid collapse of the bubble led to the theoretical examination of sonoluminescence as an effect of quantum vacuum radiation [12]. The lack of an afterglow suggested a cooperative optical emission, like that of an optical laser or superradiance [13,14]. A model of the flash resulting from a two-component plasma in the bubble (containing a low-density halo and high-density core) was developed to explain the measured properties [15]. The actual process may be a combination of any of the above [16–18]. A simplified schematic of the current model of the process is shown in Fig. 1.

Even as these theories are being explored, applications for the effect are taking shape, from fusion containment [9,19–22] to thin film deposition systems [23] and other useful applications [24–27]. Recently, claims put forward for the generation of fusion reactions have caused controversy in the scholarly and popular media [28–37]. However, if realized, harnessing the high-energy release in safe, emission-free ultrasonic processes would lead to the development of revolutionary power systems for in-flight use for both aircraft and spacecraft. The benefits of an onboard fusion power system will result in reduced fuel consumption, lower emissions, and reduced noise for many types of aircraft.

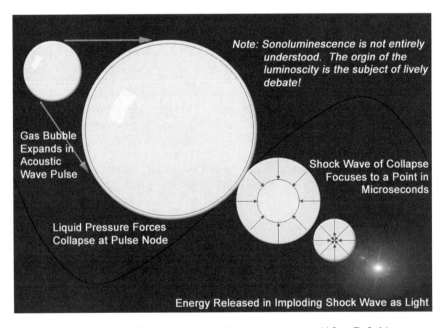

Fig. 1   Schematic of the sonoluminescence process. (After Ref. 1.)

A practical fusion power source would replace the conventional gas turbine auxiliary power units and electrical generators in aircraft and fuel cells and batteries in spacecraft, improving flight and mission capability. Longer-term use of the power source will have both an environmental benefit and act as a positive contributor to the country's energy diversification, as well as enable new missions for both air and space.

II. Approaches

A common approach for the generation of power from ultrasonic acoustic waves is through the attempted initiation of fusion reactions using the high energy of cavitation. Various methods have been referred to early on as "cavitation fusion" and more technically as "acoustic inertial confinement fusion" (AICF), but popularly it is "sonofusion" or "bubble fusion." Confusion with "cold fusion" is understandable, as the respective test cells in "sonofusion" and "cold fusion" appear at first glance to be similar bubbling glass beakers. The mechanisms are very different; "sonofusion" utilizes the ultra-high temperature and pressure conditions of the sonoluminescence to initiate fusion, whereas "cold fusion" supposedly utilizes the boson-like properties of deuterium gas in a metal lattice at room temperature [38]. "Sonofusion" is hot fusion on the pico-scale.

One of the early concepts of a fusion generator was by Hugh Flynn of the University of Rochester, who patented the cavitation fusion reactor (CFR) in 1982 [19]. The CFR design had six acoustic horns to cavitate liquid lithium metal with hydrogen, deuterium, or helium gas added. The fusion reactions initiated by the intense cavitation of the liquid metal fuses the gas with the liquid metal. Liquid metal was used due to the high speed of sound, and thus higher energy cavitations. The case would heat up from the fusion reactions, allowing energy transfer to a heat exchanger. The CFR was never built.

After the isolation of single bubble sonoluminescence, Seth Putterman's research group at the University of California Los Angeles (UCLA) in 1994 [9] and William Moss' research group at Lawrence Livermore National Laboratory in 1996 [20] addressed the possibility that conditions in sonoluminescence in water allows the fusion of two deuterium nuclei (D-D) and a deuterium and tritium nuclei (D-T). In 1997, UCLA patented Putterman's apparatus for converting acoustic power to other useful forms of energy, including D-T fusion reactions by introducing D-T gas into a flask filled with water [21]. The apparatus is a flask with two piezoelectric transducers attached on opposite sides to produce resonances to cavitate the fluid. Unlike Flynn's apparatus, Putterman's apparatus was in use in their lab. Also unlike Flynn's apparatus, no method of extracting the energy from fusion was outlined, but the flask is enclosed in a temperature-controlled box for calorimetric calculations.

As part of research sponsored by the Electric Power Research Institute, Russ George and Roger Stringham of E-Quest Sciences (now D2Fusion, Inc.) reported the generation of molten "ejecta craters" and vents along with anomalous heating of titanium, nickel, copper and palladium metal foils under cavitation in heavy water in 1997 and 1998 [29,30]. The craters and vents were compared to similar nanoparticles formed by hydrogen and deuterium implantation in palladium targets reported by Okuyama et al. of the Nagoya Institute of Technology [39]. The apparatus used two "sonicators" with disks for acoustic input to an aluminum container. George and the Stringham's calorimetric calculations and the trace amounts of helium-4 that were detected suggested to the researchers that the formation of the nanoparticles were the result of fusion reactions in the cavitation. No report was given of any results using cavitations in light water. As the research was purely to determine if cavitation in heavy water could result in fusion reactions in metals, no method of extracting the resulting energy was outlined.

The most publicized claim of sonofusion was in 2002 by Rusi Taleyarkhan et al. of Purdue University [31,32]. Taleyarkhan's group reported the generation of tritium and neutron flux from cavitation of deuterated acetone. Regular "light" acetone was used in a control run. The selection of deuterated acetone as the liquid medium was due to the high accommodation coefficient of dissociated acetone compared to water; that is, more of the vapor in the dissociated acetone is stuck to the bubble walls than water vapor bubbles. Their models indicate that the temperatures in the resulting implosion of cavitating bubbles will result in temperatures of over 1 million Kelvin for acetone, but under one-half million Kelvin for water. The results imply that higher cavitation temperatures are found in liquids with higher vapor pressure rather than liquids with higher speed of sound [9,19], surface tension, or viscosity [40]. The apparatus was a cylinder with a piezoelectric transducer ring producing resonances to cavitate the acetone. The cavitation sites were initiated with a burst of neutrons from a 14-MeV pulsed neutron source. They note that the sonoluminescence was not single bubble, but formed clusters of a thousand bubbles at the cavitation site. The 2002 Taleyarkhan experiment was repeated by Oak Ridge National Laboratory (ORNL), but no coincidence of neutrons with sonoluminescence flash was distinguishable from the neutron background [33].

In 2006, the experiment was modified by Taleyarkhan's group using alpha particle production from uranyl nitrate in a mix of benzene, tetrachloroethene, and deuterated acetone [34]. Experimental runs with light acetone, as well as cells using light and heavy water mixed with uranyl nitrate, were performed as controls. The group reported an increase in neutron and gamma ray flux, using the deuterated acetone mixture that was not seen in the other mixtures, including heavy water. Edward Forringer and his group at LeTourneau University were able to reproduce the experiment [35]. Critical analysis of the published data has led some to conclude that the neutron flux is too weak to provide definitive evidence of nuclear reactions [36], or that the recorded neutron energies are consistent with ^{252}Cf emission [37]. It is not clear if the experiment ruled out fission reactions from uranium.

Though not having any claims of nuclear fusion, the work of Ken Suslick's group at the University of Illinois at Urbana-Champaign is of note since they concentrate on the chemistry at work in ultrasonic cavitation (sonochemistry) [41–43]. Suslick's appratus consists of a transducer horn in a test cell immersed in a temperature-controlled bath [41]. Suslick's group found localized point heating due to cavitation in liquids and the resulting metal clustering and liquid decomposition can be attributed to processes with temperatures between 2900 K and 5200 K that do not include fusion processes [42,43]. Because the high temperatures generated that make sonochemistry possible may also be used for power generation, an effort for harvesting power from sonoluminescence should not be focused solely on sonofusion results.

III. Challenges for Application

The application of sonoluminescence as a means of energy harvesting requires a clear understanding of the heat generation process involved. In order for sonofusion claims to be verifiable, detection of their nuclear products is a necessity. The reactions cited are those that fuse two deuterium atoms to form: 1) helium-3 and a neutron, 2) tritium and a proton [D(d,p)T], or 3) helium-4 and a gamma ray. These reactions are reported to occur either in a sonoluminescent bubble alone or interacting with a palladium catalyst. Detection of these products will verify whether or not fusion is involved in the heating process and discrimination of the products will verify the specific process involved.

In order for a system to be applicable in flight hardware, the system needs to be scaled down to become as small as possible. Auxiliary power units in aircraft have a current design point of 440 kW with a mass of 1396 kg, the equivalent of 315 W/kg specific power [44]. The International Space Station (ISS) electrical power system contains 76 Ni-H battery cells to supply the 8-kW peak power required when the station cannot use the power from its solar cell arrays [45]. The size of a typical cell is 350 ml in volume, approximately 1/3 kg in mass, providing the equivalent of 316 W/kg specific power [46]. For a substantial improvement over state-of-the-art, a factor of 10 improvement in overall performance (power, mass, and volume) is desired. The ultimate goal then is to provide a power source capable of better than 1 kW/kg, with the volume less than 100 ml and mass less than 0.1 kg. For a sonoluminescence-based power source, most of the mass is in the transducers and support equipment. Minimizing the test cell to 20 ml or less is a realistic criterion for achieving these goals.

As an example of the scaling challenges, a first order estimate on the minimum size of a sonoluminescence test cell can be derived from the role of harmonic resonances in bubble oscillations [47]. A gas bubble in a liquid has a free oscillation frequency given by Eq. (1):

$$\omega = \frac{1}{a}\sqrt{\frac{3\gamma P}{\rho}} \qquad (1)$$

Here ω is the angular frequency at resonance ($\omega = 2\pi f$), a is the bubble radius, P is the initial liquid pressure, γ is the heat capacity ratio for the air in the bubble ($\gamma = 5/3$ for monatomic gasses, $\gamma = 7/5$ for diatomic gasses), and ρ is the liquid density at initial pressure. The forced oscillations are driven by the resonance frequency of the test cell given in Eq. (2):

$$\omega = \frac{k\pi c}{r} \qquad (2)$$

Again, ω is the angular frequency at resonance ($\omega = 2\pi f$), k is the harmonic, r is the radius of the cell, and c is the speed of sound in the liquid. This can be set equal to the free oscillation frequency to determine the maximum response. Sonoluminescence takes place in bubbles with initial radius less than 10 μm, so, theoretically, the smallest test cell (with $k = 1$) for water ($\rho = 1000$ kg/m³, $c = 1500$ m/s) at atmospheric pressure ($P = 10^5$ N/m²) can be calculated by combining Eqs. (1) and (2):

$$\frac{r}{a} = \sqrt{\frac{k^2 \pi^2 c^2 \rho}{3\gamma P}} = 230 \qquad (3)$$

With the result of Eq. (3), a test cell can theoretically be sized to 4.6 mm in diameter with a resonance frequency of $f = 326$ kHz, making the goal of scaling the test cell to under 20 ml appear realistic.

IV. Sonoluminescence at NASA

Initial examination of the phenomenon has been performed by several researchers sponsored by NASA via the NASA Research Announcement (NRA) process. The research efforts involved measurements of single and/or multibubble sonoluminescence in microgravity (0 g vertical acceleration) and hypergravity (2 g vertical acceleration) performed on a KC-135 parabolic research flight. One such experiment was performed by researchers from the University of Washington. The researchers discovered that light emission brightened promptly by 20% and increased continually under microgravity, suggesting buoyancy-driven instabilities are a critical limitation to the effect [48]. This research was selected under the NRA process for a flight experiment. Flight hardware for an ensuing experiment on ISS studying buoyancy-driven instabilities in single bubble sonoluminescence was under development in 2003 (Fig. 2), but was canceled following a redirection in NASA's space exploration efforts.

As part of our mission to conduct basic and applied research on advanced instrumentation technologies, NASA Glenn Research Center began an examination of sonoluminescence for instrumentation and measurement technique development [49]. The objective of the effort was to investigate claims and theories of power generation based on sonoluminescence. The approach would initially compare the emission from bubbles in light water, heavy water, and other solvents. Conclusions on sonoluminescence-based power generation concepts would be formulated utilizing the data from our experiments.

A. Apparatus

The basic equipment for sonoluminescence consists of a flask containing the liquid, ultrasonic transducers, a piezoceramic amplifier, and a function generator [4]. Schematics of our test apparatus are shown in Figs. 3 and 4. The transducers are driven by an amplified signal from the function generator to saturate the fluid with ultrasonic waves. The voltage and current applied to the transducers can be monitored and instrumentation (such as a microphone,

Fig. 2 Single bubble sonoluminescence flight hardware testing in 2003.

spectrometer, and photomultiplier tube) can be positioned to monitor the resulting sonoluminescence.

If the ultrasonic energy from the transducers can be concentrated into a small enough uniform spot, and the liquid does not contain a large amount of dissolved gas, the liquid can be ruptured in cavitation that is controlled. The cavitation can cause a single glowing bubble to appear that is reasonably stable, known as single-bubble sonoluminescence (SBSL) [4]. If the liquid is nearly saturated with dissolved gas, the cavitation forms filamentary patterns that are chaotic as the input energy is distributed throughout the flask. A cloud of glowing bubbles can result, which is referred to as multi-bubble sonoluminescence (MBSL) [50]. Either way, the sonoluminescence is located at or near regions of maximum pressure [50].

B. Imaging

The initial apparatus for our tests was a resonating test apparatus as shown in Fig. 3, consisting of a 250-ml round borosilicate glass flask filled with refrigerated ($\sim$17°C) distilled water with a pair of piezoceramics attached with epoxy on opposite sides. Patterns of MBSL were produced at various frequencies. Although most of the patterns were chaotic, two stable geometric patterns were reproducible. The patterns of interest were rings having four and eight nodules at 68.5 kHz and 93.0 kHz, respectively. The MBSL patterns were

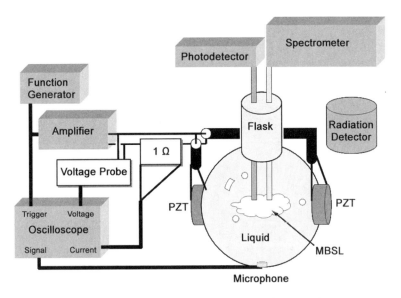

Fig. 3 Resonating test apparatus and associated equipment.

faint, requiring 20 minutes of dark adaptation in the lab for observers to see them. To determine the placement of fiber-optic instrumentation for future investigations, the position of the MBSL in relation to the sides of the flask needed to be recorded.

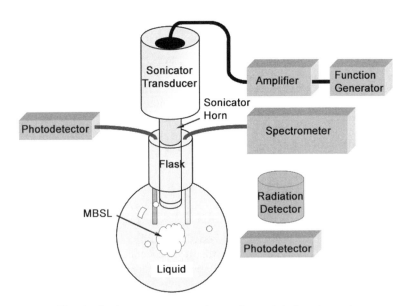

Fig. 4 Sonicator test apparatus and associated equipment.

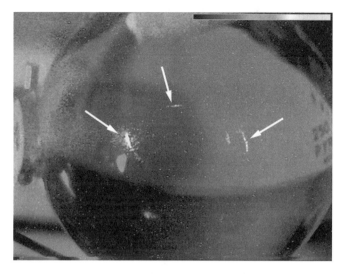

Fig. 5 Enhanced image from video of a ring of MBSL. The arrows indicate the three visible nodules.

A low lux astronomical video camera was purchased to image the patterns. The camera was set to maximum sensitivity and images were recorded using a 2-s frame integration setting. Images were recorded with and without the frequency generator on, and with and without room lights on. The resulting compilation of these is seen in Fig. 5. Only three of the nodules of the four-nodule ring are seen in the image. The eight-nodule ring was too faint to be resolved by this method.

To help increase visibility of the MBSL, a sonoluminescence setup using a high power Sonicator test apparatus as shown in Fig. 4 was assembled. The test cell used initially was a 100-ml borosilicate glass beaker filled with refrigerated distilled water, but later a 50-ml quartz flask was used. Each used a titanium high-intensity ultrasonic transducer horn probe in line with the signal generator and amplifier instead of piezoceramics. As seen in the figure, the transducer horn was inserted into the open top of the beaker and flask. Besides being capable of delivering high acoustic power, the Sonicator test apparatus has the added benefit that the transducer is not physically attached with epoxy to the test cell as it is in the resonating test apparatus.

With the resonating test apparatus, the previously reported effect [10] of brighter sonoluminescence glow in cooler water at the same driving voltage was observed. In order to keep the water temperature cool and the cells stable, a Peltier cooler plate was modified from a commercial kit and aluminum support rings were machined for use with the Sonicator test apparatus. The aluminum support rings allowed thermal contact of the round flasks with the Peltier cooler plate. The Peltier cooler was not used in the resonating apparatus, as it would interfere with the vibrations of the flask. The MBSL patterns were generated in both the beaker and quartz flask cells, with and without support rings, by this method. However, the patterns were not as defined as in the 250-ml flask, and the video camera was not able to detect the patterns as well as the ring patterns generated by the resonating test apparatus.

A 16-bit grayscale charged coupled device (CCD) imaging system for terrestrial still deep-space astrophotography was purchased to image the MBSL. Unlike the compiled image (Fig. 5) from three separate images, only brightness, contrast level enhancement, and the application of a noise filter were needed to achieve the images shown in Figs. 6 to 10. The MBSL filament feature on the bottom of the flask in Fig. 6b is approximately 25-mm wide by 1-mm thick, making the resolution approximately 100 μm/pixel. The filament structure is revealed with a gradient map of sonoluminescence intensity as false-color, rendered in black-and-white, in Figs. 7b to 10b.

The sonoluminescence filaments seen in Figs. 7 and 8 follow a conical pattern from the horn to the base of the beaker and flask. This filament pattern has been reported elsewhere [46], but in our images, the container is clearly visible, and no brightening agent was introduced into the water. The 50-ml flask produced the brightest MBSL (Fig. 8), possibly due to the small volume and thus increased concentration of the applied ultrasonic power.

The CCD imaging system was then used to look at the rings formed by the resonating test apparatus using the 250-ml flask, which had been at room temperature for several months. The ring patterns were reproduced (Figs. 9 and 10), but dimmer and at higher frequencies than in Fig. 5 due to the room temperature water. The rings in the images are smeared due to the variation of the nodules positions about the pressure maximums. The images reveal the rings as "wavy" rings, with nodules near the piezoceramics.

The nodules of the eight-nodule ring are revealed as pairs of nodules shifting about the four maximums of the four-nodule ring (only six of the eight nodules are visible in Fig. 10). In addition, when comparing Figs. 9 and 10, the four-nodule ring forms in the center of the flask, but the eight-nodule ring forms slightly lower. The shift in position is not apparent to the dark-adapted observer because the flask is not clearly visible in the darkened lab. Thus, this

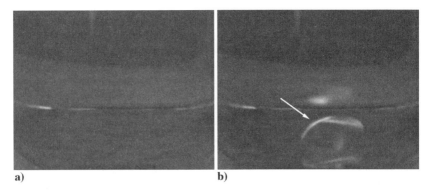

Fig. 6 Bottom of a 100-ml beaker with the Sonicator transducer horn a) off and b) on. The MBSL filament appears centered at the bottom of the beaker (*arrow*). Both pictures are 1-min exposures at f/1.2.

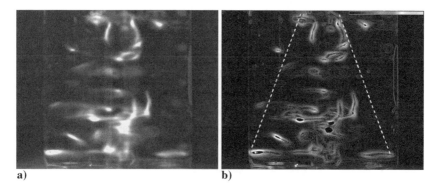

Fig. 7 Enhanced image of MBSL at 108.8 kHz in the 100-ml beaker with Sonicator. Exposure time was 2 min at f/2.8. The field of view is 6.6 × 5.1 cm. Image shown in a) grayscale and b) false-color (shown here in black-and-white), highlighting filament structure of the MBSL. Dashed lines outline the conical area beneath the Sonicator where most of the sonoluminescence occurred.

ability to image the flask and the sonoluminescence effect together is demonstrated as necessary if any precision is desired for the placement of in situ instrumentation.

A 16-bit three-color CCD imaging system for terrestrial still deep-space astrophotography was purchased for color imaging the MSBL. The camera was calibrated using a color rendition chart. The resulting true-color image of the MBSL in a 50-ml flask is shown in black-and-white in Fig. 11 with contrast enhanced. Red light from LEDs on the support equipment scattered by the flask and transducer horn help reveal their presence.

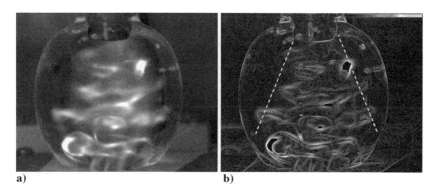

Fig. 8 Enhanced images of MBSL at 106.5 kHz in the 50-ml flask with Sonicator. Exposure time was 3 min at f/2.8. The field of view is 7.1 × 5.4 cm. Image shown in a) grayscale and b) false-color (shown here in black-and-white), highlighting filament structure of the MBSL. Dashed lines outline the conical area beneath the Sonicator where most of the sonoluminescence occurred.

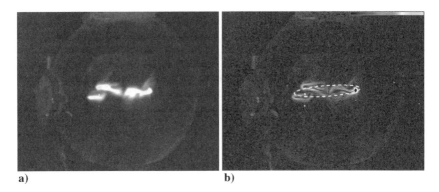

Fig. 9 Enhanced images of MBSL at 68.76 kHz in the resonating 250-ml flask. Exposure time was 3 min at f/2.8. The field of view is 11.4 × 8.8 cm. Image shown in a) grayscale and b) false-color (shown here in black-and-white), highlighting filament structure of the MBSL. Dashed line outlines the position of the ring of MBSL within the flask.

C. Sonoluminescence in Solvents

Previous studies of sonoluminescence brightness in various solvents give empirical relationships of the brightness varying with the liquid's viscosity, surface tension, inverse of the vapor pressure, or a combination of these properties [40]. Properties of several solvents that have been identified as producing some form of sonoluminescence are reported in Table 1 [40,51]. The stability of the sonoluminescence light output with temperature is also reported to be dependent on the molar heat of vaporization of the liquid divided by the boiling point. Table 1 shows that the brighter sonoluminescence should be seen in solvents with higher boiling points (>100°C); however, a higher boiling point also

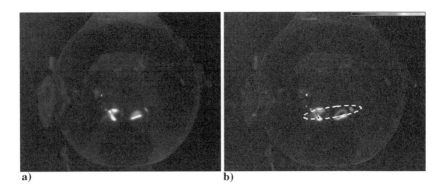

Fig. 10 Enhanced images of MBSL at 93.28 kHz in the resonating 250-ml flask. Exposure time was 5 min at f/2.8. The field of view is 11.4 × 8.8 cm. Image shown in a) grayscale and b) false-color (shown here in black-and-white), highlighting filament structure of the MBSL. Dashed line outlines the position of the ring of MBSL within the flask.

Fig. 11 True-color picture of sonoluminescence in light water (contrast-enhanced) using the 50-ml flask with Sonicator, shown here in black-and-white. Exposure time was 2 min at f/2.8.

indicates a greater sensitivity to temperature. Note that this is consistent with Flynn's use of higher density liquids for more violent cavitation, but contradictory to Taleyarkhan's use of higher vapor pressure liquids (see Sec. II).

From the information in Table 1, pure glycerin appears to be the best solvent for use in sonoluminescence studies. Like the other, higher boiling point solvents, glycerin is notoriously hydroscopic, eventually absorbing water from the ambient air to stabilize as the 80% glycerin to 20% water mixture listed in Table 1. As glycerin is relatively safe and more readily available than the other solvents, NASA Glenn began sonoluminescence studies using glycerin. Note that acetone and other low boiling point solvents, which are being explored elsewhere [34,35], were rejected based on low viscosity and high vapor pressure properties that are considered detrimental to sonoluminescence intensity.

A cloud of cavitation was generated with a Sonicator setup (Fig. 4) that corresponded to sonoluminescence in the liquid as seen under dark conditions, as shown in Fig. 12. The cavitation was particularly localized, allowing for a promising target for the spectroscopy and radiation studies that are planned.

V. Indications of High Temperature Generation

Cavitation bubbles are formed by the vaporization of the liquid due to changes in pressure. The liquid transitions to its vapor phase around nucleation points that arise from a breakdown in surface tension of the liquid. This breakdown can be initiated by the random mechanical motion of the liquid molecules, container geometry, impurities (liquid or gaseous), and cosmic radiation. The vaporization can translate into high temperatures. The highly localized nature

Table 1 Properties important to sonoluminescence for different solvents at 20°C [40,51]

Solvent	Boiling point (°C)	Density (ρ) (g/cm^3)	Mol. wt. (g/mole)	Viscosity (μ) (cP)	Surface tension (γ) (dynes/cm)	Vapor pressure (P_v) (Torr)	Molar heat of vaporization (ΔH_v) (kJ/mole)	Calculated MBSL intensity (γ^2/P_v relative to water)	Intensity stability with temperature $dI/(IdT)$ (°C^{-1})
Methanol	64.7	0.415	32.04	0.597	22.6	93.3	35.2	0.0180	−0.072
Ethanol	78.5	0.789	46.07	1.2	22.8	44	39.3	0.0386	−0.078
Acetone	56.5	0.792	58.08	0.326	23.7	186	30.3	0.0099	−0.064
Cyclohexane	81.4	0.779	84.16	1.02	25.5	84.8	29.9	0.0251	−0.059
Carbon tetrachloride	76.8	1.595	153.84	0.969	27.0	91.3	29.9	0.0261	−0.059
Benzene	80.1	0.879	78.11	0.652	28.9	74.3	30.8	0.0369	−0.060
Light water	100	1.00	18.02	1.00	73.1	17.5	40.7	1	−0.076
Heavy water	101.4	1.11	20.03	1.25	73.1	16.6	42.1	1.13	−0.078
80% Glycerin + 20% water	121	1.209	77.28	60.1	66.6	3.5	85.7	4.16	−0.151
Ethylene glycol	197	1.116	62.07	19.9	47.7	0.08	49.6	93.3	−0.073
Sulfuric acid	290	1.788	184.15	25.4	55.1	0.00006	94.1	166,000	−0.116
Glycerin	290	1.261	92.09	1490	63.4	0.001	87.9	13,200	−0.108

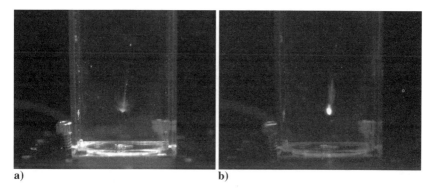

Fig. 12 Cavitation in glycerin with room lights a) on and b) off (contrast enhanced) showing localized sonoluminescence in a 40-ml container. The Sonicator transducer was set to 56 kHz.

of sonoluminescence makes a direct measurement of the bubble temperature extremely difficult with thermocouples or RTD probe. One method of comparing the temperatures of two different environments is to compare the surface modification of materials exposed to the different environments. Leveraging our expertise in thin films for high temperature sensing applications, the effect of sonoluminescent bubbles on thin films was investigated to determine differences in the temperature of sonoluminescence in light water (H_2O) and heavy water (D_2O).

A. Platinum Films

The effect of exposure of thin films to sonoluminescence in light water and heavy water was compared using several 1-mm thick, 6.1 × 6.1 mm substrates of alumina (Al_2O_3) coated with thin films. The first two samples had 3 μm of platinum deposited on the alumina substrates and were not annealed. One sample was exposed to MBSL in light water, and another was exposed to MBSL in heavy water. Images taken by a scanning electron microscope (SEM) are shown in Fig. 13, and both samples have similar features. Each sample had the general appearance of as-deposited platinum film. However, larger and more frequent grain clusters were seen in the sample exposed to sonoluminescence in heavy water. These features are typically observed after annealing platinum films at high temperatures. If the larger grains were created from exposure to sonoluminescence, their appearance would be an indication that the sonoluminescence in heavy water generated higher temperatures than the sonoluminescence in light water.

B. Palladium–Chromium Films

The three new thin film samples were fabricated by first patterning a 1-μm thick sputter-deposited platinum (Pt) resistance temperature detector (RTD) on 1-mm thick, 6.1 × 6.1 mm substrates of alumina and annealing at 1000°C for 8 hours. The RTD pattern was to be used for applications of miniature instrumentation to characterize sonoluminescence. The samples were then over-coated with

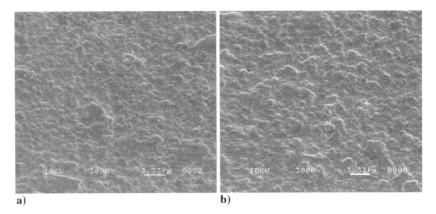

Fig. 13 SEM image comparison of as-deposited, 3-μm thick films of pure platinum coating alumina substrates exposed to sonoluminescence with a) H_2O and b) D_2O.

a 3-μm layer of palladium alloy with 13% chromium (PdCr). The PdCr alloy film is typically used as a strain gauge in high temperature environments. For this instance, the PdCr alloy was used because of the affinity of hydrogen to palladium to allow the localization of sonoluminescence, which may contain hydrogen ions [40,52]. Though the Pt itself has some limited affinity to hydrogen, the film does not adhere well to alumina at high temperatures, and some delamination of the Pt film was seen after annealing prior to PdCr deposition.

One of the new samples was exposed to MBSL in light water and another to MBSL in heavy water. Viewed under SEM, the sample exposed to MBSL in light water showed no significant modification of the film. The sample exposed to MBSL in heavy water showed 4- to 5-μm diameter craters in the PdCr film overcoating the Pt film when viewed under SEM. As the Pt film was already delaminating, the third sample was exposed to MBSL in both light water and heavy water. Images of the exposures (Fig. 14) show that the sample was exposed to sonoluminescence directly at the tip of the Sonicator horn. Again, the sample

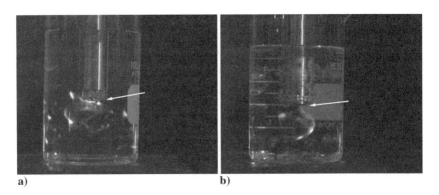

Fig. 14 Film sample on Sonicator horn tip (*arrows*) exposed to SL in a) H_2O and b) D_2O at 56 kHz in 100-ml beakers.

exposed to light water showed no modification of the film, but after the heavy water MBSL exposure, the 4- to 5-μm diameter craters in the film were apparent when viewed under SEM. No craters in the PdCr film directly deposited on the alumina substrate were seen. Figure 15 gives a side-by-side comparison of the film at different

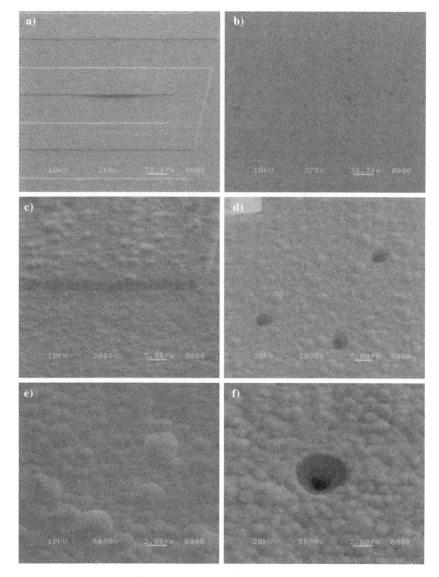

Fig. 15 Surface of the 3-μm thick PdCr films on 1-μm thick Pt patterns seen by SEM under increasing magnification. The black surface at the bottom of the craters is nonconducting alumina. The films were exposed to MBSL in a) H_2O, 260× magnification (mag), b) D_2O, 275× mag, c) H_2O, 2000× mag, d) D_2O, 2000× mag, e) H_2O, 5000× mag, and f) D_2O, 5000× mag.

Table 2 Summary of film modification by MBSL in light and heavy water

Film on alumina	MBSL in H_2O	MBSL in D_2O
Pt (as-deposited)	No damage	No damage
PdCr (as-deposited)	No damage	No damage
PdCr (as-deposited) over Pt (annealed) (some Pt delamination after annealing)	No further damage	Craters through film; isolated PdCr damage

magnifications after the light water exposure and after the heavy water exposure. A summary of the damages seen in the films is given in Table 2.

Table 3 gives a summary of some properties of the materials used in the samples [53,54]. Delaminating of films is considered a zero-order effect of mismatches between the two materials' coefficient of thermal expansion (CTE). The fact that grain failures are seen in the film with PdCr on Pt and not the PdCr suggests that the expansion of the PdCr on Pt is greater than that of the PdCr compared to alumina. However, Table 3 indicates that the adhering PdCr has a greater CTE mismatch to the alumina substrate than the Pt film that was delaminating.

A first-order estimate of the relative adhesion strength of films on oxides is the energy involved in forming an oxide of the film (or heat of oxide formation) [55]. Table 3 shows that the PdCr alloy film has a more negative heat of oxide formation, and is thus more favorable to adhesion on the alumina than platinum. The energy imparted to the film at the sonoluminescence point locations could be enough to vaporize or melt the film when not adhering to the substrate, but not enough to vaporize or melt the film when it is adhering to the substrate.

The adhesion of the film to the substrate may be allowing the substrate to absorb the energy via phonon vibrations, and the delaminating film does not allow the energy to be transferred directly into the bulk substrate material. Alternatively, the loose islands may have heated up and "popped" off due to thermal expansion. This effect was seen on two separate samples in the heavy water runs but were not seen in the light water runs. The craters did not occur in the as-deposited Pt films in either light or heavy water.

Some failures of the PdCr film that did not involve failure of the Pt film were observed after exposure to heavy water. The SEM image of the largest volume ($14 \times 8 \times 3$ μm) is shown in Fig. 16. Analysis by EDX of the failure area indicates that remnants of the PdCr film remain at the exposed Pt surface, presumably in the form of several 0.375-μm radius globules observed in the SEM image. From Table 3, at least $6.9 \text{ kJ}/\text{cm}^3$ are required to melt PdCr from 20°C, and $7.1 \text{ kJ}/\text{cm}^3$ for platinum. The craters and the PdCr film damage may be due to the same heating process of the Pt film. The lack of damage induced in the as-deposited Pt film suggests the heating is not directly from excessively high energy density of the MBSL in heavy water as compared to light water.

The primary result of these tests revealed modification of PdCr films by MBSL generated in heavy water but not light water. The film modification indicates high energy densities are generated, though not high enough to indicate net energy generation by the MBSL itself. No modification was observed of platinum films by MBSL in heavy or light water.

Table 3 Sample substrate and film properties [53,54]

Material	Density (g/cm^3)	Melting point (K)	Heat of fusion (J/g)	Specific heat (J/gK)	Thermal conductivity (W/mK)	Thermal diffusivity (mm^2/s)	Coefficient of thermal expansion (CTE) (10^{-6}/°C)	Heat of oxide formation (kJ/mol)
Al$_2$O$_3$	3.92	2300	16,435	0.880	30.0	8.70	8	—
Pt	21.5	2141	113.6	0.113	71.6	29.5	10	−173
Pd – 13%Cr	11.39	>1828	189.4	0.271	71.8	23.3	15	−468

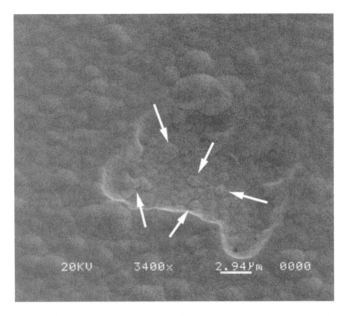

Fig. 16 A large 20 × 10 μm failure of the PdCr film over Pt film. Globules of PdCr are seen (*arrows*) on the 1-μm thick Pt film that is still adhering.

C. Comparison to Other Claims

The formation of submicron globules and film failures are consistent with the results of George and Stringham [29,30], as well as Suslick [42,43]. The film modifications suggest that sonochemical processes in the sonoluminescence in heavy water reached temperatures of at least 1830 K. The lack of similar modification in adherent platinum suggests that the temperatures of the cavitation did not reach over 2140 K. Suslick reported the temperatures of cavitations in hydrocarbon slurries can be as high as 5200 K. The globules as a remnant of larger PdCr failure suggest two possibilities: 1) The film failed due to high temperature melting and the globules are "residue," or 2) the film was not fully adhering when exposed and the globules are bits of the failing film forced into the platinum undercoat. The former interpretation is consistent with George and Stringham, and the latter consistent with Suslick. The point film failures are similar to the void formation seen by George and Stringham, but without the vent cones associated with ion implantation. The large grain failures seen in our films after high temperature exposure suggests the point failures are due to thermal expansion mismatch of the alumina and the film. The lack of similar failures in the light water runs suggests that the heavy water cavitations did get warmer.

In comparing our results with those of Suslick, George, and Stringham, we cannot claim fusion processes are at work in our sonoluminescence runs based on the film modification. However, the increased temperature in the heavy water runs is not easily explainable. As noted earlier, the development of an energy harvesting device is not dependent on the detection of fusion in sonoluminescence. The determination of the nature of the film heating, and understanding the role of light and heavy water in temperature generation, will influence the design and the estimation of the performance characteristics of an energy harvesting device.

VI. Energy Harvesting

The application of sonoluminescence as a means of energy harvesting requires a clear understanding of the heat generation process involved. There have been several claims of fusion reactions occurring in sonoluminescence [24–33]. A direct method to verify the reaction processes is by sampling in situ with radiation detectors. The ultimate application of sonoluminescence for this effort is the energy harvesting of extremely hot bubbles at point locations. New instrumentation is under development at NASA Glenn for the energy harvesting application.

A. Fusion Claims

The reactions cited in sonofusion claims are those that fuse two deuterium atoms to form: 1) helium-3 and a neutron [D(d,n)^{3}He], 2) tritium and a proton [D(d,p)T], or 3) helium-4 and a gamma ray [D(d,γ)^{4}He]. These reactions either can occur in a sonoluminescent bubble alone or interacting with a palladium catalyst [Pd:D(d,b)Y reactions]. See Table 4 for the calculated energies of the products of these reactions. In order for sonofusion claims to be verifiable, detection of their nuclear products is a necessity. Detection of the reaction products in Table 4 will verify whether or not fusion is involved in the heating process and discrimination of the products will verify the specific process involved.

B. Radiation Detection

The detection of the fusion reactions in the sonoluminescence test cells is made difficult by the nature of the liquids used to generate the reactions. The low level ionizing radiation levels are difficult to detect through water and heavy water due to their nuclear properties. The liquids can be analyzed for tritium production or the detection of helium-3 and helium-4 out-gassing

Table 4   Calculated energies of reactions of interest

Reaction	Product	Rest mass (MeV/c^2) (D = 1875.6128 MeV/c^2)	Kinetic energy (keV)
D(d,n)^{3}He	^{3}He	2808.3914	822
	n	939.5536	2458
D(d,p)T	T	2807.9027	1265
	P	938.2720	3786
D(d,γ)^{4}He	^{4}He	3727.3792	75.7
	γ	–	23,771

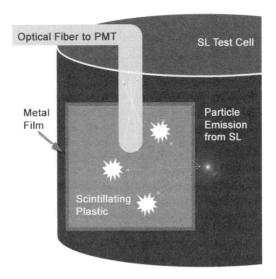

Fig. 17 Schematic of thin film coated scintillating detector.

during the tests, but background impurities and the diffusive nature of the gases makes such analysis slow and difficult. However, some success has been reported using liquid scintillators and track detectors [30,31].

A direct method to verify the reaction processes is by sampling the ionizing radiation in situ with scintillation detectors. A thin film coated scintillating detector is under development to identify fusion reactions occurring in sonoluminescence. The detector consists of a coated scintillating cube or fiber, a wave-shifting optical fiber, a fiber-optic connector, and a photomultiplier (PMT) module. A schematic is shown in Fig. 17.

The polyvinyltoluene (PVT)-based scintillator¶ generates a pulse of light if an ionizing particle interacts inside it, depositing energy. The intensity of the light is directly proportional to the energy deposited by the interacting particle with the ionization potential based on the particle's atomic number. The scintillator used is reported to be insensitive to photons with energies over 100 keV as well as neutral particles. A thin film coating of rhodium (Rh), copper (Cu), or palladium (Pd) on the scintillator functions as either an attenuator or a convertor to allow possible fusion products to react with the scintillator, and also prevents the sonoluminescence light from generating false readings. The nuclear properties of the scintillator and the coatings are given in Table 5 [56,57].

¶Bicron BC-408 scintillator was used in the prototype detector design. "Bicron" is a registered trademark of Saint-Gobain Ceramics & Plastics, Inc. for their inorganic and organic scintillators. This usage is for identification only, and does not constitute an official endorsement, expressed or implied, by NASA or AIAA.

Table 5 Nuclear properties of scintillator and coatings [56,57]

Material	Density (g/cm³)	Atomic mass (amu)	Thermal neutron absorption cross section (barns)	76 keV alpha particle range (μm)	3.75 MeV proton range (μm)	2.5 MeV electron range (mm)
Scintillator	1.032	6.25	0.35	0.7914	208.6	12.34
Rh	12.41	102.91	144.8	0.2724	42.34	1.480
Cu	8.96	63.546	3.78	0.2076	48.68	1.953
Pd	12.02	106.4	6.9	0.1797	44.83	1.552

The relative response of the scintillator with coatings for each reaction was calculated using the Monte Carlo program SRIM [58]. For each coating (Rh, Cu, Pd) and ion (p, T, ^{3}He, ^{4}He), 10,000 input particles were used, and the balance of the energy of the particles transmitted through the coatings was assumed deposited into the scintillator. The sum of the deposited energies was scaled based on the reported scintilator response to determine the light output.

For each fusion reaction scenario, the ouput of the detectors is normalized to the detector with the largest signal. The results are shown in Table 6. Thus, comparing the response of each detector will allow a reasonable identification of the reaction occuring in the cell. In this way, the fusion reactions that appear (or not) in sonoluminescent conditions can be determined, limited by the statistics of the actual counts and background radiation.

The output for the $D(d,n)^3$He reaction was determined from the helium-3 ion transport in SRIM, as well as utilizing a 3-μm thick rhodium film as a neutron convertor. Rhodium emits 2.44 MeV electrons in the beta decay of caputured neutrons, and has an excellent thermal neutron absorption cross section of 145 barns/atom, which will capture about 0.1% of the neutron flux per micron thickness of film. The emitted electron will be easily detectable as the scintillator sensitivity to electrons is reported to be five times that for protons of equivalent energy.

The use of palladium in ultrasonic systems as well as our observations in Figs. 13 and 14 suggest that the metal should be considered in radiation detection. All three of the deuterium-deuterium reactions will be tested with and without a palladium film included on the detectors. If palladium has a catalytic role in fusion reactions, then the detectors with the palladium will be more sensitive as the reactions are occurring inside the palladium film. The sensitivity was modeled in SRIM assuming the reactions occur at the bottom of the 0.3 μm Pd layer, based on the 0.375 μm radius globules seen in Fig. 15, with all the particles deposited into the scintillator or rhodium.

The current assembly of the detectors is shown in Fig. 18. This assembly is too sensitive to background counts to provide useful data at this time and improvements in the assembly are proceeding. With modifications, the concept could potentially be used in other applications, such as radiation monitoring at high altitudes or in space environments. This effort was successfully leveraged in radiation detector development for the application of dosimeters on surface suits for lunar extravehicular activities [59].

Table 6 Relative sensitivities of detectors to reactions of interest

Detector film coating	$D(d,p)T$	$D(d,n)^3$He	$D(d,\gamma)^4$He	Pd:$D(d,p)T$	Pd:$D(d,n)^3$He	Pd:$D(d,\gamma)^4$He
3.0 μm Rh	0.89	0.04	—	—	—	—
0.3 μm Cu	0.86	1.00	1.00	—	—	—
0.3 μm Pd	1.00	0.94	—	1.00	1.00	1.00
0.3 μm Pd/ 3.0 μm Rh	0.84	0.04	—	0.87	0.03	—

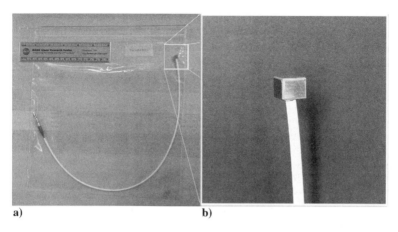

Fig. 18 Prototype thin film coated scintillating detector a) assembly and b) close-up of coated 1/4-inch scintillator cube with a thin film of Pd.

C. Energy Harvesting Concept

The common goal of energy harvesting devices is to generate electricity from the environment. In the case of energy harvesting from sonoluminescence, the environment has point locations of extremely hot bubbles. The most direct method of energy harvesting of such localized heating is by thermoelectric conversion to electricity.

Thermoelectric voltage generation uses the Seebeck effect, which is the generation of a potential difference between two ends of a conductor or semiconductor that are at different temperatures. The effect involves the movement of charge carriers from the hot end of the conductor, where the carrier density is forced lower, to the cold end, where the carrier density is allowed to be greater. The electric field due to the new distribution of charge carriers leads to a thermoelectric voltage. The voltage generation per degree difference is referred to as the Seebeck coefficient. To complete the electrical circuit a different conductor is used with its own Seebeck coefficient, making the net thermoelectric voltage generated the difference of the two with a corresponding relative Seebeck coefficient. The simplest example of this effect is a thermocouple used for temperature measurement [60].

If the thermoelectric voltage generated is high and the losses due to the thermal conduction and electrical resistance of the conductors are low, then the voltage generated by such a device can be used to generate power. Thermoelectric power generation is used on deep space probes (such as the Viking, Voyager, and Galileo missions) where solar power, fuel cells, and batteries are not practical. As their name indicates, the heat source for radioisotope thermoelectric generators (RTGs) on spacecraft is a radioactive pellet heating a ceramic block; the cold side is the darkness of space.

To take advantage of the high temperature robustness of ceramics and the miniature nonintrusiveness of thin films, NASA Glenn is investigating thin film ceramic thermocouples for high-temperature environments. Initial results revealed a chromium silicide film with a Seebeck coefficient to be high enough to be considered for use in thermoelectric generators [60]. These results suggest that ceramic thin films can be tailored for thermoelectric energy harvesting applications.

A schematic of the energy harvesting concept is shown in Fig. 19. A 6-mm diameter thermopile, originally used for fabrication of heat flux sensors (Fig. 20), will be used in fabricating devices for the initial test of a thermoelectric generator for sonoluminescence. The thermoelectric elements will be made of high-temperature ceramics, covered with electrical insulation for use in the liquid. The inner junction will be covered with a high-temperature insulator embedded with a palladium catalyst based on the indications of high temperature generation seen in Figs. 14 and 15.

The output of the generator can be estimated using the properties of the chromium silicide film [60]. A thermoelectric voltage of 200 mV per junction results, assuming sonoluminescent bubbles react on the inner junctions of the thermopile with a 2000°C temperature difference from the surrounding liquid. The thermopile resistance should be about 200 Ω per junction, giving 0.2 mW of power per junction. About half of the generated power is expected be lost to the resistance of the thermopile. Thus, a 40-pair thermopile as shown in Fig. 20 should generate 4 mW of power under sonoluminescence. Because this 28 mm^2 generator is a fraction of the MBSL area seen in Fig. 13, an array of generators can conceivably output enough electrical power to match the acoustic power input to the system. Clearly, the improvement of the Seebeck coefficient, thermal resistance, or electrical conductivity is needed for the high-performance thermoelectric generators needed to allow practical applications for energy harvesting of sonoluminescence.

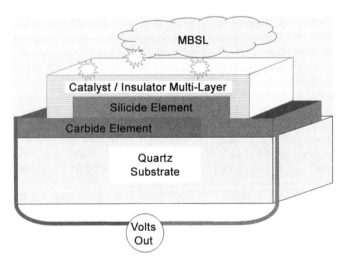

Fig. 19 Cross-section concept of a high-temperature thermopile for energy harvesting sonoluminescence.

Fig. 20 A 6-mm diameter thermopile for initial test of ceramic thin film thermoelectric device.

VII. Summary and Future Directions

Instrumentation techniques were explored at NASA Glenn to measure optical, radiation, and thermal properties of the phenomenon of sonoluminescence, the light generated using acoustic cavitation. The objective of the effort was to investigate claims and theories of power generation based on sonoluminescence, particularly from fusion reactions in the glowing bubbles. The approach was to determine whether there is any difference in the emission from bubbles in light water and heavy water and then from bubbles in other solvents to formulate conclusions of sonoluminescence-based power generation concepts utilizing the data from the experiments.

More generally, there remains much to learn about sonoluminescence. Efforts to characterize sonoluminescence and thereby form a complete theoretical description of the phenomenon have produced such a wide variety of information that much more research needs to be done. Characterization efforts—such as searching for byproducts of chemical and nuclear reactions, completely mapping the temperature distribution in a sonoluminescence cell, and spectroscopy—will all contribute to a more complete understanding of the nature of sonoluminescence. Because the results of any type of characterization measurement seem to change with reported changes in cell size and shape, temperature, pressure, liquid, dissolved gasses, and oscillator frequency and power, there is ample room for careful detailed characterization to be made.

Optical emission spectra need to be recorded for many different sonoluminescence setups, for SBSL and MBSL, as near and far from the bubble(s) as is practically achievable. For different liquids, differences in spectra, power, and whether there are single or multiple bubbles should be carefully noted. Temperature maps should be made of as much of the liquid region as possible, with the smallest probes that are practical to use. Evidence for chemical or nuclear reactions and the specifics of the condition that produced the evidence must be carefully noted.

High-speed timing of bubble flashes, and high-speed photography should be employed to understand better the mechanics of the bubble behavior. Additionally, high-speed photography or light scattering studies may help map bubble motion in the fluid. Currently, widely disparate claims about the nature and behavior of sonoluminescence are made by people who are not considering the same conditions under which the phenomenon of sonoluminescence is achieved. Until many detailed characterizations are complete, a detailed theoretical explanation of the phenomenon remains an elusive hope. The listing of measurement conditions and sonoluminescence characterizations to be varied in experiments on both SBSL and MBSL is summarized in Table 7.

A more detailed understanding of the nature of sonoluminescence will help define its practical uses, although even with our current limited understanding of the phenomenon some practical uses have arisen already. The field of sonochemistry is already using the localized heating provided by ultrasound to drive chemical reactions [24], modify surfaces [25], produce interparticle melting in liquid–solid chemical reactions [26], and produce quantum dots [27]. These techniques have already produced useful applications—even without a complete understanding of the phenomenon.

After all this, two issues must yet be resolved. The first is whether fusion actually occurs in the bubbles. This has been discussed above and is the subject of active research in laboratories around the world. One possibility for increasing the likelihood of fusion is to take advantage of the fact that D-T fusion is roughly two orders of magnitude more likely than D-D fusion. This could be done using mixtures of suitable deuterated and tritiated solvents. D-D fusion, if it can be made to work, does have the advantage that the supply of fuel is almost unlimited, whereas the tritium required for D-T fusion does not occur in nature, but must be manufactured from lithium in a reactor according to the reaction:

$$n + {}^6Li \rightarrow {}^4He + T + 4.8 \text{ MeV} \qquad (4)$$

The second is whether, even if fusion does occur, enough power is produced for sonoluminescence to be a net energy source. After all, production of

Table 7 A list of testing variables for SBSL and MBSL sonoluminescence experiments

Measurement conditions	Sonoluminescence characterizations
Test cell geometry	Flash intensity
Ambient conditions	Three-dimension temperature map
Test liquid and dissolved gasses	Emission spectrum
Applied power and frequency	High-speed and low-light photographs
Bubble location and motion	Light scattering of bubble
Bubble creation method	Precision timing of flash duration
	Chemical and nuclear reaction products

sonoluminescence is a power intensive process; only a fraction of the electrical power is converted into vibrational power in the piezo crystals, and only a fraction of the vibrational power is transmitted into the liquid. Intuitively, the power required for sonoluminescence should be less in small vessels; more power would be required to initiate sonoluminescence in Lake Erie than in a 100-ml flask. One would think that if the wave amplitude diminished with distance, perhaps because of dissipation within the liquid, that the power required would scale as some characteristic dimension r. The frequency is inversely proportional to r, so if the dissipation is frequency dependant the overall question of how the power required varies with size needs to be investigated. From the result in Eq. (3), the minimum container diameter is of the order of a few millimeters. In this case, one could envision a piezo-walled honeycomb arrangement, with each small cell holding a bubble; some arrangement such as this is probably the most efficient.

To date, a complete understanding of the nature of sonoluminescence remains elusive. Sonoluminescence is often difficult to create, difficult to see (in many cases), difficult to sustain, and difficult to characterize. Reports of days spent acquiring one data set from a single event abound. However, difficulty should not deter us from pursuing and attempting to understand the unknown. Facing the difficulties of characterizing sonoluminescence and attempting to overcome them may lead to more practical applications of this phenomenon with the acquisition of new data. For "It is likely that those doing absurd experiments will be the ones to make discoveries" [61].

VIII. Conclusions

Sonoluminescence, the generation of light from cavitation in fluids, has been associated with claims of energy production. Instrumentation techniques were explored at NASA Glenn Research Center to measure properties of the phenomenon of sonoluminescence.

A resonating test cell and a Sonicator test cell were built to generate sonoluminescence. Multi-bubble sonoluminescence in the tests cells was imaged with low-light cameras in a variety of containers. Indications of high temperature generation were observed in palladium–chromium alloy films when exposed to sonoluminescence in heavy water but not in light water. No indication of high-temperature generation was observed in platinum films exposed to sonoluminescence in heavy or light water, though the platinum film was estimated to require about the same amount of energy to be modified as the palladium–chromium film.

Localized bright sonoluminescence was generated in glycerin saturated with water, allowing future spectroscopic and other optical investigations. A design was presented for in situ radiation monitoring of the sonoluminescence, with plans for future improvements. A concept was presented for harvesting the effects seen in this study as electricity, however there is clear need for improving the thermoelectric properties of the thin films for practical energy harvesting.

Characterization efforts, such as searching for byproducts of chemical and nuclear reactions, completely mapping the temperature distribution in a sonoluminescence cell, and spectroscopy, all contribute to a more complete understanding of the nature of sonoluminescence. From these efforts, a clear understanding of the heat generation process involved may lead to the successful application of sonoluminescence as a means of energy harvesting.

Acknowledgments

This work was sponsored over the past several years through the Green Propulsion and Cryogenic Advanced Development, the Low Emissions Alternative Power, and the Breakthrough Propulsion Physics projects at the NASA Glenn Research Center. The authors thank Kenneth Weiland (retired) and James Williams of the R&D Labs Technical Branch for their optical hardware and electronics support in this effort. We also thank Drago Androjna of Sierra Lobo, Inc. (retired) and José Gonzalez of Gilcrest Electric as members of the NASA Glenn Research Center Test Facilities Operation, Maintenance, and Engineering (TFOME) organization for laboratory support. Finally, we are thankful to Jonathan Wright of University of Florida for his input in this work, and Carl Chang of ASRC Aerospace Corporation at the NASA Glenn Research Center for his technical review.

References

[1] Putterman, S. J., "Sonoluminescence: Sound into Light," *Scientific American*, February 1995, pp. 32–37.

[2] Frenzel, H., and Schultes, H., "Luminescenz im ultraschallbeschickten Wasser," *Zeitschrift für Physikalische Chemie*, Vol. B27, 1934, p. 421.

[3] Gaitan, D. F., Crum, L. A., Church, C. C., and Roy, R. A., "Sonoluminescence and Bubble Dynamics for a Single, Stable, Cavitation Bubble," *Journal Acoustics of Society of America*, Vol. 91, 1992, pp. 3166–3183.

[4] Moran, M. J., Haigh, R. E., Lowry, M. E., Sweider, D. R., Abel, G. R., Carlson, J. T., Lewia, S. D., Atchley, A. A., Gaitan, D. F., and Maruyama, X. K., "Direct Observation of Single Sonoluminescence Pulses," *Nuclear Instrumentation and Methods in Physical Reseach B*, Vol. 96, No. 3–4, May 1995, pp. 651–656.

[5] Hiller, R., Putterman, S. J., and Barber, B. P., "Spectrum of Synchronous Picosecond Sonoluminescence," *Physical Review Letters*, Vol. 69, No. 8, August 1992, pp. 1182–1184.

[6] Hiller, R., Weninger, K., Putterman, S. J., and Barber, B. P., "Effect of Noble Gas Doping in Single-Bubble Sonoluminescence," *Science*, Vol. 266, 1994, pp. 248–250.

[7] Hiller, R. A., and Putterman, S. J., "Observation of Isotope Effects in Sonoluminescence," *Physical Review Letters*, Vol. 75, No. 19, Nov. 1995, pp. 3549–3551.

[8] Didenko, Y. T., McNamara, W. B., and Suslick, K. S., "Temperature of Multibubble Sonoluminescence in Water," *Journal of Physical Chemistry*, Vol. 103, No. 50, Dec. 1999, pp. 10783–10788.

[9] Barber, B. P., Wu, C. C., Lofstedt, R., Roberts, P. H., and Putterman, S. J., "Sensitivity of Sonoluminescence to Experimental Parameters," *Physical Review Letters*, Vol. 72, No. 9, Feb. 1994, pp. 1380–1383.

[10] Prosperetti, A., "A New Mechanism for Sonoluminescence," *Journal of the Acoustical Society of America*, Vol. 101, 1997, pp. 2003–2007.

[11] Prosperetti, A., "Old-Fashioned Bubble Dynamics," *Sonochemistry and Sonoluminescence*, Crum, L. A., Mason, T. J., Reisse, J. L., and Suslick, K. S. (eds.), Kluwer, 1999, pp. 159–164.

[12] Eberlein, C., "Sonoluminescence as Quantum Vacuum Radiation," *Physical Review Letters*, Vol. 76, No. 20, May 1996, pp. 3842–3845.
[13] Brodsky, A. M., Burgess, L. W., and Robinson, A. L., "Cooperative Effects in Sonoluminescence," *Physics Letters A*, Vol. 287, Sep. 2001, pp. 409–414.
[14] Schiffer, M., "Sonoluminescence: The Superradiance Paradigm," 14 Oct. 1997. URL: http://arxiv.org/pdf/quant-ph/9710039 [cited October 2002].
[15] Moss, W. C., Clarke, D. B., and Young, D. A., "Star in a Jar," *Sonochemistry and Sonoluminescence*, Crum, L. A., Mason, T. J., Reisse, J. L., and Suslick, K. S. (eds.), Kluwer, 1999, pp. 159–164.
[16] Yasui, K., "Effect of Liquid Temperature on Sonoluminescence," *Physics Review E*, Vol. 64, June 2001.
[17] Yasui, K., "Temperature in Multibubble Sonoluminescence," *Journal of Chemical Physics*, Vol. 115, No. 7, Aug. 2001, pp. 2893–2896.
[18] Margulis, M. A., and Margulis, I. M., "Contemporary Review on Nature of Sonoluminescence and Sonochemical Reactions," *Ultrasonics Sonochemistry*, Vol. 9, 2002, pp. 1–10.
[19] Flynn, H. G., "Method of Generating Energy by Acoustically Induced Cavitation Fusion and Reactor Therefor," U.S. Patent 4,333,796, June 1982.
[20] Moss, W. C., Clarke, D. B., White, J. W., and Young, D. A., "Sonoluminescence and the Prospects for Table-Top Micro-Thermonuclear Fusion," *Physics Letters A*, Vol. 211, No. 2, Feb. 1996, pp. 69–74.
[21] Putterman, S. J., Barber, B. P., Hiller R. A., and Lofstedt, R. M. J., "Converting Acoustic Energy into Useful Other Energy Forms," U.S. Patent 5,659,173, Aug. 1997.
[22] Crum, L. A., "Sonofusion: Star in a Jar?" *Science Forum Colloquium Series*, University of Washington, 21 Nov. 2003. URL: http://www.washington.edu/research/scienceforum/pdfs/Crum.pdf [cited December 2003].
[23] Nomura, S., and Toyota, H., "Sonoplasma Generated by a Combination of Ultrasonic Waves and Microwave Irradiation," *Applied Physics Letters*, Vol. 83, No. 22, Dec. 2003, pp. 4503–4505.
[24] Didenko, Y. T., and Suslick, K. S., "The Energy Efficiency of Formation of Photons, Radicals and Ions During Single-Bubble Cavitation," *Nature*, Vol. 418, July 2002, pp. 394–397.
[25] Schewe, P., and Stein, B., "Detecting Megasonic Bubbles on Computer Chips," *Physics News Update*, No. 710, American Institute of Physics, Nov. 24, 2004.
[26] Doktycz, S. J., and Suslick, K. S., "Interparticle Collisions Driven by Ultrasound," *Science*, Vol. 247, March 1990, pp. 1067–1069.
[27] Dhas, N. A., Zaban, A., and Gedanken, A., "Surface Synthesis of Zinc Sulfide Nanoparticles on Silica Microspheres: Sonochemical Preparation, Characterization, and Optical Properties," *Chemistry of Materials*, Vol. 11, No. 3, Feb. 1999, pp. 806–813.
[28] Jorné, J., "Ultrasonic Irradiation of Deuterium-Loaded Palladium Particles Suspended in Heavy Water," *Fusion Technology*, Vol. 29, Jan. 1996, pp. 83–90.
[29] George, R., "Photographic Evidence for Micronuclear Explosions in Thin Metal Foils Initiated by Intense Ultrasonic Cavitation," *Scanning*, Vol. 19, No. 3, 1997, p. 196.
[30] Stringham, R., "Anomalous Heat Production by Cavitation," *Proceedings of the 1998 IEEE Ultrasonics Symposium*, IEEE, 1998, pp. 1107–1110.
[31] Taleyarkhan, R. P., West, C. D., Cho, J. S., Lahey, Jr., R.T., Nigmatulin, R. I., and Block, R. C., "Evidence for Nuclear Emissions during Acoustic Cavitation," *Science*, Vol. 295, March 2002, pp. 1868–1873.

[32] Taleyarkhan, R. P., Cho, J. S., West, C. D., Lahey, Jr., R. T., Nigmatulin, R. I., and Block, R. C., "Additional Evidence of Nuclear Emission during Acoustic Cavitation," *Physical Reviews E*, Vol. 69, No. 3, March 2004, 036109.
[33] Shapira, D., and Saltmarsh, M., "Nuclear Fusion in Collapsing Bubbles—Is It There? An Attempt to Repeat the Observation of Nuclear Emissions from Sonoluminescence," *Physical Review Letters*, Vol. 89, No. 10, Sept. 2002, 104302.
[34] Taleyarkhan, R. P., West, C. D., Lahey, Jr., R. T., Nigmatulin, R. I., Block, R. C., and Xu, Y., "Nuclear Emissions during Self-Nucleated Acoustic Cavitation," *Physical Review Letters*, Vol. 96, No. 3, Jan. 2006, 034301.
[35] Forringer, E. R., Robbins, D., and Martin, J., "Confirmation of Neutron Production during Self-Nucleated Acoustic Cavitation," *Transactions of the American Nuclear Society*, Nov. 2006, pp. 736–737.
[36] Lipson, A. G., "Comment on 'Nuclear Emissions during Self-Nucleated Acoustic Cavitation,'" *Physical Review Letters*, Vol. 97, Oct. 2006, 149401.
[37] Naranjo, B., "Comment on 'Nuclear Emissions during Self-Nucleated Acoustic Cavitation,'" *Physical Review Letters*, Vol. 97, Oct. 2006, 149403.
[38] Chubb, S., "An Overview of Cold Fusion Theory," *Thermal and Nuclear Aspects of the Pd/D_2O System*, SSC TR 1862, Szpak, S., and Mossier-Boss, P. A. (eds.), Spawar Systems Center, San Diego, CA, Feb. 2002, Vol. 1, pp. 91–111.
[39] Okuyama, F., Muto, H., Tsujimaki, H., and Fujimoto Y., "Palladium Nanoparticles Grown by Hydrogen and Deuterium Ion Bombardment," *Surface Science*, Vol. 355, March 1996, pp. L341–L344.
[40] Young, F. R., "Multibubble Sonoluminescence," *Sonoluminescence*, CRC Press, Boca Raton, FL, 2005, pp. 27–66.
[41] Flint, E. B., and Suslick, K. S., "Sonoluminescence from Nonaqueous Liquids: Emission from Small Molecules," *Journal of the American Chemistry Society*, Vol. 111, No. 18, 1989, pp. 6987–6992.
[42] Suslick, K. S., and Doktycz, S. J., "The Sonochemistry of Zn Powder," *Journal of the American Chemistry Society*, Vol. 111, No. 61989, pp. 2342–2344.
[43] Suslick, K. S., Didenko, Y., Fang, M. M., Hyeon, T., Kolbeck, K. J., McNamara, III, W. B., Mdleleni, M. M., and Wong, M., "Acoustic Cavitation and its Chemical Consequences," *Philosophical Transactions of the Royal Society of London, Series A*, Vol. 357, No. 1751, Feb. 1999, pp. 335–353.
[44] Freeh, J. E., Steffen, Jr., C. J., and Larosiliere, L. M., "Off-Design Performance Analysis of a Solid-Oxide Fuel Cell/Gas Turbine Hybrid for Auxiliary Aerospace Power," NASA TM-2005-213805, Dec. 2005.
[45] Thaller, L. H., and Zimmerman, A. H., "Overview of the Design, Development, and Application of Nickel-Hydrogen Batteries," NASA TP-2003-211905, June 2003.
[46] Cohen, F., and Dalton, P., "Update on International Space Station Nickel-Hydrogen Battery On-Orbit Performance," NASA TM-2003-212542, Aug. 2003; AIAA paper 2003-6065, Aug. 2003.
[47] Nigmatulin, R. I., Akhatov, I. Sh., Vakhitova, N. K., and Lahey, Jr., R.T., "Hydrodynamics, Acoustics and Transport in Sonoluminescence Phenomena," *Sonochemistry and Sonoluminescence*, Crum, L. A., Mason, T. J., Reisse, J. L., and Suslick, K. S. (eds.), Kluwer, 1999, pp. 127–138.
[48] Matula, T. J., "Single-Bubble Sonoluminescence in Microgravity," *Ultrasonics*, Vol. 38, No. 1–8, March 2000, pp. 559–565.

[49] Wrbanek, J. D., Fralick, G. C., and Wrbanek, S. Y., "Development of Techniques to Investigate Sonoluminescence as a Source of Energy Harvesting," NASA TM-2007-214982, Sep. 2007; AIAA 2007–5596, July 2007.
[50] Lauterborn, W., and Ohl, C. D., "Acoustic Cavitation and Multi Bubble Sonoluminescence," *Sonochemistry and Sonoluminescence*, Crum, L. A., Mason, T. J., Reisse, J. L., and Suslick, K. S. (eds.), Kluwer, 1999, pp. 97–104.
[51] Hodgman, C. D., Weast, R. C., and Selby, S. M. (eds.), *Handbook of Chemistry and Physics*, The Chemical Rubber Publishing Co., Cleveland, OH, 1960.
[52] Beckett, M. A., and Hua, I., "Impact of Ultrasonic Frequency on Aqueous Sonoluminescence and Sonochemistry," *Journal of Physical Chemistry A*, Vol. 105, No. 15, April 2001, pp. 3796–3802.
[53] Hulse, C. O., Bailey, R. S., Grant, H. P., Anderson, W. L., and Przybyszewski, J. S., "High Temperature Static Strain Gauge Development," NASA CR-189044, Aug. 1991.
[54] Kubaschewski, O., and Hopkins, B. E., *Oxidation of Metals and Alloys*, Butterworths, London, 1967, pp. 4–18.
[55] Campbell, D. S., "Mechanical Properties of Thin Films," *Handbook of Thin Film Technology*, Maissel, L. I., and Glang, R. (eds.), McGraw-Hill, New York, 1970, pp. 12-3–12-50.
[56] Berger, M. J., Coursey, J. S., and Zucker, M. A., "ESTAR, PSTAR, and ASTAR: Computer Programs for Calculating Stopping-Power and Range Tables for Electrons, Protons, and Helium Ions (version 1.2.2)," 2000; URL: http://physics.nist.gov/Star [cited 15 September 2004]; NIST, Gaithersburg, MD.
[57] Sears, V. F., "Neutron Scattering Lengths and Cross Sections," *Neutron News*, Vol. 3, No. 3, 1992, pp. 26–37; URL: http://www.ncnr.nist.gov/resources/n-lengths/list.html, NIST, Gaithersburg, MD [cited 02 January 2004].
[58] Ziegler, J. F., and Beirsack, J. P., *SRIM, The Stopping and Range of Ions in Matter*, Software Package, Ver. 2003.26, SRIM.org, Annapolis, MD, 2004.
[59] Wrbanek, J. D., Wrbanek, S. Y., Fralick, G. C., and Chen, L. Y., "Micro-Fabricated Solid-State Radiation Detectors for Active Personal Dosimetry," NASA TM-2007-214674, Feb. 2007.
[60] Wrbanek, J. D., Fralick, G. C., Farmer, S. E., Sayir, A., Blaha, C. A., and Gonzalez, J. M., "Development of Thin Film Ceramic Thermocouples for High Temperature Environments," NASA TM-2004-213211, Aug. 2004, also AIAA 2004-3549, July 2004.
[61] Lane, J. W., "Our Children Will Not Be Us," Letter to *APS News*, Aug./Sept. 2007, Vol. 16, No. 8, p. 4.

Chapter 20

Null Tests of "Free Energy" Claims

Scott R. Little*
EarthTech International, Austin, Texas

I. Introduction

THERE IS no question that our hydrocarbon-based energy system is slowly breaking down. The supply of fossil fuel is finite and combustion products are steadily polluting our environment. Conventional nuclear energy also faces challenges: undesirable waste products are produced, there is significant public concern about the safety of nuclear power plants, the supply of fuel is finite, and the technology is not suitable for most transportation applications. But these terrestrial problems pale in comparison to the likely energy requirements for interstellar travel. For example, consider a hypothetical mission to Alpha Centauri in a ship the size of a Boeing 757 (i.e., 10^5 kg). Acceleration is 1 g to the midpoint and deceleration is 1 g until arrival at the destination 4.3 light-years from Earth. At the midpoint, the ship is moving at 0.95 c. For the passengers the trip takes only 3.6 years whereas 5.9 years go by for those who stay on Earth. A yet-to-be-developed engine that converts its fuel entirely to energy and beams it out the tailpipe drives the ship with the maximum possible fuel efficiency. Despite this efficiency, the trip requires a staggering 3.8×10^6 kg of fuel: 38 times the weight of the ship. The energy required for this ideal one-way trip to Alpha Centauri is about 700 times the present (2008) annual energy consumption of the world. (See Chapters 2 and 3 for additional information on spaceflight energy requirements.)

Faced with such monumental problems humans naturally devise schemes to solve them. Often, the scheme involves a device that is purported to produce more energy than it takes to run it. Sometimes the inventor naively thinks the device is simply creating the extra energy. More often the inventor believes that the device is tapping a new source of energy.

Copyright © 2008 by the American Institute of Aeronautics and Astronautics, Inc. All rights reserved.
*Experimentalist.

Breakthrough energy claims are nothing new. Recorded history of perpetual motion claims begin in the 13th century with a simple mechanical device attributed to Wilars de Honecort [1]. Leonardo da Vinci toyed with the idea of a hydraulic perpetual motion machine [2]. In the 1700s Johann Bessler (a.k.a. Orffyreus) developed a mechanical perpetual motion machine [3] that was widely witnessed yet remains shrouded in mystery to this day. There are numerous other devices, primarily invented by charismatic individuals with no particular scientific training. As science progressed toward formal recognition of the laws of thermodynamics in the mid 1800s, claims to perpetual motion became considerably less acceptable but hardly less frequent. In 1870, Henry Dircks eloquently described claimants to perpetual motion as follows, "A more self-willed, self-satisfied, or self-deluded class of the community, making at the same time pretension to superior knowledge, it would be impossible to imagine. They hope against hope, scorning all opposition with ridiculous vehemence, although centuries have not advanced them one step in the way of progress" [4].

Fortunately, things have changed. In our present age of science and technology almost everyone accepts the laws of thermodynamics. The majority of new energy claims are therefore based upon the idea of tapping a new source of energy. It is these claims that are the primary focus of this chapter.

II. Testing of Energy Claims

Testing of energy devices is conceptually straightforward. The output energy of the device is measured, the apparent input energy is measured, and the two quantities are compared. If the device is working as claimed, that is, extracting energy from some unexpected source and delivering it to the output, the measured output energy will exceed the apparent input energy. Otherwise, the First Law of Thermodynamics requires that the output energy, including any losses such as heat, must be precisely equal to the input energy.

In the case of an electrical input, measurement of the input energy is relatively straightforward, particularly if the input is steady DC. If the input has temporal variations, a sophisticated power analyzer may be required to achieve satisfactory measurement accuracy. These instruments sample the voltage and current being delivered to a device rapidly and simultaneously. Each pair of samples is multiplied together to obtain a measure of the instantaneous power flowing into the device. These values are then integrated over time to obtain the electrical energy consumed by the device. In the case of mechanical input like a rotating shaft, a dynamometer, which measures torque and angular velocity, is required to directly measure the mechanical input energy.

Measurement of the output energy is often more challenging. Devices that produce an electrical output can be handled by the methods outlined above for input energy measurement. Devices that output heat energy require some form of calorimetry. Calorimetry is conceptually simple but, in practice, a great effort is required to reduce errors, primarily systematic, to acceptable levels. It is especially difficult to achieve accuracy levels better than 1% relative. Compared to a calorimeter with 1% relative accuracy, at least an order of magnitude more effort is required to achieve 0.1% accuracy [5].

A frustrating situation arises when investigating energy claims. To the claimant, a single null test often proves almost nothing. Regardless of the circumstances one can always say that the test was not conducted properly—that the right materials were not used, the apparatus was not assembled correctly, or the planets were not properly aligned. Strictly speaking, from the claimant's viewpoint, an infinite number of null tests are required to disprove a claim. A relatively small number of positive tests suffice to prove a claim.

With this in mind, the ideal condition under which to test an energy claim is with the full cooperation of the claimant using the original apparatus. The testing proceeds rapidly and, in the event of null results, the claimant can ensure as far as possible that the tests were conducted properly. If the original apparatus is not available, a new apparatus must be constructed. In this case, even with the full cooperation of the claimant, there is more room for excuses should the tests yield null results. If the claimant is noncooperative and the apparatus must be constructed from various documents and records, very little can be proved if the tests are null.

III. Some Tests of Breakthrough Energy Claims

This section contains accounts of some tests of breakthrough energy claims conducted at Earthtech International (http://www.earthtech.org). Although the results were negative, we present these cases primarily to promote better understanding of the problems involved in such testing. They are not to be taken as indisputable refutation of the claims.

A. Zero-Point Energy Devices

Quantum physics predicts the existence of an electromagnetic zero-point field whose energy density is so large that many physicists designate it as virtual. Others treat the zero-point field as real energy that surrounds and pervades everything leaving ordinary matter as an almost negligible foam riding on this vast sea of energy. According to John A. Wheeler *et al.* [6], "... elementary particles represent a percentage-wise almost completely negligible change in the locally violent conditions that characterize the vacuum." This viewpoint has led to intense speculation about the possibility of utilizing zero-point energy. (See Chapter 18 for broader discussions of this possibility.)

There is a real force that can be attributed to the zero-point field: the Casimir force [7]. Experimentally confirmed [8], this force arises when conductive surfaces are placed in very close proximity, thus creating a cavity that eliminates certain electromagnetic modes between the plates. The result is an imbalance in the radiation pressure on the two sides of each plate [9], which produces a net force that pushes the plates together. Some physicists, including Robert Forward [10], have proposed that the Casimir force provides a means of extracting energy from the zero-point field. Julian Schwinger [11] provided further stimulus by suggesting that the energy released in sonoluminescence was due to Casimir forces acting on the collapsing bubbles. (Chapter 19 discusses additional sonoluminescence tests and testing methods.)

Our first and most extensive campaign to extract energy from the zero-point field was an effort to replicate the energy claims associated with Ken Shoulders' charge clusters [12], or EVs (electrum validum; i.e., strong electron) as they are popularly known. Shoulders believes that at least one EV is formed in every spark discharge. EVs are supposed to contain at least 10^9 electrons and exist only during the transit from cathode to anode. Shoulders claims that sharply pointed cathodes and a fast rise time for the applied high voltage pulse promote EV formation. The connection with zero-point energy comes from the hypothesis that the compression of electrons into a charge cluster is due to attractive Casimir forces overpowering repulsive Coulomb forces at very short range. Shoulders did a great deal of experimentation with EVs and claimed in US Patent 5,018,180 (21 May 1991) to have observed 96 times more energy released by an EV than required to produce the EV. We pursued this claim for years and were never able to reproduce Shoulders' results. We experimented with numerous configurations in an effort to observe the direct electrical energy output that Shoulders had claimed in his patent. Failing that, we attempted calorimetric measurements that were of compounded difficulty because of the low energy levels involved, the difficulty of accurately measuring the input energy delivered to the spark discharge, and the overall difficulty of making sensitive measurements in such an electrically noisy environment. Despite these problems we eventually managed to obtain reasonable accuracy and reliability in our calorimetric measurements of EVs and the results were uniformly negative. (See Chapter 18 for additional information on EVs.)

Schwinger's hypothesis led us to give serious consideration to several cavitation-based energy claims. The Potapov device, invented by Yuri Potapov in Moldavia, is an example. The Potapov device consisted simply of a swirl chamber through which water was pumped vigorously to create a vortex and cavitation. Potapov claimed that his device imparted up to 3 times more heat energy to the water than the mechanical energy required to pump the water through it. In this case we were able to obtain a genuine Potapov device for testing and we had limited cooperation from Potapov himself. We constructed a batch calorimeter system in which the device would be operated for a certain period of time to heat up the water contained in a large insulated reservoir. Water was pumped from the reservoir, through the device, and returned to the reservoir. The pump was driven by an electric motor. For the input energy we simply measured the electrical energy required to drive the motor using a 3-phase watt-hour meter. For the output energy we measured the increase in water temperature and used the total weight of water in the tank to compute the heat energy delivered to the water. Instead of the $\sim 300\%$ efficiency claimed for the Potapov device we observed only 80% at best. As a control, we also measured the heating efficiency of a simple gate valve inserted in the flow path in place of the Potapov device and adjusted to provide about the same flow restriction. The valve heated the water just as efficiently as the Potapov device. The testing went on for months as we struggled to communicate with Potapov. He did not think we were operating his device properly and we made numerous modifications at his request. The test results remained uniformly negative (see http://www.earthtech.org/experiments).

Another device (whose specific identity is protected by a nondisclosure agreement with the developer) involved a motor-driven rotor in a close-fitting housing. Water was forced through the gaps around the rotor where intense cavitation was supposed to occur. This device was claimed to impart up to 50% more heat energy to the water flowing through it than the mechanical energy required to drive it. We tested this device using a larger version of the same batch calorimeter described above. In this case we constructed a cradle dynamometer to directly measure the mechanical input energy. With a 30-hp electric motor and copious generation of steam by the device, this was a very exciting and sometimes dangerous experiment. However, our measurements never showed any sign of excess energy. Furthermore, by comparing mechanical input energy to heat output energy we were able to obtain a near-perfect energy balance in our measurements, typically $99 \pm 1\%$. We were fortunate to have significant cooperation from the developer of this device and, during a visit to our facility, we accidentally discovered the source of most of his anomalous readings: improper usage of his electrical power analyzer.

A related claim is that of sonofusion made primarily by Roger Stringham [13]. We first investigated this claim by constructing our own apparatus without requesting any cooperation from Stringham. The apparatus consists of an ultrasonic transducer immersed in heavy water with a palladium target in close proximity. According to Stringham, more heat energy is produced in the apparatus than the acoustic energy delivered by the transducer. We employed a water-flow calorimeter to measure the heat output energy and we spent a great deal of effort to learn how to accurately measure the high-voltage, 20-kHz, low power-factor electrical input energy being delivered to the piezoelectric transducer. We evaluated several power analyzers and selected the Clarke-Hess 2330, which we found to be significantly superior to other instruments with similar accuracy specifications. We conducted a total of 48 runs with our apparatus, 12 of which used palladium targets. We never saw any sign of excess heat. After communicating our results to Stringham we arranged to visit his laboratory with a portable version of our calorimeter and our Clarke-Hess 2330. In other words, we were given a chance to test his claim using his apparatus with his full cooperation. We succeeded only in demonstrating that his input power measurements were understating the actual input power. That was the cause of his apparent excess heat on that day in 1999. More recently, Stringham has explored this phenomenon with considerably reworked apparatus and claims even higher excess energy production [14].

H. E. Puthoff [15] has calculated that the ground state of the hydrogen atom can be explained as a dynamic balance between energy lost by the electron due to acceleration radiation and energy absorbed from the zero-point field. The fact that the space between Casimir plates is a region where the zero-point field is reduced in energy density led to speculation that hydrogen might lose some of its ground state energy when placed in such a cavity. If that were the case, then that energy release would constitute zero-point energy conversion and a circulation of hydrogen into and out of a Casimir cavity might produce a continuous extraction of energy from the zero-point field. We have designed and constructed several experiments to explore this hypothesis but without success so far. Most of these experiments were attempts to detect heat energy being released by hydrogen flowing through some form of Casimir cavity. We first tried constructing cavities out of precision optical flats. For the ground state of molecular hydrogen, the optimum cavity spacing is about 1 μm. We monitored the gas temperature at the entrance and exit of the cavity looking for signs of heating due to the release of ground state energy. Although a small temperature increase in this experiment was observed, it turned out to be due only to the Joule–Thompson effect that, for hydrogen, results in a warming of the gas as it flows through a restriction. We also tried using finely powdered metals such as platinum and palladium to create a dense labyrinth of Casimir scale passages. These experiments tended to produce an exciting initial burst of heat when the flow of H_2 was started. But the burst always faded away after a minute or two and it could not be readily repeated. We finally tracked this down to $H_2 + O_2$ combustion catalyzed by the finely powdered metal. It was not readily repeatable because the apparatus was nearly sealed. Only after the apparatus had been sitting overnight or had been disassembled was there sufficient O_2 present for another heat burst.

In a different approach to testing this hypothesis we put hydrogen molecules in a Casimir cavity and used absorption spectroscopy to look for a shift in their ground state energy, which follows from the assumption that a depression of the ground state energy would produce a corresponding increase in the dissociation energy. Early attempts at this experiment were conducted at the Synchrotron Radiation Center of the University of Wisconsin-Madison. Following in the footsteps of molecular spectroscopy pioneer Gerhard Herzberg [16], extreme ultraviolet radiation was used to probe the dissociation energy of H_2 molecules in an appropriate Casimir cavity. Unfortunately we did not find any evidence of a ground state shift in this work. (See http://www.earthtech.org/experiments/src/srcreport.htm for full details of our synchrotron experiment.) Further effort along this line by a consortium of researchers is planned [17]. The theory and operating principles behind this approach are discussed in more detail in Chapter 18.

B. Electromagnetic Devices

A number of electromagnetic energy claims have been made over the past 150 years. Some of them are no more than a continuation of the quest for perpetual motion but with magnets and coils replacing the weights and levers of the earlier devices. Others, particularly the more recent claims, are not so easy to dismiss and deserve to be investigated.

A relatively simple device called the Motionless Electromagnetic Generator (MEG) (US Patent 6,362,718 26 March 2002) has been widely publicized on the Internet. We constructed our own MEG using detailed construction plans from an independent lab that reportedly had successfully replicated the excess power results. The initial results with our MEG also apparently showed excess power but we soon learned why. The MEG operates at high audio frequencies and delivers several hundred volts and a few milliamperes to a load resistor. This current was being measured using a 10-ohm current-viewing resistor in series with the load resistor. With only a few milliamperes of current through this resistor, the voltage developed is only a few 10s of millivolts.

At these frequencies it is almost impossible to accurately measure this small voltage while in intimate proximity to the output voltage, which is four orders of magnitude higher. Capacitive coupling between the voltmeter or scope leads significantly elevated the observed voltage across the current viewing resistor resulting in overestimation of the output power. Supporting this conclusion was the lack of heating in the load resistor that should have occurred had the apparent output power been correct. When we changed the current viewing resistor to 1000 ohms, which did not significantly affect the load impedance, the apparent excess power disappeared.

The essence of another case, again covered by nondisclosure agreement, can be presented as a good example of an axiom that we have come to embrace: the fact that your instruments cost a lot of money does not guarantee that their results will be accurate. A simple device was being driven by 60 Hz AC power. The output of this device was also 60 Hz AC but at a different voltage. We employed a Clarke-Hess 2330 power analyzer for the input and output power measurements and obtained a mundane efficiency of $\sim 90\%$ for the device. However, the claimant was using a state-of-the-art high-bandwidth digital oscilloscope with extensive waveform math capabilities (including power calculation) and a sophisticated clamp-on AC/DC current probe. Surprisingly, this \$30,000 collection of modern equipment, while capable of accurate power analysis on a wide variety of waveforms over an impressive range of frequencies, made significant errors in the measurement of power in these 60 Hz signals.

C. Cold Fusion

In March of 1989 Martin Fleischmann and Stanley Pons of the University of Utah announced that they had succeeded in making D-D fusion occur in an electrochemical cell near room temperature [18]. Compared to the ordinary conditions required for this reaction, this claim was aptly named "cold fusion." The announcement of cold fusion generated intense interest as it promised to solve most if not all of the energy problems on Earth. However, it also received intense scrutiny as it appeared to violate known principles of nuclear physics. With widespread failure to replicate the experiment, support for continued experimentation rapidly waned.

Despite diminished support, a number of scientists continue to investigate cold fusion. Hundreds of papers reporting positive results have been published (see the cold fusion library at http://www.lenr-canr.org) and international conferences are held every couple of years. However, to this day (2008), there exists no cold fusion demonstration experiment. In other words, the cold fusion phenomenon is not sufficiently reproducible that it can be demonstrated on demand. This situation greatly hampers cold fusion research because it makes application of the scientific method almost impossible. It is extremely difficult to test hypotheses when the experimental results are nearly random.

The primary signature of cold fusion is excess heat, which means that the electrochemical cell produces more heat power than the electrical power used to stimulate it. Thus calorimetry is often involved in testing cold fusion experiments. In our laboratory we have expended a great deal of effort on the development of calorimeters suitable for cold fusion experiments. Over the years we have had the opportunity to test a relatively small number of cold fusion cells, some that we constructed ourselves and some that were brought to our laboratory by other investigators who had seen positive signs of excess heat in their own labs. None of these cold fusion experiments have shown any convincing evidence of excess heat in our calorimeters. We cannot say that we have never seen *any* signs of excess heat in our laboratory because all calorimeters drift somewhat and, inevitably, that drift sometimes goes in a positive direction and looks just like a low level of genuine excess heat. When that occurs we strive to check the calorimeter's calibration as quickly and thoroughly as possible. Usually the drift in calibration is evident and its magnitude matches, and thus explains, the apparent excess heat signal. In a few cases the calibration check did not explain the apparent excess heat signal, but when we returned the cell to the calorimeter after the calibration check, the excess heat signal did not reappear. This tantalizing behavior either means that the cell did produce low levels of excess heat for a while or the calorimeter was simply drifting up and down in unfortunate synchrony with our observations.

In our laboratory we have about 70 years of combined experience designing, building, and operating various measuring systems. From this perspective, we find calorimetry to be unusually susceptible to subtle systematic errors. Furthermore we have found it nearly impossible to anticipate the causes of these errors. Their elucidation usually occurs only after the instruments have been constructed and tested extensively.

The culmination of our efforts to build an accurate and reliable calorimeter for cold fusion experimentation is an instrument we call MOAC (Mother Of All Calorimeters). This instrument operates on a simple and fundamental principle. Flowing water is used to extract the heat from the cell. The flow rate is measured and the temperature rise of the water is measured. The product of the temperature rise, the flow rate, and the specific heat of water yields the heat power being extracted from the cell. Despite its simple concept, MOAC is not a simple instrument. Two independent computer-based data acquisition systems monitor a total of 45 parameters, including 22 separate temperatures. Fourteen analog outputs, driven by proportional-derivative feedback algorithms, control various critical parameters. Figure 1 shows a simplified block diagram of the system.

The cold fusion cell (marked CELL in Fig. 1) and heat exchanger are located in a chamber whose walls are made almost perfectly insulating by a system that heats the outer surface of each of the six wall panels so that its temperature matches that of the corresponding inner surface. The inner and outer surface of each wall panel is composed of a thick aluminum plate for thermal uniformity. Temperatures are sensed by thermistors embedded in the center of each plate and the outer plates are heated by electrical heating elements distributed over the surface (shown schematically as long resistors in Fig. 1). Each wall panel is independently controlled by a servo algorithm in the software. This active insulation ensures that virtually all of the heat dissipated by the cell leaves the chamber via the flowing water.

A three-stage Peltier temperature regulator (which can add or remove heat as needed) controls the temperature of the water entering the heat exchanger to $\pm\ 0.0003°C$. A positive-displacement pump driven by a synchronous motor powered by a crystal-based oscillator produces an exceedingly stable flow of about 2.5 gm/s. A flowmeter consisting of an automated batch weighing

NULL TESTS OF "FREE ENERGY" CLAIMS

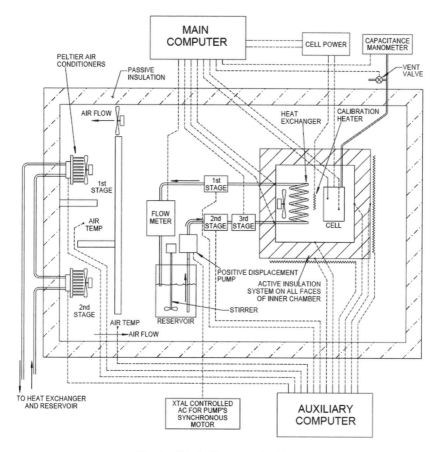

Fig. 1 Block diagram of MOAC.

system measures the flow rate periodically and typically reports a standard deviation of only ± 0.0005 gm/s (i.e., 0.02% relative). A large well-insulated enclosure houses the entire system. Air circulates over the calorimetry apparatus and then is ducted to a two-stage Peltier air-conditioner where its temperature is regulated to ± 0.001°C before it re-enters the enclosure. Figure 2 is a photograph of the entire system.

MOAC was designed to achieve ± 0.1% relative accuracy at the typical input power level of 10 watts (i.e., equivalent to ± 0.01 watts). Accuracy is tested using a standard electrolysis cell constructed of "inactive" materials (i.e., without palladium, lithium, or deuterium). The measured electrical input power is compared to the measured heat output power. When freshly calibrated and operating normally, the design accuracy of 0.1% relative is actually achieved. However, performance can degrade over time, typically drifting up to 0.03 watts a month after calibration. We believe this drift originates primarily in the thermistors used to measure inlet and outlet water temperatures.

Fig. 2 Exterior of MOAC and controlling computers.

The space available for the cell is relatively large (about 10 cm × 25 cm × 25 cm). The cell sits in a stirred air environment where it is not thermally clamped to a specific temperature. MOAC exhibits excellent specimen versatility by producing precisely the same reading regardless of the size, shape, or location of the heat source.

IV. Conclusions

Testing of breakthrough energy claims is simple in concept but often difficult in practice. It is most effectively done with the full cooperation of the claimant. Systematic errors are common. Considerable diligence is required to ensure that the measurement techniques employed are acceptably free from such problems. The specific examples discussed here can guide other researchers in avoiding similar experimental errors.

References

[1] Dircks, H., *Perpetuum Mobile*, E. & F.N. Spon, London, 1870, p. 1.
[2] Heaton, Mrs. C. W., *Leonardo da Vinci and His Works*, Kessinger Publishing, 2004, pp. 154–155.
[3] "Perpetual Motion," *Encyclopedia Britannica*, 1971 ed.
[4] Dircks, H., *Perpetuum Mobile*, E. & F.N. Spon, London, 1870, p. 354.
[5] McCullough, J. P., and Scott, D. W., *Experimental Thermodynamics, Calorimetry of Non-reacting Systems*, Butterworth, London, Vol. 1, 1968, p. 9.
[6] Misner, C. W., Thorne, K. S., and Wheeler, J. A., *Gravitation*, W.H. Freeman & Co., New York, 1973, p. 1202.

[7] Casimir, H. B. G., "On the Attraction Between Two Perfectly Conducting Plates," *Proceedings of the Koninklijke Nederlandse Akademie ran Wetenschappen*, Vol. 51, 1948, pp. 793–796.
[8] Harris, B. W., Chen, F., Mohideen, U., "Precision Measurement of the Casimir Force Using Gold Surfaces," *Physical Review A*, Vol. 62, 2000, 052109/1-5.
[9] Milonni, P. W., Cook, R. J., and Goggin, M. E., "Radiation Pressure from the Vacuum: Physical Interpretation of the Casimir Force," *Physical Review A*, Vol. A38, 1988, p. 1621.
[10] Forward, R. L., "Extracting Electrical Energy from the Vacuum by Cohesion of Charged Foliated Conductors," *Physical Review B*, Vol. 30, No. 4, 1984, p. 1700.
[11] Schwinger, J., "Casimir Light: The Souce," *Proceedings National Academy Science USA*, Vol. 90, March 1993, pp. 2105–2106.
[12] Shoulders, K. R., *EV: A Tale of Discovery*, Jupiter Technologies, Austin TX, 1987.
[13] Stringham, R. S., George, D. R., Tanzella, F. L., Williams, M., "Cavitation-Induced Heat in Deuterated Metals," *EPRI Report TR-108474*, March 1998.
[14] Stringham, R., "1.6 MHz Sonofusion Measurement and Model," *Proceedings of the American Physical Society*, Denver, CO, 2007.
[15] Puthoff, H. E., "Ground State of Hydrogen as a Zero-Point-Fluctuation-Determined State," *Physical Review D*, Vol. 35, 1987, p. 3266.
[16] Herzberg, G., "The Dissociation Energy of the Hydrogen Molecule," *Journal of Molecular Spectroscopy*, Vol. 33, 1970, pp. 147–168.
[17] Davis, E. W., Teofilo, V. L., Haisch, B., Puthoff, H. E., Nickisch, L. J., Rueda, A., and Cole, D. C. "Review of Experimental Concepts for Studying the Quantum Vacuum Field," *Proceedings of the STAIF-2006: 3rd Symposium on New Frontiers and Future Concepts, AIP Conference Proceedings*, Vol. 813, El-Genk, M. S. (ed.), AIP Press, New York, pp. 1390–1401.
[18] Browne, M. W., "Fusion in a Jar: Announcement by Two Chemists Ignites Uproar," *New York Times*, March 28, 1989.

Chapter 21

General Relativity Computational Tools and Conventions for Propulsion

Claudio Maccone*

I. Introduction

TO ADDRESS the technical challenges of breakthrough spaceflight within the sciences of General Relativity (GR) and Quantum Field Theory (QFT), computer computational tools are required. Calculations in GR and QFT are extensive and involve complex notational conventions. The time it takes to do these calculations by hand is prohibitive as is the risk of inducing transcription errors. In this chapter we make a comparative review of the main tensor calculus capabilities of the three most advanced and commercially available "symbolic manipulator" tools. We also address the challenge of the different conventions in tensor calculus that make it difficult or impossible to compare results obtained by different scholars in GR and QFT. To proceed, conventions that would be useful for space propulsion research are suggested, and then reviews of the software that take these options into account follow.

A. Computational Conventions

Mathematical physicists, experimental physicists, and engineers have each their own way of customizing tensors, especially by using the different metric signatures, different metric determinant signs, different definitions of the basic Riemann and Ricci tensors, and by adopting different systems of physical units. This inconsistency hampers progress when trying to apply the advances from one area to another.

To understand this point a little better, a comparison can be made with the situation in Europe before the French revolutionaries adopted (and imposed) the metric system; that is, each country had its own system of units, and a

---

Copyright © 2008 by Claudio Maccone. Published by the American Institute of Aeronautics and Astronautics, Inc., with permission.
*Retired scientist.

considerable amount of time had to be spent just to make the various results numerically comparable.

In this chapter we examine the major convention choices and suggest which choices are preferred for further exploring propulsion physics.

B. History of Computational Tool Development

NASA inaugurated the use of symbolic manipulators (called computer algebra codes, or symbolic mathematics codes) in the early 1960s. NASA's goal at that time was to check the validity of a number of analytical results in celestial mechanics that had been found by hand calculation in the previous 300 years. NASA decided to solve the problem by creating a new code from scratch called Macsyma, a lisp-written symbolic manipulator developed by the Artificial Intelligence Laboratory of MIT from 1965 through 1982, and later taken over by private corporations for further developments.

In the 1980s Macsyma was endowed with a tensor calculus package, the first software in history capable of handling long analytical expressions for the Riemann, Ricci, and stress-energy tensors required to compute analytical solutions of both the Einstein and combined Einstein–Maxwell equations. Other products followed and these options are examined here in the context of how well they match the recommended propulsion conventions.

When "C" became the standard programming language, two new research companies produced new codes, which have since been commercially available: "Mathematica," developed by Wolfram Research, Inc., and "Maple," developed by a team of the University of Waterloo (Canada).

Initially, both these codes did *not* have a tensor package, but several tensor add-on packages were created in the 1990s:

1) "MathTensor," written by Leonard Parker and Steven M. Christensen, which runs on Mathematica

2) "GRTensorII," written by Kayll Lake and coworkers Peter Musgrave and Denis Pollney of Queens University in Kingston, Ontario, which runs on Maple, currently under development (a smaller version of it runs on Mathematica also)

3) "Ricci," developed by John M. Lee of Washington University for Mathematica 3.0 (more details are included in Ref. 1)

This chapter examines the above-mentioned codes in the context of their utility to propulsion science research. The comparison is based on the experience of explicit calculations in the field of GR and QFT that the author has performed over the last 15 years when the latest releases of the above codes became available.

II. Recommended Propulsion Computational Conventions

In both GR and QFT, there are different conventions on the metric signature, the definition of which indices are contracted in the Riemann tensor to yield the Ricci tensor, and the use of different systems of units. The options are so complex that a number of excellent results in GR and QFT obtained by either pure mathematicians, theoretical physicists, applied technologists, or engineers are hardly comparable to each other, thus hampering their applications to research on spaceflight propulsion.

A good example of these difficulties is offered by the two pages printed on red paper that opened the famous book *Gravitation* by Misner, Thorne, and Wheeler in its first 1973 edition [2]. In these two pages the authors list 34 authoritative textbooks of GR that were published between 1922 and 1973 and, for each book, they neatly specify: 1) the metric signature; 2) the definition of the Riemann tensor; 3) the sign found by contracting Riemann to Ricci; and 4) whether the spacetime four-dimensional indices are denoted by Greek or Latin letters.

These two pages give a good idea about the "mess" that freedom of arbitrary conventions created in a century of GR.

Because propulsion science is in the midst of changes, now is a good time to implement a set of standard conventions. To that end, here are suggested choices for conventions and why each is recommended. The key convention choices are as follows: sytem of units, sign convention for the metric, sign convention for the determinant of the metric, definition of Riemann tensor in terms of Christoffel symbols and their derivatives, and definition of which indices are used when contracting the Riemann tensor into the Ricci tensor.

A. System of Units

While different branches of physics have their preferred units (e.g., esu, emu, Gaussian, Heaviside–Lorentz, geometrical, and Planck systems), using a mix of these would impede comparisons. When exploring science to seek operating principles for new space propulsion technology, it would be useful to adopt the system of units most prevalent in engineering professions and ensure that the equations explicitly list the pertinent natural constants.

Although it is common practice in mathematical physics to treat the fundamental constants of the speed of light, c, Newton's gravitational constant, G, and Plank's constant, h, as unity ($c = G = h = 1$), this is not desired for propulsion research. Although such "geometrized" units ($c = G = h = 1$) help draw attention to the power of the mathematical tools, it has the disadvantage of masking the role played by these fundamental properties of nature. For spaceflight research it is recommended to explicitly include the natural constants in the equations.

Regarding the system of units, it is recommended to use the MKS system (meter, kilogram, and second), also called SI for Système International.

B. Sign Convention for Metric

Another convention choice in Special and General Relativity is where to assign the negative sign − to the *spatial* terms or to the *temporal* term:

$$(ds)^2 = (c\,dt)^2 - dx^2 - dy^2 - dz^2 \quad \text{vs} \quad (ds)^2 = -(c\,dt)^2 + dx^2 + dy^2 + dz^2 \quad (1)$$

The version on the left is the convention recommended for spaceflight considerations, while the convention on the right is often used in General Relativity textbooks (e.g., Ref. 3). The reason that the version on the left is preferred is because all time-like displacements, $(ds)^2$, will be positive. "Time-like" is the terminology for motion in spacetime that behaves according to the normal causal relations. Conversely, "space-like" displacements are those that cover more space in a given time than light-speed propagation will allow. It is convenient for causal displacements to have a positive value while causality-violating displacements have a negative value.

C. Sign Convention for Metric Determinant

When proceeding to investigate a particular metric it is necessary to calculate the inverse of that metric. A step in that process is to find the metric determinant for which there are two sign conventions, shown below.

$$\begin{vmatrix} g_{00} & g_{01} & g_{02} & g_{03} \\ g_{10} & g_{11} & g_{12} & g_{13} \\ g_{20} & g_{21} & g_{22} & g_{23} \\ g_{30} & g_{31} & g_{32} & g_{33} \end{vmatrix} < 0 \text{ vs } \begin{vmatrix} g_{00} & g_{01} & g_{02} & g_{03} \\ g_{10} & g_{11} & g_{12} & g_{13} \\ g_{20} & g_{21} & g_{22} & g_{23} \\ g_{30} & g_{31} & g_{32} & g_{33} \end{vmatrix} > 0 \qquad (2)$$

The version on the left is the convention recommended for spaceflight considerations. The convention on the right is used in some advanced QFTs. The reason that the version on the left is preferred for spaceflight calculations is simply because this is more common in ordinary GR and its immediate generalizations.

D. Definition of Riemann Tensor in Terms of Christoffel Symbols

The Riemann curvature tensor is the mathematical tool for determining the curvature of a spacetime metric by combining terms (Christoffel sysmbols) obtained from moving a vector (using parallel transport) around a small closed path in spacetime. There are two conventions to represent the indices in these treatments:

$$R^{\lambda}_{\nu\rho\sigma} \equiv \Gamma^{\lambda}_{\sigma\nu,\rho} - \Gamma^{\lambda}_{\rho\nu,\sigma} + \Gamma^{\alpha}_{\sigma\nu}\Gamma^{\lambda}_{\alpha\rho} - \Gamma^{\alpha}_{\rho\nu}\Gamma^{\lambda}_{\alpha\sigma}$$

vs

$$R^{\lambda}_{\nu\rho\sigma} \equiv -\Gamma^{\lambda}_{\sigma\nu,\rho} + \Gamma^{\lambda}_{\rho\nu,\sigma} - \Gamma^{\alpha}_{\sigma\nu}\Gamma^{\lambda}_{\alpha\rho} + \Gamma^{\alpha}_{\rho\nu}\Gamma^{\lambda}_{\alpha\sigma} \qquad (3)$$

The version below is the convention recommended for spaceflight considerations, quite simply because it is the most widely adopted convention to define the Riemann tensor in the various tensor packages (such as those running on Macsyma and Mathematica). The convention above is simply the version below with all the signs reversed.

E. Index Convention when Contracting Riemann to Ricci

Other options in GR deal with the indices on the Ricci tensor, which result from contracting the Riemann tensor. There are two ways to represent this, with the preferred choice for propulsion research on the left:

$$R_{\nu\rho} = R^{\lambda}_{\nu\lambda\rho} \text{ vs } R_{\nu\rho} = R^{\lambda}_{\nu\rho\lambda} \qquad (4)$$

Again, the reason that the version on the left is preferred for spaceflight calculations is because it is the most widely adopted convention in the various symbolic manipulators.

III. Representative Problems in Propulsion Science

Another factor to address when considering existing and future computational tools is the nature of the calculations that need to be performed. Some representative examples include calculating the energy conditions needed to provide warp drive metrics and calculating the energy conditions needed to provide traversable wormhole metrics.

IV. Review of Existing Computational Tools

Before discussing the three major available computational tools, it must be stressed that these tools continue to undergo changes. While the comparisons here, compiled on 9 September 2007, might change later, the basis of the comparisons for their applicablity to space propulsion will remain valid for some time to come and as new products become available. For example, some of the newer codes such as *Cadabra, Tensorial*, and *Cartan*, and codes related to Casimir effects, have not yet been assessed for their propulsion research utility. The following website presented a comparison of various computer algebra codes in 2007, and might continue to offer up-to-date information: <http://en.wikipedia.org/wiki/Comparison_of_computer_algebra_systems>.

To compare the three major software packages with the features desired for space propulsion science research, Table 1 is offered. Following the table are more detailed comments on each code. Note that within the major codes, *Maxima, Mathematica*, and *Maple*, there are additional codes that are also addressed.

A. Maxima

Maxima is a free (i.e., noncommercial) version of Macsyma that was recovered by the late Professor William Shelter (1947–2001). Maxima is based on a 1982 version of Macsyma and is capable of a wide range of mathematical applications including basic tensors for GR. Volunteers are now further developing its basic tensor package.

Macsyma, the initial embodiment that lead to *Maxima*, was an "elementary math" package written around 1965 at the Artificial Intelligence Laboratory of the MIT. The codes were designed to provide NASA and the then-ongoing Apollo program with software capable of checking the analytical results in celestial mechanics that had been piling up in the previous 300 years with no practical possibility of checking them manually again. For instance, the French astronomer Charles E. Delaunay (1816–1872) had spent 10 years of his life calculating new and more accurate analytical results for the motion of the moon. He then spent

Table 1 Summary of computational tools compared to propulsion research preferences (valid as of Sept. 2007)

Propulsion preferences Product	Units: MKS (SI)	Metric sign convention: time-like is positive	Sign of metric determinant: negative	Riemann convention: Lower version of Eq. (3)	Ricci convention: $R_{\nu p} = R^\lambda_{\nu\lambda p}$	Cost estimate (circa 2007)
Maxima (formerly Macsyma)	Yes (load "units" package)	No, metric is -+++	Yes	Yes	Yes	Free[a]
Mathematica						$1800
MathTensor	Yes (user set)	Yes (user set)	Yes (user set)	Yes (user set)	Yes (user set)	$536
Ricci	No	(user set)	(user set)	(user set)	(user set)	Free[b]
Maple						$1995
GRtensorII (for Maple)	No	No, metric is -+++	Yes	Yes	Yes	Free[c]
Cadabra	Not assessed	Not assessed	Not assessed	Not assessed	Not assessed	Free[d]

[a]Available from: <http://sourceforge.net/project/showfiles.php?group_id=4933>.
[b]Available from: <http://www.math.washington.edu/~lee/Ricci/>.
[c]Available from: <http://grtensor.phy.queensu.ca/>.
[d]Available from: http://www.aei.mpg.de/~peekas/cadabra/; does not run on Windows.

10 additional years checking them, and finally published them in 1867. His results were taken for granted for over a century, basically because no one was willing to devote 20 years to check the results. But, in 1970, the advent of Macsyma allowed this daunting task to be performed: in just 20 hours of computer time André Deprit, Jacques Henrard, and Arnold Rom, at the Boeing Scientific Research Labs in Seattle, checked all Delaunay's results. They found only one analytical mistake; on page 234 of the second volume of Delaunay's book titled *Théorie de la Lune* he incorrectly wrote one fraction as 13/16, whereas the correct fraction is 33/16. Two further errors were just a consequence of this one. Happily, this error in Delaunay's calculations turned out not to be vital for the Apollo flights to the moon. This success made Macsyma the preferred worldwide symbolic computational tool from 1965 onward.

Though the Macsyma source files are written in *lisp*, the user does not need to be proficient in *lisp*. Instead, the user must learn a special "Macsyma language" to perform the requested calculations. This is described in both the paper manual and online help form. This programming language may not be immediately helpful to the user because the names for the commands are invented a bit "randomly," without strict logical rules that the programmer's mind may memorize immediately (see the Macsyma Mathematics Reference Manual and Macsyma System Reference Manual [4]). This situation is somewhat similar in Maple but not in Mathematica, where strict name rules are enforced.

B. Mathematica

When "C" became the stardard programming language, Wolfram Research, Inc. developed and commercially sold Mathematica. This remains a widely used computational tool for various applications. Specific notebooks and even more specialized codes are available to use with Mathematica, some may be downloaded without cost and others are add-on commercial products.

In Mathematica, variables are defined using the convention of long, compound input words, where uppercase letters denote the beginning of each new word within the compound word. This convention is not used in Maxima or Maple. This convention has pros and cons:

Pro: The meaning of the very long words is self-explanatory.
Con: It forces the user to input very long, case-sensitive words.

1. MathTensor

Leonard Parker of the University of Wisconsin at Milwaukee and Steven M. Christensen of the University of North Carolina at Chapel Hill created MathTensor in 1994, the first tensor symbolic manipulator running on Mathematica. They also published a book [5] as the user manual (http://smc.vnet.net/mathtensor.html). Like Mathematica, MathTensor also uses the same case-sensitive, compound word convention.

A main advantage of MathTensor is that the user can set all of the following conventions independently, by just assigning four parameters, respectively: MetricSign (signature of the metric) = 1, DetgSign (sign of the determinant of the metric) = −1, RmSign (definition of the Riemann tensor in terms of the

Christoffel symbols and their derivatives) = 1, and RcSign (definition of the Ricci tensor contraction from the Riemann tensor) = 1.

Another unique advantage of MathTensor is that the user can freely adopt the preferred unit system: one just has to set the variable "units" equal to either of the following names in order to have all equations computed in the relevant units system, respectively: emuUnits, esuUnits, GaussianUnits, HeavisideLorentz Units, RationalizedMKSUnits or SIUnits. This is a nontrivial feature of Math-Tensor, because the definition itself of some tensors, like the stress-energy tensor of the electromagnetic field, depends on conventions adopted with the units system. The amount of time saved by the user changing results from one unit system to another one is remarkable.

When it comes to research utility, MathTensor is able to derive the set of tensor equations (like the Einstein equations) as the Euler–Lagrange equations from a suitable variational principle (like the Einstein–Hilbert action). This is accomplished by MathTensor first by partial integration, and then computation of the Euler-Lagrange equations. The result is that, even for Lagrangians that are quadratic in the Riemann tensor, obtaining the relevant differential equations is virtually immediate. Based on experience, however, it is not yet clear if these operations are bug-free.

Regarding disadvantages, MathTensor is not currently capable of handling Greek letters, so symbols like the electric permittivity, ε_0, and magnetic permeability, μ_0, have to be written "eps0" and "mu0." Worse still, the user interface of MathTensor is very primitive. For example, one has to edit the "Conventions.m" file of MathTensor (i.e., not a file of Mathematica itself) in order to set up all the above conventions.

2. Ricci

Ricci is free software created by John M. Lee of the Mathematics Department of the University of Washington. It performs general differential geometry calculations and tensor calculus. The relevant Web site is <http://www.math.washington.edu/~ lee/Ricci/>.

Just one example of application of Ricci: the Beltrami (or Laplace-Beltrami) operator is the wave propagation operator in Riemannian manifolds. Ricci computes the Beltrami operator immediately, a feature not often found in other codes. (For more features see Ref. 1.)

3. GRtensorII (for Mathematica)

Kyall Lake of Queen's University in Kingston, Ontario, Canada, and coworkers Peter Musgrave and Denis Pollney, created a tensor package specifically designed to tackle research problems in GR [6]. This package is called "GRTensor II" and was originally developed to run with Maple. A more recent version of it also runs on Mathematica. Because it was first developed for Maple, its features are discussed more in the Maple section. It can be downloaded for free from <http://grtensor.phy.queensu.ca/>.

4. Others

Other recently developed packages that have not yet been evaluated for their propulsion utility are Tensorial (version 3 and higher) and Cartan. Tensorial is a new tensor package of Mathematica. Cartan is a package for conducting differential geometry and Lie groups and runs under Mathematica.

C. Maple

The other product that was introduced after "C" became the standard programming language is Maple, developed by a team at the University of Waterloo (Canada). It is a main competititor to Mathematica and is also widely used as a computational tool for a variety of scientific and engineering applications.

1. Maple's "Standard" Tensor Package

Raymond McLenaghan of the University of Waterloo in Canada created the "standard" tensor package of Maple. McLenaghan is known among GR experts as co-discoverer (with the Australian, John Carminati) of the Carminati–McLenaghan invariants (i.e., all possible invariants that one can construct by contracting the Riemann and Weyl tensors in all possible modalities; Web site <http://en.wikipedia.org/wiki/Carminati-McLenaghan_invariants>.

The tensor package of Maple is endowed with all the facilities necessary in GR. It includes a symbolic computing code to immediately list the Carminati–McLenaghan invariants for a given four-dimensional Riemannian manifold (i.e., when the metric tensor is assigned). A similar capability also is included in the GRTensorII package, described next.

2. GRTensorII (For Maple)

GRTensor II was specifically designed to tackle research problems in GR by Kyall Lake of Queen's University in Kingston, Ontario, Canada, and coworkers Peter Musgrave and Denis Pollney. It was originally developed to run with Maple (Lake had originally contributed to the tensor package of Macsyma).

To operate GRTensorII one has to learn Maple's operating language and a dedicated "GR" language to handle tensors, both old and "new" (e.g., tensors at research level like the Bel-Robinson tensor).

Many advanced features of GRTensorII are quite remarkable, for example:

1) Automatic Petrov classification of Einstein spaces
2) Immediate solutions of the full set of Einstein–Maxwell equations
3) Numerous and detailed exact solutions (Bondi metric, Stephani solution for perfect fluids, Tomimatsu-Sato solution, Cosmological metrics, etc.). Five-dimensional metrics of the Kaluza-Klein type are also starting to show up among the worked examples, and this paves the way to future generalizations to $N = 11$ dimensions as typical of supergravity, string theories, etc.
4) "GRJunction" is an innovative tool within GRTensorII that allows the symbolic computations of exact solutions made up by "cutting, pasting, and joining together" already known, simpler exact solutions. This is a unique feature of GRTensorII that no other tensor package has, including MathTensor.

The disadvantages of GRTensorII are that the user cannot select:

1) Signature of the metric (assumed to be space-like)
2) Sign of the determinant of the metric (assumed to be negative)

3) Definition of the Riemann tensor in terms of the Christoffel symbols and their derivatives
4) Definition of the Ricci tensor from the Riemann tensor
5) MKS (or SI = Système International) system of units

The inability to adjust these settings is serious as one may get a confusing factor of -1 when comparing the same exact solutions of the Einstein or Einstein–Maxwell equations by using different tensor packages, like GRTensorII and Macsyma, or GRTensorII and MathTensor. Furthermore, the inability to use MKS units is disappointing and limits its utility for space propulsion considerations.

In conclusion, GRTensorII is another available code that has advantages and disadvantages relative to its utility for space propulsion science research.

D. Other Codes

In addition to the codes examined, there are a number of other codes that might be useful but whose features for space propulsion research have not yet been assessed. A couple of examples are described next.

1. Reduce (Web site: http://en.wikipedia.org/wiki/Reduce_computer_algebra_system)

Reduce is an existing code commonly used in both quantum physics (e.g., for Dirac equations) and GR. Recently a spin-off of this code was created devoted to GR called "GRG," Web site <http://sal.linet.gr.jp/A/1/GRG.html>.

2. Casimir-Effect Codes

Jordan Maclay developed some codes within Mathematica for studying the Casimir force for numerous geometries, not just the traditional parallel plate configuration. The codes address estimating the forces on the faces of three-dimensional cavities, whose results vary depending on the aspect ratio of the cavity and can sometime become repulsive.

Similar codes were probably also written in Europe when the space mission to measure the Casimir effect in space was proposed to ESA back in 2000. These have recently become available through Dr. Martin Tajmar of ARC Seibersdorf Research, in Seibersdorf, Austria [7].

It would be interesting to "match" these codes with GR codes to find possible breakthrough propulsion physics links between GR and the Casimir effect.

V. Conclusions

As computer codes advance and as more research is conducted in propulsion physics, it would be advantageous for future propulsion researchers to begin to follow common conventions in GR equations. To that end, standardization of sign and notational conventions are recommended as follows:

1) MKS (SI) sytem of units, with all natural constants (c, G, h) explicitly shown in the equations.

2) Sign convention for the metric where causal conditions result in a positive spacetime displacement (i.e., a signature + - - -).
3) Sign convention where the determinant of the metric is negative.
4) Definition of Riemann tensor as $R^\lambda_{\nu\rho\sigma} \equiv -\Gamma^\lambda_{\sigma\nu,\rho} + \Gamma^\lambda_{\rho\nu,\sigma} - \Gamma^\alpha_{\sigma\nu}\Gamma^\lambda_{\alpha\rho} + \Gamma^\alpha_{\rho\nu}\Gamma^\lambda_{\alpha\sigma}$.
5) Definition of contracting the Riemann tensor into the Ricci tensor as $R_{\nu\rho} = R^\lambda_{\nu\lambda\rho}$.

Next, the ability of existing codes to support these conventions and their ability to support propulsion-relevant calculations is assessed. Presently there is no single, ideal product. It is the intent of this chapter that the examination of software options and the use of standardized conventions will assist future researchers in making their work easier to conduct and easier to share with their colleagues.

Acknowledgments

The late John Anderson of NASA was the first to recognize the key role that these symbolic codes might play in the development of breakthrough space propulsion. Subsequently, Marc G. Millis of NASA Glenn Research Center at Lewis Field adapted John Anderson's work into the Breakthrough Propulsion Physics project. The author of this paper acknowledges Marc G. Millis for his interest in a question as vital as "how will breakthrough propulsion research benefit from symbolic tensor manipulators?" The support and friendship of the co-editor of this book, Eric W. Davis of the Institute for Advanced Studies at Austin, is also gratefully acknowledged. Finally, this author is personally indebted to one of the most farsighted researchers in visionary fields, H. E. Puthoff, Director of the Institute for Advanced Studies at Austin, for sharing support and enthusiasm! Relativistic interstellar flight is difficult to model and investigate mathematically, but someone has to start the process. In the centuries to come we will show that we have not been working in vain.

References

[1] Lee, J. M., *Ricci—A Mathematica Package for Doing Tensor Calculations in Differential Geometry*, User Manual. URL: http://www.math.washington.edu/~lee/Ricci/.
[2] Misner, C. W., Thorne, K. S., Wheeler, J. A., *Gravitation*, W. H. Freeman & Co., New York, 1973.
[3] Hartle, J. B., *Gravity: An Introduction to Einstein's General Relativity*, Addison Wesley, San Francisco, 2003.
[4] *Macsyma Mathematics Reference Manual* and *Macsyma System Reference Manual*, Macsyma Inc., Arlington, MA, 1993.
[5] Parker, L., and Christensen, S. M., *MathTensor—A System for Doing Tensor Analysis by Computer*, Addison Wesley, Reading, MA, 1994.
[6] Lake, K., Musgrave, P., and Pollney, D., *GRTensor II User Manual*. URL: http://130.15.26.62/NewDemo/frame.html.
[7] Sedmik, R., and Tajmar, M., "CasimirSim—A Tool to Compute Casimir Polder Forces for Nontrivial 3D Geometries," *Proceedings of the STAIF-2007: 4th Symposium on New Frontiers and Future Concepts*, El-Genk, M. S. (ed.), AIP Conference Proceedings Vol. 880, AIP Press, New York, 2007, pp. 1148–1155.

Chapter 22

Prioritizing Pioneering Research

Marc G. Millis*
NASA Glenn Research Center, Cleveland, Ohio

I. Introduction

A TYPICAL challenge of any research project is to decide how best to disburse limited resources to the variety of competing options. When it comes to seeking propulsion-specific advances from science, the situation is even more challenging. First, research aimed at such *revolutionary* advancements is different from the more common work of improving technology. Balancing the vision required to extend beyond established knowledge, along with the rigor required to make genuine progress, presents additional challenges beyond just the science itself. Next, there is the unfamiliarity of the topic. This includes introducing the aerospace community to emerging science as well as introducing scientists to the specific queries behind revolutionary spaceflight. Additionally, existing research proposals span multiple disciplines of science and span different levels of progress and applicability toward the goals. To use a colloquial expression, this presents the challenge of comparing apples to oranges. This difficulty is compounded by the fact that not all of the important questions are yet represented by proposed approaches. There are likely to be unaddressed issues that are more important than the approaches proposed so far. And lastly, on topics that appear so far from fruition, the available resources are minimal. This compounds the challenge of partitioning resources in portions sufficient to ensure progress.

This chapter addresses these challenges primarily by examining the methods and lessons learned from the NASA Breakthrough Propulsion Physics (BPP) Project. This project, founded by this author, was designed to manage research on gravity control and faster-than-light travel. It assessed 10 different research approaches, produced 16 peer-reviewed journal articles [1–16], an

---

This material is a work of the U.S. Government and is not subject to copyright protection in the United States.
*Propulsion Physicist, Propellant System Branch, Research and Technology Directorate.

award-winning website [17], and covered this plus discretionary efforts thereafter for a total cost of $1.6 M spread over 1996 to 2002 [18]. The findings of the sponsored research are described in Chapter 1. Regarding other evidence of the project's effectiveness, here is a quote from an independent review panel about the BPP Project [19]:

> [The Breakthrough Propulsion Physics Project] approach was unanimously judged to be well thought out, logically structured and carefully designed to steer clear of the undesirable fringe claims that are widespread on the Internet. The claim that the timing is ripe for a modest program of this sort was agreed to be justified: Clues do appear to be emerging within mainstream science of promising new areas of investigations. The team concurred that the 1997 BPP kickoff workshop did identify affordable candidate research tasks which can be respectably pursued, and that the proposed research prioritization criteria were a valid way to select from amongst these (and future) proposals. The program approach was deemed to be sound: emphasizing near-term progress toward long-term goals; supporting a diversity of approaches having a credible scientific basis; holding workshops to exchange ideas; solicit constructive criticism and assess progress; aiming toward testable concepts [19].

The project's operating principles, research solicitation process, evaluation criteria, and other lessons learned are detailed. Introductory comparisons to the larger NASA Institute of Advanced Concepts (NIAC) [20,21] and the still larger Defense Advanced Research Projects Agency (DARPA) [22] are also provided.

The first section of this chapter examines a variety of historic lessons about revolutionary advances and then sets them in the context of seeking revolutionary propulsion advances from science. These details are included to convey the rationale behind the NASA project's methods and also pertain to both NIAC and DARPA.

Another section of this chapter is devoted to the challenge of combining vision with rigor. Because it is common that novel ideas are reflexively dismissed and that noncredible concepts are abundant on provocative topics, this specific challenge warrants special attention. Methods are suggested to avoid reflexive dismissals, to efficiently filter out noncredible concepts, and to respond to nonrigorous research submissions in a constructive manner.

All of these lessons are presented largely in the context of management, but some lessons apply to individual research as well. The intent is to provide managers and researchers insights on how to conduct research that is both visionary and productive.

For readers who want to jump ahead to the recommendations without needing their rationale, the specific recommendations can be found in the following sections. Most of these are in the form of lists:

II.H. Recommendations Deduced from Historical Perspectives
III.C. Summary of Recommendations on Combining Vision and Rigor
IV. Research Project Operating Principles

V.	Devising Prioritization Criteria (refer to the entire discussion, not just the bullet list)
VII.C	Project Metrics of Performance
VII.D	Mitigating Revolutionary Research Risks (see Table 3)

II. Historical Perspectives

To better assess the conditions necessary to elicit technological revolutions, a variety of historical perspectives are examined and then set in the context of seeking the scientific discoveries that could lead to revolutionary space propulsion. This includes Fosters's "S-Curve" of technological advancements, Henderson's "architectural innovations," Dyson's "tool-driven" perspectives, Kuhn's "paradigm shifts," Hamming's lecture on great researchers and important problems, Sir Clarke's Three Laws, the role played by science fiction, and John Anderson's "Horizon Methodology."

The literature on the history of technological advancements is abundant, and the coverage here is only a small sample of the depth of available information that reflects similar patterns. Interested readers are encouraged to examine the cited references for further insights. Following the explanations of these perspectives, a list of recommendations is offered to help researchers, managers, and advocates of revolutionary research improve the prospects for success. These recommendations are based on the recurring patterns observed in history.

A. Institutional Patterns of Technological Revolutions

History has shown that simply improving existing technology is not sufficient to sustain competitive advantage [23]. As illustrated by the recurring S-curve pattern of technological advancement (Fig. 1), it is time to look for revolutionary approaches when the existing methods have reached the point of diminishing returns. For example, jet aircraft did not result from mastering piston-propeller aircraft. Transistors were not invented by mastering vacuum tubes. Photocopiers did not result from mastering carbon paper. The recurring theme is that entirely different operating principles were pursued to sustain competitive advantage rather than refinements of known methods.

The S-curve evolution shown in Fig. 1 is typical of any successful technology. The pattern begins with the initial efforts where little advancement results until a breakthrough occurs. The breakthrough, at the lower knee of the curves, is where the technology has finally demonstrated its viability. After this point significant progress is made as several embodiments are produced and the technology becomes widely established. Eventually, however, the physical limits of the technology are reached, and continued efforts result in little additional advancement. This upper plateau is the "point of diminishing returns." To go beyond these limits, a new alternative (with its own S-curve) must be created. Shifting to the new alternative is what is meant by pursuing "revolutionary" advancements.

It has been found that it is most difficult for incumbent organizations to consider such alternatives when their familiar approaches are at the point of diminishing returns [23,24]. By then, the institutions have become too uniquely adept

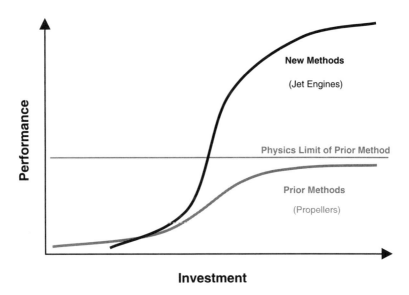

Fig. 1 S-curve pattern of technology advancement.

at their accrued technology to consider alternatives. They are also tied so closely with their existing customers that it is difficult to explore new opportunities. Because new approaches emerge in a still-developing state and have unfamiliar principles, it is also difficult for the incumbent to properly assess their merit. This difficulty is compounded because the incumbents use their prior values to judge the new approach, values that are rooted in the evaluation criteria for the different, prior technology. When at the point of diminishing returns, established institutions prefer to modify or add new features to their technology rather than to search for ways to go beyond their technology or to find new applications for their technology. Sometimes this takes the form of changing the emphasis away from performance to improving reliability and cost effectiveness.

The term for reconfiguring existing technology to address a new opportunity is "architectural innovation" [25]. Even here, the incumbent organizations will typically dismiss such innovations because the new opportunity is seemingly irrelevant when viewed per their prior values. Additionally for architectural innovations, their value is even harder to appreciate because the *technical* aspects of the innovation do not appear like noteworthy advancements.

In the case of spaceflight, the space tourism entrepreneurs are examples of such architectural innovations. They are taking existing technology and applying it in new configurations and to reach new markets [26,27]. Although this example is not directly relevant to frontiers of propulsion science, it is relevant in the context of identifying obsolescing values. As evidenced by the emergence of such firms outside the incumbent aerospace organizations, it is clear that the original values that drove the emergence of spaceflight are no longer complete. In other words, the criteria against which early spaceflight was judged are no longer the only drivers of future progress. Significant changes have occurred in societal values, technological options, and emerging science. Following historical patterns, it is likely that the incumbent spaceflight organizations will have difficulty recognizing and adapting to the contemporary opportunities and constraints [23–25].

Considering these patterns, it is not surprising that the emergence of revolutionary advances often come from outside the established organizations [28, p. 30]. A classic aerospace example is how the Wright brothers (bicycle mechanics) succeeded in heavier-than-air manned flight well in advance of the government funded (Smithsonian Institution) aerospace research of Samuel P. Langley.

B. Tool-Driven Revolutions

In Freeman Dyson's *Imagined Worlds* (1997) scientific revolutions are cast as the byproducts of new tools [29]. As examples, Dyson cites how the telescope led to Galileo's insights and how X-ray diffraction led Crick and Watson to understand DNA structure. The recurring theme is that when new observational tools become available, new phenomena are observed, which then leads to revolutionary science.

Considering our contemporary situation, this list shows tools and observations that are relevant to seeking revolutionary spaceflight along with citations for where these are discussed in this book:

1) Hubble space telescope and other advanced ground telescopes are revealing further details about the constituents of our universe, such as the anomalous redshifts from the most distant supernovae, and the apparent existence of "dark matter" (discussed in Chapters 3, 4, and 15).

2) *COBE* and *WMAP* satellites are revealing new details of the cosmic microwave background radiation [30] (discussed in Chapter 3).

3) Improved fidelity of laboratory instrumentation makes it possible to conduct ever-more accurate experiments and correspondingly requiring extra vigilance in their proper use (Chapters 5 and 20).

4) Symbolic mathematical computational tools make it easier to apply complex mathematical techniques (Chapter 21).

5) Ultrahigh-intensity tabletop lasers can momentarily produce extreme electric and magnetic fields suitable for testing space-warping theory (Chapter 15).

6) Superconductivity provides the means to explore the intriguing phenomenon of absolute-zero electrical resistance and its secondary phenomena (see Chapter 5).

7) Micro- and nano-structure engineering make it easier to tangibly explore the physics of the very small, such as Casimir forces (discussed in Chapters 12 and 18).

8) Internet communication makes it easier to access a wider network of data and collaborators. This includes those that were previously separated by geography or disciplines.

C. Paradigm Shifts

The next perspective, articulated in 1962 by Thomas Kuhn in *The Structure of Scientific Revolutions* [31], posits that scientific revolutions are sudden paradigm shifts that occur after the accumulation of physical data that do not coherently blend with the prior paradigms. Examples include the Copernicus model of planetary orbits (heliocentric model), Newton's laws of motion, Darwin's evolution, Einstein's special relativity, and quantum mechanics.

In a 1970 edition to his book [31, see Postscript], Kuhn distinguishes between two meanings of *paradigm* that were too closely woven together in earlier editions, leading to misinterpretations. One meaning is related to the scientific community as a whole, whereas the other is specific to sets of interpretations held for natural phenomena. In the context of this discussion, the latter meaning is used, specifically paradigms as sets of interpretations held for natural phenomena. In this sense, it is time to more critically challenge old paradigms and suggest new ones when accumulating evidence does not clearly fit the established paradigms.

The organizational challenge when dealing with paradigms is the implicit value system used to judge emerging possibilities. This is similar to the tendencies discussed earlier regarding S-curves, where old values are used to judge new situations. With paradigms there are implicit commitments within incumbent organizations for setting work priorities. This results in a tendency to reflexively dismiss novel approaches that are inconsistent with the established paradigm. This issue is discussed in more detail in Sec. III: Combining Vision and Rigor.

For now, to offer starting points for considering how the perspective of paradigm shifts pertain to space propulsion, a brief list of physical observations or lingering unknowns that have not yet been resolved follows. Many of these are revealed by the new tools discussed previously and their relevance to space propulsion is discussed in the cited chapters.

1) "Dark matter" as the contemporary paradigm to explain gravitational lensing and anomalous binding of rotating galaxies (Chapters 3, 4, and 15).

2) "Dark energy" as the contemporary paradigm to explain the anomalous redshifts of distant supernovae (Chapters 3, 4, and 15).

3) Incompatibility of General Relativity (large scale) and Quantum Mechanics (small scale). As one example, there is discrepancy of about 120 orders of magnitude between the General Relativistic energy estimate for the quantum vacuum and that from astronomical observations with Quantum interpretations (Chapters 3, 12, 13, and 18).

4) Causality and the nature of time (Chapters 14–16).

It is curious to contrast these lingering unknowns to the notion presented in the 1996 book, *The End of Science* [32], where its premise is that all the important questions have already been raised and are well on their way to being answered.

D. Great Researchers and Important Problems

In a 1986 lecture, Richard Hamming (a pioneer in error-correcting codes) reflected on the distinctions between *good* and *great* researchers [33]. According to Hamming, *good* researchers are those who make competent, incremental advancements, while *great* researchers are those who achieve pioneering advancements beyond their peers. The two major distinguishing characteristics are the problems those researchers choose to tackle, and the self-critical drive with which those problems are pursued.

Regarding the choice of problem, it was noted that the great researchers have the courage to tackle the "important problems." Important problems are defined as the grand challenges that will make a significant difference if solved, and where enough progress has been made to enable these problems to finally be pursued. These are the problems that their peers will not attempt. Instead, the *good* researchers opt to pursue objectives that are already well established in their field and where there is little chance of failure. In terms of the S-curve analogy discussed previously, the good researchers work to improve existing methods, while the great researchers seek the new S-curves, the alternatives to surpass the existing methods.

The choice of problem is not the only distinction. Hamming discussed numerous characteristics that great researchers tend to possess. The recurring theme is that the great researchers have both the confidence to approach the problem, with enough self-doubt and awareness of their shortcomings to sustain their objectivity. Here is an abbreviated list of the major characteristics identified by Hamming:

1) Have the courage to tackle the *important problems*:
   a) These are grand challenges that will make a significant difference, not just the "safe" research.
   b) Attackable, meaning that there is a way to begin solving the problems.
2) Start with independent thoughts and then collaborate with others.
3) Make steady progress; be driven and focused.
4) Redirect what is difficult into something easier.
5) Are open to learn things beyond their own field—"Knowledge is like compound interest."
6) Tolerate ambiguity:
   a) Believe enough in self to proceed.
   b) Doubt self enough to honestly see flaws.
7) Sell themselves well.
   a) Write well.
   b) Present well.
   c) Able to translate work into executive-level communications.
8) Honest with personal flaws and work toward overcoming them (converting liabilities into assets).

E. Sir Arthur C. Clarke's Three Laws

Evolving over the course of revisions to the book, *Profiles of the Future*, 1962–1973, Arthur C. Clarke posited the following three "laws" regarding the reaction to revolutionary advancements [34]:

1) When a distinguished but elderly scientist states that something is possible, he is almost certainly right. When he states that something is impossible, he is very probably wrong.

2) The only way of discovering the limits of the possible is to venture a little way past them into the impossible.

3) Any sufficiently advanced technology is indistinguishable from magic.

While these oft-cited laws are not laws in the strictest sense, they do serve to summarize key recurring themes and are consistent with many of the perspectives already introduced.

Regarding the first law, which is in reference to such infamous quotes as: "Space travel is utter bilge" uttered by Dr. Richard van der Riet Wooley, one year before 1957's Sputnik; and "The secrets of flight will not be mastered within our lifetime, not within a thousand years" credited to Wilbur Wright, two years before their 1903 Kitty Hawk flight, it is equally important to remember that history includes many errant ideas that were indeed critically flawed. Distinguishing these is easy in retrospect. The viable ideas survive (along with the infamous dismissive quotes), while the errant ideas tend to be forgotten.

To assess revolutionary ideas *as* they emerge is difficult. Revolutionary ideas, by their very nature, break from the familiar and can look as nonsensical at first as the more common errant ideas. Here again the notions of paradigm shifts and incumbent values pertain. Because this situation is complex and is a major challenge of revolutionary research, Sec. III: Combining Vision and Rigor is devoted to addressing this issue.

When it comes to deliberately seeking revolutionary advances in spaceflight, Clarke's second law is more opportune because it suggests where to look for the next revolutions. An oft-cited source of such inspirational impossibilities is science fiction, which is examined in the next section.

Regarding Clarke's Third Law, Robert Forward applied this perspective in his book *Indistinguishable from Magic*, which included projections of interstellar flight technology [35]. While many of the specific topics in Forward's book are addressed in more up-to-date terms here in this book, Forward's book still serves as an illustration of Clarke's third law.

F. Science Fiction

It is generally recognized that science fiction is inspirational. Just to cite one study, the American Astronautical Society (AAS) found that the rocketry pioneers, Tsiolkovsky, Goddard, von Braun, etc., were all inspired by the science fiction of their day [36]. The captivating idea of rocket travel and its implications for a better world spurred these pioneers to investigate how to make rocketry real.

There is also a common misperception that science fiction reliably predicts future technologies. Although some fiction became real (nuclear submarines, men on the moon, portable palm-sized phones, etc.) there is much fiction that has not happen or did not happen as envisioned. An excellent example is when men first walked on the moon [37, p. 142]. Although the Apollo crew size and the launch location matched Jules Verne's fiction, rockets were used instead of a giant cannon and, most importantly, when it happened for real, people around the whole world were watching on television. This unforeseen event of the entire world being able to witness this historic event is profoundly significant unto itself.

Regarding the scientific discoveries that could revolutionize spaceflight, many science fiction stories have already broadcast what interstellar flight might look like. Famous examples include *Star Trek* [38] and *Star Wars* [39,40]. This fiction, however, is not sufficient for conducting research. Although entertaining, inspiring, and thought provoking, the science behind such fiction was cast to provide a setting for drama. Accordingly, it is often incomplete or compromised to create dramatic tensions and expedient plots. As much as it might inspire, to follow its fictional images too closely would limit possibilities, as it would constrain one to only consider the approaches suggested in the fiction.

When examined in the context of its thought-provoking nature and without taking its visions too literally, science fiction is akin to "brainstorming." Brainstorming is a step in explicit processes for enhancing the problem-solving ability of teams [41]. In such processes, the focal problem is defined in the first step. Next, through brainstorming, numerous ideas are collected to address the problem. At this stage the ideas do not need to be correct, but merely provocative. This is completely consistent with science fiction visions. Once a suite of provocative ideas is collected, analytical rigor is applied to filter through the ideas and refine a workable set of approaches. For the details of such techniques, the reader is referred to Miller's *The Creative Edge* [41] or any of the other numerous books that document the human creative process and how to apply that process explicitly in organizational settings.

Following along these lines, the next section summarizes a technique developed and applied to deliberately seek revolutionary research within NASA.

G. Horizon Mission Methodology

In 1996, John L. Anderson published his 'Horizon Mission Methodology' in the *Journal of the British Interplanetary Society* and set it in the context of interstellar flight [42]. This method is a systematic approach for provoking revolutionary research within organizations—the kind of approaches indicative of the alternative S-curves described earlier. The method employs lessons from prior revolutions and deliberately uses science fiction for its inspirational value.

The process can be used by individuals or teams. Its first step is to impose a general goal that is impossible to achieve with projected technology. The goal of timely interstellar flight is an example. The use of impossible goals is deliberately intended to counteract the habit of researchers to extrapolate their familiar technologies to address problems. To use a colloquial expression, it forces the researchers to think "out of the box." Along with the seemingly impossible goals is the requirement that the team considers that the goals are achievable by some undefined, far-future technology. This is an application of Clarke's Second Law. Next, through brainstorming, the notions of science fiction are used as placeholders for the embodiments of the ultimate successes. From those provisional "solutions" the team is then asked to "look back from this future" to identify the limiting assumptions. In other words, the team is asked to determine the specific make-or-break issues that would have to be solved to make such a future plausible. In short, this means defining the "grand challenges" around which to aim research objectives. With those goals set, the next step is to identify the knowledge gaps. By comparing the grand challenges in detail with the accumulated knowledge to date, the key unknowns and critical issues are identified. From there, the researchers are asked what steps could be taken to begin addressing those unknowns and issues. In terms of Hamming's lecture, this means articulating the "important problems."

The NASA BPP Project employed this Horizon method. It was used to devise the Project's grand challenges, specifically its three goals: 1) nonpropellant propulsion, 2) hyper-fast travel, and 3) the energy breakthroughs related to those two goals. In a subsequent exercise with geographically dispersed participants, the foundational knowledge related to these challenges was collected [43]. And finally, still following the spirit of this Horizon method, a workshop was convened to deliberately identify relevant research tasks [2,44].

The next section lists recommendations based on the lessons just described. These recommendations are intended to help researchers, managers, and advocates of revolutionary research improve their prospects for success.

H. Recommendations Deduced from Historic Lessons

From historically based perspectives, it is clear that pioneers who seek support from established organizations face both technical and organizational challenges. The Horizon Mission methodology [42] is one approach to begin soliciting revolutionary approaches within established organizations, but this by itself appears insufficient when considering organizational impediments. To help overcome organizational impediments, it is recommended that advocates of pioneering research also perform the following programmatic tasks:

1) Compare the goals of the organization to the ultimate performance limits of the organization's technology. Make it clear which revolutionary advancements are required to fully satisfy those goals. In the case of viable interstellar flight, for reaching other habitable worlds, this goal cannot be achieved by merely improving technology. Alternatives from undiscovered science must be sought (see Chapters 2, 3, 14, and 17).

2) Familiarize the organization with the emerging possibilities, building on scholarly publications and impartially identifying both their strengths and weaknesses. Offer foundational information that will help the organization comprehend the opportunities and challenges. In the case of the frontiers of propulsion science, this book is a step to address this need.

3) To enable the organization to properly assess the emerging alternatives, develop the criteria against which the emerging alternatives can be compared to one another and to the organization's broader goals. This avoids the pitfall of assessing fledgling alternatives in terms of criteria that are specific to prior, more-developed methods. In the case of propulsion science, this chapter offers examples of criteria and the selection processes used by the NASA BPP project.

4) Using the foundational information in comparison to the revolutionary goals, identify the "important problems"; those problems that are both highly relevant and approachable. John Anderson's Horizon Mission methodology is one approach to identify such problems within organizations [42].

5) Closely related to identifying important problems is to place emphasis on physical observations rather than on contemporary paradigms of those observations. Paradigms are interpretations to explain observations of nature.

Empirical data are nature (partially revealed within the constraints of the given observation or experiment). For example, "dark energy" is a paradigm—a working hypothesis—to explain the anomalously large redshifts of light from distant supernovae. The redshifts and luminosity data are the empirical observations. Conversely, the liability of emphasizing empiricism is that data are more limited than a *reliable* theory. A reliable theory (which is more substantial than a paradigm) can predict observations that have not yet been made. At the current stage of seeking revolutionary spaceflight, where the theories still have not been fully formulated, it is advantageous to place more emphasis first on empirical observations, and then derive theories to fit those data, independent of such paradigms as dark matter and dark energy.

6) Tailor the earliest research proposals to fit within the relatively minor resources available to the far-future options, and then build on the progress from those steps to demonstrate the value of continuing research toward revolutionary advancements.

7) And lastly, realizing that many revolutions come from outside the expected organizations, look to other organizations and disciplines for opportunities. In terms of revolutionary propulsion science, it is prudent to approach the undiscovered physics from the *aerospace* perspective and seek the visions for the next aerospace advances in terms of emerging *physics*.

In subsequent sections of this chapter, the specific programmatic methods of the NASA BPP Project are described. Before proceeding, it is necessary to examine yet another set of lessons from dealing with revolutionary ideas—the challenge of balancing vision and rigor.

III. Combining Vision and Rigor

Exploring the edge of knowledge for profound discoveries evokes special challenges. In addition to the normal challenges of scientific research—discovering how nature truly works—the provocative nature of the edge of knowledge can encumber research. First, by pursuing truly profound improvements in the human condition, the stakes are higher and accordingly emotions run higher. Second, by operating on the genuine edge of knowledge instead of exploring refinements of established knowledge, controversial ideas are encountered. This combination of heightened emotions and controversy can encumber the normal, productive discourse of scientific study. Both skeptics and optimists can prematurely reach conflicting conclusions and, in their zeal, fail to communicate with the impartiality needed to rigorously identify, test, and resolve the real issues.

Considering that most historic breakthroughs originally sounded like fringe ideas, it is not surprising that many of the proposals for revolutionary spaceflight might sound *too* visionary at first, or at least unfamiliar. It is therefore difficult to sort out the fringe ideas that may one day evolve into tomorrow's breakthroughs from the more numerous, erroneous ideas.

This section examines this challenge with the in-going premise that a *combination* of vision and rigor is needed to make genuine progress on such topics. Vision, where one extends beyond the known using imagination and opportunistic optimism, is necessary to break from mere extrapolations of existing solutions. It is a necessary characteristic of venturing in search of new S-curves and to break free from legacy paradigms, and to seek out the important problems. It is a necessary characteristic to apply Clarke's Second Law. Perhaps this is what Einstein meant when he said "Imagination is more important than knowledge." Conversely, vision by itself is insufficient, because it lacks the critical analysis to move forward. By itself, vision is just akin to science fiction. To make genuine progress, analytical rigor is also needed. Just as vision by itself is inadequate, rigor alone is insufficient to make revolutionary progress. Rigor by itself does not take one beyond the known. By combining rigor with vision, reliable contrasts between the known and unknown become possible. Make-or-break issues can be identified. Progress is made with the resolution of these issues.

This section examines several facets of how to enable this combination of vision and rigor. This includes avoiding the tendency to reflexively dismiss unfamiliar ideas and the challenge of dealing with nonrigorous concepts that are endemic to this topic and that taint serious research. Lessons from psychological studies and from the NASA BPP project are offered.

A. Absence of Vision: Reflexive Dismissals

As reflected by the investigations of Foster [23], Utterback [28], Kuhn [31], and Clarke [34], it is common to have established experts summarily dismiss emerging possibilities. Statistically speaking, one is likely to be correct by dismissing all unfamiliar assertions. Viable revolutionary ideas are rare while errant ideas are easily generated. Such reflective dismissals are not the same as rigorously applying skepticism to identify critical issues.

To *reliably* determine feasibility of new ideas requires openness to consider the possibility and to rigorously apply skepticism to check for weak points and to judge how well these weaknesses are addressed. To apply skepticism in the assessment of a new idea is time consuming, because the unfamiliarity inherent in revolutionary ideas requires more time to assess. With familiar approaches there are already precedents with which to compare. With unfamiliar approaches, it takes time to check the citations and ensure that the assertions are logically constructed. It is analogous to assessing a topic outside of one's normal discipline. This is compounded because emerging ideas are inherently less refined, which can make them appear nonrigorous when compared to well-established knowledge. Conducting a full assessment of an unfamiliar approach is comparable to a full research task unto itself.

In the context of managing research solicitations for pioneering ideas, this presents a challenge; if a rigorous proposal review is potentially as intensive as the proposed research tasks themselves, how does one efficiently filter out errant approaches while still allowing the potentially viable ideas that might otherwise be hastily dismissed? In short, one technique is to scrutinize the *rigor* of the approach rather than trying to judge its *feasibility*. This technique is described in more detail in the section "NASA Breakthrough Propulsion Physics Project Operating Principles." Before detailing that technique, it is necessary to examine more closely the characteristics of nonrigorous work.

B. Vision Without Rigor

Individuals who dabble in vision without rigor are common. Regrettably, such nonrigorous enthusiasts can taint the serious pursuit of revolutionary research; hence it needs to be addressed here. The grand claims typical of nonrigorous work can attract undue media attention [45]. When nonrigorous approaches are reported in the press they exacerbate the difficulty of focusing attention on more promising approaches. Some works specifically related to the pursuit of spaceflight breakthroughs have appeared in such books as *Voodoo Science* [46]. One of the first lectures on this situation is the 1953 colloquium by Dr. Irving Langmuir who coined the term "pathological science" [47]. Other examples appear in a short paper "Some Observations on Avoiding Pitfalls in Developing Future Flight Systems" [48]. To address this issue, this section examines the extent of this situation, corresponding psychological considerations, techniques to identify noncredible approaches, and then techniques for responding to noncredible submissions.

1. Statistics on the Extent of Nonrigorous Submissions

In the course of the NASA BPP Project, two years of unsolicited correspondence (2000 and 2001) were analyzed to determine the relative proportion of nonrigorous enthusiasts [49]. The intent at the time was to provide management an estimate of the labor required to reply to these submissions. In the context of this chapter, these statistics convey the magnitude of nonrigorous activities. No statistics are available to reflect the proportion of researchers who possess rigor without vision.

The findings are presented in Table 1. These statistics counted the number of *messages* rather than the number of submitters. Multiple messages from a single sender, for example, were all counted. Roughly one-third of all the submissions to the NASA project were from nonrigorous individuals requesting reviews or other support for their ideas. Given that an objective review and response can take roughly three days to prepare, and considering the rate of incoming submissions, it is estimated that a fulltime staff of three or four researchers would be needed just to respond to the nonrigorous individual requests.

Within that group of nonrigorous individuals, there is another subset that warrants mention, the "lunatic fringe." The term lunatic fringe is a colloquial expression for those people whose ideas are accompanied by delusions of grandeur and/or paranoia. Please be advised that this discussion is not meant to be glib, trivial, or taken lightly. Roughly one-third of the nonrigorous correspondences (specifically 9% of total unsolicited correspondence) displayed such characteristics. Delusions of grandeur are evidenced when the submitter states that their device or theory is *the* answer to solve the world's problems, or other statements to that effect. Paranoia is easier to recognize, where the submitter expresses fear about their safety or about the theft of their ideas. Often these characteristics coexist, such as when the submitters express concern that their submission is so valuable that it will evoke suppression by evil conspirators.

2. Techniques to Identify Nonrigorous Submissions

Research managers who are scouting for potentially revolutionary ideas need to find an efficient way to screen submissions. As already shown, the proportions

Table 1 Unsolicited correspondence statistics from the NASA Breakthrough Propulsion Physics Project

1000 unsolicited submissions per year (2000–2001)	100%
Professional researchers:	32%
• News of recent peer-reviewed publications • Inquiries about future research solicitations • Employment inquires	
Amateur researchers requesting feedback	31%
• "Here is my breakthrough ..." • "Please evaluate my idea ..." • "Please help me advance my idea ..." (About one-third of amateur requests, or 9% of all submissions, display paranoia or delusions of grandeur)	
Public inquiries:	30%
• "Please tell me more about ..." • "What is your assessment of ...?" • "Please help me with my homework."	
Invitations to conferences/workshops	4%
Press interview requests	2%
Public speaking requests	1%

of nonrigorous enthusiasts are high and there is a tendency in reviews to prematurely dismiss unfamiliar concepts, concepts that might actually be viable. To conduct thorough, rigorous reviews has already been shown to be as intensive as conducting the research task itself.

The recommended tactic is to focus on the *rigor* with which the concepts are articulated rather than on the nature of the proposed concept. A lack of rigor is easier to detect. Classic symptoms of nonrigorous work are reflected in Langmuir's lecture on "pathological science" [47], Carl Sagan's "baloney detector" [50], John Baez's "Crackpot Index" [51], and lessons from the NASA BPP Project.

One of the earliest examinations on the phenomenon of noncredible science occurred in a 1953 colloquium by Dr. Irving Langmuir who coined the term "pathological science" [47]. In this lecture, Langmuir identified the following six symptoms of unproductive and misleading work (quoted directly from R. N. Hall's transcription of the lecture [47]):

> 1) The maximum effect that is observed is produced by a causative agent of barely detectable intensity, and the magnitude of the effect is substantially independent of the intensity of the cause.
> 2) The effect is of a magnitude that remains close to the limit of detectability; or, many measurements are necessary because of the very low statistical significance of the results.

3) Claims of great accuracy.
4) Fantastic theories contrary to experience.
5) Criticisms are met by ad hoc excuses thought up on the spur of the moment.
6) Ratio of supporters to critics rises up to somewhere near 50% and then falls gradually to oblivion.

Of these largely self-explanatory symptoms, the first bears further explanation. In the cases that Langmuir studied (N-rays, Davis–Barnes effect, mitogenic rays, etc.), the proponents avoided tests to increase the magnitude of their effect by increasing the magnitude of its hypothesized cause. Instead, the phenomena were treated as if they were independent of controllable causes.

Another list comes from Carl Sagan's *The Demon-Haunted World* [50]. Here Sagan offers a list of roughly 30 characteristics for identifying fallacious or fraudulent arguments, sometimes referred to as "baloney detectors." Although the list is too long to quote here in its entirety, the following seven items reflect many of its key points. (The full list has been posted on the Internet at <http://www.xenu.net/archive/baloney_detection.html>.)

1) Wherever possible there must be independent confirmation of the facts.
2) If there is a chain of argument every link in the chain must work.
3) Is the hypothesis testable? Can it, at least in principle, be falsified (shown to be false by some unambiguous test)? Can others duplicate the experiment and get the same result?
4) Was there observational selection (counting the hits and forgetting the misses)?
5) Are the statistics of small numbers (such as drawing conclusions from inadequate sample sizes)?
6) Is there a misunderstanding of the nature of statistics (e.g., President Eisenhower expressing astonishment and alarm on discovering that fully half of all Americans have below average intelligence!)?
7) Is there confusion of correlation and causation?

A similar list is the Crackpot Index from John Baez. It provides a scoring system for rating potentially revolutionary contributions to physics, but where the higher scores refer to less-rigorous assertions. The list is based on common symptoms of nonrigorous submission, such as offers for prize money to anyone who proves their theory wrong (item 13) or that the "scientific establishment" is engaged in a "conspiracy" to prevent their work from gaining its well-deserved fame (item 34). The full list of 37 items can be found at <http://math.ucr.edu/home/baez/crackpot.html> [51].

From the experience of the NASA BPP Project, these behaviors were similarly observed. It is common for the nonrigorous proponents to blame their lack of success on the phenomenon of reflexive dismissals, without considering the possibility that their work has flaws. This lack of self-criticism was observed as a recurring theme among unsuccessful researchers. Another consistent behavior of nonrigorous submissions was the lack of relevant reference citations. This is a particularly easy characteristic to use as a checklist item in research solicitations.

3. Additional Psychological Considerations

It has been found in psychological studies that those who are most incompetent also lack the competence to realize their incompetence [52]. This is not meant to be a glib comment; this is a real psychological characteristic that compounds the difficulty of responding to nonrigorous research.

The cited study covered many different tests, including humor, grammar, and logic; the trends were similar throughout. The study divided the participants into quartiles based on their test scores and then compared how the participants perceived their abilities without knowing the scores. The trend is that the poorest performers are the least aware of their limitations. There was, however, a crossover point typically with the third quartile who tended to accurately judge their ranking. The most competent quartile had the opposite perspective. They tended to underestimate how well they fared compared to their peers. The self-assessments of their *test scores* tended to be accurate, but the top quartile tended to overestimate the abilities of others. In other words, they tend to think others were similarly competent to themselves.

The study also found that raising the skill level of the less competent helped them better realize their limits. By teaching the less competent how to improve their skills, they become more aware of their deficiencies. In the context of dealing with nonrigorous submissions, this means that there is a benefit from teaching the submitters how to take the next step with their work rather than to just dismiss their work.

4. Responding to Nonrigorous Submissions

Because of the possibility that researchers and managers of provocative research will occasionally be asked to respond to nonrigorous submissions, the following recommendations are offered. These recommendations are offered with the observation (from the NASA BPP project [49]) that poorly crafted feedback can result in additional time-consuming and unproductive consequences. Recall that the poorest performers are the least aware of their limitations. If not dealt with constructively, they can counter with emotionally charged correspondence that also brings in others into their dispute. This short list of recommendations is followed with explanations and examples.

1) Do not reply to submissions that display delusions of grandeur or paranoia.

2) Give the submitters a task to perform that is aimed at improving the rigor of their work and at a level commensurate with their abilities.

3) Make the successful completion of that task a condition of continued correspondence.

4) When stating that the submitter's claims run contrary to known physics, use terms such as "*appears* to violate a well-established law of *nature*."

Based on the advice of a psychologist, it is recommended to not send any response at all to the *lunatic fringe*, those whose submissions display delusions of grandeur and/or paranoia (author's discussion with Dr. Joseph R. Wasdovich, Employee Assistance Program Manger and psychologist, NASA Glenn Research Center, 11 December 2002). The reason is that a technical response will not provide the submitter with the kind of help they need and will only encourage more unproductive correspondence. Delusions of grandeur are evidenced when the submitter states that their device or theory is *the* answer to solve the world's problems, or other statements to that effect. Paranoia is easier to recognize, where the submitter expresses fear about their safety or about the theft of their ideas. Often these characteristics coexist, such as when the submitters express concern that their submission is so valuable that it will evoke suppression by competing conspirators.

The option to not reply to any submissions is a viable option. Even though reviews of unsolicited submissions were occasionally conducted early in the NASA BPP Project, eventually the project adopted the policy of not reviewing any unsolicited submissions due to the volume of correspondence. For formal solicitations, however, where the submitter has followed the rules of submission, debriefings are often required.

Most of the nonrigorous submissions are curiosity-driven and the individuals simply do not know where to turn for guidance. Normal venues for assessments (journal peer-reviews or analytical assessments) are excessively difficult for nonrigorous individuals; it has been found to be more effective to give the submitters a task that is within their ability to perform and that will educate them to the critical details. Sometimes this can be as simple as showing them an example of a similar, previously submitted idea that was proven not to work as claimed. This is also consistent with the psychological study from Kruger and Dunning [52] where it was found that poor performers are more apt to realize their actual level of capabilities if taught to improve. A quintessential example is where mechanical oscillators are claimed to produce net thrust (see Chapter 6). In this case, the experimenter can be asked to perform further experiments using a level pendulum and instructed on how to distinguish vibrations from net thrust. This gives the submitters a task that is both within their ability and that will illustrate that their device does not operate as they originally interpreted. It is an exercise they can learn from.

When drafting the response letter, it is *not* effective to simply dismiss the ideas as violating known physics. This does not give the submitters a path to improving themselves or for learning the real operation of their device, plus it tends to evoke inflammatory, emotionally charged reactions. Remember, according to the study by Kruger and Dunning [52], they are likely to be unaware of the extent of their limitations. It is necessary in the response to raise the issue of a violation of physics, but in a way that leaves them an option to prove otherwise. Also, rather than using the phrase "laws of physics," it is more constructive to use phrases such as "laws of nature." The distinction is that nature is inarguably impartial and *the* basis of the physics.

The following is an example of a more effective response letter:

> Your device appears to violate conservation of momentum (or insert whichever natural law applies to their case), a well-evidenced law of nature. To convince us that something other than this is occurring with your device, we require that you perform additional tests to a higher standard of proof. More detailed suggestions are included for such tests as well as explanations for why other devices, that appear similar to yours, were found not to be viable. If, after following our suggestions and ensuring that all false-positive

effects have been dismissed, you will then have more convincing evidence that your device is operating in a novel manner. At that point we will gladly reconsider your submission. Until such conditions are met, we regret that we do not have the resources to maintain correspondence or provide further assistance in your investigations. We wish you the best in your endeavors.

This response puts the burden of proof back onto the submitter, setting clear and reasonable steps to move ahead to the next level of legitimacy. In cases where the submitters have built devices, it is recommended that the response dictate further experimental tests as a condition of deeper inquiry. By giving the submitter a reasonable next step and advice on what pitfalls to avoid, it gives them the means to learn from their own experiences. By making successful tests a condition for engaging in future correspondence, it relieves reviewers from further time consuming correspondence, yet leaves open the option of learning about possible positive results.

C. Summary of Lessons on Combining Vision and Rigor

Extracting the key points from the previous discussions, below is a short list of recommendations for successfully combining vision and rigor. These address both the tendency to reflexively dismiss novel ideas (an absence of vision), as well as to efficiently filter out nonrigorous ideas by taking advantage of their recurring critical flaws.

1) In evaluations, focus on the *rigor* with which the work is constructed rather than on trying to assess the *viability* of its approach.
2) Checklist of characteristics that reflect required rigor:
   a) The submitter is aware of the focal make–break issues related to their approach.
   b) The submitter is cognizant of the reliable relevant literature. Note, however, that it is common that credible researchers are not aware of *all* of the relevant literature. Some omissions are reasonable to expect.
   c) Any alternative and unconventional interpretations of known phenomena are accompanied by correct citations of those phenomena.
   d) The submitter cites examples of their prior publications or work to reflect their competence to conduct the proposed work.
3) Checklist of the characteristics that reflect noncredible work. The appearance of any of these is sufficient to reject the submission:
   a) The submitter does not provide any reference citations.
   b) The submitter is unaware or does not mention the critical issues underlying their concept.
   c) The submitter displays delusions of grandeur or paranoia.
   d) The submitter blames others for their lack of prior success.
4) The bottom line evaluation question is: "Has the researcher demonstrated that the results will be rigorous enough to be a reliable conclusion upon which to base future decisions, regardless if their idea works or not?"

IV. Research Project Operating Principles

Considering the historical perspectives and the need to combine vision and rigor, managing a research project devoted to revolutionary advances presents challenges beyond managing nearer-term research. As shown in Fig. 2, the manager faces the challenge of balancing a number of opposites. In addition to balancing vision and rigor, the manager must demonstrate progress in the near-term on goals whose solutions are not immediately conceivable. Further, the research approaches are likely to be numerous and diverse, while the manager can only focus on a few. And lastly, the supported research has to be within budgets, yet in portions sufficient to make genuine progress.

The NASA BPP Project devised the operating strategies described below to deal with these challenges [18]. These strategies are offered as an example for others who are responsible for managing revolutionary research. Later in this chapter, the specific research selection criteria are also detailed. The items in this introductory list are subsequently discussed in detail.

1) Reliability: Define success as acquiring *reliable* knowledge, rather than as achieving a breakthrough.

2) Immediacy: Focus research on the *immediate* unknowns, make-or-break issues, or curious effects, rather than extrapolating beyond reliable foundations.

3) Measured: Measure the relative maturity of the research using recognized steps of scientific and technological progress.

4) Iteration: Achieve overall progress by repeating cycles of solicitation, research, and reassessment, where the cycle duration is roughly two to four years.

5) Diversified: Explore multiple, divergent research approaches simultaneously.

6) Impartial: Judge selections based on *credibility* and *relevance*, rather than trying to predict *feasibility*.

7) Empirical: Preference is given to experiments and empirical observations over purely analytical studies.

8) Published: Results are published, regardless of outcome.

A. Reliable Knowledge Gained as the Success Criteria

Although it is a common practice when advocating research to emphasize the ultimate technical benefits, this practice is not constructive on topics as visionary and provocative as seeking revolutionary advances in spaceflight. Instead, it is more constructive to emphasize the *reliability* of the information to be gained. Compared to other propulsion research, the current frontiers of propulsion science are at their infancy. It is expected, therefore, that any practical embodiment is years, perhaps decades, away, if not impossible. Although breakthroughs, by their very definition, happen sooner than expected, no breakthrough is genuine until it has been *proven* to be genuine. Hence, the reliability of the information is a paramount prerequisite to the validity of any conclusions.

To place the emphasis where it is needed, no research approach should be considered unless it is sufficiently rigorous, regardless of the magnitude of claimed benefit. Success is defined as acquiring *reliable* knowledge, rather than as achieving a breakthrough.

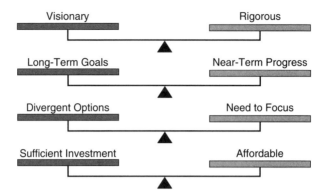

Fig. 2 Management challenges for revolutionary research.

This success criterion even means that a *failed* concept (test, device, etc.) is still a *success* if the information gleaned from that failure provides a reliable foundation for future decisions. This is a departure from the more common notion of judging success by how closely outcomes match expectations. In the more common approach, a task fails if the device does not work as desired, regardless of the lessons gained from the attempt. While such expectation-specific success criteria are appropriate for manufactured goods, they can be detrimental to fundamental research. By placing the emphasis on the fidelity of the findings, it encourages researchers to apply rigor to their work and to take the risks necessary to discover what others have overlooked. It also makes it easier to accept the results as they are, rather than to be tempted to skew the findings to match the expectations.

B. Immediate Research Steps

Another technique to shift the emphasis away from provocative situations and toward constructive practices is to focus the research on the *immediate* questions at hand. These immediate unknowns, issues, and curious effects can be identified by comparing established and emerging science to the ultimate propulsion goals. A subsequent section in this chapter, "Identifying Important Problems," deals with methods to assist with this step. The scope of any research task should ideally be set to the minimum level of effort needed to resolve an immediate "go/no-go" decision on a particular issue. This near-term focus for long-range research also makes the tasks more manageable and more affordable. Specifically, it is recommended that any proposed research be configured to reach a reliable conclusion in one to three years. Should the results be promising, a sequel can be proposed in the next solicitation cycle.

This is a departure from the Phase I and Phase II practice of both the Small Business Innovative Research (SBIR) [53] and the NIAC [20,21] programs. In these systems, phase-I awards are 6 to 12 month feasibility studies, whereas Phase II awards are larger and longer-term awards to move from feasibility toward an application embodiment. While this may be a prudent strategy for *technology* that is approaching fruition, such a two-stage approach is premature for basic research, even for *applied* basic research. For example, the Phase I and Phase II approach is based on a system where *success* is based on the *feasibility* of the concept as opposed to revolutionary research where success is based on the lessons learned.

C. Measured Progress

To help identify a suitable research increment and to provide managers a means to measure progress, the Scientific Method has been adapted as a readiness scale in a manner similar to how the technology readiness levels are used to measure *technological* progress [54]. The readiness scale developed for the BPP project consists of three stages that gauge the *applicability* of the work (reflecting how research can evolve from the more general, to the more specific application), and within each of these three stages, the five steps of the *scientific method* are repeated (from recognizing the problem, through testing the hypothesis). This equates to 15 levels of relative maturity, with the most advanced level being equivalent to Technology Readiness Level 1 (i.e., basic principles observed and reported).

1. Three Stages of Applicability

The scope of scientific research can span from the very general, broad-sweeping considerations all the way down to specific details of a given application. The more focused that a research increment is toward a desired application, the greater its applicability, as reflected in the following three stages:

Stage 1: General Physics. The research topic deals with general underlying physics related to the desired application.
Stage 2: Critical Issues. The research topic deals with an immediate unknown, critical make–break issue, or curious effect relevant to the desired application.
Stage 3: Desired Effects. The research topic deals with a specific effect or device for answering the goals of the application, such as force-inducing effects or the conversion of energy from one form to another.

2. Five Levels of the Scientific Method

Within each of these ranges of applicability, the Scientific Method can gauge interim progress. The following definitions for the steps of the Scientific Method have been tailored to reflect *applied* research progress. Because applied research implies a tangible product, these steps distinguish between *empirical* approaches (those based on the emergence of empirical effects) and *theoretical* approaches (those based on theory). The most significant distinction is with the final stage, where the hypotheses are tested. For applied science, again emphasizing a tangible product, this final stage can only be satisfied with an *empirical* test. Another noteworthy distinction from the common definitions of the Scientific Method is the inclusion of the zeroth step. This gives a placeholder for emerging opportunities that have not yet been addressed.

Step 0: Pre-Science
Empirical: Observations of an unconfirmed anomalous effect have been reported (includes observations of natural phenomena or claims of unverified devices), *or*
Theory: A correlation between a desired goal (or unsolved problem) and the existing knowledge base has been articulated.

Step 1: Problem Formulated
Empirical: An experiment has been defined that can collect the data required to isolate and characterize the anomalous effect, *or*
Theory: A goal (or problem) has been defined specifically enough to identify the remaining knowledge gaps toward achieving the goal (or solving the problem).

Step 2: Data Collected
Empirical: Data have been collected and analyzed from experiment to isolate and characterize the anomalous effect, *or*
Theory: The relevant data to fill the critical knowledge gaps, identified in the previous step, have been collected through experiment, observation, or mathematical proof (this level includes assessments of theory using mathematical analysis).

Step 3: Hypothesis Proposed
Empirical: A mathematical representation of the physical principles underlying an effect has been offered to explain the effect and predict additional (testable) effects, *or*
Theory: A mathematical representation of the relation between physical phenomena has been offered that addresses the goal (or problem) formulated previously.

Step 4: Hypothesis Tested and Results Reported
The hypothesis has been tested by comparison to observable phenomena or by experiment sufficiently to determine if it appears viable, and the results reported.
Note: In the context of applied research, testing of a hypothesis must be empirical; that means it must be done by comparison to observable phenomena or by experiment, rather than just by mathematical proof. Although mathematical proof can be used to test the consistency of a theory against known science, such a mathematical test alone is not sufficient to warrant achievement of Step 4. Instead, a mathematical test of a theory reflects achieving Step 2.

After a given research objective has been ranked relative to this 15-level scale, the next logical increment of research would be to advance that topic to the next readiness level. This is consistent with the *incremental* research strategy. A "Research Summary Form" is presented in the appendix.

D. Iterated Research

To accumulate progress over the long term, it is recommended to solicit a suite of proposals every two to three years, and to let the findings of the prior suite influence the next round of selections. This provides an opportunity for new approaches, sequels to the positive results, and redirections around null results. At any point, if a research task leads to the discovery of a new propulsion or energy effect, it can be pulled out of this process into its own advancement plan. This strategic approach is recommended for high-gain/high-risk research, where cycles of peer-reviewed solicitations can examine a diverse portfolio of options, and where the decisions build on the lessons learned from prior cycles of research.

Again, for basic research aimed at revolutionary advances, this iterative strategy is recommended over the Phase I and Phase II strategy of SBIRs and NIAC projects. The distinction is that revolutionary research requires taking risks beyond testing seemingly feasible approaches. This is also tied to shifting the definition of success from that of feasibility to that of lessons learned.

E. Diversified Portfolio

It is far too soon, in the course of seeking spaceflight breakthroughs, to downselect to just one or two hot topics. Instead, a variety of research approaches should be investigated in each review cycle. In simple terms, this is to diversify the research portfolio. This is different than the more common practice with advanced propulsion research where further advancements are primarily sought on the technical approaches already under study. Although this more common strategy can produce advances on the chosen topics, it faces the risk of overlooking potentially superior emerging alternatives and the risk that support will wane unless the chosen topics produce unambiguous positive results.

F. Impartial Reviews

When inviting research on the edge of knowledge, controversial ideas are encountered. Considering that most historic breakthroughs originally sounded like fringe ideas, it is not surprising that many of the proposals for breakthrough spaceflight might sound too visionary at first, or at least unfamiliar. It is therefore difficult to sort out the fringe ideas that may one day evolve into tomorrow's breakthroughs from the more numerous, erroneous ideas. During proposal reviews, it is common to have some reviewers reflexively assume that unfamiliar ideas will not work. To reliably determine technical feasibility, however, is beyond the scope of a proposal review, constituting a full research task unto itself. Instead of expecting proposal reviewers to judge technical feasibility, it is recommended to have reviewers judge if the task is leading to a result that other researchers will consider as a reliable conclusion upon which to base future investigations. This includes the possibility of learning from null results. This posture of judging credibility rather than prejudging feasibility is one of the ways of being open to visionary concepts while still sustaining rigor.

This posture also eases the burden on the reviewers. By asking them to focus on rigor rather than feasibility, it is easier for them to review work beyond their immediate area of expertise. The hallmarks of rigor, and the absence thereof, are relatively easy to spot and are similar across a wide range of disciplines.

G. Empirical Emphasis

When seeking advancements that can eventually lead to new technology, there is a decided preference toward tangible observations over purely analytical studies with all other factors being equal (cost, technical maturity, etc.).

Experiments, being hardware, are considered closer than theory to becoming technology. Also, experiments are considered a more direct indicator of how nature works. Theories are interpretations to explain observations of nature, while the empirical data *are* nature, partially revealed within the constraints of the given experiment. This tactic also helps to break from the limitations implicit in dominant paradigms.

H. Publishing Results

The final recommendation is to ensure that the research findings are published, regardless of outcome. Results, pro or con, set the foundations for guiding the next research directions. Although there can be a reluctance to publish null results when a given approach is found not to work, such dissemination will prevent other researchers from repeatedly following deadends. Again, by defining success as gaining reliable knowledge, such dissemination of lessons-learned becomes easier.

V. Devising Prioritization Criteria

Before detailing the selection criteria developed by the NASA BPP Project, this section describes the process by which those criteria were developed, including the scoring equation. These methods are offered for other managers who are tasked to develop scoring criteria tailored to their own projects.

The BPP evaluation process was adapted from a procedure developed by Bruce Banks during his tenure at the NASA Glenn Research Center and evolved over several years and among a variety of teams. A documented example of this process is the selection of the replacement thermal control materials for the Hubble Space Telescope [55]. This decision-making process is designed for prioritizing options where there are many issues of varying importance and influence that make such selections complex. A *multiplicative* scoring method is employed, based on principles of probability and statistics, which is significantly more sensitive than *additive* scoring methods.

The process has the following advantages and characteristics:

1) It is ideal for making decisions when many complex and often conflicting issues must be considered.

2) The process is applicable to prioritizing choices and making decisions on almost any topic such as:
    a) Strategic planning (determining which of several potential plans is best)
    b) Personnel selections or promotions
    c) Career options
    d) Contractor selections
    e) Major purchases

3) Every issue and every opinion is considered.

4) One individual or team can perform the process, but the team approach is recommended on those decisions where a team consensus on the final decision is paramount.

5) The decision-making process results in excellent "buy-in" by those using the process, because all issues and everyone's opinions can be taken into account.

6) The process employs multiplicative scoring that is significantly more sensitive to critical issues than additive scoring methods.
7) The decisions are prioritized in a quantified manner.
8) The resulting decisions are highly defendable.
9) The process minimizes skewing from overly assertive people.
10) The process can be performed via e-mail.
11) The mathematical methods of the process can be automated in software.

A. Participants

The process starts by assembling a team of representative experts and customers of the desired technology. "Customers" are those who are sponsoring the research solicitation, and "experts" are representative practitioners who are capable of conducting the research. Through brainstorming and voting, the team defines the relevant evaluation criteria, and then narrows these criteria down to a minimal list with weighting factors for each. The group also must distinguish among those criteria that are *mandatory* (criteria that *must* be met), and those criteria that are just *enhancing*. It is essential that the customers for the research concur with the criteria, and it is crucial that the other participants concur before applying the criteria to actually evaluate the options.

B. Required Characteristics of Evaluation Criteria

To determine the criteria to be used in the selection, it is helpful to have a suite of illustrative options (i.e., a sample of items whose prioritization is sought). With these up for discussion, it is easier to start collecting a list of evaluation factors. When the committee discusses the evaluation criteria, every proposed criterion should be listed, regardless of how many people feel it is meritorious. After all the criteria are listed, individual or group voting will be done to determine the relative importance of each.

Evaluation criteria should be:

1) Phrased in a positive sense (express as desired characteristics or freedom from undesirable characteristics)
2) Phrased such that there is majority acceptance of the wording
3) Independent of each other (no duplication of issues)
4) Address single issues (issues should not contain the word "and")
5) Able to be numerically scored (or graded)
6) Include all relevant issues

C. Relative Weighting of Criteria

Once the criteria have been selected, the committee decides, by consensus or majority vote, which criterion is the most important. The "relative importance" of this criterion is assigned a value of 1. Voters decide their values for relative importance of each evaluation criterion knowing that:

1) The most important criterion is rated 1, and is not changeable.
2) Because the most important criterion has been rated 1, the relative importance of the remaining criteria will have proportional importance values between

0 and 1. For example, a criterion half as important as the most important criterion should be rated 0.5.

3) Values can be entered that have up to three decimal places to the right of 0.

D. Multiplicative Scoring Principles

To quickly filter out substandard submissions, it is desired to have a feature whereby *any* failure to meet a *mandatory* criterion will eliminate the entire submission from competition. To provide this feature as an integral part of an evaluation system, the total score is determined by *multiplying* together, rather than by *adding*, the individual criterion scores. In this manner, any zero score (failing grade) on any *mandatory* criterion will result in a total score of zero.

To implement such a system, there are three details to take into account: 1) how to handle nonmandatory criteria, 2) how to handle weighting functions, and 3) how to normalize scores. The sample equation below illustrates a multiplicative system for two mandatory criteria and one nonmandatory criterion:

$$\text{Total Score} = \left(\frac{A}{N_A}\right)^a \left(\frac{B}{N_B}\right)^b \left(\frac{C + C_{\min}}{N_C}\right)^c \qquad (1)$$

where:

$A, B, C =$ criteria scores

$a, b, c =$ weighting factors, where 1 is the maximum value, and lower priorities are fractions of 1

$N_A, N_B, N_C =$ normalizing values or functions

$C_{\min} =$ a preset, non-zero value to prevent the parenthetical term from equaling zero, in the event that $C = 0$, thereby making criterion C nonmandatory

To allow nonmandatory criteria into a multiplicative system, two different approaches can be employed. The easiest, and the way employed with the NASA BPP process, is to just assign a score range for that criteria where the lowest possible score is not zero. The alternate approach, shown in the equation above, is to include a non-zero value in the criterion's equation. This second approach, however, complicates how normalizing functions are included.

To accommodate weighting factors, *exponents* are used analogous to the way *coefficients* are used for additive systems. It is recommended to use positive values equal to or less-than 1 for these exponents, where an exponent of 1 represents the highest priority. An exponent of 0.5 represents a criterion that is half as important.

In practice, the effect of the weighting functions also is tied to the maximum-point-value that each criterion can attain. Therefore, it is necessary that each criterion be normalized to the same maximum-point-value (the terms within the parentheses). For normalization, which means equalizing each criterion prior to applying its weighting exponent, a simple fractional coefficient is applied, so that the maximum possible values of all the criteria are equal.

Although a generic set of equations can be derived for how to implement a multiplicative system that accommodates all possibilities of mandatory and nonmandatory criteria, and accommodates criteria with differing scoring ranges, it is far simpler to implement the system with constraints on the scoring ranges. If all criteria have the same maximum point value, no normalization is required. If all nonmandatory criteria have a non-zero value as their minimum possible score, then no additional constants or associated normalization functions are required.

E. Scholastic Grading Standard

Experience has shown that an evaluation depends not only on the perceived merit of the idea, but also on the evaluators' interpretations of how to *score* the idea. For example, if the scoring range is 0 to 25 on a given criterion, such as with the Small Business Innovative Research evaluations [53], two different evaluators may use significantly different point values to mean the same grade. To avoid this problem, it is recommended that a familiar and limited grading system be used, such as the scholastic four-point scale:

A (4 points) = Excellent or outstanding, meeting the criteria to the maximum amount
B (3 points) = Good, or well above average
C (2 points) = Average, or the score to use if there is no reason to score high or low
D (1 point) = Poor or well below average
F (0 points) = Fails to meet the criteria

In those cases where these discriminators do not fit, it is still recommended to have the scoring range limited to about 5 gradations where possible, and to have clear text explanations to accompany each gradation. Because the final scores combine several criteria, it is possible to get sufficient distinctions with the total final scores even with such limited gradations.

VI. NASA Breakthrough Propulsion Physics Research Solicitation

This section is based on the 1999 research solicitation conducted by the NASA BPP Project [56]. It covers the selection criteria, the composite scoring equation, the two-stage review process, and the lessons learned. A "Research Summary Form," with instructions, is included in the appendix to help convey pertinent details.

A. NASA Propulsion Physics Criteria and Scoring Equation

Using methods very similar to those just described, the NASA BPP Project created the following prioritization criteria. In the early stages of the project, a Product Definition Team helped devise the first set of criteria and their weighting factors [43]. This team included a mix of both the customers of the research (NASA, DOD, DOE) and practitioners of research (physicists from a variety of organizations). A revised set of criteria were tested during a workshop in 1997 [44], and a further revised set [18] was finally adapted for use in the formal solicitation that spanned 1999 through 2000 [56]. The criteria listed below incorporate revisions from the lessons learned through the solicitation process.

1. Criteria

There are three categories of selection criteria. Technical Relevance relates directly to the project's Grand Challenges, the Credibility criteria judge the rigor of the research, and the Resource criteria address affordability and timeliness. The total composite score is achieved by multiplying the individual criteria scores as illustrated in Eq. (2) in the next section.

Technical Relevance:

1) Gain: Magnitude of performance improvement relative to all three of the Grand Challenges, assuming the approach under consideration ultimately reaches fruition.

2) Empiricism: Does the topic deal with tangible physical effects or just theory?

3) Readiness: The present maturity of the topic/concept under study as measured using the Applied Science Readiness Levels.

4) Progress: Magnitude of progress to be achieved by the research task, as measured by the difference in the readiness now (criteria 3), and the anticipated readiness level to be reached upon completion of the task, as measured using the Applied Science Readiness Levels.

Credibility:

5) Foundations: Based on credible references.

6) Contrasts: Compared to current credible interpretations.

7) Tests: Leading toward a discriminating test.

8) Results: Probability that the task will result in knowledge that will be a reliable foundation for future decisions.

Resources:

9) Triage: Will it be done anyway or is it unique to this project?

10) Cost: Funding required (reciprocal scoring factor).

11) Time: Time required to complete task (reciprocal scoring factor).

Some revisions were needed based on the lessons learned during the actual solicitation. The version presented above is the *revised* version. In the original solicitation, the Technical Relevance group of criteria (called Directness in the NRA solicitation) and the Scientific Method Readiness Scale were separate. When submitters and reviewers were scoring readiness, they often overlooked how readiness and applicability were linked, sometimes leading to contradictory assessments of readiness and progress. This is why the revised readiness scale has these two criteria explicitly interwoven. The other problem was that there was a "lineage" criteria that was found to be redundant to the "Probability of Successful Completion" criteria. Lineage has been deleted and is now integral to "Credible Results" (criteria 8).

2. Composite Equation

The total composite score is achieved by multiplying the individual scores, as illustrated in Eq. (2). This has the feature whereby a failure to meet any mandatory

Table 2 Evaluation criteria correlations to composite scoring equation

Criteria	Variable name	Score range	Equation	Normalizing variable	Weighting variable
1: Gain	G	0–4	G = (GM + GS + GE)/3	a	WG
Gain, Goal 1—Mass	GM	0–4	—	—	—
Gain, Goal 2—Speed	GS	0–4	—	—	—
Gain, Goal 3—Energy	GE	0–4	—	—	—
2: Empiricism	E	1–4	—	—	WE
3: Readiness (now)	RN	0–15	—	RNV = 15/4[b]	WRN
4: Progress	P	0–15	P = RA−RN	PNV = 1/2[c]	WP
Readiness (after)	RA	0–15	—	—	—
5: Credible Foundations	CF	0–4	—	—	WCF
6: Credible Contrasts	CC	0–4	—	—	WCC
7: Credible Tests	CT	0–4	—	—	WCT
8: Credible Results	CR	0–4	—	—	WCR
9: Triage	TR	1–4	—	—	WTR
10: Cost (discrete bands)	C	0–4	—	d	WC
11: Time (discrete bands)	TI	1–4	—	d	WTI

[a]The 1/3, embedded in the Total Gain equation, is to normalize the total gain to a maximum value of 4.
[b]The Readiness Normalization Value (RNV) is set to 15/4 so that a ranking of Technology Readiness Level 1 (score = 15) equates to a score of 4.
[c]The Progress Normalization Value (PNV) is set to 1/2, so that the typical progress of incrementing up one level of the Applied Science Readiness Levels equates to a score of 2 (an average condition).
[d]For Cost and Time, the scoring gradations are set to discrete bands rather that directly entering time or cost, and where the higher scores refelct lower cost and time. Specific scoring gradations examples are in the appendix.

criterion (zero score on criteria) will result in a total score of zero. Table 2 lists the characteristics of each criteria for traceability to the equation.

$$\text{Total Score} = G^{WG} E^{WE} (RN/RNV)^{WRN} (P/PNV)^{WP}$$
$$\times CF^{WCF} CC^{WCC} CT^{WCT} CR^{WCR} TR^{WTR} C^{WC} TI^{WTI} \qquad (2)$$

3. Research Summary Form

To streamline the review process, the proposal submitters were asked to encapsulate the key points of their proposal onto a Research Summary Form. This form is configured to pull together the information that pertained most directly to the evaluation criteria, and to present this information in a standard format to make it easier for the reviewers to find the key information. This form, along with its instructions, is provided in the appendix. Separate instructions are included for both the submitters and the reviewers to illustrate the difference between the collection of information and how the reviewer examines that information. Note how the inputs on the form trace directly to the evaluation criteria.

On a real proposal, the summary form would be accompanied by a 10- to 20-page proposal document. To better convey the meaning and application of selection criterion, the instructions for the reviewers that are presented in the appendix deal with each criteria in detail. For example, one area of expanded detail instructs the reviewers how to apply the four credibility criteria to accommodate the following five possible proposal situations:

1) New unreported effect
2) Known, unconfirmed effect (such as the Tajmar observations of frame dragging in the vicinity of very low temperature rotating rings [57])
3) Known, confirmed effect
4) New theory/hypothesis
5) Known theory

From the lessons of the actual 1999 solicitation, the use of such a proposal summary sheet was found to be effective. This made it very easy to quickly filter out noncompliant proposals and to focus on the key points needing further scrutiny. This might also become a useful tool for pre-proposal screening, and submissions of just the summary sheet could be used as a first screening, and only those submissions with acceptable summary sheets are invited to submit a full proposal. With the 1999 solicitation, a similar form was required from the reviewers, which made it easy to compile the scores. The version presented in the appendix combines both the submitter and reviewer inputs onto one sheet.

Another possibility that was under development when the project funding was deferred, was to adapt such a form to allow Internet submissions where appropriate information fields would be configured to automatically reject fringe submissions. This format was also to serve as the outline for archiving BPP research findings into an electronic database.

B. Inviting Relevant Research Proposals

The NASA BPP solicitation used a NASA Research Announcement (NRA) format that gave submitters the flexibility to suggest how their work could contribute to the solicitation's general goals. A lesson learned from the solicitation is that not all the important issues were covered by a viable proposal. If the project does not have another venue through which to direct work, such as an in-house lab, then the broader success of the project is entirely dependent on the proposals.

Before research proposals should be solicited, it is prudent to identify focal issues to help guide submissions. Workshops are one way to address this issue, provided that the workshop is designed for that purpose [44]. The Horizon Mission methodology, already discussed, is yet another approach. In addition, the NASA BPP Project used a "traceability map" to help contrast the foundational science to the propulsion goals. This map and its methods are discussed in Sec. VII.A.

C. Review and Selection Process

The research prioritization and selection process employed by the NASA BPP project followed a two-stage peer-review process. In the first stage, proposals were numerically scored relative to the criteria. In the second stage, a diverse suite of proposals were selected from the top-ranking candidates.

Having a two-stage review process was found to be effective. The first stage allowed each proposal to receive a thorough review, and the second stage allowed a smaller team to effectively compare the proposals. The first stage produced a large amount of data that was, by itself, not sufficient to identify the best proposals. Even after these data were analyzed to obtain an average score for each proposal and the value for the standard deviation of the reviewers' scores (Fig. 3), further discussion was required to effectively review and distill these results.

In the first stage of the review process, proposals were subject to scientific review by discipline specialists consistent with the topic matter of the proposal. On topics as visionary as breakthrough propulsion physics, it is prudent to have numerous reviewers to provide well-rounded assessments. For the 1999–2000 solicitation, at least four reviewers per proposal were secured for all 60 proposals, with virtually all proposals being scored by five reviewers. Proposals were reviewed by a combination of in-house and selected external reviewers, with due regard for conflict-of-interest and protection of proposal information. These external reviewers included individuals from government labs, universities, or industry.

Although it was challenging to secure the volunteer services of the more than 50 reviewers required for this coverage, having this many reviewers allowed

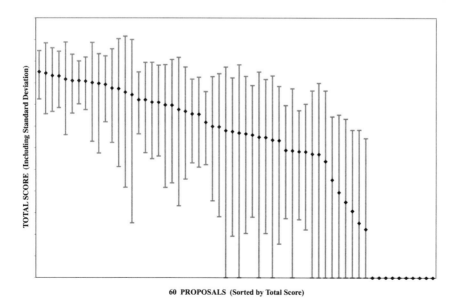

Fig. 3 Scores of proposals from the 1999 BPP Project solicitation.

several important functions to be employed. First, it was easy to avoid conflicts of interest by not having any proposals scored by people from their own or competing institutions. Next, it was possible to match the areas of expertise of the proposals and reviewers, which varied greatly because there is a wide span of physics underlying the BPP Project's ambitions. As already noted under the operating principles section, having the reviewers focus on rigor rather than feasibility made it easier for them to review topics that were not identical to their area of expertise.

With multiple reviewers per proposal, it was possible to calculate the standard deviations of scores to flag any disparate reviews for further scrutiny. This feature, of assessing the relative agreement among the reviewers of a given proposal, was found to be an effective tool for screening out questionable proposals and to identify problems with any reviews. Figure 3 reflects the scoring results using these techniques. In particular, note how the standard deviations are lower for the highest-scoring proposals whereas the mid range proposals have larger deviations, and many of the failing proposals have zero deviation. This reflects how the best and worst proposals are more obvious (less disagreement among the reviewers) and the medium-range proposals are more ambiguous.

The second stage of the review process involved the compilation and review of these scores by a smaller team of reviewers, again with due regard for conflict-of-interest and protection of proposal information. This team consisted of 10 government employees from multiple government labs, including NASA, the Department of Energy, and the Department of Defense. With a smaller but still diverse group of reviewers, it was relatively easy to sort through the total scores, evaluate the disparate reviews, and then select a diverse suite of research from top-ranking proposals. This second-stage review required about 2.5 days to complete.

A suite of different approaches is desired to "diversify the portfolio" because it is not possible to tell which research paths will lead most directly to revolutionary advancements. This means that the proposals selected for award were not necessarily selected contiguously from the highest-ranking set. For example, if the top two ranking proposals are both to perform an experimental test of theory A, and the third-ranking proposal is to test theory B, then it is the prerogative of this selecting team to award the best proposal on A and the third-ranking proposal on B, while skipping the second-ranking proposal on A if this supports diversification of research.

Five proposals were selected for award and their final results were presented at the 2001 Joint Propulsion Conference [58–62] along with one in-house supported task [63]. Some of these went on to have publications in peer-reviewed journals as well [2–14]. The intent was to use the conference sessions as a step toward preparing for the next solicitation around which additional work could be proposed toward promising areas and the null results dispensed.

VII. From Individual to Overall Progress

This last group of recommendations address building from the individual task progress into overall progress. This covers three aspects: 1) plotting the task progress relative to the goals, 2) providing metrics for upper level managers, and 3) mitigating the risks associated with revolutionary research.

A. Task Progress in the Big Picture

As stated in the Operating Principles, overall progress results from the accumulation of findings from individual research tasks. When managing a project, however, it is helpful to more specifically convey how such individual items compare to the general state of art. One method discussed is to measure progress with the Applied Science Readiness Scale (Sec. IV.C "Measured Progress"). Another tool toward this objective is the Traceability Map, which helps show how the individual tasks are connected to both the founding science and the project's goals. Figure 4 presents an abbreviated version to introduce the concept. In the left-most column, the various disciplines of physics are listed, which then branch out into specific items of interest relative to propulsion science. In the right-most column, the desired goals are listed, which then branch out into a variety of hypothetical concepts for achieving the desired effects. These concepts help define the critical questions. Between these two extremes reside the individual tasks that address the relevant effects, unknowns, and issues.

Figure 5 presents a more substantive example showing work supported by the NASA BPP project. In the figure, the shaded blocks represent work supported by the project, some of which encompass further details. Numbers in brackets refer to the resulting publications (listed with this chapter's references). Of these, the items having Refs. 3 to 13 and 58 to 62 were from the 1999–2000 competitive research selection; Ref. 63 is a small in-house task; Ref. 14 is a small grant; and Refs. 1, 16, and 49 represent work accomplished on discretionary time. For a version of a traceability map that shows all the items known to the BPP Project (144 blocks total), the reader is referred to page 16 of the project's management report [18]. As further illustration, a traceability

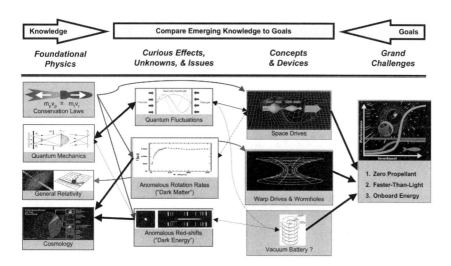

Fig. 4 Introduction to the concept of the traceability map.

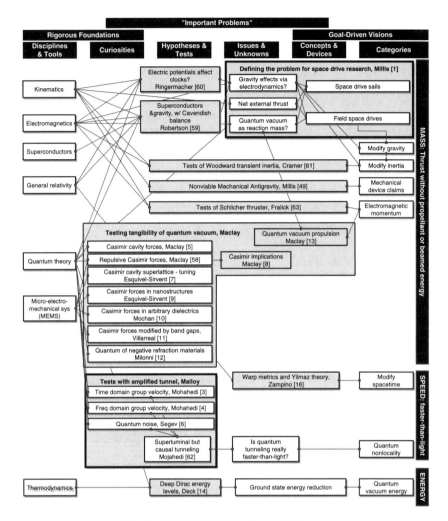

Fig. 5 NASA BPP task traceability map.

map for this book is provided in the preface. Unfortunately, printing limitations prevented showing the numerous interconnections between the items on the book's traceability map.

It is important to stress that items shown in Fig. 5 do not represent the full range of work or even the most critical issues, but rather the opportunities that were available to the project at the time. Many critical questions and unknowns still remain unaddressed. Likewise, the traceability map for this book presented in the Preface only covers the items discussed in this book. Neither is a complete set of possibilities.

To help identify required research, John Anderson's Horizon Methodology [42] can be employed. This method was introduced earlier in the context of historic lessons, and now it will be illustrated here in the context of populating a traceability map with representative research task options. Also, this is identical to identifying the "important problems" as in the context of Hamming's lecture [33].

B. Horizon Mission Methodology Applied to Traceability Map

1) Define impossible goals. These appear in the right-most column.

2) Apply Clarke's second law and brainstorming to expand those goals into notional concepts. These concepts appear just to the left of the goals, and connectors indicate which goals are addressed by which concepts.

3) Contrast these notional concepts to our established knowledge base, by:
   a) First listing the relevant foundational knowledge in the left-most column, and then
   b) Branch out from those disciplines to identify active areas of research that pertain to the notions emerging from the right-hand side of the chart. This includes curious effects and open hypotheses and tests.

4) "Look back from the future" of the notional concepts to identify the critical issues and unknowns that prevent their creation. These issues and unknowns connect the notional concepts with items on the left of the diagram.

5) Complete drawing the linkages between the individual items on the diagram, linking established disciplines to the desired revolutionary advances.

Additionally, this map could assist in prioritizing those options. In principle, because an individual task is linked to the concepts it addresses and the foundational science upon which it is based, the number of those connections is a measure of its relative importance. The research options that are linked to the most concepts are more important, and the ones linked to more foundational physics are likelier to be easier to study. The notion of applying the numerical techniques of fault tree analysis for such prioritizations has been introduced by Zampino [64], but has not been advanced further. This is a potential area for improvement.

C. Project Metrics of Performance

In addition to conveying how individual tasks progress maps to the general state of art, it is also useful to address how to measure progress and risks of the project as a whole. To that end, the following project metrics are provided:

1) Number of incremental unknowns resolved. When numerically counting these, devise a system whereby further advanced approaches are weighted with higher point values to reflect their greater importance (as gauged by the readiness levels).

2) Number of findings published in peer-reviewed literature with the goal being 100% of submissions accepted for publication.

3) Number of citations of the published works.

4) Number of visionary notions converted into new research approaches.
5) Number of students inspired (can only count those that send comments).
6) Number of spin-offs or educational materials produced.

To measure the overall performance of such a pioneering research project, all of the productivity measures above can be tallied, and then divided by the amount of resources (funding and time) consumed to achieve them:

$$\text{Relative performance} = \frac{\sum \text{Project metrics}}{(\$) \cdot (\text{Duration})} \qquad (3)$$

By including resources, it provides a means to gauge efficiency over time and to compare the effectiveness of this project to others in an organization (provided that the other projects are judged according to similar metrics).

D. Managing Revolutionary Project Risks

Risk management takes on a different meaning than that typically associated with technology development. In contrast to the issues of hardware reliability, research projects deal with knowledge. Therefore, the risks are associated with the reliability and relevance of the knowledge. Table 3 describes those risks along with the mitigation strategy for each.

Table 3 Mitigating revolutionary research risks

Risks	Mitigations
Credibility damaged by nonrigorous research reporting	Emphasize building rigorous, incremental advancements in knowledge, rather than requiring breakthroughs, and by collaborating with academia and other institutions, including peer reviews.
Leadership stature damaged by missing relevant breakthrough or by being too conservative	Pursue visionary research, extending well beyond the known and into areas that other organizations choose not to address. Sustain active scouting for ongoing research, forging widespread collaborations to find the most reliable advanced research. Support as much research as possible.
Competitive advantage weakened from premature disclosures (for individuals or commercially sponsored research)	Threshold of attention is when devices can be engineered. Disclosure strategy is to publish only enough for others to independently confirm the validity of the operating principles, followed by the marketing of improved versions once independent confirmation is reliably published.

VIII. Final Lessons and Comparisons

The BPP Project was found to be effective, although areas of improvement were identified and discussed in the 2004 project management report [18]. Some improvements, particularly refinements to the research criteria themselves, had already been incorporated. Other planned improvements included shifting the operation of the project to a nonprofit consortium of government, university, and industry. This shift was intended to improve the ability of the project to engage with both academia and industry, respond to the large number of interested researchers and research approaches, create and maintain a database of research findings and future options, and improve continuity to better manage progress. These revisions were underway when funding for the project was terminated in 2003 [65, p. 325].

The BPP Project is not the only government-funded effort to seek revolutionary advances. Others include the larger NASA Institute of Advanced Concepts (NIAC) and the still larger Defense Advanced Research Projects Agency (DARPA). For an introduction to the accomplishments produced by NAIC and DARPA, the reader is referred Ref. 20 for NIAC and Ref. 22 for DARPA.

Because both of these other efforts have a different scope and operating budget, direct one-to-one comparisons are inappropriate, but some features merit mention. First, regarding the differences in scope: The BPP Project only addressed emerging *science* relevant to space propulsion and had an effective budget of approximately \$0.2M/yr (total budget divided by years of operation) [18]. NIAC's scope was broader, seeking revolutionary concepts on "architectures or systems" of space exploration in general, and had an effective budget of more than 10 times the BPP effort, specifically about \$3.4M/yr [20]. NIAC began in 1998 and was terminated in 2007. DARPA has been in existence since 1958 and has an even larger scope and budget, seeking revolutionary advances on any aspects related to defense. In 2007, DARPA's portion of the budget that was devoted to space activities was almost \$500M [22], which is roughly two orders of magnitude higher than NIAC funding. With these differences, the mechanisms of how work is selected and supported are quite different and would be difficult to objectively compare.

What can be examined are some of the operating principles followed by both DARPA and NIAC. For example, DARPA rotates its staff so they are in positions no longer than six years to avoid the "we tried that, it didn't work" syndrome. This tactic is one way to avoid institutional paradigms or the hesitation to venture beyond current approaches. Another DARPA practice is that they have "the freedom to fail." This tactic allows researchers to take the risks necessary to extend beyond the known. For NIAC, one of its founding characteristics addressed the issue of incumbent limitations. NIAC was directed to seek innovations from *outside* NASA. NASA researchers were not allowed to compete for NIAC funding. Also, one of the criteria was to judge if the work was revolutionary instead of evolutionary. And finally, to reflect its operating attitude, the following quote is prominently displayed on their 2005 brochure [21]: "Don't let your preoccupation with reality stifle your imagination..."

These practices are indicative of the same historical lessons discussed at the beginning of this chapter.

IX. Conclusions

Although pioneering research is difficult, enough lessons have accumulated from history to guide the management of projects devoted to revolutionary research. These historic lessons plus the precedents set with the NASA BPP Project are offered as guidance to future managers and researchers of revolutionary concepts.

A key recommendation is to combine vision with rigor. Vision is needed to extend beyond existing knowledge, whereas rigor is needed to impartially compare those visions to accrued knowledge. The intent from that contrast is to identify the critical issues, make/break questions, and curious effects related to the desired goals. Once articulated, these become the important problems for future, pioneering research.

Another key recommendation is to not attempt to judge technical feasibility during proposal review, because that would constitute a research task unto itself. Instead, focus attention on judging if the proposed work will reach a reliable conclusion upon which other researchers and managers can make sound decisions for the future. To make such judgments, focus on the following four credibility criteria: foundations, contrasts, testability, and results. *Foundation* refers to the source material from which the proposal was based. *Contrast* is how the proposed concepts are compared to the accrued knowledge: Does the work possess the self-criticality that is characteristics of great researchers? Does the work reflect a working knowledge of accrued science and technology? *Testability* asks if the research is advancing toward a discriminating test. And finally, *results* refers to how likely the results, pro or con, will be reliable enough to become a basis for future management and research decisions. This, in large part, can be reflected by the track record of the research team.

From the experiences of the NASA project over its seven years, some additional useful strategies have been:

1) Breaking down the long-range goals into near-term immediate "go/no-go" research objectives that can each be assessed within one to three years.

2) Devising a numerical means to impartially compare research options and inherently reject nonrigorous submissions.

3) Addressing a diversified portfolio of research approaches.

Other insights into revolutionary research have been offered that demonstrate that revolutionary research faces implementation challenges in addition to the challenges of the research itself, and that these challenges can be met. It is hoped that by articulating these leading-edge project challenges and mitigating strategies, that other leading-edge research projects can improve their prospects for success.

Appendix

I. Research Summary Form

II. Instructions for the Research Summary Form

A. Instructions for the Submitter

The submitter only fills in blocks that are labeled with a capital letter. The reviewers will fill in the blocks in the "Review Column." The entries onto this summary form should only be an *introduction* to the proposed work as opposed to a thorough explanation of the work. Reference citations are required parts of this form that are intended to give interested readers leads to further study.

BLOCK A: Submitted by

This block contains three separate fields: Submitter's first name with middle initial (as one field), submitter's last name, and the date of submission (mm/dd/yy format).

BLOCK B: Title of Submission

For published results, enter the full reference citation in Block B. For proposed tasks or suggested inquiries, enter an appropriate title. The use of familiar keywords is encouraged to help convey context.

BLOCK C: Submission Type

Check one box to reflect the type of submission (research findings, proposed tasks, or suggested areas of inquiry), and note the other associated data blocks to skip or include.

Note: Adding this block allows the form to also be used for cataloging results or provisional research suggestions into a database.

BLOCK D: Central Issue, Unknown, or Observation Under Study

Provide a short paragraph to reflect the key focus of the completed (or proposed) work. What is the central, single question that this work aims to resolve? Use Block N to list the primary references that gave rise to this increment of work.

BLOCK E: Findings

This block should only be filled out if *results of published research* are being submitted. In that case, provide a short paragraph reflecting the conclusions of the research.

BLOCK F: Sequels Expected

This block should only be filled out if *results of published research* are being submitted. In that case, indicate whether continued investigations are expected. Note that null-findings, where no sequels are expected, are still considered valuable results worth disseminating.

BLOCK G: Relevance to Project

Provide a brief paragraph explaining how the research is relevant to the ultimate goals of the project. For general physics research, some traceability to the final goals needs to be articulated.

BLOCK H: Ultimate Improvement

This is a continuation of the relevance question, specifically addressing the degree to which the research is relevant to the ultimate three goals of the project. For these questions, assume that the work reaches a successful conclusion and that all subsequent works building on its result ultimately lead to a new propulsion or power device. For each of the three project goals (mass, speed, energy), check the box pertaining to the statement that best describes the ultimate envisioned performance of this new propulsion or power device. Granted, it may be difficult to predict this because the concept may be far from fruition, but these questions provide a way to distinguish among different levels of ambition.

MASS: Discover new propulsion physics that eliminates or dramatically reduces the need for propellant.
0   Not applicable to this goal.
1   Applicability doubtful.
2   Applicable, but potential impact unknown.
3   Intended to significantly reduce propellant requirement.
4   Intended to eliminate the need for propellant and the need for directed energy.

(The term "directed energy" means any form of energy sent from a central location such as from the Earth or sun.)

SPEED: Discover how to circumvent existing speed limits to dramatically reduce transit times.
0   Not applicable to this goal.
1   Applicability doubtful.
2   Applicable, but potential impact unknown.
3   Intended to eliminate speed constraints caused by limits of propellant or energy supply.
4   Intended to circumvent the light speed limit.

ENERGY: Discover new energy physics sufficient for interstellar flight.
0   Not applicable to this goal.
1   Applicability doubtful.
2   Applicable, but potential impact unknown.
3   Better energy conversion physics, but still limited to a consumable onboard supply.
4   Intended to provide energy sources and conversion methods accessible in flight.

BLOCK I: Increment of Work Proposed or Reported
Provide a brief paragraph describing the work performed (or proposed) to resolve the critical issue, unknown, or observation under study that was specified in Block D.

BLOCK J: Type of Research
Use this block to specify whether this work is theoretical or experimental, by checking the one box that best describes the type of research, using the options offered below:
  Study Comparative study, data collection, or literature search.
  Theory Theoretical work only, without empirical investigations.

Expmt Experimental tests or empirical observations only, not aimed at verifying or extending a specific theory.

Ex&Th Experiment or empirical observations coupled with theory. (This implies that the theory is sufficiently advanced to numerically predict effects or assess the utility and scalability of the effects beyond just a single demonstration experiment.)

BLOCKS K and L: Readiness and Progress[†]

Using the readiness levels described below, specify the level of progress of the work by checking the corresponding boxes. Note that two columns are required to be checked: one specifying the readiness **before** starting the task, and the other for the readiness **after** completion of the task (actual completion level for reported results, and *predicted* completion level for proposed research).

Answer this question **within the limits of the specific increment of research being addressed**. Note that "successful completion" is defined as completing the proposed work and learning more about reaching the breakthrough, rather than actually achieving the breakthrough.

BLOCK M: Contrasting or Skeptical Challenges to the Proposed (or Reported) Approach

Provide a brief paragraph summarizing the key skeptical challenges against the work. For *proposed work*, mention what is being done to answer the challenges. For the *results of completed work*, the response to the challenges can be addressed in the *Findings* section (Block E), and/or the *Increment of Work* section (Block I). Use Block Q to list the primary references for the skeptical challenges. In those cases where specific challenges have not been published, at least describe the weak points of the work and cite textbooks where the basic foundations behind such issues are found. For the BPP Project, researchers are required to explicitly cite and respond to the skeptical challenges to the work.

BLOCK N: Founding References

List the primary references upon which the work is based. These are the references from which the central issue, described previously in Block D, can be found.

BLOCK O: Representative Graphic (Optional)

This block is optional. The intent of this block is to provide a quick visual icon representing the essence of the work. In the final compiled document where many

---

[†]The Applied Science Readiness Levels were devised to provide a measure for the progress of *applied science* in an analogous manner to how the *Technology Readiness Levels* are used to rank relative maturity of *engineering* developments. Specifically, these Applied Science Readiness Levels consist of three stages for *applicability* (reflecting how applied research evolves from the more general to the more specific), and within each of these three levels, five steps of the *Scientific Method* are repeated. This equates to 15 levels of relative maturity, with the most advanced level being equivalent to Technology Readiness Level (TRL) 1: Basic Principles Observed and Reported. The form also has a placeholder for progress beyond this level ($\geq$TRL2). A more thorough description of these levels in provided in Sec. IV.C.

summary sheets are presented, such visual icons are useful to quickly find items of interest. Provide a graphic that represents some key feature of the work such as graphs of important data, photographs of hardware, significant equations, diagrams, etc.

BLOCK P: Related Disciplines
Check all boxes that apply. This block is used for identifying those disciplines that would be needed in a review and identifying which disciplines might benefit from the findings of the described research.

BLOCK Q: Contrasting or Skeptical References
List the primary references that describe the skeptical challenges to this work, as previously discussed in Block M. In those cases where specific challenges have not been published, at least cite textbooks where the basic foundations behind such issues are found.

BLOCK R: Prior Publications to Reflect Qualifications of Proposed Researchers
For *proposed work*, list examples of prior publications from the principal Investigator that demonstrate the ability to conduct and complete the proposed research.

BLOCK S: Resources
In addition to specifying the cost (in 1000s of U.S. dollars), and the anticipated duration of the proposed research in their appropriate boxes, also specify the "triage" situation. Triage is used here to predict whether the proposed work will be conducted anyway, or whether it is uniquely suited to the BPP project. The explanations accompany the checkbox options are as follows:
    Elsewhere   Certain to be credibly done without the support of the project.
    ?   Unknown. May or may not be done without the support of the project.
    Only-BPP   Exclusively suited to the BPP Project.

BLOCK T: Performing Organization
Specify the affiliation of the Principal Investigator under which the work has been conducted or is being proposed. Only the name and general location of the organization is required, not necessarily the full address. It is assumed for proposed work that the Submitter is the Principal Investigator or a representative of the Principal Investigator. In the case of published findings, it is assumed the first author cited in Block C is the Principal Investigator.

BLOCK U: Other Sponsors (Optional)
If the proposed (or completed) work is (or was) cosponsored by another affiliation, please list this here. Only the name and general location of the organization is required.

B. Instructions for Evaluators

The submitter only fills in blocks that are labeled with a capital letter. Reviewers will fill in the blocks in the "Review Column." The entries on this summary form are only an *introduction* to the proposed work as opposed to a thorough explanation of the work. See the full text of the proposal before evaluating. The instructions that follow explain how the entry blocks correspond to the criteria. More information is provided to help interpret these criteria.

Reviewer ID#
Self-explanatory

SCORES
The first and last rows of this block are part of the automated scoring equations that will register data once all the information has been entered. The middle row is where you enter your subjective assessment of the proposal, using the following familiar scholastic gradations. Unless otherwise specified, these gradations will be used for other criteria scores:

A (4 points) = Excellent or outstanding, meeting the criteria to the maximum amount.
B (3 points) = Good, or well above average.
C (2 points) = Average, or the score to use if there is no reason to score high or low.
D (1 point) = Poor or well below average.
F (0 points) = Fails to meet the criteria.

The bottom row, "Automated Pre-Score," is an automatic score generated from the inputs of the submitter as they select gradations from their own input blocks. The equation that supports this function is set up such that failure to meet any mandatory criteria—by virtue of the information on the form—will render a zero score. This includes looking for null entries for required inputs (especially Blocks M, N, Q, and R).

SEQUELS JUSTIFIED
In the case where results are reported instead of being a proposal (see Block C), judge if the issues (Block D), findings (Block E), the submitter's assessment (Block F), and relevance (Block G) merit further inquiry. Select one of the three stoplight options provided.

RELEVANCE: 1. Gains Toward Goals
This criterion grades how the proposal relates to all three of the BPP Grand Challenges, assuming that the concept behind the proposed research ultimately reaches fruition. Each Grand Challenge is graded separately and the final Gain criteria involves the sum of these three subset scores.

It is mandatory that the proposed work seek advances in science that are in some way relevant to the Project's three propulsion challenges or any critical issues or unknowns related to these goals. The scope is limited to further advances in science from which genuinely new technology can eventually emerge, technology to surpass the limits of existing methods as opposed to further developments of known technology. This means that if the proposed work only builds on known technology, then it fails this criterion.

For each of the three Grand Challenges, specify which of the statements best describes the ultimate achievable performance of the concept being addressed by the proposal, while entertaining the assumption that a final embodiment ultimately functions as desired. Granted, it may be difficult to predict this ultimate impact because the concepts may be far from fruition or the concept may appear not to be viable. For grading this criterion, assume for the moment, that the concept is viable and will reach fruition. Other criteria will grade readiness and credibility.

Note that this is a mandatory criterion, which means that a failure to meet this criterion (a zero score on all three subcriteria) will result in a total score of zero. Because the scores for all three Grand Challenges will be added, it is only mandatory that one of the three goals be addressed.

Grand Challenge 1—MASS: Discover new propulsion physics that eliminates or dramatically reduces the need for propellant. Scoring Gradations:

0   Not applicable to this goal (default answer if no answer specified).
2   Applicable, but potential impact unknown.
3   Intended to significantly reduce propellant requirement.
4   Intended to eliminate the need for propellant and the need for directed energy.

(The term "directed energy" means any form of energy sent from a central location such as from the Earth or sun.)

Grand Challenge 2—SPEED: Discover how to circumvent existing speed limits to dramatically reduce transit times. Scoring Gradations:

0   Not applicable to this goal (default answer if no answer specified).
2   Applicable, but potential impact unknown.
3   Intended to eliminate speed constraints caused by limits of propellant or energy supply.
4   Intended to circumvent the light speed limit.

Grand Challenge 3—ENERGY: Discover new energy physics to power these propulsion devices at levels sufficient for interstellar flight. Scoring Gradations:

0   Not applicable to this goal (default answer if no answer specified).
2   Applicable, but potential impact unknown.
3   Better energy conversion physics, but still limited to a consumable onboard supply.
4   Intended to provide energy sources and conversion methods accessible in flight.

RELEVANCE: 2. Empiricism

Does the topic deal with tangible physical effects or just theory? Because this project is interested in advancements that can eventually lead to new technology, and because empiricism is necessary to validate theories, there is a decided preference toward empirical observations over purely analytical studies, all other factors being equal. Experiments, being hardware, are considered closer than theory to becoming technology. Also, experiments are considered a more direct indicator of how nature works. Theories are interpretations to explain observations of nature, while the empirical data are nature. The most desired research task is an experiment that is coupled with theory. Experiments that are backed by a sound theoretical foundation provide a means to numerically assess the utility and scalability of the effects beyond just a single demonstration experiment. The next preference is experimental work by itself; for example, to independently test a claimed anomalous effect. The next preference is theoretical work by itself. Lowest on this priority scale is work that only involves comparative studies of existing approaches or literature searches. Scoring Gradations:

1 Comparative study, data collection, or literature search.
2 Theoretical work only, without empirical investigations.
3 Experimental tests or empirical observations only.
4 Experiment or empirical observations coupled with theory.

RELEVANCE: 3–4. Readiness and Progress

Specify the level of readiness and progress of the work by checking the corresponding boxes. Note that two columns are required to be checked: one specifying the readiness at the **before** starting the task, and the other for the readiness **after** completion of the task (actual completion level for reported results, and *predicted* completion level for proposed research). Note that item 4: Calculated Progress will be automatically filled in by the difference of the two columns, including application of its normalization function.

Answer this question **within the limits of the specific increment of research being addressed**. Note that "successful completion" is defined as completing the proposed work and learning more about reaching the breakthrough, rather than actually achieving the breakthrough.

CREDIBILITY: 5. Foundations

This criterion grades how well the proposed work is grounded in credible foundations. The proposed work must be based in some way on data or theories that are in the peer-reviewed literature. *Note:* Requiring reference citations is one of the techniques to filter out "fringe" submissions. This is a mandatory criterion, which means that failure to meet this criterion (zero score) will result in a total score of zero. Grade this criterion on how well the author identifies the most relevant references for the topic of investigation. A variety of specific guidelines are provided below on how this criterion maps to different proposed situations. The scoring gradations follow the previously described scholastic grading scale.

New, Unreported Effect: In cases where an unconfirmed anomalous effect is being investigated (where the effect has not yet been independently reported or confirmed in the peer-reviewed literature), the author must cite peer-reviewed references to indicate why the newly observed phenomenon would be considered anomalous. For example, in the case of anomalous thrust observations, where no reaction mass is readily apparent, it would be appropriate to cite references on momentum conservation. *Note*: Requiring an admission that the effect does not match the physics known to date is one of the techniques to filter out "fringe" submissions. If the author assumes that the effect is genuine, despite not having been independently confirmed, or despite the appearance of contradicting known physics, then the author fails this criterion. For consistency in this case, it is expected that the readiness level specified under criteria 3 should match Scientific Method Step 0 (Pre-Science) at whatever Applicability Stage fits the proposed work.

Known, Unconfirmed Effect: In cases where an unconfirmed anomalous effect is being investigated that has been reported in the literature, then these references must be cited. It is expected, in this case, that the existing literature would already list suspect causes for a "false positive" to help guide independent confirmation or refutation. For consistency in this case, it is expected that the readiness level specified under criteria 3 should match Scientific Method Step 1 (Problem Formulated), at whatever Applicability Stage fits the proposed work. If the suspect causes have not yet been articulated in the literature, then the readiness level specified under criteria 3 should match Scientific Method Step 0 (Pre-Science). If the author assumes that the effect is genuine, despite not having been independently confirmed, or despite the appearance of contradicting known physics, then the author fails this criterion.

Known, Confirmed Effect: In the case where the proposed work builds on an effect that has already been confirmed in the peer-reviewed literature, then the author must cite those references. For consistency in this case, it is expected that the readiness level specified under criteria 3 should match Scientific Method Step 2 (Data Collected) at whatever Applicability Stage fits the proposed work.

New Theory: In the case where work involves a theory that is not yet in the peer-reviewed literature, then the author must cite peer-reviewed references of the data or phenomena with which they are claiming consistency. It is not necessary that the author agree with current interpretations of these data, but it is mandatory that the theories are consistent with credible empirical evidence. For consistency in this case, it is expected that the readiness level specified under criteria 3 should match Scientific Method Step 2 (Data Collected) at whatever "Applicability Stage" fits the proposed work.

Known Theory: In the case where the proposed work builds on a theory that is already in the peer-reviewed literature, then the author must cite those references. For consistency in this case, it is expected that the readiness level specified under criteria 3 should match Scientific Method Step 3 (Hypothesis Proposed) at whatever Applicability Stage fits the proposed work.

CREDIBILITY: 6. Contrasts

This criterion grades how well the author articulates how the proposed work compares to existing credible interpretations, relative to the BPP Grand Challenges. This is to ensure that an idea is oriented toward the goals of the Project, and to ensure that the author has done the requisite homework on the existing literature. This not only checks for relevance to BPP, but also positions the work to address the next criterion of a discriminating test. *Note:* Requiring reference citations is one of the techniques to filter out "fringe" submissions. This is a mandatory criterion which means that a failure to meet this criterion (zero score) will result in a total score of zero. Grade this criterion by how well the author identifies the most relevant literature and on their understanding of this literature. Again, the usual scholastic scoring gradations apply.

Also, recall from the objectives of the BPP project, that the perceived *correctness* of the author's alternative interpretations is **not** being judged with this criterion. Unless there is some obvious error, it is considered too difficult to reliably determine such feasibility during a proposal review. Such an assessment would constitute a full research task unto itself. The burden of addressing feasibility, via a discriminating test, is addressed by criteria 7. Instead, judge how well the author demonstrates an understanding of the current, credible interpretations that are cited, and the author's ability to contrast this prior knowledge to the approach they are offering. Recall that the BPP Project examines emerging physics in the context of propulsion and power, and as such, there is latitude to consider alternative perspectives beyond that from *general* physics. Even though the current credible interpretations have already passed their own rigorous tests, this does not imply that such interpretations are a complete or best representation of the actual underlying physics, especially in the context of propulsion and power. Conversely, however, if the proposed interpretation has already been raised and dismissed in the open literature, then the author must cite and address these references.

A variety of specific guidelines are provided below on how this criteria maps to different proposed situations.

New, Unreported Effect: In cases where an unconfirmed anomalous effect is being investigated, where the effect has not yet been reported or confirmed in the peer-reviewed literature, then the author must focus on comparing the effect with other, credibly known effects that might lead to a false-positive conclusion. In the prior criteria ("foundations"), the author had to acknowledge that the effect was anomalous. In this criterion, the author must demonstrate that they are astute to the conventional interpretations that must be tested to determine if the effect is genuinely new. References that cover these conventional interpretations must be cited. Also, the author must explain why the effect (if genuine) might be advantageous to the BPP challenges. If the author does not address the issue of ruling out the suspect causes, then the author fails this criterion.

Known, Unconfirmed Effect: In cases where an unconfirmed anomalous effect is being investigated, that has already been reported in the literature, two different scenarios can apply. If the existing published report (that the author had to cite under "foundations") did *not* list a well-rounded set of suspect causes for a "false positive," then judge adherence to this criterion in the same manner stated for the case of *new unreported effects*. On the other hand, if the existing published report did sufficiently list suspect causes for a "false positive," then the original report will suffice for the required "contrast" citation, but the author must still demonstrate that they are astute to the conventional interpretations necessary to determine if the effect is genuinely new. Also, the author must explain why their proposed investigation is more applicable than the existing or past investigations into the effect. In the case where *null* results were previously published, the author must cite these and explain why these prior tests were incomplete, in error, or why a reinvestigation is warranted. Reference citations for these other investigations are required. Also, the proposal must explain why the effect (if genuine) might be advantageous to the BPP challenges. If the author does not address the issue of ruling out the known, suspect causes, then the author fails this criterion.

Known, Confirmed Effect: In the case where the proposed work builds on an effect that has already been confirmed in the peer-reviewed literature, then the author must explain why the effect might be relevant or advantageous to the propulsion challenges and why the investigation is more applicable to BPP than the prior or ongoing investigations into the effect. Reference citations for the confirmation publication, and for the prior or ongoing investigations, are required. If the author is challenging the current interpretations of the effect, then also judge this criterion by the guidance offered under "new theory," below.

New Theory: In the case where a theory is proposed that is not yet in the peer-reviewed literature, then it is mandatory that the new theories be compared to the contemporary theories that address the same phenomenon. Reference citations for the contemporary theories are required. The comparison must explain why the new theory would be more advantageous to the propulsion challenges than the contemporary theories. Judge this criterion by how well the author demonstrates an understanding of the existing theories, and on the author's ability to identify the unresolved issues of both theories with respect to the goals of breakthrough propulsion or power. The author must also demonstrate a willingness to consider that the theory might be in error, by identifying its weak points. If the author assumes that the theory is correct, without it having been confirmed with rigorous empirical tests, then the author fails this criterion.

Known Theory: In the case where the proposed work builds on a theory that is already in the peer-reviewed literature, then the author must describe how the work will be more applicable to BPP than the prior or ongoing work on the same theory. Reference citations for the contemporary theories are required. If the theory is still under debate in the open literature, then the author must acknowledge its potential weaknesses, and cite references that highlight these issues. Judge this criterion by how well the author demonstrates an understanding of the known theory, its debated issues, and on the author's ability to identify how the theory applies to the goals of breakthrough propulsion or power.

CREDIBILITY: 7. Credible Tests

This criterion grades how well the research advances the topic toward a discriminating test. It is required that the proposed work be leading toward a discriminating test or actually be a discriminating test. If a discriminating test can be completed within the budget and time guidelines requested of proposals, it is necessary that the test actually be proposed. Otherwise, it is sufficient to propose the design of an experiment for a make-or-break test, or to further advance a theory toward testable predictions. Again, the usual scholastic scoring gradations apply.

This requires that the author must identify the critical make-or-break issues for their immediate area of investigation. Also, the proposed next-step must be consistent with the scientific method, with due consideration for the current status of the topic as specified by the author. Further note that, depending on the status of the proposed task, independent verification may be warranted. In such a case, the vested interests of the Principle Investigator must be taken into account. This is a mandatory criterion, which means that a failure to meet this criterion (zero score) will result in a total score of zero. A variety of specific guidelines are provided below on how this criterion maps to different proposed situations.

Unconfirmed Effect (Reported or Not): In cases where an unconfirmed anomalous effect is being investigated, a discriminating test must be suggested that could distinguish between possible conventional explanations or whether this is a genuine new effect. The task should propose to at least design a discriminating experiment, or to actually conduct an experimental test. The work will be considered more credible if the proposal concentrates only on the experimental methods rather than on speculating on a new cause for the effect. For consistency in this case, it is expected that the completion readiness level (the level anticipated after the task is completed), will be at least at Scientific Method Step 1 (Problem Formulated) for an experiment design, or Scientific Method Step 2 (Data Collected) if an experimental test is actually planned.

Known, Confirmed Effect: In the case where the proposed work builds on an effect that has already been confirmed in the peer-reviewed literature, a logical next step would be to develop a theory to describe the anomaly. It would also be appropriate to propose a reconfiguration of the effect so that its propulsive or energy implications could be assessed. For consistency in this case, it is expected that the completion readiness level could be anywhere between Scientific Method Step 2 (Data Collected) through Scientific Method Step 4 (Hypothesis Tested), depending on the breadth of the proposed work.

Theory: In the case where the proposed work deals with theory, it is mandatory that the new theories are at least matured to the point where mathematical models are offered (this is one of the "fringe" filters). Then, either further mathematical analysis to predict testable effects, comparison to credible empirical observations, or experimental tests must be proposed that can bring the theory closer to a correctness resolution. An actual empirical test is preferred. For consistency in this case, it is expected that the completion readiness level could be anywhere between Scientific Method Step 1 (Problem Formulated) through Scientific Method Step 4 (Hypothesis Tested), depending on the breadth of the proposed work.

CREDIBILITY: 8. Results

This criterion grades the expected fidelity of the conclusions to be reached at the end of the proposed task. Will the task result in knowledge that will be a reliable foundation for future research decisions? Again, the usual scholastic scoring gradations apply.

Successful completion of the research task is defined as learning more about reaching the breakthrough, rather than actually achieving the breakthrough. Negative test results are considered progress. What is required, for successful completion, is that the work reaches a credible resolution that is clearly communicated. If it is likely that the work can be completed within the funding and time allocations specified, and that the results will be accepted by other researchers as a credible foundation for future work, then a high score is warranted. Base this assessment on a combination of the realism of the proposed work, its cost and schedule, and on the credentials of the proposed research team and their facilities. If cost-sharing is mentioned in the proposal, judge this criterion on the total resources to be devoted, not just the amount to be charged to NASA. Consider the clarity and quality of the proposal and any prior publications from the authors as a good reflection of the clarity and quality of the final product. Note too that, depending on the status of the proposed task, independent verification may be warranted. In such cases the vested interests of the Principle Investigator must be taken into account to ensure that there is no conflict of interest in the outcome of the device, phenomenon, or theory under test. This is a mandatory criterion, which means that a failure to meet this criterion (zero score) will result in a total score of zero.

RESOURCES: 9. Triage

Will this research be done anyway or must this project support it? This criterion addresses the possibility that the BPP project can save its resources if the topic is likely to be explored without support of the BPP project. Specify which statement best describes the situation. Note that this is not a mandatory criterion. A minimum score here will only result in demoting an overall "A" grade to a "C" grade. Scoring gradations:

1. (D) Certain to be credibly done without the support of the BPP project.
2. (C) Unknown.
3. (A) Exclusively suited to the BPP project.

RESOURCES: 10. Cost

This is a reciprocal scoring factor that addresses practical resource concerns. The more costly the work, the lower the overall score, all other factors being equal. Scoring gradations follows. (Note that the threshold values are based on 1999 costs.)

1. 0 (F) If the cost is outrageous, then assign a failing grade.
2. (D) Cost $\geq$ \$400K: Below average.
3. (C) Cost $=$ \$200K: Average.
4. (B) Cost $=$ \$100K: Good or well above average.
5. (A) Cost $\leq$ \$50K: Excellent (but verify that this is realistic for the work offered).

RESOURCES: 11. Time

This is a reciprocal scoring factor that addresses practical resource concerns. The longer to reach a reliable conclusion, the lower the overall score, all other factors being equal. Scoring Gradations:

1. (D) Duration $\geq$ 3 years: Below average.
2. (C) Duration $=$ 2 years: Average.
3. (B) Duration $=$ 1.5 years: Good or well above average.
4. (A) Duration $\leq$ 1 year: Excellent (but verify that this is realistic for the work offered).

References

[1] Millis, M. G., "Challenge to Create the Space Drive," *Journal of Propulsion and Power*, Vol. 13, No. 5, 1997, pp. 577–582.

[2] Millis, M. G., "NASA Breakthrough Propulsion Physics Program," *Acta Astronautica*, Vol. 44, No. 2–4, 1999a, pp. 175–182.

[3] Mojahedi, M., Schamiloglu, E., Hegeler, F., and Malloy, K. J., "Time-Domain Detection of Superluminal Group Velocity for Single Microwave Pulses," *Physical Review E*, Vol. 62, 2000, pp. 5758–5766.

[4] Mojahedi, M., Schamiloglu, E., Agi, K., and Malloy, K. J., "Frequency Domain Detection of Superluminal Group Velocities in a Distributed Bragg Reflector," *IEEE Journal of Quantum Electronics*, Vol. 36, 2000, pp. 418–424.

[5] Maclay, G. J., "Analysis of Zero-Point Electromagnetic Energy and Casimir Forces in Conducting Rectangular Cavities," *Physical Review A*, Vol. 61, 2000, 052110-1–052110-18.

[6] Segev, B., Milonni, P. W., Babb, J. F., and Chiao, R. Y., "Quantum Noise and Superluminal Propagation," *Physical Review A*, Vol. 62, 2000, 0022114-1–0022114-15.

[7] Esquivel-Sirvent, R., Villarreal, C., and Cocoletzi, G. H., "Superlattice-Mediated Tuning of Casimir Forces," *Physical Review A*, Vol. 64, 2001, 052108-1–052108-4.

[8] Maclay, G. J., Fearn, H., and Milonni, P. W., "Of Some Theoretical Significance: Implications of Casimir Effects," *European Journal of Physics*, Vol. 22, 2001, pp. 463–469.

[9] Esquivel-Sirvent, R., Villarreal, C., Mochan, W. L., and Cocoletzi, G. H., "Casimir Forces in Nanostructures," *Physica Status Solidi (b)*, 230, 2002, pp. 409–413.

[10] Mochan, W. L., Villarreal, C., and Esquivel-Sirvent, R., "On Casimir Forces in Media with Arbitrary Dielectric Properties," *Revista Mexicana de Fisica*, Vol. 48, 2002, p. 339.

[11] Villarreal, C., Esquivel-Sirvent, R., and Cocoletzi, G. H., "Modification of Casimir Forces due to Band Gaps in Periodic Structures," *International Journal of Modern Physics A*, Vol. 17, 2002, pp. 798–803.

[12] Milonni, P. W., and Maclay, J., "Quantized-Field Description of Light in Negative-Index Media," *Optics Communications*, Vol. 228, 2003, pp. 161–165.

[13] Maclay, J., Forward, R., "A Gedanken Spacecraft that Operates Using the Quantum Vacuum (Dynamic Casimir Effect)," *Foundations of Physics*, Vol. 34, 2004, pp. 477–500.

[14] Deck, R., Amar, J., and Fralick, G., "Nuclear Size Correction to the Energy Levels of Single-Electron and -Muon Atoms," *Journal of Physics G*, Vol. 38, 2004, pp. 2173–2186.

[15] Millis, M. G., "Assessing Potential Propulsion Breakthroughs," *New Trends in Astrodynamics and Applications*, Belbruno, E. (ed.), Annals of the New York Academy of Sciences, New York, Vol. 1065, 2005, pp. 441–461.

[16] Zampino, E. J. "Warp-Drive Metrics and the Yilmaz Theory," *Journal of the British Interplanetary Society*, Vol. 59, 2006, pp. 226–229.

[17] Millis, M. G., "Warp Drive, When?," 2006. URL: <http://www.nasa.gov/centers/glenn/research/warp/warp.html> [cited 27 Feb. 2008].

[18] Millis, M. G., "Breakthrough Propulsion Physics Project: Project Management Methods," NASA TM-2004-213406, 2004.

[19] Merkle, C. L. (ed.), "Ad Astra per Aspera, Reaching for the Stars," Report of the Independent Review Panel of the NASA Space Transportation Research Program, Jan. 1999.

[20] Cassanova, R., Jennings, D., Turner, R., Little, D., Mitchell, R., and Reilly, K., NASA Institute of Advanced Concepts, 9th Annual & Final Report, 2006-2007, 12 July 2006–31 August 2007.
[21] Bradley, A. (ed.), *The First Five Years and Beyond*, NASA Institute of Advanced Concepts. URL: <http://www.niac.usra.edu> [cited 29 Feb. 2008].
[22] Wislon, J. R., "Fifty Years of Inventing the Future, 1958–2008 (DARPA)," *Aerospace America*, Vol. 46, No. 2, 2008, pp. 30–43.
[23] Foster, R. N., *Innovation: The Attacker's Advantage*, Summit Books, New York, 1986.
[24] Shepherd, D. A., and Shanley, M., "Common Wisdom on the Timing of the Entry," *New Venture Strategy*, Sage, Thousand Oaks, CA, 1998, pp. 1–14.
[25] Henderson, R. M., and Clark, K. B., "Architectural Innovation: The Reconfiguration of Existing Product Technology and the Failure of Established Firms," *Administrative Science Quarterly*, Vol. 35, 1990, pp. 9–30.
[26] "Bezos in Space," *Newsweek on MSNBC News* (5 May). URL: <http://www.msnbc.com/news/9o4842.asp> [cited 29 May 2003].
[27] "Race to Blast Tourists into Space," CNN.com (21 March). URL<http://www.cnn.com/2006/TECH/space/03/20/space.tourism.ap/index.html> [cited 23 March 2006].
[28] Utterback, J. M., "Dominant Designs and the Survival of Firms," *Mastering the Dynamics of Innovation*, Harvard Business School Press, Cambridge, MA, 1994.
[29] Dyson, F., *Imagined Worlds*. Harvard Univ. Press, Cambridge, MA, 1997.
[30] *Legacy Archive for Microwave Background Data Analysis (LAMBDA)*, NASA Goddard Space Flight Center (7 Oct.). URL: <http://lambda.gsfc.nasa.gov/> [cited 13 March 2006].
[31] Kuhn, T. S., *The Structure of Scientific Revolutions*, 2nd ed., Univ. Chicago Press, Chicago, 1970.
[32] Horgan, J., *The End of Science*, Addison-Wesley, Reading, MA, 1996.
[33] Hamming, R., "You and Your Research," Morris Research & Engineering Center Lecture, 7 May 1986. URL: < http://www.cs.virginia.edu/~robins/YouAndYourResearch.pdf> [cited 17 Feb. 2006].
[34] Clarke, A. C., *Profiles of the Future: An Inquiry into the Limits of the Possible*, Bantam Books, New York, 1972.
[35] Forward, R. L., *Indistinguishable from Magic*, Baen Books, New York, 1995.
[36] Emme, E. M. (ed.), *Science Fiction and Space Futures Past and Present*, American Astronautical Society, San Diego, CA, 1982.
[37] von Puttkamer, J., "Reflections on a Crystal Ball: Science Fact vs. Science Fiction," *Science Fiction and Space Futures Past and Present*, Emme, E. M. (ed.), American Astronautical Society, San Diego, CA, pp. 137–150.
[38] *Star Trek*, Gene Roddenberry, NBC, 78 episodes, 8 Sept. 1966–3 June 1969.
[39] *Star Wars* movies, Lucasfilm Ltd, 1997, 1980, 1983, 1999, 2002 and 2005.
[40] Rodley, E. (ed.), *Star Wars—Where Science Meets Imagination*. National Geographic, Washington, DC, 2005.
[41] Miller, W. C., *The Creative Edge: Fostering Innovation Where You Work*, Addison-Wesley, Reading, MA, 1987.
[42] Anderson, J. L., "Leaps of the Imagination: Interstellar Flight and the Horizon Mission Methodology," *Journal of the British Interplanetary Society*, Vol. 49, 1996, pp. 15–20.

[43] Millis, M. G., *Breakthrough Propulsion Physics Research Program*, NASA TM 107381, 1997.
[44] Millis, M. G., and Williamson, G. S., *NASA Breakthrough Propulsion Physics Workshop Proceedings*, NASA CP 1999-208694, Cleveland, OH, 12–14 Aug. 1997.
[45] Hartz, J. and Chappell, R., *Worlds Apart: How the Distance Between Science and Journalism Threatens America's Future*, 2nd ed., First Amendment Center, Nashville TN 1998. URL: <http://www.firstamendmentcenter.org/about.aspx?id=6270> [Accessed May 2005].
[46] Park, R. L., *Voodoo Science: The Road from Foolishness to Fraud*. Oxford University Press, New York, 2000.
[47] Langmuir, I., "Pathological Science, Colloquium at The Knolls Research Laboratory," 18 Dec. 1953, transcribed and edited by Hall, R. N; URL: <http://www.cs.princeton.edu/~ken/Langmuir/langmuir.htm> [cited 27 Feb. 2008].
[48] Bennett, G. L. "Some Observations on Avoiding Pitfalls in Developing Future Flight Systems," *33rd AIAA/ASME/SAE/ASEE Joint Propulsion Conference*, AIAA 97-3209, Seattle, WA, July 1997.
[49] Millis, M. G., "Responding to Mechanical Antigravity," NASA/TM–2006-214390, 2006.
[50] Sagan, C., and Druyan, A., *The Demon-Haunted World: Science as a Candle in the Dark*, Ballantine Books, New York, 1997.
[51] Baez, J., Crackpot Index, 1998. URL: <http://math.ucr.edu/home/baez/crackpot.html> [cited 27 Feb. 2008].
[52] Kruger, J., and Dunning, D., "Unskilled and Unaware of It: How Difficulties in Recognizing One's Own Incompetence Lead to Inflated Self-Assessments." *Personality and Social Psychology*, Vol. 77, No. 6, 1999, pp. 121–1134.
[53] (SBIR) evaluations. URL: http://sbir.nasa.gov [cited May 2004].
[54] Hord, R. M., *CRC Handbook of Space Technology: Status and Projections*, CRC Press, Boca Raton, FL, 1985.
[55] Townsend, J. A., Hansen, P.A., McClendon, M. W., de Groh, K. K., and Banks, B. A., "Ground-Based Testing of Replacement Thermal Control Materials for the Hubble Space Telescope," *High Performance Polymers*, Vol. 11, 1999, pp. 63–79.
[56] "NASA Research Announcement: Research and Development Regarding 'Breakthrough' Propulsion," NRA-99-LeRC-1, NASA Lewis Research Center, 9 Nov. 1999.
[57] Tajmar, M., Plesescu, F., Seifert, B., Schnitzer, R., and Vasiljevich, I., "Search for Frame-Dragging-Like Signals Close to Spinning Superconductors," *Proceedings of the Time and Matter 2007 Conference*, Bled, Slovenia, World Scientific Press, 2007.
[58] Maclay, J., Hammer, J., George, M., Sanderson, L., and Clark, R., "First Measurement of Repulsive Quantum Vacuum Forces," *Joint Propulsion Conference*, AIAA Paper 2001-3359, July 2001.
[59] Roberson, T., "Exploration of Anomalous Gravity Effects by rf-Pumped Magnetized High-T Superconducting Oxides," *Joint Propulsion Conference*, AIAA Paper 2001-3364, July 2001.
[60] Ringermacher, H., Conradi, M., Browning, C., and Cassenti, B., "Search for Effects of Electric Potentials on Charged Particle Clocks," *Joint Propulsion Conference*, AIAA Paper 2001-3906, July 2001.
[61] Cramer, J. "Tests of Mach's Principle with a Mechanical Oscillator," *Joint Propulsion Conference*, AIAA Paper 2001-3908, July 2001.

[62] Mojahedi, M., and Malloy, K., "Superluminal but Causal Wave Propagation," *Joint Propulsion Conference*, AIAA Paper 2001-3909, July 2001.
[63] Fralick, G., and Niedra, J., "Experimental Results of Schlicher's Thrusting Antenna," *Joint Propulsion Conference*, AIAA Paper 2001-3357, July 2001.
[64] Zampino, E. J., and Millis, M. G., "The Potential Application of Risk Assessment to the Breakthrough Propulsion Physics Project," *Annual Reliability and Maintainability Symposium 2003 Proceedings*, 2003, pp. 164–169.
[65] Budget of the United States Government, Fiscal Year 2003; Executive Office, Office of Management & Budget. GPO, Washington, DC, 2002; URL:<http://www.gpoaccess.gov/usbudget/fy03/pdf/bud27.pdf> [cited 18 April 2006].

Subject Index

Abraham-Lorentz constant, 425
Abraham-Minkowski controversy, 22, 341–342
  Rontgen interaction, 355
  theoretical history, 348–352
Abraham's formulation, 346–347
  Ashkin and Dziedzic, 349–350
  Balasz, 348
  Brevik's analysis, 351–352
  Gordon's analysis, 350
  Halpern, 348
  Jones and Richards experiment, 349
  photon drag experiment, 352
  Shockley, 349
  Walker experiment, 350–351
Absence of vision, characteristics of, 674
Absolute reference frame, 135
Accelerated matter, gravitomagnetic field and, 245
Acceleration increase, Casimir drive and, 413–415
Acceleration, effects on mission trip time, 61–63
ACO. *See* anharmonic Casimir oscillator.
Acoustic inertial confinement fusion (AICF), 607
ACT. *See* Advanced Concepts Team.
ACT. *See* asymmetrical capacitor thrusters.
Active gravitational mass, 161, 162
Additive scoring method, 686
Advanced Concepts Team (ACT), 19–20
  nonviable concepts, 19–20
Advanced nuclear electric propulsion (NEP), 65
Advanced Stirling radioisotope generator (ASRG), 555
AFM. *See* atomic force microscope.
AICF. *See* acoustic inertial confinement fusion.
Alcubierre warp drive, 8–9, 487–489
Alpha Centauri mission, 88–89
ALSEPs. *See* Apollo Lunar Surface Experiments Packages.
Alzofon's antigravity propulsion, 221

Anderson, John J., 671–672
Angular momenta, 256
Anharmonic Casimir oscillator (ACO), 401
Anomalous forces, 231
Anomalous redshifts, 131
Antigravity
  cosmological, 192–198
  exact relativistic antigravity propulsion, 189–190
  gravitomagnetic foundations, 184
  negative energy induced, 190–192
  unrelated devices
    frame dragging, 260
    gyroscopic antigravity, 254–259
    oscillation thrusters, 249–254
    reaction/momentum wheel, 259–260
    spinning superconductors, 260
    torque wheel, 259–260
Antimatter
  fusion propulsion and, 78–80
  production of, 77
Antimatter annihilation propulsion, 42
Antimatter catalyzed nuclear fission/fusion, 80
Antimatter propulsion, 107, 108
Antimatter rocket issues, 103–104
Antimatter rockets
  light sails vs, 93–94
  propulsion system, 86
Antimatter space power, 564–565
Apollo Lunar Surface Experiments Packages (ALSEPs), 545
Areal density, 55
Ashkin and Dziedzic, 349–350
ASRG. *See* advanced Stirling radioisotope generator.
Asymmetric capacitors lifters (Biefeld-Brown effect), nonviability of, 16
Asymmetrical capacitor thrusters
  ablative material, 337
  corona discharge regime, 315

Asymmetrical capacitor thrusters
(Continued)
  corona wind, 324–325
  current flow, 335
  DC measurements, forces generated, 303–304
  dielectrophoretic forces, 324
  early development of, 329–330
  early history, 293–294
  electrogravitic coupling, 295
  electrohydrodynamic forces, 324
  electromagnetic manipulation, 295
  experiments, 333–335
    atmospheric pressure, 333–334
    DC measurements, 297–304
    partial vacuum, 334
    results, 297–304
    set ups, 295–297
    vacuum chamber, 339
  force power law, 312, 313–314
  Fowler-Nordheim, 314, 315
  geometric variations, 310–317
    linear arrangement, 310
  geometries of, 331–332
  grounding issues, 333
  high voltage, 295, 330
  image charges, 337
  induced gravitational field, 295
  ion drift, 337–339
    thrust mechanism, 324
  Kelvin forces, 324
  lifter geometries of, 330–331
  magnetic field measurements, 317–319
    corona discharge, 321–322
    Fowler-Nordheim emission, 322–323, 325
    Townsend equation, 322
  numerical calculations, 336
  Sinusoidal excitation measurements, 304–309
  theories, 294–295
    corona wind, 295
  thermodynamic forces, 324
  voltage gradients, 535
Atmospheric pressure, asymmetrical capacitor thruster experiments and, 333–334
Atomic force microscope (AFM), 400

Balance response, Micro-Newton thrust balance and, 378
Balasz, 348

Batteries, 537
Beamed-energy propulsion, 80–81
  laser electric, 81
  solar thermal, 81
Beamed-momentum light sails, 108
Beamed-momentum propulsion, 81–85
  electromagnetic catapults, 85
  sails
    electromagnetic, 84–85
    light, 82–83
    solar, 81–82
Beginning of life (BOL) values, 534
Beginning of mission (BOM) values, 534
Bell inequality tests, 511
Bias drive, 165–167
Biefeld-Brown effect. *See* asymmetric capacitor lifters.
Bonnor and Piper, 201
Bonnor and Swaminarayan, 181
Bose-Einstein condensate experiment, 354–355
Boyer's correlation function, 446–451
BPP. *See* Breakthrough Propulsion Physics (BPP) Project.
Breadboard device Micro-Newton thrust balance, 386–388
Breakthrough Propulsion Physics (BPP) Project, 11–17, 663–692
  goals/grand challenges, 12
  kickoff workshop, 13
  research proposal invitations, 692
  review and selection process, 693–694
  scoring equation, 689–692
  tasks, 13–14
Breakthrough propulsion studies, 1–2
  Abraham-Minkowski debate, 22
  advance concepts, 3, 7
    Sanger Study, 7
  Breakthrough Propulsion Physics Project, 11–17
  coupling of electromagnetism and gravity, 1
  early efforts, 2
  Einstein's Special Relativity, 1
  ESA General Studies Programme, 19–20
  Institute for Advanced Studies at Austin, 11
  Podkletnov disk, 9–10
  Project Greenglow, 17–19
  quantum tunneling, 1, 9
  quantum vacuum, 10–11

reality based, 2–3
rotating superconductors, 21–22
timeline of, 3–6
transient inertia phenomena, 22
vacuum fluctuation energy, 1
Vision-21, 7–8
warp drive, 1
  physics, 8–9
  wormholes, 8–9
Brevik's analysis, 351–352
Brito, 362–366
Buoyancy effects, superconductor experiments and, 232–233
Bussard interstellar ramjet, 59–61, 69–72, 151
  cylindrical-geometry magnetic confinement fusion reactor, 60–61
  extraterrestrial resource utilization, 69–72
  momentum drag, 60–61
  ram-scoop design, 60–61

Calorimetry, 640
  design, 645–646
Cartan, 658–659
Casimir cavity, 213–214
Casimir drive
  applications of, 416–417
  testing, 416
  unresolved physics issues, 415–417
  vibrating mirror, 405–415
Casimir dynamic effect, 397–398
Casimir effect, 391, 395–397, 481–484, 570–572, 583–585, 589–590, 641
  alternative theories, 398–399
  analytical codes, 660
  fluctuating matter fields, 590
  measuring of, 399–402
    anharmonic oscillator, 401
    atomic force microscope, 400
    microelectromechanical systems, 400–401
  moving mirrors, 483
  plenum, 589
  semiconductor surfaces, 401–402
  traversable wormholes, 483–484
Cassini project, 553–554
Catalyzed nuclear fission, 78–80
Catapults, electromagnetic, 85
Cavitation energy claims, 642
  Potapov device, 642

Cavitation fusion reactor (CFR), 607
Cerenkov radiation, 464
CFR. See cavitation fusion reactor.
Charge clusters, Shoulders, Ken, 642
Charge leakage, superconductor experiments and, 236
Charge pooling, superconductor experiments and, 235
Chemical rocket propulsion, 65
  high energy density matter, 65
  space shuttle main engine, 65
Chemical space power, 536–537
  batteries, 537
  fuel cells, 537
Christoffel symbols, Riemann Tensor, 654
Clarke, Arthur C, three laws, 669–670
Classical interaction, ZPF and, 445
Classical oscillator, 451–452
Classical rocket equations, 43–45, 108–110
  propellant loss, 46–48, 114–116
Closed-cycle gas-core nuclear rocket, 560
Closed timelike curve (CTC), 496
CMB. See cosmic microwave background.
Coherence-entanglement complementarity, 518–520
Cold fusion, 645–648
  calorimetry design, 645–646
Conservation Laws, negative matter propulsion and, 182–183
Conservation of momentum and energy, 128–130, 167, 359
  inertia, 129
  traversable wormholes, 501
  warp drives, 501
Constant acceleration rocket, 465–467
Continuous machine, 590–591
Control volume analysis, 259
Cooper pair mass anomaly, 243
Corona discharge, 321–322
  regime, 315
Corona plasmas, 323
Corona wind, 296, 324–325
Corum, 360
Cosmic microwave background (CMB), 195
  radiation, 132
Cosmological antigravity, 192–198
  dark energy, 195–197
  dark/vacuum energy, 198
  vacuum energy, 193–195

Cosmological inflation, 197–198
Coupled Maxwell-Einstein fields, 204–207
  gravitons, 204–207
Coupled phenomena, 20
Cruise velocity, effects on mission trip time, 61–63
CTC. *See* closed timelike curve.
Current flow, 335
Curvature of spacetime, 137
Cyclindrical-geometry magnetic confinement fusion (MCF) reactor, 60–61

Daedalus fusion rocket, 69, 70–71
Damping oscillation motion, 425–426
  Abrahm-Lorentz constant, 426
Damping system, 377
Dark energy, 20, 131–132, 195–197, 668
  anomalous redshifts, 131
  cosmic microwave background, 195
Dark matter, 3, 20, 132, 668
  galaxy rotation, 132
  gravitational lensing, 132
Dark/vacuum energy, 198
  warp drive, 198
DC measurements, asymmetrical capacitor thrusters and, 297–304
De Matos, 243
Deep Dirac energy assessment, 14
Deep space travel energy, 143–149
  Earth to orbit energy, 146–147
  General Relativity, 146
  gravitational dipole toroid, 146
  Kinetic Energy Equation, 144
  Krasnikov Tubes, 146
  levitation energy, 147–149
  Newtonian equations, 146
  Rocket Equation, 144
  type comparison, 145, 146
  warp drives, 146
  wormholes, 146
Degradable quantum vacuum, 594–597
  gravitational squeezing, 594–595
  melting, 596–597
  redshifting, 595–596
  vacuum field energy, 597
  zero-point fluctuation modes, 597
DeWitt, 236–237
Diametric drive, 160–162
  gravitational mass, 161
  inertial mass, 161

Dielectric media
  Abraham's formulation, 346–347
  electromagnetic fields, 345–347
  Minkowski's formulation, 346
Dielectrophoretic forces, 324
Differential sail, 154–155
Diode sail, 155–156
Dipole gravitational field generator, 185–187
Disjunction drives, 162–164
  active gravitational mass, 162
Displacement current, 358
Distances, interstellar flight technology and, 32
Doppler shifts, 464
Drop tests, 258
Dynamic Casimir effect, 397–398
Dynamic friction, 251
Dynamic space propulsion, 405

Earth to orbit energy, 146–147
  Rocket Equation, 147
  Space Shuttle, 147, 148
Einstein's General Theory of Relativity, 184–196
  antigravity, 184–186
  cosmological, 192–198
  exact relativistic antigravity propulsion, 189–190
  computational tools, 651–661
  Field equation, 136
  negative energy induced, 190–192
Electric coupling, superconductor experiments and, 234
Electric power system (EPS), 531
  components, 531
  power management and distribution, 531
Electric propulsion, 65–66
  advanced nuclear, 65
  low specific mass, 534
  radioisotope thermoelectric generator, 66
  solar, 65
Electrogravitational theory, 263–264
Electrogravitic coupling, 295
Electrohydrodynamic (EHD) forces, 324
Electromagnetic catapults, 85
Electromagnetic coupling, superconductor experiments and, 235

SUBJECT INDEX

Electromagnetic device energy claims, 644–645
  Motionless Electromagnetic Generator, 644
Electromagnetic effects, superconductor experiments and, 234–235
Electromagnetic field momentum, 363–364
Electromagnetic fields, dielectric media and, 345–347
Electromagnetic inertia manipulation (EMIM) propulsion, 362–366
  Brito, 362–366
  electromagnetic field momentum, 363–364
  electromagnetic momentum density, 365
  experiments, 365–366
  mass-inertia tensor, 364
Electromagnetic manipulation, asymmetrical capacitor thrusters and, 295
Electromagnetic momentum density, 365
Electromagnetic momentum exchange, 356–366
  Slepian's space ship, 357–360
Electromagnetic sails, 84–85
  MagOrion, 84–85
  particle beam driven, 84–85
Electromagnetic stress-tensors, 360–362
  Corum, 360
  Hartley oscillator, 360
  Heaviside force density, 360–362
Electromagnetic vortex (EV) phenomenon, 585–587
Electrostatic effects, superconductor experiments and, 235–236
Electrostatic forces, asymmetrical capacitor thrusters and, 337
EM torsion theory testing, 15
EMIM. *See* electromagnetic inertia manipulation propulsion.
Empirical observations, 685–686
End of mission (EOM) values, 534
Energy conditions, 472–484
  violating field, 474
Energy density function, 426
Energy storage systems, 534
  high energy density, 534
  high specific energy, 534

Energy-momentum stress tensor, 137
EPS. *See* electric power system.
Equivalence Principle, 140
ESA General Studies Programme, Advanced Concepts Team, 19–20
Escape velocity, 148
ETRU. *See* extraterrestrial resource utilization.
EV. *See* electromagnetic vortex.
Exact relativistic antigravity propulsion, 189–190
Exotic matter, 472–484
Extraterrestrial resource utilization (ETRU), 69–72

Faster-than-light (FTL) spacetimes
  closed timelike curve, 496
  conservation of momentum, 499–500
  General Theory of Relativity and, 471–501
  motion on causality, 497–499
  time machines, 496–499
  traversable wormholes, 484–487
  warp drives, 487–489
Faster-than-light travel paradoxes, 461
Feigel hypothesis, 366–367
Fermi energy, 188
Feynman diagrams, 433, 437
  external lines, 437
Field drives, 159–168
  bias, 165–167
  conservation of momentum, 167
  diametric, 160–162
  disjunction, 162–164
  gradient potential, 164–165
  net external force requirements, 164, 167
Fission fragment propulsion, 42, 67–68
  rocket, 6
  sail, 69
Fission space power, 556–561
  nuclear electric propulsion, 556
  nuclear thermal propulsion, 556
  Soviet Union, 560–561
  United States, 556–558
Fission thermal propulsion, 66–67
  nuclear engine for rocket vehicle applications, 66
Fluctuating plenum, 588–589
Fluctuation matter fields, Casmir effect and, 590
Force beams, Podkletnov, 242

Force power law, 312, 313–314
Forward, dipole gravitational field generator, 185–187
Forward, Robert, 2, 7
Forward's negative matter propulsion analysis, 181
Forward's Six-Mass Compensator, 177–179
Fowler Nordheim, 314, 315
Fowler-Nordheim emission, 322–323, 325
Frame dependent considerations, 135–136
Frame dependent effects, 21
  Mach's principle, 21
Frame dragging, 185–187, 243–245, 260
Fraudulent submissions, 675–677
Free energy, testing claims of, 639–648
  cold fusion, 645–648
  electromagnetic devices, 644–645
  output energy measurement, 640
  zero-point energy devices, 641–644
Free radicals, 533
Friction, 250–251
  dynamic, 251
  static, 251
Fuel cells, 537
Fusion, 78–80
Fusion propulsions, 69–72
  all-onboard types, 69
  antimatter-catalyzed fission fragment sail, 80
  Bussard interstellar ramjet, 69–72
  combinations, 78–80
    antimatter catalyzed nuclear fission/fusion, 80
  Daedalus, 69, 70–71
  inertial confinement, 69
  magnetic confinement, 69
  matter–antimatter annihilation, 72–78
  muon catalyzed fusion, 79–80
Fusion ramjet issues, 105–106
Fusion reactions, 562
  inertial confinement, 562
  magnetic confinement, 562
Fusion space power, 561–564
  fusion reactions, 562
Fusion, sonoluminescence energy harvesting and, 625

Galactic hydrogen, 133
Galaxy rotation, 132
Galileo spacecraft, 551–552

Gamma ray bursts (GRB), 495
Gas-core nuclear rocket (GCNR), 560
GCNR. *See* gas-core nuclear rocket.
General purpose heat source radioisotope thermoelectric generator (GPHS-RTG), 551
General Relativity, 146, 184–196, 219, 230, 394
  Quantum Mechanics vs, 668
General Relativity computational tools, 651–661
  existing types, 655–660
    Casimir-Effect codes, 660
    history, 652
    Macsyma, 652, 655, 657
    Maple, 659–660
    Mathematica, 657–659
    Maxima, 656, 657
    Reduce, 660
  propulsion computational conventions, 652–656
  summary, 656
General Theory of Relativity, 184–186
  energy condition, 472–484
    violating field, 474
  exotic matter, 472–484
  faster-than-light spacetimes, 471–501
    wormholes, 471–472
  negative energy, 474
  theorems, 473
Generational ships, 94–97
Geometric variations, asymmetrical capacitor thrusters and, 310–317
Gertsenshtein effect, 209–210
Gertsenshtein waves (GW), 209
Gliese mission, 89–92
Gordon's analysis, 350
  pseudo-momentum, 350
GPHS-RTG. *See* general purpose heat source radioisotope thermoelectric generator.
Gradient effects, superconductor experiments and, 235
Gradient potential drive, 164–165
Gravitational bremsstrahlung radiation, 208
Gravitational dipole toroid, 147
Gravitational lensing, 132
Gravitational mass, 161
  active, 161
  passive, 161
Gravitational scalar potentials, 167

Gravitational squeezing, 594–595
Gravitational synchrotron radiation, 208
Gravitational wave rockets,
  coupled Maxwell-Einstein fields, 204–207
  Gertsenshtein effect, 209–210
  photon rocket spacetime metric, 201–203
  realistic use of, 210–211
Gravitational wave transducers, 242–243
Gravitational waves, 231
  high-frequency, 231
Gravitationally altered launch pad, 140–143
  analyses of, 143
Gravitationally squeezed electromagnetic zero-point fluctuations, 479–481
Gravitomagnetic field
  accelerated matter and, 245
  rotating superconductors and, 21–22
Gravitomagnetic foundations, 184
  Fermi energy, 188
  Forward's dipole gravitational field generator, 185–187
  Lense-Thirring effect, 185
  Maxwellian electrodynamics, 185
  reality of, 187–188
Graviton production, 204–207
  particle accelerators, 207–209
Gravitophoton interaction field, 219–220
Gravity
  electromagnetism and, 1
  modification of, 231
Gravity control
  Alzofon's antigravity propulsion, 221
  Einstein's General Theory of Relativity, 184–198
  field drives, 159–168
  gravitational wave rockets, 201–211
  Heim's quantum theory, 218–221
  Levi-Civita effect, 198–200
  negating, 177
  negative matter propulsion, 180–184
  Newtonian Levitation Energy Estimate, 179
  Newtonian physics, 176–184
  nonretarded quantum interatomic dispersion force, 215–218
  orbital stability, 177–179
    Six-mass Compensator, 177–179
  quantum antigravity propulsion, 211–221

Gravity shielding, Podkletnov, 239–242
Greenglow, Project, 17–19
Ground state energy, 644
  reduction, 579–582
Grounding issues
  asymmetrical capacitor thrusters, 333
  superconductor experiments, 235
GRtensorII, 658, 659–660
Gyroscopic antigravity, 254–259
  angular momenta, 256
  testing, 257–259
    control volume, 259
    drop tests, 258

Haisch model, 429–431
Haisch, Rueda and Puthoff (HRP), 424
Halpern, 348
Hamming, Richard, 668–669
Harmonic oscillator, 427
Hartley oscillator, 360
Heaviside force density, 360–362
HEDM. *See* high energy density matter.
Heim's quantum theory, 218–221
  General Relativity, 219
  gravitophoton interaction, 219–220
Heisenberg uncertainty principle, 574
Helicopter analogy, 148
HFGW. *See* high-frequency gravitational waves.
Hidden momentum, 375
High energy density, 534
High energy density matter (HEDM), 65
  propellant, 98, 99
High specific power, 534
High-frequency gravitational wave producer, 203
High-frequency gravitational waves (HFGW), 231
Hooper antigravity coils, nonviability of, 16
Horizon Mission Methodology, 697
  Anderson, John L., 671–672
HSMs. *See* hydrogen switchable mirrors.
Human generational issues, 94–97
Hydrogen switchable mirrors (HSMs), 584

ICF. *See* inertial confinement fusion.
ICP. *See* inertial confinement propulsion.
Image charges, 337
  electrostatic forces and, 337
Imaging, multi-bubble sonoluminescence (MBSL) and, 611–615

SUBJECT INDEX

Impossible missions, concept of, 36–40
Impulse representation, 252
Impulse treatment, 148
Indexing convention, contracting Riemann to Ricci, 654–655
Indigenous reaction mass, 130–137
  cosmic microwave background radiation, 132
  dark energy, 131–132
  dark matter, 132
  galactic hydrogen, 133
  quantum vacuum fluctuations, 132–133
  spacetime, 134–137
  virtual particle pairs, 133
Indigenous space phenomena, 131
Induced charges, superconductor experiments and, 235
Induced gravitational field, 295. *See also* gravity control.
Induction sail, 155
Inertia, 129
Inertia control, 156–159, 404–405
Inertia modified rocket, 156
Inertial confinement fusion (ICF), 562
Inertial confinement propulsion (ICP), 68–69
Inertial frame, 134
Inertial mass, 161
Inertial mass, stochastic electrodynamics and, 432–452
Inertial space drive modifications, 156–159
  inertia modified rocket, 156
  oscillatory inertia thruster, 157–159
In-FEEP (indium field emission electric propulsion) thrusters, 376
Infinite specific impulse perspective, 138
Input current/rotational speed equation, 269, 271
Instant messaging paradoxes, 461–463
Institute for Advanced Studies at Austin, 11
Internal drive, 249
Interstellar dust grains, 107
Interstellar dust particles, 63–65
Interstellar flight technology
  antimatter, 106, 107
  antimatter rocket issues, 103–104
  beamed-energy, 80–81
  beamed-momentum light sails, 108
  beamed-momentum propulsion, 81–85
  challenge of, 32–42
    distances, 32
  critical nonpropulsion technologies, 99–100
  dust grains, 106
  engineering vs physical feasibility of, 101–105
  evaluation, 40–53
    infrastructure, 42
    technology, 42
    velocity change capability, 40–41
  fundamentals, 43–65
    Bussard interstellar fusion ramjet mission performance, 59–61
    mission trip time, 61–63
    relativistic light sail equations, 53–59
    rocket equation, 43–53
  fusion ramjet issues, 105 106
  human and cultural effects, 95–96
  impossible missions, 36–40
  interstellar dust, 63–65
  laser light sails, 106–107
    issues, 104–105
  light sails vs antimatter rockets, 93–94
  minimum communications system power, 101
  mission requirements, 33–35
    robotics, 34
    Vision Mission, 33
  multiple generations, 94–96
  possible missions, 85–97
    Alpha Centauri, 88–89
    antimatter rocket propulsion, 86
    Gliese, 876, 89–92
    light sail relativistic effects, 94
    single-stage light sail, 86–87
    Vision Mission, 92–93
  reliability vs reparability, 100–101
  rocket-based propulsion, 65–80
  societal investment in, 101
  specific energy density, 98–100
  specific impulse, 36–40, 97
  specific power, 98–100
  stage dry mass, 36–40
  wet mass, 36–40
Interstellar fusion ramjet mission performance, 59–61
Interstellar jet propulsions, 151–152
  Bussard ramjet, 151
Interstellar travel, barriers to, 12
Ion drift, 324, 337–339
Ion effects, superconductor experiments and, 235
Iterated research, 684–685

SUBJECT INDEX 727

Jones and Richards experiment, 349
Journal of the British Interplanetary Society, 2

Kelvin forces, 324
Kinetic Energy Equation, 144
Kowitt, 239
Krasnikov Tubes, 146

Lamb, Willis, 393, 394
Langrangian, 352, 436
Laser beam power, 54–55
Laser electric propulsion (LEP), 81
Laser light sails, 107
  issues, 104–105
Lense-Thirring effect (rotational frame dragging effect), 185–187, 243–245, 260
Leonhardt's analysis, 355
  Abraham-Minkowski controversy Rontgen interaction, 355
LEP. *See* laser electric propulsion.
Levi-Civita effect, 198–200
  quantum electrodynamic vacuum breakdown field, 200
Levitation energy, 147–149
  damped oscillation, 148
  escape velocity, 148
  helicopter analogy, 148
  impulse, 148
  normal accelerated motion, 148
  thermodynamics, 148
Li, 237, 239
Lifter geometries, 330–331
Lifters. *See* asymmetrical capacitor thrusters.
Light sails, 82–83
  antimatter rockets vs, 93–94
  diameter, 54
  hypothetical sail drives, 152–156
  mass, 55
  relativistic effects, 92
Lightspeed limit, 134
Lightweight radioisotope heater units (LWRHUs), 553
Lincoln satellites, 545–546
Linear arrangement, asymmetrical capacitor thrusters and, 310
Liquid core reactor, 560
Liquid effects, superconductor experiments and, 233–234
Lisp, 657

Lorentz force law, 353–354, 428–429
Lorentz Transformation, 456–459
Loudon, 354
Low specific mass, 534
LWRHU. *See* lightweight radioisotope heater units.

Mach's principle, 21, 135, 167, 373
  absolute reference frame, 135
  frame dependent considerations, 135–136, 515–516
Mach-5C Micro-Newton thrust balance, 381–382
Mach-6C Micro-Newton thrust balance, 382–385
Mach-6CP Micro-Newton thrust balance, 385–386
Macsyma, 652, 655, 657
  lisp, 657
  tensor calculus package, 652
Magnetic confinement fusion (MCF), 562
  reactor, 60–61
Magnetic confinement propulsion (MCP), 69
Magnetic coupling, superconductor experiments and, 234
Magnetic fields, 475–576
  measurements, asymmetrical capacitor thrusters and, 317–319
  changes in, 320
Magnetic Orion (MagOrion), 84
Mallove, Eugene, 2
Mansuripur, 353–354
  Lorentz force law, 353–354
  Minkowski-Abraham controversy, 353–354
Maple, 659–660
  GRTensorII, 659–660
  tensor package, 659
Mars Science Laboratory, 555–556
  Multi-Mission Radioisotope Thermoelectric Generator, 555
Mass-inertia tensor, 364
Mass renormalization, 434
Mass, characteristics of, 161
Mathematica, 657–659
  Cartan, 658–659
  GRtensorII, 658
  MathTensor, 657–658
  Ricci, 658
  Tensorial, 658–659
MathTensor, 657–658

Matter–antimatter annihilation propulsion, 72–78
 antimatter, 78
 vehicle sizing, 78
Matter–antimatter annihilation reaction, 48, 117, 118, 119
Maxima, 656, 657
Maxwell's equation, 342–345
Maxwellian electrodynamics, 185
MBSL. See multi-bubble sonoluminescence.
MCF. See magnetic confinement fusion.
MCP. See magnetic confinement propulsion.
MEG. See motionless electromagnetic generator.
Melting the vacuum, 596–597
MEMS. See microelectromechanical systems.
Metric determinant sign convention, 654
Metric sign convention, 653–654
Michelson–Morely experiment, 463
Microelectromechanical systems (MEMS), 21, 392, 400–401
Micro-Newton thrust balance
 balance response, 378
 damping system, 377
 experiment setup, 378–380
  2-MHz breadboard device, 386–388
  Mach-5C, 381–382
  Mach-6C, 382–385
  Mach-6CP, 385–386
 hidden momentum, 375
 In-FEEP thrusters, 376
 sensors, 376–377
 theoretical considerations, 374–376
 Woodward effect, 373–388
Minkowski's formulation, 346
Minkowski–Abraham controversy. See Abraham-Minkowski controversy.
Mission trip time, 61–63
 cruise velocity, 61–63
 effects of acceleration, 61–63
MKS system, 653
MMRTG. See Multi-Mission Radioisotope Thermoelectric Generator.
Molecular effects, superconductor experiments and, 235

Momentum, 128–130
 conservation of, 167, 499–500
Momentum domain entanglement, 514–518
Motionless Electromagnetic Generator (MEG), 644
Moving mirrors, 483
Multi-bubble sonoluminescence (MBSL), 611
 imaging, 611–615
 testing variables, 632
Multi-Mission Radioisotope Thermoelectric Generator (MMRTG), 555
Multiplicative scoring method, 686, 688–689
Muon catalyzed fusion, 79

NASA, sonoluminescence and, 610–617
Negative energy, 474
 drives, 404–405
 generation of, 475–484
  Casimir effect, 481–484
  gravitationally squeezed electromagnetic zero-point fluctuations, 479–481
  magnetic fields, 475–476
  squeezed quantum vacuum, 476–479
  static radial electric, 475–476
 induced antigravity, 190–192
 observation of, 495–496
  gamma ray bursts, 495
 quantum inequalities, 492–495
 spatial distributions, 492–495
 traversable wormholes, requirements of, 489–490
 warp drive, requirements of, 490–492
Negative matter propulsion, 180–184
 actual existence of, 183–184
 analysis of, 180
  Bonnor and Swaminarayan, 181
  Forward, 181
 Conservation Laws, 182–183
 diametric drive, 160–162
 Principle of Causality, 184
 Second Law of Thermodynamics, 184
 Totalitarian Principle, 184
Nelson, 352
 Lagrangian approach, 352
NEP. See nuclear electric propulsion.
NERVA. See nuclear engine for rocket vehicle applications.

SUBJECT INDEX

Net external force requirements, gradient potential drive and, 164, 167
New Horizons project, 554
Newton's gravitational constant, 166, 167
Newtonian Levitation Energy Estimate, 179
Newtonian physics, 128, 146, 176–184
Nieminen, R., 237–239
Nimbus-3 meterological satellite, 544
Nonlinear quantum mechanics, 524–525
Nonlocal communication, 520–522
Nonlocal signaling, 520–522
Non-nuclear space power, 535–537
  chemical, 536–537
  photovoltaic, 535–536
Nonretarded quantum interatomic dispersion force, 215–218
  challenge to, 218
  Pinto, 215–218
Nonrigorous submissions, 675
  responding to, 678–680
Normal accelerated motion, 148
NTP. *See* nuclear thermal propulsion.
Nuclear electric propulsion (NEP), 65, 556
Nuclear engine for rocket vehicle applications (NERVA), 66
Nuclear space power, 537–565
  radioisotopes, 537–556
Nuclear thermal propulsion (NTP), 556

Onboard fusion propellants, 69
Orbital stability, gravity control and, 177–179
Oscillation thrusters, 249–254
  friction, 250–251
  impulse representation, 252
  internal drive, 249
  nonviability of, 16
  single plane motion, 260
  slip-stick drive, 249
  sticktion drive, 249, 250
Oscillatory inertia thruster, 157–159
Output energy measurement, 650
  calorimetry, 640

Palladium-chromium films, 619–623
  resistance temperature detector, 619
Paradigm shifts, research and, 667–668
Paradoxes, 523–524

Particle accelerators, 207–209
Particle beam driven electromagnetic sails, 84
Particle bed reactor (PBR), 560
Passive gravitational mass, 161
PBR. *See* particle bed reactor.
PDF. *See* probability distribution function.
Photon band gaps, 355–356
Photon drag experiment, 352
Photon implications, 345–347
  dielectric media, 345–347
Photon momentum
  Abraham–Minkowski controversy, 341–342
  theoretical history, 348–352
  Abraham's formulation, 346–347
  electromagnetic inertia manipulation propulsion, 362–366
  electromagnetic momentum exchange, 356–366
    Slepian's space ship, 357–360
  electromagnetic stress-tensors, 360–362
  Feigel hypothesis, 366–367
  Maxwell's equation, 342–345
  Minkowski's formulation, 346
  Nelson, 352
    Lagrangian approach, 352
  photon band gaps, 355–356
  theoretical history
    Bose-Einstein condensate experiment, 354–355
    Leonhardt's analysis, 355
    Loudon, 354
    Mansuripur, 353–354
Photon rocket, 467–468
  spacetime metric, 201–203
    Bonnor and Piper, 201
    high-frequency gravitational wave producer, 203
Photovoltaic space power, 535–536
  solar arrays, 536
Pinto, 215–218
Pioneer projects, 547–548
Piper and Bonner, 201
Planck mass, 429
Platinum films, 619
Plenum, 588
  Casimir effect and, 589
  fluctuating, 588–589
PMAD. *See* power management and distribution.

Podkeltnov gravity shield, 239–242
  nonviability of, 16
Podkletnov disk, 9–10
  testing of, 14–15
Podkletnov, Eugene, 237–239
Podkletnov, force beams and, 242
Polarization correlation condition, 510–511
Pole vaulter paradox, 459–460
Polyvinyltoluene-based scintillator, 626
Potapov device, 642
Power management and distribution (PMAD), 531
Poynting vector, 432, 446–551
Principle of Causality, 184, 496, 522
Probability distribution function (PDF), 426
Project Greenglow, 17–19
Project metrics of performance, 697–698
Project risk management, 698
Propellant loss, 46–51
  classical rocket equations and, 114–116
  relativistic rocket equations and, 116–122
Propulsion computational conventions, 652–656
  Christoffel symbols, Riemann Tensor, 654
  indexing, 654–655
  system of units, 653
    metric determinant sign convention, 654
    metric sign convention, 653–654
    MKS system, 653
Pseudo-momentum, 350
Publishing, research findings and, 686
Pulsed propulsion, 66–67

QED. *See* quantum electrodynamics.
QFT. *See* quantum field theory.
QI. *See* quantum inequalities.
Quantum antigravity propulsion, 211–221
  quantum vacuum zero-point fluctuation force, 213–215
Quantum chromodynamic, 569
Quantum electrodynamics 391, 392, 433–438
  alternatives to, 592–594
    stochastic electrodynamics, 437, 438–441, 593–594

Feynman diagrams, 433
  less second-quantized fields, 594
  quantum field theory, 433
  theory, 574–575
    Heisenberg uncertainty principle, 574
  vacuum, 588
    breakdown field, 200
    Casimir effect, 589–590
    fluctuation effects, neoclassical theories, 592
    plenum, 588
    fluctuating, 588–589
Quantum entanglement, 509–525
  Bell inequality tests, 511
  coherence-entanglement complementarity, 518–520
  definition of, 509
  early experiments, 510–511
  glossary, 526–528
  momentum domain entanglement, 514–518
  nonlinear quantum mechanics, 524–525
  nonlocal communication, 520–522
  nonlocal signaling, 520–522
  paradoxes, 523–524
  polarization correlation condition, 510–511
  quantum nonlocality, 509–510
  quantum no-signal theorems, 512–513
  retrocausal nonlocal communication, 522
  superluminal communication, 522
Quantum field theory (QFT), 433, 651
  Lagrangian, 436
  looping of, 435
  mass renormalization, 434
  Unruh-Davies temperature, 437
  vacuum fluctuation, 434
Quantum inequalities (QI), negative energy and, 492–495
Quantum mechanics, 393
  General Relativity vs, 668
  Lamb, Willis, 393, 394
Quantum nonlocality
  definition of, 509–510
  early experiments, 510–511
  Special Relativity vs, 513–514
Quantum no-signal theorems, 512–513
Quantum tunneling, 1, 9
  testing, 15
  nonviability of, 16

SUBJECT INDEX

Quantum vacuum
　Casimir forces, 391, 392, 395–397
　　alternative theories, 398–399
　　dynamic effect, 397–398
　　measuring, 399–402
　degradable, 594
　effects, 21
　　microelectromechanical structures, 21
　extracting energy from, 569–598
　　Casimir force, 570–572, 641
　　degradability, 591–592
　　early concepts, 570–574
　　experiments, 577–587, 642
　　　electromagnetic vortex phenomenon, 585–587, 642
　　　ground state energy reduction, 579–582, 644
　　　tunable Casimir effect, 583–585
　　　ZPF voltage fluctuations, 577–579
　　quantum chromodynamic, 569
　　quantum electrodynamics vacuum, 588
　　type I transient machine, 590–591
　　type II continuous machine, 590–591
　　zero-point field, 570
　　zero-point field energy, 574–576
　fluctuations, 132–133
　General Relativity, 394
　history, 393–394
　　quantum mechanics, 393
　microelectromechanical systems, 392
　quantum electrodynamics, 391, 392
　space propulsion, 402–405
　　dynamic systems, 405
　　negative energy drives, 404–405
　　sails, 402–404
　　vacuum energy density, 404–405
　studies, 10–11
　　zero-point energy, 10–11
　theoretical calculation limitations, 399
　vibrating mirror Casimir drive, 405–415
　zero-point field energy, 394–395
　zero-point fluctuation force, 213–215
　　Casimir cavity, 213–214

Radiation detection, 625–628
Radioisotope power
　sources, 537–556
　　thermoelectric generator, 537–538

used by United States, 539
　advanced Stirling radioisotope generator, 555
　Apollo Lunar Surface Experiments Packages, 545
　Cassini project, 553–554
　Galileo spacecraft, 551–552
　general purpose heat source radioisotope thermoelectric generator, 551
　lightweight radioisotope heater units, 553
　Lincoln satellites, 545–546
　Mars Science Laboratory, 555–556
　New Horizons project, 554
　Nimbus-3 meterological satellite, 544
　Pioneer projects, 547–548
　transit satellites, 540–544
　Ulysses project, 553
　Viking space probes, 548–550
Radioisotope thermoelectric generator (RTG), 66, 537–538, 629
Ram-scoop design, momentum drag, 60–61
Reaction/momentum wheel, 259–260
Redshifting vacuum, 595–596
Reduce, 660
Reflexive demissals, 674
Relativistic effects on sail acceleration, 57–59
Relativistic light sail equations, 53–59
　areal density, 55
　laser beam power, 54–55
　light sail diameter, 54
　light sail mass, 55
　relativistic effects, 57–59
　thermally limited light sail acceleration, 55–57
Relativistic particle half-lives, 463–464
Relativistic rockets, 465–468
　constant acceleration, 465–467
　equation, 45–46, 111–114
　　propellant loss, 48–51, 116–122
　　matter–antimatter annihilation reaction, 48
Relativistic transformation, uniformly accelerated frame, 442–445
Research, prioritizing of, 663–713
　absence of vision, 674
　Breakthrough Propulsion Physics project, 663, 689–694

Research, prioritizing of *(Continued)*
  criteria development, 686–689
    additive scoring method, 686
    multiplicative scoring method, 686, 688–689
    participants, 687
    scholastic grading standards, 689
    weighing of, 687–688
  fraudulent work, 675–677
  historical perspectives, 665–673
    Arthur C. Clarke, 669–670
    Horizon Mission Methodology, 671–672
    lessons learned, 672–673
    paradigm shifts, 667–668
    researchers, 668–669
    science fiction, 670–671
    technology revolutions, 665–667
    tool-driven revolutions, 667
  Horizon Mission Methodology, 697
  operating principles, 681–694
    diversified portfolio, 685
    empirical emphasis, 685–686
    immediate research steps, 682–683
    impartial reviews, 685
    iterated, 684–685
    measured progress, 683
  publishing, 686
  reliable knowledge, 681–682
  Scientific Method, 683–684
  performance project metrics, 697–698
  project risk management, 698
  Richard Hamming, 668–669
  sample procedure, 701–714
  task progress, 695–697
  vision without rigor, 675–680
Resistance temperature detector (RTD), 619
Rest mass, ZPF and, 445
Retrocausal nonlocal communication, 522
Reviews, impartially of, 685
Ricci, 658
Ricci, Riemann from, 654–655
Riemann Tensor, 654
Riemann, Ricci to, 654–655
Rigor
  importance of, 675–680
  nonrigorous submissions, 675–678
Robotic interstellar missions, 34
Rocket-based propulsion, 65–80
  chemical, 65–66
  electric, 65–66
  fission fragment, 67–69

fission thermal, 66–67
fusion, 68–73
pulsed, 66–67
upper performance limits, 65
Rocket equations, 43–53, 108–122, 144, 147,
  classical, 43–45, 108–111
  comparison of, 51–53
  relativistic, 45–46, 110–113
Rocket inertia, modification of, 138–143
Rotating superconductors, gravitomagnetic fields in, 21–22
Rotational frame dragging effect, 185, 243–245
Rotational speed/input current equation, 269, 271
RTD. *See* resistance temperature detector.
RTG. *See* radioisotope thermoelectric generator.

Sails
  acceleration, relativistic effects on, 57–59
  beamed-momentum light, 108
  drives, 152–156
    differential, 154–155
    diode, 155–156
    induction, 155
  electromagnetic, 84–85
  fission fragment propulsion for, 68–69
  issues in use of, 104–105
  laser light, 107
  light, 81–82
  propulsion, quantum vacuum and, 402–404
  solar, 81–82
Sanger Study, 7
SBSL. *See* single-bubble sonoluminescence.
Schlicher Thruster testing, 14
Scholastic grading standards, 689
Science fiction, 670–671
Scientific Methods, 683–684
Scintillator, 626
S-curve evolution, 665–666
Second Law of Thermodynamics, 184
Second quantized electrondynamics, 594
SED. *See* stochastic electrodynamics.
Seismic effects, superconductor experiments and, 233
Semiconductor surfaces, Casimir forces and, 401–402
Sensors, 376–377

SEP. *See* solar electric propulsion.
Shockley, 349
 simplest case theory, 349
Shoulders, Ken, 642
SI, 653
Simplest case theory, 349
Single plane motion, 260
Single-bubble sonoluminescence (SBSL), 611
 testing variables, 632
Single-stage light sail mission, 86–87
Single-stage to geosynchronous Earth orbit (SSTO-GEO), 98
Sinusoidal excitation measurements, 304–309
Sinusoidal high voltage excitation, corona plasmas, 323
Six-mass Compensator, Forward, 177–179
Slepian's electromagnetic space ship, 357–360
 conservation of momentum and energy, 359
 displacement current, 358
Slip-stick drive, 249
Society's investment, interstellar flight technology and, 101–102
Solar arrays, 536
Solar electric propulsion (SEP), 65–66, 533
Solar sails. *See* light sails.
Solar thermal propulsion (STP), 80–81
Solvents, sonoluminescence and, 616–617, 618
Sonofusion, 606–610, 617–633, 643
 Roger Stringham, 643
Sonoluminescence, 605–633
 approaches, 607–609
  acoustic inertial confinement fusion, 607
  applications, 609–610
  cavitation fusion reactor, 607
  Suslick, Ken, 608–609
  Yaleyarkhan, Rusi, 608
 energy harvesting, 625–630
  concepts, 629–639
  fusion, 625
  polyvinyltoluene-based scintilator, 626
  radiation detection, 625–628
  radioisotope thermoelectric generators, 629
 high temperature generation, 617, 619–625

 palladium-chromium films, 619–623
 platinum films, 619
 multi-bubble, 611
 NASA, 610–617
  apparatus, 610–611
  single-bubble, 611
  solvents, 616–617, 618
Soviet Union fission space power, 560–561
Space drives
 conservation of momentum, 128–130
 deep space travel energy, 143–149
 Equivalence Principle, 140
 gravitationally altered launch pad, 140–143
 hypothetical mechanisms, 149–168
  field drives, 159–168
  inertial modifications, 156–159
  interstellar jet propulsions, 151–152
  listing of, 150
  sail drives, 152–156
 indigenous reaction mass, 130–137
 infinite specific impulse perspective, 138
 net external force requirements, 164, 167
 Newtonian physics, 128
 physics of, 168
 practicality of, 168
 rocket inertia, 138–143
 strategy, 14
 sustainability of, 168
Space power, 531–565
 antimatter, 564–565
 electric power system, 531
 electric propulsion, 534
 energy storage systems, 534
 fission, 556–561
 fusion, 561–564
 high power density, 534
 high specific power, 534
 non-nuclear, 535–537
 nuclear, 537–565
 performance characteristics, 534
 solar electric propulsion, 533
 specific impulse, 532–533
 values at
  beginning of life, 534
  beginning of mission, 534
  end of mission, 534
Space shuttle main engine (SSME), 65
Space Shuttle, energy used, 147, 148

Spaceflight, relativistic limits of, 455–469
  photon, 467–468
  rockets, 465–468
  Special Theory of Relativity, 455–459
Spacetime, 134–137
  inertial frame, 134
  lightspeed limit, 134
  Mach's principle, 135
  Young's modulus, 136
Spatial distributions, negative energy and, 492–495
Spatial terms, 653–654
Special Relativity, quantum nonlocality vs, 513–514
Special Theory of Relativity
  experiments re, 463–465
    Cerenkov radiation, 464
    Doppler shifts, 464
    Michelson–Morley, 463
    relativistic particle half-lives, 463–464
    Tachyon searches, 464–465
  paradoxes, 459–463
    faster-than-light travel, 461
    instant messaging, 461–463
    pole vaulter, 459–460
    twin, 460
  principles, 456
    Lorentz Transformation, 456–459
    postulates, 456–459
Specific energy density, 98–99
Specific impulse, 36–40, 532–533
  free radicals, 533
  metastable, 533
  technology, 97–98
Specific power, 98–99
Spinning superconductors, 260
Squeezed quantum vacuum 476–479
SSME. *See* space shuttle main engine.
SSTO-GEO. *See* single-stage to geosynchronous Earth orbit.
Stage dry mass, 36–40
Stargate, 471
Static Casimir effect, 407–409
Static friction, 251
Static radial electric, 475–476
Stiction drive, 249, 250
Stochastic electrodynamics (SED)
  background, 425–432
  damping oscillation motion, 425–426
  disadvantages of, 442
  energy density function, 426
  Haisch model, 429–431
  harmonic oscillator, 427
  inertial mass, 432–452
  Lorentz force, 428–429
  modeling, 593–594
  Planck mass, 429
  probability distribution function, 426
  quantum electrodynamics (QED) vs, 433–438, 437, 438–441
  semi-quantized, 432
  theory, 424, 574
  zero-point fields, 424
  ZPF Poynting vector, 432
STP. *See* solar thermal propulsion.
Stringham, Roger, 643
Superconductors
  anomalous forces, 231
  experiment considerations, 231–236
    buoyancy effects, 232–233
    charge leakage, 236
    charge pooling, 235
    electric coupling, 234
    electromagnetic coupling, 235
    electromagnetic effects, 234–236
    gradient effects, 235
    grounding issues, 235
    induced charges, 235
    ion effects, 235
    liquid effects, 233–234
    magnetic coupling, 234
    molecular effects, 235
    seismic/vibration effects, 233
    temporal effects, 234
    thermal effects, 232
    vacuum effects, 233
  General Relativity, 230
  gravitational
    force, 231
    waves, 231
  gravitomagnetic field and accelerated matter, 245
  gravity shielding, Podkletnov, 239–242
  history, 236–245
    DeWitt, 236–238
    force beams, Podkletnov, 242
    gravitational wave transducers, 242–243
    Kowitt, 239
    Li and Torr, 237, 239
    Podkletnov and Nieminen, 237–239
    Tajmar experiments, 243–245
  modifying of gravity, 231

SUBJECT INDEX 735

spinning, 260
yttrium barium copper oxide ceramics, 229
Superluminal communication, 522
Suslick, Ken, 608–609
Swaminarayan and Bannor, 181

Tachyon searches, 464–465
Tajmar experiments, 185, 243–245
  Cooper pair mass anomaly, 243
  De Matos, 243
Task progress, 695–697
Technology revolutions
  institutional patterns, 665–667
  S-curve evolution, 665–666
Temporal effects, superconductor experiments and, 234
Temporal terms, 653–654
Tensor calculus package, 652
Tensorial, 658–659
Thermal effects, superconductor experiments and, 232
Thermally limited light sail acceleration, 55–57
Thermodynamic forces, 324
Thermodynamics, 148
Thrust balance, 376–378
Time machines, 496–499
Tool-driven revolutions, 667
Torque wheel, 259–260
Torr, 237, 239
Totalitarian Principle, 184
Townsend equation, 322
Transient inertia phenomena. *See* Woodward effect.
Transit satellites, 540–544
  TRIAD, 542–544
Traversable wormholes, 472, 483–487
  conservation of momentum and, 499–500
  construction of, 494
  physical requirements, 485–486
  requirements of, 489–490
TRIAD satellites, 542–544
Tunable Casimir effect, 583–585
  hydrogen switchable mirrors, 584
Twin paradoxes, 460
Type I transient machine, 590–591
Type II continuous machine, 590–591

Ulysses project, 553
Uncertainty Principle, 132
Uniformly accelerated frame, relativistic transformation, 442–445

United States fission space power, 556–558
  rockets, 559
  closed-cycle gas-core nuclear rocket, 560
  gas-core nuclear rocket, 560
  liquid core reactor, 560
  particle bed reactor, 560
Unruh-Davies, 451–452
  temperature, 437

Vacuum chamber, 339
Vacuum effects, superconductor experiments and, 233
Vacuum energy degradability, 591–592
Vacuum energy density propulsion, 404–405
  inertia control, 404–405
Vacuum energy, 193–195
Vacuum field energy, 597
Vacuum fluctuation, 434
  energy, 1
Vacuums, asymmetrical capacitor thruster experiments and, 334
Van de Graaf generator, 271, 273
Velocity change capability, 40–41
Vibrating mirror Casimir drive, 405–415
  acceleration increase, 413–415
  modeling, 409–413
  static, 407–409
  unresolved physics issues, 415–417
Vibration effects, superconductor experiments and, 233
Viking space probes, 548–550
Virtual particle pairs, 133
Vision Mission, 33, 91–94
Vision-21, 7–8
Voltage gradients, 535

Walker experiment, 350–351
Warp drives, 1, 22–23, 146, 198, 472, 487–489
  Alcubierre, 487–489
  conservation of momentum and, 499
  physics, 8–9
    Alcubierre, 8–9
  requirements of, 490–492
  York extrinsic time, 487
Wet mass, 36–40
Woodward effect, 22, 373–388
  Mach's principle, 373
  oscillatory inertia thruster, 157–158

Woodward's transient inertia,
  testing of, 15, 378–388
Wormholes, 8–9, 146, 471–472
  stargate, 471
  traversable, 472
  warp drive, 472

Yaleyarkhan, Rusi, 608
Yamashita electrogravitational patent,
    263–290
  experiment background, 264–267,
      278–287
    input current/rotational speed
        equation, 269, 271
    Van de Graaf generator, 271
  experiment conclusions, 290–291
  experiment results, 273–278,
      287–290
  theoretical background, 263–264
YBCO. *See* yttrium barium copper
    oxide ceramics.
York extrinsic time, 487
Young's modulus analogy to,
    136–137
Yttrium barium copper oxide ceramics
    (YBCO), 229

Zero-point energy (ZPE) devices,
    641–644
  Casimir force, 570–572, 641
  cavitation, 642
  charge clusters, 585–587, 642
  ground state energy, 579–582, 644
  sonofusion, 643
Zero-point energy (ZPE) history, 10–11
Zero-point fields (ZPF), 424, 570
  classical interaction, 445
  Haisch, Rueda and Puthoff, 424,
      425–444
  Poynting vector, 432
  rest mass, 445
  voltage fluctuations, 577–579
Zero-point fields energy, 394–395,
    574–576
  quantum electrodynamics theory,
      574–575
  stochastic electrodynamics theory, 574
Zero-point fluctuation
    modes, 597
ZPE (ZPF energy), 570
ZPE. *See* zero-point energy.
ZPF. *See* zero-point field.
ZPF energy. *See* ZPE.

Author Index

Bennett, G. L. 531
Buldrini, N. 373
Cambier, J.-L. 423
Canning, F. X. 329
Cramer, J. G. 509
Cassenti, B. N. 455
Davis, E. W. 175, 471, 569
Drummond, T. J. 293
Fralick, G. C. 605
Frisbee, R. H. 31
Gilster, P. A. 1
Hall, N. R. 605
Hathaway, G. D. 229

LaPointe, M. R. 341
Lawrence, T. J. 263
Little, S. R. 639
Maccone, C. 651
Maclay, G. J. 391
Miller, P. B. 293
Miller, W. M. 293
Millis, M. G. 127, 249, 663
Puthoff, H. E. 569
Siegenthaler, K. E. 263
Tajmar, M. 373
Wrbanek, J. D. 605
Wrbanek, S. Y. 605

Supporting Materials

Many of the topics introduced in this book are discussed in more detail in other AIAA publications. For a complete listing of titles in the Progress in Astronautics and Aeronautics Series, as well as other AIAA publications, please visit http://www.aiaa.org.